OXYGEN TRANSPORT
TO TISSUE XVII

ADVANCES IN EXPERIMENTAL MEDICINE AND BIOLOGY

Recent Volumes in this Series

OXYGEN TRANSPORT TO TISSUE XVII

Edited by

C. Ince
Academic Medical Center
Amsterdam, The Netherlands

J. Kesecioglu
Academic Medical Center
Amsterdam, The Netherlands

L. Telci
University of Istanbul
Istanbul, Turkey

and

K. Akpir
University of Istanbul
Istanbul, Turkey

PLENUM PRESS • NEW YORK AND LONDON

Library of Congress Cataloging-in-Publication Data

Oxygen transport to tissue XVII / edited by C. Ince ... [et al.].
        p.   cm. -- (Advances in experimental medicine and biology ; v.
    388)
        "Proceedings of the 22nd annual meeting of the International
    Society on Oxygen Transport to Tissue, held August 22-26, 1994, in
    Istanbul, Turkey"--T.p. verso.
        ISBN-13:978-1-4613-8002-3   e-ISBN-13:978-1-4613-0333-6
        DOI: 10.1007/978-1-4613-0333-6

        1. Oxygen--Physiological transport--Congresses. 2. Tissue
    respiration--Congresses. 3. Oxygen therapy--Congresses.   I. Ince,
    C. (Can) II. International Society on Oxygen Transport to Tissue.
    Meeting (22nd : 1994 : Istanbul, Turkey) III. Series.
        [DNLM: 1. Oxygen Consumption--physiology--congresses. 2. Oxygen-
    -metabolism--congresses. 3. Biological Transport--physiology-
    -congresses.   W1 AD559 v.388 1995 / WF 110 098 1995]
    QP99.3.0909397 1995
    599'.012--dc20
    DNLM/DLC
    for Library of Congress   95-50023
                                                                    CIP

Proceedings of the 22nd annual meeting of the International Society on Oxygen Transport to Tissue, held August 22–26, 1994, in Istanbul, Turkey

© 1996 Plenum Press, New York
Softcover reprint of the hardcover 1st edition 1996

A Division of Plenum Publishing Corporation
233 Spring Street, New York, N. Y. 10013

10 9 8 7 6 5 4 3 2 1

INTERNATIONAL SOCIETY ON OXYGEN TRANSPORT TO TISSUE

1993-1994

Officers

Presidents:	C. Ince, The Netherlands
	K. Akpir, Turkey
President-Elect:	E.M. Nemoto, U.S.A.
Past President:	P.D. Wagner, U.S.A.
Secretary:	A.G. Hudetz, U.S.A.
Treasurer:	S.M. Cain, U.S.A.

Executive Committee

D. Delpy, U.K.	K. Groebe, Germany
K. Rakusan, Canada	W. Erdmann, The Netherlands
A. Eke, Hungary	P. Vaupel, Germany
N.S. Faithfull, U.S.A.	D.F. Wilson, U.S.A.

Other Official Functions

Archivist:	I.S. Longmuir, U.K.
Award Committee:	D.F. Bruley, U.S.A.

Organizing Committee

K. Akpir
C. Ince
J. Kesecioğlu
B. Sta van Uiter
L. Telci
L. Visser

SPONSORS

We are gratefully for the financial of the following for the 1994 ISOTT meeting:

- Alliance Pharmceutical Corp.
- Baxter Healthcare Corp., U.S.A.
- Baxter NV/SA, Belgium
- Biomedical Sensors Ltd/Pfizer, U.K.
- CRTICON/ Johnson and Johnson Professional Prod. Ltd., U.K.
- Eppendorf-Netheler-Hinz Gmbh, Germany
- Hamamatsu Photonics, Japan
- Physio Medical Systems BV, The Netherlands
- Radiometer International, Denmark

PREFACE

The 22nd meeting of the International Society on Oxygen Transport to Tissue (I.S.O.T.T.) of which this volume is the scientific proceedings, was held in Istanbul, Turkey on August 22-26, 1994. It was a historical occasion in that it was almost 200 years to the day that one of the founding fathers of oxygen research, Antoine Lavoisier, on May 8, 1794 found his early demise at the hands of the guillotine. This spirit of history set the tone of the conference and in the opening lecture the contribution that this part of the world has given to the understanding of oxygen transport to tissue was highlighted. In particular, the contribution of Galen of Pergamon (129-200) was discussed who for the first time demonstrated that blood flowed through the arteries and whose view on the physiology of the circulation dominated the ancient world for well over a millennium. A forgotten chapter in the history of the circulation of the blood is the contribution made by Ibn al Nafis of Damascus (1210-1280) who for the first time described the importance of the pulmonary circulation by stating that all venous blood entering the right ventricle of the heart passes to the left ventricle, not through pores in the septum of the heart as had been postulated by Galen, but through the circulation of the lungs. This important advancement in the insight into the importance of the pulmonary circulation had been generally attributed to Michael Servetus (1511-1553) who made the same observation some 300 years later. These historical aspects of research into oxygen transport to tissue were followed by a lecture by Dr. Sugioka who gave an overview of oxygen measuring techniques developed this century and which lay the foundations of our society. In this spirit of history, the introduction to this year's conference was thereby characterized by highlighting both ancient and modern historical aspects of the molecule of our common interest in its passage from the lungs to the tissues.

The Istanbul meeting was well attended with participants representing the technical, physical, physiological and clinical sciences. The success of our meeting was in large part due to the inexhaustible organization of Betty Sta van Uiter and the conference organizers, Ortur Turizm. They made the conference run smoothly and gave us all an unforgettable stay in Istanbul. We were treated each evening to a different social event resulting in maximal interaction between the participants of the meeting while at the same time enjoying the magic of Istanbul. The social program culminated in our conference banquet. Here, for the first time, the Lübbers Award was presented in honour of Dr. Lübbers' contribution to the science of oxygen supply and his professional contributions to the society. Its aim is to encourage young scientists to participate in the I.S.O.T.T. conference. The recipient of the Lübbers Award was M.V. Dubina (Russia) for his work on the effect of adaptation to hypoxia. The annual Melvin H. Knisely Award for outstanding scientific achievement was awarded to Dr. K. van Rossem (Belgium) for his work on brain oxygenation. Finally, Dr. Sugioka was honoured by the organizing committee for his appreciation of Kemal Atatürk, the founding father of modern Turkey. The presidentship of the society was then handed over to Dr.

Nemoto who called on the membership to attend the 1995 meeting of the I.S.O.T.T. in Pittsburgh Pennsylvania, U.S.A.

It is worth noting that 1994 marked the twenty-second meeting of the I.S.O.T.T. and that this edition of our proceedings is already the seventeenth volume. It is indeed a credit to our society and its members that a multidisciplinarian scientific society such as our own has maintained its high standard of excellence throughout this time. This volume of the proceedings indeed reflects this high level of scientific work being carried out over a wide and complete range of issues relating to oxygen transport to tissue. It is both the depth and range of topics covered in this series of proceedings that has made them into standard reference works on oxygen in (patho)physiological systems.

The contents of this year's proceedings reflect the four main themes covered each year in our meeting: A) Methods for assessment of oxygenation, in which new technology and the underlying physics of these techniques are discussed, B) The physiology of oxygen transport, where fundamental issues related to the passage of oxygen from the lungs to the mitochondria are investigated, C) Organ systems in disease, which concerns the dysfunction of oxygen handling in various types of pathologies, and D) Restoration of oxygenation, which mainly concerns approaches to restoring tissue oxygenation once it has become impaired following disease. The latter theme is largely clinically oriented and reflects a growing trend of the society to address clinical issues on oxygen transport to tissue. This trend was reflected by the sponsorship of this year's meeting by the European Society on Intensive Care Medicine.

The editors of this volume wish to thank Laraine Visser–Isles for her excellent language editing of the present proceedings, and to Silvia Reus for her tireless effort in the final editing proceedures. Thanks are also due to Lies Pruin for her secretarial assistance in the handling of the manuscripts. Reviewing and editing the manuscripts submitted to these proceedings was a real pleasure. We hope that you the reader will enjoy them as much as we did.

For the editors
Can Ince

CONTENTS

THE PHYSIOLOGY OF OXYGEN TRANSPORT

ORGAN SYSTEM IN DISEASE

RESTORATION OF OXYGEN IN DISEASE

INDICES

FROM GERMANY TO BRITTON

A Personal View of Experiences in Almost 50 Years of Measuring Oxygen

Kenneth Sugioka

Duke University Medical Center
Department of Anesthesiology
PO Box 3094, Durham, North Carolina 27710

From the title of this discussion, you can see that I consider Germany the cradle of oxygen measurement, even though I was not there when in 1897 Danneel[1] at the University of Gottingen devised the first platinum oxygen electrode. However, I did work in Gottingen (1962-1963) with Kurt Kramer. My acquaintanceship with Kramer began in 1946 while I was a medical student. James Elam, an investigator at Washington University, and I, with advice from Earl Wood, Professor of Physiology at the Mayo Clinic, were trying to develop an improved ear oximeter. Efforts to develop a good war time oximeter were triggered by the fact that the Luftwaffe was equipped with oximeters long before the allied air forces, and with a hypoxia warning device, their pilots were able to fly safely at higher altitudes. One of the first things we did was to obtain, with some difficulty, a reprint of the original paper on oximetry by Kramer and Matthes in Zeitschrift Fur Biologie in 1935.[2] Illustrations of his original work can be seen in Figures 1 and 2. Based on Glen Millikan's[3] ear oximeter (Figure 3) we developed a twin beam oximeter using selenium photo cells and Wratten Filters with some, but not notable success. In the midst of our work we were pleasantly surprised to find that the Army was sending us Kramer. He had been "rescued" from Russian hands by the U.S. Army who had dressed him as an American soldier in order to bring him back through the Russian lines. With his help we were able to gain new insights into the principles of oximetry.[4] While I was working on the ear oximeter, I worked also as a research technician in a laboratory which measured oxygen content in arterial blood, pre and post pneumonectomy. This was done by obtaining arterial samples, considered a highly dangerous procedure, and analyzing them in the Van Slyke apparatus.[5] (Figure 4) The analyses were performed by transferring a sample of blood into the closed end of a mercury manometer along with a reagent to free any bound oxygen and then observing the nanometric change that occurred as a result of the free gas. Though cumbersome and time consuming, since each analysis took 45 minutes, and a repeat analysis was always done to check on the first, the values obtained were so accurate that this method persists to this day as a way of obtaining the oxygen content of blood. The principles of this method were developed in part by J.P. Peters[6] Professor of Biochemistry at Yale University, who became the forgotten man in this

Oxygen Transport to Tissue XVII, Edited by Ince et al.
Plenum Press, New York, 1996

1

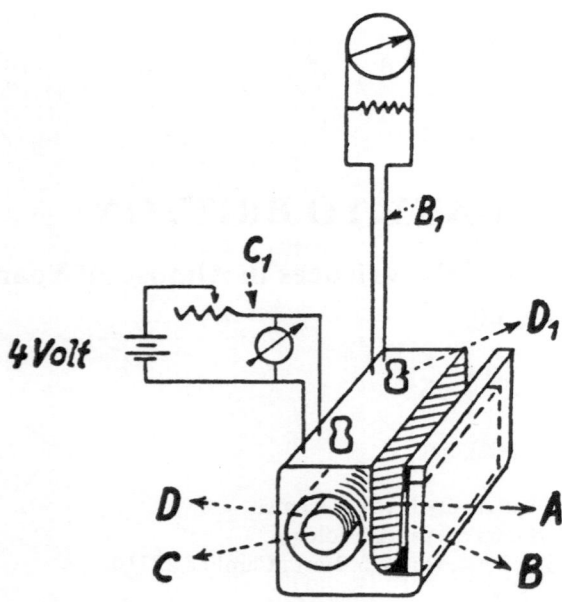

Figure 1. Schematic drawing of one of Kramer's oximeters. A. photocell. B. wire from photocell to galva-nometer. C. light source. C. battery and rheostst. D. water jacket. D_1. coupings for water hoses.

remarkable development in oxygen measurement. I had the pleasure some years later of spending some time in his laboratory and found him to be most helpful.

Because of frustration with developing an accurate ear oximeter using selenium cells of our own manufacture, and trying to obtain Wratten filters that would give us accurate wave lengths of 600 and 800 nanometers, I gave up on it as ever being a clinical tool. Proof of my devotion to the ear oximeter is evident by the scars on my earlobes which came about as a result of burns inflicted by the incandescent lamp used for ear oximetry at that time. In 1975 a leading instrument company asked me what I thought of the future of oximetry, especially one which was pulsed. My advice to them was not to waste any time on it. This is perhaps an early example of my not so farsighted "wisdom", since pulse oximetry is now widely used, and profitably so.

Thanks to Leland Clark[7], one of the kindest of all men, I obtained one of his oxygen electrodes in 1957 and decided that the future of tissue oxygen measurement lay with this instrument. Fascinated by the work of Kety and Schmidt[8], who measured the oxygen tension in brain tissue with recessed platinum electrodes, I obtained the so-called needle electrode (Figure 5) and obtained some of the early quantitative measures of brain tissue oxygen tension[9] (Figure 6).

In 1960 I was approached by Tom Rosse who was forming a company called Instrumentation Laboratories. He was interested in having his engineers develop a clinical instrument for measuring oxygen and carbon dioxide tensions using the Clark and Stowe electrodes and based on the system developed by Severinghaus. At the University of North Carolina we performed the first clinical trials with this instrument which is now used world wide. My greatest financial mistake was not to invest in his company when he made an offer at that time. Instrumentation Laboratories then went on to develop the Co-Oximeter[10] (Figure 7). This instrument, which is widely used, was one of the first to use the visible light spectrum

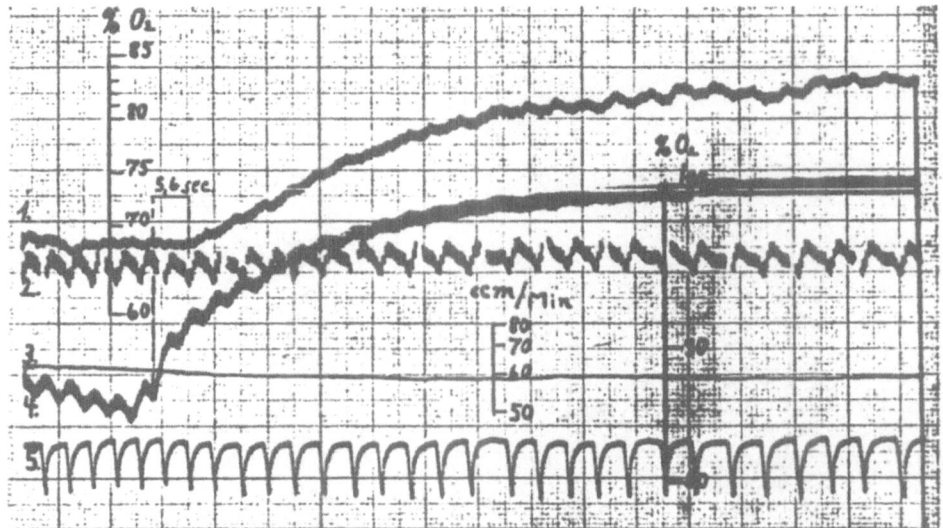

Figure 2. Recording of oximeter output during normoxia and hyperoxia in the experimental animal. 1. changes in saturation in the femoral vein. 2. blood pressure recording. 3. temperature recording. 4. changes in saturation in the femoral artery. 5. resporation.

for analysis of oxygen content and other components of hemolyzed blood. We also performed the early clinical trials with this instrument in our laboratory.

In 1961 I had an unexpected reunion with Kurt Kramer, when I went to Gottingen to work with Hans Loeschcke and found that Kramer was the Professor at that institute. At that time he was attempting to measure the oxygen saturation of the kidney cortex with a modified

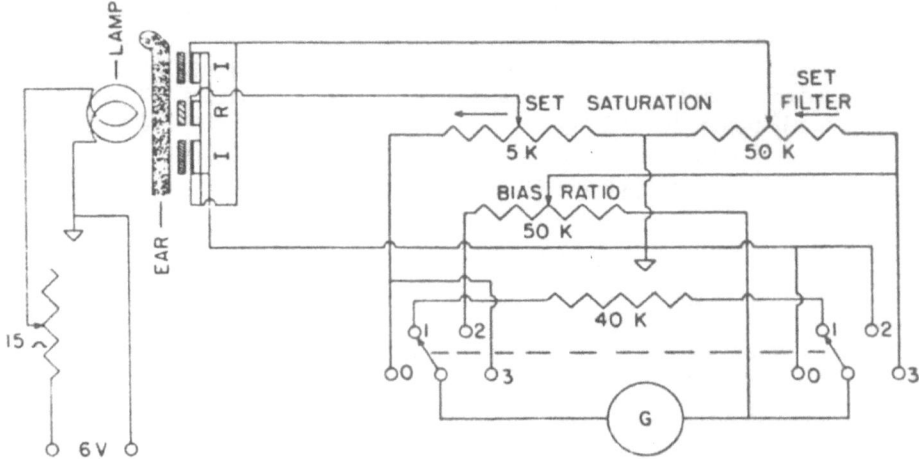

Figure 3. Circuit diagram of Millikan oximeter. G, galvanometer (sensitivity such that 0.0006 μamp produces deflection of 1 mm); R, "red" photoelectric cell covered by Wratten 29F filter; I, "green" photoelectric cell covered by Wratten 61N filter. K = 1000 ohms.

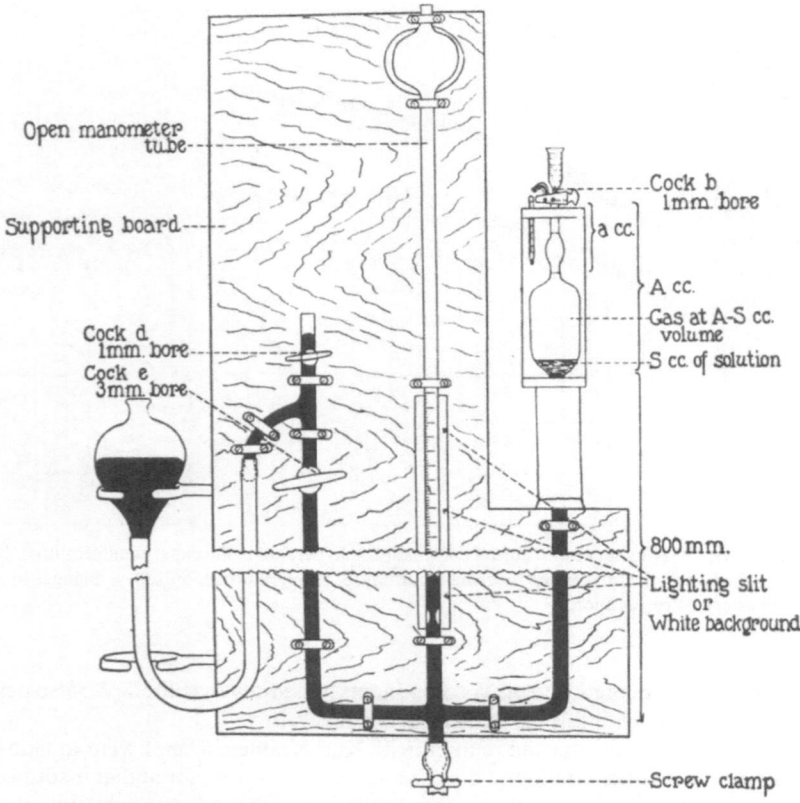

Figure 4. Apparatus, with solution in position for extraction of gases. Manometer os of open type.

oximeter and invited me to work with him. We used the needle electrode, which has been previously described, to measure oxygen tension in the kidney cortex, and a paper was published on this subject in the German literature. At that time, I learned that Kramer, at an early age, had to choose between being a concert pianist and a physician. When I last spoke with him shortly before he died, he informed me that he had just published a book on the life of J.S. Bach. He continued to play the piano, and institute members would frequently gather outside his window when he would play the piano in the evening.

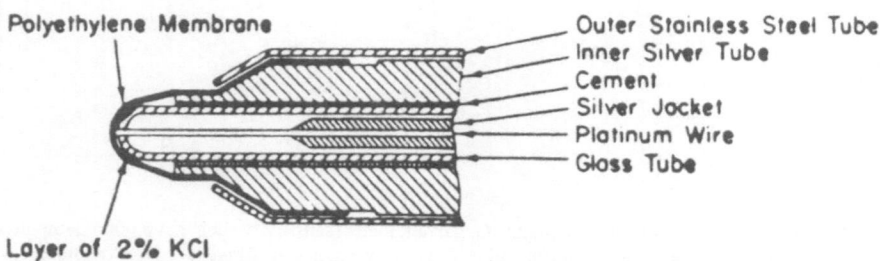

Figure 5. Section of the electrode tip of the oxygen analyzer.

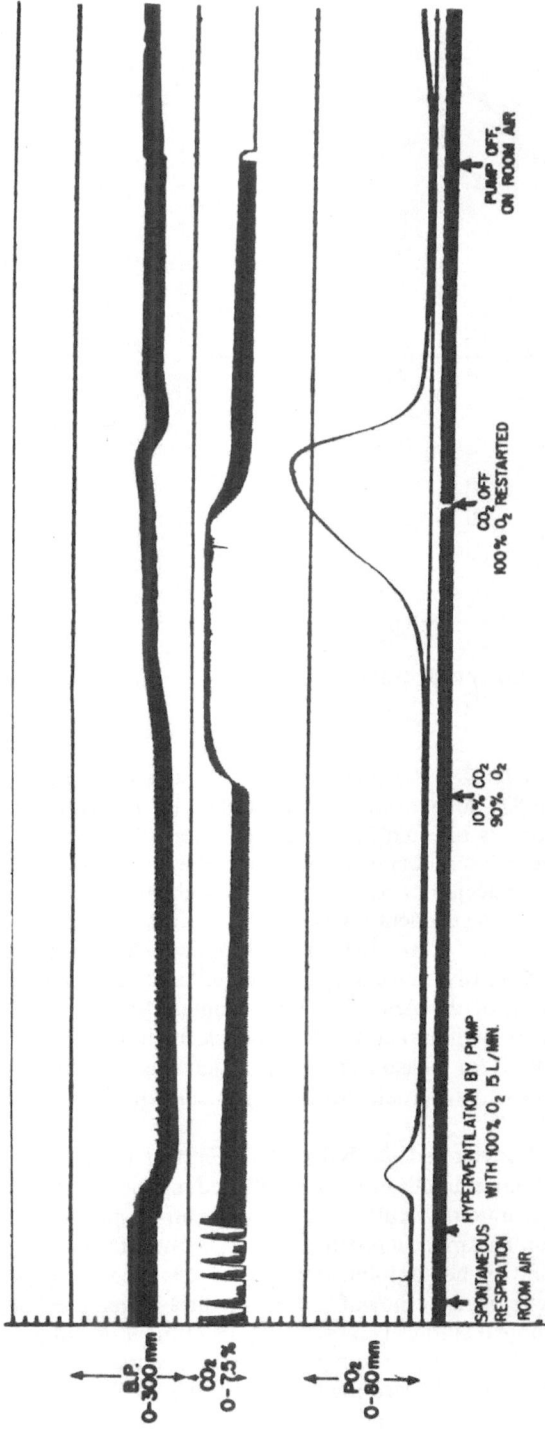

Figure 6. Record taken when an animal was hyperventilated first with 100 per cent oxygen, then with a 90 per cent oxygen 10 per cent carbon dioxide mixture, then with 100 per cent oxygen again.

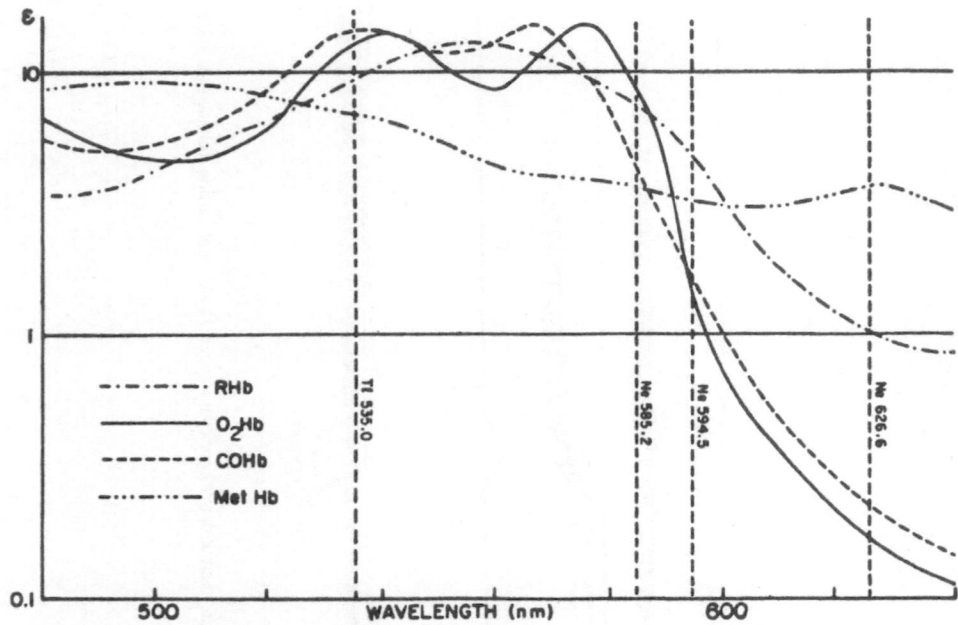

Figure 7. Extinction spectra of RHb, O₂Hb, COHb, and MetHb, with spectral lines superimposed.

Upon my return to the U.S., I was asked by the manufacturers of an instrument called the Lex-O₂-Con (Figure 8)[11] to perform clinical trials with it. With this instrument a 20 microliter sample of blood is injected into distilled water which hemolyses the cells and releases the bound oxygen. A scrubber gas then carries the oxygen to a galvanic cell, where it is reduced. Since each molecule of oxygen requires 4 electrons to reduce it, by counting the number of electrons, the instrument gives a highly accurate measure of oxygen content in the blood sample. The results correlated very well with the Van Slyke method and is one of the most accurate methods of determining the oxygen content of whole blood.

It was at a meeting of the American Physiological Society in 1974 that Dietrich Lubbers invited me to work with him at the Max-Planck Institute in Dortmund. He felt that the needle electrode we used for measuring tissue oxygen was inaccurate and wanted me to use the Lubbers "Multiple Wire Microelectrode". The results of our work with this electrode was published in 1978[12].

Jobsis, in 1977,[13] developed the NIROSCOPE (near infrared oxygen sufficiency scope) which utilized four laser diodes in the near infrared region. (Figure 9) Photons emitted at these wave lengths penetrate the skull sufficiently to permit spectrophotometric analysis of the cerebral gray matter. This made it possible to measure non-invasively and qualitatively, changes in hemoglobin and oxyhemoglobin in the brain. After years of experimentation with animal models, Jobsis developed a clinical instrument and I joined his laboratory where we used the instrument in clinical studies (Figure 10)[14]. Its clinical usefulness is limited by its inability to obtain quantitative values of oxygen in blood and tissue.

In 1984 Frank and Kessler[15] described the use of the EMPHO (Erlangen micro-light guide spectrophotometer) (Figure 11) in the beating dog heart and were able to demonstrate changes in the oxygenation of myocardial tissue. The light source for this instrument is from a high pressure xenon lamp. The light is transmitted to the tissue through a fiber optic light

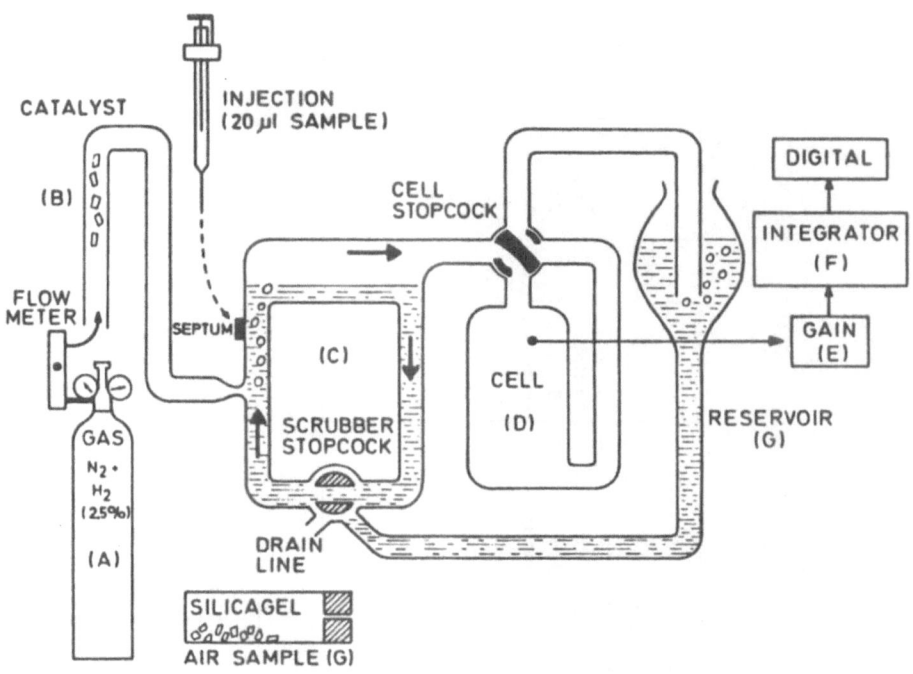

Figure 8. Diagram of the equipment: *A*, a cylinder (1 to 3 per cent hydrogen in nitrogen); *B*, a catalyst; *C*, a glass circuit containing distilled water through which the gas mixture passes continuously; *D*, an oxygen-sensitive galvanic cell; *E*, an ajustable amplification system; *F*, an integrator with a digital display ; and *G*, a tank filled with distilled water free of oxygen and a tube containing a dessicating agent (silicagel).

guide, and the reflected light collected by several fiber optic bundles. The collected light is then subjected to spectral analysis by the use of special filter discs turned at high speed by a micromotor. The various spectra in the visible light range can be analyzed at the rate of 70 or more per second, and 64 points per revolution of the disc can be plotted on each spectral curve. As a result, detailed analysis of the spectrum of oxy and reduced hemoglobin can be performed. This instrument, combined with high speed computers permits on line, near quantitative, analysis of oxygenation in the intact tissue. It is possible to calibrate this instrument in vitro using blood-containing cuvettes. By using this method of calibration, combined with a pattern recognition system of oxygenation in various tissues, it is possible to obtain near quantitative values of oxygenation in some clinical situations.

Some of us working in Kessler's laboratory attempted to correlate the EMPHO signals with data acquired from a pulse oximeter, a transcutaneous oxygen electrode, and arterial blood samples. As expected, the correlations were not precise but were certainly in line with what we expected the EMPHO to demonstrate. That is, the arterial pO_2 and the transcutaneous pO_2 bore no correlation to the EMPHO data, but the pulse oximeter did roughly correlate with the EMPHO (Figure 12). The work was done on 5 normal volunteers, using the skin of the back of the hand, which is subject to wide thermal and neural influences on changes in blood flow. In studies carried out on exposed tissue, such as that of the liver or of the heart, during surgical procedures, the correlation between readings from the EMPHO and oxygen content as analyzed by established methods were quite close. It is extremely difficult to calibrate in absolute quantitative terms the EMPHO or any other

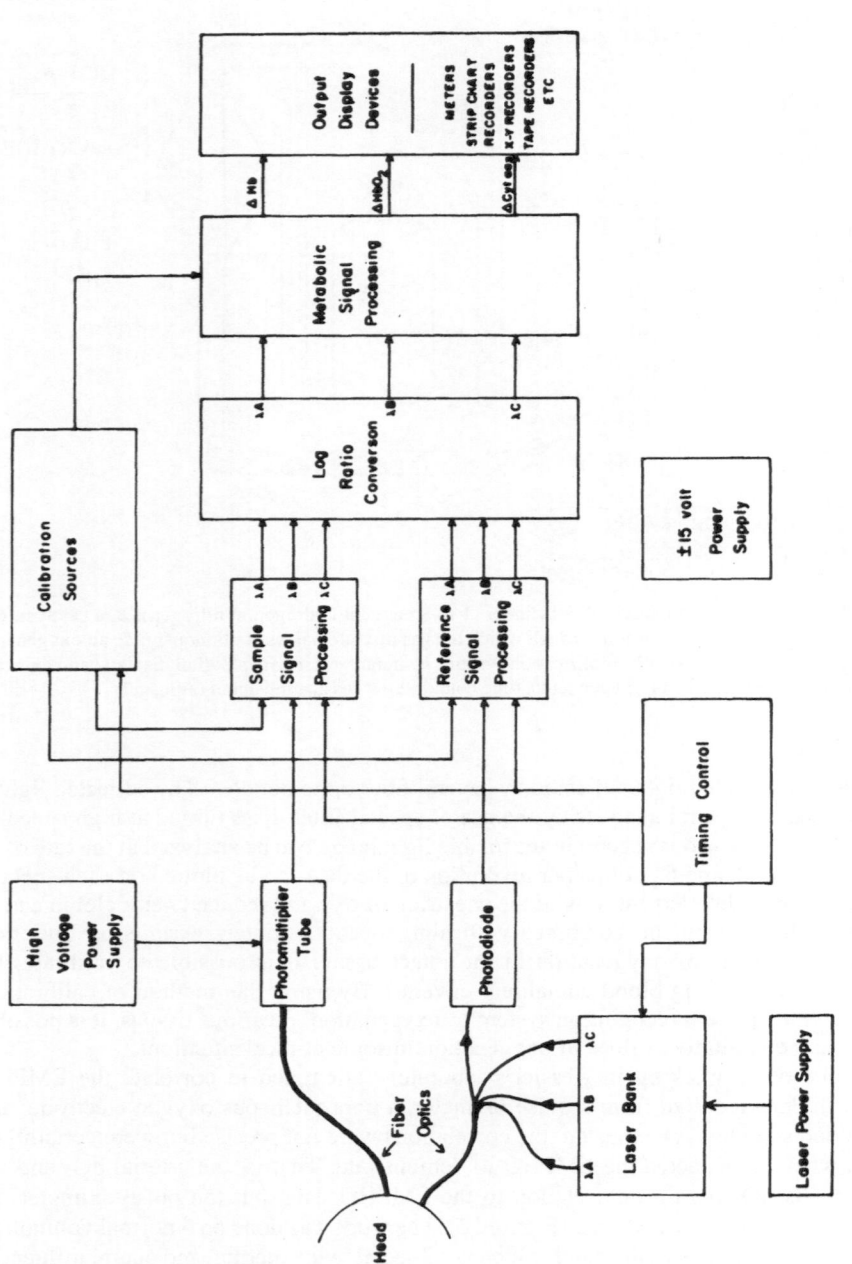

Figure 9. A schematic of NIROSCOPE.

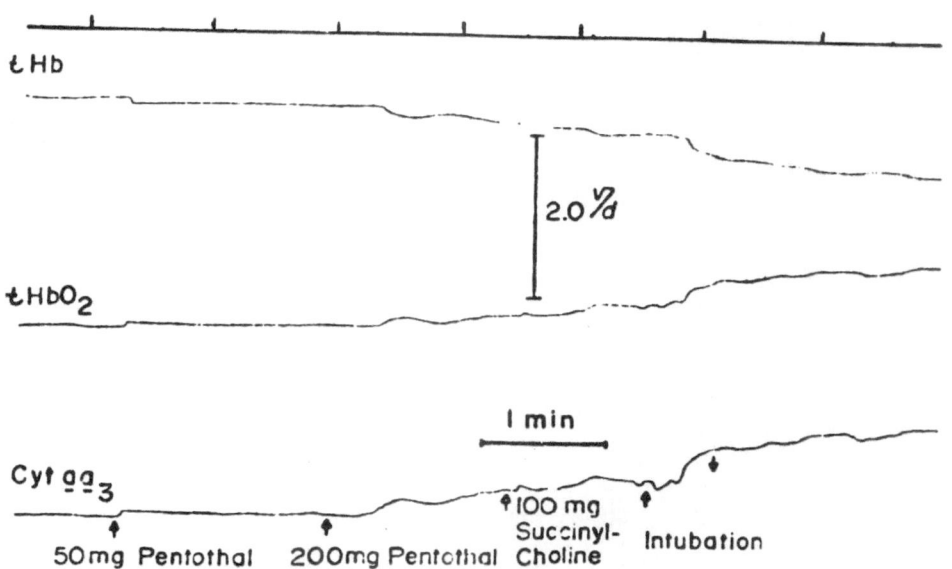

Figure 10. Induction of anesthesia with pentothal. After the 50 mg. test dose, the sleepdose (200 mg) produces a marked oxidation of cytochrome c oxidase and a shift toward more arterially colored blood within the regional, cerebral circulation. Intubation and ventilation with 100% O_2 produced a further oxidaton of cyt aa₃ and a shift of the microcirculation toward greater oxygenation.

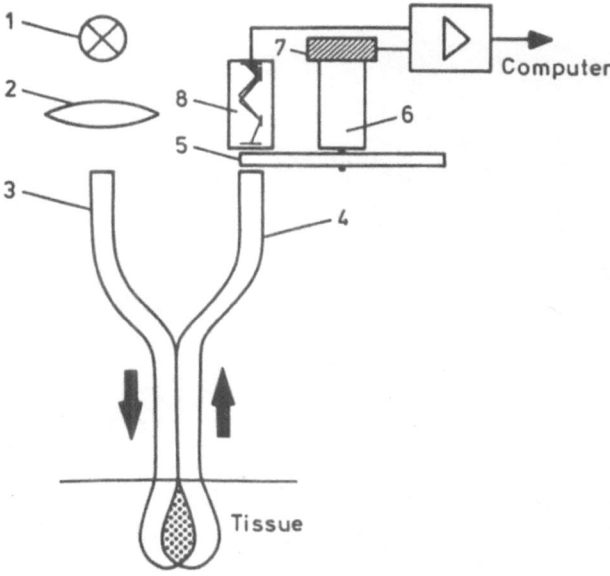

Figure 11. Schematic drawing of the EMPHO I. 1. Xenon high-pressure arc lamp. 2. Lens system. 3. Illuminating micro-lightguide fiber. 4. Detecting micro-lightguide. 5. Interference filter disk.6. Driving micromotor. 7. Decoder wheel. 8. Photomultiplier tube.

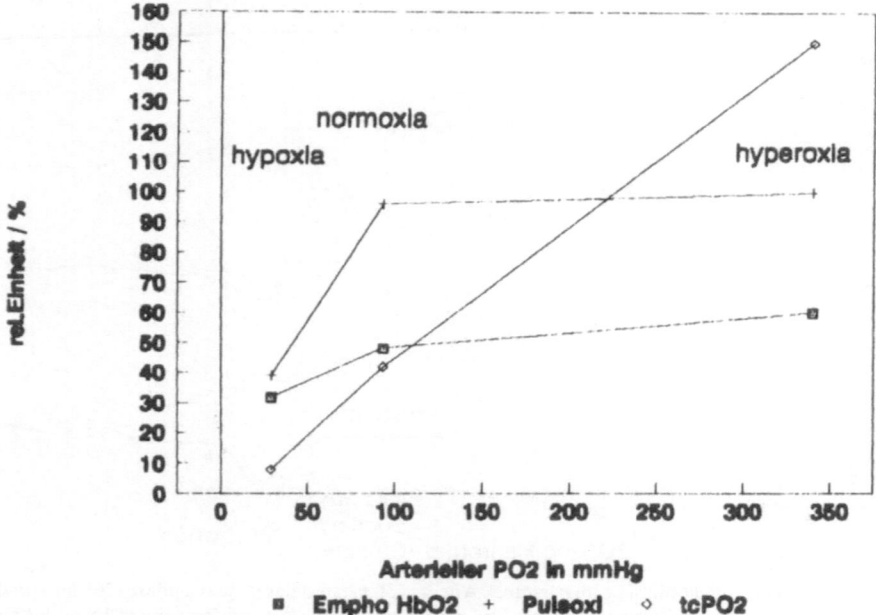

Figure 12. Recordongs of simultaneous measurements of oxygenation in a human subject during normoxia and hyperoxia using a transcutaneous electrode, a pulse oximeter, and the EMPHO.

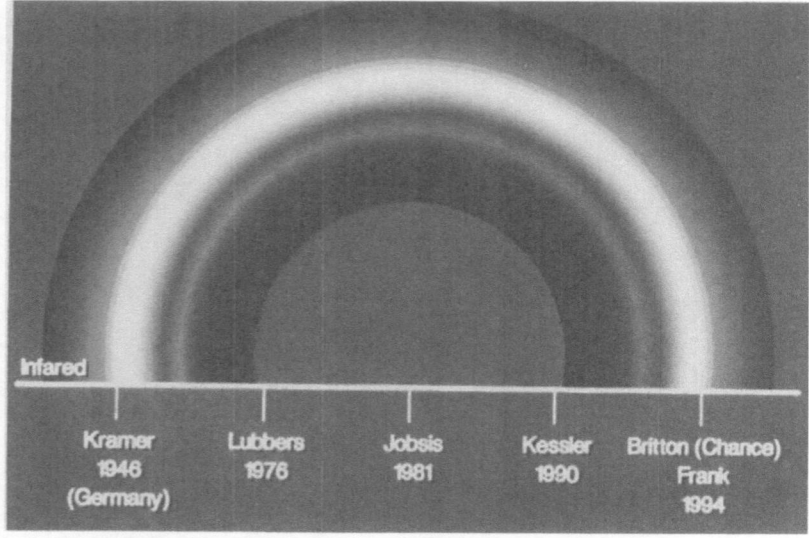

Figure 13. A rainbow of spectra from infrared to violet light with the names and dates of investigators with whom the author collaborated.

spectrophotometer when dealing with living tissue or blood. However, with increasingly sophisticated techniques of blood and tissue sampling and analysis, especially the use of phase modulation spectroscopy currently being developed by Britton Chance and Klaus Frank, quantitative analysis of tissue oxygen content seems highly possible.

In summary, I have had the privilege of working in the laboratories of the men listed in Figure 13 and this figure illustrates the spectrum of my almost 50 years of experiences in oxygen measurement. "The rainbow" of oxygen measurement will be, I am sure, extended until a near perfect method is developed. This would be analogous to finding the proverbial "pot of gold" at the end of the rainbow.

REFERENCES

1. Danneel, H.L. Uber den durch diffundierende Gase her vorqerufenen Restom. Z Electrochemie 1897/1898; 4: 227-242.
2. Kramer, K Ein Verfahren zur fortlaufenden Messung des Saverstoffgehaltes in stromenden Blute an uneroffneten Gefassen. Z fur Biologie 1935, 96: 61-75.
3. Millikan GA, Pappen heimer JR, Rawson AS, Hervey, J Continuous measurement of oxygen saturation in man. American Journal of Physiology 133:390, 1941.
4. Elam Jo, Neville JF, Sleator W, Elam WN Sources of error in oximetry. Ann. Surg. 130:755-773, 1949.
5. Var Slyke DD, Neill JM The determination of gases in blood and other solutions by vacum extraction and manometric measurement. J. Biol chem 61:523-573, 1924.
6. Peters J.P., et. al Studies of gas and electrolyte equilibria in blood. Technique for collection and analysis of blood for its saturation with gas mixture, of known composition. J. Biol Chem 54: 121-147, 1922.
7. Clark LC, Wolf R, Granger D, Taylor Z Continuous recording of blood oxygen tensions by polarography. J. Appl. Physiol G: 189-193, 1953.
8. Kety SS, Schmidt CF Effects of alterations in arterial tension of carbon dioxide and oxygen on cerebral blood flow and cerebral oxygen consumption of normal young men. Fed. Proc. 5:55, 1946.
9. Sugioka K, Davis D.A. Hyperventilation with oxygen - a possible cause of cerebral hypoxia. Anesthesiology 21: 135-143, 1960.
10. Brown, L.J. A new instrument for the simultaneous measurement of total hemoglobin, 70 oxyhemoglobin, % Carboxyhemoglobin % Methmoglobin, and oxygen content in whole blood. IEE, Trans. Biomed. Eng. 3: 132-138, 1978.
11. Clerbaux, G. Gerets G., Frans A. Oxygen content determination using a new analyzer: The Lex 02 - Con. J. Lab. Clin. Med. 82:342-348, 1973.
12. Sugioka K, Lubbers DW: Effect of changes in PaC_2 on the blood microflow and on PO_2 in liver tissue. Fed Proc 37:851, 1978.
13. Jobsis, F.F. Non invasive monitoring of cerebral and myocardial oxygen sufficiency and circulatory parameter. Science198. 1264-1267, 1977.
14. Jobsis, F.F., Fox E., Sugioka, K. Monitoring of cerebral oxygenation and cytochrome aa_3 redox state. Internat. Anes. Clinics 19:85-122, 1981.
15. Frank K.H., Kessler M., Friedl A., Brunner M., Ellerman R., Kerl G., and Hoper J. Measurements of intracapillary hemoglobin spectra in the beating heart, skeletal muscle and the liver, using the Erlangen micro-light guide spectrophotometer. Pfluger's Arch. ges Phsiol. 400 (suppl 219) R55, 1988.

2

OXYGEN ELECTRODES AND OPTODES AND THEIR APPLICATION *IN VIVO*

D. W. Lübbers

Max Planck Institut für Molekulare Physiologie
Rheinlanddamm 201
44139 Dortmund
Germany

INTRODUCTION

In the 1940s it was well known that the transport of oxygen within the tissue occurs by diffusion. Since by diffusional transport oxygen pressure gradients develop, the state of tissue oxygen supply can be characterized by the distribution of local pO_2. At that time, however, it was only possible to measure mean tissue pO_2. This was done by introducing e.g. a small gas bubble into the tissue[1]. After equilibration the bubble was withdrawn and analyzed by gas analysis. It was tried to understand these measurements by applying the cylindrical tissue model of Krogh[2,3], but these measurements did not help much to understand the physiological presuppositions of anoxia or hypoxia because local tissue measurements were missing. The situation was changed when in 1942 P.W. Davies and F. Brink published their paper on "Microelectrodes for Measuring Local Oxygen Tension in Animal Tissue"[4]. This work was done in the laboratory of D.W. Bronk at the Johnson Foundation in Philadelphia, USA. The authors describe in great detail manufacturing and application of polarographic pO_2 electrodes with a sensor surface of only 25 µm. Such a small sensor surface was needed because their aims were to measure "variations of oxygen tensions over small distances in tissues" to locate "sites of oxygen consumption by mapping concentration gradients" or "to measure the oxygen tension in blood of superficial arterioles and venules of the cat cerebral cortex, in the cortical substance itself, at the surface of muscle cells and at various distances from the surface of single celled organs". About the development of this sensor Roseman, Goodwin and McCulloch mention in a footnote of their paper,[5] that qualitative and quantitative polarographic analysis "was first employed in Dr. Bronk's laboratory. By this method Brink and Davies determined the metabolism of the excised nerve. But it is another of Dr. Brink's collaborators, Dr.

Oxygen Transport to Tissue XVII, Edited by Ince et al.
Plenum Press, New York, 1996

13

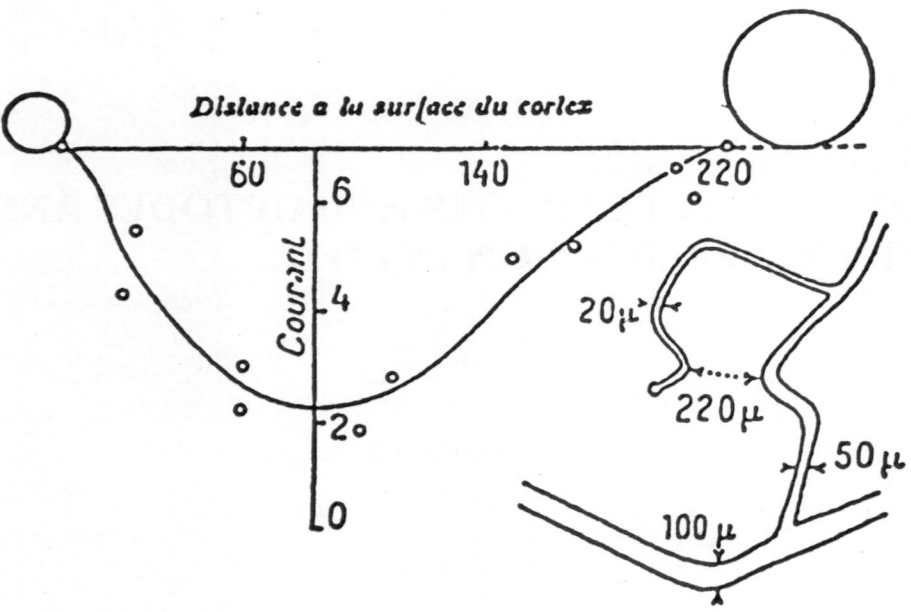

Figure 1. pO₂ profiles on the cortical surface of the cat recorded in 1948 by A. Rémond[8].

Glenn Millikan, who thrust the bare Pt electrode into the cortex to determine the O_2 tension *in vivo*, that the authors are immediately indebted, for it was he who suggested the method to Dr. Nims and to the two present authors". Glenn Millikan had studied in the 1930s the O_2 supply of the skeletal muscle by measuring continuously the O_2 saturation of myoglobin claiming that the oxygenation of myoglobin can be used as an indicator of tissue metabolism and of tissue oxygen tension[6]. Neverthless, the above mentioned statements of Davies and Brink prove clearly that at this time the authors were aware of possible applications and of the importance of the new pO_2 microsensor to study tissue oxygen supply.

In their experiments Davies and Brink demonstrated that the open bare platinum microcathode fused in glass showed in NaCl solutions and in dead tissues a sufficiently good polarographic plateau (see below). To obtain more stable diffusion conditions they constructed a microelectrode with a recess as proposed by Laitinen and Kolthoff[7]. The recess could be filled with agar and covered by a collodion membrane. For absolute measurements a pulsing technique was used so that the length of the pulse was adapted to the length of the recess. Such a collodion protected recessed electrode had a drift of only 2 % in 2 hours. Rémond[9] who had worked together with Davies and Bronk was the first who published 1948 quantitative graphic presentations of pO_2 profiles on the brain surface. Fig.1 shows such pO_2 profiles between a small artery and an arteriole. Between the two vessels a clear pO_2 minimum is formed demonstrating that individual O_2 supply zones exist within the tissue. It was estimated that the pO_2 in the artery amounted to ca. 95 Torr and the pO_2 at the minimum to about 30 Torr. The pO_2 profiles could be influenced by respiration of oxygen and by flow stop. They demonstrated that also veins can supply the tissue with oxygen. These measurements can be considered as the basis of all later measurements of the Oxygen Transport to Tissue, the last three letters "OTT" of the name of our society, ISOTT.

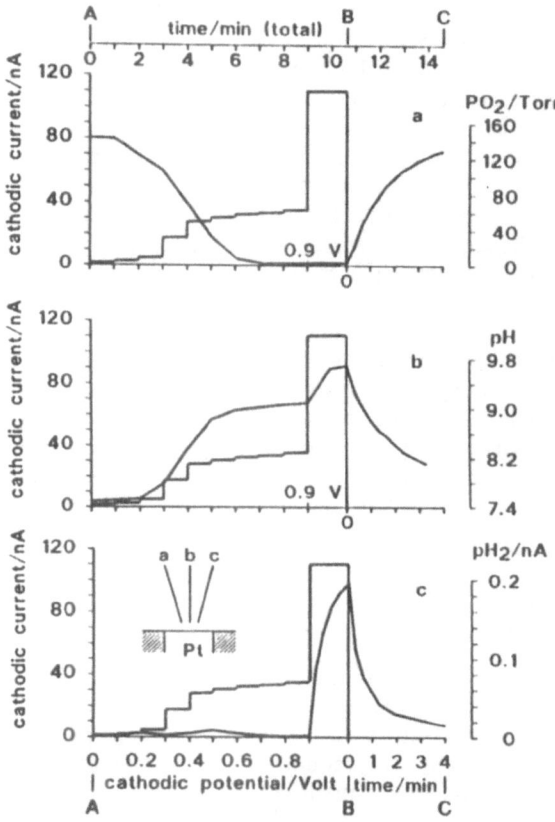

Figure 2. Polarogram of a stationary polarographic pO_2 electrode. The cathodic reduction current (left ordinate) increases with the stepwise increase of the negative cathodic potential up to 0.5 V, but then the polarographic plateau is formed. Under these conditions the pO_2 at the cathode surface (measuring setup see insert in c) becomes zero showing that each O_2 molecule reaching the cathode becomes reduced (a, right ordinate). The polarographic reaction produces OH^- ions changing the local pH (b, right ordinate). At a potential above 0.9 V H_2 is generated (c, right ordinate). Data reconstructed from[16].

MEASURING METHODS

1. Polarographic Oxygen Sensors

Basic Phenomena. In a polarographic sensor molecular oxygen is reduced at the surface of a cathode by applying a negative voltage between cathode and anode[9-15]. In neutral or alkaline electrolytes the following overall reaction occurs

at the cathode $O_2 + 4\ e^- + 2H_2O = 4\ OH^-$
and at the anode $4\ Ag + 4\ Cl^- = 4\ AgCl + 4\ e^-$

i. e. the reduction current e^- is proportional to the number of oxygen molecules being reduced. The polarographic reaction can be tested by measuring the so-called polarogram[16]. First, by increasing the negative voltage an increase of the reduction current is produced, but then inspite of a further voltage increase the reduction current remains constant and a plateau is formed. This is demonstrated in Fig.2.

By a stepwise increase of the cathodic potential from 0.0 - 0.5 V the cathodic reduction current increases from zero to 30 nA, but the further increase of the potential up to 0.9 V results in an increase of the current of only 5 nA, i.e. in the polarographic plateau. During this reaction the local pO_2 (Fig.2a) as well as the local pH (Fig.2b) and the local pH_2 is measured in front of the platinum cathode and close to its surface as is shown in the insert (Fig.2c). With increasing reduction current the local pO_2 starts to decrease. Within the plateau the negative voltage becomes high enough to reduce each oxygen molecule reaching the cathode so that the O_2 concentration at the cathode surface becomes zero. At the same time OH^- ions are produced changing the local pH in front of the cathode. If the potential increases above 0.9 V hydrogen is developed and the local pH_2 increases. To establish a fixed relationship between the oxygen concentration of the solution and the reduction current, defined conditions for the oxygen transport to the cathode must exist. This can be achieved by putting an oxygen permeable membrane in front of the cathode[4]. To avoid side reactions and to improve selectivity Leland Clark[17] made the very important proposal to place cathode as well as anode behind the same membrane. In the Clark electrode a suitable membrane achieves:

1. separation of the measuring space in front of the electrode from the medium to be measured,
2. defined transport of oxygen between medium and cathode, and
3. improved selectivity for the analyte oxygen.

The diffusive O_2 transport through a membrane is determined by its permeability and the pO_2 difference across the membrane. Since within the polarographic plateau the pO_2 at the cathode surface is zero (see Fig.2a), the pO_2 of the medium at the membrane surface determines the reduction current. Thus, such a sensor measures directly the oxygen pressure of a medium; to obtain its oxygen concentration $[O_2]$ the measured pressure has to be multipied by the oxygen solubility coefficient α: $[O]_2$ $= \alpha \cdot pO_2$. However, since during continuous measurements oxygen is steadily consumed and has to be supplied from the medium to be measured, the pO_2 at the membrane surface depends on the oxygen transport within the medium[18]. Therefore, during continuous measurements the influence of the medium can not be eliminated. It can only be minimized, for example by choosing a membrane having a much smaller oxygen conductivity than the medium. A complete elimination is only possible in non-continuous measurements. For such measurements pulses of polarizing voltage are used[4,7,19,20]. Using sufficiently short pulses it can be achieved that the expanding diffusion field remains in a space of defined diffusion conditions, for example within the membrane or within a defined unstirred layer of the solution. In this case the same relationship between reduction current and pO_2, i.e. the same calibration curve of the electrode holds for measurements in different media.

Due to technical developments it is now possible to construct continuously measuring polarographic pO_2 electrodes using the Clark principle by which reliable pO_2 values can be measured supposed that calibration was done in the medium to be measured. They show a linear relationship between pO_2 and reduction current and only a small drift. Technical developments and measuring results up to 1976 were reviewed by I. Fatt[11]. There are many publications about construction and application of microelectrodes. Already in 1980 F. Kreuzer[12] listed in a review more than 100 technical papers about pO_2 microelectrodes used in tissue measurements. Therefore, it is impossible to describe all the different electrodes in this contribution in detail. However, there are still methodological limitations which have to be considered to obtain reliable results. In the following we will discuss these limitations and some special problems which arise with the construction of microelectrodes and their application to measure the distribution of local pO_2 within tissue.

a) Material, Size and Insulation of the Cathode. For the polarographic cathode a material should be selected which guarantees a surface at which a homogeneous and constant reduction can occur. The predominant cathode material used to construct polarographic microelectrodes is platinum[4,12,21-23], but other materials as gold or silver are also proposed[12,23-25]. Platinum wires can be etched in a conical shape with a tip size down to only 0.1 µm[22]. Such small cathode surfaces can only be obtained with a reasonable output if physically pure platinum is used; spectroscopically pure platinum did not give better results, probably because the remaining impurities are not evenly distributed, but form clusters. To insulate the platinum wire down to the tip melting glass is available with the proper expansion coefficient and sufficient electrical resistance. Additional dc-polishing of the wire produces a smooth surface which is needed to guarantee a tight contact to the insulating glass. The tip of the electrode is cleaned and polished. It can be ground to facilitate puncturing of tissue. To produce a recess the platinum wire is etched[23]. Gold cathodes show a rather stable and broad polarogram[24-27], platinum[22] or metal alloy filled in a micropipette[28] has been plated with gold.

b) Space in Front of the Cathode. Since during O_2 reduction OH^- ions are steadily produced the pH in the space in front of the cathode is changing (see Fig.2b). These changes mostly do not affect the electrode reaction, but if this space is not completely separated OH^- ions may penetrate into the tissue and disturb the measurements. There are many different ways to membranize the tip of a microelectrode and many different materials have been proposed, but if the tip size becomes smaller than about 5 µm then it can not be excluded that such a thin membranization is incomplete[22]. Furthermore, the thin membrane is often damaged if the tip is inserted into the tissue. Table 1 demonstrates that the polarographic reduction current of a bare, circular cathode (diameter d_i = 20 µm) varies considerably in different media; e.g. using for evaluation of measurements in muscle (3.2 nA/20 µm d_i) the calibration curve obtained in saline (9.5 nA/20 µm d_i) the evaluated pO_2 values would be

Table 1. Polarographic reduction current of a bare, flat, circular cathode in different media[23]

Medium	Polarographic reduction current of a cathode with a diameter of 20 µm at 150 Torr and 37°C (nA)
Water	9.8
Saline	9.5
1 mol KCL	7.5
Serum	7.3
Plasma	6.3
Brain	6.1
2 mol KCL	5.2
Heart	4.9
3 mol KCL	4.3
Muscle	3.2
Saturated KCL	2.4

d_i = diameter of the electrode.

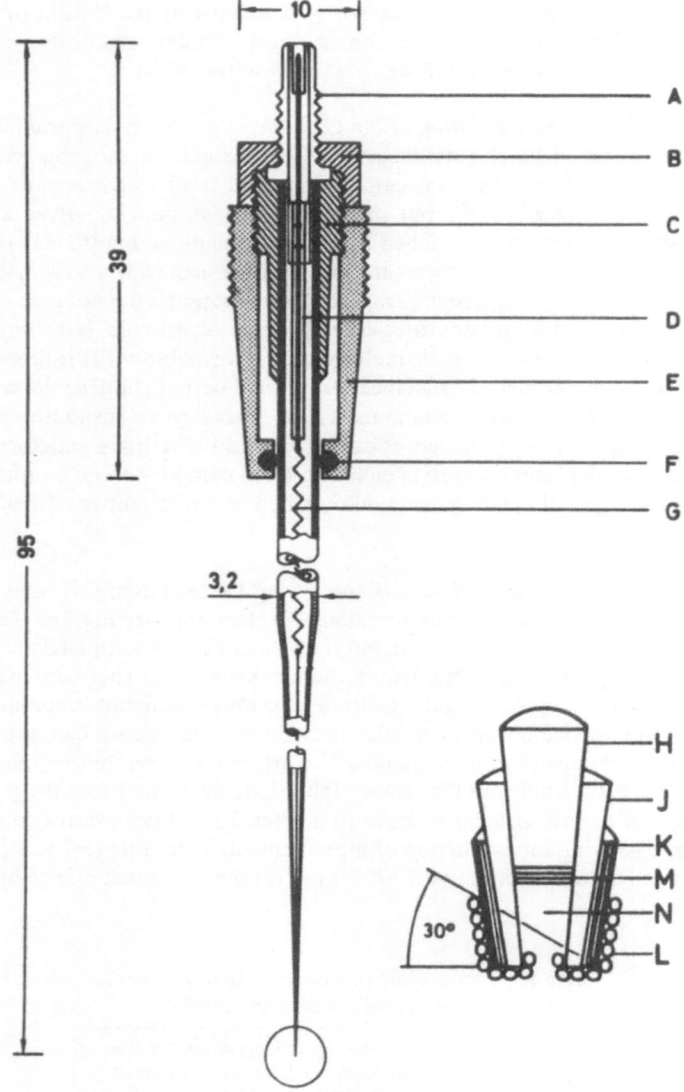

Figure 3. Schematic drawing of the pO$_2$ microcoaxial needle electrode. A = HF chassis socket, B = fixing nut, C = metallic contact to reference electrode, D = polyethylene coated Cu wire, E = exchangeable lucite housing in NS-10 form, F = siliconized O-ring, G = metallic contact to the Pt wire, H = Pt wire etched by ac-current, J = glass insulation, K = sputtered three-layered reference electrode (Ta/Pt/Ag), L = oxygen permeable membrane, M = electrochemically polished Pt surface coated galvanically with Au, N = recess[22]

ca. 3 times too small. Since the calibration curve of the electrode changes if the properties of the membrane change, erroneous measurements may result. These problems can be partly overcome if recessed electrodes are used. The recess forms a space of a defined size. It can be filled with a suitable material and its wall serves as a support for the membrane[4,22,29].

c) Reference Electrodes. Mostly Ag/AgCl electrodes are used as reference electrodes (e.g.[11-15]). They are tolerated by the tissue and can be miniaturized. However, if the reference electrode is separated from the pO_2 electrode the reduction current has to pass the tissue and depends on its properties. Using microelectrodes this is especially dangerous because of their small reduction currents. Therefore, a technique has been developed by which the reference electrode is placed directly on the shaft of the microelectrode as close as possible to the cathode. Thin silver layers alone can not be used in a medium containing chloride ions since such layers become completely chloridized in a rather short time. Systematic experiments showed that this can be avoided by applying a thin platinum layer under the silver layer. Then, a part of the silver remains unchanged so that a stable Ag/AgCl potential is obtained[22]. To produce the thin layers on the glass shaft we use the RF sputtering technique. To achieve a good sticking of the layers to the glass surface, first a thin tantalum layer (only 8 nm thick!) is sputtered, followed by a layer of platinum (ca. 30 nm thick) and finally by a layer of silver (ca. 62 nm thick). The reference electrode is isolated by sputtered dielectric layers (e.g. SiO_2) only allowing a small area at the tip to have contact with the medium to be measured[22]. At high pO_2 values reference electrodes of silver oxide may be of advantage[30]. In Fig. 3 a schematic drawing of the microelectrode is shown which was developed in our laboratory together with H. Baumgärtl and was used in most of our experiments[22,23]. Puncturing can be facilitated by grinding the tip for example in an angle of 30^0.

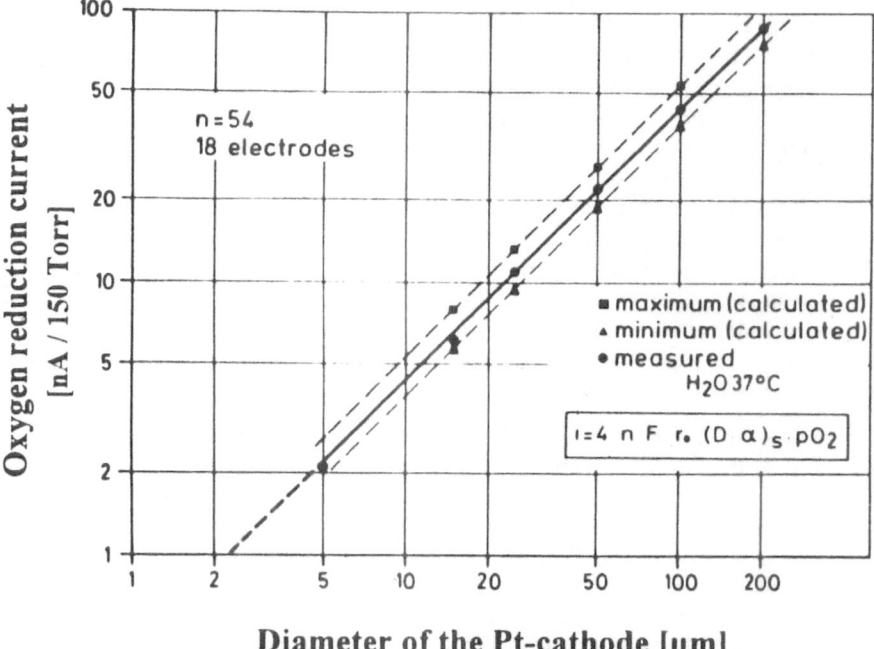

Figure 4. Relationship between polarographic O_2 reduction current and diameter of the cathode. Using a double logarithmic scale there is a linear relationship between the polarographic O_2 reduction current (measured in slightly alkalic water at 37^0 C) and the diameter of a bare, flat, circular cathode. The maximum and minimum values are calculated using $(D \cdot \alpha)$ values reported in the literature[23]

Test of the Electrodes. To obtain reliable results by using a microelectrode the function of each electrode should be carefully tested.

1. Test of the polarogram. Test measurements should be performed in air saturated saline. Electrodes with a diameter of the cathode smaller than 20 µm rarely show a plateau of the current parallel to the abcissa, but there must be a clearly reduced increase of the reduction current, i.e. a suffiently flat polarogram when the negative polarization voltage becomes larger than -550 mV (see Fig. 2). If, after a certain testing time, such a polarogram is not observed the cathode has to be cleaned, electrochemically polished or etched to obtain a suitable surface. As already mentioned the polarogram can often be improved by plating the platinum cathode with a thin layer of gold[22].

2. Functional test of glass insulation. The glass insulation can be tested by comparing the reduction current measured in a N_2 saturated solution to the current in an air saturated one. The current in the N_2 saturated solution should be less than 5 % of the other current, otherwise it is highly probable that the glass insulation is insufficient (microcracks, splits in the fusion zone between platinum and glass). Such electrodes are unstable and should not be used.

3. Test of the the reduction current and of the linearity of the calibration curve. According to the Faraday law the reduction current of a flat, circular cathode is given by

$$i = 2 \cdot n \cdot F \cdot d_i \cdot (D \cdot \alpha) \cdot pO_2 \tag{1}$$

where n = number of involved electrons, F = Faraday constant, d_i = diameter of the cathode, D = diffusion coefficient, α = solubility coefficient.

Varying the radius between 2 and 200 µm it was experimentally found that in a double logarithmic scale there is a linear relationship between the reduction current at a fixed pO_2 (e.g. air = 150 Torr) and the diameter of the cathode (Fig. 4). The reduction current of a flat, bare, circular cathode normalized for a diameter of 20 µm and measured in water after addition of a small amount of NaOH at 37°C amounts to a value between $8.9 \cdot 10^{-9}$ and $9.8 \cdot 10^{-9}$ A / (150 Torr $\cdot$ 20 µm) corresponding to the range of values of oxygen conductivity $c_v = (D \cdot \alpha)$ reported in the literature. Considering the actual $(D \cdot \alpha)$ the measuring current of an electrode should not exceed this value.[23] Higher values could mean that not only oxygen, but also other substances are being reduced falsifying the results. A properly constructed pO_2 microelectrode has a linear calibration curve (Fig. 5)[22].

4. Test of the membranization. Because of the small dimensions a microscopic test is insufficient. Membranization can be tested by measuring the change of reduction current in a solution with and without convection. Flow sensitivity (stirring effect) is usually tested during the calibration by measuring the signal change which occurs with and without a strong stream of gas bubbles. With a proper membranization, the stirring effect should be smaller than 5%[22,31]. Larger values are obtained when the membrane is relatively thin or incomplete. The intactness of the membrane can be tested by adding substances to the test solution which change the polarographic reduction current if they reach the cathode. As test substances we use e.g. bicarbonate, magnesium and calcium ions or cystein[22,23]. The response time of the electrode t_{90} is mainly determined by the thickness d_m and the diffusion coefficient D_m of the membrane $t_{90} = [(0.3 \cdot d_m^2) / D_m]$[23].

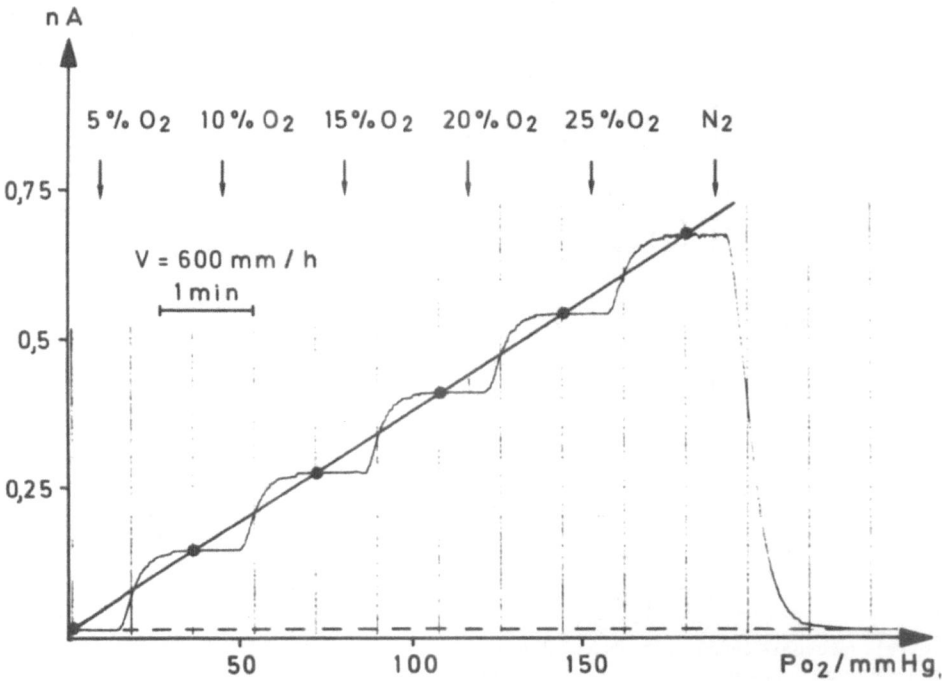

Figure 5. Calibration curve of a pO_2 microelectrode. Plotting the reduction current (ordinate) vs. O_2 partial pressure (abcissa) results in a linear calibration curve. V = recording speed. Arrow = start of a new gas mixture.

2. Oxygen Sensors Using Optical Indicators

Basic Phenomena. For optical monitoring of oxygen two different optical reactions are used: absorbance and luminescence[32,33]; luminescence comprises fluorescence and phosphorescence. In both cases the indicator molecule absorbs parts of the incident light so that by measuring the intensity of the absorbed light at different wavelengths an absorbance spectrum is obtained by which the state of the indicator molecule can be characterized. According to Lambert-Beer's law the absorbance (A) of a clear indicator solution is given by

$$A = -\log (I / I_0) = \varepsilon \cdot C \cdot L \tag{2}$$

where I_0 = intensity of incident light, I = intensity of the transmitted light, ε = molar extinction coefficient, C = molar concentration of the indicator, L = optical pathlength.

The system is not linear since there is a logarithmic dependence between the measured intensity I and the changes of the concentration C of the indicator.

An absorbance indicator changes its absorbance spectrum by binding oxygen. For example, in a hemoglobin solution equilibrated by an oxygen containing gas mixture there is the following equilibrium

$$([Hb] \cdot [O_2]) / [HbO_2] = K_D$$

where K_D = equilibrium constant, [ ] = molar concentration

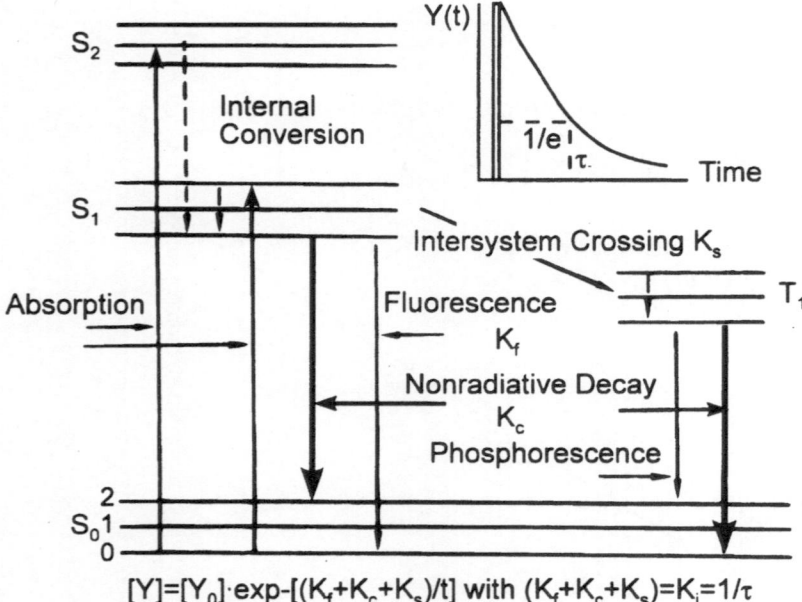

$$[Y]=[Y_0]\cdot exp-[(K_f+K_c+K_s)/t] \text{ with } (K_f+K_c+K_s)=K_i=1/\tau$$

Figure 6. Jablonski diagram. Molecular energy levels after absorption of light. The energy levels are not stable, but decay with different rate constants: K_f = energy loss by emission of a photon: fluorescence, K_c = non radiative energy loss, K_s = transition from the excited singlet state S to the excited triplet state T. $K_i = K_f+K_c+K_s$. Insert: curve of the decay of fluorescence intensity after a light pulse describable by the mean lifetime τ. (see text for details).

There are some wavelengths at which the absorbance changes are large, whereas at other so-called isosbestic wavelengths the absorbance remains constant. The relative oxygenation of hemoglobin, the oxygen saturation $HbSO_2 = [HbO_2] / ([HbO_2] + [Hb])$ is obtained by measuring the total concentration of hemoglobin ($[HbO_2] + [Hb]$) at an isosbestic wavelength and $[HbO_2]$ e.g. at a maximum of the absorbance spectrum. The evaluation method using two wavelengths already introduced in the 1930s by Matthes[34] is applied very often. In transparent and not to strongly scattering media it gives reliable results.

A luminescence indicator does not form a chemical compound with oxygen. By absorbing light the electronic ground state of the luminescence indicator changes to different excited singlet states which have a higher energy level than the ground state as it is shown in the Jablonsky diagram (Fig.6)[33,35-37]. The electronic states can have different vibrational levels (0, 1, 2) of relatively small energy differences (arrows with broken lines) whereas energy differences between excited singlet states and ground state are much larger (arrows with solid lines). The excited singlet states are not stable, but they decay by losing their energy. The concentration of the excited molecules [Y] decreases exponentially in time with the rate constant K_i

$$[Y] = [Y]_0 \cdot exp(-K_i \cdot \tau) \tag{3}$$

The reciprocal value of K_i denotes the time at which [Y] has decreased to $1/e$ of $[Y]_0$, the so-called mean lifetime of the excited state, τ. Absorption and emission. The energy loss can occur by three different ways described by three different rate constants: 1.) by an emission of a photon, i.e. by fluorescence (K_f), 2.) by a non-radiative energy loss by strong collisions (K_c) and/or 3.) by an energy loss from the excited singlet state to an excited triplet state of lower energy (intersystem crossing), (K_s). The emission of photons occurring by the transition from the excited triplet state to the ground state is called phosphorescence. The overall transition from the excited singlet state to the ground state is decribed by the sum of the three rate constants $K_i = (K_f + K_c + K_s)$ and by the mean lifetime $\tau_0 = 1 / (K_f + K_c + K_s)$. Absorption and emission are symmetric processes dependent on the structure of the molecule. Therefore, excitation and emission spectra are symmetric in appearance, but the emission spectrum is shifted to longer wavelengths (red shift, Stoke's shift).

Oxygen influences this sum because the non-radiative energy loss (thick arrows) increases by collisions of oxygen molecules with indicator molecules[38]. Therefore the mean lifetime becomes shorter and since [Y] is proportional to the fluorescence intensity F, also the fluorescence intensity decreases, i.e. the fluorescence is quenched. The rate constant for this collisional quenching is determined by the product of the collisional quenching constant K_q and concentration of the quencher oxygen $[O_2]$. The life time with the quencher τ is given by

$$\tau = 1 / (K_f + K_c + K_s + K_q \cdot [O_2]) \tag{4}$$

Under normal conditions the collisional quenching can be described by

$$K_q = 1 / (\gamma \cdot 4p \cdot D \cdot R \cdot N \cdot 10^{-3}) \tag{5}$$

where γ = efficiency factor, D = sum of the diffusion coefficients of the involved molecules, R = sum of the radii of the involved molecules, $N \cdot 10^{-3}$ = number of molecules per millimol

This equation demonstrates clearly that the quenching process is determined by molecular properties of the indicator molecule and local diffusion conditions in its environment. The collision of oxygen molecules with indicator molecules in the excited triplet state influences the mean lifetime of phosphorescence in a similar way.

The oxygen concentration is measured by the relation of the lifetimes or of the fluorescence intensities with and without quencher

$$\tau / \tau_0 = F / F_0 = (K_f + K_c + K_s) / (K_f + K_c + K_s + K_q \cdot [O_2]) = 1 / (1 + K_q \cdot \tau_0 \cdot [O_2]) \tag{6}$$

or the Stern Volmer equation[39]

$$\tau / \tau_0 = F / F_0 = 1 / (1 + K_{sv} \cdot [O_2]) = 1 / (1 + K_{sv} \cdot \alpha \cdot pO_2) = 1 / (1 + K_t \cdot pO_2) \tag{7}$$

where $K_{sv} = K_q \cdot \tau_0$ and the overall quenching constant $K_t = K_q \cdot \tau_0 \cdot \alpha$

The relationship can be linearized by taking the reciprocal value τ_0 / τ

$$\tau_0 / \tau = F_0 / F = 1 + K_q \cdot \tau_0 \cdot \alpha \cdot pO_2 = 1 / (1 + K_t \cdot pO_2) \tag{8}$$

Luminescence signals can be evaluated by measuring the luminescent intensity or, in the case of collisional quenching, by lifetime measurements. Lifetime can be measured by the pulse method or by the phase modulation. When the exciting light is sinusoidically modulated, the emitted light has the same frequency, but it is delayed in phase by the phase angle Φ. The phase angle depends on the life time of the indicator

$$\tau = (2pf)^{-1} \cdot \tan \Phi \tag{9}$$

where f = modulation frequency (Hz)

Since τ depends on f in a non-linear way the optimal modulation frequency for an indicator has to be selected.

A) Luminescence Based Oxygen Sensors: O_2 Optodes

1) Indicators and Immobilization. Many luminescent compounds are quenched by oxygen, but it was difficult to find a compound which could serve as an indicator in the pO_2 range of 0-150 Torr[40,41]. A useful indicator should have a high quantum efficiency, a good light stability and a suitable quenching constant. Furthermore, its absorbance and luminescence should be in a wavelength range for which light sources and detectors are easily available. In 1972 Knopp and Longmuir[42] used pyrene butyric acid[43] as oxygen indicator and measured the changes of fluorescence intensity after application to tissue. However, with this technique it was difficult to obtain absolute pO_2 values since the indicator molecule was directly dissolved within the tissue, i.e. in a heterogeneous environment having different unknown collisional quenching constants and different values of α (see Equs.5 and 7). To obtain better results we proposed to construct a sensing element in which the indicator is enclosed so that a stable environment is guaranteed. We called such an element optrode or later -etymologically more correct(!)- optode and could show that with such an optode reliable measurements of the pO_2 in different media are possible[44]. As O_2 fluorescence

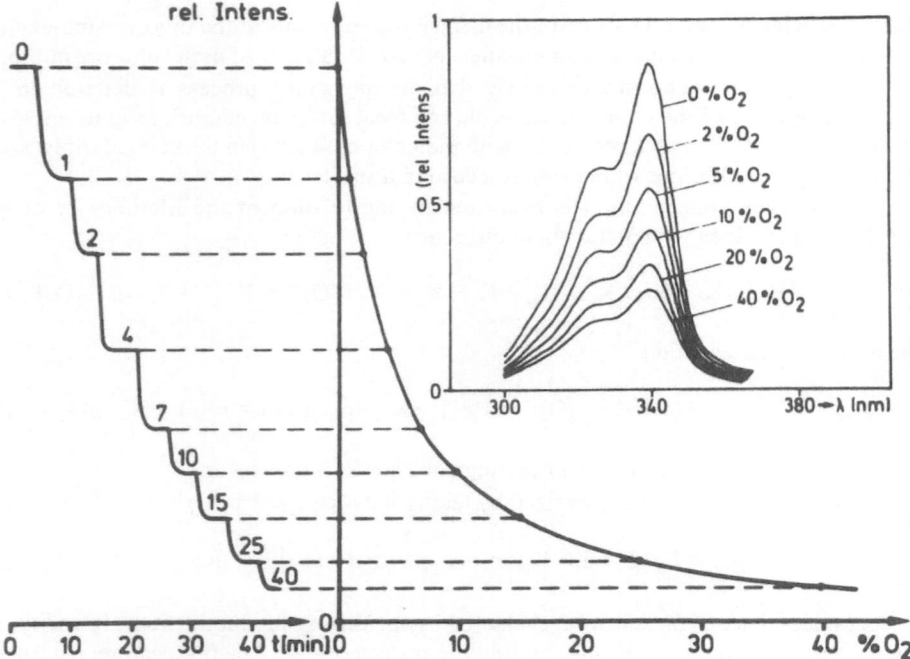

Figure 7. Changes of the fluorescence intensity of an oxygen optode (indicator PBA) caused by changing O_2 contents of the calibration gas. Right: direct record of the changes of the relative fluorscence intensities equilibrating the optode with gas mixtures of different O_2 content (0 - 40 % O_2 in N_2). Left: the corresponding hyperbolic calibration curve. Insert: changes of the the excitation spectrum of PBA (λ_{em} = 395nm) at different O_2, but concentrations. It changes only the amplitude, but not the form of the spectrum.

indicators have been proposed e.g. fluoroanthene[41], perylene dibutyrate[45], pyrene, fluorescence yellow, decacyclene, diphenyl anthracene, several ruthenium compounds as e.g. tris-(2,2'dipyridine)ruthenium(II)chloride[46] and as phosphorescence indicators fluorescein, porphin derivates[47,48] and bromonaphtalenes[49] (for further information see[50-52]). The indicator can be immobilized, for example in a fixed fluid layer[44], in a polymeric membrane as silicone[53], in polymeric particles or capsules[54] (for further information [50,52,55]).

2) Calibration Curve and Accuracy. Figure 7 shows an O_2 calibration curve of an optode consisting of a 50 μm silicone rubber membrane in which pyrenebutyric acid (PBA) was embedded[53]. Such a membrane can be fixed for example on the front of a fiberoptic lightguide connected to a fluorescence photometer. To avoid optical influences from the surroundings the sensor membrane was protected by a thin black silicone membrane. The left side of Fig. 7 shows how the fluorescence intensity decreases with increasing O_2 concentrations of the equilibrating gas. Since the exchange between gas and optode membrane depends on the O_2 diffusion into the membrane the optode measures directly oxygen pressure and only indirectly oxygen concentration. The insert demonstrates that the amplitude of the excitation spectrum changes, but not its form. The hyperbola in Fig. 7 (right) shows that the quenching process follows the Stern-Volmer equation (Equ.6). The sensitivity of the sensor S′ is described by the changing slope of the hyperbola dF / dpO_2

$$S' = dF / dpO_2 = (K_t \cdot F_0) / (1+K_t \cdot pO_2)^2 \tag{10}$$

With $\Delta_n F$ being the noise of the fluorescence signal, the quotient $F / \Delta_n F$ describes the signal to noise ratio (SNR) of the fluorescence signal. For the uncertainty of the pO_2, $\Delta_n pO_2$ follows

$$\Delta_n F = F / SNR = S' \cdot \Delta_n pO_2 \tag{11}$$

Assuming that the photodetector noise is the predominant noise we obtain for $\Delta_n F$

$$\Delta_n F = (2g \cdot e \cdot \Delta f \cdot F) \tag{12}$$

where e = elementary charge, Δf = frequency bandwidth and g = amplification factor and

$$\Delta_n pO_2 = F / (S \cdot SNR) \tag{13}$$

For the accuracy A (i.e. the reciprocal value of the uncertainty) follows

$$A = 1 / \Delta_n pO_2 = (SNR_0 \cdot K_t) / (1+K_t \cdot pO_2)^{1.5} \tag{14}$$

where SNR_0 = SNR without quencher.

Since accuracy (Equ.14) and sensitivity (Equ.10) have different exponents they have different maxima. The reason for this difference is the fact that the photonoise depends on the fluorescence intensity F of the signal (see Equ.12). Equ.14 shows that the accuracy of the pO_2 measurement by the described oxygen optode is determined by the signal to noise ratio without quencher and the overall quenching constant. With Equs. 11 and 14 the accuracy of the measurement for a selected pO_2' can be calculated by

$$pO_2' = \{(K_t \cdot \Delta_n pO_2 \cdot SNR_0)^{2/3} - 1\} / K_t \tag{15}$$

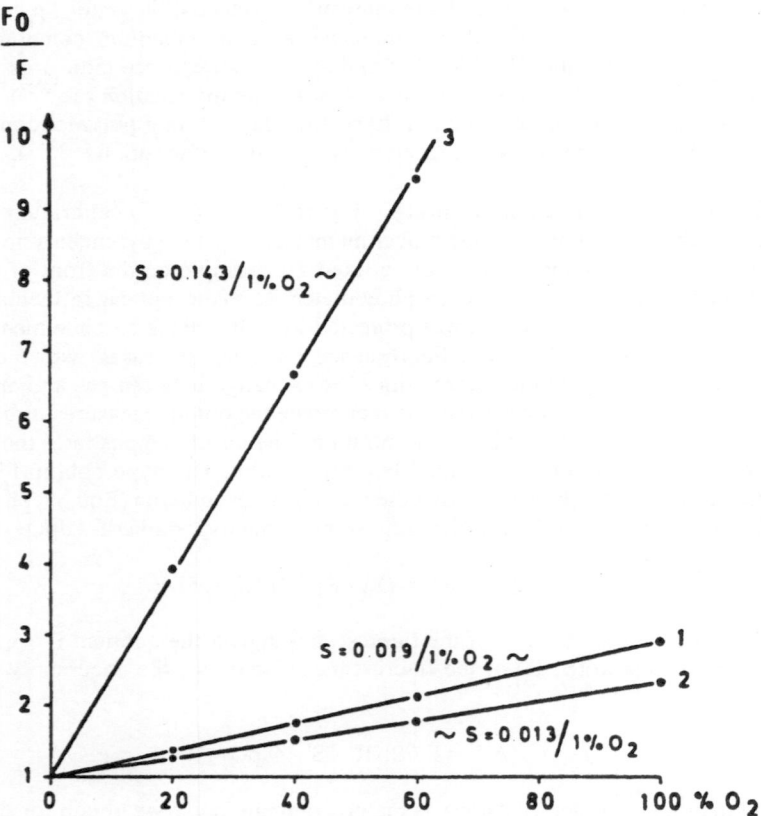

Figure 8. Linearized calibration curves of polyurethane microcapsules doped with PBA in different environments. 1.) in dioctylphtalate, 2.) in water, 3.) in a silicone rubber membrane. s = steepness of the calibration curve.

Using Equ.15 with $K_t = 1.61 \cdot 10^{-2}$ Torr^{-1} and $SNR_0 = 235$, we calculate for a pO_2' of 109 Torr a $\Delta_n pO_2 = \pm 0.8$ Torr and found experimentally $\Delta_n pO_2 = \pm 1.1$ Torr (n = 15)[56]. K_t can be manipulated by choosing a suitable membrane, SNR_0 by selecting the proper technical equipment and by increasing the size of the sensing area as large as possible.

It has been already mentioned that the luminescence optode has the advantage that the measuring process does not consume oxygen. Therefore, in contrast to the polarographic pO_2 electrode the medium to be measured should not influence the calibration curve. However, Equs.5 and 8 show that the fluorescent signal can be influenced by the properties of the environment if the indicator layer becomes very thin. Within the indicator layer the indicator molecules are statistically distributed. Since the mean diffusion paths of O_2 molecules, r_D, is

$$r_D = (6D \cdot t)^{1/2} \tag{16}$$

an excited indicator molecule with the mean lifetime t_0 can be quenched by O_2 molecules which are in a sphere with the radius $r_D = (6D \cdot \tau_0)^{1/2}$. To demonstrate a possible environmental influence we used nanocapsules (diameter 150-250 nm) doped with PBA suspended in 1) dioctylphtalate, 2) in water and embedded in a silicone membrane[54]. Fig. 8 shows that there is a strong influence of the environment on the steepness of the linearized calibration curve. The steepness for a change of oxygen concentration of 1% amounts in water to s = 0.013 and in silicone to s = 0.143, i.e. s is about 10 times larger. Under such conditions optodes have also to be calibrated in the medium to be measured. Furthermore, stray light of the surroundings or of the indicator layer may disturb the Stern-Volmer behaviour of the sensor. By embedding the indicator in a polymeric membrane multiexponential decay curves may originate. Therefore, in praxis often non-linear calibration curves are obtained.

B) Absorbance Based Oxygen Sensors. Absorbance O_2 indicators have been proposed as fiberoptic sensors using e.g. immobilized hemoglobin or other compounds[57,58,59].

APPLICATION *IN VIVO*

A) Invasive Measurement of Local Tissue pO_2

The pO_2 field measured with pO_2 microelectrodes mirrors the O_2 flux within the tissue caused by the O_2 consumption of its cells. To obtain reliable measurements the O_2 flux produced by the O_2 consumption of the electrode should be small as compared to the local O_2 flux within the tissue. The O_2 flux towards the electrode [$mlO_2 / (cm^2 \cdot min)$] is obtained by dividing the oxygen consumption of the electrode, m_e by the area of its cathode, a_e

$$m_e / a_e = \{(M \cdot i \cdot t) / (n \cdot F)\} / a_e \qquad (17)$$

where m = amount of oxygen reduced (in g), M = molecular weight (in g), i = electrical current (in A), t = time (in s), n = number of electrons involved and F = Faraday constant (in $A \cdot s \cdot mol^{-1}$).

A reduction current of i = 1 nA corresponds to an O_2 consumption of 4.99 µg O_2 / min = 0.156 · 10^{-6} µmol O_2 / min =3.5 µl · $10^{-6}O_2$ / min. To avoid misreadings

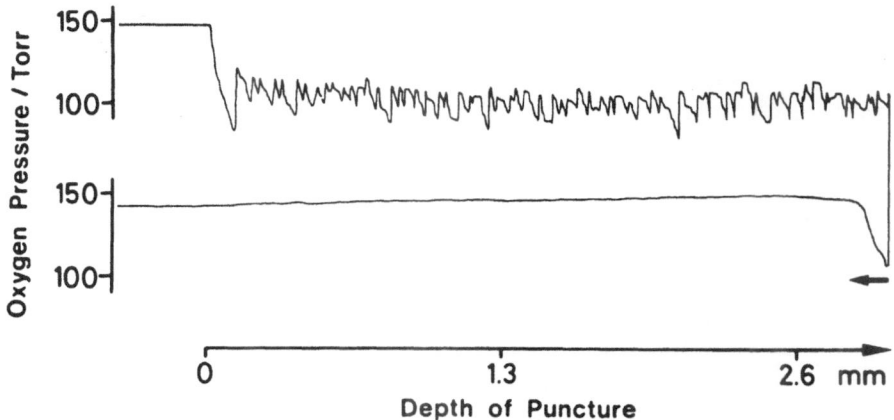

Figure 9. The effect of the puncturing on the polarographically measured pO_2. Upper trace: A layer of 3% air saturated gelatine is punctured by continously inserting a pO_2 microelectrode (tip size ca 1µm) with a continuous speed of 25 µm/min. The measured pO_2 is lower than the air value and varies characteristically. Lower trace: When the electrode is withdrawn the air value is measured. (for futher explanations see text).

the microelectrode should be calibrated in a solution with an oxygen conductivity similar to that of the tissue. To simulate the O_2 conductivity, c_v of the tissue we are using buffer solutions of ethyleneglycol-borax-biphosphate[23]. As already mentioned there are many reports about measurements with similar microelectrodes.

To measure the pO_2 distribution in the interior of the tissue the tip of the electrode has to be inserted into the tissue. A proper insertion technique is very important since the insertion produces at the tip and the shaft of the electrode strong mechanical forces which may cause structural changes resulting in a distinctly changed local pO_2 (see[60]). Such insertion artifacts can be demonstrated by puncturing defined polymeric materials[23].

In Fig. 9 a pO_2 microelectrode (tip diameter ca. 1 µm, cathode diameter 0.7 µm, length of the recess 0.7 µm) is inserted perpendicularly into the surface of a 3% gelatine layer (equilibrated by the oxygen of the surrounding air) and then continuously moved on with a speed of 25 µm/min (upper trace). After insertion pO_2 decreases continuously, but then, the decrease is followed by a sudden increase. This decrease/increase pattern recurs with different amplitudes up to the end of the puncture. By withdrawing the electrode (lower trace) the pO_2 increases slowly to about the air value and remains at this value. This demonstrates that puncturing changes the structure of the polymer so that its oxygen conductivity is dramatically altered; possibly, at this slow movement first the large polymer molecules become concentrated in front of the tip forming a layer with altered oxygen conductivity. By the continuously moving electrode this layer will be pierced through since the main part of the molecules will be not displaced during puncturing, but remains fixed in the polymer. Because of the environmental influences on the electrode signal the reduction currents measured during the slow insertion can not be evaluated because the actual oxygen conductivity of the gelatine at the tip of the electrode is unknown. Evaluating the data by using the calibration curve for a normal oxygen conductivity would give erroneous pO_2 values; but the trace can be used to demonstrate the relative changes of the local oxygen conductivity during the puncture. Other experiments showed that the decrease/increase pattern as shown in the upper trace is mainly determined by the structure of the polymer, by

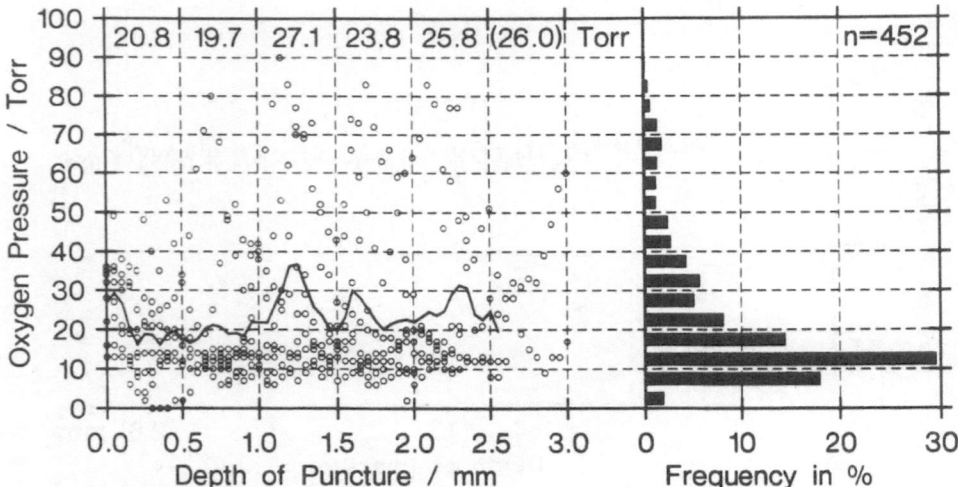

Figure 10. Local pO_2 values of the brain cortex of the guinea pig in dependence of the depth of the puncture and the corresponding frequency histogram. A single pO_2 value is marked by a circle. The solid line describes the mean pO_2 values. The figures in the upper row denote the mean pO_2 values in the correponding tissue layer. n = number of measuring values[62].

the diameter of tip and shaft of the electrode and by the speed of the puncture. This demonstrates that in order to obtain reliable results the structural alterations have to be relieved before taking a reading. Since the behaviour of living tissue is difficult to predict we use the following technique of puncturing: the puncturing is carried out by stepwise inserting and withdrawing of the electrode. We apply a nanostepper (Co. Bachofen, Reutlingen, Germany) which allows to vary the speed of the insertion and the direction as well as the length of the steps. For example, to measure the distribution of the local pO_2 at different depths in the gray and white matter a pO_2 microelectrode (tip size ca. 1 µm, length of the recess about 2 µm, collodion-polystyrene double membrane, Ag/AgCl reference electrode sputtered on the shaft, t_{90} <1 s (with recorder)) was pushed in by 2 steps of 50 µm and then immediately withdrawn by 1 step of 50 µm. One step was performed in 10 ms. The pO_2 value stabilized in 10 s. Fig. 10 shows the result of 9 punctures[62]. The values show the expected strong heterogeneity. It is interesting that the mean values (solid line) do not vary much. A pO_2 frequency histogram (Fig.10, left side) can be used to present quantitatively the measured pO_2 distribution (see[63,64]).

An interesting variation of the puncturing technique has been worked out by Fleckenstein and Weiss[26]. The idea is that a correct local pO_2 value should be obtained when after a very quick puncture the reading is taken before tissue reactions may occur. The tissue is punctured e.g. by a pO_2 microelectrode mounted in an injection canula having an outer diameter of ca. 300 µm and a response time t_{90} smaller than 500 ms. The electrode has a gold cathode of a diameter of 12.5 µm in a recess. The polarographic circuit is completed by a separate reference electrode. The electrode setup is commercially available (GMS, Kiel-Mielkendorf, Germany and Fa Eppendorf, Hamburg, Germany) and used in clinical research e.g. to analyze the oxygen supply of muscles[27] or of human tumors[65].

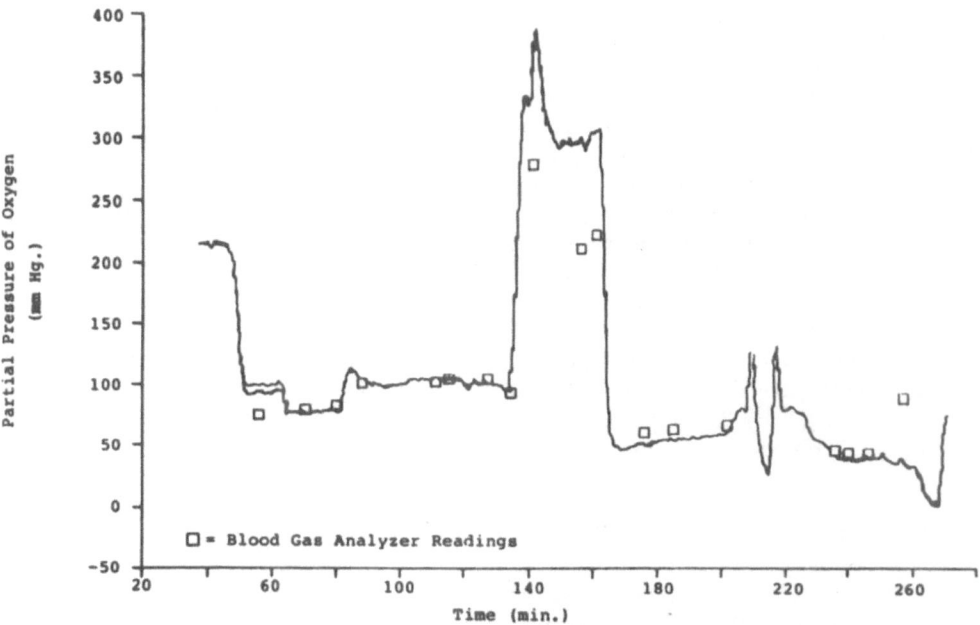

Figure 11. Continuous measurement of arterial pO_2 by an intravascular fiberoptic pO_2 optode. pO_2 vs. time. Squares: analyses by a conventional blood gas analyzer. Insert: Diagram of the tip of the optical fiber with indicator membrane and an overcoat for optical insulation (anaesthetized dog)[71].

The described electrodes are constructed to measure the heterogeneity of tissue pO_2 in space, but there are also electrodes described by which the heterogeneity in time can be monitored (see[66,67]) taking the risk that inflammatory reactions may develop. Continuous mean tissue pO_2 can be measured by introducing an oxygen permeable tube (e.g. a silastic catheter) filled with a suitable solution into the tissue. After equilibration with tissue pO_2, the pO_2 of the solution can be measured e.g. with a Clark electrode (e.g.[68]). The setup has a relatively long time constant, but it allows an easy control of the function of the measuring unit. It has been used, for example to study the oxygen supply of healing tissue[69].

For continuous measurements of blood pO_2 flexible catheters with small Clark electrodes are used. To minimize the danger of blood clotting membranes with an improved biocompatibility are being developed[70]. A combined intra-arterial analysis of pO_2, pCO_2 and pH can be performed by fiber optic sensors since the single sensor can be manufactured rather small[71-74] (for review[75]). It has been also shown that indicators with different spectra can be combined in the same optode membrane or in a membrane sandwich allowing the simultaneous measurement of different parameters[76]. As a practical example of a fiber optic analysis Fig.11 shows a continuous measurement of arterial pO_2 by an intravascular pO_2 optode[71]. The values have been checked by conventional blood gas analysis (squares). There is a good agreement between the measuring values except at high pO_2, but this may be caused by the errors involved in the conventional analysis (e.g. by the O_2 consumption of the blood). Since clotting is reduced at high local flow values a three component fiber optic catheter has been developed which produces an oscillatory blood flow[77]. There are indicators and polymeric materials which tolerate temperatures at least up to 200^0C[78]. About the possibility to measure intra-arterial pO_2 after injection of a phosphorescence indicator will be reported by D.F Wilson during this conference.

B) Non-Invasive Measurement of the pO_2

To obtain a pO_2 histogram on the surface of an organ a multiwire pO_2 electrode[79] can be used. In future, the measurement of local pO_2 distributions on an organ surface will be facilitated by using disposable, membrane-shaped, flexible pO_2 optodes covering the whole part to be studied. With the help of a camera a regional image of the local pO_2 distribution can be obtained. By combining such optodes with a diffusion test membrane it is for the first time possible to measure directly the amount of oxygen which is taken up through the skin surface, the O_2 flux. Since the local O_2 flux mirrors the O_2 supply of the skin the O_2 flux optode promises to become an interesting diagnostic tool[80,81]. A pO_2 surface electrode combined with a device for local heating is used to measure the trancutaneous pO_2. By heating up the skin to a constant temperature blood flow can be standardized. Since there is a hyperbolic relationship between $tcpO_2$ and blood flow, at large hyperemia the $tcpO_2$ can be used to check continuously arterial pO_2[82]. The non-invasively measured conjunctival pO_2 has been used to investigate its relationship to the arterial pO_2 as well as to the oxygen supply of the brain (e.g.[12,83]). To study the pO_2 changes during the respiratory gas exchange fast recording pO_2 optodes are recommended ($t_{90} = 30$ ms)[84].

CONCLUSION

In recent years the technical presuppositions have been developed so that it is possible to measure invasively the local pO_2 distribution within the tissue. However, to obtain reliable results environmental influences on the electrode signal as well as effects of the electrode (O_2 consumption and insertion artifact) on the O_2 supply and pO_2 distributon of the tissue have to be considered. Less invasive surface measurements are easier to perform, but environmental influences and effects of the O_2 consumption of the electrode must be minimized. In principle, optical pO_2 measurements have some advantages: 1.) They do not consume oxygen. 2.) Only the signal of very thin indicator layers can be disturbed by environmental influences. 3.) There is no reference electrode. 4.) The optode membrane can cover a larger area and monitor its regional pO_2 distribution. 5.) Indicator layer and measuring unit can be separated so that only the optode membrane is in contact with the tissue. 6.) Optical indicators with suitable absorbance spectra allow multi-indicator sensors. Hopefully, proper instruments will be available soon.

I like to thank my coworkers who have helped to develop proper sensors and to understand the tissue measurements: U. Grossmann, W. Grunewald, R. and A. Huch, M. Kessler, R. König, D. von Wulfen, G. Weidemann, but I like especially to mention the contributions of H. Baumgärtl to the development of the pO_2 microelectrode and of N. Opitz to the development of fluorescence based chemical sensors.

REFERENCES

1. J.A. Campbell, Gas tensions in the tissues, *Physiol.Rev.* 11:1 (1931).
2. A. Krogh, The number and distribution of capillaries in muscles with calculations of the oxygen pressure head necessary for supplying the tissue, *J. Physiol.* 52:509 (1919).
3. F. Kreuzer, Oxygen supply to tissues: The Krogh modell and its assumptions, *Experientia* 38:1415 (1982)
4. P.W. Davies. and F. Brink jr., Microelectrodes for measuring local oxygen tensions in animal tissue, *Rev. Scient. Instr.* 13: 524 (1942).
5. E. Roseman, C.W. Goodwin and W.S. McCulloch, Rapid changes in cerebral oxygen tension induced by altering the oxygenation and circulation of the blood, *J. Neurophysiol.* 9:33 (1946).
6. G.A. Millikan, Experiments on muscle haemoglobin *in vivo*: the instantanious measurement of muscle metabolism, *Biol. Science* 123:218 (1937).
7. H.A. Laitinen and I.M. Kolthoff, A study of diffusion processes by electrolysis with microelectrodes, *J. Am. Chem. Soc.* 61:3344 (1939).
8. A. Rémond, Aspects physiologiques de l'oxygene cortical, *Rev. Neurol [Paris]* 80:579 (1948)
9. P.W. Davies, The oxygen cathode, *Phys. Techn. Biol. Res.* 4:137 (1962).
10. D.W. Lübbers, H. Baumgärtl, H. Fabel, A. Huch, M. Kessler, K. Kunze, R. Riemann, D. Seiler, and S. Schuchhardt, Principle of construction and application of various platinum electrodes, *Prog. Resp. Res.* 3:136 (1969).
11. I. Fatt. "Polarographic Oxygen Sensor", CRC Press Inc. Cleveland, Ohio (1976).
12. F. Kreuzer, H.P. Kimmich and M. Brezina, Polarographic determination of oxygen in biological materials, *in:* "Medical and Biological Applications of Electrochemical Devices", J. Kroyta, ed., John Wiley&Sons Ltd., New York (1980).
13. E. Gnaiger and H. Forstner. "Polarographic Oxygen Sensors", Springer-Verlag, Berlin (1983).
14. V. Linek, J. Sinkule, and V. Vacek, Dissolved oxygen probes, *in:* "Comprehensive Biotechnology", M. Moo-Young, ed., Pergamon Press, Oxford (1985).
15. E.A.H.Hall, The electrochemical sensor, *in:* "Medical Applications of Microcomputers", W.A. Corbett, ed., John Wiley&Sons Ltd., New York (1987).
16. H. Baumgärtl, W. Zimelka, D.W. Lübbers, pH changes in front of the hydrogen generating electrode during measurements with an electrolytic hydrogen cleareance sensor, *Adv.Exp.Med.Biol.* 277:107 (1990).
17. L. Clark, Monitor and control of blood and tissue oxygen tension, *Trans. Am. Soc. Artif. Intern. Organs* 2:41 (1956).
18. J.S. Lundsgaard, J. Grønlund, and H. Degn, Error in oxygen measurements in open systems owing to oxygen consumption in unstirred layer, *Biotechnol. Bioengin.* 20:809 (1978).
19. E.A.H. Hall, The pulsed membrane gas electrode, *in:* "Neonatal Physiological Measurements", P. Rolfe, ed., Butterworth (1986).
20. D.J. Gavaghan, J.S. Rollet and C.W. Hahn, Numerical simulation of the time-dependent current to membrane-covered oxygen sensors, *J. Electroanal. Chem.* 348:15 (1993).
21. D.B. Cater, I.A. Silver and G.M. Wilson, Apparatus and technique for the quantitative measurement of oxygen tension in living tissues, *Proc. R. Soc. Lond. [Biol.]* 151:256 (1959).

22. H. Baumgärtl and D.W. Lübbers, Microcoaxial needle sensor for polarographic measurement of local O_2 pressure in the cellular range of living tissue. Its construction and properties, in: "Polarograpic Oxygen Sensors", E. Gnaiger and H. Forstner, eds, Springer-Verlag, Berlin (1983).

23. H. Baumgärtl, Systematic investigations of needle electrode properties in polarographic measurements of local tissue pO_2, in: "Clinical Oxygen Pressure Measurement", A.M. Ehrly, J. Hauss and R. Huch, eds., Springer-Verlag, Berlin (1987).

24. W.J. Albery, W.N. Brooks, S.P. Gibson and C.E.W. Hahn, An electrode for PN_2o and Po_2 analysis in blood gas, J. Appl. Physiol. 45:637 (1978).

25. I. Bergmann, Amperometric oxygen sensors: problems with cathodes and anodes of metals other than silver, Analyst 110:365 (1985).

26. W. Fleckenstein and C.H. Weiss, A comparison of pO_2-histograms from rabbit hindlimb muscles obtained by simultaneous measurements with hypodermic needle electrodes and with surface electrodes, Adv. Exp. Med. Biol. 169:447 (1984).

27. P. Boekstegers, M. Weiss and W. Fleckenstein. The effect of hypercapnia on the distributuion of pO_2 values in resting muscle, in: "Clinical Oxygen Pressure Measurement II", A.M. Ehrly, W. Fleckenstein, J. Hauss and R.Huch, eds., Blackwell Ueberreuther Wissenschaft, Berlin (1990)

28. W.J. Whalen, J. Riley and P. Nair, A microelectrode for measuring intracellular pO_2, J. Appl. Physiol. 23:789 (1967).

29. D.G. Buerk and T.K. Goldstick, Analysis of the oxygen barrier in the arterial wall from recessed pO_2 microelectrode measurements, Microvasc. Res. 17:69 (1979).

30. B.Yu, H. Baumgärtl and D.W. Lübbers, An improved polarographic multiwire pO_2 electrode, particularly for measurement of high pO_2 values, Adv. Exp. Med. Biol. 169:877 (1984).

31. G. Gust, K. Booij, W. Helder and B. Sundby, On the velocity sensitivity (stirring effect) of polarographic oxygen microelectrodes, Netherland. J. Sea Res. 21(4):255 (1987)

32. G Kortüm, "Reflexionsspektroskopie", Springer Verlag, Berlin (1969)

33. J.R. Lakowicz, "Principles of Fluorescence Spectroscopy", Plenum Press, New York (1983).

34. K. Matthes, Untersuchungen über den Verlauf der Oxyhämoglobin Reduktion in der menschlichen Haut, Pflügers Arch. 246:70-91 (1942).

35. G.G. Guilbaut, "Practical Fluorescence", M. Dekker, New York (1973).

36. D.W. Lübbers, Fluorescence based chemical sensors, Adv. Biosens. 2:215-260 (1992).

37. O.S. Wolfbeis, Fiber optical fluorosensors in analytical chemistry, in: "Molecular Luminescence Spectroscopy: Methods and Applications", Part II, S.G. Schulman, ed., Wiley, New York (1984).

38. H. Kautsky, Quenching of luminescence by oxygen, Trans. Faraday Soc. 35:216-219 (1932).

39. O. Stern and M. Volmer, Über die Abklingzeit der Fluoreszenz, Physikal Z. 20:183 (1919).

40. I.B. Berlman, "Handbook of Fluorescence Spectra of Aromatic Molecules", Academic Press, New York (1971)

41. I. Bergman, Rapid-response atmospheric oxygen monitor based on fluorescence quenching, Nature 218:396 (1968).

42. J.A. Knopp and I. Longmuir, Intracellular measurement of oxygen by quenching of fluorescence of pyrenebutyric acid, Biochim. Biophys. Acta 279:393 (1972).

43. W.M. Vaughan and G. Weber, Oxygen quenching of pyrenebutyric acid fluorescence in water: a dynamic probe of the microenvironment, Biochemistry 9:464 (1970).

44. D.W. Lübbers and N. Opitz, The pO_2-"optode", a new tool to measure pO_2 of biological gases and fluids by quantitative fluorescence photometry, Pflügers Arch. Suppl. 359: R145 (1975).

45. J.I. Peterson, R.V. Fitzgerald and D.V. Buckhold, Fiber-optic probe for in vivo measurement of oxygen partial pressure, Anal. Chem. 56:62-67 (1984).

46. M.E. Lippitsch, J. Pusterhofer, M.J.P. Leiner, and O.S. Wolfbeis, Fiber-optic oxygen sensor with the fluorescence decay time as the information carrier, Anal. Chim. Acta 205: 1 (1988).

47. J.M. Vanderkooi, G. Maniara, T.J. Green and D.F. Wilson, An optical method for measurement of dioxygen concentration based upon quenching of phosphorescence, J. Biol. Chem. 262:5476 (1987).

48. D.B. Papovsky, Luminescent porphyrins as probes for optical (bio)sensors, Sensors and Actuators B 11:293 (1993).

49. E.D. Lee, T.C. Werner and W.R. Seitz, Luminescence ratio indicators for oxygen, Anal. Chem. 59: 279 (1987).

50. O.S. Wolfbeis, Oxygen sensors, in:"Fiber Optic Chemical sensors and Biosensors", O.S. Wolfbeis, ed., CRC Press, Boca Raton.USA (1991).

51. A. Sharma and O.S. Wolfbeis, Fiber-optic oxygen sensor based on fluorescence quenching and energy transfer, Appl. Spectrosc. 42:1009 (1988).

52. W.R. Seitz, Chemical sensors based on immobilized indicators and fiber optics. *CRC Critical Reviews in Analytical Chemistry* 19:135 (1988).

53. N. Opitz and D.W. Lübbers, Theory and development of fluorescence-based opto-chemical oxygen sensors: oxygen optodes, *in:* "Advances in Oxygen Monitoring", K.K. Tremper and S.J. Barker, eds., Little Brown and Company, Boston (1987).

54. D.W. Lübbers, N. Opitz, P.P. Speiser, and H.J. Bisson, Nanoencapsulated fluorescence indicator molecules measuring pH and pO_2 down to submicroscopical regions on the basis of the optode- principle. *Z. Naturforsch. C.* 32:133 (1977).

55. D.W. Lübbers and N. Opitz, Optical fluorescence sensors for continuous measurement of chemical concentrations in biological systems, *Sensors and Actuators* 4:641 (1983).

56. N. Opitz and D.W. Lübbers, Increased resolution power in pO_2 analysis at low pO_2 levels via sensitivity enhanced pO_2 sensors (pO_2 optodes) using fluorescence dyes, *Adv. Exp. Med. Biol.* 180:261 (1985).

57. Z. Zhujun and W.R. Seitz, Optical sensor for oxygen based on immobilized hemoglobin, *Anal. Chem.* 38:220 (1985).

58. R.A. Wolthuis, S. McCrea, J.C. Hartl, E.Saaski, G.L. Mitchell, K. Garein and R. Willard, Development of a medical fiber-optic sensor based on optical absorption, *IEEE Trans. Biomed. Eng.* 39:185 (1992).

59. H.Y. Ebril and B.M Baysal, A new colorimetric method for the determination of the "dissolved" oxygen permeability coefficients of polymeric membranes, *J. Membr. Science* 26:199 (1986).

60. W. Müller, A. Winnefeld, O. Kohls, T. Scheper, W. Zimelka and H. Baumgärtl, Real and pseudo oxygen gradients in Ca-alginate beads monitored during polarographic pO_2 measurements using Pt- microelectrodes, *Biotechnol. Bioeng.* 44:617 (1994).

61. Y. Okada, K. Mückenhoff, G. Holtermann, H. Acker, and P. Scheid, Depth profiles of pH and pO_2 in the isolated brain stem-spinal cord of the neonatal rat, *Resp.Physiol.* 93:315-326 (1993).

62. D.W. Lübbers, H. Baumgärtl and W. Zimelka, Heterogeneity and stability of local pO_2 distribution within the brain tissue, *Adv.Exp.Med.Biol.* 345:567 (1994).

63. D. Jamieson and H.A.S. van den Brenk, Comparison of oxygen tensions in normal tissue and yoshima sarcoma of the rat breathing air or oxygen at 4 atmospheres, *Brit. J. Cancer* 17:70 (1963).

64. K. Turek,. K. Rakusan, J. Olders, L. Hoofd, and F. Kreuzer, Computed myocardial pO_2 histograms: effects of various geometrical and functional conditions, *J.Appl.Physiol.* 70:1845-1853 (1991).

65. M. Höckel, K. Schlenger, C. Knoop and P. Vaupel, Oxigenation of carcinomas of the uterine cervix: evaluation by computerized O_2 tension measurements, *Cancer. Res.* 51:6098 (1991).

66. K. van Rossem, H. Vermarien and R. Bourgain, Construction, calibration and evaluation of pO_2 electrodes for chronicle implantation in rabbit brain cortex. *Adv.Exp.Med.Biol.* 316:85 (1992)

67. P. Boekstegers, J. Diebold and Ch. Weiss, Selective ECG synchronized suction and retroinfusion of coronary veins: first results of studies in acute myocardial ischemia in dogs, Cardiovasc. Res. 24:456 (1990)

68. T.K. Hunt, A new method of determining tissue oxygen tension, *Lancet* 26:1370 (1964).

69. K. Jonsson, J.A. Jensen, W.H. Goodson, H. Scheuenstuhl, J. West, H. Williams Hopf, and T.K. Hunt, Tissue oxygenation, anemia and perfusion in relation to wound healing in surgical patients, *Annals of Surg.* 214:605 (1991).

70. A.P. Murphy and P. Rolfe, Intravascular oxygen sensor with polyetherurethane membrane: *in vitro* performance, *Med.&Biol.Eng.&Comput.* 30:121 (1992).

71. L. Gehrich, D.W. Lübbers, N. Opitz, D.R. Hansmann, W.W. Miller, J.K. Tusa and M. Yafuso, Optical fluorescence and its application to an intravascular blood gas monitoring system. *IEEE Trans. Biomed. Eng., BME* 33:117 (1986).

72. S.J. Barker, K.K. Tremper, J. Hyatt, J. Zaccari, H.A. Heitzmann, B.M. Holman, K. Pike, L.S. Ring, M. Teope, and T.B. Thaure, Continuous fiberoptic arterial oxygen tension measurements in dogs, *J.Clin.Monit.* 3:48 (1987).

73. B.A. Shapiro, R.D. Cane, C.M. Chomka, L.E. Bandala and W.T. Peruzzi, Preliminary evaluation of an intra- arterial blood gas system in dogs and humans, *Crit.Care Med.* 17:455 (1989).

74. B.E. Slain, P.H. King and L. Schlain, Clinical evaluation-Continuous real-time intra-arterial blood gas monitoring during anesthesia and surgery by fiber optic sensor, *Int. J. Clin. Monit. Comput.* 9:45 (1992).

75. A. Gottlieb, S. Divers and H.K. Hui, *in vivo* applications of fiberoptic chemical sensors, *in:* "Biosensors with Fiberoptics", D.L. Wise and L.B. Wingard ,eds.,, Humana Press, Lifton, NJ (1991).

76. D.W. Lübbers and N. Opitz, Die pCO_2/pO_2-Optode: Eine neue pCO_2- bzw. pO_2-Messonde zur Messung des pCO_2 oder pO_2 von Gasen und Flüssigkeiten. *Z. Naturforsch. C.* 30:532 (1975).

77. J.S. Barker and J.H. Hyatt, Continuous measurement of intra-arterial pHa, $paCO_2$ and paO_2 in the operating room, *Anesth. Analg.* 73:43 (1991).

78. N. Opitz, H.-J Graf and D.W. Lübbers, Oxygen sensor for the temperature range 300 to 500 K based on fluorescence quenching of the indicator-treated silicone membranes, *Sensors and Actuators* 13:159 (1988).

79. M. Kessler and D.W. Lübbers, Aufbau und Anwendungsmöglichkeiten verschiedener pO_2 Elektroden, *Pflügers Arch.* 291:82 (1966).

80. G.A. Holst, D.W. Lübbers and E. Voges, O_2-flux-optode for medical application, *SPIE Adv.Fluoresc.Sens. Technol.* 1885:216 (1993).

81. D.W. Lübbers, Chemical in *vivo* monitoring by optical sensors in medicine, *Sensors and Actuators B* 11:253 (1993).

82. R. Huch, A. Huch and D.W. Lübbers, "Transcutaneous pO_2", Georg Thieme, Stuttgart (1981).

83. H. Haljamäe, I. Frid, J. Holm and S. Holm, Continuous conjunctival oxygen tension ($pcjO_2$) monitoring for assessment of cerebral oxygenation and metabolism during carotid artery surgery, *Acta Anaesthesiol. 6* Scand. 33:610 (1989).

84. H.Karpf, H.W. Kroneis, H.J. Marsoner, H. Metzler and N. Gravenstein, Fast responding oxygen sensor for respiratorial analysis, *SPIE Chem. Biochem. Environmen. Sens* 1172: 296 (1989).

3

RETINO-CHOROIDAL OXYGEN IMAGING THROUGH A FUNDUS CAMERA

S. Blumenröder, A. J. Augustin, M. Spitznas, F. Koch, and F. Grus

University Eye Hospital
University of Bonn
Sigmund-Freud-Strasse 25, D-53105 Bonn, Germany

INTRODUCTION

An adequate oxygen supply is essential to the retina. Without oxygen, vision vanishes within seconds. Retinal hypoxia is an important factor in the origin and the development of retinal disorders, i.e. diabetic retinopathy, occlusion of retinal vessels and glaucoma.

Microelectrodes have been used to investigate the pathogenesis of such disorders. In recent years, aided by advances in computing and electronics, the oxygen dependent quenching of the phosphorescence of certain intravenously administered metallo-porphyrins has been employed as a valuable means of determining the oxygen tension in retinal vessels that is non-invasive except for the injection of the porphyrin.

Hitherto, these investigations have been carried out using a microscope with an epifluorescent attachment and a zero-diopter contact lens. We report on the adaptation of this set-up to a commercially available fundus camera. We present measurements of the oxygen concentration in retinal vessels and describe how it is influenced by a rise in intraocular pressure.

THEORETICAL BACKGROUND

Following excitation with a flash of light, the phosphorescence of porphyrins can be used to determine the oxygen concentration of a tissue in vivo. Phosphorescence is the emission of light via a spin-forbidden transition from an excited triplet to a singlet ground state.

It is most convenient to select a porphyrin with a reasonably long lifetime of the phosphorescence that can be easily measured using a gated and intensified CCD camera. After intravenous administration of the porphyrin, blood oxygen interacts with the excited porphyrin molecule and thereby increases the transition probability from the triplet to the ground state. The characteristic lifetime of the excited state is inversely related to the transition probability.

Oxygen Transport to Tissue XVII, Edited by Ince et al.
Plenum Press, New York, 1996

The form of the oxygen that interacts with the porphyrin is the physically dissolved oxygen rather than that bound to hemoglobin. Temperature and pH influence the decay rate to a lesser degree [Shonat et al., 1992].

METHODS

Animals were treated according to the ARVO Resolution on the Use of Animals in Research. Two pigs weighing 11 and 12 kg were used. Anesthesia was induced with an i.m. injection of pentobarbital and maintained with subsequent i.v. injections of pentobarbital 60 mg/h. The phosphorescent oxygen probe was administered in a dose of 21 mg/kg body weight of palladium meso-tetra-(4-carboxyphenyl) porphine (Porphyrin Products, Logan, Utah, USA). This corresponds to 230 mg/kg of a lyophilized powder (Medical Systems Corp., Greenvale, NY, USA) which is more convenient to prepare. The solution of the powder in sterile water was injected intravenously in an ear vein. The animals were mechanically ventilated with oxygen and N_2O, normoxia was maintained. The hemoglobin saturation as measured by reflection oximetry was between 95% and 98% during the experiments. The arterial oxygen tension was measured to be 100 mmHg, the pH was 7.6. We set the quenching constant to 300 $Torr^{-1}sec^{-1}$ and the lifetime at zero oxygen tension to 640 μsec. These are probably not the exact values [Wilson, 1991], but the possible error in oxygen tension should be less than 15%.

The intraocular pressure was regulated with a Ringer solution consisting of Na^+ 130 mmol, K^+ 5.4 mmol, Ca^{2+} 1.85 mmol, Cl^- 112 mmol and lactate 27 mmol.

The OXYMAP^R system used by our group is a commercially available system that is supposed to be used with an epifluorescence attachment. We adapted this set-up to a Zeiss FF4 fundus camera (Zeiss, Oberkochem, Germany). We removed the U-shaped flash lamp of the FF4 camera and replaced it at the proper position with an EG&G 45-watt flash tube with a flash duration of about 10 μsec. A constant illumination of the fundus by a halogen lamp was possible with a beam-splitting mirror in the excitation beam path. As an excitation filter we used a green interference filter with maximum transmission at 537 nm and a bandwidth at half-maximum of 45 nm [Shonat et al., 1992; Wilson et al., 1992].

The observation (i.e. emission) filter had a cutoff at 630 nm and was positioned next to the CCD camera, whereas the excitation filters replaced the excitation filter of the FF4 used for conventional fluorescein angiography. By flipping a mirror it was possible to switch between camera observation and direct observation of the fundus through the eyepiece of the fundus camera.

The adaptation of the new flash tube to the FF4 fundus camera results in a certain loss of performance of the fundus camera. In order to avoid corneal and other reflections, the fundus camera separates the illumination and the observation ray path. The camera illuminates the retina by projecting a ring in the pupil plane, whereas the observation is made through the center of the ring. This prevents the corneal reflection of the illumination ring from entering the observation path. By reversing the path of the light, the ring in the pupil plane corresponds to a similar image ring in the plane of the flash spark with 3 mm outer and 1.3 mm inner diameter [Sümmerer, 1994]. Therefore it is desirable that the flash tube generates a spark of this relatively large size. On the other hand, the flash duration has to be short, necessitating a small area of the flash spark. As a result, the requirement of filling the ring image with the spark area was not fully met by the EG&G flash. On the other hand, the retina illumination was sufficiently homogeneous.

The electronic and computing equipment was not altered and is described elsewhere in detail [Shonat et al., 1992]. The fitting algorithm of the OXYMAP^R software is based on the Levenberg–Marquardt method. Additionally, the fit was checked with another fitting program.

RESULTS AND DISCUSSION

As the porphyrin is bound to bovine albumin during preparation, it is limited to the retinal and choroidal vessels. We thus measure the intravascular oxygen tension rather than the tissue oxygen tension. Because of the much slower blood flow in veins, most of the retinal blood volume is accounted for by the veins. Exact values of the ratio of arterial volume versus venous volume are not known. However, in the bat wing it is known that this ratio is about 0.18 (Wiedeman, 1984). Our sampling volume is therefore a mixture of arterial and venous volume with preponderance of the latter.

Without intravenously given porphyrin, the penetration depth of the exciting light is greater than the thickness of the retina. Therefore, the image intensifier samples light from different levels of the retina and possibly also from inner parts of the choroid. With porphyrin, the situation is more complicated, as absorption is increased and the penetration depth is decreased.

There are some arguments that the choroid should not contribute too much to the observed light. The low oxygen tension in retinal veins demonstrates that the highly oxygenated choroid does not contribute to the measured intravascular oxygen tension, at least in veins. It should be noted that this argument does not rule out a substantial contribution of the choroid to retinal tissue oxygenation.

Figure 1 shows a phosphorescence intensity image taken 40 μsec after the flash (left side) and the oxygen image of the same area (right side). Two major veins are seen; additional smaller veins extend orthogonally into the capillary regions. At the bottom of the oxygen image, the full scale corresponds to an oxygen tension of 128 mmHg (1.7 kPa). We thus find an oxygen tension of about 10 to 20 mmHg (1.3 to 2.6 kPa) in the major veins. These values are lower than the systemic oxygen tension in the vena cava and are also slightly lower than those published for the cat retina [Shonat et al., 1992].

The low values are due to the high oxygen consumption of the retina and the consequently high oxygen extraction of the retina. This high extraction is also known from other experiments [Linsenmeier, 1986] and is especially pronounced in the dark-adapted

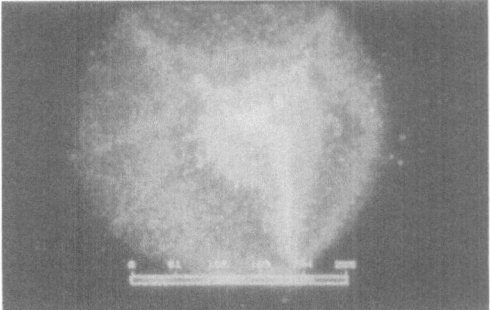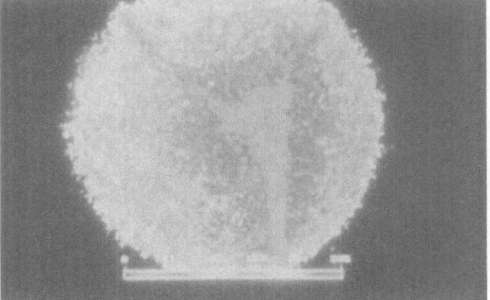

Figure 1. On the left side, an intensity image taken 40 μsec after the flash. Veins and capillary region can be clearly distinguished. On the right side, the oxygen image under normoxic conditions calculated from a sequence of images similar to the left image.

retina [Linsenmeier, 1990]. The extraction seems to be even higher than in the brain [Linsenmeier, 1986].

In the capillary region we found the lowest values close to the major veins. In the area between the two major veins we measured oxygen tensions of 20 to 50 mmHg (2.7 to 6.7 kPa) with an average of about 40 mmHg (5.3 kPa). A feature common to all our experiments is a spatial and temporal variation of the oxygen tension in the capillary bed, if the oxygen tension is measured in a small area. As these unexplained oscillations were also observed with other methods [Linsenmeier, 1986], an artifactual origin is improbable.

An increase in intraocular pressure of 30 mmHg (4.0 kPa) resulted in a decrease in the oxygen tension in the veins of about 5 mmHg (0.67 kPa). This small change could be intensified by inducing hypoxia by means of reducing the ventilation rate. When the intraocular pressure was changed back to normal, we found an overshoot of the oxygen tension beyond the starting value. This is thought to be due to the oxygen autoregulation of the retina, probably via regulation of blood perfusion. As it is known to take some time [Alder, 1990] for the retina to achieve a steady state after the pressure release, the overshoot can be attributed to the retinal vessels remaining dilated for some minutes after the release of the intraocular pressure elevation.

At an intraocular pressure of 90 mmHg (12.0 kPa) the pulsation of the central retinal artery vanished, indicating an occlusion of the arterial blood supply to the retina. In the oxygen image we found an oxygen tension of 0 mmHg (0 kPa) in veins. In the capillary region the oxygen tension was 10 mmHg (1.3 kPa) shortly after the occlusion. Seven minutes after the occlusion the oxygen tension in the capillary bed was 0 mmHg (0 kPa). In the reperfusion phase we observed very high oxygen tension of about 100 mmHg (13.3 kPa) in veins and in the capillary regions of the retina. This increase was probably not due to delayed constriction, i.e. delayed autoregulation. Because the retina suffered hypoxic damage, the oxygen consumption of the photorector cells in the early reperfusion phase was very low. For this reason we found little or no oxygen extraction from the arterial blood. The venous oxygen tension was then similar to the oxygen tension of the arteries entering the retina.

Our experiments showed that the use of a fundus camera in phosphorescence imaging ensures good image quality as the fundus camera guarantees reflection-free images because illumination and observation ray path are separated.

Additionally, the fundus camera simplifies the experiments. With the fundus camera set-up it is possible to observe parts of the retina that lie off the optical axis and to center these areas on the screen, because the fundus camera can be rotated around a vertical axis. With a newer commercially available fundus camera, rotation of the camera around a horizontal axis is also possible. With respect to future application of the method in clinical investigations, the adaptation to the fundus camera is of great advantage to patients, as it makes the investigation easier and less strenuous for them. However, for clinical applications of the method a safe and non-toxic phosphorescent dye is needed.

CONCLUSIONS

The adaptation of a phosphorescence imaging system to the fundus camera facilitates measurement and ensures good image quality. Our findings support the concept of retinal autoregulation of perfusion after a change in retinal oxygenation. The adaptation to the fundus camera enables clinical research application of phosphorescence imaging if a non-toxic dye is developed.

ACKNOWLEDGMENT

We would like to thank Synthelabo Co., Puchheim, Germany, for donating the computing and imaging system.

REFERENCES

Alder, V. A., Cringle, S. J. (1990), Vitreal and retinal oxygenation. Graefe's Arch. Clin. Exp. Ophthalmol. *228*, 151-157

Linsenmeier, R. A. (1986), Effects of light and darkness on oxygen distribution and consumption in the cat retina. J. Gen. Physiol. *88*, 521-542

Linsenmeier, R. A. (1990), Electrophysiological consequences of retinal hypoxia. Graefe's Arch. Clin. Exp. Ophthalmol. *228*, 143-149

Shonat, R. D., Wilson, D.F., Riva, C. E., Pawlowski, M. (1992), Oxygen distribution in the retinal and choroidal vessels of the cat as measured by a new phosphorescence imaging method. Appl. Opt. *31*, 3711-3718

Sümmerer, G., personal communication, Zeiss Corp., Oberkochem, Germany

Wiedeman, M. P., in : Handbook of Physiology, Vol. IV, ed. E. M. Renkin and C. C. Michel, American Physiological Society, Bethesda, Maryland 1984

Wilson, D. F., Pastuszko, A., DiGiacomo, J. E., Pawlowski, M., Schneiderman, R., Delivoria- Papadopoulos, M. (1991), Effect of hyperventilation on oxygenation of the brain cortex of newborn piglets. J. Appl. Physiol. *70*, 2691-2696

Wilson, D. F., Cerniglia, G. J. (1992), Localization of tumors and evaluation of their state of oxygenation by phosphorescence imaging. Cancer Res. *52*, 3988-3993

4

CEREBRAL OXYGENATION DURING CARDIOPULMONARY BYPASS

D. N. F. Harris

Department of Anaesthesia
Royal Postgraduate Medical School
London, United Kingdom

INTRODUCTION

Despite the low mortality of coronary bypass surgery (CABG) there is still a high incidence of neuropsychological defects at 1 and 6 months after surgery(1,2). The introduction of alpha stat CO_2 regulation for hypothermic cardiopulmonary bypass (CPB) preserves auto regulation of cerebral blood flow if mean arterial blood is maintained above 50 mm Hg(3), and it was thought that cerebral perfusion was likely to be adequate for the reduced cerebral oxygen requirements during hypothermia.

The introduction of arterial line filtering and membrane oxygenators has significantly reduced the incidence of cerebral emboli detected by Trans-cranial Doppler (TCD), but there has not been a corresponding fall in the incidence of neuropsychological defects. This has led to a re-examination of the suggested adequacy of cerebral perfusion. Near Infra Red Spectroscopy provides a non-invasive, continuous assessment of the amount of oxidised and reduced cerebral haemoglobin, and this paper presents the initial data from patients undergoing hypothermic cardiopulmonary bypass (CPB).

MATERIAL AND METHODS

After approval from the Hospital Ethics Committee and with informed consent, 10 male patients aged 58-72 years requiring routine CABG were studied. Following induction of anaesthesia and insertion of the standard monitoring lines, two optodes (NIR 1000 Hamamatsu Photonics UK) were placed high on the frontal region, avoiding the sagittal sinus and the temporalis muscle and were covered with a light-tight bandage. An optode separation of 5-6 cm was used, depending on the hairline, to minimise the contribution from the external carotid territory. Measurement of [Hb], [HbO] and Cytochrome caa3 were made every 15 seconds and stored on disk. Heart rate, oxygen saturation, arterial and central venous pressures and nasopharyngeal temperature were recorded continuously.

Oxygen Transport to Tissue XVII, Edited by Ince et al.
Plenum Press, New York, 1996

Cardiopulmonary Bypass

This was carried out at 28°C with pulsatile perfusion through a membrane oxygenator with a 40 micron arterial line filter at 2.2 litres per m², with a mean perfusion pressure maintained above 50 mm of Hg. The pump circuit was primed with 2 litres of crystalloid, and blood cardioplegia was used.

Anaesthesia

Following induction with 10 mcg per kg fentanyl and 2-4 mg midazolam, anaesthesia was maintained with 50% nitrous oxide in oxygen and < 0.5 MAC Isoflurane until 10 minutes before bypass. A propofol infusion at 100-250 mg per hour was used during and after bypass with ventilation using O_2 / Air.

RESULTS

Figure 1 shows a representative trace from patient 5. At the start of bypass there was an immediate fall in [HbO] and [Hb] due to the haemodilution by the crystalloid bypass prime from a haematocrit of c. 35 to around 25%. In four patients the haematocrit was measured every 10 seconds: the value stabilised within 2 minutes, and did not change significantly over the duration of bypass as blood cardioplegia was used. After 4 +/- 2 minutes the aortic cross clamp was applied and the patient cooled rapidly to 28°C (see fig. 1). There was no change in oxidised or reduced haemoglobin for a further 5-12 minutes, after which

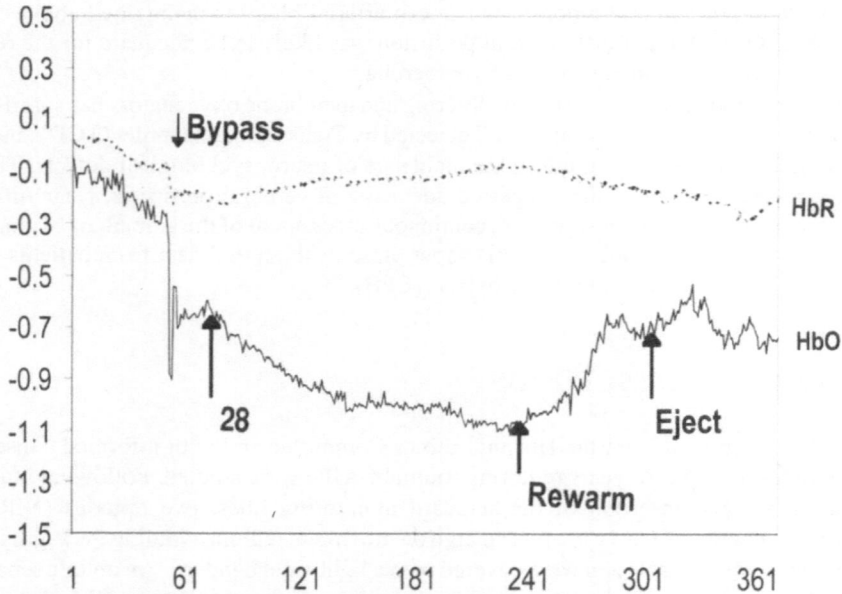

Figure 1. NIR changes in mmol.cm during hypothermic cardiopulmonary bypass. The upper trace is oxidised hemoglobin (HbO) and the lower reduced Haemoglobin (Hb). 'Bypass' indicates the start of bypass, '28°C' the time at which full hypothermia was reached, 'Rewarm' the start of rewarming, and 'Eject' the end of bypass with normal cardiopulmonary circulation and myocardial ejection.

[HbO] fell steadily for 10 minutes and then remained constant until the start of rewarming. There was no change in [Hb] at this time, so the sum of [HbO] + [Hb], indicating total blood volume, fell. The fall in [HbDiff] ([HbO]-[Hb]) indicates a fall in oxygen saturation of the mixed blood pool in the region under the optodes. During rewarming there was a rapid increase in [HbO] with a small increase in [Hb] to the levels at the start of bypass, followed by a further sharp increase in both [HbO] and [Hb] when the heart started to eject as bypass was discontinued. This lasted for 15 minutes and resembled a classic hyperaemic response. [Hb] Diff increased throughout rewarming, indicating a relative increase in regional cerebral oxygen saturation.

DISCUSSION

During hypothermic CPB oxygen consumption falls due to the drop in temperature. It was thought that coupling of flow and metabolism was maintained, and if adequate cerebral perfusion pressure was maintained (greater than 50 mm Hg) that cerebral oxygenation was adequate. The NIR data in this study show that for the first 15 minutes oxygen supply and demand appear to be balanced, despite the rapid cooling of the brain as measured by nasopharyngeal temperature. This agrees with Doppler data showing a fall in the middle cerebral artery velocity (MCAV) during cooling. The subsequent fall in [HbDiff] 15 minutes after cooling suggests that cerebral oxygen supply was less than that required, which would imply an uncoupling of flow and metabolism.

A previous study using jugular bulb measurements of cerebral venous haemoglobin saturation showed a fall in saturation during rewarming, indicating relative cerebral hypoxia(4). The NIR data in this study however, suggests an increase in saturation at this time, returning to the level at the start of bypass. Jugular bulb saturation was not measured in this study, and the apparent increase in haemoglobin saturation cannot be directly confirmed. In a similar group of patients MCAV has been shown to rise in parallel with the rise in [HbO] and to show a further increase when the heart starts to eject. The jugular bulb reflects global cerebral perfusion while NIR reflects a small region underneath the optodes and it was possible that regional variations in oxygenation could account for the discrepancy between jugular bulb and NIR measurements, but the variation would have to be very large and are unlikely to explain the apparently contradictory results.

All the NIR light recorded must pass through scalp and skull on the way both to and from the cerebral tissue, and thus always reflects extra-cranial oxygenation to some extent: the amount of this influence is at present disputed. It is possible that hypothermia and the abnormal perfusion of cardiopulmonary bypass may produce different effects on the perfusion of extra-cranial tissues to those on intra-cerebral perfusion. Also the lower mean perfusion pressure used during bypass (50 v. 90 before bypass) may specifically affect the extracranial circulation as it shows little auto regulation of blood flow. If so, these factors could explain some of the discrepancy between NIRS and Jugular Bulb indices of cerebral oxygen saturation. However this would imply that NIRS is significantly affected by the extracranial circulation and could not reliably be used to study truly intra-cerebral oxygenation. This question clearly needs to be resolved.

REFERENCES

1. Shaw-PJ; Bates-D; Cartlidge-NE et al Neurologic and neuropsychological morbidity following major surgery: comparison of coronary artery bypass and peripheral vascular surgery. Stroke. 1987 Jul-Aug; 18(4): 700-7

2. Treasure-T; Smith-PL; Newman-S et al Impairment of cerebral function following cardiac and other major surgery. Eur. J. Cardiothorac. Surg. 1989; 3(3): 216-21
3. Patel-RL; Turtle-MR; Chambers-DJ et al Hyperperfusion and cerebral dysfunction. Effect of differing acid-base management during cardiopulmonary bypass. Eur. J. Cardiothorac. Surg. 1993: 7(9): 457-63
4. Croughwell-ND; Frasco-P; Blumenthal-JA et al Warming during cardiopulmonary bypass is associated with jugular bulb desaturation. Ann. Thorac. Surg. 1992 May; 53(5): 827-32

5

LIGHTGUIDE SPECTROPHOTOMETRY FOR THE ASSESSMENT OF SKIN HEALING VIABILITY IN CRITICAL LIMB ISCHAEMIA

D. K. Harrison,[1] D. J. Newton,[1] P. T. McCollum,[2] and A. S. Jain[3]

[1] Department of Medical Physics
[2] Department of Surgery
 Vascular Laboratory, Ninewells Hospital and Medical School
 Dundee DD1 9SY, Scotland, United Kingdom
[3] Department of Orthopaedic Surgery
 Dundee Royal Infirmary
 Dundee DD1 9ND, Scotland, United Kingdom

INTRODUCTION

Techniques in vascular surgery and medicine have advanced considerably over the last decade. Whereas, in the past, critical ischaemia of the lower limb frequently resulted in amputation, nowadays, even lower limbs subject to occlusion of distal arteries can frequently be salvaged (Shah et al, 1992). Nonetheless, currently, of about 5000 patient visits per year to our Vascular Laboratory, some 100 of those patients with critical limb ischaemia (CLI) ultimately still need to undergo an amputation. Whether a below knee amputation is viable or not is currently predicted in our laboratory by means of skin blood flow measurements made at the site of the proposed skin flap (McCollum et al, 1985).

However, whilst this method is successful in predicting the outcome of a below knee amputation in 93% of cases (McCollum et al, 1988), despite the development of many new techniques for the assessment of skin blood flow and metabolism there is still no reliable method for predicting the outcome of amputations in patients with critical lower limb ischaemia (CLI) any more distal than the below knee (BKA) level. This applies particularly in patients who might be suitable for an amputation through the foot, which would allow greater mobility than one at the below knee level.

Furthermore, whilst skin blood flow may be an indicator of rate of delivery of those substrates required for the healing process, in particular oxygen, it is not a direct indicator of the adequacy of that rate of delivery and hence a more direct, non-invasive measure of skin oxygen supply at the skin flap would be desirable.

We have previously used the technique of lightguide spectrophotometry to investigate haemoglobin saturation (SO_2) in inflamed human skin (Harrison et al, 1992) and in the skin of claudicant patients after treadmill exercise (Hickman et al, 1992). In this study a

Oxygen Transport to Tissue XVII, Edited by Ince et al.
Plenum Press, New York, 1996

45

Photal MCPD-1000 (Otsuka Electronics, Osaka) was used to measure SO_2 in the skin of patients with CLI about to undergo amputation. The aim of the investigation was to determine whether lightguide spectrophotmetry could be used to predict amputation level in terms of mean oxygen saturation or critical level of oxygen supply in skin and to compare the technique with the "gold standard" skin blood flow method. In comparing the techniques, a further aim was to investigate relationships between skin blood flow and oxygen supply in critical ischaemia.

METHODS

The MCPD-1000 employed a Y configuration lightguide consisting of randomly orientated transmitting and receiving fibres (each 200 μm diameter). Analysis of the spectra was carried out in the visible range (500-586 nm) using a 6 wavelength technique described in detail elsewhere (Harrison et al. 1992). The saturation (SO_2) values thus obtained enabled an accuracy of ±5% to be achieved.

41 consenting patients undergoing regular amputation level assessment were investigated. Measurements were taken at 1 cm intervals starting medial to the tibial tuborosity (TT) and progressing to the big toe. In addition, measurements were made at 2 points, 10 cm distal from and 3 cm medial and lateral to the TT, which lie along the line of a potential BKA skin flap. The (I^{125}) 4-Iodoantipyrine (IAP) clearance technique was also used to measure skin blood flow (SBF) at these points following local injection of the isotope (McCollum et al. 1985).

Further investigations were carried out along the legs of twelve healthy adult volunteers (age range 21-42 yr). However, it was not considered appropriate to inject a radioactive substance in young, healthy volunteers for comparisons with the photometric measurements. For these investigations, therefore, photometric measurements only were carried out, as described above.

The mean SO_2 values measured at the two injection sites were recorded together with the mean value measured along the limb. The measurements along the limb were also analysed in terms of frequency histograms representing the real distribution of tissue SO_2.

Statistical analysis of the results, carried out with the aid of Lotus 123 and SPSS for Windows software, involved the Mann Whitney U and Wilcoxon tests for non-parametric data. Values of $p < 0.05$ were considered to be statistically significant.

RESULTS AND DISCUSSION

The 41 patients studied were classified into two groups according to the level of amputation they actually received: transtibial, below knee amputation (BKA) and transfemoral, above knee amputation (AKA). Of these 32 were recommended for BKA on the basis of the SBF measurements. However due to wound infection or for other clinical reasons, four of these 32 BKAs were revised to AKA. These patients were thus excluded from the analysis. 9 patients were recommended for AKA, again on the basis of SBF measurements, but one actually underwent a successful BKA. This individual was reclassified accordingly. The total population investigated thus consisted of 29 BKAs, 8 AKAs and 12 normal subjects.

Figure 1 shows the summary cumulative frequency histograms of the SO_2 values along the leg in each of the three groups. It can be seen that very few values (<2%) are to be found in the normal group within the class 0-10% SO_2. The BKA histogram is similar but shifted to the left and demonstrated a significantly greater proportion (10.8%) of SO_2 values

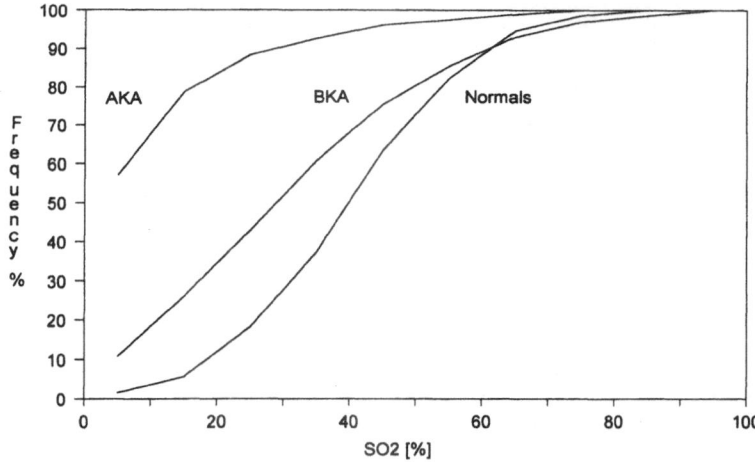

Figure 1. Cumulative frequency histograms of SO_2 summed from the two groups of patients (AKA and BKA) and the normal subjects.

in the 0-10% class. The histogram for the AKA group is shifted even more markedly to the left with 57.6% of SO_2 values less than 10%. The summary statistics for the SO_2 values along the leg are presented in the form of box-plots in figure 2. There was a significant difference in the mean SO_2 between the BKA and AKA groups (p<0.001). Also, the BKA group tended to be lower than the normals, although the difference just failed to reach significance at the 95% level.

The degree of hypoxia in the limb was defined as the percentage of SO_2 values along the leg which were below 10% saturation. Figure 3 shows a box-plot of this data. Differences

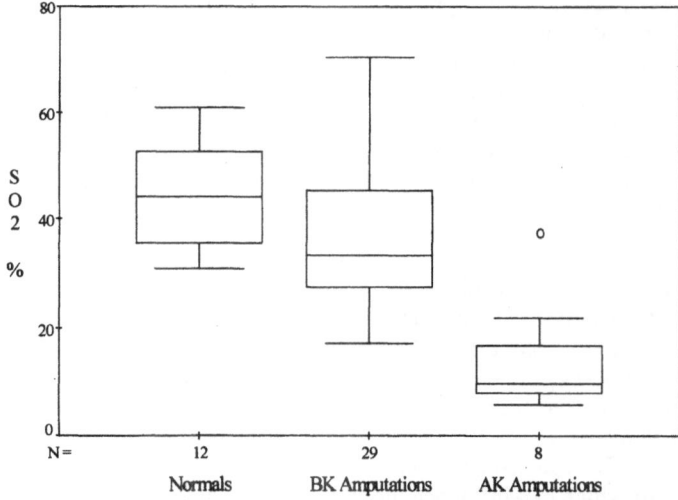

Figure 2. Box-plot summarising the results of SO_2 measurements along the leg for each group studied. Outliers are also marked (o).

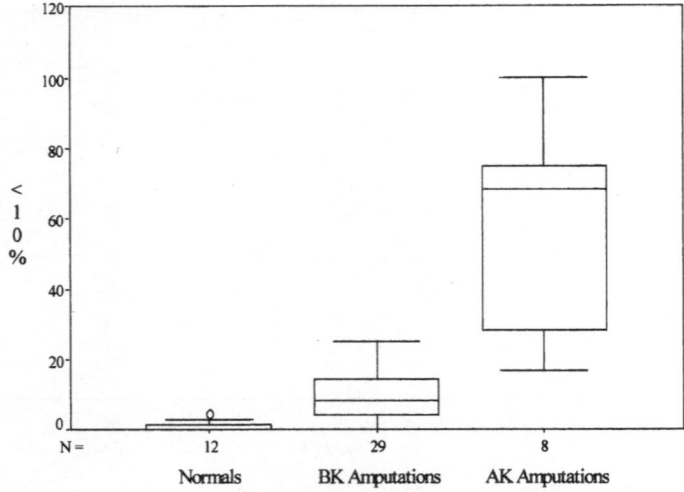

Figure 3. Box-plot summarising the results of the proportion of SO_2 values <10% (degree of hypoxia) along the leg for each group studied. Outliers are also marked (o).

between each of the three groups are significant at the p<0.001 level. It should be borne in mind that any tissue with >5% anoxia locally is already in a state of critical oxygen supply. Given also the large catchment volume of the lightguides used (Egglishaw 1994), it can be assumed that the degree of anoxia represented in figures 1-3 is substantially underestimated in this study. The oxygen supply to the skin can thus be considered as highly pathological in both patient groups.

The mean values for the SBF measurements were 6.0 and 9.8 ml/100g/min for the BKA group (medial and lateral sites respectively), and 1.6 and 1.5 ml/100 g/min for the AKA group; the differences between the two groups were significant (p<0.0001). The difference between the two sites was only significant (p<0.001) in the BKA group. These results are consistent with those previously reported from this laboratory (McCollum et al. 1985).

The mean SO_2 values for both patient groups together were compared with the SBF data in a series of linear and non-linear regresssions in order to investigate the existence of any relationship between the two parameters. Figure 4 shows a log-linear regression of mean SO_2 against SBF. It can be seen that a relationship was found between these parameters with a correlation coefficient (r) of 0.63, p<0.0001 such that:

$$SO_2 = 14.4 + 11.8 \times \ln(SBF).$$

The pattern of relationship between SBF and skin SO_2 is similar to that reported between skin SO_2 and laser Doppler flux in normal and inflamed human skin (Harrison et al. 1993). Any control of blood flow, however, is most likely to be related directly to tissue pO_2 (Kessler et al. 1984) rather than intravascular SO_2. Thus, no attempt is made to form any conclusions here as to causal relationships with regard to local regulation of flow as no information was available on local temperature, pCO_2 and pH, all of which are changed greatly in skin in CLI (Harrison & Walker 1979), each of which influences the dissociation curve, and any of which can influence blood flow independently.

The data were similarly investigated for a relationship between degree of hypoxia (% of values along the leg < 10% SO_2) and SBF. Figure 5 shows that an inverse relationship

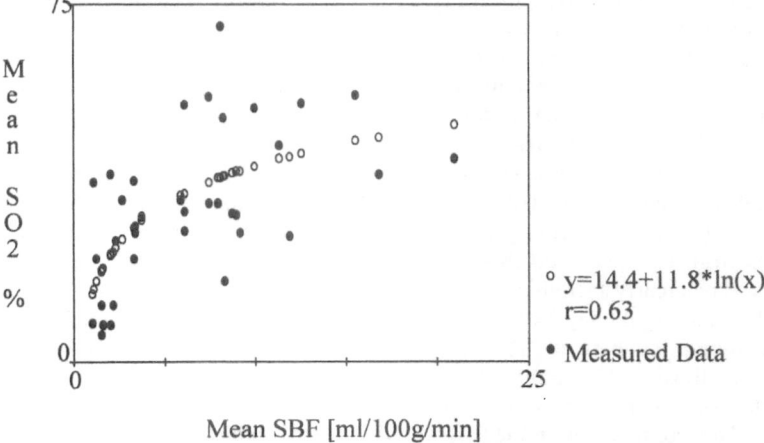

Figure 4. Relation between mean SO_2 along the leg and skin blood flow, together with the best-fit regression curve, for the two groups of patients.

was found between these two parameters, again for both patient groups together, with r=0.74 and p<0.0001:

$$\% \text{ of values} < 10 \ \%SO_2 = 2.0 + (62.1/SBF).$$

In previous work in skeletal muscle, animal experiments have shown that the degree of coupling between "local" and "central" control of capillary blood flow and tissue pO_2 is dependent on the degree of local tissue hypoxia itself (Harrison et al. 1990). However, in individual experiments when an animal was in a pathological circulatory state any element

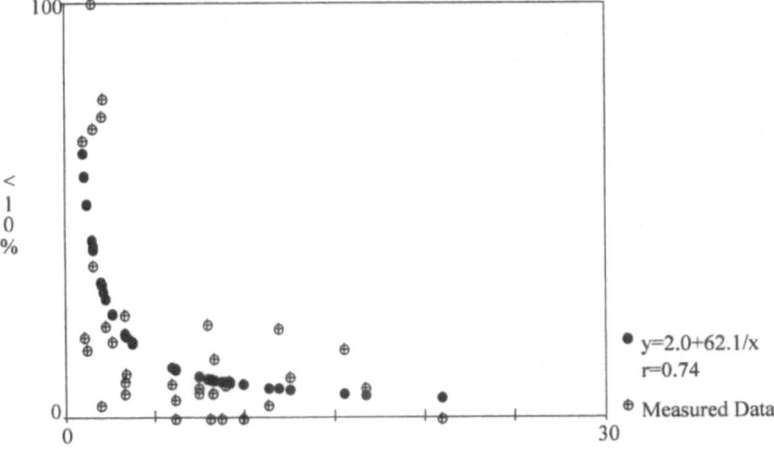

Figure 5. Relation between proportion of SO_2 values <10% (degree of hypoxia) along the leg and skin blood flow, together with the best-fit regression curve, for the two groups of patients.

of local regulation of flow was absent (Harrison et al. 1989). It is thus not surprising that, under the highly pathological flow conditions present in CLI, the degree of tissue anoxia is dependent upon local flow available rather than the existence of a degree of local physiological regulation which would normally ensure an adequate local supply (Harrison et al. 1990).

The mean values of SO_2 at the two injection sites (lateral and medial respectively) were 33.9% and 35.4% in the BKA group, and 20.6% and 10.4% in the AKA group. In the AKA group the difference between the medial and lateral sites was significant ($p<0.0001$) as was the difference from the BKA group ($p<0.04$) at the medial site.

Figure 6 shows a scatter plot of degree of hypoxia along the leg against the mean SO_2 values from the measurements at the two injection sites for each patient in the two groups. It can be clearly seen that all patients with more than 30% hypoxia (% of SO_2 values less than 10%) along the leg had an AKA, and all with less than 15% hypoxia had a BKA. Of those with hypoxia between 15 and 30%, those with a mean site $SO_2 < 30\%$ all went for AKA whereas those $> 30\%$ had a BKA. If these were to be taken as selection criteria for amputation level, the present study would yield values of 1.0 for sensitivity and specificity for the test. This compares with the SBF "gold standard" which, again for the present study, gave a selectivity of 1.0 and specificity of 0.93. The apparent high reliability of tissue spectrophotometry must be judged against an accuracy of measurement of skin SO_2 of ±5% (Harrison et al. 1992) which, with close examination of figure 6, would give some scope for possible erroneous classification of a patient. In the present study, however, the pathological differences were greater than the apparative errors.

CONCLUSIONS

The results of the study demonstrate that lightguide tissue spectrophotometry can be used for predicting the outcome of amputation in critical limb ischaemia using the criteria

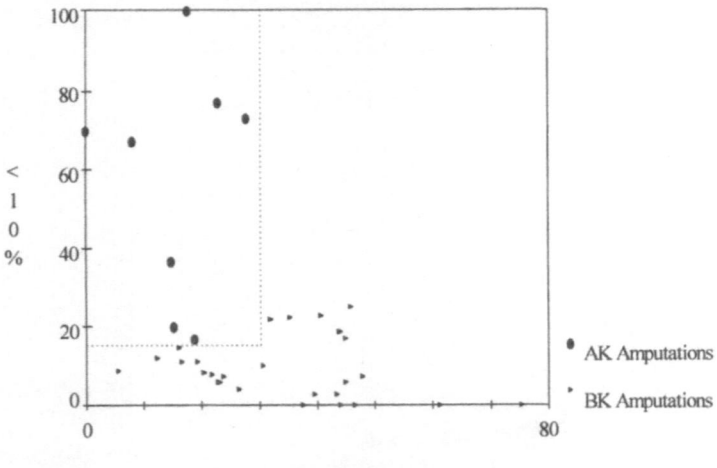

Mean SO2 at Site [%]

Figure 6. Scatter diagram, for both AKA and BKA groups, of the proportion of SO_2 values <10% (degree of hypoxia) along the leg against the mean SO_2 value from the two sites of measurement of SBF. It can be seen that the AKA values all lie within a top left hand-corner quadrant with the lower boundary intercepting the ordinate at 15% hypoxia and the right hand boundary intercepting the abscissa at 30% mean site SO_2 .

that more than 30% hypoxia (SO_2 <10%) along the lower limb is indicative of non-viability of skin flap healing and less than 15% is consistent with successful healing. Further differentiation can be achieved by measuring the SO_2 along the line of the proposed BKA skin flap. A mean value greater than 30% *in combination with* a degree of hypoxia <15% along the limb appears to be reliably consistent with a successful outcome to a BKA.

ACKNOWLEDGMENTS

Purchase of the spectrophotometer was made possible by a grant from the Research Initiatives Fund, University of Dundee. David Newton is supported by the Medical Research Council.

REFERENCES

Egglishaw, C., 1994 Quantification of catchment volumes of MCPD and EMPHO spectrophotometers, for visible wavelengths, using an in vivo model of the skin, *Hons. Disstn.*, Dept. App. Phys. Elect. and Manuf. Eng., Univ. Dundee.

Harrison, D. K., and Walker, W. F., 1979, Microelectrode measurement of skin pH in humans during ischaemia, hypoxia and local hypothermia, *J. Physiol. (Lond.)* 291:339-350.

Harrison, D. K., Birkenhake, S., Hagen, N., and Kessler, M., 1989, Capillary blood flow and local oxygen supply in skeletal muscle during acute haemodilution with hydroxyethyl starch, *Adv. Exp. Med. Biol.* 248:693-698.

Harrison, D. K., Kessler, M., and Knauf, S. K., 1990, Regulation of capillary blood flow and oxygen supply in skeletal muscle in dogs during hypoxaemia, *J. Physiol. (Lond.)* 420:431-446.

Harrison, D. K., Evans, S. D., Abbot, N. C., Swanson Beck, J., and McCollum, P. T., 1992, Spectrophotometric measurements of haemoglobin saturation and concentration in skin during the tuberculin reaction in normal human subjects, *Clin. Phys. & Physiol. Meas.* 13:349-363.

Harrison, D. K., Abbot, N. C., Swanson Beck J., and McCollum, P. T., 1993, Preliminary assessment of laser Doppler imaging for measurement of skin perfusion using the tuberculin reaction in human skin as a model, *Phys. Meas.* 14:241-252.

Hickman, P., Walker, M.A., Harrison, D., Belch, J., and McCollum P.T,, 1992, Use of lightguide reflectance spectrophotometry in the assessment of peripheral arterial disease, *Br. J. Surg.* 79:361.

Kessler, M., Höper, J., Harrison, D. K., Skolasinska, K., Klövekorn, W. P., Sebening, F., Volkholz, H. J., Beier, I., Kernbach, C., Rettig, V., and Richter, H., 1984, Tissue O_2 supply under normal and pathological conditions, *Adv. Exp. Med. Biol.* 169:69-80.

McCollum, P.T., Spence, V. A., and Walker, W. F., 1985, Circumferential skin blood flow measurements in the ischaemic limb, *Br. J. Surg.* 72:310-312.

McCollum, P. T., Spence, V. A., and Walker, W. F., 1988, Amputation for peripheral vascular disease: the case for level selection, *Br. J. Surg.* 75:1193-1195.

Shah, D. M., Darling, R. C. III., Chang, B. B., Kaufman, J. L., Fitzgerald, K. M., and Leather, R. P., 1992, Is long vein bypass from groin to ankle a durable procedure? An analysis of a ten-year experience, *J. Vasc. Surg.* 15:402-407.

6

IN VIVO MEASUREMENT OF OXYGEN PRESSURE USING 19F-NMR IMAGING

J. Lutz[1], U. Nöth[2], S. P. Morrissey[3], H. Adolf[2], R. Deichmann[2], and
A. Haase[2]

[1] Department of Physiology, University of Würzburg
Röntgenring 9, D-97070 Würzburg, Germany
[2] Department of Biophysics, University of Würzburg
Am Hubland, D-97074 Würzburg, Germany
[3] Department of Neurology, University of Würzburg
J.-Schneider-Str. 11, D-97080 Würzburg, Germany

INTRODUCTION

Nuclear magnetic resonance (NMR) of the fluorine nucleus is an *in vivo* method to measure non-invasively oxygen pressures. The measurement is based on the effect that the longitudinal relaxation rate ($R_1 = 1/T_1$) increases linearly with the partial oxygen pressure (pO_2) when the temperature is kept constant [1,2]:

$$R_1 = 1/T_1 = A + B \cdot pO_2 \tag{1}$$

The calibration constants A and B depend strongly on temperature, magnetic field strength, the fluorinated substance, the resonance line used for imaging, and on the tissue type.

Since fluorine does not occur in the organism except in small quantities in bones and teeth it has to be administered as contrast agent. Thus, no disturbing background signals arise as e.g. in 1H NMR. On the other hand, the signal to noise ratio (SNR) is limited by the amount of the fluorinated contrast agent which can be injected.

Emulsions of perfluorocarbons (PFCs) initially developed as artificial oxygen carriers [3-5] are suitable contrast agents for this purpose. Their molecular mass is of about 460 to 520 g/mole, and in most cases all hydrogen atoms are substituted by fluorine atoms. Since not all fluorine atoms in these compounds are chemically equivalent, they normally exhibit NMR spectra with several resonance lines resulting in chemical shift artifacts in 19F magnetic resonance imaging (MRI). We used only the resonance line with the strongest intensity to perform quantitative imaging, all others were suppressed.

Oxygen Transport to Tissue XVII, Edited by Ince et al.
Plenum Press, New York, 1996

MATERIALS AND METHODS

The PFC used in this study was bis-perfluorobutylethylene (F-44E). This substance has two chemically equivalent CF_3 groups resulting in the strongest line of the spectrum used for imaging and three further lines from the CF_2 groups which were suppressed.

The calibration curve R_1 versus pO_2 was determined by measuring the T_1 values of the CF_3 line for neat F-44E bubbled with six different gas mixtures of oxygen and nitrogen at atmospheric pressure (760 Torr) and 36°C.

All NMR experiments were performed on a 7.05 Tesla Bruker Biospec System with a 21 cm horizontal bore, operating at a fluorine frequency of 282.54 MHz.

The high resolution 19F-images of the T_1 distribution, i.e. the "T_1 maps" were acquired in the following way: Directly after the inversion of the magnetization 8 T_1-weighted 19F-NMR images were acquired using a fast NMR-imaging sequence [6,7] which suppresses all off resonance lines. From these images the T_1 map is calculated pixelwise as described by Deichmann and Haase [8] and converted into a pO_2 map by using the calibration curve (eq. 1).

For *in vivo* studies female Lewis rats with a body weight of about 180g received an i.v. bolus injection via the tail vein of a 90% w/v emulsion of F-44E. The dose was 10 g/kg body weight. Anesthesia was introduced with 5% and maintained at 1.4% isoflurane in oxygen flowing at a rate of 1.5 l/min. NMR studies were performed three days post injection of F-44E when the concentration in the liver and spleen is near its maximum [9].

RESULTS AND DISCUSSION

The calibration constants A and B were obtained from a linear fit of R_1 versus pO_2 according to equation 1 (Fig. 1). The partial oxygen pressure is given in percent; 100% correspond to a partial pressure of 760 Torr. The values are (0.319 ± 0.026)/sec for A and (0.0165 ± 0.0005)/(sec%) for B. The calibration was done in a phantom, i.e. in vitro, and not in the tissue.

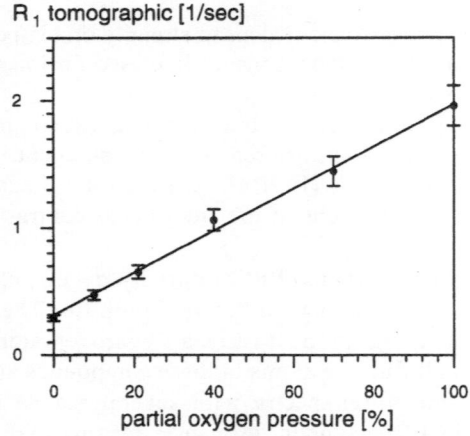

Figure 1. Calibration curve for F-44E at 36°C and 7.05 Tesla. Partial oxygen pressures are given in percent. 100% corresponds to 760 Torr.

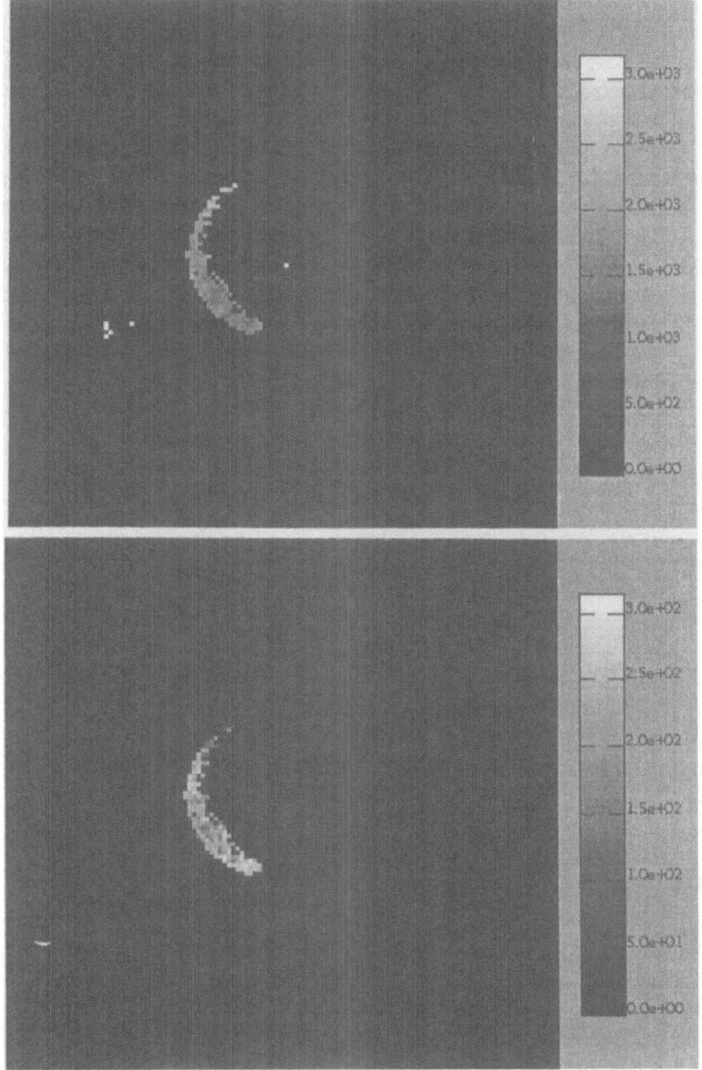

Figure 2. (a) Quantitative T_1 map of the spleen with a resolution of 0.6 mm x 0.6 mm and a slice thickness of 4 mm. The image was acquired 3 days post injection of a dose of 10 g/kg F-44E in form of a 90% w/v emulsion. T_1 values are given in msec. The average T_1 value is 1.9 sec. (b) Quantitative pO_2 map of the spleen calculated pixelwise from the T_1 map of Fig. 2a using the calibration constants A and B. pO_2 values are given in 10^{-1} %. The average pO_2 value is 15% or 114 Torr.

The T_1 and pO_2 maps of the liver and spleen have a field of view of 8 cm x 8 cm with a resolution of 0.6 mm x 0.6 mm and a slice thickness of 4 mm, i.e. a voxel resolution of 1.45 mm^3. The total acquisition time of one pO_2 map amounted to about 20 minutes. Fig. 2 shows a representative T_1 (Fig. 2a) and pO_2 map (Fig. 2b) three days post injection of F-44E of the spleen of a rat which is breathing pure oxygen. The average pO_2 value from 3 rats was (38 ± 11) Torr in the liver and (170 ± 50) Torr in the spleen. The pO_2 values in the liver are lower than in the spleen. This is likely to be due to a higher metabolism in the liver, and, therefore, resulting in a higher oxygen consumption in this organ.

During the first 36 hours after i.v. administration when most of the PFC is still in the blood, intravascular pO_2 can be measured in large vessels [10]. When all PFC is removed from the blood and sequestered by the cells of the reticuloendothelial system, pO_2 maps of the liver and spleen represent the intracellular pO_2 values [10]. During the time in between, the pO_2 maps in the liver and spleen reflect a weighted average of the blood and intracellular pO_2 value. When the rats are breathing a mixture of 10% oxygen and 90% nitrogen, i.e. under hypoxic conditions, the calculated pO_2 values are negative, an effect also observed by other groups [11,12]. We suppose that especially the calibration constant A, determined in vitro, depends on the tissue, and that this is the reason for the negative pO_2 values. However, when measuring pO_2 values in the same tissue under different conditions, it is possible to obtain the pO_2 change, which does not depend on the calibration constant A, and will be useful as an *in vivo* parameter.

CONCLUSION

In conclusion, we described a method to acquire high resolution 19F-NMR images without chemical shift artifacts. From the acquired data quantitative T_1 and pO_2 maps can be calculated. Furthermore, by performing pO_2 measurements at different oxygen concentrations in the breathing gas and at different time points post injection of the PFC, changes in pO_2 and intravascular and intracellular pO_2 values can be determined [10]. This method is of potential clinical interest, e.g. as an *in vivo* parameter to assess tumor therapy. Macrophages scavenge PFC and are a histological feature of inflammation and tumor tissue. Thus, pO_2 measurements in tumors [12-15] are another important topic, since the state of oxygenation of the tumor is an important parameter in order to evaluate the efficacy of tumor treatment, e.g. by radiotherapy.

ACKNOWLEDGMENT

This work was supported by the Wilhelm Sander–Stiftung (grant 93.001.1). Many thanks to Dr. J.G. Riess and Dr. M.P. Krafft for providing the F-44E emulsion and Dr. J.L. Howell for supplying with neat F-44E. U. Nöth and R. Deichmann acknowledge a postgraduate scholarship of the Studienstiftung des deutschen Volkes.

REFERENCES

1. Parhami P, Fung BM. Fluorine-19 relaxation study of perfluoro chemicals as oxygen carriers. J. Phys. Chem. *87*: 1928-1931 (1983)
2. Mason RP, Shukla H, Antich PP. *in vivo* oxygen tension and temperature: Simultaneous determination using 19F NMR spectroscopy of perfluorocarbon. Magn. Reson. Med. *29*: 296-302 (1993)
3. Naito R, Yokoyama K. Perfluorochemical blood substitutes Fluosol-43, Fluosol-DA 20% and 35%. Technical Information Ser. No. 5. Green Cross Corp., Osaka p. 1-177 (1978)
4. Riess JG, Le Blanc M. Solubility and transport phenomena in perfluorochemicals relevant to blood substitution and other biomedical applications. Pure & Appl. Chem. *54*: 2383-2406 (1982)
5. Long DM, Long DC, Mattrey RF, Long R, Burgan AR, Herrik WC, Shellhamer DF. An overview of perfluoroctylbromide - application as a synthetic oxygen carrier and imaging agent for x-ray, ultrasound and nuclear magnetic resonance. Biomat. Art. Cells Art. Organs *16*: 411-420 (1988)

6. Haase A. Snapshot FLASH MRI. Applications to T_1, T_2, and chemical-shift imaging. Magn. Reson. Med. *13*: 77-89 (1990)

7. Nöth U, Deichmann R, Adolf H, Schwarzbauer C, Haase A. Fast acquisition of pO_2 maps using 19F MRI and a new method for the suppression of chemical shift artifacts. J.Magn. Reson. B *105*, in press (1994)

8. Deichmann R, Haase A. Quantification of T_1 values by SNAPSHOT-FLASH NMR ima-ging. J. Magn. Reson. *96*: 608-612 (1992)

9. Nöth U, Morrissey SP, Deichmann R, Schwarzbauer C, Lutz J, Haase A. Retention half life of the perfluorocarbon F-44E in the reticuloendothelial system measured by fast 19F-MRI. 2nd Meeting of the SMR, San Francisco, Proc. Soc. Magn. Reson. Vol. *2*: 721 (1994)

10. Nöth U, Morrissey S, Deichmann R, Adolf H, Schwarzbauer C, Lutz J, Haase A. *in vivo* measurement of partial oxygen pressure in large vessels and in the reticuloendothelial system using fast 19F-MRI (submitted). Magn. Reson. Med. 34 (1995)

11. Mason RP, Jeffrey FMH, Malloy CR, Babcock EE, Antich PP. A noninvasive assessment of myocardial oxygen tension: 19F NMR spectroscopy of sequestered perfluorocarbon emulsion. Magn. Reson. Med. *27*: 310-317 (1992)

12. Hees PS, Sotak CH. Assessment of changes in murine tumor oxygenation in response to nicotinamide using 19F NMR relaxometry of a perfluorocarbon emulsion. Magn. Reson. Med. *29*: 303-310 (1993)

13. Mason RP, Antich PP, Babcock EE, Constantinescu A, Peschke P, Hahn EW. Non-invasive determination of tumor oxygen tension and local variation with growth. Int. J. Radiation Oncology Biol. Phys. *29*: 95-103 (1994)

14. Mason RP, Antich PP. Tumor oxygen tension: Measurement using Oxygent™ as a 19F NMR probe at 4.7 T. Art. Cells Blood Subs. Immob. Biotech. *22*: 1361-1337 (1994)

15. Dardzinsky BJ, Sotak CH: Rapid tissue oxygen tension mapping using 19F inversion-recovery echo-pla-nar imaging of perfluoro-15-crown-5-ether. Magn. Reson. Med. *32*: 88-97 (1994)

7

O₂ FLUX OPTODE

A New Sensing Principle to Determine the Oxygen Flux and Other Gas Diffusions

D. W. Lübbers, T. Köster, and G. A. Holst

Max-Planck-Institut für Molekulare Physiologie
Rheinlanddamm 201
D-44139 Dortmund
Germany

INTRODUCTION

By applying a transparent test membrane of defined diffusion properties between two layers of optical chemical oxygen sensors, O_2 optodes, the measurement of oxygen diffusion, the oxygen flux, across the membrane becomes possible.[1] The optode has the inherent advantage towards the established method of electrochemical oxygen measurements that the sensor is permeable for oxygen and does not consume the analyte. With this new sensing principle (Fig. 1) the O_2 flux is calculated as the product of the oxygen partial pressure (pO_2) difference between both sides of the membrane and the diffusion properties of the membrane. This sensor can be used to measure the O_2 flux in different applications, e. g. the O_2 flux into human tissue, into technical compartments as bioreactors or into biological systems. The selectivity of the proposed sensor is strongly influenced by the choice of an appropriate test membrane. With suitable indicators it is also possible to measure other gas fluxes. So, this principle opens up a new field of flux measurements

Measurements of the pO_2 in skin tissue with a needle electrode[2] have shown the presence of a pO_2 gradient in the upper layers of the skin. The gradient between the pO_2 of the atmosphere and the pO_2 minimum in the living tissue causes a diffusion of oxygen, an O_2 flux, into the skin. Although this oxygen uptake represents only 1-2% of the basal human O_2 consumption, it is an important factor for the O_2 supply of the upper layers of the skin. The O_2 uptake through the skin surface strongly depends on local skin blood flow and therefore can be used as an important parameter to qualify and quantify local microcirculation. For this reason the presented work is focussed on the measurement of the O_2 flux into human skin.

Oxygen Transport to Tissue XVII, Edited by Ince et al.
Plenum Press, New York, 1996

59

MATERIAL AND METHOD

Sensor Membrane

The O_2 flux optode consists of a test membrane, permeable to oxygen, and two optodes that measure the pO_2 gradient across the membrane[3,4,5]. So in this multilayer sensor the pO_2 is measured with two immobilized oxygen sensitive fluorophores in the indicator layers. As the diffusion properties K (Fig. 1) of the membrane are known, the oxygen flux JO_2 across is calculated by the following equation:

$$JO_2 = C \cdot (pO_2(\text{optode}_2) - pO_2(\text{optode}_1)) \qquad \text{with } C = \frac{K}{d} \tag{1}$$

$K = \alpha \cdot D$, $\alpha = O_2$ solubility coefficient, $D = O_2$ diffusion constant, d = thickness of the membrane.

Many procedures of constructing optical pO_2 sensors or sensing layers are published[6-10,11-17]. For the O_2 flux measurement the proposed multilayer sensor was modified as follows.

If the upper pO_2 is the atmospheric pO_2 it is measured by other means optode1 in Fig. 1 can be omitted. The diffusion membrane of the sensor is made of perfluoralkoxy-co-polymer - PFA (Teflon, DuPont de Nemur, Bad Homburg, Germany or Hostaflon, Hoechst, Wiesbaden, Germany) or polypropylene - PP (Trespaphan, Hoechst, Wiesbaden, Germany). As indicator a ruthenium complex: tris (1,10-phenantroline) ruthenium (II) chloride hexa-hydrate (λ_{ex} = 470 nm, λ_{em} = 605 nm, Aldrich–Chemie, Steinheim, Deutschland) was adsorbed on silica gel particles (Merck, Darmstadt, Germany). These dyed particles are embedded in a 50 µm silicone layer (Wacker–Chemie, Burghausen, Germany). The optical insulation is made of black silicone (Wacker–Chemie, Burghausen, Germany) and prevents high amounts of straylight from disturbing the optical detector. Technical uncertainties in

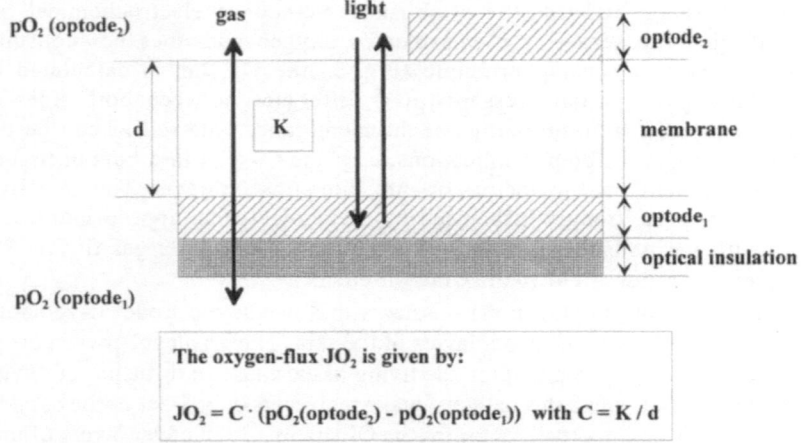

Figure 1. Principle of the O_2 flux optode (schematic drawing). The diffusion of O_2 through the test membrane causes a pO_2 difference between optode1 and optode2 that is proportional to the O_2 flux.optode1, optode2 = indicator layers, membrane = test or diffusion membrane, optical insulation = layer to separate the signal of the optode from optical background, d = thickness of the membrane, K = diffusion properties of the membrane.

the process of manufacturing cause a variability in thickness of the optical insulation (for more details about the production process [8,11,13,18])

To test the reproducibility of the method and the measuring system an additional sensor membrane was used for transcutaneous pO_2 ($tcpO_2$) measurements because of the better reproducibility of the manufacturing process. The layer setup of the $tcpO_2$ membrane is similar to the above described (Fig. 2):

a. a transparent polyester foil (Mylar D, DuPont de Nemur, Bad Homburg, Germany), impermeable for oxygen, thickness = 100 μm,
b. an indicator layer as described with an embedded oxygen sensitive fluorophore, thickness = 50μm,
c. an optical insulation layer of black silicone, thickness = 50 μm.

Sensor Head

The sensor head should be easily applicable to the skin surface and should contain a thermal regulation to keep the temperature constant at the place of measurement because nearly all fluorescence measurements are strongly temperature dependent.

The developed sensor head (Fig. 3) contains the holding for the sensor membrane. It can be applied to the skin surface by pressing it slightly with a rubber band that is fixed around the sensor head and, for example, the forearm. The light for excitation of the fluorophore and the emitted light are transmitted via a bended bifurcated glass fibre bundle (bifb, NB68, fibre diameter = 50 μm, active diameter = 2 mm, number of fibres = 1300, Volpi, Schlieren, Switzerland) where the fibres at the sensor head end are statistically mixed. The bifb is held by the cylindrical sensor housing (diameter = 25 mm and height = 18 mm) which is divided in three parts: the upper sensor housing (ush, black polyvinyl chloride, PVC), the heating ring (hr, stainless steel) and the lower sensor housing (lsh, black PVC). The heating ring has several functions in this setup. It contains a small gas test chamber (k), which allows different gas mixtures to be present on the upper side of the flux-optode. The feeding of the chamber

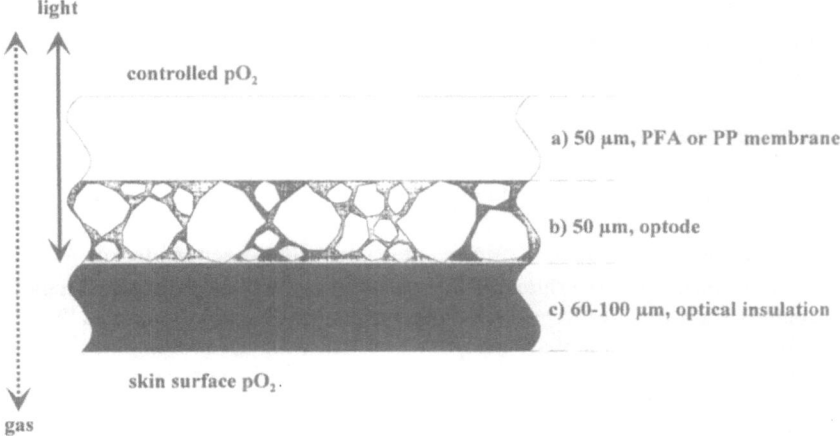

Figure 2. Structure of the O_2 flux optode (schematic drawing). a) test or diffusion membrane, thickness = 50 μm, b) oxygen sensitive indicator layer, thickness = 50 μm, c) optical insulation, variable thickness = 60-100 μm.

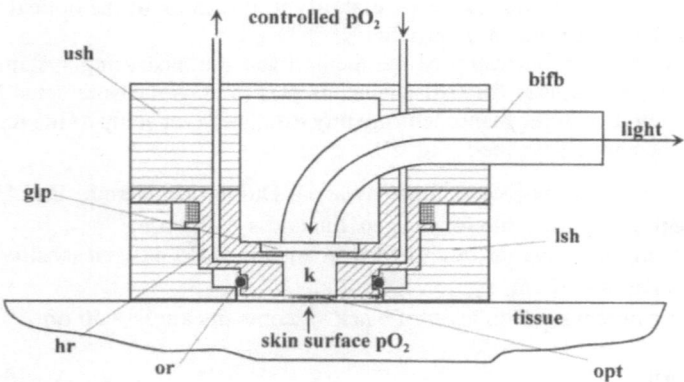

Figure 3. O_2 flux sensor head (schematic drawing). Ush = upper sensor housing, glp = glass plate, hr = heating ring, or rubber ring, k = test chamber, opt = O_2 flux optode, lsh = lower sensor housing, bifb = bifurcated glass fibre bundle (for explanation see text).

k is done by small stainless steel tubes (outer diameter = 1.6 mm, inner diameter = 1 mm). Towards the bifb the chamber is closed by a glass plate (glp) which is glued on hr. Additionally hr carries a winding of a heating wire and a temperature sensor (not drawn) for the necessary thermal regulation. Then the sensor membrane (opt) is held at the lower end of hr by straining it with a rubber ring (or). This technique proved to be the simplest way of mounting the flexible sensor membranes (opt) while closing the chamber k gas-tight. The tcpO$_2$ sensor head is similar but it does not contain a test chamber k.

The Measuring System - FLOX

For the evaluation of the oxygen dependent fluorescence signal a special measuring system was developed. The dynamic quenching of fluorescence changes the fluorescence intensity and life time. As the more stable parameter the fluorescence lifetime was chosen to measure the desired pO$_2$ in the layer between test membrane and skin surface.

If the exciting light of the fluorophore is sinusoidally intensity modulated then there exists the following well known relation[6,7] between fluorescence lifetime τ and phase angle Φ between exciting and emitted light:

$$\tan(\Phi) = 2\pi f_{mod}\tau \tag{2}$$

with Φ = phase angle, f_{mod} = modulation frequency, τ = fluorescence lifetime

The lifetime of the used fluorophore lies in the range of a few hundred nanoseconds which demands short time resolved measuring systems or - using concepts of the transmission of information signals - a phase measuring system with a modulation frequency range of a few hundred kilohertz. The developed system belongs to the phase measuring systems.[19] The measuring system, FLOX (FLux and OXygen), can be divided into three main units: 1) the already described sensor with sensor head and sensor membrane that is connected via a bifurcated glass fibre bundle to 2) the analog signal processing unit FLOX, where the phase signals are detected, and 3) the digital signal processing FLOX-Control, where the phase signals are stored and evaluated.[20]

The analog signal processing FLOX uses a light emitting diode (LED) (λ(peak) = 470 nm, GaN, Ledtronics, Torrance, CA, USA) with an optical bandpass filter (FITCA 40, Schott, Mainz, Germany) as light source for the excitation and a photo multiplier tube (PMT) (HC125, Hamamatsu, Hersching, Germany) with an optical longpass filter (KV550, Schott, Mainz, Germany) for the detection of the emission. Additionally it contains all detection, thermal regulation and safety circuits for the measurement.

The digital signal processing FLOX Control is a personal computer with input/output boards to control the measuring system, store the data and display the measured signals and calculated values.

Experimental Setup

Two different setups were used for the presented results as shown in Figure 4.

a. O₂ flux measurements at the skin surface (Fig. 4, a). One flux sensor was connected to the measuring system, FLOX and FLOX-Control, by the bifurcated glass fibre bundle (bifb) and the electrical leads (not drawn) of the temperature control. The sensor head (sh) and the cables were fixed with a rubber band. The rubber band keeps the sensor head with a constant slight pressure on the skin surface and prevents any gas diffusion from the side. During the measurement there was room air present in the test chamber above the sensor membrane with the in- and outlet tubings just left open.

b. Calibration measurements (Fig. 4, b). For this purpose the sensor head was clamped on a calibration chamber (cch). Then both chambers, the one above the sensor membrane and the calibration chamber below, were perfused with the same gas mixtures (gm). These mixtures were generated by a gas mixing pump (gmp, 1M100/a-F, precision +/- 0.01%, Wösthoff, Bochum, Germany) from two gases, oxygen (O₂) and nitrogen (N₂) and preheated (prh) before they enter the chambers.

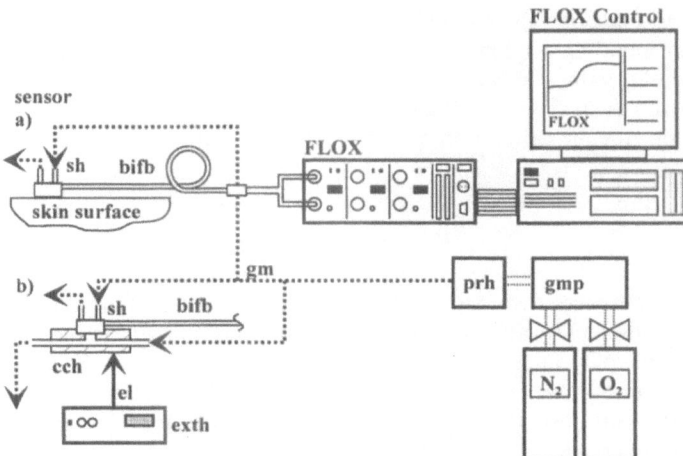

Figure 4. Experimental setup with the FLOX measuring system and the O₂ flux sensor (schematic drawing). Sh = sensor head for O₂ flux or tcpO₂ measurements, bifb = bifurcated glass fibre bundle, gm = gas mixtures, cch = calibration chamber, el = electrical leads, extc = external temperature control, prh = preheating water bath for the gas mixtures, gmp = gas mixing pump, N₂ = gas cylinder containing nitrogen, O₂ = gas cylinder containing oxygen (for explanation see text).

The gas mixing pump enables a control in the one percent volume range. The supply tubings were made from stainless steel (outer diameter = 1.6 mm, inner diameter = 1 mm) except for the short connections and the supply tubes to the sensor head that were made of silicone. All gas mixtures were pumped against a water column of around 100 mm level to keep constant pressure conditions in the test chambers. This setup was identical for both sensor heads except the $tcpO_2$ sensor has no test chamber above the sensor foil to be perfused.

Test Procedures

a. O_2 flux measurements - at the beginning of each measurement the pO_2 of the surrounding air was measured for 5 minutes. Then the flux sensor head was applied to the skin surface at the lower forearm of a test person and fixed with the rubber band. Before the application a few drops of a physiological sodium chloride solution were given on the skin. When the signal reached a steady state, a cuff around the arm was inflated to stop the blood circulation of the arm (occlusion) for 4 to 5 minutes. After opening the occlusion the measurement was continued until a steady state was reached. Then the sensor head was removed and the pO_2 of the surrounding air was measured for 10 minutes. During this time an ointment (Finalgon forte, Thomae, Biberach, Germany) was rubbed in at the measured spot to increase the blood circulation. Then the described procedure was repeated. Finally when the sensor head was removed the second time the pO_2 of the surrounding air was measured for 5 minutes and the measurement was stopped.

b. Calibration measurements. For every sensor membrane both chambers were perfused with the same gas mixtures. The calibration cycles: 20-30-100-room air-10-0-5 % O_2 (with the corresponding percent values of N_2) were repeated three times. Each mixture was kept for 4-5 minutes to reach a steady state. The measured and stored signals were calculated, converted into ASCII data and imported into a commercial scientific program (Sigma Plot, Jandel Scientific, Erkrath, Germany), where they were further processed. From the steady part of the measured phase angle of a gas mixture a 51 s time window was copied and the values of the corresponding window were averaged. For the conversion of the percent O_2 values into pO_2 values in Torr dry gases were assumed because the gases were taken directly from the gas cylinders. The following equation was used for a curve fit to describe the calibration data and convert the flux measurement:

$$\Phi_{mess} = A \cdot \frac{(1 + C \cdot pO_2)}{1 + B \cdot pO_2)} \tag{3}$$

The fit parameters A, B and C were calculated for the measured phase angles Φ_{mess} and the corresponding pO_2 values using the graphic software. It was found that by fitting the values in the whole pO_2 range of pO_2 = 0 - 760 Torr (0 - 100 % O_2, ABC100-Fit) a systematic overestimation of lower pO_2 values occurred. Therefore the fits were calculated for a pO_2 range of 0 - 160 Torr (0 - 20 % O_2, ABC20-Fit). The phase angles of the O_2 flux measurement were converted with these ABC20 parameters:

$$pO_2 = \frac{(A - \Phi_{mess})}{9B \cdot \Phi_{mess} - C \cdot A)} \tag{4}$$

RESULTS

To prove the performance of the measuring system and of the sensor membranes calibration curves were measured with different sensor spots of the same sensor foil (see Tab. 1, 4 spots at different days).

The calibration measurements were done using the tcpO$_2$ sensor foil, because of the better reproducibility in the production process of the sensor (AVL, Graz, Austria). The best reproducibility was found in the lower pO$_2$ range. This is caused by the better signal to noise ratio in this range. In the range of 0 - 160 Torr the phase angle changed by a mean difference of $\Delta\Phi_{0-160}$ = 26.38 ° while between 160 - 760 Torr there is only a mean difference of $\Delta\Phi_{160-760}$ = 17.63 °. For the used sensor membrane the frequency was adjusted to f_{mod} = 300 kHz. Table 1 shows at different pO$_2$ values the phase angle, its standard deviation in degree and the corresponding pO$_2$ value in Torr. The last column gives the deviation of the measured phase angle from the ABC20-fit. In the range of 0 - 160 Torr the average resolution of the determination of the pO$_2$ amounts to a ΔpO$_2$ between 0.65 - 2.27 Torr.

A typical O$_2$ flux measurement is shown in Fig. 5. The pO$_2$ value before the test measurement amounted to 172.7. +/- 0.3 Torr (n = 10). By conditioning the membrane the steady state value for the pO$_2$ of air is 161.2 +/- 0.3 Torr (n = 51; t5) and 160.6 +/- 0.3 (n = 51; t7). At t1 the flux sensor head is applied to the left lower forearm of a male test person as described in "Test Procedures". The sudden drop of the pO$_2$ signal is caused by temperature changes occurring when the sensor surface comes into contact with the skin. The FLOX reacts with heating up the sensor head until the adjusted temperature is reached (37 °C for the actual flux measurement). A higher heating power is necessary because of the active cooling of the blood circulation. It takes from t1 to t2 to reach nearly (in this example) a steady state of 114.8 +/- 0.1 Torr in the layer between diffusion membrane and skin surface (Fig. 2, optode2). At t2 the cuff is inflated (occlusion) so that oxygen is no longer delivered by the circulatory system. The upper layers of the skin react in taking up more oxygen from the outside which can directly

Table 1. Reproducibility measurements using the tcpO$_2$ sensor
(four different spots of the same sensor foil): Resolution and accuracy

Date	pO$_2$ [Torr]	Phase angle [°]	Standard deviation +/- [°]	Standard deviation +/- [Torr]	Deviation from fit [°]
13.01.94	15.2	80.66	0.24	0.8	-0.07
14.01.94	15.2	80.87	0.19	0.7	-0.10
17.01.94	15.4	81.01	0.15	0.5	-0.10
20.01.94	15.4	79.73	0.17	0.6	-0.09
13.01.94	76.0	68.05	0.19	1.2	-0.02
14.01.94	75.9	68.16	0.17	1.1	0.19
17.01.94	77.2	68.41	0.20	1.3	0.20
20.01.94	77.2	67.07	0.24	1.6	0.16
13.01.94	152.0	59.39	0.29	3.3	0.11
14.01.94	151.8	59.42	0.17	1.9	0.11
17.01.94	154.4	59.21	0.13	1.5	-0.03
20.01.94	154.4	58.00	0.20	2.3	-0.07
13.01.94	760.0	41.77	0.18	17.7	0.24
14.01.94	759.0	41.77	0.14	13.5	0.25
17.01.94	772.0	41.28	0.15	14.1	0.32
20.01.94	772.0	40.67	0.16	15.9	0.22

Measured with a tcpO$_2$-sensor foil.

pO$_2$ [Torr]

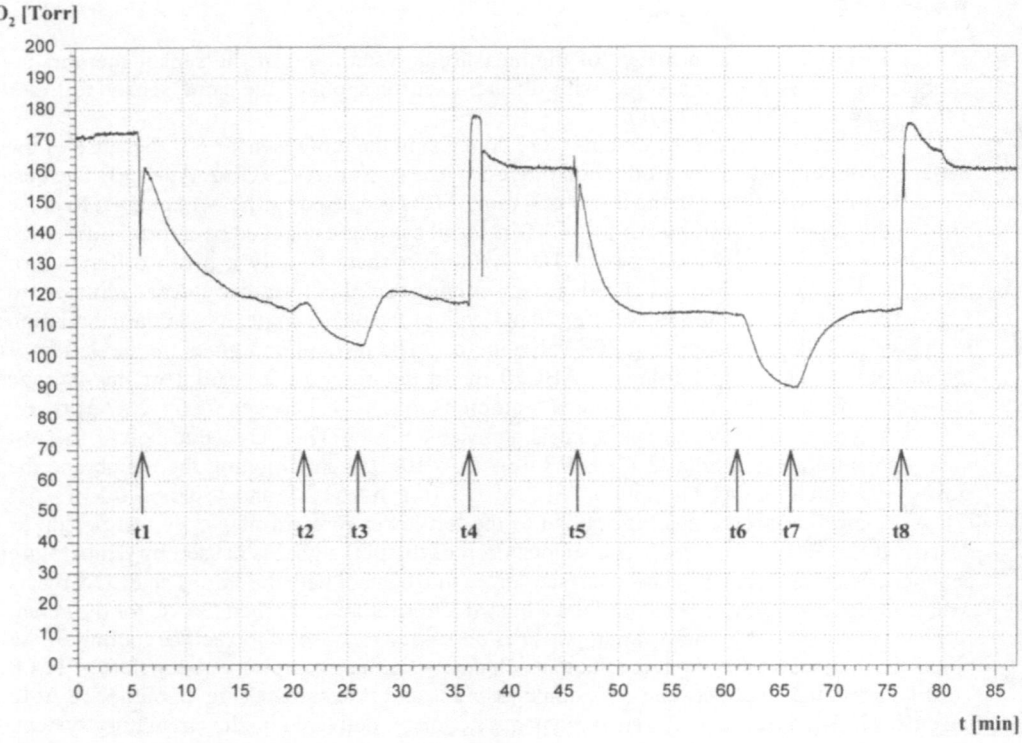

Figure 5. O$_2$ flux measurement on the human skin (pO$_2$ vs. time). After skin contact the sensor pO$_2$ decreases from 172.7 +/- 0.3 Torr to 114.8 +/- 0.15 Torr and after occlusion to 103.8 Torr. t1 = application of the sensor head to the skin surface, t2 = inflation of a cuff to stop the blood flow (occlusion), t3 = opening of the cuff, t4 = removal of the sensor head, t5 = second application of the sensor head after applying Finalgon at the measured spot, t6 = inflation of a cuff, t7 = opening of the cuff, t8 = final removal of the sensor head.

react in taking up more oxygen from the outside which can directly be seen by the decrease of the pO$_2$ to = 103.8 +/- 0.10 Torr between t2 and t3. This decrease of pO$_2$ means an increase of the pO$_2$ difference to the steady state value of the pO$_2$ of the air of ΔpO$_2$ = 44.0 Torr to a ΔpO$_2$ = 57.4 Torr which corresponds to an increase of the oxygen flux by 30.3 %. At t3 the cuff is opened again and the pO$_2$ value reaches a steady state of 117.7 +/- 0.2 Torr. At t4 the sensor head is removed and measures room air (161.2 +/- 0.3 Torr). After a short time the foil was pressed against the sensor housing which could be seen in the signal drop after t4. In the time between t4 and t5 the skin surface was rubbed with Finalgon. The provoked increase of the blood circulation could be seen in a slight redness of the skin. At t5 the sensor head was applied a second time to the same measuring spot. Between t5 and t6 first the temperature caused signal drop can be seen while the signal reaches its steady state value of pO$_2$ = 114.1 +/- 0.2 Torr much faster than before. The time in which the signal reaches 90% of the steady state value t_{90} amounts to 138 s in the first measurement and after application of Finalgon it decreases to t_{90} = 249 s.

At t7 the cuff was opened again and the value reaches a pO_2 = 115.4 +/- 0.2 Torr. Finally at t8 the sensor head was removed and for 10 minutes the pO_2 of room air was measured and the measurement was stopped thereafter.

CONCLUSION

The developed sensors and the measuring system FLOX showed a sufficient performance to measure for the first time an oxygen flux e.g. into human tissue. Reproducibility and resolution of the oxygen measurement have been checked by using tcpO₂ sensor foils (see Tab. 1). The relationship between phase angle and pO_2 is nonlinear, therefore, sensitivity and accuracy depend on the size of the pO_2 value. In the range of 0 - 160 Torr the average resolution amounts to a ΔpO_2 between 0.65 and 2.27 Torr. Sufficient calibration measurements are mandatory and have to be performed after each flux measurement.

The presented O_2 flux measurement show the expected close relation between local microcirculation and local O_2 flux into human skin. The change in time constants t_{90} shows furthermore that the ointment Finalgon has distinct effects on the gas exchange of the skin. This should be further investigated. These first results of measurement with the O_2 flux optode and the measuring system FLOX offer the basis for clinical evaluation using this new method to diagnose microcirculatory disturbances. This type of sensor membrane has an inherent advantage as it could be manufactured in a bigger size and still remains flexible to cover larger areas. With the imaging techniques it would be possible to receive two-dimensional, topographic pictures of the O_2 flux into the skin.

With these first O_2 flux measurements the proposed new principle could be proved. Consequently, similar sensors for other gas diffusion processes through surfaces are possible and can be used in other biomedical and biotechnical applications.

ACKNOWLEDGMENTS

We gratefully acknowledge the financial and technical support of the AVL List GmbH, Graz, Austria and the financial support of the Lipha Arzneimittel GmbH, Karlsruhe, Germany. We would also like to thank H. Baumgärtl, R. König, L. Teckhaus and their groups at the Max-Planck-Institut für Molekulare Physiologie, Dortmund, Germany for their excellent collaboration.

REFERENCES

1. D.W. Lübbers, Fluorescence Based Chemical Sensors, *Adv. Biosens.*, 2:215-260 (1992).
2. H. Baumgärtl, A.M. Ehrly, K. Saeger-Lorenz and D.W. Lübbers, Initial Results of Intracutaneous Measurements of pO_2 Profiles, in A.M. Ehrly, J. Hauss and R. Huch (ed.), *Clinical Oxygen Pressure Measurement*, Springer Verlag, Berlin (1984) pp. 121-128.
3. D.W. Lübbers, Optical Monitoring of Oxygen, in A.M. Ehrly, W. Fleckenstein and M. Landgraf (ed.), *Clinical Oxygen Pressure Measurement*, Vol. III, Blackwell Wissenschaft, Berlin (1992), pp. 1-14.
4. D.W. Lübbers, Chemical *in vivo* Monitoring by Optical Sensors in Medicine, *Sens. Actuat.B*, 11:253-262 (1993).
5. G.A. Holst, E. Voges and D.W. Lübbers, O₂-Flux-Optode for Medical Application, Advances in Fluorescence Sensing Technology, *SPIE Biomedical Optics, Los Angeles, CA, USA, Jan. 16-22 (1993), Vol. 1885,*pp. 216-227.
6. J.R. Lakowicz, *Principles of Fluorescence Spectroscopy*, Plenum Press, New York andLondon (1983).
7. K.W. Berndt and J.R. Lakowicz, Electroluminescent Lamp-Based Phase Fluorometer and Oxygen Sensor, *Analyt. Bioch.*, 201:319-325 (1992).

8. M.E. Lippitsch, J. Pusterhofer, M.J.P. Leiner and O.S. Wolfbeis, Fibre-Optic Sensor with the Fluorescence Decay Time as the Information Carrier, *Analyt. Chim., 205:*1-6 (1988).

9. O.S. Wolfbeis, Oxygen Sensors, in O. S. Wolfbeis (ed.), *Fiber Optic Chemical Sensors and Biosensors*, Vol. II, CRC Press, Boston and London (1991), pp. 19-52.

10. J.M.Vanderkooi and D.F. Wilson, A New Method for Measuring Oxygen Concentration in Biological Systems, *Adv. Exp. Med., 200:*189-193 (1986).

11. J.R. Bacon and J.N. Demas, Determination of Oxygen Concentrations by Luminescence Quenching of a Polymer Immobilized Transition-Metal Complex, *Analyt. Chem.. 59:*2780-2785 (1987).

12. D.B. Papkovsky, J. Olah, I.V. Troyanovsky, N.A. Sadovsky, V.D. Rumyantseva, A.F. Mironov, A.I. Yaropolov and A.P. Savitsky, Phosphorescent Polymer Films for Optical Oxygen Sensors, *Biosens. & Bioelectr.,7:*199-206 (1991).

13. M.J.P. Leiner, Luminescence Chemical Sensors for Biomedical Applications: Scope and Limitations, *Analyt. Chim., 255:*209-222 (1991).

14. X.-M. Li and K.-Y. Wong, Luminescent Platinum Complex in Solid Films for Optical Sensing of Oxygen, *Analyt. Chim., 262:*27-32 (1992).

15. J.I. Peterson, R.V. Fitzgerald and D.K. Buckhold, Fiber-Optic Probe for In Vivo Measurement of Oxygen Partial Pressure, *Analyt. Chem., 56:*62-67 (1984).

16. B.D. MacCraith, C.M. McDonagh, G. O'Keeffe, E.T. Keyes, J.G. Vos, B. O'Kelley and J.F. McGlip, Fibre Optic Oxygen Sensor Based on Fluorescence Quenching of Evanescent-Wave Excited Ruthenium Complexes in Sol-Gel Derived Porous Coatings, *Analyst,118:*385-388 (1993).

17. J.L. Gehrich, D.W. Lübbers, N. Opitz, D.R. Hansmann, W.W. Miller, J.K. Tusa and M. Yafuso, Optical Fluorescence and its Application to an Intravascular Blood Gas Monitoring System, *IEEE Trans. Biomed. Eng., BME-33/2:*117-132 (1986).

18. G.A. Holst, T. Köster, E. Voges and D.W. Lübbers, FLOX - an Oxygen-Flux-Measuring System Using a Phase-Modulation Method to Evaluate the Oxygen Dependent Fluorescence Lifetime, *Sens. & Act. B,,* (1994) *(in press)*.

19. H.D. Lüke, *Signalübertragung*, Springer-Verlag, Berlin, 3rd edn. (1985).

20. G.A. Holst, *Entwicklung und Erprobung einer Sauerstoff-Flux-Optode mit einem Sauerstoff-Sensor nach dem Prinzip der dynamischen Fluoreszenzlöschung*, VDI Verlag, Düsseldorf, 1994.

8

REFLECTANCE PULSE OXIMETRY

Accuracy of Measurements from the Neck of Fetal Lambs

R. Nijland,[1] H. W. Jongsma,[1] J. J. M. Menssen,[1] J. G. Nijhuis,[1]
P. P. van den Berg,[1] and B. Oeseburg[2]

[1] Perinatal Research Group and Department of Obstetrics and Gynaecology
[2] Perinatal Research Group and Department of Physiology
Faculty of Medical Sciences
University of Nijmegen
The Netherlands

INTRODUCTION

Continuous fetal heart rate (FHR) monitoring is used to assess the fetal condition during labour. Unfortunately, FHR patterns are not always easy to interpret, with low sensitivity and specificity as consequence. Other continuous methods for fetal surveillance have been proposed during labour (e.g. transcutaneous pO_2 and pCO_2, and pH-monitoring), but are not widespread used.

With the development of reflection pulse oximetry (RPOX), fetal arterial oxygen saturation (SaO_2) can be estimated continuously and noninvasively and this method may become an additional monitoring technique during labour.[1,2] Tissue is transilluminated by red (R) and infrared (IR) light and from the alternating back-scattered light intensities, caused by the pulsating blood volume in the tissue, an R to IR ratio can be calculated. This R to IR ratio is empirically calibrated with blood sample SaO_2 values. Since the fetus is inaccessible for arterial blood sampling and fetal SaO_2 values are predominantly below 70%, the accuracy of RPOX is usually evaluated with *in vitro* models or animal models. Studies with *in vitro* models often use homogeneous media,[3] which is a simplification of the inhomogeneous medium of the skin and underlying tissue structures.

The accuracy of a prototype reflection fetal pulse oximeter (Nellcor) was assessed by comparing measurements on the fetal lamb scalp with arterial blood samples.[4] We used an identical sensor and questioned whether measurements on the fetal neck would give the same calibration line and a comparable precision. Additionally, we studied the influence of wet hair on the calibration.

Oxygen Transport to Tissue XVII, Edited by Ince et al.
Plenum Press, New York, 1996

69

MATERIAL AND METHODS

Under general anaesthesia (2% enflurane in 50/50 O_2 and N_2O) six ewes of 126-141 days of gestation were operated on. After hysterotomy, a catheter was placed in the left fetal brachial artery for arterial sampling. A fiberoptic catheter (Abbott Opticath®, U440, 4 French, Oximetrix Inc, Mt. View, CA) was inserted into the fetal carotid artery. Both catheters remained with the tip in a pre-ductal position. ECG-electrodes were sutured subcutaneously and two RPOX sensors (Nellcor, Pleasanton, CA) were placed on the neck of the fetal lamb. The RPOX sensor is developed for fetal monitoring and consists of two Light Emitting Diodes (LED's), one for R (660 nm) and one for IR light (890 nm). A single photodiode is placed 10 mm adjacent to the LED's and receives the back-scattered light.

The sensors were placed on the shaved skin of the neck, except in 3 fetal lambs in which one of the sensors was placed on the wet, unshaved skin. The sensors were fixed by an elastic band, without applying excessive pressure to prevent compression of circulation. The sensors made appropriate contact with the skin, preventing direct shunting from R and IR light to the detector.[5] To avoid optical shunting from one sensor towards the photodiode of the other sensor, sensors were placed at appropriate distance of each other. Lack of optical shunting was verified, prior to start of the experiment (disconnection of one RPOX system did not change the R and IR pulses of the second system). The fetal head and neck remained exposed during the experiment and cooling of the fetal lamb was prevented by a thermostatic heating pad under the ewe and a warming lamp above the fetus. The sensors were covered by a warm saline soaked towel and the operation lights were turned off.

Both sensors, together with the amplified ECG-signal, were connected to prototype N-400 oximeters (Nellcor, Pleasanton, CA). The ECG-signal is used to cardiosynchronise the R and IR pulses (C-Lock, Nellcor). The calibration of the prototype N-400 oximeter is based on measurements on human volunteers (SaO_2 range 50-100%) and fetal sheep (SaO_2 range 10-50%). The signal quality of the N-400 is expressed as a score (range 0-100%) and is based on various factors such as pulse amplitude, synchrony of R and IR waveform and synchrony with the fetal ECG. All signals with a quality of $\geq$ 50% resulted in a pulse oximeter saturation (SpO_2) display and were accepted for analysis. The fiberoptic catheter was connected with an Oximetrix computer (Oximetrix Inc, Mt. View, CA) and the operation mode was set to SaO_2. The fiberoptic catheter was calibrated *in vitro* according to the manufacturer's instruction and connected to an infuser containing heparinized saline (5 I.U./ml at 1 ml/h). Fiberoptic SaO_2 values were off line linear calibrated with sample SaO_2 values. All continuous signals were stored and analyzed on a personal computer. Signals were averaged over a 5-second period. The catheter in the brachial artery was used for sampling of blood, avoiding changes of Oximetrix SaO_2 display due to the sampling. Heparinized samples were analyzed within 5 minutes to assess SaO_2 (Instrumentation Laboratory 482®, corrected for fetal sheep blood[6]), pH and blood gases (Instrumentation Laboratory 1312®, corrected to 39° C).

Different fetal desaturation levels were achieved by a stepwise change of the maternal F_IO_2 from 30% to 9%. In each lamb one or two desaturation experiments were performed.

Continuous SpO_2 readings were compared to continuous SaO_2 measurements of the fiberoptic oximeter. Paired measurements on shaved and unshaved skin were analyzed with the Student's t-test. Simultaneous measurements of sample SaO_2 were compared with SpO_2 readings by calculation of the mean difference (bias) and the standard deviation of the differences (precision). With linear regression analysis (least square method) the calibration line and the standard deviation of residuals (SD_{res}) were calculated. Significance was held at a p-value <0.05.

RESULTS

An example of a desaturation experiment is given in Fig. 1. One sensor was placed on the shaved neck and one on the unshaved neck. With both RPOX sensors a continuous recording could be obtained, but SpO_2 readings differed substantially from sample and fiberoptic SaO_2 values. This difference in saturation increased from 5-10% at the beginning towards 15-20% at the lowest point of hypoxia. The pH at the beginning and end of this period was 7.37.

In Fig. 2 all simultaneous measurements with the sensor placed on the shaved or unshaved skin versus the sample SaO_2 values are shown in 3 fetal lambs. The mean difference between both sensors was 0.1% (SD=3.2, n=17; p> 0.05, Student's paired t-test).

In Fig. 3 another example is given of a recording with one RPOX sensor on the shaved skin. A perfect relation was observed between both continuous tracings and the sample values until the lowest point of hypoxia, but an increasing difference between fiberoptic values and SpO_2 values was observed as SaO_2 returned to control values.

Since no difference was observed between shaved and unshaved measurements all paired data of SaO_2 values and SpO_2 readings were analyzed (Fig. 4). The bias±precision was 4.7±7.3% over a SaO_2 range of 16-81%. The SpO_2 readings were statistically higher than the SaO_2 values (p< 0.001, Student's paired t-test). Bias was in general larger after the first desaturation experiment. Linear regression gave the following equation:

$$SpO_2 = 1.02 \cdot SaO_2 + 3.63 \ (SD_{res}=7.3, \ n=82).$$

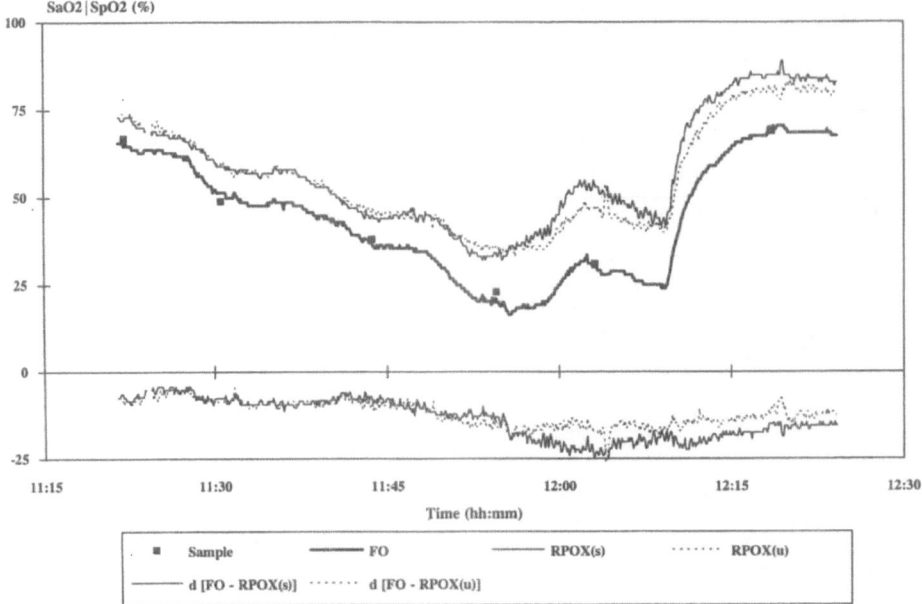

Figure 1. A desaturation experiment with continuous $SaO_2|SpO_2$ recordings of fiberoptic (FO) oximetry, RPOX sensors on shaved (s) and unshaved (u) skin, sample values and the difference between fiberoptic SaO_2 and RPOX SpO_2.

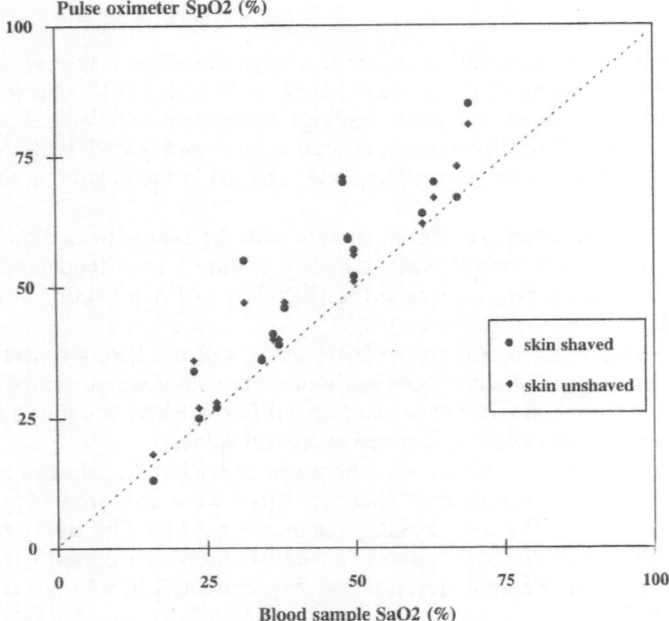

Figure 2. Paired measurements of the RPOX sensors on shaved and unshaved skin (wet hair) versus the sample SaO_2 values in 3 fetal lambs.

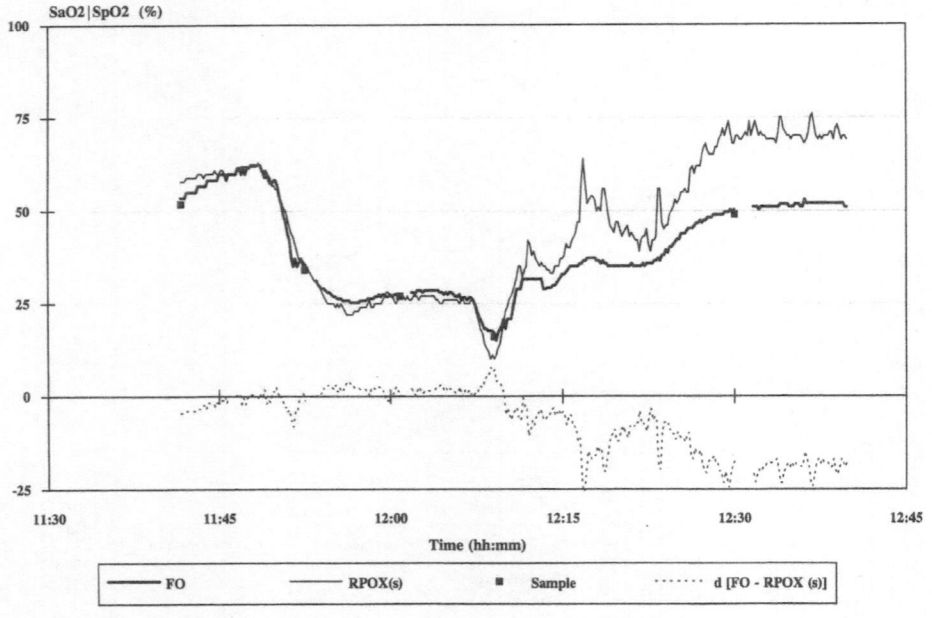

Figure 3. A desaturation experiment with continuous $SaO_2|SpO_2$ recordings of fiberoptic (FO) oximetry, RPOX sensor on shaved (s) skin and sample values and the difference between fiberoptic SaO_2 and RPOX SpO_2.

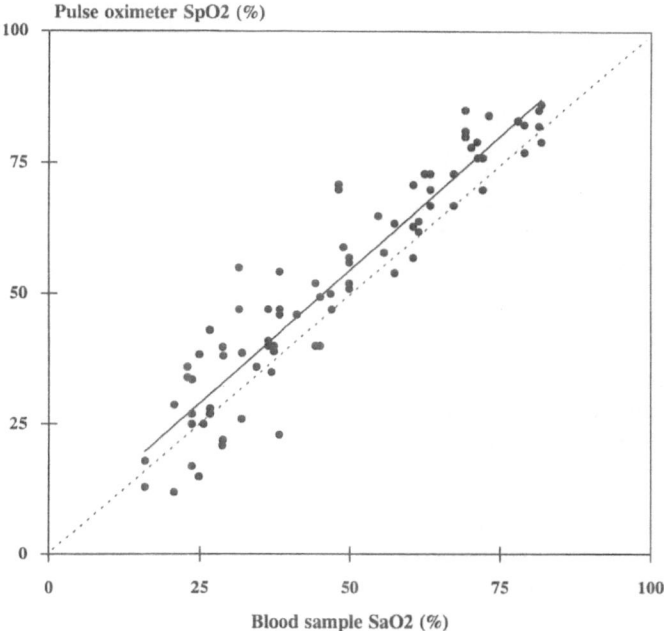

Figure 4. Paired measurements of SaO$_2$ and SpO$_2$ in 6 fetal lambs. Dashed line is line of identity. Solid line is calibration line with equation: SpO$_2$=1.02 · SaO$_2$ + 3.63 (SD$_{res}$=7.3, n=82).

DISCUSSION

Bias and precision are used to describe the accuracy of one method relative to another method.[7] In this study sample SaO$_2$ values are considered as the "gold standard" and the bias is thus an estimation for a systematic error, that is an overestimation or underestimation relative to the sample SaO$_2$ value. We found a slightly different calibration line of the precalibrated oximeter with the RPOX sensor on the fetal neck, with an overestimation in saturation of 4.7%. Mendelson et. al. used in pigs an experimental sensor with two pairs of 660 and 930 nm LEDs and a concentric array of 6 photodiodes and also found a systematic difference between scalp and neck measurements.[8] Neck measurements overestimated the SpO$_2$ with 4% at 100% SaO$_2$ and underestimated the SpO$_2$ with 5% at 30% SaO$_2$, compared to scalp measurements. It is not clear whether the difference between their[8] and our study at low saturation is caused by differences in sensor design, species or other reasons.

The precision is an estimation for the random error of the RPOX system. Our results showed a precision of 7.3 (neck) which is worse than the 5.5% (scalp) reported for 3 fetal lambs with an identical sensor.[4] Other studies using a RPOX sensor with a comparable LEDs/photodiode configuration reported a precision for fetal sheep scalp measurements of 4.7 and 6.6%,[9,10] with respectively 4 and 2 fetal lambs. Mendelson et. al.[8] found a precision of 3.5% and 4.1% for scalp and neck, respectively. Their study showed a large difference between the number of paired sample points for the scalp (n=37) and the neck (n=321); the total number of animals was not mentioned.

Various RPOX systems may differ in their performance. However, the results of the various systems are difficult to compare because of differences in: species, numbers of animals, sample points and acceptance criteria for data. We like to comment on some factors which contribute to the overall precision of RPOX systems:

1. The pulse oximeters used in the cited studies are provided with a general calibration line based on the pooled measurements of different subjects (humans and/or animals); this calibration is considered applicable for individual subjects. In our study the precise continuous SaO_2 recording by the fiberoptic oximeter made it possible to distinguish between short term deviations (artifacts) and systematic errors. We found as well systematic differences in the calibration between individual animals, as an animal in which the calibration changed over time during the experiment (Figs. 1 and 3). These 2 examples show systematic errors in individual animals which are not apparent in an overall bias and precision. None of the previous mentioned studies reported systematic differences in individual calibrations, however from a figure in one study this also can be concluded[10].

2. Acceptance criteria of signals (visual or small SD of R to IR ratio) were set at the moment of analysis,[4,9,10] however it would have been better if those criteria were set beforehand to avoid observer bias. We used the quality scale of the N-400 and set it beforehand to 50%.

3. Several avoidable circumstances may lead to a reduced accuracy. Inappropriate sensor contact may lead to an inaccurate estimation of the SpO_2[5]. We therefore verified sensor placement before and after the experiments. Optical shunting through hair is also raised as a possible cause of inaccurate measurements. We could not observe any difference between shaved and unshaved wet hair. Since the colour of our fetal lamb hair was white, this observation may not hold for other colours or dry hair. Furthermore, fetal pulses are small[9] resulting in a smaller signal to noise ratio and the possibility of a decreasing accuracy. In this study most pulse signals were visually verified and no artifacts were observed which could explain the deviation from the samples or fiberoptic tracing.

4. Improvement of the data processing may result in more accurate estimation of the SpO_2 as far as an increased signal to noise ratio results in a reduced random error. However, the part of the (im)precision of the calibration which is due to systematic differences between subjects can not be improved by a better data processing. Theoretical models and *in vitro* models predict that several factors e.g. blood volume, haematocrit[3,11] affect the accuracy of RPOX systems, specially at low SaO_2 levels. Differences in skin structure, subcutaneous tissue and local differences in vascular density may add to a decreased accuracy.

More studies have therefore to be performed with a sufficient amount of animals, before obstetricians can rely on SpO_2 readings of RPOX systems for intrapartum fetal monitoring.

ACKNOWLEDGMENTS

We acknowledge the help of Jane Crevels, Ineke Verbruggen, Theo Arts and Biny Ringnalda for there technical help during the experiments.

REFERENCES

1. Johnson N, Johnson VA, Fisher J, Jobbins B, Bannister J, Lilford RJ. Fetal monitoring with pulse oximetry. Br J Obstet Gynaecol 1991;98:36-41.

2. Dildy CA, Clark SL, Loucks CA. Preliminary experience with intrapartum fetal pulse oximetry in humans. Obstet Gynecol 1993;81:630-635.

3. de Kock JP, Tarassenko L. *In vitro* investigation of the factors affecting pulse oximetry. *J Biomed Eng 1991;13:61-66.*

4. Harris AP, Sendak MJ, Chung DC, Richardson CA. Validation of arterial oxygen saturation measurements in utero using pulse oximetry. Am J Perinat 1993;10:250-254.

5. Gardosi JO, Damianou D, Schram CM. Inappropriate sensor application in pulse oximetry. Lancet 1992;340:920.

6. Nijland R, Ringnalda B, Jongsma HW, Oeseburg B, Zijlstra WG. The measurement of oxygen saturation by multi-wavelength analyzer; influence of interspecies differences. Clin Chem (in press).

7. Bland JM, Altman DG. Statistical methods for assessing agreement between two methods of clinical measurements. Lancet 1986;i:307-310.

8. Mendelson Y, Yocum BL. Noninvasive measurement of arterial oxyhemoglobin saturation with a heated and a non-heated skin reflectance pulse oximeter sensor. Biomed Instrum Technol 1991;25:472-480.

9. Dassel AC, Graaff R, Aarnoudse JG, Elstrodt JM, Heida P. Koelink MH. Greve J. Reflectance Pulse Oximetry in Fetal Lambs. Pediatric Research 1992;31:266-269.

10. Jongsma HW, Crevels J, Menssen JJM, Arts THM, Mulders JGM, Nijhuis JG, Oeseburg B 1991 Application of transmission and reflection pulse oximetry in fetal lambs. In: Lafeber HN, Aarnoudse JG, Jongsma HW (eds) Fetal and Neonatal Physiological Measurements, Excerpta Medica, Amsterdam, pp 123-128.

11. Schmitt JM. Simple photon diffusion analysis of the effects of multiple scattering on pulse oximetry. IEEE Trans Biomed Eng 1991;38:1194-1203.

9

A NOVEL METHOD TO DEMONSTRATE CAPILLARY GEOMETRY AND PERFUSION PATTERNS IN RAT BRAIN FROM THE SAME HISTOLOGICAL SECTION

Sanjay Batra and Martin F. König

Institute of Anatomy
University of Berne
Bühlstrasse 26, CH-3012 Berne, Switzerland

INTRODUCTION

The geometry of the microvascular bed as well as its perfusion pattern are important determinants of oxygen transport to tissues. In the present study, we analyzed capillary geometry in the rat parietal cortex using discriminative morphometric methods. Specifically, we labeled arteriolar capillary (i.e. the proximal portion of the capillary bed, AC) and venular capillary (i.e. the distal portion of the capillary bed, VC) endothelium with different colors. While the geometry of the capillary bed sets the surface area for diffusion of oxygen, the pattern of perfusion within the capillary bed under *in vivo* conditions provides information as to regional conditions for oxygen diffusion. Accordingly, we used a highly concentrated colloidal gold solution as a plasma label, to identify capillaries perfused with plasma under controlled conditions. While gold particles may be directly visualized on electron micrographs, they require enhancement with silver to make them visible at the light microscopic level. To our knowledge, this is the first study where these methods have been combined within the same histological section, i.e. enzymatic labeling of the capillary endothelium as well as delineation of capillary perfusion with silver enhanced colloidal gold particles.

MATERIAL AND METHODS

Preparation of BSA-gold complexes

A volume of 750 ml of 8 nm diameter gold particles were prepared as previously described (10). Briefly, a 600 ml solution (containing 593 ml H_2O plus 7.5 ml 1% w/v $HAuCL_4$·aq (from Fluka No 50800) in water) was heated to 60°C and quickly mixed with another solution (containing 113 ml H_2O, 120 ml 1% trisodium citrate, 1.5 ml 1% tannic

Oxygen Transport to Tissue XVII, Edited by Ince et al.
Plenum Press, New York, 1996

77

acid, also heated to 60°C). This yielded ~10^{16} gold particles (15). The gold colloid was then concentrated by gently boiling down to approximately 50% of its original volume. A 5% w/v solution of bovine serum albumin (BSA) in water was prepared and serial dilution's of this were used to determine the minimal amount of RSA that would stabilize the gold against electrolyte induced aggregation (6.25 μl of 5% w/v BSA per ml of concentrated gold solution). The crude complex was then prepared by rapidly mixing the concentrated gold solution with 4.7 ml BSA (5% w/v) added with ~3 ml 10 mM sodium phosphate buffer. For further concentration, batches of the BSA-gold suspension were centrifuged for 90 minutes at 35'000 g until the final concentration of 1 ml colloidal gold suspension was obtained. Before use, the gold solution was passed through 0.25 μm Millipore filters, previously flushed with 0.1% BSA-Ringer.

Experimental Method

Six male Wistar rats with a body weight of ~300 g were premedicated with heparin i.p. (1000 U), and anaesthetized by 50 mg/kg pentobarbital i.p. During the surgical proce-dure, further pentobarbital was given as necessary. The gold concentrate (1 ml) was warmed to body temperature and injected within 2 seconds through a femoral catheter. One minute after the end of gold injection, the circulation was stopped by rapidly (~2 sec) injecting a bolus of 1 ml KCl-gold solution (0.9 ml half saturated KCl added with 0.1 ml gold concentrate). Blood pressure measurements in the femoral artery revealed that heart arrest occurred within 1 second of the end of KCl injection. Immediately after cardiac heart arrest, the brain was carefully removed. For enzymatic staining, the parietal cortex was separated, covered in OCT cryostat embedding media and frozen in liquid nitrogen. Small tissue blocks from the contralateral parietal cortex were fixed in half concentrated Karnovsky fixative for electron microscopy.

Histochemical Method

The frozen tissue samples of the parietal cortex were sectioned at -20°C (section thickness = 25 microns). All tissue sections were exposed to a two-step histochemical method to identify the microvasculature. The method is a modification of the method first introduced by Lojda (12). Tissue sections were first incubated in a solution for the demonstration of dipeptidyl peptidase IV (DPP IV) in the capillary endothelium. In contrast to our previous description of this method for heart (2) in brain, 0.1 M phosphate buffer (pH 7.4) was used in place of acetate buffer and the incubation time was increased to 150-180 minutes. This procedure was specific to the distal portion of the capillary bed staining venular capillaries (VC) red. Next, sections were transferred to a solution for the demonstration of alkaline phosphatase (AP) in the capillary endothelium for 25 minutes, which was specific to the proximal arteriolar portion of the capillary bed, staining all arteriolar capillaries (AC) blue. As a result of this histochemical technique, the cerebral microvasculature was labeled in a bivariate manner. Arterioles and the proximal portion of the capillary bed stained blue in color; venules and the distal portion of the capillary bed stained red in color. The transitional zone that demonstrated both DPP IV and AP activity stained a purple color. This zone was relatively small, and for the purpose of consistency was considered to be AC.

Silver Enhancement of Gold Particles

To visualize the gold label on light microscopic sections, particles were enlarged using silver enhancement. The sections were placed for 5 minutes in citrate buffer containing 0.25% w/v hydroquinone (pH 3.8) and transferred to the silver developer (0.1% silver acetate in 0.25% hydroquinone citrate buffer). After 15 minutes, the sections were rinsed briefly in distilled water and fixed with Ilford fix (1:9 diluted in H_2O).

Tissue Processing for Electron Microscopy

Standard electron microscopic techniques were followed to dehydrate and embed the tissue in Epon resin (8,13). Sections were cut at 60 nm and mounted on uncoated 200 mesh grids. In order to improve the visibility of the gold particles, the sections were only weakly counterstained (with uranyl acetate and lead citrate each for 5 minutes). The sections were analyzed at a final magnification of X50'000. The principle was to identify capillary profiles with an unbiased counting frame to assess whether the plasma contained gold particles or not. The ratio of marked capillaries over all capillary profiles estimates the fraction of the capillary path that has been perfused at least once during the time interval between label injection and stop of blood flow prior to fixation. This can be expanded to estimate the fraction of capillary volume or surface perfused by using point and intersection counting methods (17).

RESULTS

At the light microscopic level (Fig. 1) it was possible to clearly identify capillaries in the parietal cortex. Randomly oriented sections gave rise to capillary profiles in transverse, longitudinal and oblique orientation. Arteriolar and venular capillary regions were easily distinguished on the basis of color. Arteriolar capillaries appeared blue where venular capillaries were stained bright red. It was interesting to note that in brain, very few transitional capillaries were observed that stained a purple color. The overall architecture of the microvasculature in parietal cortex may be described a highly anatomosing network with relatively straight capillary segments (the capillary length between two branch points) but sharp branching angles.

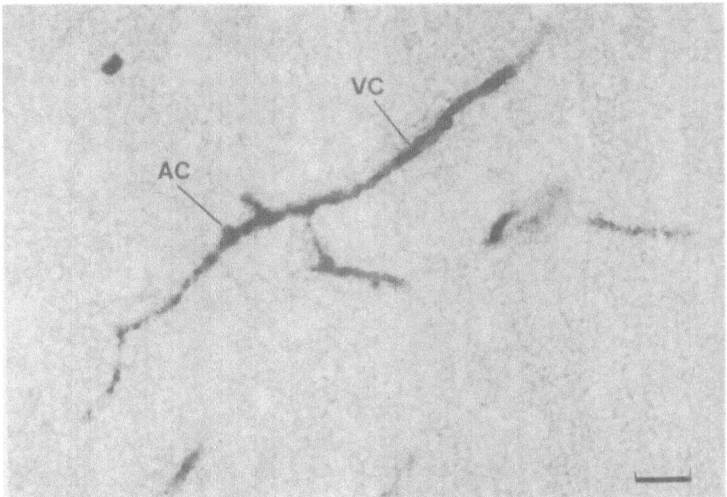

Figure 1. Cryosection illustrating an arteriolar capillary (AC) branching into a venular one (VC). (bar = 20 µm).

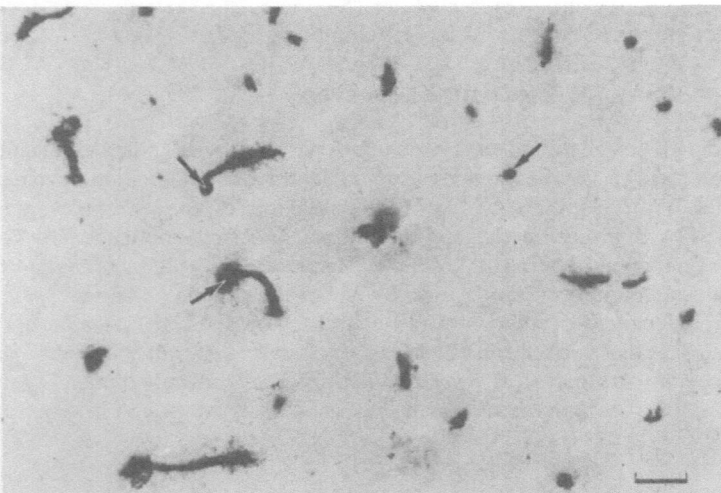

Figure 2. After silver enhancement of cryosections, blood plasma is intensely labeled (arrows); compared with Figure 1, this results in a higher contrast of capillary profiles in black and white prints. (bar = 20 μm).

Following silver enhancement of tissue sections, gold particles appeared as black plasma filled within vascular lumen (Fig. 2). Even through stained capillary endothelium it was possible to observe the black plasma tracer. At the microscope, with transillumination and the ability to focus at various planes within the section, it was even possible to detect minute quantities of the plasma tracer. Red blood cells appeared as negative images within the black labeled plasma.

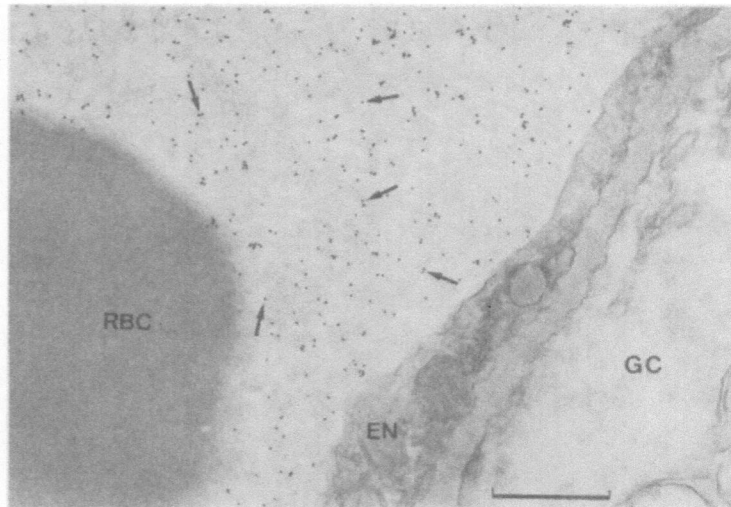

Figure 3. In electron micrographs, colloidal gold particles (arrows) are clearly identifiable between red blood cells (RBC) and endothelium (EN). GC=glia cell. (bar = 0.5 μm).

The higher magnification of electron microscopy allows to identify all cells and all intracellular structures (Fig. 3). Colloidal gold particles appear as black spheres homogeneously dispersed within the plasma of capillary profiles. Due to the high resolution of electron microscopy, arterioles and venules are clearly distinguished from capillaries by their smooth muscle layer. At the EM level it was not possible to distinguish the proximal and distal portions of the capillary bed. However, all capillary profiles contained gold nanospheres in plasma. In accordance, after AP/DPP IV reaction and silver enhancement all capillaries that were examined showed black labeled plasma. Following one minute of circulation, all capillaries in brain appear to be perfused, at least with plasma.

DISCUSSION

The question of capillary reserve and perfusion pattern in brain is of fundamental importance as this organ is vulnerable to even short periods of ischemia. To this end, we developed a new method which enabled us to differentiate arteriolar and venular capillaries and to visualize the capillary perfusion pattern at the same time from a single histological section. This approach was validated by electron microscopy.

Identification of Plasma Tracer and Capillaries

So far, experiments demonstrating capillary perfusion pattern in brain required several steps in series to combine the identification of all anatomically present capillaries with the recognition of plasma tracer. Many studies have used fluorescent dyes as perfusion tracer in combination with alkaline phosphatase as capillary marker (e.g. 5,18). Since the intensity of fluorescence may be affected by the alkaline phosphatase reaction, as a first step, fluorescent emission of plasma tracer containing capillaries is examined and photographically recorded. Then the same section is exposed to the enzymatic alkaline phosphatase reaction. The fields previously photographed for fluorescence are relocated and a new photograph is obtained. This "overlay" method allows a quantification of perfused vs. unperfused capillaries. Newer methods using the combination of two different fluorescent dyes, i.e. FITC as plasma tracer and lissamine-rhodamine as capillary marker (7,16) may allow tracer and capillary identification on the same tissue section. Nonetheless, this method still requires an intermediate recording step, where the capillaries containing the first fluorescent tracer are recorded and compared to the capillaries marked by the second tracer which is visualized after changing the filter system. Since in our study, both capillary and plasma marker are simultaneously visible by simple transluminal lightmicroscopy, no recording or intermediate step is required. We believe that our approach will reduce the errors associated with overlaying different photographs and trying to find the same capillary fields.

Differentiation of Capillaries

The alkaline phosphatase method alone is known to be limited by the fact that venular capillaries might remain unstained (1,12,14). Therefore we modified the combination method of alkaline phosphatase and dipeptidyl peptidase IV, as we originally applied for heart capillaries (3). Since this method allows to distinguish arteriolar from venular capillaries by different colors, its combination with colloidal gold particles used as perfusion marker could reveal differences in the perfusion pattern along the capillary bed from arteriole to venule. This would have very interesting implications for oxygen offloading to tissue.

Capillary Recruitment

One major question in brain perfusion is whether there exists a capillary reserve which is recruited under certain conditions. While some studies report a reserve of ~50% of the capillary bed (4,6), others find no evidence for a significant capillary recruitment (9,11,16). Vetterlein et al. (16) recently have shown that within 2 seconds ~85% of the capillaries and after 10 seconds the whole capillary bed of parietal cortex, hypocampus and basal ganglia contained plasma tracer. From their findings, they suggest that differences in plasma flow velocities within the capillary bed are the major reason for the findings of partial perfusion in the initial time interval after tracer infusion. In addition, it should be noted that controversial results respecting the microvasculature perfusion pattern may be largely based upon differences in methodological approach or experimental design (16).

CONCLUSION

We conclude that the combination of AP/DPP IV capillary staining with colloidal gold particles as a plasma tracer is a reliable method to examine microvascular perfusion in brain simultaneously from the same histological section. These results were confirmed with analysis at the electron microscopic level. Following one minute of circulation, all capillaries in parietal cortex were perfused, atleast with plasma. To address the question of capillary recruitment, further experiments using shorter circulation times are required.

REFERENCES

1. Batra, S., Koyama, T., Gao, M., Horimoto, M., and Rakusan, K. , 1992, Microvascular geometry of the rat heart: arteriolar and venular capillary regions, *Jpn. Heart J.* 33:817-828.
2. Batra, S., Kuo, C., and Rakusan, K. , 1989, Spatial distribution of coronary capillaries: A-V segment staggering, *Adv. Exp. Med. Biol.* 248:241-247.
3. Batra, S. and Rakusan, K. , 1991, Geometry of capillary networks in volume overloaded rat heart, *Microvasc. Res.* 42:39-50.
4. Buchweitz, E. and Weiss, H. R. , 1986, Alterations in perfused capillary morphometry in awake vs. anesthetized brain, *Brain Res.* 377:105-111.
5. Buchweitz-Milton, E and Weiss, H. R. , 1988, Perfused microvascular morphometry during middle cerebral artery occlusion, *Am. J. Physiol.* 255:H623-H628.
6. Francois-Dainville, E., Buchweitz, E., and Weiss, H. R. , 1986, Effect of hypoxia on percent of arteriolar and capillary beds perfused in the rat brain, *J. Appl. Physiol.* 60:280-288.
7. Göbel, U., Theilen, H., and Kuschinsky, W. , 1990, Congruence of total and perfused capillary network in rat brains, *Circ. Res.* 66:271-281.
8. Hoppeler, H., Mathieu, O., Weibel, E. R., Krauer, R., Lindstedt, S. L., and Taylor, C. R. , 1981, Design of the mammalian respiratory system. VIII. Capillaries in skeletal muscles, *Respir. Physiol.* 44:129-150.
9. Klein, B., Kuschinsky, W., Schröck, H., and Vetterlein, F. , 1986, Interdepency of local capillary density, blood flow, and metabolism in rat brains, *Am. J. Physiol.* 251:H1333-H1340.
10. König, M. F., Lucocq, J. M., and Weibel, E. R. , 1993, Demonstration of pulmonary vascular perfusion by electron and light microscopy, *J. Appl. Physiol.* 75:1-7.
11. Kuschinsky, W., Klein, B., Schröck, H. and Vetterlein, F. ,1987, The lokal density of perfused capillaries in rat brain. In Stroke and Microcirculation. J. Cervos–Navarro and R. Ferszt, editors. Raven, New York. 167-170.
12. Lojda, Z., 1979, Studies on dipeptidyl (amino) peptidase IV (glycyl-proline naphthylamidase). II. Blood vessels, *Histochem.* 59:153-166.
13. Mermod, L., Hoppeler, H., Kayar, S. R., Straub, R., and Weibel, E. R. , 1988, Variability of fiber size, capillary density and capillary length related to horse muscle fixation procedures, *Acta Anat.* 133:89-95.
14. Romanul, F. C. A. and Bannister, R. G. , 1962, Localized areas of high alkaline phosphatase activity in the terminal arterial tree, *J. Cell Biol.* 15:73-84.

15. Slot, J. W. and Geuze, H. J. , 1985, A new method of preparing gold probes for multiple-labeling cytochemistry, *Eur. J. Cell Biol.* 38:87-93.

16. Vetterlein, F., Demmerle, B., Bardosi, A., Göbel, U., and Schmidt, G. , 1990, Determination of capillary perfusion pattern in rat brain by timed plasma labeling, *Am. J. Physiol.* 258:H80-H84.

17. Weibel, E.R, 1979, Stereological methods. Vol. 1. Academic, London.

18. Weiss, H. R., Buchweitz, E., Murtha, T. J., and Auletta, M. , 1982, Quantitative regional determination of morphometric indices of the total and perfused network in the rat brain, *Circ. Res.* 51:494-503.

10

COMPARISON BETWEEN CONVENTIONAL HEMODYNAMIC MONITORING AND POLAROGRAPHIC TISSUE PO$_2$-MONITORING OF THE LIVER IN EARLY SEPTICEMIA OF THE PIG

K. Wagner,[1] R. Schäfer,[1] W. Gerling,[1] A. Michelsen,[2] and P. Schmucker[1]

[1] Department of Anesthesiology
[2] Department of Surgery
Medical University Lübeck
Germany

INTRODUCTION

The common therapeutic approach to septicemia - aside from elimination of its cause and antibiosis - consists of the attempt to keep tissue perfusion within physiological limits. A frequently employed therapeutic option is catecholamine administration. Since the reduction of peripheral vascular resistance in septicemia is at least partly due to a TNF induced vascular NO-synthase, the inhibition of this system does affect tissue perfusion as well. While there are controversial results as to the benefit of inhibition of NO-synthase on mean arterial pressure (MAP) and cardiac index (CI) (1) the effect of NO-synthase inhibition on liver circulation has not been described. In addition we were interested in the effect of NO-synthase inhibition on liver pO$_2$.

Therefore the aim of this experimental study was to measure liver perfusion and liver pO$_2$ in a pig septicemia model and compare the effects of catecholamine treatment with those of NO-synthase inhibition.

MATERIAL AND METHODS

Female pigs of German landrace breed (weight 19-25 kg) were anesthetised with intramuscular ketamine (10 mg/kg bw), benzodiazepam (1 mg/kg), and 0.5 mg atropine. The animals were tracheotomized and anesthesia was maintained with i. v. thiopental (100-200 mg/hr), fentanyl (0.2 - 0.5 mg/hr), and pancuronium (5 mg/hr). The animals were normoven-

Oxygen Transport to Tissue XVII, Edited by Ince et al.
Plenum Press, New York, 1996

85

tilated with an EVA ventilator (Dräger, Lübeck, Germany) and the FIO_2 adjusted to keep an arterial pO_2 between 80 and 150 mmHg.

Surgical Preparation

An appropriately sized electromagnetic flow probe was attached to the portal vein (8 - 12 mm). The common hepatic artery was exposed and all branches by-passing the liver parenchyma were carefully ligated. An ultrasonic flow probe was attached to the common hepatic artery.

The abdominal superior vena cava was punctured in cranio ventral to dorso caudal direction and a plastic cannula advanced into a hepatic vein. The position of the cannula was verified in all cases by post mortem examination.

Methods

Blood flow of the portal vein was measured with an electromagnetic flow meter (Narcomatic RT 500, Narco Bio-Systems, Housten, Texas, USA). Blood flow of the hepatic artery was measured with an ultrasonic flow meter (principle of transit-time measurement, Transonic System Inc, Ithaca, USA).

Tissue pO_2 was measured with a computer controlled pO_2 histograph (Eppendorf-Netheler-Hinz GmbH, Hamburg) . The tip diameter of the pO_2 probe was 250 μm. Between 20 and 40 single pO_2 values were measured in sequence. At the end of each measuring period tissue temperature was determined with a thermocouple (Philips, type K, tip diameter 260 μm). Hemoglobin saturation and hemoglobin content were measured with an OSM3 (Radiometer Copenhagen). Standard pressure transducer systems (Sirecust, Siemens, Germany) were used for measurement of MAP (measured in the femoral artery), pulmonary artery pressure (PAP), pulmonary capillary wedge pressure (PCWP), portal vein pressure, central venous pressure (CVP). CI was measured according the thermodilution principle. From these primary data the following derived parameters were calculated: total peripheral resistance, pulmonary vascular resistance, cardiac index, hepatic artery resistance

Experimental Protocol

The study comprised two experimental groups and one time-control group (control group, n = 7). After induction of hyperdynamic septicemia, group one was treated with norepinephrine alone (NE group, n = 11) while the second was treated with norepinephrine and L-NAME (L-NAME group, n = 5). The control group received no LPS and no treatment.

All three groups had an identical protocol for the first two measurements. After the surgical preparation measurement 1 (baseline data, 0 minutes) was recorded. Next septicemia was induced by administration of LPS (0.025 μg/kg/min). LPS was stopped when systolic pulmonary artery pressure had risen above 60 mm Hg or MAP had fallen below 80 mm Hg which ever came first (2). The animals were stabilised by fluid administration and norepinephrine infusion. On average 20 minutes after induction of septicemia the second measurement was taken. From now on the treatment of the experimental groups differed. In the NE group only a norepinephrine infusion was given to keep the animal stable. In the L-NAME group after measurement 2 had been completed an infusion of L-NAME (1 mg/kg/min for 4 minutes) was given and the norepinephrine infusion stopped.

Measurement 3 was 40 minutes, measurement 4 was 60 minutes, measurement 5 was 90 minutes and measurement 6,120 minutes after induction of hyperdynamic septicemia.

For each measurement the following set of data was obtained: blood samples were drawn from the femoral artery, pulmonary artery, portal, and hepatic vein. Each of these

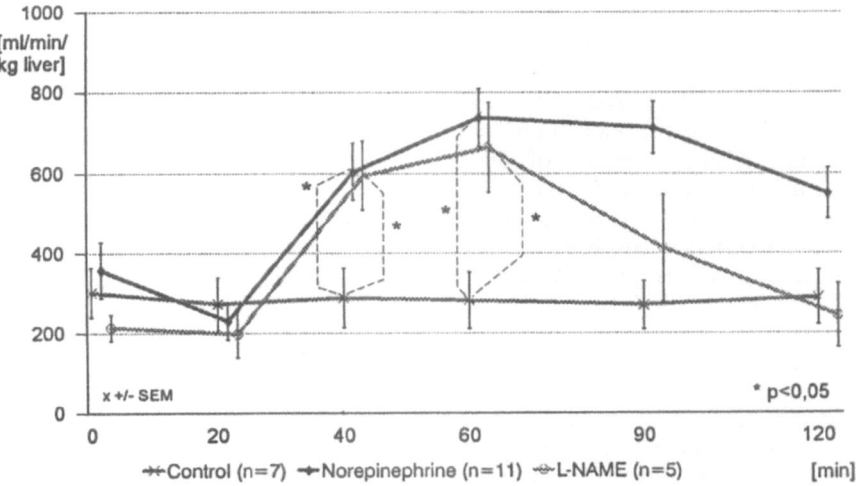

Figure 1. Specific hepatic artery flow.

samples was analysed for Hb content and Hb saturation. A liver pO_2 histogram was recorded and the circulatory parameters described above were read and cardiac output averaged from three measurements. Flow and flux readings of the liver were recorded. Data labelled as 'specific' are normalised for kg liver weight. The data presented here are those specific to the subject of this paper. Results concerning general circulation and additional aspects of the liver metabolism and circulation will be published elsewhere.

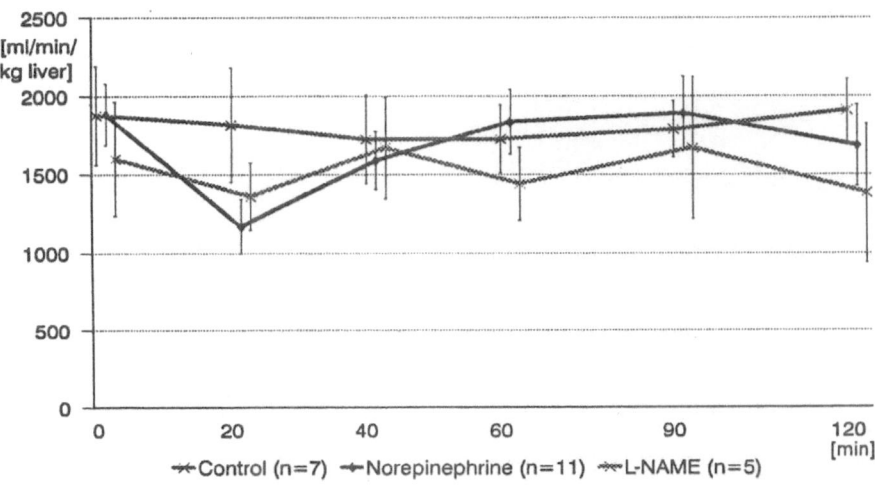

Figure 2. Specific portal vein flow.

Statistics

Although we strongly assume that all data measured during this study are normally distributed and we therefore use the mean and standard error of the mean to describe the results, this could not be proven. As a consequence we applied nonparametric statistics not to overestimate the significance of differences within our data. As a first step a global test was performed followed - in case of significance - by a second detailed one. The level of significance was set at $p < 0.05$. To test for differences within each group the Friedman-test was used as a global test followed by the Wilcoxon–Wilcox test if the p value of the global test indicated a significant difference. The tests for differences between the groups were the H-test of Kruskal–Wallis (global) and the Nemenyi-test as the detailed one.

RESULTS

Specific hepatic artery flow (Figure 1) decreased - compared to baseline values - at 20 minutes. Just prior to this measurement the infusion of LPS had been stopped. Thereafter specific hepatic artery flow increased by approximately 100%. At 40 and 60 minutes flow readings from controls compared to the NE and L-NAME group were significantly increased (p<0.05). While specific hepatic artery flow tended to stay elevated (with a small decrease at 120 minutes) in the NE group, administration of L-NAME reduced specific hepatic artery flow to baseline levels at the end of the experiment. For the control group specific hepatic artery flow remained constant over the whole experimental time.

Specific portal vein flow (Figure 2) of the NE and the L-NAME group were reduced at measurement 20 minutes and no hyperemia during the latter course of the experiment was found.

Hepatic O_2 delivery (Figure 3) of the NE and the L-NAME group decreased between measurements at 0 and 20 minutes due to the LPS infusion. Afterwards in the NE group flow did increase significantly (p<0.01) between measurements at 20 minutes and at 60 and 90

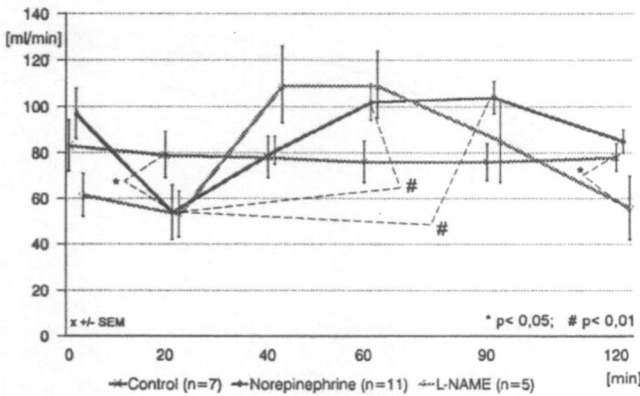

Figure 3. Hepatic oxygen delivery.

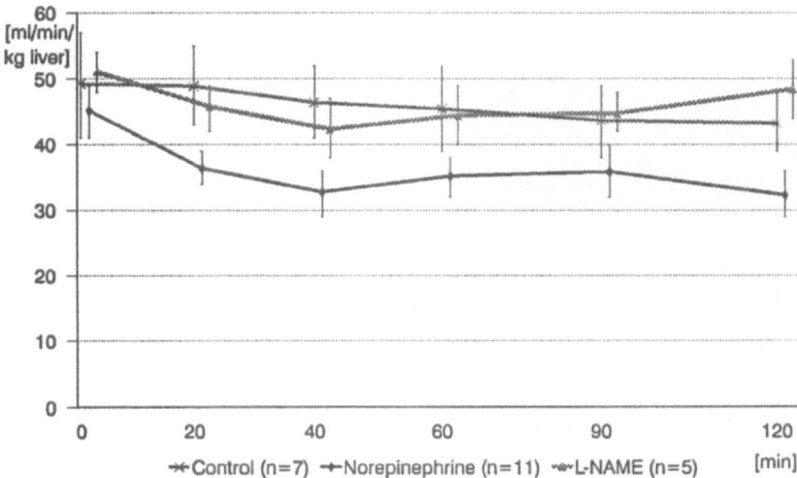

Figure 4. Hepatic oxygen consumption.

minutes. A comparable tendency was found for the L-NAME group for measurements at 20 minutes compared to 60 minutes but hepatic O_2 delivery decreased continuously thereafter and fell below baseline values at the end of the experiment.

Hepatic O_2 consumption (Figure 4) in the L-NAME group tended to increase towards the end of the experiment, but the differences were not significant between the groups.

Hepatic O_2 extraction ratio (VO_2/DO_2, Figure 5) of the NE group was reduced to about half of the baseline value as from 40 minutes until the end of the experiment.

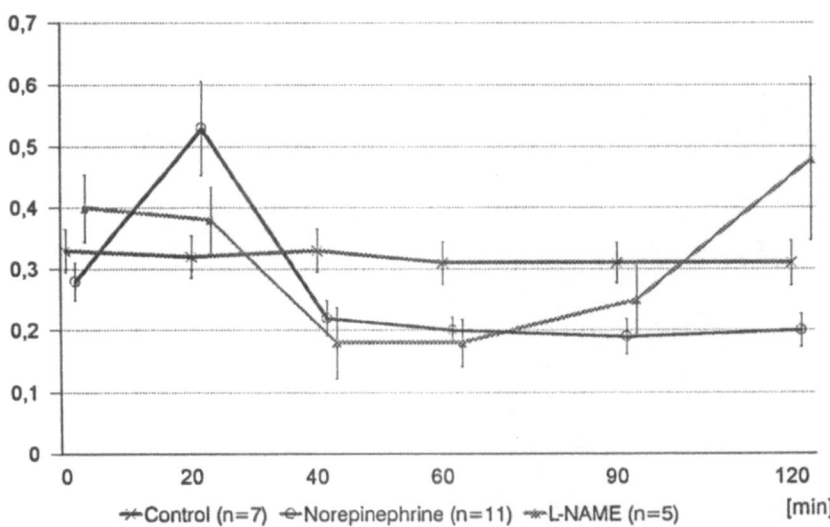

Figure 5. Hepatic oxygen extraction ratio (VO_2/DO_2).

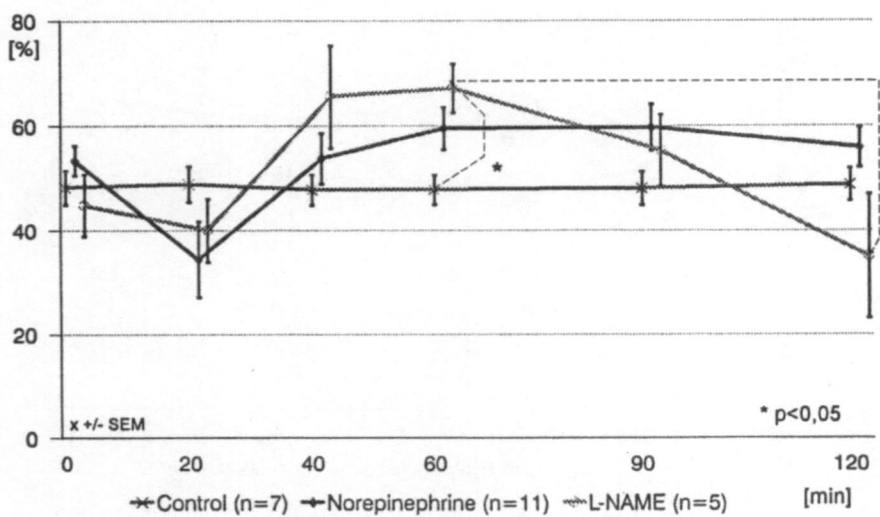

Figure 6. Liver vein hemoglobin saturation.

In the L-NAME group hepatic O_2 extraction ratio (VO_2/DO_2) was half compared to the baseline value at 40 and 60 minutes but increased to well above baseline values at 120 minutes. This is reflected by corresponding changes of the liver vein saturation (Figure 6), which was significantly reduced at 120 minutes compared to 60 minutes ($p<0.05$).

Despite of the marked changes in hepatic artery flow, O_2 delivery and O_2 extraction ratio, mean liver tissue pO_2 (Figure 7) remains stable over the whole experimental time

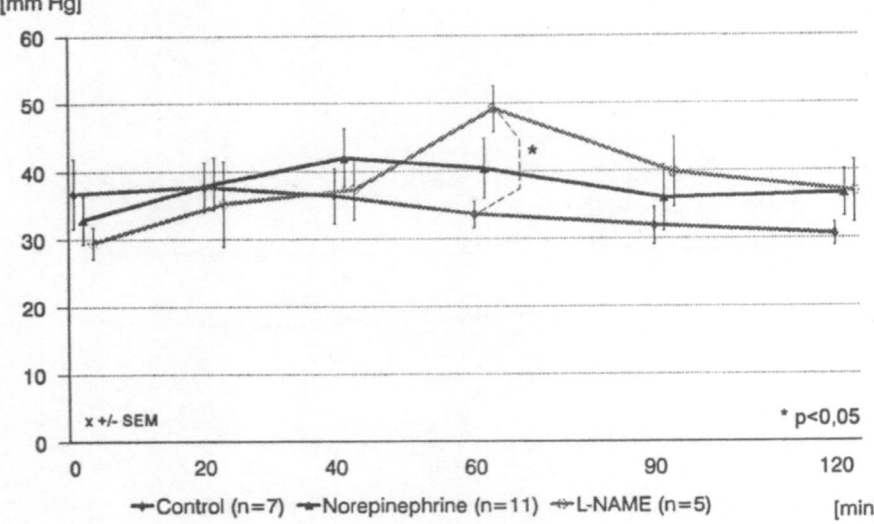

Figure 7. Mean liver tissue pO_2.

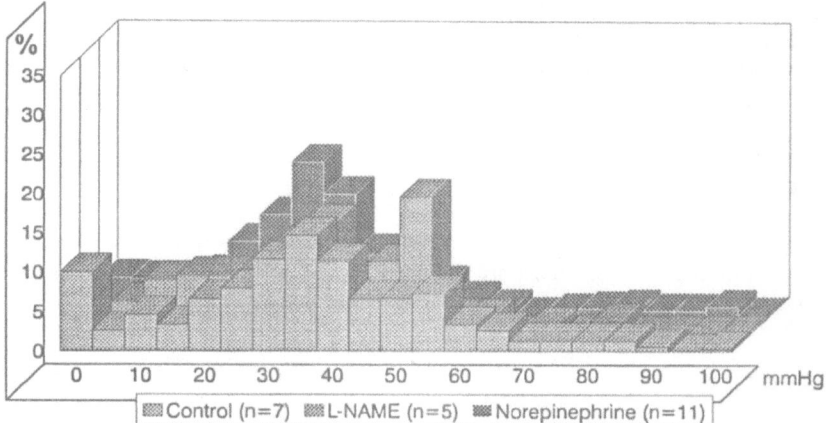

Figure 8. Liver pO$_2$ histograms measured at 120 minutes.

without significant differences between the groups (except a chance significance between the control and the L-NAME group at 60 minutes). The liver tissue pO$_2$ values from measurement 120 minutes were physiologically distributed for the NE and the L-NAME group as shown in Figure 8.

DISCUSSION

In this study some data of the liver circulation from a clinic-oriented porcine model of endotoxin shock were presented. The animal model was characterised by hyperdynamic state, i.e. markedly reduced peripheral resistance and elevated cardiac output, PCWP-oriented compensation of fluid imbalance and pressure-stabilisation.

After administration of LPS a reduction of hepatic artery, portal vein blood flow and the derived parameters developed. This was the effect of the LPS infusion which had been started just after measurement 0 minutes and lead to a circulatory depression. When cut off criteria were reached LPS was stopped and the animals resuscitated by restoring MAP to 80 mm Hg.

In the NE group norepinephrine alone was given for pressure stabilisation. Here the hepatic artery flow doubled and remained elevated. These data reflect the hyperemic state that developed. The results are in contradiction to the data of Wang et al. (3), who found an increased portal vein blood flow in early septicemia. With regard to the oxygen metabolism the oxygen supply was increased. The hyperemia of the hepatic artery blood flow and the oxygen-delivery cannot be explained by a higher oxygen demand, because liver oxygen consumption did not increase. The shift of the oxygen balance towards an increased supply did not lead to an increase of the mean liver tissue pO$_2$ and the liver pO$_2$ histogram remained normally distributed.

A different pattern was found in the L-NAME group, where norepinephrine was changed for L-NAME between measurements 20 and 40 minutes. Still initially a marked hyperemia (flow increase 270 %) developed but after measurement 60 minutes a steady decline of the hepatic artery flow was found. During hyperemia the oxygen supply was increased as in the NE group but at the end of the experiment the supply had returned to baseline values and the oxygen demand had moderately increased. This situation led to an increased oxygen extraction of the liver blood. The initial increase in oxygen supply was not associated with an increase of the mean liver tissue pO_2 as in the NE group. But the shift of the oxygen balance to a reduced oxygen availability at the end of the experiment did not lead to a decline of the mean liver tissue pO_2 either and the liver pO_2 histogram was still normally distributed. Apparently the regulatory ability of the liver kept tissue pO_2 at physiological levels and distribution although oxygen consumption was moderately increased and oxygen delivery was not adequately increased.

In conclusion the administration of the NO-synthase inhibitor L-NAME led to a decrease of hepatic oxygen delivery which had to be compensated for by an increase of the oxygen extraction of the liver. Therefore the use of L-NAME in hyperdynamic septicemia should be questioned.

REFERENCES

1. Amir S, English AM: An inhibitor of nitric oxide production N-nitro-L-arginine-methyl ester, improves survival in anaphylactic shock. Eur J Pharmacol, 203, 125-127, 1991
2. Breslow MJ et al: Effect of vasopressors on organ blood flow during endotoxin shock in pigs. Am J Physiol 253, H291-300,1991
3. Wang P, Ba ZF, Chaudry IH: Increase of hepatic blood flow during early sepsis is due to increased portal blood flow. Am J Physiol 261, R1507-R1512, 1991

11

ESTIMATION OF CEREBRAL BLOOD VOLUME AND TRANSIT TIME IN NEONATES FROM QUICK OXYGEN INCREASES MEASURED BY NEAR-INFRARED SPECTROPHOTOMETRY

M. Wolf, H. U. Bucher, M. Keel, K. von Siebenthal, and G. Duc

Clinic for Neonatology
University Hospital
8091 Zurich, Switzerland

INTRODUCTION

Quick oxygen increases are used to estimate the cerebral blood flow (cbf) or cerebral haemoglobin flow (cHbf) in neonates by near infrared spectrophotometry (NIRS) (Edwards et al, Skov et al., Bucher et al.). One of the essential assumptions for the correct estimation of cbf by the Fick principle is, that the time a bolus of tracer needs to pass the cerebral compartment - the so-called transit time (tt) - exceeds 6s to 8s. This paper describes a way of estimating the tt.

If the mean tt is known, the suggested one compartment model will also supply estimates for the cerebral blood volume (cbv) or the cerebral haemoglobin weight (cHbw).

THEORY

For the analysis a one compartment model is applied (Fig. 1). The following assumptions are made:

1. The oxygen saturation (SaO_2) of the haemoglobin and the haemoglobin concentration of the blood flowing into the brain are known.
2. The brain corresponds to one compartment. In this compartment all of the blood takes the same time tt to flow through.
3. NIRS measures the global changes in cerebral oxygenated haemoglobin (O_2Hb) and deoxygenated haemoglobin (HHb).
4. The brain is well oxygenated before each measurement ($SaO_2 > 85\%$). Giving additional oxygen does not change the cerebral oxygen consumption, which is

Oxygen Transport to Tissue XVII, Edited by Ince et al.
Plenum Press, New York, 1996

93

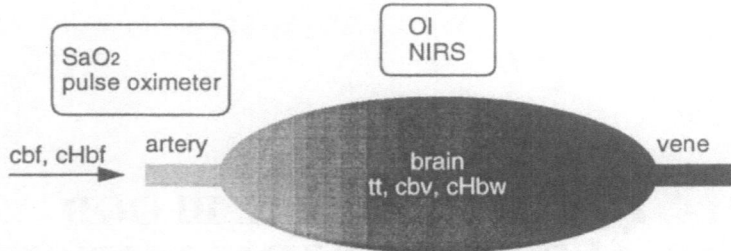

Figure 1. The one compartment model. The arterial oxygen saturation (SaO_2) is measured by a pulse oximeter and the cerebral oxygen index (OI) is measured by near-infrared spectrophotometry (NIRS). This allows calculation of the cerebral blood flow (cbf), the transit time (tt) and the cerebral blood volume (cbv).

unknown. Only the additional oxygen is taken into account for all the following calculations.

The SaO_2 is adjusted to a lower normal level (aim 90%) and should be in a steady state for about one minute. A step increase in oxygen is given. These measurements are restricted to ventilated infants needing additional oxygen. Fig. 2 shows a typical measurement with an increase in SaO_2 followed by an increase in the oxygen index (OI), which equals to OI = (O_2Hb - HHb)/2. OI gives a better signal to noise ratio than O_2Hb as long as the total haemoglobin volume (THb) remains the same during the measurement.

$$cbf(ip) = k * \frac{\Delta OI(ip)}{Hb * \int_0^{ip} SaO_2 * dt}$$

(1)

For the calculation of cbf as a function of the integration period (ip), equation 1 is applied to a quick increase in oxygen. k is a conversion constant and Hb corresponds to the haemoglobin content in arterial blood.

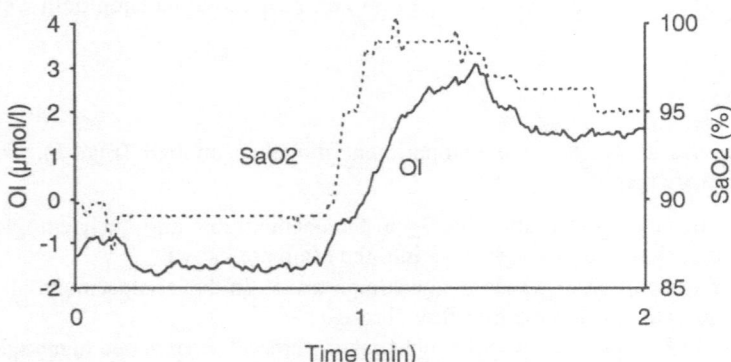

Figure 2. A typical example of a quick oxygen increase. A rapid increase of arterial oxygen saturation (SaO_2) is followed by a slower increase of oxygen index (OI).

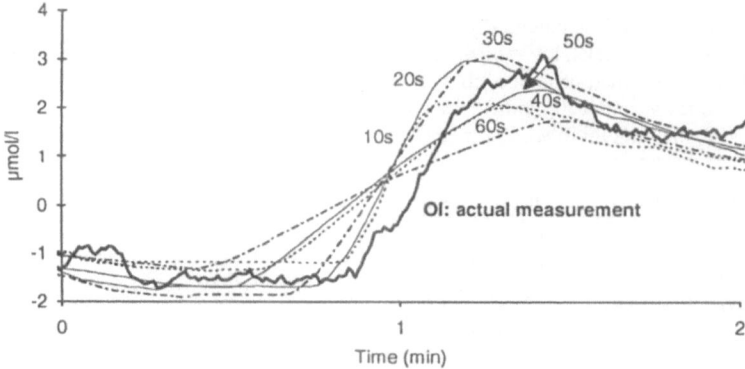

Figure 3. Comparison of simulated oxygen index (OIsim) curves for different integration periods (10s, 20s, 30s, 40s, 50s, 60s) to the measured oxygen index (OI) curve. The longer the integration period the flatter is the slope of the OIsim curve. The curve with 30s integration period corresponds best to the actual measurement.

$$OIsim(t, ip) = \frac{cbf(ip) * Hb * \int_{t-ip}^{t} SaO_2 * dt}{k} \tag{2}$$

Knowing cbf, OI can be simulated as a function of time (t) and ip (equation 2). The slope of OIsim during the time of the oxygen increase correlates inversely to ip (Fig. 3). The ip, where the slope of OIsim corresponds best to the slope of OI is determined numerically by minimising the mean square distance between the OIsim - and the measured OI curve. tt is equal to this ip.

$$cHbf(ip) = k * \frac{\Delta OI(ip)}{\int_{0}^{ip} SaO_2 * dt} \tag{3}$$

$$OIsim(t, ip) = \frac{cHbf(ip) * \int_{t-ip}^{t} SaO_2 * dt}{k} \tag{4}$$

By an analogue procedure, it is possible to determine the cHbf (equations 3 and 4). Assuming a one compartment model cbv = tt * cbf or cHbw = tt * cHbf.

MATERIAL AND METHODS

Eighty-two measurements were obtained from nine mechanically ventilated babies, who needed additional oxygen.

A Nellcor N-200 pulse oximeter in beat to beat or 2s sampling time mode was used to measure the SaO_2. The changes in O_2Hb and HHb were recorded by a Critikon Cerebral Oxygenation Monitor 205 at a sampling time of 0.57s.

Table 1. The results of the calculations. The values in this table are given in: Mean (SEM) ga = gestational age in weeks, bw = birthweight in g, pa = postnatal age in days, N = Number of measurements, tt = transit time in s, cbf = cerebral blood flow in ml/(100g*min), cHbf = cerebral haemoglobin flow in g/(100g*min), cbv = cerebral blood volume in ml/100g, cHbw = cerebral haemoglobin weight in g/100g, hmd = hyaline membrane disease

ga	bw	pa	N	tt	cbf	cHbf	cbv	cHbw	clinical condition
26 (0/7)	980	19	5	16.6 (1.8)	5.9 (0.5)	0.7 (0.06)	1.59 (0.15)	0.20 (0.02)	severe hmd
26 (1/7)	980	3	10	14.4 (1.0)	15.1 (2.3)	2.1 (0.31)	3.44 (0.38)	0.48 (0.06)	severe hmd
27 (5/7)	1020	0	12	20.0 (1.4)	4.0 (0.3)	0.8 (0.06)	1.30 (0.12)	0.27 (0.05)	hmd
28 (2/7)	1160	1	16	21.5 (2.5)	5.2 (0.8)	0.7 (0.11)	1.48 (0.13)	0.25 (0.03)	severe hmd
28 (3/7)	920	7	7	13.6 (1.9)	10.4 (2.0)	1.3 (0.25)	1.98 (0.05)	0.30 (0.04)	oesophageal atresia, severe hmd
30 (6/7)	1520	1	12	10.6 (0.9)	5.8 (0.6)	1.0 (0.11)	0.98 (0.10)	0.18 (0.03)	severe hmd
32 (0/7)	1775	2	7	15.0 (3.3)	5.8 (2.9)	0.9 (0.43)	1.63 (0.12)	0.21 (0.03)	hmd
36 (5/7)	2570	0	10	23.7 (3.5)	7.5 (1.1)	1.0 (0.14)	2.60 (0.33)	0.39 (0.05)	sepsis, hypotension
37 (2/7)	3090	1	3	23.1 (3.5)	3.7 (1.3)	0.5 (0.22)	1.36 (0.40)	0.21 (0.04)	diaphragm. hernia, severe asphyxia

RESULTS

The detailed results are in Table 1. The mean tt was 17.60 (SD = 4.36) s, the cbf 6.98 (3.48) ml/(100g*min), the cHbf 1.00 (0.42) g/(100g*min), the cbv 1.82 (0.71) ml/100g and the cHbw 0.28 (0.09) g/100g.

DISCUSSION

Pulse oximetry may be a source of considerable error. Most pulse oximeters only have a resolution of 1% in SaO$_2$ and 2s in time. The pulse oximeter we used, was capable

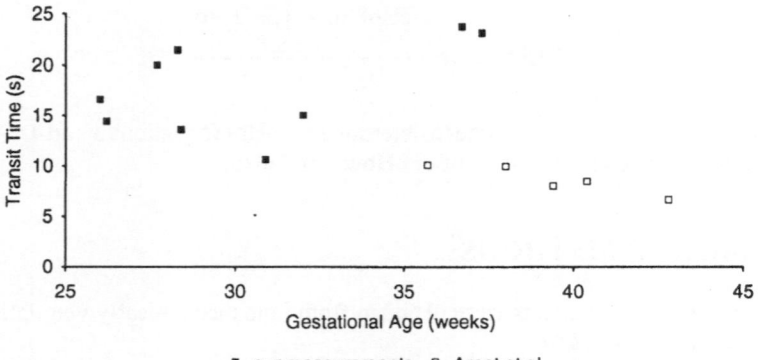

Figure 4. The mean transit times of the nine infants versus the gestational age. In addition the values measured by Arnot et al. are shown.

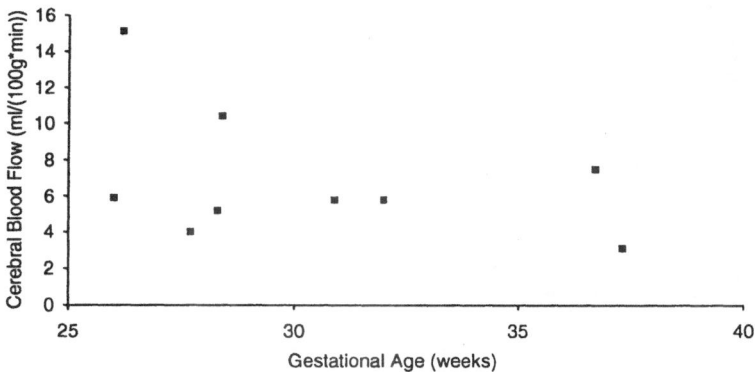

Figure 5. The cerebral blood flow versus the gestational age.

of giving one value per heart beat and a resolution of 0.1%, having a higher noise level in return.

The haemoglobin content in the blood (Hb) may vary depending on the diameter of the vessel (Lammertsma et al.). Therefore it is more reliable and physiologically more relevant to give cHbf and cHbw instead of cbf and cbv. In this paper cbf and cbv are only used to compare the values to the ones previously published.

Considering the blood circulation, the head of the newborn infant contains three compartments: grey matter, white matter and skin. From the 133Xenon clearance method it is known, that the grey matter compartment is very small in neonates (10%) and has a very high flow (8 to 10 times higher than white matter), so that the transit time must be very small (around 1s to 3s). On the one hand the time resolution of the current measuring equipment is not good enough to get stable estimates. On the other hand such a small compartment can be neglected. The skin compartment is small and has a low blood flow. Therefore it can be neglected as well.

The principles of NIRS have been discussed extensively by Cope et al.

Our results for tt are higher than those reported by Arnot et al., who found values between 6.6s and 10s in 5 patients with gestational age between 42 and 36 weeks. The

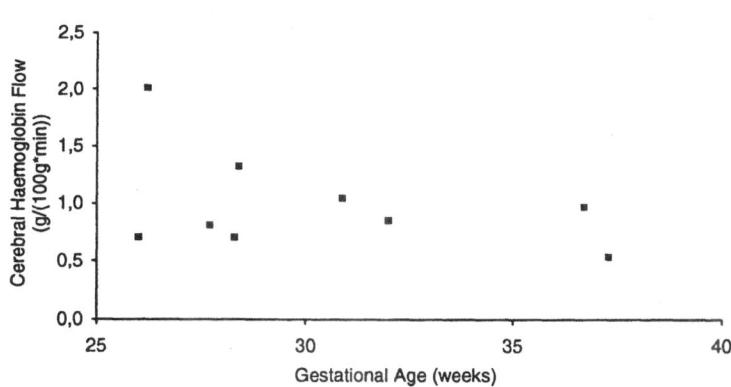

Figure 6. The cerebral haemoglobin flow versus the gestational age.

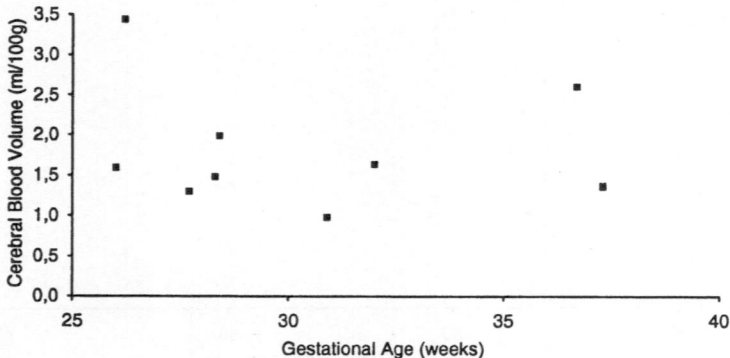

Figure 7. The cerebral blood volume versus the gestational age.

difference can be explained by the lower gestational age of our infants except for the two infants with a gestational age above 35 weeks. These two infants were both severely ill (Table 1). It is impossible to get healthy infants at this age, which are intubated. Our remaining tt fit well into the extrapolation of the values of Arnot et al.

The values for cbf are in the lower range of previous publications (Edwards et al, Skov et al., Bucher et al.). They used shorter ip between 6s to 8s, which result in higher estimates. The ip used for the calculation of cbf in this paper equals to the tt, which is only correct under the strict assumption, that all the blood takes the same time to pass the compartment. What ip should be taken? There are two effects: 1. Due to the grey matter compartment, the cbf estimates decrease with increasing integration period. 2. If some of the additionally oxygenated blood leaves the brain before the ip is elapsed, this results in underestimation of the cbf. The advantages of choosing an ip in relation to the tt would be, that a maximum ip can be adjusted individually to each infant, which will give less noisy results and reduce the influence of effect 1. To avoid effect 2 the ip must be shorter than the tt. Therefore we suggest to use an ip, which is around 70% of the tt. This issue needs further consideration.

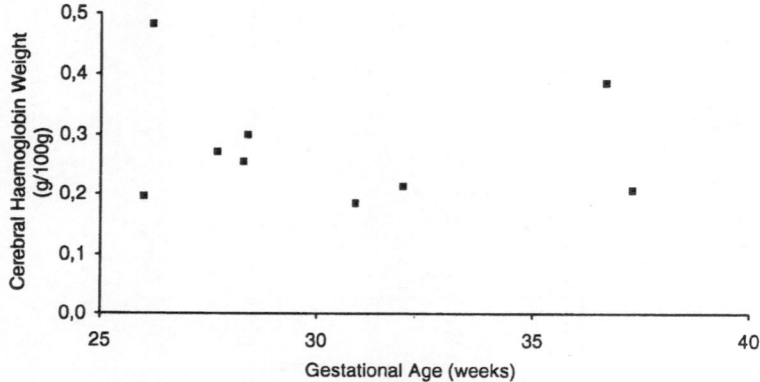

Figure 8. The cerebral haemoglobin weight versus the gestational age.

Our values for cbv are in the lower range of the 2.22 (SD 0.40) ml/100g published by Wyatt et al., that are also based on near-infrared spectrophotometry (NIRS), but analysed in a different way.

CONCLUSION

We report a simple mathematical model using NIRS data, which gives reasonable results for cerebral blood and haemoglobin flow, blood volume, haemoglobin weight and transit time. Further studies are needed to show its accuracy and clinical value.

REFRENCES

Arnot, R.N., Glass, H.I., Clark, J.C., Davis, J.A., Schiff, D. and Picton-Warlow, C.G., 1970, Methods of Measurement of Cerebral Blood Flow in the Newborn Infant using Cyclotron Produced Isotopes, *Radioaktive Isotope in Klinik und Forschung* 9:60-74.

Bucher, H.U., Edwards, A.D., Lipp, A.E. and Duc, G., 1993, Comparison between near infrared spectroscopy and 133Xenon clearance for estimation of cerebral blood flow in critically ill preterm infants, *Pediatr.Res.* 33:56-60.

Cope, M. and Delpy, D.T., 1988, System for long-term measurement of cerebral blood and tissue oxygenation on newborn infants by near infra-red transillumination, *Med.Biol.Eng.Comput.* 26:289-294.

Edwards, A.D., Wyatt, J.S., Richardson, C., Delpy, D.T., Cope, M. and Reynolds, E.O., 1988, Cotside measurement of cerebral blood flow in ill newborn infants by near infrared spectroscopy, *Lancet* 2:770-771.

Lammertsma, A.A., Brooks, D.J., Beaney, R.P., et al, 1984, In vivo measurement of regional cerebral haematocrit using positron emission tomography, *J.Cereb.Blood.Flow.Metab.* 4:317-322.

Skov, L., Pryds, O. and Greisen, G., 1991, Estimating cerebral blood flow in newborn infants: comparison of near infrared spectroscopy and 133Xe clearance, *Pediatr.Res.* 30:570-573.

Wyatt, J.S., Cope, M., Delpy, D.T., et al, 1990, Quantitation of cerebral blood volume in human infants by near-infrared spectroscopy, *J.Appl.Physiol.* 68:1086-1091

12

OXYGEN DEPENDENT QUENCHING OF PHOSPHORESCENCE

A Status Report

David F. Wilson, Sergei Vinogradov, Leu–Wei Lo, and Lu Huang

Department of Biochemistry and Biophysics
University of Pennsylvania
Philadelphia, Pennsylvania 19104

INTRODUCTION

Oxygen dependent quenching of phosphorescence has now been in use as a method for measuring oxygen for about 10 years. As expected for a new technique, these years have seen a rapid evolution of every aspect of the method, and as it matures the full power of the technique is becoming apparent. Several properties of oxygen measurements by phosphorescence which are of great value compared to other available method for measuring oxygen were established early, including; its very rapid response time (msec), accuracy to low oxygen pressures (< 0.1 Torr), wide dynamic range, and applicability to measurements in tissue *in vivo*. The limits of the method have been determined in part by the available phosphorescent oxygen probes, the best of which has been a series of Pd-porphyrins. These probes provide a good measure of oxygen, but the excitation light needed to be in the part of the spectrum which is strongly absorbed by pigments in tissue and this restricted oxygen measurements to the surface about 1 mm of tissue.

Improvements have occurred in every aspect of the application of phosphorescence quenching to oxygen measurement. We will focus on those improvements which are most relevant to measurements of oxygen *in vivo*, dividing them into improvements in probe design, data analysis and instrument performance.

NEW OXYGEN PROBES

Phosphors have been synthesized which absorb in the near infra-red window of tissue, i.e. the region of the spectrum where the absorbance by natural pigments of tissue is low. This near infra red "window" of tissue begins at about 620 nm and extends to about 1,000 nm. These phosphors with both absorbance and emission in this window of tissue absorption allow oxygen measurements in substantial thickness of tissue and

Oxygen Transport to Tissue XVII, Edited by Ince et al.
Plenum Press, New York, 1996

101

substantially increase the sensitivity of the measurements. Several new phosphor structures have been synthesized in our laboratory and the properties of two of the most
promising are briefly summarized in Table 1. Phosphors based on the same basic structure
as Green 2W provide the first generation of oxygen probes absorbing in the near infra
red and are nearly ready for general use in research. Green 2W is nearly ideal for research
in which oxygen measurements in the blood are desired. It is very water soluble and
therefore can be injected into the blood without previous binding to albumin. This
decreases the amount of foreign material and the volume of solution which must be
injected prior to making the oxygen measurements. There are several additional factors
which contribute to the decreased amount of phosphor which must be injected to obtain
oxygen measurements *in vivo* but three of these are:

First, the absorbance by the natural chromophors in tissue is much less at the longer
wavelengths (about 540 nm vs about 640 nm). As a result, the excitation light is less
attenuated as it passes through the tissue and it illuminates a much larger volume of tissue.
This means a larger volume of blood is illuminated and there is a higher probability that the
light will be absorbed by phosphor.

Second, the absorption coefficient for Green 2W at 636 nm is approximately 51,000
$M^{-1}cm^{-1}$, more than twice that for the absorption of the Pd-porphyrins in the 520 to 560 nm
wavelength range. As a result, for a flash of excitation light of any particular intensity, a
substantially greater number of photons will be absorbed by the phosphor and there will be
a proportionate increase in the amount of phosphorescence emitted. Conversely, less phosphor must be added to obtain the same phosphorescence intensity.

Third, the quantum efficiency for phosphorescence of Green 2W in aqueous solution
and bound to albumin is about 8%, similar to that of the Pd-meso-tetra (4-carboxyphenyl)
porphyrin. The latter is currently the phosphor most commonly used for *in* vivo measurements. These are high quantum efficiencies and as a result the emitted phosphorescence is
comparable to the emission of the best fluorophors. The phosphorescence decay can
therefore be measured with an excellent signal to noise.

Unfortunately, in its current chemical form, Green 2W does not stay in solution in
the blood. Preliminary data indicates it binds to the vascular lining where it retains its
sensitivity to oxygen but with heterogeneity in τ^o and quenching constant. Thus although
the phosphorescence lifetimes accurately indicate changes in oxygen pressure in the vasculature, the absolute values cannot be considered accurate because the probe environment is
no longer that of the albumin binding site. Currently efforts are underway to "redesign" the
phosphor such as to overcome this limitation.

Table 1. Basic phosphors for physiological oxygen measurements

Phosphor	Absorption maximum (nm)	Phosphorescence maximum (nm)	Lifetime (μsec)	k_Q (Torr^{-1} sec^{-1})	Quantum efficiency
Pd-porphyrin	524	695	650	320	6%
Green 2W	636	790	290	210	8%

Pd-porphyrin values are for Pd-meso-tetra (4-carboxyphenyl) porphyrin, the oxygen probe
most widely used at the present time. The values for Pd-porphyrin and Green 2W are for 38°
and neutral pH, with the phosphor bound to albumin in aqueous solution. Quantum efficiency
is the ratio of photons absorbed to phosphorescence photons emitted in the absence of oxygen.

PHOSPHORESCENCE IMAGING THROUGH SUBSTANTIAL THICKNESS OF TISSUE

As noted above, the thickness of tissue penetrated by 636 nm red light is much greater than that by green light. We have used mice to demonstrate that oxygen distribution in tissue can be imaged through the body of a mouse. As shown in Figure 1, 0.3 mg of Green 2W was injected into a 32 g mouse, the mouse was anesthetized and laid on its stomach in a plastic petri dish. The excitation light was introduced from below the mouse the lower abdomen and the phosphorescence imaged from above. Thus the light, excitation and/or emission, passed completely through the abdomen of the mouse before being imaged by the camera. A series of 8 images were taken, using delay times after the flash of 30, 50, 80, 120, 180, 240, 420, and 2,500 μsec after the flash of excitation light. The phosphorescence lifetimes were then calculated for each pixel of the image set followed by conversion of the lifetime maps into a maps of apparent oxygen pressure (see above) using the Stern–Volmer relationship (for more description see Pawlowski and Wilson, 1992). In this particular mouse, subcutaneous tumors had been planted on each flank, and these can be seen as the regions of lower oxygen pressure. The larger tumor was on the left side of the oxygen map and the smaller on the right side. The tumor regions can be easily seen, although the oxygen pressures appear to be closer to that of normal tissue than is the case for the tumor tissue *per se*. Trans illumination means the measurements include both tumor and normal tissue, giving a weighted average which shows the presence of hypoxic regions but not the minimal oxygen pressure values in those hypoxic regions.

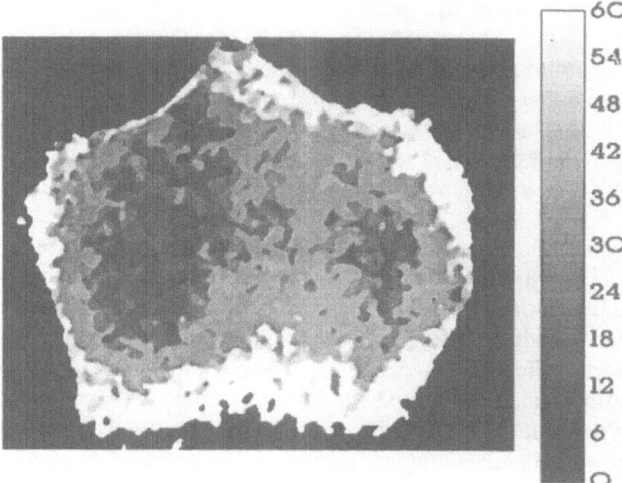

Figure 1. Oxygen pressure maps measured through a mouse using trans-illumination. The mouse was injected with 0.3 mg of Green 2W, anesthetized, and laid on its stomach in a clear petri dish. The excitation light (635 ± 15 nm) was introduced from below and the phosphorescence (> 715 nm) imaged from above using a Xybion intensified CCD camera. The phosphorescence images were taken with delay times after the flash of 30, 50, 80, 120, 180, 320, 420, and 2,500 μsec. The image with a delay of 2,500 μsec (background) was subtracted from the other images and then the phosphorescence decay constant calculated for each pixel of the phosphorescence image set by fit to a single exponential. The tail of the mouse is toward the top of the map. Subcutaneous tumors had been implanted in each flank and had grown to a diameter of about 8 mm (left side) and 4 mm (right side). The calculated oxygen pressures in Torr are given in the grey scale provided on the right. These values are approximate.

TOXICITY OF THE PHOSPHORESCENT PROBES

The toxicity of the Pd-porphyrins is extremely low. In our laboratory the Pd-meso-tetra-(4-carboxyphenyl) porphyrin has been used for oxygen measurements in many experiments using newborn piglets, adult cats and adult rats. There has been no evidence of toxicity due to the phosphor as determined by the measured physiological parameters (blood pressure, oxygen levels in the brain cortex, metabolic status of the brain, heart rate or blood pH, $PaCO_2$ or PaO_2). All of these experiments have been acute, however, and the animals were not allowed to recover from anesthesia. Thus, although the physiological parameters of the animals were carefully monitored for up to 6 hours and showed no effect of phosphor, the longer term effects were not tested. On the other hand, a newer phosphor, Green 2W, has been injected into the blood of 10 mice in amounts up to more than 10 times that needed for oxygen measurements and the mice followed for 10 days. In this case, the mice appeared completely normal throughout the 10 day period. Thus, there appears little reason for concern about toxic effects of the phosphors, although as new phosphors are developed each will need to be tested before being extensive used *in vivo*.

IMPROVEMENTS IN DATA ANALYSIS

As blood passes through the vasculature of the tissue, the oxygen pressure decreases continuously, falling from arteriolar to venous end capillary values. Thus, the phosphor in the tissue sample being illuminated is in the presence of a range of oxygen pressures, from arterial values to values below that in the veins. The initial intensity of the phosphorescence is dependent only on the volume of blood in the tissue, the concentration of phosphor and the intensity of the illuminating light. The time course of phosphorescence decay, however, is dependent on the oxygen pressure. Each phosphorescence decay measurement from tissue therefore contains information on the complete distribution of oxygen pressures in the blood as well as the fraction of the blood with each oxygen pressure. This is all of the information necessary to construct a histogram of the oxygen distribution in the blood of that tissue sample.

Calculation of the oxygen histogram require deconvolution of the phosphorescence decay data into the distribution of exponentials which gives rise to the measured decay data, determining both the phosphorescence lifetime and the initial intensity of the component of the phosphorescence with that lifetime. The phosphorescence lifetime is then converted into oxygen pressure while the initial intensity is a measure of the blood volume in the illuminated tissue. Computational methods suited for deconvolution of phosphorescence decay (Vinogradov and Wilson, ISOTT 1993; Biophys. J., in press) have been successfully implemented. The algorithm described in the Biophys. J. paper, based on quadratic programming principles, can deconvolute a phosphorescence decay into a base of 200 different oxygen pressures in about 2.0 minutes using a personal computer with a 60 MHz Pentium processor. Good resolution requires a signal to noise of greater than 400, but with current instrumentation values greater than this are readily attainable. Thus, application of this algorithm can rapidly and significantly increase the biochemical and physiological information obtained with each measurement of phosphorescence decay in tissue. Preliminary application of the deconvolution method to *in vivo* phosphorescence decays provide reasonable oxygen histograms but time will be required to establish that these histograms fully reflect the distribution of oxygen in the vasculature and their relationships to histograms determined by other oxygen measuring techniques.

IMPROVEMENTS IN PHOSPHORESCENCE MEASURING INSTRUMENTATION

Most of the currently available systems for measuring phosphorescence utilize photomultipliers or photodiodes. In general, photomultipliers are more much sensitive than photodiodes but, unlike the latter, their sensitivity usually decreases rapidly at wavelengths greater than about 800 nm. Thus it has not been possible to measure phosphorescence in the spectral region from 800 to 1,000 nm with as high a sensitivity as can be routinely achieved in the visible wavelengths. New detectors, such as avalanche photodiodes, are currently being developed which have excellent sensitivity extending to above 1,000 nm. These new detectors are allowing phosphorimeters to be constructed with improved performance in the wavelength range of interest to oxygen measurement in tissue. Phosphorimeters can now readily determine phosphorescence decay with signal to noise of greater than 400, sufficient for accurate deconvolution into distributions of lifetimes.

ACKNOWLEDGMENTS

Supported by grants NS-10939 and NS-31465 from the U.S National Institutes of Health.

REFERENCES USING PHOSPHORESCENCE TO MEASURE OXYGEN

Alcala, J.R., Yu, C. and Yeh, G.J. (1993) Digital phosphorimeter with frequency domain signal processing: Application to real-time fiber-optic oxygen sensing. *Rev. Sci. Instrum.* 64, 1554-1560.

Green, T.J., Wilson, D.F., Vanderkooi, J.M. and DeFeo, S.P. (1989) Phosphorimeters for analysis of decay profiles and real time monitoring of exponential decay and oxygen concentrations. *Analytical Biochem.* 174, 73-79.

Ince, C., Ashruf, J.F., Avontuur, J.A., Wiering, P.A., Spaan, J.A., and Bruining, H.A. (1993) Heterogeniety of the hypoxic state in rat heart is determined at capillary level. *Am. J. Physiol.* 264(2), H294-H301.

Iturriaga, R., Rumsey, W.L., Lahiri, S., Spergel, D., and Wilson, D.F. (1992) Intracellular pH and oxygen chemoreception in the cat carotid body in vitro. *J. Appl. Physiol.* 72(6), 2259-2266.

Lahiri, S., Rumsey, W.L., Wilson, D.F., and Iturriaga, R. (1993) Contribution of *in vivo* microvascular PO_2 in the cat carotid body chemotransduction. *J. Appl. Physiol.* 75(3): 1035-1043.

Lee, W.W.S., Wong, K.Y., Li, X.M., Leung, Y.B., Chan, C.S., and Chan, K.S. (1993) Halogenated platinum porphyrins as sensing materials for luminescence-based oxygen sensors. *J. Mat. Chem.* 3: 1031-1035.

Pawlowski, M., and Wilson, D.F. (1992) Monitoring of the oxygen pressure in the blood of live animals using the oxygen dependent quenching of phosphorescence. *Adv. Exptl. Med. Biol.* 316, 179-185.

Plant, R.L. and Burns, D.H. (1993) Quantitative, depth-resolved imaging of oxygen concentration by phosphorescence lifetime measurement. *Applied Spectroscopy.* 47, 1594-1599.

Pastuszko A. (1994) Metabolic response of dopaminergic system during hypoxia-ischemia and reoxygenation in the immature brain. *Review, Biochem. Med. and Metabolic Biology.* 51, 1-15.

Pastuszko, A., Saadat-Lajevardi, N., Chen, J., Tammela, O., Wilson, D.F., and Delivoria-Papadopoulos, M. (1993) Effects of graded levels of tissue oxygen pressure on dopamine metabolism in the striatum of newborn piglets. *J. Neurochem.* 60, 161-166.

Robiolio, M., Rumsey, W.L., and Wilson, D.F.: (1989) Oxygen diffusion and mitochondrial respiration in neuroblastoma cells. *Am. J. Physiol.* 256, C1207-C1213.

Rumsey, W.L., Iturriaga, R., Wilson, D.F., Lahiri, S., and Spergel, D. (1990) Phosphorescence and fluorescence imaging: new tools for the study of carotid body function. in: "Chemoreceptors and chemoreceptor reflexes" (H. Acker et al, eds) Plenum Press, New York, pp. 73-79.

Rumsey, W.L., Iturriaga, R., Spergel, D., Lahiri, S., and Wilson, D.F. (1991) Optical measurements of the dependence of chemoreception on oxygen pressure in the cat carotid body. *Amer. J. Physiol.* 261, C614-C622.

Rumsey, W.L., Lahiri, S., Iturriaga, R., Mokashi, A., Spergel, D., and Wilson, D.F. (1992) Optical measurements of oxygen and electrical measurements of oxygen chemoreception in the cat carotid body. *Adv. Exptl. Med. Biol.* 317, 387-395.

Rumsey, W.L., Robiolio, M. and Wilson, D.F.: (1989) Contribution of diffusion to the oxygen dependence of energy metabolism in human neuroblastoma cells. *Adv. Exptl. Med. Biol.* 248, 829-833.

Rumsey, W.L., Schlosser, C., Nuutinen, E.M., Robiolio, M. and Wilson, D.F. (1990) Cellular energetics and the oxygen dependence of respiration in cardiac myocytes isolated from adult rat. *J. Biol. Chem.* 265, 15392-15399.

Rumsey, W.L., Schlosser, C., Nuutinen, E.M., Robiolio, M., and Wilson, D.F. (1992) The oxygen dependence of mitochondrial oxidative phosphorylation and its role in regulation of coronary blood flow. *Adv. Exptl. Med. Biol.* 316, 279-284.

Rumsey, W.L., Vanderkooi, J.M. and Wilson, D.F.: (1988) Imaging of phosphorescence: A novel method for measuring the distribution of oxygen in perfused tissue. *Science,* 241, 1649-1651.

Shonat, R.D., Wilson, D.F., Riva, C.E., and Cranston, S.D. (1992) Effect of acute increases in intraocular pressure on intravascular optic nerve head oxygen tension in cats. *Inv. Ophthalmology & Visual Science* 33, 3174-3180.

Shonat, R.D., Wilson, D.F., Riva, C.E., and Pawlowski, M. (1992) Oxygen distribution in the retinal and choroidal vessels of the cat as measured by a new phosphorescence imaging method. *Appl. Opt.* 33, 3711-3718.

Tammela, O., Pastuszko, A., Lajevardi, N., Delivoria-Papadopoulos, M. and Wilson, D.F. (1993) Activity of tyrosine hydroxylase in the striatum of newborn piglets in response to hypocapnic hypoxia. *J. Neurochem.* 60, 1399-1406.

Torres, I.P. and Itaglietta, M. (1993) Microvessel PO_2 measurements by phosphorescence decay method. *J. Appl. Physiol.* 265, 1434-1438.

Torres, I.P., Leunig, M., Yuan, F., Intaglietta, M., and Jain, R.K. (1994) Noninnvasive measurement of microvascular and interstitial oxygen profiles in a human tumor in SCID mice. *Proc. Natl. Acad. Sci. U.S.* 91, 2081-2085.

Vanderkooi, J.M., Maniara, G., Green, T.J., and Wilson, D.F. (1987) An optical method for measurement of dioxygen concentration based on quenching of phosphorescence. *J. Biol. Chem.* 262, 5476-5482.

Vanderkooi, J.M., and Wilson, D.F. (1986) A new method for measuring oxygen concentration in biological systems. *Adv. Exptl. Med. Biol.* 200, 189-193.

Vanderkooi, J.M., Wright, W.W. and Erecinska, M. (1990) Oxygen gradients in mitochondria examined with delayed luminescence from excited-state triplet probes. *Biochem.* 29(22), 5332-5338.

Vinogradov, S.A. and Wilson, D.F. (1993) Recovery of the distribution of oxygen in tissue from phosphorescence decay data. *Adv. Exptl. Med. Biol.* in press.

Vinogradov, S.A. and Wilson, D.F. (1994) Phosphorescence lifetime analysis for determination of oxygen distribution in heterogenous systems. *Biophys. J.* in press.

Vinogradov, S.A. and Wilson, D.F. (1994) New phosphorescent probes for oxygen measurement. *J. Chem. Soc. Perkin Trans. II,* in press.

Wilson, D.F. "Tissue energy metabolism and its dependence on oxygen pressure" In: *The Adaptations.* (J. R. Sutton, G. Coates and J.E. Remmers, eds.) B.C. Decker, Inc. Toronto, 1990, pp. 106-111.

Wilson, D.F. (1990) Contribution of diffusion to the oxygen dependence of energy metabolism in cells. *Experientia* 46, 1160-1162.

Wilson, D.F. (1992) Oxygen dependent quenching of phosphorescence: a perspective. *Adv. Exptl. Med. Biol.* 317, 195-201.

Wilson, D.F. and Cerniglia, G.J. (1992) Localization of tumors and evaluation of their state of oxygenation by phosphorescence imaging. *Cancer Research,* 52, 3988-3993.

Wilson, D.F., Rumsey, W.L., Green, T.J., and Vanderkooi, J.M.: (1988) The oxygen dependence of mitochondrial oxidative phosphorylation measured by a new optical method for measuring oxygen. *J. Biol. Chem.* 263, 2712-2718.

Wilson, D.F., Gomi, S., Pastuszko, A., and Greenberg, J.H. (1992) Oxygenation of the cortex of the brain of cats during occlusion of the middle cerebral artery and reperfusion. *Adv. Exptl. Med. Biol.* 317, 689-694.

Wilson, D.F., Gomi, S., Pastuszko, A., and Greenberg, J.H. (1993) Microvascular damage in the cortex of the cat from middle cerebral artery occlusion and reperfusion. *J. Appl. Physiol.* 74(2), 580-589.

Wilson, D.F., Pastuszko, A., DiGiacomo, J.E., Pawlowski, M., Schneiderman, R., Delivoria-Papadopoulos, M. (1991) Effect of hyperventilation on oxygenation of the brain cortex of newborn piglets. *J. Appl. Physiol.* 70(6), 2691-2696.

Wilson, D.F., Rumsey, W.L. and Vanderkooi, J.M.: (1989) Oxygen distribution in isolated perfused liver observed by phosphorescence imaging. *Adv. Exptl. Med. Biol.* 248, 109-115.

13

EMPIRICAL MODELING FOR OXYGEN TRANSPORT PROCESSES AND RELATED PHYSIOLOGICAL AND BIOPROCESS SYSTEMS

Duane F. Bruley

Bioengineering Program
University of Maryland Baltimore County
College of Engineering
5401 Wilkens Ave. ECS 202A
Baltimore, Maryland 21228

INTRODUCTION

Engineering analysis is based upon quantification of systems via mathematical modeling and computer simulation. The most fundamental approach is to begin with the basic principles of physics, chemistry and biology and formulate phenomenological models that are based upon mass, energy and momentum balances. This procedure evolves equations with specific system parameters such as, diffusivity, velocity, rate constants, etc., that can be examined to determine system sensitivity to various inputs. The theoretical modeling approach has many advantages because it yields a primary understanding of system behavior at the process parameter level.

Another powerful approach however, **empirical modeling,** can be employed to quantitate a process. The method involves disturbing the process with a known input function and recording the response function.[1,2,3,4,5] Two common approaches include direct frequency forcing and pulse forcing. The input/output information can then be reduced via computer and empirical transfer functions (Laplace transform of the output variable divided by the Laplace transform of the input variable) can be determined. The transfer functions do not contain explicit system parameters but do identify gross system parameters (steady-state gain, time constant, system order, and pure time delay) that can be important for discerning basic process mechanisms and system response trends for given disturbances.

The empirical transfer function or frequency response results are also valuable in verifying theoretical models derived from basic principles. The two plots, magnitude ratio (MR) vs. forcing frequency and phase shift (ϕ) vs. forcing frequency, can be used to cross-check the validity of phenomenological models because both curves must be satisfied to yield a valid set of process equations. Curve fitting in the time domain is not as effective as in the frequency domain because only one curve is available for comparison.

Oxygen Transport to Tissue XVII, Edited by Ince et al.
Plenum Press, New York, 1996

109

Arbitrary pulse testing is especially useful for empirical modeling because of the simplicity in application. Using an efficient, reliable data reduction procedure[9] the technique will theoretically yield MR and ϕ to infinite forcing frequency.

THEORY

The application of pulse testing to obtain process dynamics treats the system as a black box. Once the system is at steady-state or equilibrium, a disturbance is added, and system response is recorded until it returns to steady-state or to re-equilibration.

It is essentially an art to design an appropriate input function for any empirical process identification attempt. While sinusoidal inputs are the classical disturbance, these are difficult to generate for many processes[4,5]. The use of a single input pulse can provide a frequency response determination over a wide range of forcing frequencies. With an appropriate sharp pulse, approximating a dirac delta function, this is especially true. Although the pulse must be small enough not to force the system into non-linear operation, it must be large enough to produce a discernible response.

The system response is recorded as a function of time and then, using Fourier transforms[2,3,6], the data is reduced to the frequency domain,

$$@EQ = F(\omega) = \int_{-\infty}^{+\infty} f(t)e^{-j\omega t}dt$$

were f(t) is the time domain pulse data, $F(\omega)$ is the pulse data in the Fourier domain, t is the time, and ω is the frequency. In theory, the integral would be over an interval from positive to negative infinity. For closed pulses, however, the interval begins and ends at zero. The system response in the Fourier domain is defined as the transfer function G(s), and when s equals the imaginary part of the complex number, $s = j\omega$, a transfer function in the frequency domain results:

$$@EQ = G(\omega) = \frac{Y(\omega)}{X(\omega)} = \frac{\int_{0}^{Ty} y(t)e^{-j\omega t}dt}{\int_{0}^{Tx} x(t)e^{-j\omega t}dt} .$$

When reducing experimental pulse data, the normalized input frequency content (NFC) is useful for discerning when the transformed data has become too noisy for use. The NFC is defined as:

$$@EQ = NFC = \frac{\left| \int_{0}^{Tx} x(t)e^{-j\omega t}dt \right|}{\left| \int_{0}^{Ty} x(t)dt \right|} .$$

NFC varies between values of 0 and 1, and decreases with increasing forcing frequency. As a rule of thumb, when the NFC falls below 0.3, results are assumed unusable. The sharper the input pulse, the greater the range of frequencies the response will cover before the NFC falls below 0.3. The NFC of a dirac delta function is unity throughout the entire frequency domain, with zero phase shift.

In the frequency domain, the system is represented by the resulting magnitude ratio (MR) and phase shift (ϕ) parameters. These quantities are defined as: [2,3,4,7]

$$@EQ = MR = |G(\omega)| = \sqrt{Re^2(\omega) + Im^2(\omega)}$$

$$@EQ = \phi = \phi|_{G(\omega)} = \tan^{-1}\left[\frac{Im(\omega)}{Re(\omega)}\right]$$

After substitution of the Euler relationship

$$@EQ = e^{-j\omega t} = Cos(\omega t) - jSin(\omega t),$$

the real and imaginary contributions can be determined:

$$@EQ = Re(\omega) = \frac{AC + BD}{C^2 + D^2}$$

$$@EQ = Im(\omega) = \frac{AD - BC}{C^2 + D^2},$$

and the values A,B,C, and D are defined as:

$$@EQ = A = \int_0^{T_y} y(t)\cos(\omega t)dt, \qquad @EQ = C = \int_0^{T_x} x(t)\cos(\omega t)dt$$

$$@EQ = B = \int_0^{T_y} y(t)\sin(\omega t)dt, \qquad @EQ = D = \int_0^{T_x} x(t)\sin(\omega t)dt.$$

The computation of the product integrals is carried out numerically using a Filon's Method Data Reduction Code[9]. Application of various quadrature methods such as the trapezoidal rule, Simpson's rule, approximation by linear or higher order curves followed by integration of the subsequent trigonometric functions are discussed in the literature[3,4]. Problems arise due to oscillations of the trigonometric functions at high frequencies. To offer a smooth approximation of the pulse curves, Filon[8] proposed a quadrature formula for the integrals based on approximation by parabolic segments as in Simpson's rule. However, the Simpson's rule coefficients are replaced by functions of $\omega\Delta t$.[4] The transformed, reduced data is plotted in Bode diagrams. These Bode diagrams are analyzed using process control theory to yield transfer functions that empirically model the system.

To fit transfer functions to systems with a non-integer order requires modifications to standard process control formulae. For example, the time delay ($\ominus$) of the system can be calculated at defined frequency steps using the following equation:

$$@EQ = \ominus = \phi(\omega) + \sum_{l=1}^{int(n)} \tan^{-1}(\tau_l\omega) + (n-int[n])\tan^{-1}(\tau_n\omega).$$

The overall transfer function of the system can be calculated using:

$$@EQ = G(s) = \frac{K_p\exp(-\ominus s)}{\left(\prod_{l=1}^{int(n)}(\tau_l s + 1)\right)(\tau_n s + 1)^{(n-int(n))}},$$

where the contributions of each integer order to the overall transfer functions are multiplicative in the frequency domain.[10]

APPLICATIONS

Recent published examples where pulse testing has been used to determine system frequency response information include Near Infrared-Time Resolved Spectroscopy (NIR-TRS)[18,19] for tissue optical property determination and medical imaging and for the process identification of liquid chromatography systems for the separation of Protein C[10], a powerful anticoagulant. For NIR-TRS testing, process parameter sensitivity studies were done and for liquid chromatography, empirical transfer functions were determined.

An illustration that demonstrates *the pulse testing* method for quantifying very complex systems is the process of oxygen transport in the microcirculation of living tissue. For instance past investigations of tissue oxygen autoregulation[11,12,13,14,15,16,17] can be quantified via an approximate empirical transfer function by converting time domain recordings into the frequency domain[9] and backing out the transfer function. Since oxygen transport in the microcirculation is somewhat non-linear it is important to note that this approach has limitations and the results might be valid for only a narrow range of process conditions.

Anesthetized, curarized cats were ventilated with pure nitrogen for 90 seconds. Microelectrodes were positioned, with micro manipulators, at 1 micron increments into the grey matter of the brain and interstitial oxygen tensions were recorded over the closed pulse.[11,12] Input/output curves are illustrated in Figure 1.

The input/output data from this experiment is presented in Table 1.

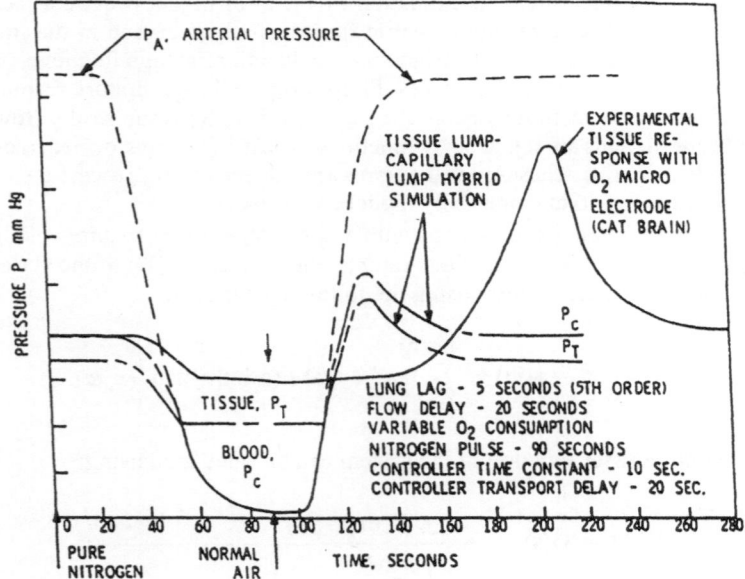

Figure 1. Krogh cylinder dynamics to step change in inspired air.

Table 1

Time (seconds)	Input data (O_2) (air, O_2)	Output data (O_2) (mm Hg)
0	20	40
10	0	40
20	0	40
30	0	40
40	0	40
50	0	35
60	0	31
70	0	30
80	0	29
90	0	28
100	20	28
110	20	29
120	20	30
130	20	30
140	20	31
150	20	32
160	20	35
170	20	40
180	20	45
190	20	55
200	20	70
210	20	80
220	20	65
230	20	55
240	20	45
250	20	43
260	20	42
270	20	41
280	20	40

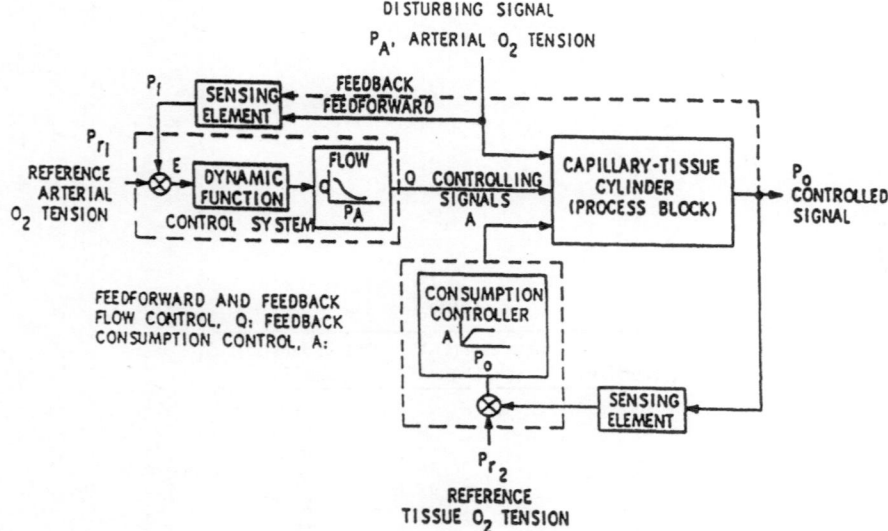

Figure 2. Lumped capillary-tissue cylinder with autoregulation.

The system is initially assumed to be at steady-state with respect to the oxygen supplied by normal air (20% oxygen). As stated earlier the input disturbance was a square pulse of pure nitrogen which resulted in a response function, with an overshoot. This behavior implies that there is inherent feedback control that is related to the oxygen tension levels in the tissue.[11,12] Blood flow rate and metabolic rate are regulated to maintain tissue oxygen levels during the insult period as shown in Figure 2. This information is captured in the resulting transfer function for the overall process.

The input/output data was fed into FORTRAN[9] programs that reduced the information to the frequency domain using Filons quadrature algorithm[8] discussed previously. Bode plots (see Figure 3) were then constructed from the frequency domain data and relevant process parameters (time constants, steady-state gain, process order and pure time delay) were extracted to construct a transfer function for the system.

The anoxic - anoxia experimental transfer function represents the input air oxygen level to the lungs of a cat and the output is the cat brain grey matter oxygen tension response. This extremely complex system that would be essentially impossible to model with phenomenological mathematical equations can be represented by a relatively simple second order transfer function with a steady-state gain and pure transport delay. The transfer function is:

$$@EQ = G(s) = \frac{7.94 \exp(-\tau\Theta s)}{(106.1\,S+1)\,(79.6\,S+1)}$$

This is not an exact transfer function because it assumes that the process was disturbed only in a linear regime. However, the empirical pulse testing technique yields a quantitative expression that can be converted to differential equations or used in the frequency domain to mathematically investigate the process. The results can also be used to validate phenomenological models or to examine process control phenomena.

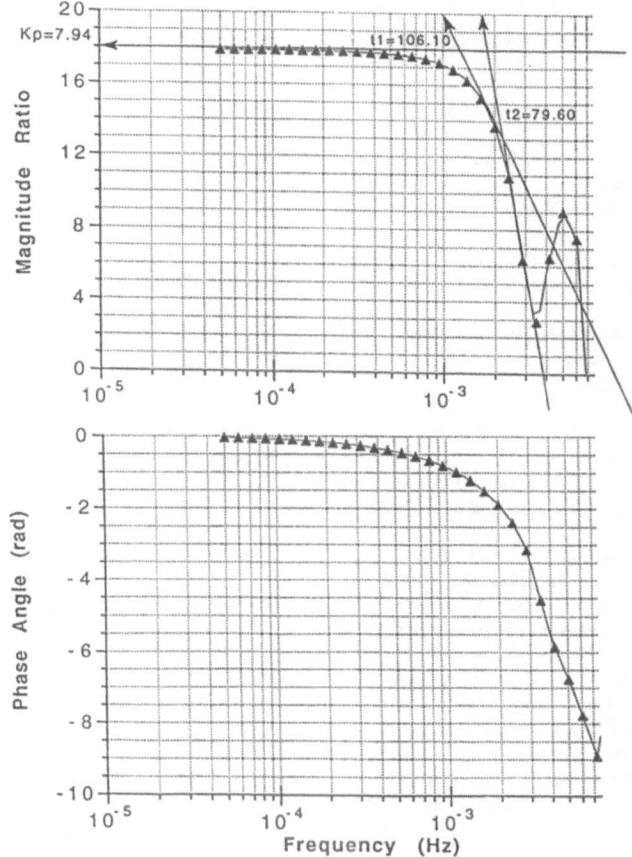

Figure 3. Bode Plots for the cat brain (O_2 data).

ACKNOWLEDGMENTS

This work was supported by The Whitaker Foundation Special Opportunities award "Symbiosis of Biomedical and Bioprocess Engineering Utilizing TQM to Enhance Health Care Quality and Delivery."

REFERENCES

1. Lees, D., and Hougen, J., Journal of Industrial & Engineering Chemistry 48, 1064-1068
2. Lewis, C.I., Bruley, D.F., and Hunt, D.H., 1967, Evaluation of temperature pulse characteristics and pulse testing for thermal dynamic analysis. *I&EC Process Des. Dev.* 6:281-286.
3. Luyben, W.L., 1990 Process Modeling, Simulation and Control for Chemical Engineers, 2nd Edition. New York: McGraw-Hill.
4. Clements, W.C., Jr., and Schnelle, K.B., 1963. Pulse testing for dynamic analysis. *I&EC Process Des. Dev.* 2(2):94-102.
5. Bruley, D.F., and Prados, J.W., 1964, The frequency response analysis of a wetted wall adiabatic humidifier. *AIChE J.* 10:612-616.

6. Kang, K.A., Bruley, D.F., Londono, J.M., and Chance, B., 1994, Highly Scattering Optical System Identification via frequency Response Analysis of NIR-TRS Spectra, *Annuals of Biomedical Engineering*. Vol.22:240-252

7. Coughanowr, D.R., 1991, Process systems analysis and control. New York, McGraw-Hill.

8. Filon, L.N.G. 1928, On a quadrature formula for trigonometric integrals., *Proc. Roy. Soc. Edinburgh*. 49:38-47.

9. Bruley, D.F., 1975, Filon's Method Data Reduction Code. Unpublished.

10. Dalton, J.C., Gupta, S., Bruley, M.D., Kang, K.A., and Bruley, D.F., Liquid Chromatographic Process Identification via Pulse Testing for Column Standardization and Scale-Up, Submitted to the Journal of Chromatography

11. Bruley, D.F., Bicher, H.I., Reneau, D.D., and Knisely, M.H., 1971, Auto-regulatory Phenomena Related to Cerebral Tissue Oxygenation, *Advances in Bioengineering. CEP Symposium Series*. 114, Vol. 67, 195-201.

12. Bruley, D.F., Reneau, D.D., Bicher, H.I. and Knisely, M.H., September 1973, Studies of Cerebral Tissue Oxygenation and Related Autoregulation, *Advances in Chemistry Series. Chemical Engineering in Medicine*. 118, 290-302

13. Bicher, H.I., Reneau, D.D., Bruley, D.F., and Knisely, M.H., 1972, Autoregulation of Oxygen Supply to Microareas of Brain Tissue Under Hypoxic and Hyperbaric Conditions, *VIIth International Meeting, European Society for Microcirculation*. Published by S. Karger. 529-531

14. Bruley, D.F., Bicher, H.I., Reneau, D.D. and Knisely, M.H., 1973, Autoregulation of Chemotransport to Brain Tissue, *Regulation and Control in Physiological Systems Symposium Proceedings*, Rochester, New York, 501

15. Bruley, D.F. and Hunt, D.H. 1974, Theoretical Studies of Brain Autoregulation; Oxygen Transport to Tissue, *Microvascular Research*, Vol. 8, 314-319

16. Bruley, D.F., Bicher, H.I., Hunt, D.H., and Flache, W.E., 1975, Autoregulatory Mechanisms Controlling the Supply of Oxygen to Microareas of Brain Tissue, *Biochemistry and Experimental Biology*, Vol. XI - N. 2, 1255-161.

17. Bicher, H.I., Hunt, D.H., and Bruley, D.F., 1976, Automatic and Pharmacological Control of Oxygen Autoregulation Mechanisms in Brain Tissue, *Oxygen Transport to Tissue - II;* Plenum Publishing Corporation, Editors, J. Grote, D. Reneau and G. Thews, 383-389.

18. Kang, K.A., Bruley, D.F., Londono, J. Chance, B., Frequency response by pulse reduction for the analysis of TRS spectra. *Proc. ISOTT XVI.* In: Hogan, M.C., Mathieu-Costello, O., Polle, D.C., Wagner, P.D., eds. Advances in experimental medicine and biology. New York: Plenum Press. (in press)

19. Kang, K.A., Bruley, D.F., Kitai, T., and Chance, B., System Parameter Analysis of NIR-TRS Spectra from Homogeneous Media with and without an Absorbing Boundary and Heterogeneous Media with a single Absorber (this volume).

14

HOW TISSUE OPTICS INFLUENCES REFLECTANCE PULSE OXIMETRY

R. Graaff,[1] A. C. M. Dassel,[2] W. G. Zijlstra,[3] F. F. M. de Mul,[4] and
J. G. Aarnoudse[2]

[1] Centre for Biomedical Technology
[2] Department of Obstetrics and Gynaecology
[3] Department of Pediatrics
  University of Groningen
  Oostersingel 59, 9713 EZ Groningen, The Netherlands
  The Netherlands
[4] Department of Applied Physics
  University of Twente
  The Netherlands

INTRODUCTION

In this paper the phenomena in tissue optics that influence reflectance pulse oximetry are described systematically for the purpose of making the subject accessible to those who are involved in pulse oximetry but less familiar with tissue optics. The results of Monte Carlo simulations have been used to show that reflectance pulse oximetry does not only depend on the differences in absorption coefficient caused by the blood volume changes, but also by changes in the reduced scattering coefficient, and by the pulse-independent parts of the absorption and reduced scattering coefficients.

PULSE OXIMETRY

It is well-known that the absorption spectra of HbO_2 and Hb are different. Optical methods can therefore be used to measure the oxygen saturation of blood. Pulse oximetry is a non-invasive optical measuring technique for the determination of S_aO_2, the arterial oxygen saturation. This method takes advantage of the intensity fluctuations that occur each heart beat, such as shown in Fig. 1. These fluctuations are caused by pulse wave induced changes in the blood volume in tissue, most likely the arterial blood volume fraction. Aoyagi et al.[1] discovered in the early seventies that the ratio of the relative changes of the pulse sizes, when measured at two different wavelengths (630 and 805 nm) after transmission through an earlobe, correlates very well with the arterial oxygen saturation, without the need for

Oxygen Transport to Tissue XVII, Edited by Ince et al.
Plenum Press, New York, 1996

117

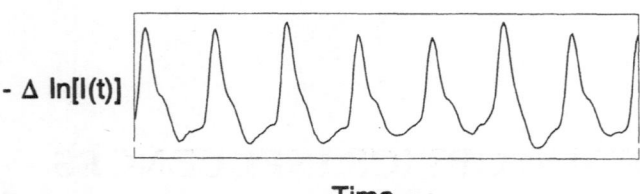

Time ⟶

Figure 1. Pulsatile variation of the logarithm of the measured intensity, $\Delta\ln[(I(t)]$, plotted as a function of time (~7 seconds).

individual calibration. This idea was the basis for the development of pulse oximetry,[2,3] which became an important breakthrough for the non-invasive monitoring of S_aO_2.

However, in some situations, pulse oximetry can only be used when the light sources and the detector are located at the same side of the skin surface. With this configuration the fluctuations in the photon flux caused by changes in the arterial blood volume can also be detected. "Reflectance pulse oximetry" is the accepted term for this way of measurement, which is currently under development for monitoring the arterial oxygen saturation of the fetus during labour, where the only available location is the fetal scalp.

THE LAMBERT–BEER MODEL OF PULSE OXIMETRY

The arterial oxygen saturation, S_aO_2, is defined as the ratio of the concentration, c, of oxygenated hemoglobin (HbO$_2$) to the sum of the concentrations of HbO$_2$ and de-oxygenated hemoglobin (Hb) of the arterial blood:

$$S_aO_2 = \frac{c_{HbO2}}{c_{Hb} + c_{HbO2}}. \tag{1}$$

The absorption spectra of the hemoglobin derivates measured by Van Assendelft[4] have widely been used to describe the relation between the absorption spectra of hemoglobin solutions and the oxygen saturation. More recently, data for fetal hemoglobins have been obtained at the same laboratory, as well as data with improved accuracy for adult hemoglobins.[5]

The theory of pulse oximetry has been based on the theory that was used for oximetry. For that reason Lambert–Beer's law has also been used as a basic theory for transmission pulse oximetry. Therefore, pulse oximetry is often explained in the literature by comparing transmission through tissue with transmission through a cuvette, neglecting the influence of light scattering in the derivation.[6-8] The "cuvette" contains only the absorbers of skin, venous blood, and arterial blood. In the absence of light scattering, the absorbers may be thought to be separated without changing the transmission, as shown in Fig. 2.

The pulsatile changes in the intensity observed in pulse oximetry, caused by arterial blood volume changes, have been included into the model by assuming a pulsatile change in the thickness of the layer with arterial blood. It has thereby been assumed as well that the arterial blood volume fluctuations do not introduce pulsatile changes of the venous (and capillary) blood volume fraction.

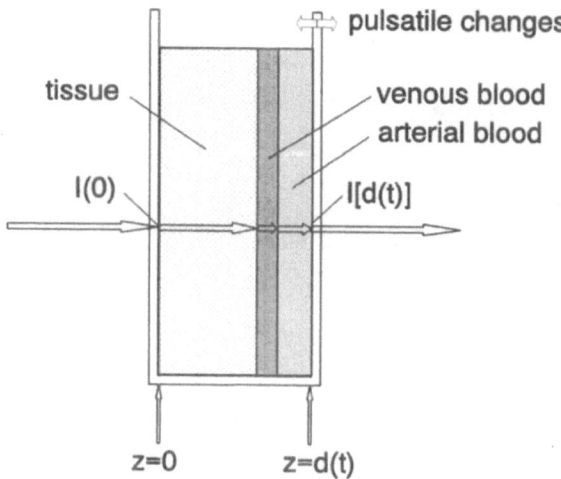

Figure 2. Lambert–Beer's law applied to pulse oximetry.

With the Lambert-Beer model the following equations can be derived for the ratio between the light intensities at $z = d$ and $z = 0$:

$$\ln \left\{ \frac{I[d(t),\lambda]}{I[0,\lambda]} \right\} = - \Sigma_{a,t}(\lambda)\, d_t - \Sigma_{a,v}(\lambda)\, d_v - \Sigma_{a,a}(\lambda)\, d_a(t), \tag{2}$$

where $\Sigma_{a,t}(\lambda)$, $\Sigma_{a,v}(\lambda)$ and $\Sigma_{a,a}(\lambda)$ are the absorption coefficients for tissue, venous blood and arterial blood, respectively, and d_t, d_v, and $d_a(t)$ are the thicknesses of the layers of tissue, venous blood, and arterial blood, respectively. Because the pulsatile changes are only caused by $d_a(t)$, the other thicknesses disappear in the expression for changes in the measured transmitted intensities. In pulse oximetry the changes are often designated by R and IR, for red and infrared light with wavelengths λ_1 and λ_2, respectively. When these changes are measured simultaneously at two wavelengths, the following relation is found for the ratio of changes:

$$R/IR = \frac{\Delta \ln[I(\lambda_1)]}{\Delta \ln[I(\lambda_2)]} = \frac{\Sigma_{a,a}(\lambda_1)}{\Sigma_{a,a}(\lambda_2)} = \frac{\epsilon(\lambda_1)}{\epsilon(\lambda_2)}, \tag{3}$$

showing that, in the Lambert–Beer approximation, R/IR only depends on the ratio of molar absorptivities of arterial blood, and not on any other optical property. With the approximation $\Delta\ln(I) \approx \Delta I/I$ for small changes in I, we obtain

$$\Delta\ln(I) = \frac{\Delta I}{I} = \frac{AC}{DC} \tag{4}$$

The intensity I corresponds with DC, the direct current or average part of the measured intensity, whereas the fluctuations ΔI correspond with AC, the alternating current or fluctuating part of the measured intensity. In several pulse oximeters, the ratio R/IR is

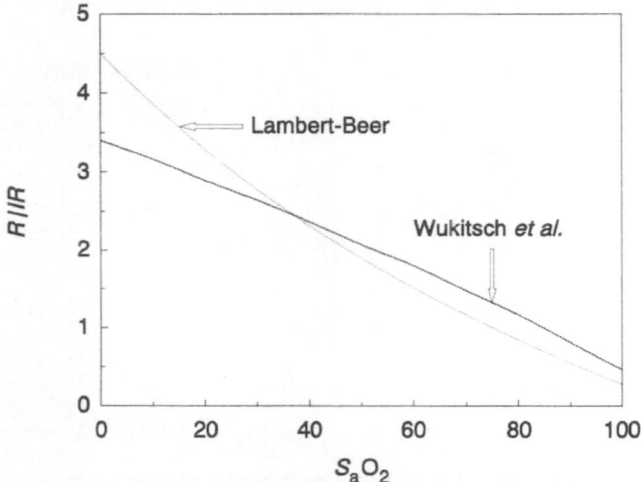

Figure 3. The relation between R/IR and the arterial oxygen saturation according to the Lambert–Beer model for pulse oximetry with $\lambda_1 = 660$ nm and $\lambda_2 = 940$ nm, compared with the relation used in transmission pulse oximetry according to Wukitsch et al.[7]

therefore obtained from the measurement of the AC and DC components of the red and infrared intensity signals, although they can also be obtained by the measurement of $\Delta\ln(I)$ for these wavelengths.

Several reasons exist to be careful with the Lambert–Beer model for pulse oximetry, because the empirical calibration curves used in transmission pulse oximetry differ from the relation given by Eq. (3), as shown in Fig. 3.[7-9] Obviously, the Lambert–Beer approach is insufficient to describe the relations that are used in transmission pulse oximetry.

When we started our research in 1986 at least two papers provided evidence that differences between reflectance pulse oximetry and transmission pulse oximetry might exist: The first paper contains the early experiments on reflectance pulse oximetry by Mendelson et al.[10] with 635 and 935 nm LEDs, which showed that large deviations between individual calibration curves occured. The second paper contains experimental observations on photo-electric plethysmography by Nijboer and Dorlas.[11] They measured the blood pressure at the same arm where the intensity fluctuations were measured. During a part of the period of cuff inflation, they observed inversion of the peaks in the reflected intensity, whereas the fluctuations in the transmitted intensity did not invert. This might imply that reflectance pulse oximetry is much more sensitive to the influence of light scattering than transmission pulse oximetry.

Because of these discrepancies, we postulated that light scattering within tissue does influence pulse oximetry, in particular reflectance pulse oximetry. The description of the influence of light scattering on reflectance pulse oximetry is the main goal of the remaining part of this paper.

CONDENSED MONTE CARLO METHOD

The present study on reflectance pulse oximetry is based on a much more reliable model of light propagation: the simulation of light propagation in tissue with the Monte Carlo technique, as depicted in Fig. 4. Monte Carlo simulations have often been used to simulate

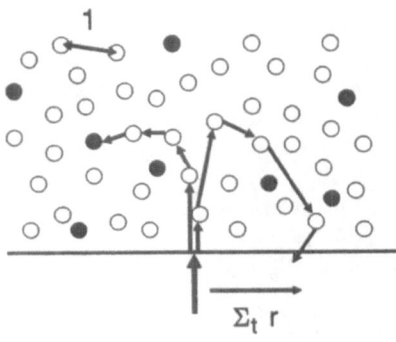

● = Absorption   ○ = Scattering

Figure 4. Monte Carlo simulations are used to predict the relative reflectance expressed per unit of mean free path squared at the surface of a semi-infinite medium with known absorption and scattering coefficients as a function of $\Sigma_t r$.

light propagation in turbid media, e.g., biological tissues. Apart from statistical noise, results with this method are more accurate than with the diffusion theory, although a considerable amount of simulation time is needed for each case.

Variables for each simulation include Σ_a, Σ_s', the geometry, the refractive index of the medium, and the phase function, where Σ_a and $\Sigma_s' \equiv \Sigma_s(1 - g)$ are the absorption and reduced scattering coefficients, and g and Σ_s are the average cosine of the scattering angle and the scattering coefficient, respectively.

In order to reduce the simulation time, the *condensed* Monte Carlo technique is used: as we have shown elsewhere,[12-13] the results of a Monte Carlo simulation can be used to replace simulations with the same value of $\Sigma_t \equiv \Sigma_a + \Sigma_s$, but with different albedos c, with $c \equiv \Sigma_s / \Sigma_t$. This can be performed when the location where each photon leaves the medium is stored as well as the number of scattering interactions for each photon. For a semi-infinite medium this implies that Σ_a and Σ_s' can both be varied independently without the need for a new simulation.

Variation of the phase function, the distribution of the scattering angles, influences light propagation in turbid media. However, it has been shown that the results with different phase functions are practically the same when the values of Σ_a and $\Sigma_s' \equiv \Sigma_s(1 - g)$ for both phase functions are equal. This also remains true for a reduced albedo with $c' \ll 1$, where $c' \equiv \Sigma_s' / \Sigma_t'$ and $\Sigma_t' \equiv \Sigma_a + \Sigma_s'$, especially for the case that two media are compared which consist of forward scattering particles within the range $0.9 < g < 1$.[12-16] This similarity rule can, therefore, be applied to scattering within biological media, e.g., the human skin, provided that the simulation is also performed with a phase function where forward scattering dominates, e.g., $g \approx 0.9$.

SIMULATION OF REFLECTANCE PULSE OXIMETRY

The intensity for detectors at different distances from the light source were obtained as a function of Σ_a and Σ_s' by means of the condensed Monte Carlo technique.[17] The simulation results of 30,000 photon paths were used by calculating the chance that a photon emitted into the medium at $r = 0$ leaves the medium within the detector area. The condensed Monte Carlo simulation was performed with a point source at the origin of a semi-infinite medium with refractive index $n_{\text{refr}} = 1.4$, with Henyey–Greenstein scatterers with $g = 0.875$, and with a smooth interface to air. The size and form of the rectangular detectors have been taken into account in the simulations by accepting only the contribution of photons that leave the medium within the detector area.

Subsequently, we fitted the results for each detector with an equation of the form

$$\ln(I/I_0) = A_0 + A_1 \Sigma_a + A_2(\Sigma_a)^{0.5} + A_3 \Sigma_s' + A_4 \ln(\Sigma_s') +$$

$$A_5 \Sigma_s'(\Sigma_a)^{0.5} + A_6 (\Sigma_s')^2 . \tag{5}$$

Equation (5) is the basis for the calculations below to derive values for the intensities. From this equation we also derived expressions to describe the pulsatile intensity fluctuations, caused by fluctuations in the absorption and reduced scattering coefficients, $\partial\ln(I/I_0)/\partial\Sigma_a$ and $\partial\ln(I/I_0)/\partial\Sigma_s'$, respectively. It should be noted that these properties are a function of Σ_a and Σ_s', the weighted optical properties of the homogeneous mixture of skin, arterial and venous blood volumes. The fluctuations can be expressed by

$$\Delta\ln[I(\lambda)] = \frac{\partial\ln[I(\lambda)/I_0(\lambda)]}{\partial\Sigma_a} \Delta\Sigma_a + \frac{\partial\ln[I(\lambda)/I_0(\lambda)]}{\partial\Sigma_s'} \Delta\Sigma_s' . \tag{6}$$

Results for the red and infrared wavelengths can be used to derive expressions for R/IR.

The changes in the absorption and reduced scattering coefficient, $\Delta\Sigma_a$ and $\Delta\Sigma_s'$, respectively, are caused by the pulsatile changes of the arterial blood volume fraction at each heart beat. Equation (6) should be applied for all wavelengths used, which may be more than two, because LEDs, which are generally used in pulse oximetry, are not monochromatically.[17]

INFLUENCE OF OPTICAL PROPERTIES IN REFLECTANCE PULSE OXIMETRY

Influence of Photon Path Lengths

In reflectance pulse oximetry as well as in transmission pulse oximetry all detected photons propagate through the medium over photon paths with different lengths. The average value and the distribution of the path lengths of detected photons depend on the scattering properties of the medium. For each of these path lengths Lambert–Beer's law can be applied by unfolding the photon path

$$I(\ell)/I(0) = \exp(-\Sigma_a \ell), \tag{7}$$

where Σ_a is the absorption coefficient within the medium, ℓ is the photon path length, and $I(\ell)/I(0)$ is the ratio between the detected photon flux and the detected photon flux when no absorption occurs, $\Sigma_a = 0$.

Table 1. The ratio of pulse sizes at adjacent detectors at 7.5 and 4.8 mm from the light sources, measured simultaneously. Measurements were taken from the distal phalanx of the index finger; four subjects[17]

$\Delta\ln(I_2)$ / $\Delta\ln(I_1)$	
660 nm LED	940 nm LED
1.92 ± 0.14	1.77 ± 0.09

By using Eq.(6), we have implicitly assumed that the pulsatile changes of Σ_a and Σ_s', which cause the fluctuations in intensity, are distributed homogeneously. In this case it can be derived for a medium with a fluctuating absorption coefficient that the contribution of photons with path length ℓ to the relative pulse size $\Delta I(\ell)/I(\ell)$ equals

$$\Delta I(\ell)/I(\ell) = \Delta\ln[I(\ell)/I(0)] = -\Delta\Sigma_a\,\ell \qquad (8)$$

Eq. (8) shows that the contribution to the pulse size by a group of photons that have propagated over a length ℓ is proportional to the change in absorption coefficient and to the photon path length. For pulse oximetry this implies that the size of the pulsations increases at larger distances between source and detector at the skin surface, which confirms the observations of Mendelson and Ochs.[18] With a reflectance pulse oximeter with several detectors at the skin surface at several distances from the light source, which permits the simultaneous measurement of pulse sizes at different detectors, shown in Table 1,[17,19] we obtained similar results.

A more important observation is that the determined ratio R/IR has generally not the same value at both detectors, and differs from $\Delta\Sigma_a(R)/\Delta\Sigma_a(IR)$.[17,19] The reason for these differences are explained by the fact that R/IR also depends on the photon path lengths through the medium for red and infrared light, which may differ. In the following paragraphs it is exposed how these photon path lengths are influenced by the mean absorption and reduced scattering coefficients of the medium. Moreover, the effects of changes in the reduced scattering coefficient during the heart beat are described below.

Influence of the Non-Pulsatile Part of the Absorption Coefficient

The measured intensity fluctuations are influenced by the non-pulsatile part of the absorption coefficient in two ways.

Firstly, Eq. (8) shows that the pulse size is proportional to the change in absorption coefficient $\Delta\Sigma_a$. This change in absorption coefficient does not only depend on the absorption coefficient of the arterial blood, but also on the absorption coefficient of the non-pulsatile part of the homogeneous medium, since $\Delta\Sigma_a$ is proportional to the *difference* between the absorption coefficients of arterial blood and the absorption coefficient of the homogeneous medium. However, the absorption coefficients of the bloodless skin are very small compared with the absorption coefficients of blood. It implies that the variation between subjects of the non-pulsatile part of the absorption coefficient has only a relatively small influence on $\Delta\Sigma_a$, unless the non-pulsatile blood volume fraction is large.

Secondly, the path length of the detected photons within the medium varies between photons, as illustrated in Fig. 5. When the non-pulsatile part of the absorption coefficient is increased, the photons that propagate over the longer photon paths have a greater chance to be absorbed somewhere at that path within the medium.

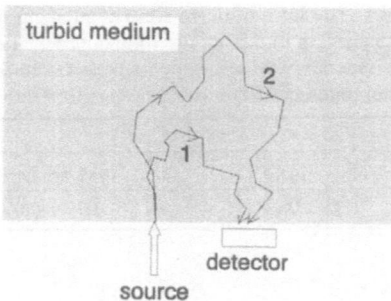

Figure 5. Two photon paths, a longer and a shorter one, that end at the detector.

In that case the average path length of detected photons decreases, which implies that a smaller pulse size is found for a given value of $\Delta\Sigma_a$, since the relative pulse size is proportional to the photon path length, as given by Eq. (8).

The effect of absorption on the pulse size can also be observed in Fig. 6, which shows the photon fraction, defined as the fraction of photons that is detected, at a detector with given geometry at the surface of the medium, after emission by a point source towards a semi-infinite medium. The fraction of detected photons in Fig. 6 was obtained from Monte Carlo simulations on a semi-infinite medium as a function of the reduced scattering coefficient Σ_s' for various values of the absorption coefficient Σ_a. When Lambert–Beer's law would be valid, the results in Fig. 6 would be given by straight lines with a slope $\Delta\ln[I/I_0]$ $/\Sigma_a = -\langle \ell \rangle$, where $\langle \ell \rangle$ is the apparent path length. The changes in the slope for increasing Σ_a confirm that $\langle \ell \rangle$ decreases for increasing absorption.

Path length distributions and apparent path lengths can be measured with time-resolved spectroscopy[20] and phase resolved spectroscopy,[21] respectively. Benaron et al.[21,22]

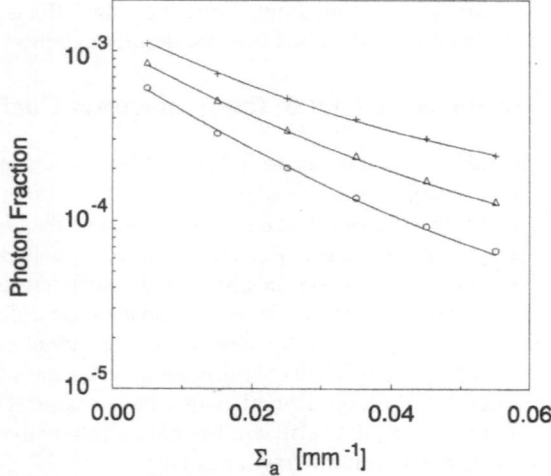

Figure 6. Fraction of photons detected after propagation in a semi-infinite medium as a function of the absorption coefficient. Results for several values of the reduced scattering coefficient of the medium, $\Sigma_s' = 0.9$ (+), 1.4 (Δ), and 1.9 (O) mm^{-1}. Henyey–Greenstein scattering, $g_{HG} = 0.875$; $n_{refr} = 1.4$. The centre of the detector (1.4 x 1.4 mm) at the surface is 7.5 mm from the source.

and others have studied optical path lengths of light emitted into the heads of infants, using phase resolved spectroscopy. They found that the apparent path length varies with the intercranial absorption coefficient, which is in agreement with the phenomena described in this section.

Influence of the Non-Pulsatile Part of the Reduced Scattering Coefficient

The influence of scattering on the photon fraction that reaches a simulated detector is more complex than the influence of absorption. Figure 7 shows the relative reflectance as a function of the distance from the light source for two different values of the reduced scattering coefficient. The relative reflectance has thereby been defined as the detected fraction of emitted photons per unit of area at a detector at a given distance between source and detector; the detector should have dimensions which are small compared to $1/\Sigma_s'$.

For photons detected at short distances from the light source, the results in Fig. 7 show that an increase in the reduced scattering coefficient increases the fraction of detected photons. However, at larger distances from the light source the situation is different: an increase in the reduced scattering coefficient decreases the relative reflectance.

The decrease in the relative reflectance at larger distances is partly caused by an increased photon path length, as can be observed from the slopes in Fig. 6, which are steeper for increased scattering. However, in the vicinity of the source an increase of the reduced scattering coefficient *decreases* the path length of detected photons. For the photons that leave the medium after only one (back)scattering event, the decrease in photon path length can be understood quite easily: the average path length of these photons is proportional to the mean depth where scattering occurred, and thus to the mean free path between scattering events.

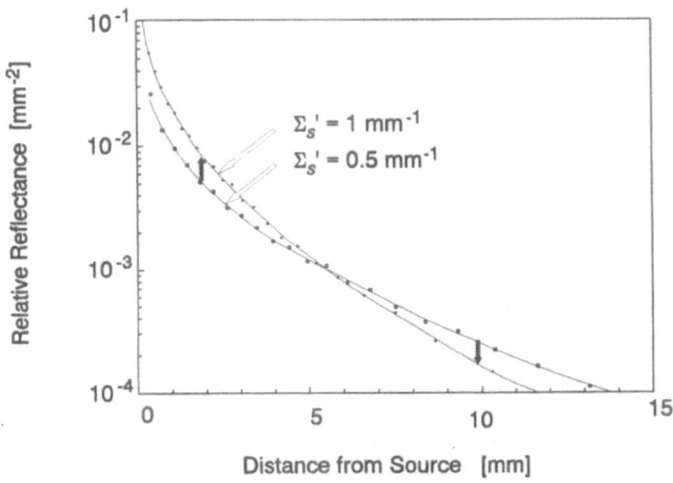

Figure 7. Relative reflectance as a function of the distance from the light source for two reduced scattering coefficients: $\Sigma_s' = 0.5$ and 1.0 mm^{-1}. Monte Carlo simulation with Henyey–Greenstein scattering ($g_{HG} = 0.875$), $n_{refr} = 1.4$ and $\Sigma_a = 0.01$ mm^{-1}. Source and detector at the surface of a semi-infinite medium.

Influence of Fluctuations in the Reduced Scattering Coefficient of Blood, Caused by Blood Flow Variations during Each Pulse

In the preceding paragraphs it has been shown that the pulse size for a given detector is proportional to the photon path length and to the change in the absorption coefficient, caused by the pulse wave. The influence of changes in the reduced scattering coefficient with the heart beat have not been considered yet. These changes in the reduced scattering coefficient can be caused by two different mechanisms:

A. *Changes in the (arterial) blood volume fraction.* When the reduced scattering coefficient of blood differs from that of the surrounding tissue, a difference in blood volume fraction caused by the pulse wave will cause pulsatile changes in the reduced scattering coefficient of the homogeneous mixture.

B. *Changes in the reduced scattering coefficient of blood.* The reduced scattering coefficient of blood depends on flow-related phenomena, such as red cell orientation, red cell aggregation, and red cell deformation. Within each heart cycle the shear rate decreases and increases, which may induce pulsatile changes in the reduced scattering coefficient of blood. It is expected that the influence of these flow-related phenomena increases when low perfusion or stasis occurs during part of the heart cycle.

Figures 8 and 9 have been added to clarify these phenomena. Figure 8 is based on the same Monte Carlo simulation data as Fig. 6 for a detector at 7.5 mm from the light source. Arrows have been added in the range where the changes for red and infrared light are assumed to occur. The arrows "A" in Fig. 8 give an example of the effect of changes in the arterial blood volume fraction from diastole to systole as described under A, although the size of the

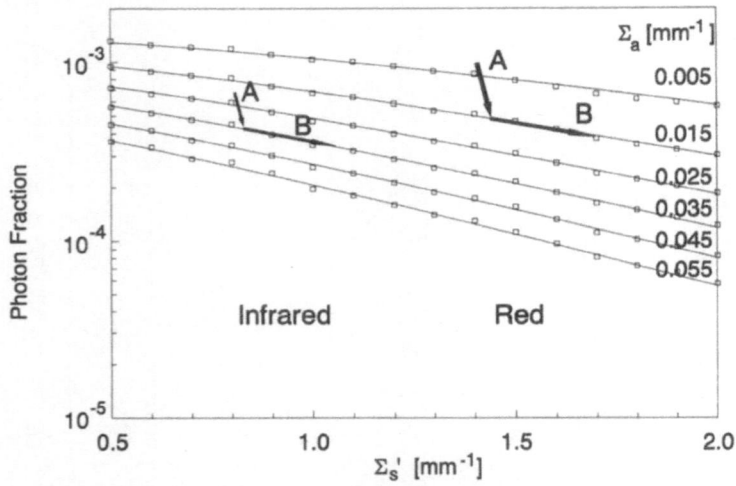

Figure 8. The fraction of photons that is detected at a semi-infinite turbid medium as a function of the reduced scattering coefficient Σ_s' of the medium for various absorption coefficients Σ_a. Data are given for a rectangular detector (1.4 x 1.4 mm) with its center at 7.5 mm from the source. Henyey–Greenstein scattering, $g_{HG} = 0.875$; $n_{refr} = 1.4$. The arrows "A" denotes the changes when the arterial blood volume is changed, while Σ_s' for arterial blood remains constant. The arrows "B" denote the changes when the reduced scattering coefficient of blood is increased, while the arterial blood volume remains constant.

pulse given by the arrow is exaggerated. The changes in absorption and reduced scattering coefficients are both proportional to the change in blood volume fraction.

The arrows "B" describe the additional changes in intensity described under B, both for red and infrared light. These changes occur when the blood volume and the absorption coefficient are constant whilst the scattering coefficient of the blood increases, e.g. when the blood flow rate increases during systole. It should be noted that the whole blood volume fraction within the measuring volume is involved in this process, not just the difference in blood volume. This implies that these changes in the reduced scattering coefficient of blood do not depend on the difference in arterial blood volume fraction, but on the whole blood volume fraction. Therefore, the changes under B might become much larger than the changes described under A.

Consequently, when the reduced scattering coefficient of blood increases during systole the pulse size for the detector of Fig. 8 may become larger than the size expected from the arterial blood volume change only, whereas the size of the pulsations would decrease when the reduced scattering coefficient would be decreased.

For detectors in the vicinity of the light source the effects for a change in the reduced scattering coefficient of blood are opposite. This inversion can be observed for small values of Σ_s' in Fig. 9, which gives results for a detector at 4.8 mm from the light source. Figure 9 also shows that inversion does not occur at this detector location for higher values of Σ_s'. The inversion in Fig. 9 is in agreement with Fig. 7 where the arrows are directed toward increased scattering, just as the arrows "B" in Figs. 8 and 9. Figure 7 shows that an increase in the reduced scattering coefficient during systole would only lead to a larger relative reflectance, e.g. for an increase of Σ_s' from 0.5 to 1.0 mm^{-1}, when the detector is in the vicinity of the light source. Since the increased absorption during systole decreases the relative reflectance, addition of both effects leads to a smaller pulse size, or even to inversion of the direction of the intensity fluctuations.

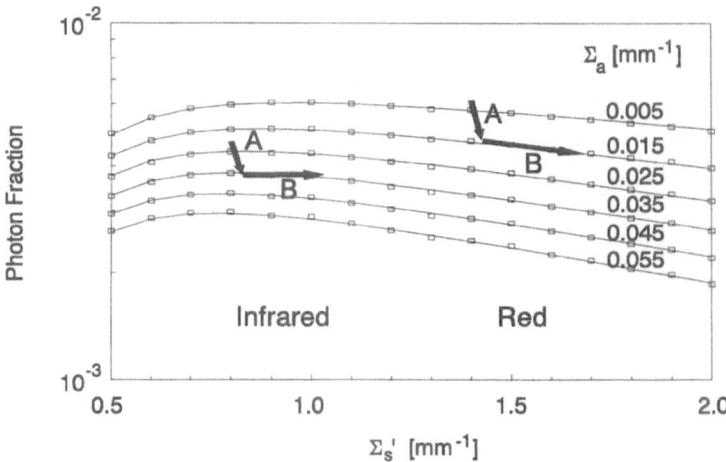

Figure 9. Detector in the vicinity of the light source. The fraction of photons that is detected at a semi-infinite turbid medium as a function of the reduced scattering coefficient Σ_s' of the medium for various absorption coefficients Σ_a. Data are given for a rectangular detector (1.4 x 1.4 mm) with its center at 4.8 mm from the source. Henyey–Greenstein scattering, $g_{HG} = 0.875$; $n_{HG} = 1.4$. For further details see Fig. 8. The arrows "B" are parallel with the curves for constant absorption.

The observations inversed waveforms by Nijboer and Dorlas[11] with photo-electric plethysmography, which proved the limitations of the Lambert-Beer approach, as mentioned above, can now be understood with the mechanisms described under B. When the cuff is released to pressures between the systolic and diastolic pressure a flow stop will occur during each diastole, which causes relatively large contribution of differences in the reduced scattering coefficient and thus to the relative reflectance.

The direction of the arrows "B" in Fig. 8 shows that we assumed that the reduced scattering coefficient of blood increases at the start of systole. The direction of the arrow "B" in Fig. 8 does not describe the inversion phenomenon in that case. However, at small distances from the source, inversion is expected, as shown by Fig. 7 and by Fig. 9 for low values of Σ_s'. The contribution of scattering and absorption to changes in the measured intensity are opposite at that location. Inversion occurs when the contribution of scattering is larger than the contribution of absorption.

APPLICATION TO CALIBRATION CURVES FOR REFLECTANCE PULSE OXIMETRY

The theory given above has consequences for reflectance pulse oximetry. Monte Carlo simulations have shown that variation in the absorption and reduced scattering coefficients of the medium, e.g. variation of the non-pulsatile part of the blood volume fraction, influences the relation between S_aO_2 en R/IR.[17]

Figure 10 gives an example of the influence of tissue optics on the calibration curves for reflectance pulse oximetry. It shows the calibration curves obtained with Monte Carlo theory for given tissue properties for the blood-free medium for two venous blood volume fractions. The solid curves were obtained from Eq. (6). Effects of variations in the reduced

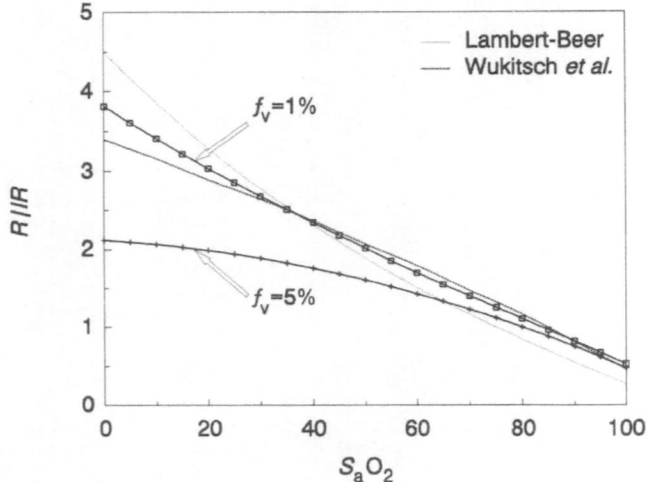

Figure 10. Calibration curves for reflectance pulse oximetry for a 660 nm LED with 14% secondary emission at 900 nm, and a 940 nm LED. The detector (1.4 mm x 1.4 mm) is located at 7.5 mm from the light sources. Variation of venous blood volume fraction f_v with $S_aO_2 = S_aO_2 - 10\%$. Tissue properties: Σ_a (660 and 940 nm) = 0.016 and 0.030 mm^{-1}, respectively; Σ_s' (660 and 940nm) = 1.2 and 0.76 mm^{-1}, respectively. Henyey-Greenstein scattering with $g_{HG} = 0.875$. Also given are the empirical calibration curve of Wukitsch et al.[7] and the calibration curve according to Lambert–Beer's law without secondary emission.

scattering coefficient of blood, as given by arrows "B" in Figs. 8 and 9, were not taken into account. The dotted curves were taken from Fig. 3. The solid curves show the advantage of the Monte Carlo simulation over the Lambert–Beer model. The deviations of Lambert-Beer theory are partly caused by the secondary emission that occurs in most red LEDs around 900 nm.[17]

It should be noted that the Monte Carlo results are very close to the calibration curve of Wukitsch *et al.*,[7] especially for the low venous blood volume fraction. The solid curves also show the influence of the increased blood volume fraction. This influence is small for high oxygen saturations, but much larger for the lower saturations. It is also noteworthy that the decrease at low oxygen saturation could be expected, since the absorption coefficient for red light is much larger in that case than for infrared light. The average path lengths for red light will decrease more in that case than the path lengths for infrared light. Consequently, the red pulse size will decrease more than the infrared pulse size, which results in a decreased value for R/IR.

APPLICATION TO REFLECTANCE PULSE OXIMETRY MEASUREMENTS

Measurements with reflectance pulse oximetry on the distal phalanx of fingers of normal breathing volunteers have shown that the values of R/IR increased for an increasing distance between light source and detector, corresponding to a saturation difference of 2%.[19] Those results were obtained with a sensor as shown in Fig. 11. In a previous study we have investigated the influence of the estimated optical properties of the tissue on Monte Carlo predictions of the ratio R/IR at the three detectors which have shown that the values of R/IR could be predicted very well when assuming $S_aO_2 = 98\%$.[17] The increase in R/IR at adjacent detectors has thereby been confirmed as well;[17] this increase is the result of the differences between Figs. 8 and 9. Also the increase in the pulse sizes, as given in Table 1, could be explained by the simulations.

Our experimental results have also shown that the variation in R/IR between the healthy normal breathing subjects depends on variations in the non-pulsatile part of the tissue properties, which was obtained from the ratios of intensities for red and infrared light at adjacent detectors.[23] A correction method could be obtained where the variation between the

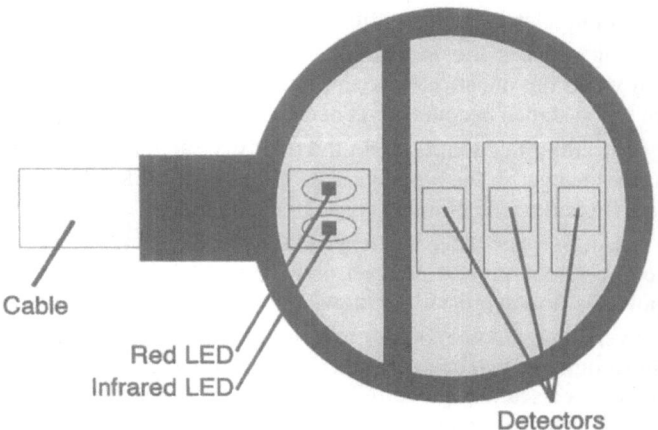

Figure 11. Multi site sensor for reflectance pulse oximetry.

Table 2. Predictions of the arterial oxygen saturation
from the simultaneously measured values of R/IR
and intensity ratios I_2/I_1 for red and infrared light,
obtained from detectors #1 and #2, at 4.8 and
7.5 mm from the light sources, respectively.
Measurements at the distal phalanx of the
index finger, four subjects

	$S_aO_2 \pm$ s.d. [%]	
	Detector 1	Detector 2
Constant tissue properties	97.5 ± 1.1	97.7 ± 0.8
Calculated tissue properties	97.4 ± 0.8	97.8 ± 0.6

predicted saturations could be further improved, as shown in Table 2, even without arterial blood gas samples. These results also confirm the findings of Dougherty and Lowry,[24] who have found empirically that the ratio of the intensities for red light at adjacent detectors can be used as a correction term for R/IR.

CONCLUSIONS

The following conclusions on the influence of light propagation can be drawn:

1. The contribution of photon paths to the intensity fluctuation is proportional to their path length through the medium. Detected photons that have propagated over a long path within the medium will therefore contribute relatively much to the detected intensity fluctuations.

2. The size of the observed pulsations in reflectance pulse oximetry depends on the difference between the absorption coefficients of the blood and the medium. When the surrounding medium has the same absorption coefficient as the blood no pulsations will be observed for a given path length, except for pulsatile changes caused by changes in the reduced scattering coefficient.

3. Increased absorption decreases the measured intensity, the apparent path length $<l>$, and the relative intensity fluctuation, $\Delta \ln(I) = \Delta I/I$.

4. Scattering reduces the measured intensity fluctuations and the apparent path length when the distance between light source and detector is small. However, when this distance becomes larger scattering increases the apparent path length.

5. The contribution of scattering to the intensity fluctuations is proportional to the pulsatile changes in the blood volume fraction, provided that the reduced scattering coefficient of the blood is constant.

6. The reduced scattering coefficient of blood also depends on red cell aggregation, red cell orientation, and red cell deformation. However, these changes are not proportional to the pulsatile changes in blood volume function.

7. The inversion phenomenon as observed by Nijboer and Dorlas in photo-electric plethysmography can be explained when the intensity increase caused by changes in the reduced scattering coefficient of blood is larger than the intensity decrease that occurs because of a higher blood volume fraction. This phenomenon can only occur when the distance between light source and detector is small.

Application of the Monte Carlo results to reflectance pulse oximetry has shown that the results can very well describe the experimental results at distal phalanx of the finger. However, more experimental research will be necessary to prove whether the homogeneous model given here will satisfy in practice, or whether future Monte Carlo studies are needed, e.g., for the interpretation of reflectance oximetry just above bone tissue, where the assumptions of a homogeneous distribution of pulsations might not be justified.

In this study the influence of light propagation on reflectance pulse oximetry has been investigated with Monte Carlo simulations instead of diffusion theory, because it is likely that diffusion theory would give less accurate results.

The disadvantage of the long simulation time needed in Monte Carlo simulations has been eliminated by using the condensed Monte Carlo technique. It is concluded that this technique provides a valuable tool to gain insight in the principles of reflectance pulse oximetry, as well as to obtaining reliable quantitative data that take the influence of tissue optics into account. Therefore, we will also use the condensed Monte Carlo technique in future studies on reflectance pulse oximetry to interpret experimental results.

REFERENCES

1. T. Aoyagi, M. Kishi, K. Yamaguchi, and S. Watanabe, 1974, Improvement of the earpiece oximeter, *Abstracts of the Jap. Soc. of Med. Electronics and Biol. Engng.*: 90-91.
2. J. W. Severinghaus and P. B. Astrup, 1986, History of blood gas analysis. VI. Oximetry, *J. Clin. Monit.* 2: 270-288.
3. J. W. Severinghaus and Y. Honda, 1987, History of blood gas analysis. VII. Pulse oximetry, *J. Clin. Monit.* 3: 135-138.
4. O. W. Van Assendelft, *Spectrophotomerty of haemoglobin derivates*, Thesis (Groningen, The Netherlands, 1970).
5. W. G. Zijlstra, A. Buursma, and W. P. Meeuwsen-van der Roest, 1991, Absorption spectra of human fetal and adult oxyhemoglobin, de-oxyhemoglobin, carboxyhemoglobin, and methemoglobin, *Clin. Chem.* 37: 1633-1638.
6. I. Yoshiya, Y. Shimada, and K. Tanada, 1980, Spectrophotometric monitoring of arterial oxygen saturation in the fingertip, *Med. & Biol. Eng. & Comput.* 18: 27-32.
7. M. W. Wukitsch, M.T. Petterson, D.R. Tobler, and J.A. Pologe, 1988, Pulse oximetry: analysis of theory, technology, and practice, *J. Clin. Monit.* 4: 290-301.
8. M. Yelderman, "Pulse oximetry," in *Monitoring in Anesthesia and Critical Care Medicine*, C.D. Blitt, ed. (Churchill Livingstone, New York, 1990), 417-427.
9. J. P. Payne, and J. W. Severinghaus, *Pulse Oximetry*, (Springer-Verlag, Berlin, 1986): 188-191.
10. Y. Mendelson, P. W. Cheung, M. R. Neuman, D. G. Fleming, and S.D. Cahn, 1983, Spectrophotometric investigation of pulsatile blood flow for transcutaneous reflection oximetry, *Adv. Exp. Med.* 159: 93-102.
11. J. A. Nijboer, and J. C. Dorlas, 1982, The origin of inverted waveforms in the reflection plethysmogram, *Br. J. Anaesth.* 54: 1289-1293.
12. R. Graaff, M. H. Koelink, F. F. M. de Mul, W. G. Zijlstra, A. C. M. Dassel, and J. G. Aarnoudse, 1993, Condensed Monte Carlo simulations for the description of light transport, *Appl. Opt.* 32: 426-434.
13. R. Graaff, A. C. M. Dassel, M. H. Koelink, F. F. M. de Mul, J. G. Aarnoudse, and W. G. Zijlstra, 1993, Optical properties of human dermis in vitro and in vivo, *Appl. Opt.* 32: 435-447.
14. H. C. van de Hulst, *Multiple Light Scattering*, Vol. 2 (Academic, New York, 1980): Chap. 12.2 and 14.1.
15. R. Graaff, J. G. Aarnoudse, F. F. M. de Mul, and H. W. Jentink, 1989, Light propagation parameters for anisotropically scattering media, based on a rigorous solution of the transport equation, *Appl. Opt.* 28: 2273-2279.
16. R. Graaff, J. G. Aarnoudse, F. F. M. de Mul, and H. W. Jentink, 1993, Similarity relations for anisotropic scattering in absorbing media, *Opt. Eng.* 32: 244-252.
17. R. Graaff, A.C.M. Dassel, M.H. Koelink, J.G. Aarnoudse, F.F.M. de Mul, W.G. Zijlstra, and J. Greve, 1993, Condensed Monte Carlo simulations applied to reflectance pulse oximetry, in *Photon Migration and Imaging in Random Media and Tissues*, B. Chance and R.R. Alfano, Eds., *Proc. SPIE* 1888: 201-212.
18. Y. Mendelson and B. D. Ochs, 1988, Noninvasive pulse oximetry utilising skin reflectance photoplethys-mography, *IEEE Trans. Biomed. Eng.* 35: 798-805.

R. Graaff, J. G. Aarnoudse, W. G. Zijlstra, P. Heida, F. F. M. de Mul, M. H. Koelink, and J. Greve, 1990, Reflection pulse oximetry depends on source-detector distance, *Intensive Care Med.* 16: S99, Abstract.

20. D. T. Delpy, M. Cope, P. van der Zee, S. Arridge, S. Wray, and J. Wyatt, 1988, Estimation of optical path length through tissue from direct time of flight measurement, *Phys. Med. Biol.* 33: 1433-1442.

21. D. A. Benaron, S. Gwiazdowski, C. D. Kurth, J. Steven, M. Delivoria–Papadopoulos, and B. Chance, 1990, Optical path length of 754nm and 816 nm light emitted into the heads of infants, *Proc. Ann. Int. Conf. I.E.E.E. Eng. in Med. and Biol. Soc.* 12: 1117-1119.

22. D. A. Benaron, C. D. Kurth, J. Steven, L. C. Wagerle, B. Chance, and M. Delivoria-Papadopoulos, 1990, Non-invasive estimation of cerebral oxygenation and oxygen consumption using phase-shift spectroscopy, *Proc. Ann. Int. Conf. I.E.E.E. Eng. in Med. and Biol. Soc.* 12: 2004-2006.

23. R. Graaff, *Tissue optics applied to reflectance pulse oximetry*, Thesis (Groningen, The Netherlands, 1993).

24. G. Dougherty and J. Lowry, 1992, Design and evaluation of an instrument to measure microcirculatory blood flow and oxygen saturation simultaneously, *J. Med. Eng. & Technol.* 16: 123-128.

15

EFFECT OF OPTODE SEPARATION ON BRAIN PENETRATION IN ADULTS

D. N. F. Harris, S.M. Bailey, F. Cowans, and D. Wertheim

Department of Anaesthesia
Royal Postgraduate Medical School
London, United Kingdom

INTRODUCTION

Near Infra-red Spectroscopy (NIRS) has been widely used to study intracerebral oxygenation and haemodynamics in neonates (1,2), but doubts have been raised over the penetration of NIR of the thick adult skull when using optode separations of 3-4 cm (3), as suggested by the classical estimates of scattering and absorption, and one commercial instrument has been shown to reflect only extra-cranial tissue perfusion. Indocyanine Green (ICG) is easily detected by NIRS and is unaffected in changes in oxygenation. This study investigates the amount of cerebral tissue seen by NIR at different optode separations using indocyanine green as a specific indicator of flow during selective clamping of the carotid arteries.

MATERIAL AND METHODS

16 patients undergoing carotid endarterectomy were studied, 14 male and 2 female aged 52-71 years. Two optodes from the Hamamatsu NIR 1000 were applied at 5 cm separation and allowed to stabilise. Transcranial Doppler was used to document the velocity in the Middle Cerebral Artery on the same side during clamping of the internal carotid artery. After re-vascularisation, the external carotid artery was clamped and 1 mg of indocyanine green (ICG) in 1 ml Normal Saline injected into the internal carotid artery. When the NIR indocyanine signal had fallen to <50% of the peak, the external clamp was released. After 5 minutes the procedure was reversed, with the internal carotid clamped and 1 mg ICG injected into the external carotid artery. The optode separation was changed to 3 cm and then 7 cm, and the 2 injections repeated. The NIR signal was sampled at one Hz and stored for subsequent analysis, when the height of the peak ICG concentration from the baseline was calculated for each injection. Absolute concentration of ICG in each territory was not determined as external and internal carotid flows could not be measured. In 3 patients

Oxygen Transport to Tissue XVII, Edited by Ince et al.
Plenum Press, New York, 1996

133

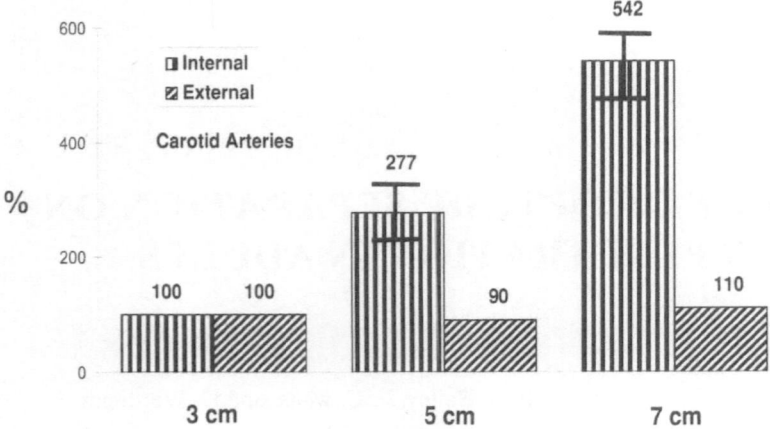

Figure 1. Indocyanine Green peak concentrations after selective carotid injection, relative to the signals at 3 cm optode separation averaged for the group.

common carotid flow was measured using continuous Doppler (Opdop–Scimed Ltd UK) during selective carotid clamping.

RESULTS

Carotid flow measured by Doppler on the common carotid by had a high variability (Figure 1), but tended to confirm that internal carotid flow was somewhat greater than external carotid flow. As the concentration of ICG in internal and external carotid arteries could not be determined accurately, the data was normalised so that the values at 3 cm were 100%. The external carotid signal was not significantly different at the 3 spacings, while the internal signal increased to 277% (5 cm) and 542% (7 cm) of the signal at 3 cm with increased optode separation (p < 0.002).

DISCUSSION

The external carotid signal was not significantly different at all 3 spacings, reflecting the constant extra-cranial tissue path. As the optode separation increased the internal carotid signal increased to 5.7 times, and thus the percentage of the Near Infra-red signal from the internal carotid increased accordingly. This confirms the relative lack of penetration when using a small optode separation, and the marked increase in the intra-cerebral signal at wide separations. This data conflicts with a previous study (4) showing a significant signal from the internal carotid injection (c. 1/10 that of the external signal) at 2.7 cm, with little internal signal at 1.0 cm, which was used as the basis for the Invos 3100 monitor (Somanetics Corporation). Subsequent tests appear to confirm the lack of a truly intracerebral signal at such a narrow optode separation, and the current model has been revised, using a 5 cm optode separation. Delpy et al. (5) have measured the NIR scattering using human skull and found it to be much greater than expected, which would explain why a greater separation is needed than that predicted by Monte Carlo modeling using classical estimates of absorption and scattering. As the concentration in the internal and external carotid arteries was not known, the absolute amounts of territory seen cannot be determined; however internal carotid flow is usually greater than external carotid flow, which would suggest that at 7 cm separation the contribution of the external carotid territory is less than 15% of the measured NIR signal

CONCLUSION

This study confirms previous suggestions that NIRS in adults reflects essentially external carotid territory with optode spacing less than 3 cm, and that the contribution from intracranial tissues increases markedly increasing separation. We suggest that optode separations of less than 5 cm should be used with caution if contamination by the external carotid territory is to be avoided.

REFERENCES

1. Wyatt-JS; Cope-M; Delpy-DT et al. Quantitation of cerebral blood volume in human infants by near-infrared spectroscopy. J. Appl. Physiol. 1990 Mar; 68(3): 1086-91
2. Edwards-AD; Wyatt-JS; Richardson-C et al. Cotside measurement of cerebral blood flow in ill newborn infants by near infrared spectroscopy. Lancet. 1988 Oct 1; 2(8614): 770-1
3. Harris-DNF; Bailey-SM Near infrared spectroscopy in adults. Does the Invos 3100 really measure intracerebral oxygenation? Anaesthesia. 1993 48(8): 694-6
4. McCormick-PW; Stewart-M; Lewis-G et al. Intracerebral penetration of infrared light. Technical note. J. Neurosurg. 1992 76(2): 315-8
5. Firbank-M; Hiraoka-M; Essenpreis-M; Delpy-DT Measurement of the optical properties of the skull in the wavelength range 650-950 nm Phys. Med. Biol. 1993 38(4): 503-10

16

THEORETICAL BASIS VERSUS CLINICAL PRACTICE OF OXYGEN PARAMETERS OF BLOOD

Rolf Zander

Institute of Physiology and Pathophysiology
University of Mainz
D-55099 Mainz, Germany

INTRODUCTION

Under optimal conditions all oxygen parameters of human blood are defined at a theoretical basis, accepted and understood by the clinicians, measured correctly by instruments of different manufacturers and, therefore, leading to clear diagnoses for the patients.

The oxygen status of arterial blood is described at least by four variables:

Oxygen partial presssure (pO_2, mmHg), oxygen saturation (sO_2, %), hemoglobin content (cHb, g/dl) and oxygen content (cO_2, ml/dl). Beside perfusion, however, the oxygen supply of all organs is decisively determined by the mean capillary pO_2 which itself is primarily dependent on the arterial cO_2. Therefore, the oxygen availability or transport (cardiac output x caO_2, ml/min) may be described by the caO_2 value in blood or those variables which determine the latter one.

In arterial blood, oxygen partial pressure is the result of O_2 diffusion rate within the lungs into the blood (lung function). Oxygen saturation describes the portion of chemically bound oxygen expressed as O_2Hb in relation to total Hb ($HHb + O_2Hb + COHb + MetHb$). Oxygen content (concentration) is the total amount of oxygen in blood chemically bound plus physically dissolved.

CLINICAL SITUATION

Under clinical conditions the diagnostic significance of the four variables becomes very clear as shown in Table 1.

Disturbances of lung function decrease all three variables, pO_2 (hypoxia), sO_2 (hypoxygenation) and cO_2 (hypoxemia), to produce hypoxic hypoxemia.

Carbon monoxide poisoning or methemoglobin formation decreases two variables, sO_2 and also cO_2, whereas the pO_2 remains normal and results in toxic hypoxemia.

Oxygen Transport to Tissue XVII, Edited by Ince et al.
Plenum Press, New York, 1996

137

Table 1. Diagnostic value and clinical consequences. Arterial O_2 partial pressure (pO_2), partial O_2 saturation (psO_2), O_2 saturation (sO_2), hemoglobin content (cHb) and O_2 content (cO_2) measured (or calculated[*]) by a blood gas analyzer (BGA), a pulse oxymeter (PO), a multi-wavelength oxymeter (OM) or the Oxystat System (OS)

Methods Disturbance	BGA pO_2 [mmHg]	PO BGA[*]/ OM[*] psO_2 [%]	OM OS* sO_2 [%]	OM / OS cHb [g/dl]	OS cO_2 [ml/dl]
Hypoxia	↓	↓	↓	→	↓
Toxemia (COHb / MetHb)	→	→	↓	→	↓
Anemia	→	→	→	↓	↓
		"functional" sO_2	sO_2	"fractional" O_2Hb	

Anemia with a decrease in the hemoglobin content lowers cO_2 only, while pO_2 and sO_2 remain normal (anemic hypoxemia).

The diagnostic significance of the single variables of the oxygen status consequently increases in the order of pO_2, sO_2 (cHb) and cO_2.

METHODS AND MEASURED VALUES

Measurement of arterial pO_2 is possible using an oxygen electrode in vitro (blood gas analyzer) or in vivo (transcutaneous O_2 electrode).

Heme oxymeters are multi-wavelength blood oxymeters for the in vitro measurement of total Hb content (cHb), all Hb derivatives (COHb, MetHb) and O_2 saturation (sO_2), together with some derived parameters, e.g. chemically bound O_2 content.

Additionally to the sO_2 value the heme oxymeter calculates the so-called partial oxygen saturation psO_2, sometimes also called functional saturation.

For some reasons, e.g. sample volume, accuracy, HbF blood or animal blood, the OSM 3 heme oxymeter (Radiometer) is proposed.

The partial saturation (functional saturation), described in Table 2, is defined as the percentual proportion of O_2Hb in relation to the sum of only O_2Hb plus HHb. The term "partial" is used in the sense that only a portion of the total hemoglobin, i.e. that available

Table 2. Definitions of oxygen saturation (sO_2) (fractional saturation) versus partial oxygen saturation (psO_2) (functional saturation) in relation to the corresponding methods

$$sO_2\,[\%] = \frac{cO_2Hb}{cO_2Hb + cHHb + cCoHb + cMetHb} \times 100 \quad \text{Photometry}$$

$$sO_2\,[\%] = \frac{cO_2 - phys.\,diss.\,O_2}{cHb \times 1.39\,(theor.\,O_2\,capacity)} \times 100 \quad \text{Gas analysis (gold standard)}$$

(normal value in arterial blood = 96%)

$$psO_2\,[\%] = \frac{cO_2HB}{cO_2Hb + cHHb} \times 100 \quad \text{Photometry}$$

(normal value in arterial blodd = 98%

for O_2 transport, is taken into consideration. For example, when the blood contains 10% of COHb or MetHb, the value of sO_2 amounts to about 86% whereas that of the psO_2 amounts to about 98%.

Pulse oxymeters try to measure arterial partial O_2 saturation (psO_2) in vivo and continuously. Using only two wavelengths they are unable to differentiate between possible Hb derivatives other than O_2Hb and HHb, e.g. COHb and MetHb. Therefore, in relation to the used algorithms they can measure only the partial O_2 saturation.

Unfortunately, the necessary in vivo calibration of pulse oxymeters until now is the most crucial problem. Obviously, some manufacturers calibrate their oxymeters according to the O_2 saturation ("sO_2 fractional") others to the partial O_2 saturation ("sO_2 functional").

An ECRI test report (1989) which presents only those data for pulse oxymeters that correspond with the calibration method used by each manufacturer is, therefore, not acceptable for clinicians.

The results of a typical test of different pulse oxymeters from different manufacturers within our laboratory (Hohmann and Zander, 1988) under normoxic and hypoxic conditions with 10 smokers (6,7 ± 2,2% COHb) and 10 non-smokers are shown in Table 3.:

While the preferred oxymeter from Critikon (Oxyshuttle) measures the psO_2 in the case of smokers and non-smokers very accurately, the oxymeter from Radiometer (Oximeter) underestimates psO_2 within the normoxic and hypoxic range as a consequence of the sO_2 calibration. The normal value of arterial blood in the first case is measured as about 98% (psO_2) and in the second case as about 96% (sO_2).

The Oxystat method (Zander and Lang, 1994) is a photometric in vitro procedure using disposable cuvettes for the determination of oxygen as well as of hemoglobin content together with a battery-operated mini-photometer. A special dosing attachment for a 10 µl blood sample is affixed to the cuvettes, which contain specific and highly sensitive reagents for both oxygen and hemoglobin. From the two measured absorbance differences the built-in calculator of the photometer calculates all necessary values of the oxygen status: cO_2 (ml/dl), cHb (g/dl) and sO_2 (%). Since a crystalline standard with a purity of 99.5% exists for both procedures, the Hb cuvette (chlorohemin) and the O_2 cuvette (KIO_3), the user can perform an optimal quality control of the entire procedure - namely, cuvette, dosing system, photometer and calculator - using concentrated solutions of defined composition, which are treated the same way as blood.

Table 3. Values of arterial O_2 saturation measured by two pulse oxymeters in relation to the expected value of the partial O_2 saturation (psO_2)

Expected psO$_2$ (%)	Measured psO$_2$ (%)			
	Critikon (Oxyshuttle)		Radiometer (Oximeter)	
	smoker	non-smoker	smoker	non-smoker
98	98.3	98.5	94.6	95.8
90	90.6	90.5	86.0	87.0
80	80.9	80.6	75.2	76.0
70	71.2	70.6	64.4	65.0

GOLD STANDARDS

The only gold standards today applicable to all values of the O_2 status of human blood are of gasometric kind, i.e. the oxygen content or concentration of primary standards (air, KIO_3) or secondary standards (equilibrated distilled water or human blood), as described before (Zander, 1991).

Another kind of gold standard might be the so-called theoretical O_2 capacity using the theoretical Hüfner number, i.e. the maximum amount of chemically bound oxygen in a blood sample of known Hb content (tab. 2): $cHb \times 1.39 = cO_2$.

But, this O_2 capacity is essentially a theoretical value since no experimental method is available that allows complete saturation of the total Hb with O_2: Traces of COHb and MetHb will always remain and, although not interfering with the determination of the O_2 capacity, will be included in the measurement of Hb as total Hb.

Therefore, the "actual O_2 capacity" of blood, termed BO_2 by Oeseburg (1994), related to the "available hemoglobin" of that blood, must be a misleading value only.

Neither primary nor secondary standards are available for calibrating the determination of sO_2 as mentioned by Müller-Plathe (1991), therefore, for the calibration of a heme oxymeter concerning the Hb concentration a "red ink" is used produced by the manufacturer.

A typical consequence of this situation is the need for an empirical correction of errors in the measurement of hemoglobin oxygen saturation if multi-wavelength heme oxymeters are used (Nijland et al. 1994).

MISLEADING VALUES

The term "available Hb" for the sum of HHb + O_2Hb, i.e. the Hb available for O_2 transport, as well as the "actual O_2 capacity" of a blood sample, are very undefined theoretical values which are not practical for clinical routine.

The reason for this is the fact that the conditions of measurement are not defined:

Depending upon the time of exposure, the magnitude of the pO_2, and the MetHb-reductase activity COHb and MetHb are converted into "available Hb" resulting in an increase of the "actual O_2 capacity" in vivo and in vitro.

As a consequence, the recommendation of Payne and Severinghaus (1986) to saturate with a "minimum volume of oxygen" to prevent removal of COHb and MetHb, is not practicable. The term "dyshemoglobin concentration (cdysHb)", as recommended by Oeseburg (1994), is superfluous in this connection.

The problems by using these misleading values may be demonstrated for example by a patient with carbon monoxide poisoning: The blood of this patient, e.g. with a COHb concentration of 30%, if monitored by a heme oxymeter, shows a sO_2 value of 67.6%, but a psO_2 value of 98%, if measured by a pulse oxymeter. The severe hypoxia of this patient is monitored by a decreased sO_2 and cO_2 only whereas pO_2, psO_2 and cHb are normal (see tab. 1).

The aim of the primary therapy in this case is the increase of arterial pO_2 (monitored by an O_2 electrode) by the administration of pure oxygen to the patient to decrease cCOHb; the success of the therapy will be indicated by the increase of sO_2 back to its normal value. In contrast, the psO_2 value alone is indifferent and, therefore, the display of the pulse oxymeter will show before, during and after therapy a normal value of 98%.

Additionally, the "available hemoglobin" as well as the "actual O_2 capacity" due to Oeseburg (1994) will increase, but both values are not measurable under clinical conditions.

REFERENCES

ECRI: Pulse Oximeters. Health Devices 18, 185 (1989).

Hohmann, C., Zander, R.: Vergleich verschiedener Pulsoxymeter unter Hypoxie bei Rauchern und Nichtrauchern. Anaesthesist 37 (Suppl.), 93 (1988).

Müller–Plathe, O.: Calibration and quality control of equipment used for determining O_2 saturation. In: Zander, R., Mertzlufft, F. (eds.): The Oxygen Status of Arterial Blood, pp 130 - 135. Karger, Basel (1991).

Nijland, R., Ringnalda, B., Jongsma, H.W., Oeseburg, B., Zijlstra, W.G.: The effect of interspecies differences on the measured oxygen saturation by a multi-wavelength oximeter. Abstracts of the 22nd ISOTT meeting, Istanbul (1994).

Oeseburg, B.: Physiological basis of definitions relating to oxygen transport in blood. Abstracts of the 22nd ISOTT meeting, Istanbul (1994).

Payne, J.P., Severinghaus J.W.: Definitions and symbols. In: Payne, J.P., Severinghaus, J.W. (eds.): Pulse Oximetry, pp xxi. Springer, London (1986).

Zander, R.: Calibration and quality control of equipment used for measuring O_2 concentration. In: Zander, R., Mertzlufft, F. (eds.): The Oxygen Status of Arterial Blood, pp 224 - 227. Karger, Basel (1991).

Zander, R., Lang, W.: Photometric determination of the O_2 status of human blood using the Oxystat system: cO_2 (ml/dl), sO_2 (%), cHb (g/dl). Adv. Exp. Med. Biol. 345, 849 - 852 (1994).

PHYSIOLOGICAL BASIS OF DEFINITIONS RELATING TO OXYGEN TRANSPORT IN BLOOD

Berend Oeseburg

Department of Physiology
Faculty of Medical Sciences
University of Nijmegen
PO Box 9101, 6500HB
Nijmegen, The Netherlands

Recently, the literature has been burdened with a futile discussion concerning which quantities are the most appropriate for being displayed by a CO-oximeter. The introduction of the first CO-oximeter by IL®, displaying fractional concentrations started all this. Since CO-oximeters are replacing the traditional VanSlyke determination as gold standard for quantities pertaining to oxygen transport the definitions of these variables have to be reconsidered.

Molecular oxygen (O_2) in blood exists in two forms: that associated with hemoglobin and that dissolved in blood but not associated with any other substance. The substance concentration of total O_2 (ctO_2) is defined as the sum of the substance concentrations of these two forms.

$$ctO_2 = cO_2(Hb) + cO_2 \qquad (1)$$

Substance concentration of total O_2 corresponds with the earlier designation O_2 content. The O_2 capacity of blood (BO_2) is defined as the maximum amount of O_2 that can be carried by the hemoglobin contained in one volume unit of blood. The O_2 saturation of blood (sO_2) is defined as the actual amount of hemoglobin-bound O_2 per unit volume of blood divided by the O_2 capacity:

$$sO_2 = cO_2(Hb)/BO_2 = (ctO_2 - cO_2)/BO_2 \qquad (2)$$

Originally, sO_2 was determined by gasometric methods such as the manometric VanSlyke–Neill procedure[1]. The O_2 content of a blood specimen was measured before and after equilibration with air and sO_2 was calculated with equation [2]. The binding of O_2 by hemoglobin is reversible and depends primarily on pO_2. The sO_2 corresponding with a given pO_2 is determined by the O_2 affinity. The O_2 affinity is commonly described with the help

Oxygen Transport to Tissue XVII, Edited by Ince et al.
Plenum Press, New York, 1996

143

of the O_2 dissociation curve (ODC), which is a graph of sO_2 vs. pO_2. For one to compare oxygen affinity data of various individuals and different species the ODC is reduced to standard conditions: plasma pH = 7.40, pCO_2 = 40 mm Hg and T = 37 °C. The position of the ODC is expressed by the pO_2 corresponding to sO_2 = 50% ($p50$).

Since the introduction of more dedicated methods in blood gas analysis a lot of confusion started about definition and notation on oxygen related quantities in blood. The (US) National Committee for Clinical Laboratory Standards (NCCLS) has published a proposed guideline in which these problems are thoroughly discussed[2]. In this presentation a consistent set of definitions will be given of the principal quantities pertaining to oxygen in blood in relation to the methods employed in the measurement of the quantities. Its core is the correct definition of oxygen saturation of hemoglobin as given in equations [2] and [4], which is in agreement with that given by NCCLS[5]. This system at least is consistent and the arguments are presented in this presentation in a number of statements.

Dyshemoglobins (dysHb) are normal hemoglobins which have permanently or temporarily lost the capability of reversible O_2 binding at physiological pO_2[3]. Well-known dyshemoglobins are methemoglobin (MetHb; Hi), carboxyhemoglobin (COHb) and sulf-hemoglobin (SulfHb; SHb). Only MetHb and COHb are frequently present in human blood[4]. The international reference method for hemoglobin in human blood measures the total hemoglobin concentration ($ctHb$)[5]. Consequently, when BO_2 is calculated from $ctHb$, the dyshemoglobin concentration ($cdysHb$) has to be taken into account:

$$BO_2 = ctHb - cdysHb \qquad (3)$$

The early one- and two-color photometric methods for the determination of sO_2 were calibrated against VanSlyke's method[1]. The usual spectrophotometric procedures are 2-λ methods analyzing the two-component system oxyhemoglobin/deoxyhemoglobin (O_2Hb/HHb). They do not measure the substance concentrations of the two hemoglobin species but only their ratio. For these methods the defining equation of sO_2 is usually written as

$$sO_2 = cO_2Hb/(cO_2Hb + cHHb) \qquad (4)$$

which is equivalent with [2]. Multiwavelength spectrophotometric methods, and automated multiwavelength photometers for routine measurement of hemoglobin derivatives ("CO-oximeters"), such as the IL282 and 482, the Corning 2500 and 270, and the Radiometer OSM 3, actually determine the substance concentrations of the chosen hemoglobin derivatives. From the concentrations various other quantities are calculated. Usually only these derived quantities are displayed. This started the confusion about fractional concentration, fractional saturation and functional saturation[6]. The use of incorrect definitions is clearly illustrated in the following example. A patient with a CO-poisoning has a $FCOHb$ = 0.5 (50%), ventilated with F^IO_2 = 1.0 the arterial pO_2 will be > 500 mm Hg. The arterial sO_2 will be 100% with a $p50 \approx 18$ mm Hg, due to the direct effect of CO on the affinity[7]. If fractional saturation was used it would have been 50% and so the $p50$ would have been > 500 mm Hg.

The crucial quantity in the O_2 supply to the tissues is the end-capillary pO_2. This quantity is on the one hand determined by perfusion parameters and capillary architecture, and on the other hand by the O_2 transport properties of the blood and the arterial pO_2. About the latter a CO-oximeter can give a wealth of information: $ctHb$ is the principal determinant of the O_2 capacity, sO_2 represents pO_2, and knowledge of both $FCOHb$ and $FMetHb$ is desirable, because they similarly affect O_2 capacity and O_2 affinity, but need different therapeutic interventions. Therefore, since all four of them are available, all four should be displayed. This does not yield a complete picture of the O_2 transport capability of blood ($p50$

is still lacking), but for practical purposes it is usually enough. Since FO_2Hb depends on pO_2 and $c$dysHb, $FO_2Hb > 95\%$ signals that pO_2 is high enough and that no appreciable amounts of COHb and MetHb are present. However, when FO_2Hb is too low, it does not convey much useful information any more, because the decrease may be caused either by a fall in pO_2 or by an increase in $c$dysHb or by both. Moreover, it should be stressed that FO_2Hb is **not** a measure of hemoglobin-bound O_2 in the blood. The misuse of FO_2Hb for oxygen saturation is clearly demonstrated in a number of clinical papers. For illustration just one sample is added. A paper by Rieder et al.[8] is an interesting case report, but the authors are not clear in what they mean with the notation of "saturation". Their patient has a large amount of methemoglobin (FMetHb = 0.64). Nevertheless they report FO_2Hb (0.36) as arterial saturation (36%). By correct definition the arterial oxygen saturation in this patient at that moment is 100%. The authors put their patient on controlled ventilation with 100% O_2, resulting in an arterial pO_2 of 53.5 kPa with still a FO_2Hb of 0.34. In this case the use of 100% O_2 might have been lifesaving due to the increase in dissolved oxygen, but of course did not lead to an increase in the arterial saturation, which was, by definition, already 100%.

REFERENCES

1. VanSlyke DD, Neill JM. Determination of gases in blood and other solutions by vacuum extraction and manometric measurement. J Biol Chem 61, 523-573, 1924.
2. Moran RF, Clausen JL, Ehrmeyer S, Feil M, Van Kessel AL, Eichhorn JH. Oxygen content, hemoglobin oxygen saturation, and related quantities in blood: terminology, measurement, and reporting. NCCLS Document C25-P. Vol 10, No 2, 1990.
3. Dijkhuizen P, Buursma A, Fongers TME, Gerding AM, Oeseburg B, Zijlstra WG. The oxygen binding capacity of human haemoglobin. Hüffner's factor redetermined. Pflügers Arch 369, 223-231, 1977.
4. Zwart A, van Kampen EJ, Zijlstra WG. Results of routine determination of clinically significant hemoglobin derivatives by multicomponent analysis. Clin Chem 32, 972-978, 1986.
5. International Committee for Standardization in Haematology. Recommendations for reference method for haemoglobinometry in human blood (ICSH standard 1986) and specifications for international haemiglobincyanide reference preparation (3rd edition). Clin lab Haemat 9, 73-79, 1987.
6. Oeseburg B, Rolfe P, Siggaard Andersen O, Zijlstra WG. Definition and measurement of quantities pertaining oxygen in blood. In: P. Vaupel, R. Zander & D.F. Bruley (Eds), Oxygen transport to tissues XV. Adv Exp. Med Biology, Vol 345, Plenum Publ Corp, New York and London, 925-930, 1994.
7. Kwant G, Oeseburg B, Zijlstra WG. Reliabilty of the determination of whole-blood oxygen affinity by means of blood-gas analyzers and multi-wavelength oximeters. Clin. Chem. 35, 773-777, 1989.
8. Rieder HU, Frei FJ, Zbinden AM, Thomson DA. Pulse oximetry in methaemoglobinaemia. Anaesthesia 44, 326-328, 1989.

18

MEASUREMENT OF THE OXYGENATION STATUS OF THE ISOLATED PERFUSED RAT HEART USING NEAR INFRARED DETECTION

Johannes H. G. M. van Beek,[*] Mary D. Osbakken, and Britton Chance

Laboratory for Physiology
Institute for Cardiovascular Research
Free University
1081 BT Amsterdam, The Netherlands
Department of Biochemistry and Biophysics
University of Pennsylvania
Philadelphia, Pennsylvania

INTRODUCTION

The oxygenation status of isolated hearts has often been measured using optical detection methods for the visible range of the spectrum (Tamura et al., 1989). To the best of our knowledge there are no reports on measurement of myoglobin oxygenation using detection methods in the near-infrared region of the spectrum. On the other hand, equipment for near-infrared measurements on tissue is widely available and is in some instances quite inexpensive. Therefore we investigated in the present study whether near-infrared (NIR) detection at two wavelengths provides a way to conveniently determine myoglobin oxygenation in an isolated organ as small as the saline-perfused rat heart.

Although sophisticated methods to quantitate hemoglobin deoxygenation making use of the time of flight of photons through tissue exist (Chance et al., 1988), a simple and inexpensive continuous NIR radiation dual wavelength system was investigated in the present study. This dual wavelength method makes use of the opposite changes in absorption of NIR radiation upon deoxygenation of myoglobin: at 760 nm NIR absorption increases while at 850 nm absorption decreases. The difference of the radiation intensity at 760 and 850 nm, respectively, gives a signal proportional to deoxygenation of hemoglobin and myoglobin (Chance et al., 1992).

[*] Address corresopndence to J. H. G. M. van Beek, Laboratory for Physiology, Institute for Cardiovascular Research, Free University, Van der Boechorststraat 7, 1081 BT Amsterdam, The Netherlands. Phone: 31-20-4448121; fax: 31-20-4448255; e-mail: VANBEEK@SARA.NL

Oxygen Transport to Tissue XVII, Edited by Ince et al.
Plenum Press, New York, 1996

147

Because the NIR absorption spectrum of myoglobin is similar to the hemoglobin spectrum, near-infrared detection may be used for both. Dual wavelength NIR tissue oxygen monitors have usually been applied to muscle tissue where the largest part of the near infrared absorption is due to hemoglobin in the blood. In the present application hemoglobin has been flushed out of the heart and changes of myoglobin oxygenation in the cardiac myocytes are detected. However, because the absorption by myoglobin in the near-infrared is considerably weaker than in the visible range it remained to be established whether a saline-perfused rat heart yields a sufficient signal-to-noise ratio to make near-infrared detection of tissue oxygenation feasible.

METHODS

Hearts were isolated from seven male Sprague–Dawley rats, weighing 250-300 gram, anesthetized with 45 mg/kg pentobarbital sodium. The isolated hearts were perfused according to Langendorff with recirculated Krebs–Henseleit buffer solution at 37°C at a perfusion pressure of about 100 mmHg (see Fig. 1).

The heart could be perfused from two vessels: buffer in the first vessel was gassed with 95% O_2 - 5% CO_2, in the second with 95% N_2 - 5% CO_2. A stopcock close to the aortic cannula allowed switching between the deoxygenated and oxygenated buffer. The tubes that ran between the vessels, which were located 140 cm above the heart, and the aortic cannula were made of glass to prevent oxygen leakage and ran through chambers with thermostatted water to maintain temperature. The left ventricle contracted against a water-filled latex balloon. The heart was paced with ~10 V 0.5 ms pulses with a Grass S48 stimulator (Grass, Quincy, MA, USA) via two thin platinum needle electrodes hooked to the atrial tissue.

A tungsten bulb connected to a RUNMAN CWS-2000 unit (NIM Inc., Philadelphia, PA, USA) was placed facing the right ventricular surface and was flashed at 0.4-1 Hz by a 2 V square wave, with 25% duty cycle. A light guide (6 mm diameter for the set of light fibers) was placed against the left ventricular surface, just touching the epicardial tissue. The light guide conducted the NIR radiation from the left ventricle to two silicon diode NIR detectors fitted with filters at 760 and 850 nm (peak accuracy 5 nm, half-width 20 nm). The signals for both wavelengths were electronically integrated during the flash after background radiation, as detected in between the flashes, was subtracted. The difference signal of both wavelengths gave the oxygenation and was displayed immediately after the flash and the signal was kept constant via a sample and hold circuit until the next flash's difference signal

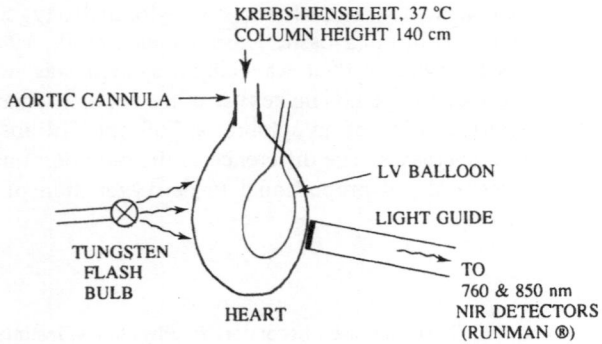

Figure 1. Schematic drawing of the setup for perfusion of the isolated rat heart with measurement of tissue oxygenation using the dual wavelength near-infrared tissue oxygen detector.

had been processed. Normally the difference signal of the sample and hold circuit is filtered with a time constant of more than 10 s (Chance et al., 1992), but in the present study the time constant of the filter was modified and was 0.12 s. Therefore the signal consists of steps that reflect the consecutive levels of the sample-and-hold circuit which can clearly be distinguished in the recording. Each step represents the difference between the 760 and 850 nm radiation having traversed the heart during the immediately preceding flash. In view of the 25% duty cycle and the 0.4-1.0 Hz flash rate the signal was thus averaged over 250-625 msec. Because the heart rate was in the range from 4-10 Hz this means that the signal was averaged over the cardiac cycle.

The sum of the 760 and 850 nm signals for the radiation transmitted through tissue is also calculated and gives the total amount of absorbing molecules in tissue irrespective of changes in O_2 saturation of myoglobin. However, in the isolated heart the amount of myoglobin inside the myocytes does not change for all practical purposes, and no use was made of the sum signal in the present study.

The response time of the equipment was tested in an *in vitro* setup. Silicone tubing of 1 cm diameter was filled with heparinized rat blood and the tungsten flash bulb was positioned on one side of the tube while the light guide was put on the other side touching the tube. By contact with nitrogen and air, two levels of hemoglobin saturation were obtained and blood with low oxygen content was layered inside the tube on top of blood with higher oxygen content. During measurement of the difference signal the blood was moved inside the tube by pushing or pulling a large syringe connected to the silicone tube exactly in between two flashes. The result was that a different layer of blood with different oxygenation was now "seen" by the light guide. Upon this step change in oxygenation in between flashes the difference signal, displayed on a recorder, changed essentially stepwise. Thus any delay in the signal after a stimulus to the isolated heart indicates a delay due to the physiological properties of the tissue and is not due to instrument delay.

When the background radiation level was suddenly changed by switching on an incandescent lamp or attenuating it with white copy paper the difference signal responded

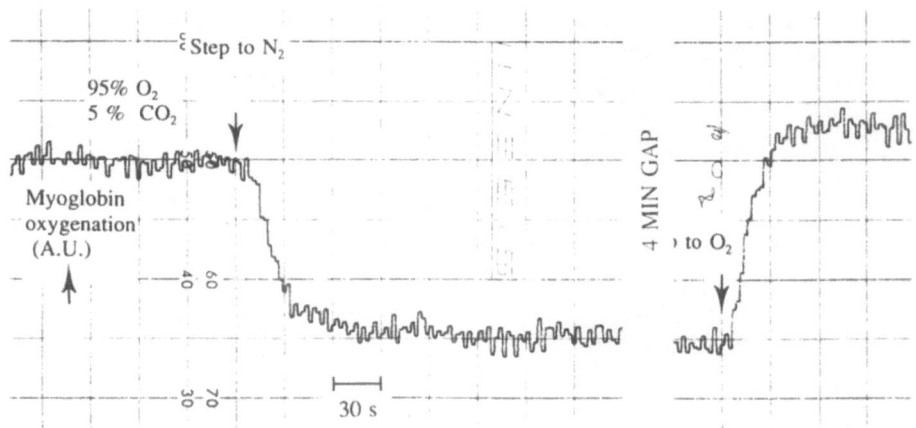

Figure 2. The difference signal from the near-infrared (NIR) tissue oxygen monitor, sensitive to myoglobin oxygenation, obtained from a saline-perfused rat heart. The flash rate of the NIR equipment was ~0.4 Hz. Each step in the tracing, occurring every 2.4 s, is a reading of the difference between the 760 and 850 nm signals obtained during a single flash of the tungsten bulb immediately before the step in the recording. At the beginning of the trace the heart was perfused with buffer gassed with O_2 and CO_2. At the arrows we switched to buffer gassed with N_2 and CO_2 and back again. The heart was paced at 4 Hz. A.U. (arbitrary units).

with a slow change that took several seconds to complete, in agreement with previous determinations of the frequency response to overall light levels (Chance et al., 1993). However, when the background radiation was kept constant the response to oxygenation changes was immediate when an oxygenation change was completed before the flash. For this reason, a black cloth was used to cover the heart-light-guide setup so that stray light did not reach the detectors, and thus distort the signal originating from the heart.

RESULTS

The perfusion flow through the hearts was 9.0 ± 3.4 ml/g wet weight/min (mean $\pm$ SD). Following transitions from perfusate gassed with 95% O_2 to perfusate gassed with 95% N_2 (always with 5% CO_2) and back to 95% O_2 there was a reproducible change in the difference signal with good signal-to-noise ratio (Fig. 2). Immediately after each flash of the bulb a reading of the tissue oxygenation was obtained. The half time of the change in the oxygenation signal was 23 ± 13 s (n=18) for steps from O_2 to N_2 and 23 ± 22 s (n=18) for steps from N_2 to O_2. The mean delay time between switching perfusion flow and the first noticeable change in oxygenation signal was 10.5 ± 11.2 s for the step from O_2 to N_2 and was 7.4 ± 10.2 s (n=18) for the step from N_2 to O_2.

At the end of the trace in Fig. 2 the signal seemed to return to an oxygenation level above the initial one. This may have been due to a 20% increase in the perfusion flow that was measured in the heart 5 min after we had switched back to buffer gassed with 95% O_2. Fifteen minutes after switching back to O_2 the perfusion flow had returned to its initial value and at that time also the oxygenation signal had returned to its initial baseline level. Upon steps in heart rate the NIR difference signal changed by up to 40% of the amplitude of the O_2 to N_2 transition (see example in Fig. 3).

For steps in heart rate from 4 to 7 Hz, see Fig. 4, the half-time of the change in oxygenation was 6.6 ± 2.2 s (mean $\pm$ SD, n=23) and for steps from 7 to 4 Hz the half-time was 7.6 ± 2.3 s (n=23). The half-time was significantly larger for the downward steps (P = 0.03, Wilcoxon signed-ranks test). Often it appeared that the oxygenation signal responded to a step in heart rate with a delay of at least several heart beats, not due to instrumental delay. This minimum delay time before a change was noticeable was 1.9 ± 2.0 s (n= 23) for steps from 4 to 7 Hz and was significantly longer (P = 0.002, Wilcoxon signed-ranks test), i.e., 3.7 ± 2.9 s (n=23) for steps from 7 to 4 Hz.

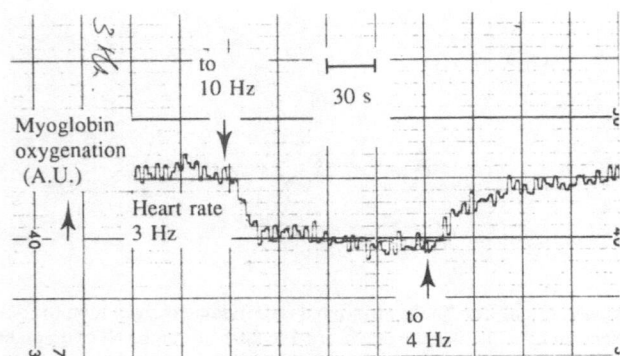

Figure 3. Difference signal from a saline-perfused rat heart. The heart was continuously perfused with buffer gassed with 95% O_2 and 5% CO_2. At the beginning of the trace the heart was paced at 3 Hz. At the arrows the pace rate was switched to 10 Hz and then back to 4 Hz. A.U.

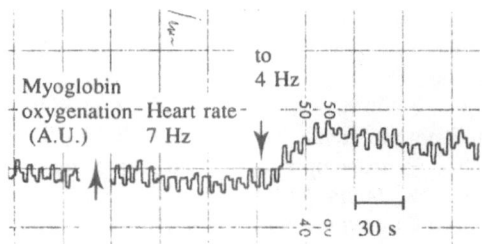

Figure 4. Difference signal from saline-perfused rat heart continuously perfused with buffer gassed with 95% O_2/5% CO_2. At the arrow the pace rate was stepped from 4 Hz to 7 Hz.

The flash rate of the NIR equipment could be increased to 1 Hz while retaining a similar signal-to-noise ratio as at 0.4 Hz.

DISCUSSION

Methodological Considerations

This study showed that the tissue oxygenation in the saline-perfused heart can be measured very conveniently and cost-effectively using dual wavelength near-infrared detection. A limitation of the present technique is that the oxygenation signal was averaged over the heart cycle. This follows from the duty cycle of 25% of the flash during which the signal was integrated and from the heart rate (3-10 Hz) and the rate of flashing of the bulb (0.4-1 Hz). Therefore the fluctuations in the signal caused by the within beat variation of myoglobin oxygenation that have been reported by Tamura et al. (1976) in the saline-perfused rat heart using visible wavelengths cannot be observed with this NIR monitor.

A stable position of the light guide on the surface of the heart is needed to keep the signal stable: when the fiber optic cable and the heart are displaced even slightly relative to each other, a permanent shift in the baseline of the signal will result. On the other hand, the signal-to-noise ratio was similar in a heart that had been arrested by a high level of KCl in the perfusate as in beating hearts, so that contraction of the heart did not deteriorate the signal's quality appreciably. This may have been partly due to the restriction of the movement of the heart due to the left ventricular balloon.

When the light level in the room is changed, the difference signal giving the apparent oxygenation changes only transiently. Because the background radiation measured via the fiber optics will also depend on the transmission of radiation from the environment through the heart tissue and because the measured background radiation might thus vary with the condition of the heart we suppressed background radiation by surrounding the isolated heart with black cloth.

The NIR equipment makes use of two wavelengths. There are two molecular species in the saline-perfused heart that absorb significant quantities of radiation at 760 and 850 nm and whose absorption may change with the oxygenation status of the tissue: myoglobin and the cytochrome a,a_3 part of cytochrome oxidase. Although the two types of absorber have clearly distinct absorption spectra, they cannot be discerned with two wavelengths. The myoglobin content of the rat heart on a molar basis is eight times that of cytochrome a,a_3 (Kennedy and Jones, 1986). On the other hand the change in extinction coefficient on a molar basis caused by a transition from the oxygenated to the deoxygenated state for myoglobin is considerably smaller than the change for a transition of cytochrome a,a$_3$ from oxidized to reduced. However, the two wavelengths, 760 and 850 nm, employed for NIR detection in the present study have been chosen such that they are on opposite sides of the isosbestic wavelength for myoglobin, ~800 nm. At the isosbestic wavelength the absorption of NIR radiation by myoglobin is not affected by its oxygenation state. Consequently, after subtracting the *negative* extinction increase upon deoxygenation of myoglobin at 850 nm from the positive extinction increase at 760 nm the result is a large positive number for the difference signal. However, cytochrome a,a$_3$ does not have an isosbestic wavelength between 760 and 850 nm. Consequently, the extinction decreases both at 760 and at 850 nm upon reduction of cytochrome a,a$_3$ and these changes are for a major part cancelled when the difference signal is obtained. It could be estimated that for a hypothetical change in myoglobin oxygenation from 100 to 0% and a simultaneous change in cytochrome oxidation state from 100 to 0% the change in cytochrome a,a$_3$ absorption accounts for less than 25% of the total difference signal. Further, in the isolated rat heart, a preparation similar to that used in the present study, changes in cytochrome oxidation are exactly proportional to the changes in myoglobin oxygenation (Tamura et al., 1978), a phenomenon that has been called coherence. Thus, in the present application a minor part of the total difference signal may be due to cytochrome a,a3, but this part varies exactly proportionally to the myoglobin signal which constitutes the major part of the signal.

The dual wavelength method is usually applied to skeletal muscle in vivo. In such applications a large amount of hemoglobin is present in the skeletal tissue which cannot be discerned from an about 25% contribution by myoglobin (Chance et al., 1992). In such applications on skeletal muscle the relative contribution by cytochrome a,a$_3$ is expected to be much less than in the isolated rat heart because of the large amount of hemoglobin that is present and because the mitochondrial density is lower in skeletal muscle than in the heart.

The geometric arrangement was such that virtually all NIR photons that are measured by the silicon diode detectors first entered the right ventricular surface and must traverse the left ventricular free wall before being collected by the light guide. Consequently, the photons, which have almost all been scattered multiple times (Chance et al., 1988), have effectively sampled the whole heart volume before contributing to the signal. Therefore, oxygen diffusion across the surface of the heart which influences the oxygenation of the outer layer of the heart (Van Beek et al., 1992) has a negligible influence on the measurement.

Good recordings of myoglobin oxygenation with the NIR equipment have also been obtained in the saline-perfused rabbit heart (unpublished results) by doubling the voltage to the flash bulb. The reading of the NIR instrument was calibrated in the rabbit experiments by inhibiting oxygen consumption at the end of the experiment with oligomycin or, preferably, with rotenone to obtain a reading at a level of myoglobin oxygenation that is virtually 100%, and by perfusion with buffer without O$_2$ to obtain the oxygenation reading at 0%.

Physiological Consequences

The balance of oxygen supply via the coronary vasculature and mitochondrial oxygen consumption is directly reflected by myoglobin oxygenation in the cardiac myocytes. Monitoring MbO$_2$ thus shows whether the balance of O$_2$ supply and consumption is disturbed. The isolated rat heart perfused with buffer is presumably vasodilated with respect to the in vivo situation and vasoregulation is probably impaired. Further, the increase in perfusion flow following an increase in metabolic demand is slow. The early changes in the oxygenation signal after a heart rate step therefore take place with little change in capillary flow and by inference the early change in myoglobin oxygenation reflects the change in mitochondrial oxygen consumption under these conditions.

The delay in the change of oxygenation in the myocytes upon a step in heart rate suggests that cardiac oxygen consumption does not change immediately following an increase in heart rate. This agrees with the time constant of about 8 s with which mitochondrial oxygen consumption changes after a step in heart rate at 37°C in the isolated rabbit heart (Van Beek and Westerhof, 1991). This gradual change in oxygen consumption is accompanied by a small but significant decrease in phosphocreatine (Eijgelshoven et al., 1994). The gradual change in oxygen consumption is therefore not caused by a delay in the change in ATP hydrolysis after the step in heart rate. The gradual change in oxygen consumption in turn also means that ATP synthesis follows ATP consumption with a certain delay, leading to an increase in inorganic phosphate and a decrease in phosphocreatine concentration (see Fig. 5). These changes are by inference accompanied by an increase in the concentration of free ADP. A gradual increase in inorganic phosphate concentration has been found using NMR spectroscopy (Eijgelshoven et al., 1994). Inorganic phosphate and ADP can thus stimulate oxidative phosphorylation. This constitutes a feedback loop (see Fig. 5) which may contribute to the regulation of cardiac oxidative phosphorylation. The gradual change in oxidative phosphorylation is therefore not due to a slow response within the mitochondria themselves, which show a very fast response (<100 ms; Chance et al., 1965), but the gradual change reflects [ADP] in the cytosol, which changes very little in the first few seconds after a step in heart rate due to the efficient resynthesis of ADP to ATP from the phosphocreatine energy reserve via the enzyme creatine kinase. We hypothesize that the change in cardiac oxidative phosphorylation has a similar time course as the change in tissue perfusion during increases in cardiac workload to prevent temporary oxygen shortage in the cardiac myocytes.

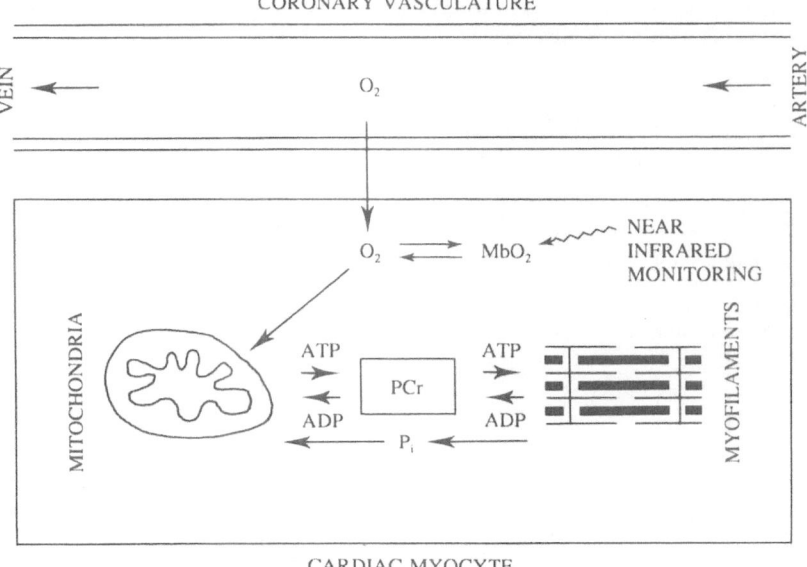

Figure 5. A schematic representation of oxygen supply from the coronary vasculature and mitochondrial oxygen consumption in the cardiac myocyte. Myoglobin oxygenation, which can be monitored with near-infrared radiation, reflects the balance of oxygen supply and consumption. Phosphocreatine (PCr) reflects the balance of ATP production and consumption. Inorganic phosphate (Pi) is a direct breakdown product of ATP. Both PCr and Pi can be monitored with NMR equipment, which is however much more expensive than NIR equipment.

SUMMARY

We have shown that an inexpensive dual wavelength near infrared tissue oxygen monitor may be very useful to detect myoglobin oxygenation in a volume of tissue as small as the isolated buffer-perfused rat heart. The time resolution can be as low as one second. Upon a step increase in heart rate the myoglobin deoxygenation changes with a half-time of about 6 s. Often there seems to be a small delay of at least several heart beats before myoglobin oxygenation clearly starts to change, suggesting that oxygen consumption does not change immediately upon a step in heart rate.

ACKNOWLEDGMENTS

This work was supported by a travel grant of the Netherlands Organization for Scientific Research NWO to J.H.G.M. van Beek and by NIH grant RO1-HL39208.

REFERENCES

Beek, J.H.G.M. van, Loiselle, D.S, and Westerhof, N, 1992, Calculation of oxygen diffusion across the surface of isolated perfused hearts, *Am.J.Physiol.* 263: H1003-H1010.
Beek, J.H.G.M. van, and Westerhof, N., 1991, Response time of cardiac mitochondrial oxygen consumption to heart rate steps, *Am.J.Physiol.* 260: H613-H625.
Chance, B., 1965, The energy-linked reaction of calcium with mitochondria, *J.Biol.Chem.* 240: 2729-2748.
Chance, B., Dait, M.T., Zhang, C., Hamaoka, T., and Hagerman, F. , 1992, Recovery of exercise-induced desaturation in the quadriceps muscles of elite competitive rowers, *Am.J.Physiol.* 262: C766-C775.
Chance, B., Nioka, S., Kent, J., McCully, K., Fountain, M., Greenfeld, R., and Holtom, G., 1988, Time-resolved spectroscopy of hemoglobin and myoglobin in resting and ischemic muscle, *Anal. Biochem.* 174:698-707.
Chance, B., Zhuang, Z., Unah, C., Alter, C., and Lipton, L., 1993, Cognition-activated low-frequency modulation of light absorption in human brain, *Proc. Natl. Acad. Sci. USA* 90: 3770-3774.
Eijgelshoven, M.H.J., Beek, J.H.G.M. van, Mottet, I., Nederhoff, M.G.J., Echteld, C.J.A. van, and Westerhof, N., 1994, Cardiac high-energy phosphates adapt faster than oxygen consumption to changes in heart rate, *Circ. Res.*, in press, scheduled for Vol. 75, No. 4, 9 pages.
Kennedy, F.G., and Jones, D.P., 1986, Oxygen dependence of mitochondrial function in isolated rat cardiac myocytes, *Am.J.Physiol.* 250: C374-C383.
Tamura, M., Hazeki, O., Nioka, S., and Chance, B., 1989, In vivo study of tissue oxygen metabolism using optical and nuclear magnetic resonance spectroscopies, *Annu. Rev. Physiol.* 51:813-834.
Tamura, M., Oshino, N., Chance, B., and Silver, I., 1978, Optical measurements of intracellular oxygen concentration of rat heart *in vitro*, *Arch. Biochem. Biophys.* 191: 8-22.

19

PLASMA MIXING IS LIKELY TO AFFECT CAPILLARY OXYGEN TRANSPORT IN HARD WORKING RAT HEART

Cees Bos,[1] Louis Hoofd,[1] Thom Oostendorp,[2] and Berend Oeseburg[1]

[1] Department of Physiology
[2] Department of Medical Physics and Biophysics
University of Nijmegen
P.O. box 9101, 6500 HB
Nijmegen, The Netherlands

INTRODUCTION

The delivery of oxygen to tissue occurs primarily in the capillaries. Therefore a substantial part of the research on oxygen transport into the tissue has been directed towards oxygen release and transport in and around capillaries. The influence of the major determinants on this transport, such as diffusion coefficient, blood flow, reaction kinetics, has been well investigated. One of the possible influences on the oxygen transport in the capillaries was, seemingly, eliminated by Aroesty and Gross (1970). They investigated the effect of plasma mixing on oxygen transport in capillaries and showed that the net effect of mixing was negligible. The enhancement near the red blood cell (RBC) downstream of a small plasma volume was cancelled out near the upstream RBC. However, in a previous model (Bos et al., in press) we showed that the results of their study were influenced considerably by their choice of boundary conditions. In fact, one could easily show a 50% increase of the oxygen flux through the plasma to the tissue by altering the boundary conditions.

Although it was shown that there could be some effect of plasma mixing, the model did not fully disclose the influence of this mixing on oxygen transport. Therefore an extended model is presented here, which now includes a thin layer of tissue around the capillary. The addition of this thin layer of tissue results in a model that is significantly less sensitive to changes in boundary conditions and that enables to estimate the effect of plasma mixing on oxygen transport in capillaries.

MODEL FORMULATION

The model encompasses a plasma volume trapped between two successive RBCs and a thin layer of adjacent tissue. A thin layer should suffice since the oxygen partial pressure

Oxygen Transport to Tissue XVII, Edited by Ince et al.
Plenum Press, New York, 1996

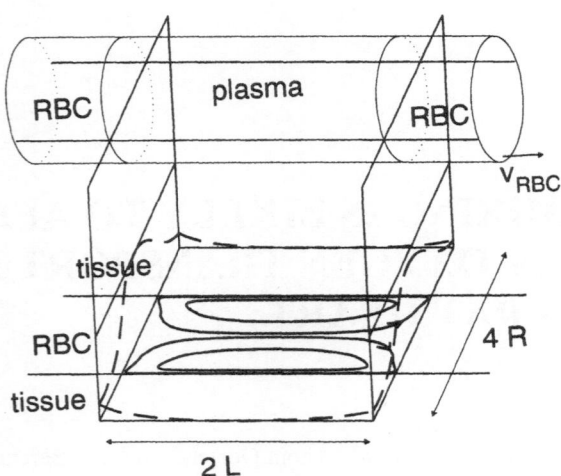

Figure 1. Model lay-out. At the top a section of a capillary with RBCs is shown, which is projected onto a 2-dimensional plane at the bottom. The vertical axis is used to indicate the pO_2 at the boundaries of the model (dashed line). Typical streamlines in the plasma fluid are depicted by the solid lines in the center of the plane at the bottom.

gradients are steep close to the capillary, and level out within a few micrometers of the capillary wall (Groebe, 1990; Hoofd, 1992). The plasma is assumed to be totally sealed off by the RBCs, which is true when the RBCs fill up the width of the capillary. The basic phenomena in this model are diffusion, forced convection, and zero-order oxygen consumption. The equations describing this model are derived for axially symmetric cylindrical coordinates (r, ϕ, z) relative to the RBCs (where ϕ can be omitted since the system is axially symmetric). The lay-out of the model is depicted in figure 1. The radius of the capillary is R, the thickness of the adjacent tissue layer also is taken to be R, the length of the plasma gap is 2L, and v_{RBC} is the velocity of the RBCs. The movement of the RBCs causes mixing of the plasma. The oxygen is transported from the RBCs through the plasma gap by forced convection and diffusion to the tissue. In the tissue oxygen is transported by diffusion where it is consumed.

 The velocity components of the plasma are the same as those in the previous model (Bos et al., in press). The equations describing the fluid mechanics are the equation of motion, derived from the momentum balance, and the equation of continuity, derived from the mass balance (Bird et al., 1960). Since blood plasma, in contrast to whole blood, is a Newtonian fluid, and the Reynolds number is low, the equations of Stokes flow can be used:

$$\begin{cases} \vec{\nabla}p = \mu\nabla^2\vec{v} \\ (\vec{\nabla}\cdot\vec{v}) = 0 \end{cases}$$

where p is the plasma pressure, $\vec{v}$ is the plasma velocity vector, $\vec{\nabla}$ is the gradient operator, ∇^2 is the Laplace operator and the dot denotes the inner product of two vectors. The streamlines can be calculated from the stream function ψ. They are defined by ψ = constant. ψ is calculated from the velocity components in such a way that the equation of continuity is automatically satisfied, by $\vec{v} = (v_r, v_z) = (\partial\psi/r\partial z, -\partial\psi/r\partial r)$, where v_z is the axial component of the velocity, and v_r is the radial component of the velocity. A depiction of typical streamlines can be found at the bottom of figure 1.

Using the velocity profile found from these equations, the transport and consumption of the oxygen is calculated from the following steady-state equation:

$$(\vec{v} \cdot \vec{\nabla}) = D\nabla^2 c + R$$

where c is the oxygen concentration, D is the diffusion coefficient, and R is the chemical production term, which is zero in the plasma and a negative constant ($-\dot{Q}$ for the zero-order oxygen consumption in the tissue. The calculations for oxygen transport are performed with a non-dimensionalized version of this equation, where the dimensionless concentration is defined as: $c^* = (c-c_0)/c_1$. Both c_0 and c_1 are irrelevant for this investigation. A frequently used dimensionless number is the Peclet number defined here as $Pe = (v_{RBC} \cdot L)/D$, which defines a relationship between the mass transfer by forced convection and the mass transfer by diffusion.

The boundary conditions investigated here are both realistic and unrealistic boundary concentrations at the plasma-tissue border and at both the up- and downstream plasma-RBC border. The realistic boundary concentrations were estimated from a model that neglects the effect of mixing, but describes a whole capillary with accompanying tissue (Hoofd et al., 1994). The simulations are evaluated by calculation of the flux ratio $\varphi(Pe)$, which is defined as the ratio of the total flux with mixing, for a certain Pe number, to the total flux without mixing. The total flux is the total oxygen flux through the capillary wall along the plasma volume. See Bos et al. (in press) for a mathematical description of $\dot{\varphi}(Pe)$.

RESULTS AND DISCUSSION

The data used here are for rat heart (Altman et al., 1958; Hoofd et al., 1990). The Pe numbers were calculated for rat heart by Bos et al. (in press). From the Pe number it can be seen that the effect of mixing depends of the RBC velocity (v_{RBC}) and the RBC spacing (2L). In this study we did calculations for two velocity ranges, a resting and a hard working rat heart, and two RBC spacings, 20% hematocrit and 40% hematocrit. A capillary radius, R, of 2.44 µm is used and Pe varies from 1.4 to 18.

In table 1 Pe numbers with the resulting flux ratio $\varphi(Pe)$ are shown, referring to figure 2 where the corresponding calculated oxygen concentration profiles are presented. A and B

Table 1. Effect of mixing on oxygen flux at the capillary wall in rest and at work, and for 20% and 40% hematocrit. The boundary concentrations (b.c.) along the four plasma borders are shown in Figure 2

b.c.	Hct	Pe		$\varphi(\ )$	
		Rest	Work	Rest	Work
Fig. 2A	40%	1.4	6.8	1.00	1.00
Fig. 2B	20%	3.7	18	1.07	1.51
Fig. 2C	40%	1.4	6.8	1.00	1.01
Fig. 2D	20%	3.7	18	1.01	1.12
Fig. 2E	40%	1.4	6.8	1.02	1.10
Fig. 2F	20%	3.7	18	1.04	1.20
Fig. 2G	40%	1.4	6.8	1.01	1.03
Fig. 2H	20%	3.7	18	1.04	1.21

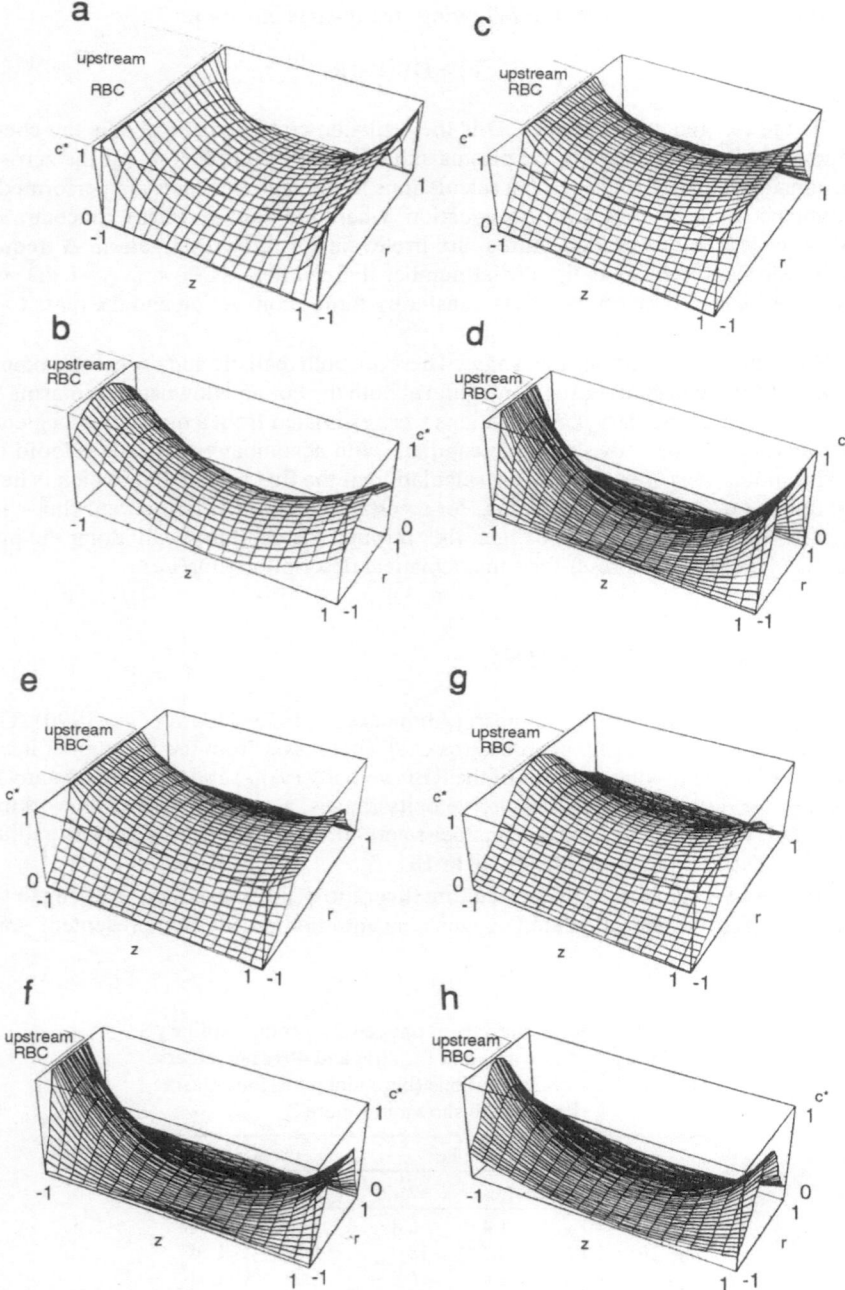

Figure 2. Dimensionless concentration gradients for Pe = 0. A and B are from the previous model (without tissue sheet). C - H are from the present model (which includes a thin layer of tissue). A, C, E and G are simulations for different boundary conditions (concentration profiles at the four plasma borders) with a 40% hematocrit, and B, D, F and H for a 20% hematocrit value. The boundary concentration profiles of B, G and H are estimated from a diffusional model (Hoofd et al., 1994).

in figure 2 were calculated using the previous model, which did not include tissue, and the other panels are calculated using the present model.

The figures 2A and 2B are added to show the difference in effect of mixing between the two models. The boundary concentrations as shown in figure 2A, did not result in an enhanced oxygen flux due to mixing. The same was true for 20% hematocrit (results not shown here). However, for the 20% hematocrit an example of calculations for more realistic boundary concentrations is presented, which shows an enhancement of the oxygen flux by 50%, caused only by application of these different boundary concentrations. This dependence on the choice of boundary concentrations is notably reduced in the present model. When case 2A is compared to 2C and 2D, it can be seen that, although 2C and 2D have similar crude boundary concentrations, the change in flux is different. If we compare cases 2C, 2E and 2G, then it seems that the flux is predominantly affected by the difference in oxygen concentration in the plasma at the two RBCs. However, from 2D it can be seen that for the 20% hematocrit, the flux is also affected without a concentration difference between the two successive RBCs. This difference in oxygen transport between the 20% and 40% hematocrit results is probably due to the difference in concentration drop between the center of the plasma gap and the RBCs. In the 40% hematocrit case, the concentration along the axis of the capillary is high everywhere and mixing will hardly effect the concentration drop from the center of the plasma gap to the capillary wall.

The physiological importance of mixing cannot easily be assessed, since this model describes only a part of the capillary. However, the increase of flux can be translated into a change in the radius of the well-known Krogh cylinder. The maximal effect can be found when the Krogh cylinder is surrounded by anoxic tissue. Then an enhancement of the flux by 20% results in an enlargement of the Krogh cylinder radius by approximately $\sqrt{1.2} = 1.095$ times. To be precise the enlargement of the tissue radius should be calculated by:

$$\frac{R_{T,new}}{R_{T,old}} = \sqrt{f + (1-f)\left(\frac{r_{cap}}{R_{T,old}}\right)^2}$$

where $R_{T,old}$ is the old Krogh cylinder radius, $R_{T,new}$ is the Krogh cylinder radius with improved flux, r_{cap} is the capillary radius, and f is the flux enhancement factor. For $r_{cap} = 2.44$ μm, $R_{T,old} = 10$μm, and f = 1.2, a $R_{T,new} = 10.90$ μm will be found.

The translation of a 20% increase of the flux into the Krogh cylinder radius shows that the effect of plasma mixing on oxygen transport is noticeable, but not overwhelming. However, it might be important when realistic distributions of RBCs are incorporated into models. It is known that the RBCs are rarely evenly spaced and that the distance between two successive RBCs can be significantly larger than the average spacing. In that case it might be worthwhile to incorporate plasma mixing in the model for the large gaps. In conclusion it can be said that mixing of plasma enhances oxygen transport, and that incorporation of plasma mixing in oxygen transport models might be useful at high red blood cell velocities combined with a low hematocrit, especially when uneven distribution of RBCs is considered.

SUMMARY

Based on a study of Aroesty and Gross (1970), mixing of plasma between two successive red blood cells (RBCs) in capillaries is usually assumed to be unimportant for oxygen transport. In a previous study we showed that their findings were highly dependent on the boundary conditions. In this study we present a model that is less subject to changes in boundary conditions. The results obtained with the present model indicate that there is an enhancement of oxygen transport, but the importance of this effect is highly dependent on the RBC velocity and the distance between two successive RBCs.

REFERENCES

Altman, P.L., Gibson, J.F., and Wang, C.C., 1958, *Handbook of Respiration*, Eds. Dittmar, D.S. and Grebe, R.M., Saunders company, Philadelphia and London

Aroesty, J., and Gross, J.F., 1970, Convection and diffusion in the microcirculation, *Microvasc. Res.* 2:247-267

Bird, R.B., Stewart, W.E., and Lightfoot, E.N., 1960, *Transport Phenomena*, Wiley & Sons, New-York.

Bos, C. Hoofd, L., and Turek, Z., Local plasma convection can be important for oxygen release in tissue capillaries, *In: Oxygen Transport to Tissue XVI, Adv. Exp. Med. Biol.*, in press.

Groebe, K. 1990, A versatile model of steady state O_2 supply to tissue. Application to skeletal muscle. *Biophys. J.* 57:485-498

Hoofd, L., 1992, Updating the Krogh model - assumptions and extensions, *In: Oxygen Transport in Biological Systems*, Cambridge University Press, pp. 197-229.

Hoofd, L., Bos, C., Turek, Z., 1994, Modelling erythrocytes as point-like sources in a Kroghian cylinder model, *In: Oxygen transport to tissue XV, Adv. Exp. Med. Biol.*, 345:893-900

Hoofd, L., Olders, J., and Z. Turek, Z., 1990, Oxygen pressures calculated in a tissue volume with parallel capillaries, *In: Oxygen transport to tissue XII, Adv. Exp. Med. Biol.*, 227:21-29

20

PRACTICAL APPLICATIONS OF MODELS OF OXYGEN SUPPLY, DIFFUSION, AND CONSUMPTION

Past, Perspectives, and Problems

K. Groebe

Institut für Physiologie und Pathophysiologie
Johannes Gutenberg-Universität Mainz
Duesbergweg 6, D-55099
Mainz, Germany

INTRODUCTION

It is the objective of this paper to describe, what models of oxygen supply, diffusion, and consumption have been and can be used for and to discuss some common problems in employing mathematical models of O_2 transport. In a first part, four typical fields for model applications are selected from the wide variety of former investigations and are illustrated with some recent examples.

In modelling oxygen transport to tissue, there has been some kind of evolution from very straightforward and greatly simplified approaches to highly elaborate and much more realistic ones. Extrapolating from this development, one could state rather heretically that it is the ultimate challenge to a modeller of a physiological system to abolish the necessity of experiments on this system in living organisms, and to simulate these experiments in the computer instead. Despite the rapidly increasing availability of computing power and the growing complexity of models, at the present time we are still far away from this goal, but it nevertheless can point out a direction to our thinking about future development of model applications. The most crucial prerequisite for replacing experiments by computer simulations is the availability of models which *quantitatively* describe the physiological system in question. In developing such models, a number of typical problems has to be overcome some of which are illustrated in a second part. Even though some encouraging results demonstrate quite good agreement between model and experiment, this part addresses many more open questions than it offers answers. Rather, some extraordinarily difficult problems are characterized which all relate to the appropriateness and acquisition of input data for modelling. Finally, requirements on models and on input data for models are formulated which are aimed at improving the applicability of models for quantitative simulation and analysis.

Oxygen Transport to Tissue XVII, Edited by Ince et al.
Plenum Press, New York, 1996

PURPOSE OF MODEL APPLICATIONS, EXAMPLES

The following list singles out four classes of model applications which, in a way, also represent stages in the above mentioned evolution of modelling oxygen transport:

- Identification of mechanisms of possible physiological significance,
- quantification of parameter values governing the function of transport mechanisms,
- assessment of "measuring values" which cannot be obtained in real experiments due to methodological limitations ("unfeasible experiments"), and
- analysis of the entirety of all processes involved in oxygen transport in their mutual dependencies and interactions ("system analysis").

In the following, these classes of model applications are illustrated with a few exapmles.

Identification of Mechanisms of Physiological Significance

Identification of physiologically relevant mechanisms has been the purpose of the majority of all model applications. Models of this class typically are comparatively simple by including only those parts of the oxygen transport system which are relevant for the mechanisms under study, and are usually of more qualitative nature. The first model that has ever been developed for describing O_2 supply to tissue is a member of this class and was

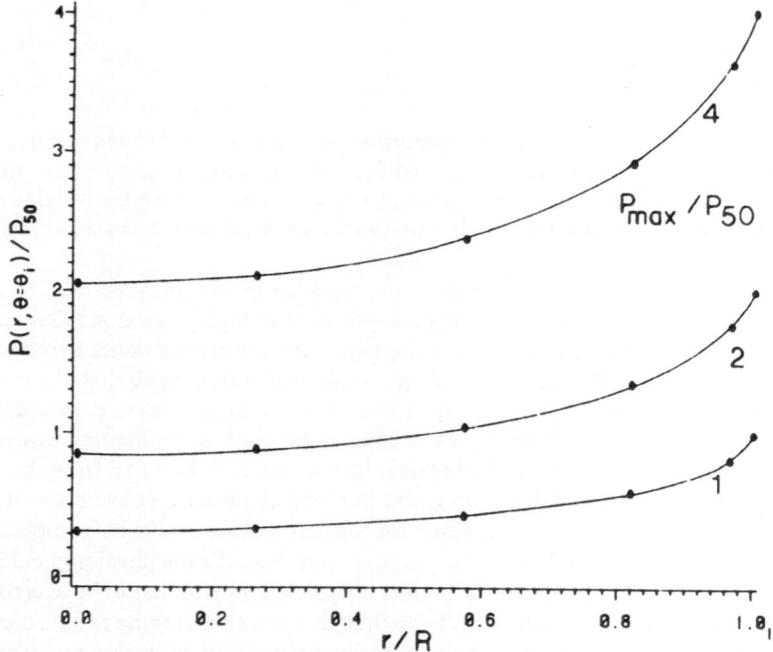

Figure 1. Calculated P_{O_2} profiles in a muscle fiber at $\dot{V}_{O_2} = 15$ ml $O_2 \cdot 100g^{-1} \cdot min^{-1}$ for different settings of P_{O_2} at the sarcolemma. Abscissa: Radial coordinate, given as fraction of muscle fiber radius R. Ordinate: P_{O_2} given as multiple of myoglobin P_{50} (which equals 5.3 mmHg). P_{O_2} drops within the muscle fiber are the smaller the lower P_{O_2} at the sarcolemma. From Federspiel (1986).

introduced by August Krogh (1918b, c) in order to demonstrate that in frog skeletal muscle oxygen transport from capillaries to the sites of consumption can be explained by diffusion of free oxygen. In a more recent example, Federspiel (1986) has simulated O_2 transport in the presence of myoglobin (Mb). He investigated the effects of Mb-O_2 reaction and diffusion on oxygen partial pressure (P_{O_2}) distributions in cross sections of heavily working muscle fibers supplied out of non-uniform sources at the sarcolemma. This model is much closer to the real situation but still concentrates on the muscle fiber and neglects capillary blood, capillary, and interstitial space. Among others, the calculations showed that in the presence of Mb-facilitated O_2 diffusion, P_{O_2} drops in the muscle fiber of down to 3.7 mmHg (or 38% the ones without Mb) may be sufficient for driving maximum O_2 flux, and that these P_{O_2} drops are the smaller the lower P_{O_2} at the sarcolemma (see Fig. 1).

Quantification of Basic Transport Parameters

There is a number of parameters which determine the functions of the various steps in the O_2 transport chain and which are important for understanding transport processes. Some of these parameters may easily be calculated from suitable experiments, *e.g.*, O_2 conductances in a flat tissue layer (*cf.* Krogh 1918a). However, there are other parameters which cannot easily be measured, like Mb diffusivities in tissue or reaction rates between oxygen and hemoglobin (Hb) under conditions of red blood cells. Finding numerical values of these parameters from appropriate *in vitro* experiments, the second class of models is designed for. These models need to give a *quantitative* description of the experimental situation which, in turn, can be controlled very precisely by manipulating geometry, boundary conditions, *etc.*, and which is typically set up in a way as to keep the model as simple as possible. Under these conditions, measuring values may be determined from the experiment as well as

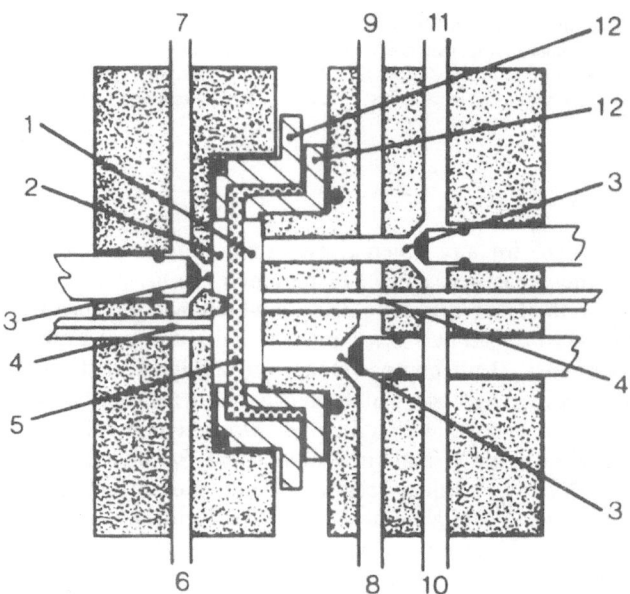

Figure 2. Cross section of diffusion chamber. 1, 2: Gas compartments. 4: P_{O_2} electrode. 5: Layer of material under study. 6, 7, 8–11: Inlet/outlet system with shutoff valves (3). From de Koning *et al.* (1981).

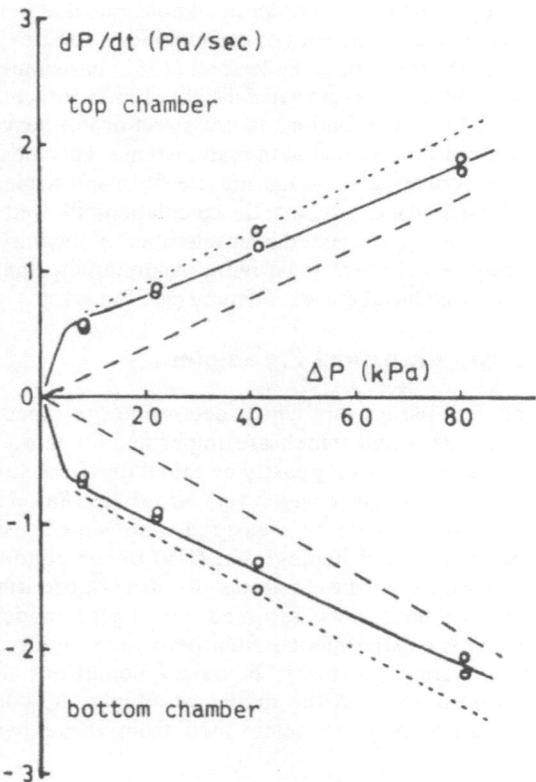

Figure 3. Results of measurements in diffusion chamber (circles), given as initial rate of P_{O_2} change, $\dfrac{dP}{dt}$, after shutting off gas flow to the gas compartments (*cf.* Fig. 2) for a membrane of Hb solution and for various P_{O_2} differences between the compartments (abscissa). From the least squares fits with an appropriate model function (solid curves), the unknown parameters, in this case Hb-O_2 reaction rates, may be determined. From Bouwer (1987).

predicted by the model for any given choice of unknown parameters. By fitting the model results to the measured data, the parameters may be calculated. Hoofd has developed many methods of this type one of which will be outlined briefly: A thin layer of the material to be studied is mounted in a diffusion chamber, by that dividing it in two compartments. Both compartments are equipped with a P_{O_2} electrode and with an inlet/outlet system through which their gas contents may be exchanged (Fig. 2). After equilibrating the system with gases of different composition in the two compartments, gas flow is stopped and time courses of P_{O_2} and hence of oxygen fluxes are recorded. Fig. 3 shows typical results of such measurements (circles), given as initial rate of P_{O_2} change, $\dfrac{dP}{dt}$, for a membrane of Hb solution separating the compartments and for various P_{O_2} differences between them (abscissa). From the least squares fits with an appropriate model function (solid curves), the unknown parameters, in this case Hb-O_2 reaction rates, may be determined. Other examples of this class include determination of O_2 consumption rate in multicellular tumor spheroids (Mueller–Klieser 1984) or assessment of diffusion coefficients by non-steady state methods (Groebe *et al.* 1994). A different approach for identifying system parameters is detailed in Bruley (1995).

Simulating "Unfeasible Experiments" in a Model

Assessment of the spatial P_{O_2} profiles in the vicinity of capillaries in heavily working red muscle – which are needed if, *e.g.*, the distribution of resistance to O_2 diffusion is to be analyzed (see below) – is an example for the class of "undoable experiments" which currently can be performed in a computer program only. At the present time, in working red muscle, experimental P_{O_2} profiles with the highest spatial resolution may be obtained by the cryophotometric method which has been used by Gayeski and Honig (1986, 1988) for collecting P_{O_2} maps in electrically stimulated dog gracilis muscles. In a muscle at maximum O_2 consumption rate ($\dot{V}_{O_2}$), P_{O_2} is very low and largely uniform (Gayeski and Honig 1986, 1988) due to carrier-facilitated O_2 diffusion (see above). Major P_{O_2} gradients are present only

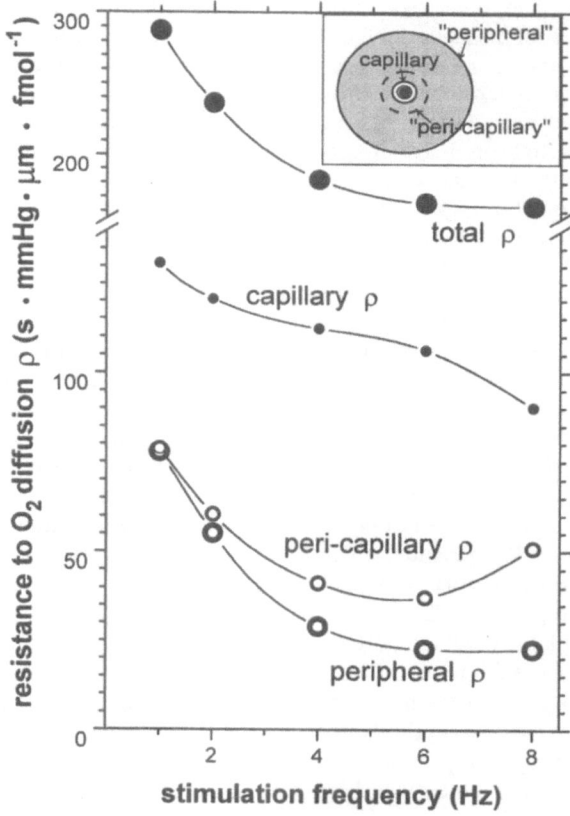

Figure 4. Total and partial resistances to O_2 diffusion, , in dog gracilis muscles stimulated at 1, 2, 4, 6, and 8 Hz (abscissa). The upper curve displays total resistance to O_2 diffusion; the curves below refer to the portions of the diffusion path which are outlined in the inset: The capillary region includes capillary blood and a carrier-free layer separating erythrocytes and sarcolemma. The peri-capillary region comprises the first one fifth of the diffusion path within the muscle fiber. The peripheral region is made up of the rest of the diffusion path. Even though the carrier-free region covers no more than 10 to 15% of total diffusion path length, capillary resistance accounts for about half of total resistance or more.

in a carrier-free layer (which separates red blood cells and muscle fiber, and which consists of a plasma sleeve, capillary endothelium, and interstitial space) and within several microns from a capillary. In this latter region, however, cryophotometry cannot furnish very accurate results because of contamination of the Mb spectra by the hemoglobin signal arising from the capillary. This gap can be bridged by using model calculations which employ the parameter values pertinent to the muscle under study and from which the P_{O_2} at locations close to capillaries may be determined.

System Analysis

If the various components of an O_2 transport system are known well enough, the system as a whole can be analyzed with respect to the mutual dependencies and interactions of its constituent subprocesses. While in the above classes of model applications the numerical results typically provide an immediate answer to the question assessed by the model, this is not true in the case of system analysis. Rather, the resulting distributions of P_{O_2}, O_2 fluxes, etc. need to be transformed in a way as to allow for a general characterization of the system and its components. To that end, one needs to devise concepts of how to approach such a characterization, one has to define quantities that are suitable to convey the desired information about the system, and one must compute their numerical values from the direct model results. Quantities of that type include parameters of the P_{O_2} frequency distribution, diffusing capacities or resistances to O_2 diffusion of the system as a whole or of ist components (which all may be defined on a macroscopic or on a microscopic scale), factors that describe the overall effect of Mb-facilitated O_2 diffusion on average P_{O_2} drops or on maximum $\dot{V}_{O_2}$, etc.

As an example, Fig. 4 shows an analysis of the distribution of O_2 transport resistances in dog gracilis muscles stimulated at 1, 2, 4, 6, and 8 Hz (abscissa; $\dot{V}_{O_2}$ = 3 to 15 ml $O_2 \cdot 100g^{-1} \cdot min^{-1}$). The upper curve displays total resistance to O_2 diffusion, . The lower tracings give partial resistances and refer to the regions which are sketched in the inset: The capillary region includes capillary blood and the carrier-free layer separating erythrocytes and sarcolemma. The peri-capillary region comprises the one fifth of the diffusion path within the muscle fiber that is located closest to the capillary, and the peripheral region is made up of the remaining 80% of the diffusion path. Even though the carrier-free region covers no more than 10 to 15% of total diffusion path length, capillary resistance accounts for about half of total resistance or more (which is due to absence of a carrier and to unfavourable diffusion geometry; see below). Similarly, resistance in a thin peri-capillary layer is larger than that in the rest of the fiber. This is because facilitation of O_2 transport by diffusion of Mb-O_2 is P_{O_2}-dependent: Facilitated transport is virtually absent at high P_{O_2}'s and may exceed diffusion of free oxygen by a factor of 5 at P_{O_2}'s close to 0 mmHg. Since in the peri-capillary region P_{O_2} is high, facilitation of O_2 transport by myoglobin is little effective which explains the large portion of total diffusion resistance located there. Moreover, P_{O_2}-dependence of facilitation renders muscle fiber resistances smaller with increasing performance because P_{O_2} drops between capillary and tissue grow larger (cf. Fig. 6) and tissue P_{O_2} decreases. An exception is peri-capillary resistance at maximum $\dot{V}_{O_2}$ which increases because capillary P_{O_2} (and hence also peri-capillary P_{O_2}) needs to be raised in order to drive the exceedingly high oxygen fluxes out of the capillary.

PROBLEMS IN QUANTITATIVE MODELLING

For performing "unfeasible experiments" or "system analysis", it is essential that the model not only qualitatively describes the physiological situation but that model results are also in reasonable quantitative agreement with reality. As a matter of fact, one could state that a physiological system may not be regarded as being fully understood unless any actually observed phenomena can quantitatively be reproduced by a mathematical model which employs all of and anything but the mechanisms known to be present and effective within the system or, in a word, unless the system is perfectly predictable.

Since any model calculations actually performed in a computer cannot include any more than a small section of the system studied, quantitative results clearly can be expected only if this section is representative for the entire system, *i.e.*, if the model is sufficiently comprehensive and if adequate geometric relations and parameter settings are chosen. This requirement poses severe problems some of which will be dealt with in the following. Topics to be addressed include:

- Identifying important mechanisms or factors that affect the function of the system,
- establishing descriptors which define the basic properties of the system and which are suitable for model input,
- accounting for heterogeneities in the system, and
- using specific data sets.

Mechanisms Affecting O$_2$ Transport

As a first example, a factor is discussed which helps to explain the extraordinarily high diffusion resistance of the carrier-free region mentioned above. In narrow muscle

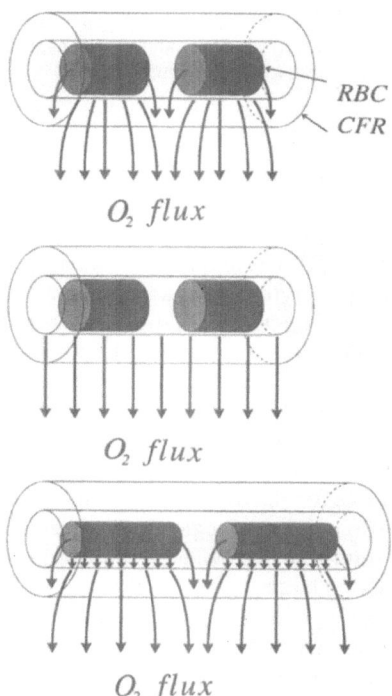

Figure 5. Model geometries for red cell O$_2$ unloading in capillaries. a: Realistic arrangement (oxygen flux (arrows) out of the erythrocyte in radial direction and partly also in longitudinal direction). b: Idealized models (O$_2$ flux out of the capillary assumed to be homogeneous and radial only). c: Geometry for halved red cell diameter (additional zone of high O$_2$ flux density (short arrows) surrounding erythrocyte).

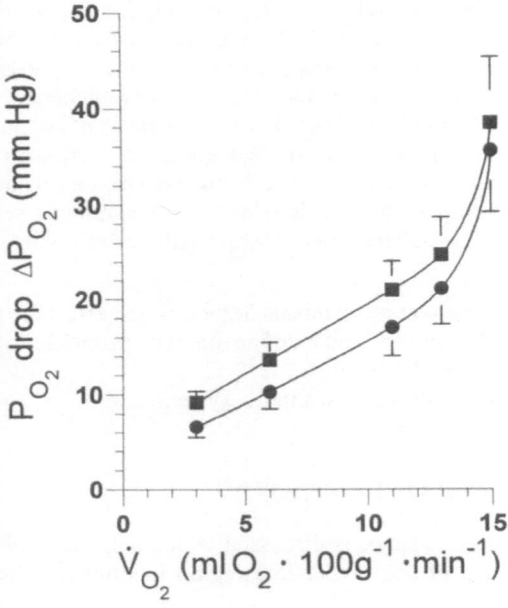

Figure 6. Effect of red cell spacing on average P_{O_2} drops between capillary and tissue (shown with standard deviations) for electrically stimulated dog gracilis muscles at O_2 consumption rates between 3 and 15 ml O_2 · $100g^{-1}$ · min^{-1}. Red cell spacing was considered for computing and neglected for . Assuming homogeneous O_2 flux may underestimate average P_{O_2} drops by up to 4 mmHg (or 28% of total P_{O_2} drop).

capillaries, the disk-shaped erythrocytes typically are folded up and take on an appearance similar to bullets or cylinders. These bullets are travelling in single file flow with interspersed plasma gaps (Fig. 5). As Hellums (1977) pointed out first, this has the consequence that only those capillary sites are maximally effective in O_2 exchange at which a red cell is present. At these locations, oxygen leaves erythrocyte and capillary mainly in radial direction. To a small portion, O_2 also diffuses in longitudinal direction across the "front" and "back" ends of the RBC, and from there through the inter-erythrocytic plasma gap out of the capillary (Fig. 5a). In former models of tissue O_2 supply, red cell spacing has mostly been neglected, and a homogeneous O_2 flux out of the capillary has been assumed, just as if the capillary were perfused with hemoglobin solution (Fig. 5b). The effects of this simplification are demonstrated in Fig. 6 which displays average P_{O_2} drops between capillary and tissue (, shown with standard deviations) for electrically stimulated dog gracilis muscles at performances ranging from very low to maximal. Neglecting red cell spacing ( ) may underestimate *average* P_{O_2} drops by up to 4 mmHg (or 28% of total P_{O_2} drop). *Maximum* differences between the two cases are almost twice as large.

Another example of an even more important mechanism in muscle O_2 transport is Mb-facilitated O_2 diffusion for which the average P_{O_2} differences between capillary and tissue are shown in Fig. 7. Surprisingly, P_{O_2} drops without facilitation (, solid, left ordinate) are hardly any larger than those with facilitation considered (, solid, left ordinate), even at highest performance. This is because P_{O_2} is very low in the bulk of the muscle fiber even with myoglobin present, so without facilitation P_{O_2} cannot decrease very much more. Rather, large regions fall anoxic. From the tracings of anoxic fiber portions, one realizes that without myoglobin (,dashed, right ordinate) anoxia starts to occur at 4 Hz stimulation rate and

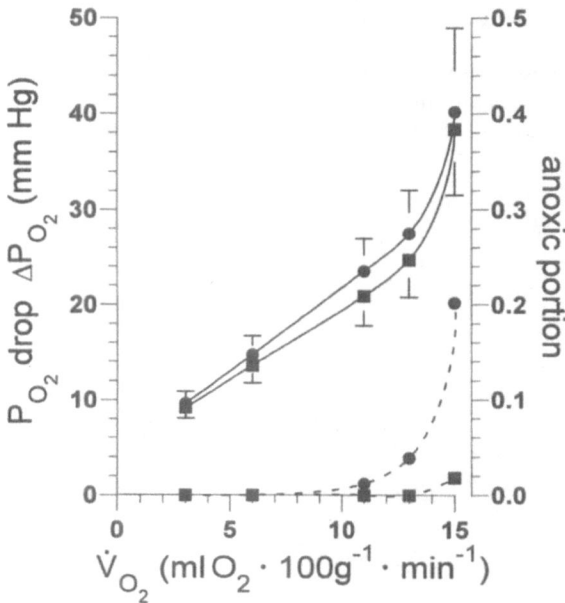

Figure 7. Effect of Mb-facilitated O_2 diffusion on average P_{O_2} drops between capillary and tissue (solid, left ordinate; shown with standard deviations) and on anoxic tissue portions (dashed, right ordinate) for electrically stimulated dog gracilis muscles at O_2 consumption rates between 3 and 15 ml $O_2 \cdot 100g^{-1} \cdot min^{-1}$. Mb-facilitated O_2 diffusion was considered for computing and neglected for . Without facilitation, anoxic tissue portions may become as large as 20% of the muscle fiber.

becomes as large as 20% of the tissue at maximum performance at which normally only 2% anoxia would be expected (, dashed, right ordinate).

Establishing Suitable Descriptors of System Properties

For quantitative modelling, precise data specifying the basic properties of the system under study are needed. Obviously, this requirement may give rise to major technical difficulties, in particular if dynamic quantities such as capillary blood flows are to be measured. In addition, there are other and potentially even more intricate problems which are less obvious: For some properties of the system, it may not be clear at all in which parameters they are adequately reflected. The capillarization in muscle tissue, *e.g.*, is classically described by functional (*i.e.* perfused) capillary density. From Potter and Groom (1983) it is known, however, that capillaries are not straight tubes but rather take a meandering course. Therefore, the capillary is longer than the line connecting capillary origin and end – by about 25% in unstretched muscles (Kayar *et al.* 1994) –, and the O_2 flux unloaded from red cells per unit length of capillary and the P_{O_2} drops necessary to drive this O_2 flux are accordingly smaller. Since functional capillary density does not reflect this phenomenon, alternatives for describing capillarisation have been proposed (a discussion of which is beyond the scope of the present article), *e.g.*, combining functional capillary density *and* tortuosity, employing aggregate capillary length or capillary surface, or disregarding capillarization altogether and applying a Hill-type solid cylinder model (*cf.* Ellis *et al.* 1983, Mathieu-Costello *et al.* 1991).

Accounting for Heterogeneities

Heterogeneities in the various input parameters present one of the most complicated unresolved problems in quantitative modelling of O_2 transport. However, this is
not primarily due to lack of suitable models, at least not in muscle tissue There are a

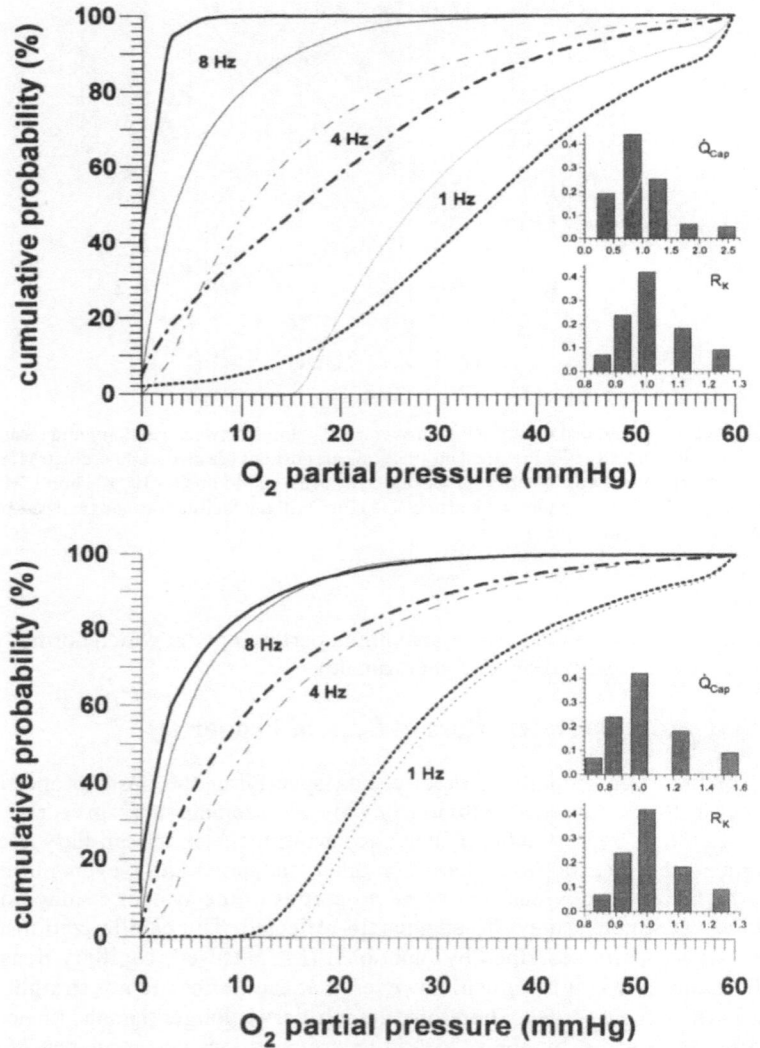

Figure 8. Calculated cumulative P_{O_2} probability distributions in dog gracilis muscle stimulated at 1, 4, and 8
Hz (curve labels) for homogeneous (hair lines) and heterogeneous (bold curves) capillary densities and
perfusions. Upper panel: Calculations for heterogeneous data are based on unrelated measured distributions
of capillary density and of capillary perfusion. Lower panel: Distribution of capillary density is the same as in
upper panel, but capillary perfusion is proportional to capillary domain size. Insets: Distributions of capillary
perfusion Q_{cap} and of capillary domain radius R_K employed for the heterogeneous cases. While P_{O_2} distributions for the homogeneous and the heterogeneous cases are entirely dissimilar in the upper panel, good
agreement is found in the lower one.

number of quite elaborate models, for example by Clark *et al.* (1989), by Hoofd *et al.* (1989), or by Groebe (1990) that can be used to assess heterogeneity effects, if appropriate data quantifying heterogeneities are available. This latter condition, though, usually is not satisfied because analogous and even more arduous difficulties apply than for determination of individual parameters: *(i)* Presently, there is no conclusive set of descriptors which are suitable for quantifying tissue heterogeneity. These descriptors first need to be developed – a task which by no means is trivial because in many cases heterogeneity cannot satisfactorily be quantified by just measuring the frequency distribution of the respective parameter values: One and the same probability distribution of functional capillary densities, *e.g.*, may have very different impacts on O_2 supply, depending on capillary arrangement. If large and small capillary domains are "well mixed" within the tissue there will be at least partial compensation of the differences by diffusion of O_2 from well to poorly supplied domains. If, *vice versa*, there is clustering of large domains and of small domains, the potential for compensation is very limited due to long diffusion distances, and patches of "oversupplied" and of "undersupplied" tissue may result. Hence, this kind of neighbourhood relationships has to be accounted for in descriptors of capillarisation heterogeneity. For quantification of tissue heterogeneities, *cf.* also Turek *et al.* 1987, Egginton and Ross 1989, Batra and Rakusan 1992. *(ii)* The technical problems in measuring *distributions* of dynamic quantities are even more serious than for mean parameter values. This is particularly true, since *(iii)* some essential information is still lacking even if the individual frequency distributions for all parameter values are known: For given distributions of capillary domain size and capillary blood flow, for example, it makes a major difference if in an arbitrary capillary domain the actual values of domain size and blood flow are the result of random combination, or if they are dependent on each other (*e.g.*, if capillary blood flow is proportional to domain size; see below). Hence, for modelling, the *joint* distributions of capillary blood flows, capillary domain cross sectional areas, *etc.* are needed instead of their individual ones.

Notwithstanding the present lack of appropriate input data, it is quite informative to study some heterogeneity effects in a model. For the calculations, a model composed of a number of individual capillary cylinders has been used which tends to overestimate P_{O_2} heterogeneity. Fig. 8 shows calculated cumulative P_{O_2} probability distributions in dog gracilis muscle stimulated at 1, 4, and 8 Hz (curve labels) for homogeneous (hair lines) and matched heterogeneous (bold curves) input data. In the upper panel, computations for heterogeneous data were based on *unrelated* measured distributions of capillary density by Honig *et al.* (1980) and of capillary perfusion by Klitzman and Johnson (1982) which are represented by the histograms in the insets. Evidently, the P_{O_2} distributions for the homogeneous and the heterogeneous cases are entirely dissimilar. According to the latter results, 2% of the tissue would be anoxic even at very light work, whereas in homogeneous tissue at the same performance, no P_{O_2} values below 15 mmHg are predicted. At maximum performance, anoxia would even be present in 38% of the heterogeneous muscle which is entirely unrealistic (Gayeski *et al.* 1987). The lower panel shows another type of extreme combination of capillary domain size and perfusion in which the same distribution of capillary density and the same mean capillary blood flow have been assumed: Local capillary perfusion was chosen in proportion to capillary domain size, so the volume related perfusion rate is the same in each capillary domain. Resulting P_{O_2} distributions for the homogeneous and heterogeneous cases resemble each others surprisingly well. These distributions are also very similar to the ones measured by Gayeski and Honig (1988) or by Gayeski *et al.* (1987). This latter observation suggests that locally capillary perfusion rate should be extraordinarily well matched to capillary domain size in working muscle.

Using Specific Data Sets

Apart from inhomogeneity present within an organ, there is considerable inter-individual variability in the data needed for model input which is the cause for another kind of complication in quantitative modelling: Depending on the nature of the respective parameter, it may be sufficient if a value is used that has been determined in a different species of the same genus or even class. For other parameters, values measured under identical experimental conditions in the same organ of two individuals of the same species may yield completely divergent model results. In order to illustrate this problem with a practical example, P_{O_2} measurements in dog gracilis muscle stimulated at 6 Hz have been simulated in the computer. The resulting P_{O_2} distributions are compared to experimental ones by Gayeski and Honig (1988) which were obtained in muscles stimulated at 4 and at 6 Hz (Fig. 9, thin dashed curves). The bold solid distribution which has been calculated using data that were specific for the latter experimental muscle approximates the measured distribution very well. For computing the bold dashed distribution, identical data were used, except for functional capillary density which had been determined earlier by the same group (Honig *et al.* 1982) and which represents the mean value for six dog gracilis muscles, that were also stimulated at 6 Hz. Evidently, for this choice of input data there is a striking difference between measured and calculated P_{O_2} profiles. As a matter of fact, the latter one is quite similar to the P_{O_2} distribution determined in a muscle which was stimulated at 4 Hz and the O_2 consumption rate of which is more than 25% less. In consequence, functional capillary density is one of the parameters that need to be chosen specifically for the individual under study in order to yield quantitative agreement between model and experiment.

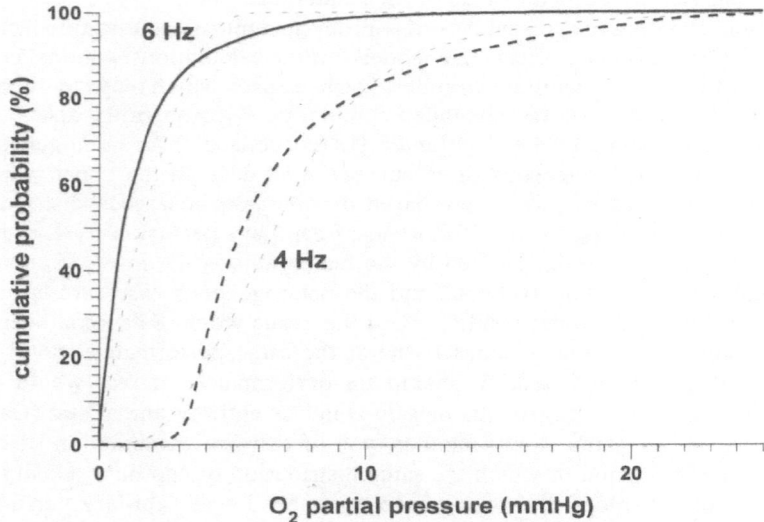

Figure 9. Comparison between experimental cumulative P_{O_2} probability distributions in dog gracilis muscles (dashed hair lines) stimulated at 4 or 6 Hz, respectively, (curve labels) and distributions calculated for muscles at a twitch rate of 6 Hz (bold curves). Bold solid curve: Data specific for the corresponding experimental muscle were employed. Bold dashed curve: Same data were used except that for functional capillary density a mean value obtained in six individuals under identical experimental conditions was substituted. Good agreement between model and experiment is found only for the calculations based upon specific data.

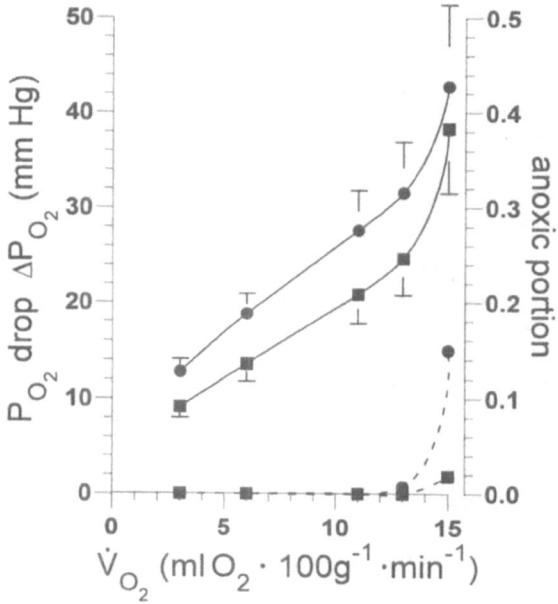

Figure 10. Effect of halving intra-capillary red cell diameter on calculated average P_{O_2} drops between capillary and tissue (solid, shown with standard deviations) and on anoxic tissue portions (dashed) for electrically stimulated dog gracilis muscles at O_2 consumption rates between 3 and 15 ml $O_2 \cdot 100g^{-1} \cdot min^{-1}$. Assumed red cell diameter was 4 m for and 2 m for . At halved red cell diameter, anoxic tissue portions may become as large as 15% of the muscle fiber.

As a final illustration of the requirement for specific input data, intra-capillary diameter of the folded, bullet-shaped red blood cells is addressed. Similar to capillary perfusion, red cell shape in capillaries is difficult to quantify since it is a dynamic parameter which is influenced by the actual microscopic pressure and shear rate distributions in the vessel. To appreciate the importance of red cell diameter, recall that in muscle the carrier-free region represents the site of major resistance to O_2 diffusion and of the largest P_{O_2} gradients along the diffusion path. The steep gradients are necessary to drive the extremely high O_2 flux densities in the vicinity of red blood cells. (Note that all of the oxygen consumed in the entire fiber must pass throug this tiny region!) This is particularly true because an O_2 carrier is lacking. If one assumes a smaller red cell diameter (Fig. 5c) one implicitly adds an extra layer to the carrier-free region in which O_2 flux density is even higher (short arrows) because of its smaller area normal to the direction of the flux. Therefore, changes in red cell diameter affect diffusion resistance at a location at which P_{O_2} drops and tissue P_{O_2} distributions are the most sensitive to it. This is demonstrated in Fig. 10 by means of calculated average P_{O_2} drops between capillary and muscle fiber for the same range of O_2 consumption rates used before (abscissa). Squares represent the results for red cell dimensions as measured by Carl Honig [personal communication], whereas for the circles half this red cell diameter has been employed. Differences in mean P_{O_2} drops are pronounced and amount to as much as 7 mmHg. More significantly, however, the anoxic tissue fraction at maximum performance is increased from 2 to 15%.

CRITERIA FOR QUANTITATIVE MODELS OF O_2 TRANSPORT AND FOR INPUT DATA

Based upon the above considerations and similar ones, in this last section some requirements are proposed which should be met in practical, tentatively quantitative applications of oxygen transport models and which are aimed at minimizing the loss of information and significance in the transfers from physiological data to model calculations and back to interpretations of the physiological system. Quantitative models should be

- *representative* of the system modelled by including a sufficiently large representative section of the tissue or, if necessary, a representative weighted sample of such sections,
- as *complete* as necessary by considering all of the factors and mechanisms that are known to play a major role in the transport processes studied,
- of as *little complexity* as possible to allow for maximum transparency and easy handling, and to facilitate interpretation and application of model results,
- of practical usefulness by presenting the findings of model calculations in terms of *applicable concepts* and *suitable derived quantities* which contribute to a deeper understanding of the physiological system under study, and
- designed to allow for their *verification* by comparison of model results to measured data.

These demands on mathematical models correspond to requirements which input data have to meet. Data sets specifying the basic properties of the system to be modelled ought to be

- *representative* by accounting for intra- and inter-individual heterogeneities as well as by covering a spectrum of functional states of physiological interest,
- *consistent* with established physiological and physical laws and regularities like Fick's principle,
- sufficiently *specific* in that they may – at least potentially – all be measured in one and the same individual at a time and under given experimental conditions,
- *redundant* by supplying data values which are expressive regarding the actual state of the system, and which are not needed as input data for modelling and therefore may be used for verifying the model.

Particularly for quantifying model input parameters and their heterogeneities, close cooperation between experimentalists and theoreticians will be needed in order to establish suitable descriptors for the basic properties of the system that extract the desired information in a form appropriate for model input on the one hand and that can be measured with acceptable effort on the other.

REFERENCES

Batra, S. and Rakusan, K., 1992, Capillary length, tortuosity, and spacing in rat myocardium during cardiac cycle, *Am.J.Physiol.* 263:H1369–H1376

Bouwer, S., 1987, Facilitated oxygen diffusion through hemoglobin solutions. Measurement of diffusion and reaction parameters, Doctoral Thesis, University of Nijmegen

Bruley, D.F., 1995, *Empirical modeling for oxygen transport processes and related physiological systems*, this volume

Clark, P.A.A., Kennedy, S.P., and Clark, A., 1989, Buffering of muscle tissue P_{O_2} levels by the superposition of the oxygen field from many capillaries, *Adv.Exp.Med.Biol.* 248:165–174

Egginton, S. and Ross, H.F., 1987, Quantifying capillary distribution in four dimensions. *Adv.Exp.Med.Biol.* 248:271–280

Ellis, C.G., Potter, R.F., and Groom, A.C., 1983, The Krogh cylinder geometry is not appropriate for modelling O_2 transport in contracted skeletal muscle, *Adv.Exp.Med.Biol.* 159:253–268

Federspiel, W.J., 1986, A model study of intracellular oxygen gradients in a myoglobin-containing skeletal muscle fiber, *Biophys.J.* 49:857–868

Gayeski, T.E.J. and Honig, C.R., 1986, O_2 gradients from sarcolemma to cell interior in red muscle at maximal V_{O_2}, *Am.J.Physiol.* 251:H789–H799

Gayeski, T.E.J., Connett, R.J., and Honig, C.R., 1987, Minimum intracellular P_{O_2} for maximum cytochrome turnover in red muscle in situ, *Am.J.Physiol.* 252:H906–H915

Gayeski, T.E.J., Honig, C.R., 1988, Intracellular P_{O_2} in long axis of individual fibers in working dog gracilis muscle, *Am.J.Physiol.* 254:H1179–H1186

Groebe, K., 1990, A versatile model of steady state O_2 supply to tissue. Application to skeletal muscle, *Biophys.J.* 57:485–498

Groebe, K., Erz, S., and Müller–Klieser, W., 1994, Glucose diffusion coefficients determined from concentration profiles in EMT6 tumor spheroids incubated in radioactively labeled l-glucose, *Adv.Exp.Med.Biol.*, in press

Hellums, J.D., 1977, The resistance to oxygen transport in the capillaries relative to that in the surrounding tissue, *Microvasc.Res.* 13:131–136

Honig, C.R., Odoroff, C.L., and Frierson, J.L., 1980, Capillary recruitment in exercise: rate, extent, uniformity, and relation to blood flow, *Am.J.Physiol.* 238:H31–H42

Honig, C.R., Odoroff, C.L., and Frierson, J.L., 1982, Active and passive capillary control in red muscle at rest and in exercise, *Am.J.Physiol.* 243:H196–H206

Hoofd, L., Turek, Z., and Olders, J., 1989, Calculation of oxygen pressures and fluxes in a flat plane perpendicular to any capillary distribution, *Adv.Exp.Med.Biol.* 248:187–196

Kayar, S.R., Hoppeler, H., Jones, J.H., Longworth, K., Armstrong, R.B., Laughlin, M.H., Lindstedt, S.L., Bicudo, J.E.P.W., Groebe, K., Taylor, and C.R., Weibel, E.R., 1994, Capillary blood transit time in muscles in relation to body size and aerobic capacity, *J.Exp.Biol.*, in press

Klitzman, B. and Johnson, P.C., 1982, Capillary network geometry and red cell distribution in hamster cremaster muscle, *Am.J.Physiol.* 242:H211–H219

de Koning, J., Hoofd, L.J.C., and Kreuzer, F., 1981, Oxygen transport and the function of myoglobin, *Pflügers Arch.* 389:211–217

Krogh, A., 1918a, The rate of diffusion of gases through animal tissues with some remarks on the coefficient of invasion, *J.Physiol.(London)* 52:391–408

Krogh, A., 1918b, The number and distribution of capillaries in muscles with calculations of the oxygen pressure head necessary for supplying the tissue, *J.Physiol.(London)* 52:409–415

Krogh, A., 1918c, The supply of oxygen to the tissues and the regulation of the capillary circulation, *J.Physiol.(London)* 52:457–474

Mathieu–Costello, O., Ellis, C.G., Potter, R.F., MacDonald, I.C., and Groom, A.C., 1991, Muscle capillary-to-fiber perimeter ratio: morphometry, *Am.J.Physiol.* 261:H1617–H1625

Müller–Klieser, W., 1984, Method for determination of oxygen consumption rates and diffusion coefficients in multicellular spheroids, *Biophys.J.* 46:343–348

Potter, R.F. and Groom, A.C., 1983, Capillary diameter and geometry in cardiac and skeletal muscle studied by means of corrosion casts, *Microvasc.Res.* 25:68–84

Turek, Z., Hoofd, L., and Rakusan, K., 1987, A comparison of the methods for assessment of the heterogeneity of myocardial capillary spacing, *Adv.Exp.Med.Biol.* 215:13–19

21

THE EFFECT OF CAPILLARY BLOOD FLOW ON THE OXYGEN RELEASE INTO RAT HEART TISSUE

Model Calculations with Point-like Sources Representing the Erythrocytes

Cees Bos,[1] Louis Hoofd,[1] and Thom Oostendorp[2]

[1] Department of Physiology
[2] Department of Medical Physics and Biophysics
University of Nijmegen
P.O. box 9101, 6500 HB Nijmegen, The Netherlands

INTRODUCTION

Modelling of oxygen transport into tissue can be considered in two steps. The first step comprises the release from the oxygen carriers, the erythrocytes, in the streaming blood up to the capillary wall. The second step models the diffusional transport in the - stagnant - tissue, including the capillary wall. Many literature models only involve the second step. The first model by Krogh (1919) used the simplified geometry of a tissue cylinder around a centrally located capillary. Since, several extensions have been made to this model establishing the basis of calculation of pO_2 in muscle tissue (Hoofd, 1992).

While modelling of oxygen diffusion in the tissue is quite straightforward, the capillary release poses considerable problems. We developed analytical methods (Bos et al., 1994; Hoofd, 1992) allowing an estimation of its effect on tissue pO_2 in various circumstances. To this end, pO_2 was calculated using two models only different in one respect, using homogeneous blood and separate erythrocytes respectively. In these two models, the boundary conditions were chosen such that tissue pO_2s at locations distant from the capillary became the same. The resulting difference in initial boundary condition, homogeneous blood pO_2 or erythrocyte pO_2, was defined as Extraction Pressure (EP). Consequently, EP quantifies the effect of the particulate nature of blood on tissue pO_2. This model is also valid when the tissue contains myoglobin, but the resulting difference in tissue pO_2 may be smaller. The pO_2 drop can be quantified using the concept of the facilitation pressure (Hoofd, 1992).

The two former models (Bos et al., 1994; Hoofd, 1992) were combined and extended to account for erythrocyte movement in the capillary (Hoofd et al., submitted). The present model only handles evenly spaced erythrocytes all moving with the same speed. Again,

Oxygen Transport to Tissue XVII, Edited by Ince et al.
Plenum Press, New York, 1996

177

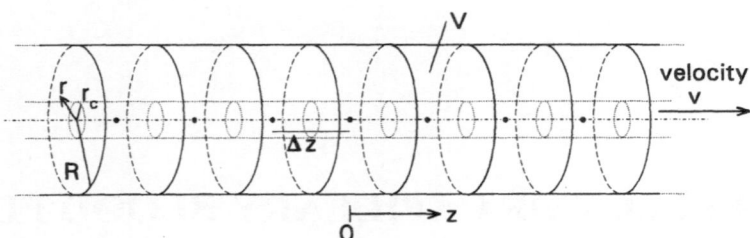

Figure 1. Cylindrical coordinate system $\vec{r}$ =(r,z) with capillary of radius r_c containing point-like oxygen sources (black dots) with equidistant spacing Δz all moving along the z-axis with constant velocity v. In a surrounding tissue cylinder of radius R, each source distributes its oxygen into an equal amount of tissue, of volume V.

results can be presented in terms of extraction pressures, where it is most interesting now to look at the differences with the former models.

Theory

Assumptions and basic equations

Diffusional transport of oxygen is considered in a cylindrical lay-out, with coordinates (r,ϕ,z). The angle coordinate ϕ will be dropped because we assume cylindrical symmetry, so $\vec{r}$ = (r,z). Along the z axis, point-like O_2 sources are located at even distances Δz moving with velocity v along the axis (Figure 1). The point sources represent erythrocytes moving in a capillary of radius r_c. Tissue oxygenation will be considered in a concentric tissue cylinder of radius R so that each point source supplies a tissue volume $V=\pi R^2\Delta z$. In the figure, such volumes are shown separated by circular cross-sections with the point sources as black dots in their centers. The origin of the z axis is chosen at one of these source locations so that the ith source is located at $\vec{r_i}$ = $(0,z_i)$ where $z_i=i\Delta z$.

The O_2 transport equations in the tissue are for combined diffusion and mass balance:

$$\frac{\partial c}{\partial t} - \wp\nabla^2 p = -\dot{Q} \tag{1}$$

where c is oxygen concentration, $\wp$ is oxygen permeability of the tissue (product of oxygen diffusion coefficient D and oxygen solubility α), p is oxygen partial pressure, and $\dot{Q}$ is oxygen consumption per tissue volume. The concentration is proportional to pressure according to Henry's law, c = αp. In former treatments (Bos et al., 1994; Hoofd, 1992), these equations were extended to account for myoglobin (Mb) in the tissue. Here, in a time-dependent system, Mb poses problems in particular because the reaction with O_2 is not infinitely fast and consequently interferes with time-dependent diffusion. This cannot easily be modelled. Close to the capillary-tissue interface, pO_2 mostly is high enough to keep Mb fully saturated, avoiding significant chemical reaction with O_2. Therefore, we will neglect the myoglobin in the treatment here for the calculation of EP. Further on in the tissue, outside this capillary-tissue boundary region, time-dependent phenomena quickly damp out and the influence of Mb might be incorporated again through EP as done before (Bos et al., 1994).

Solutions

In a previous paper (Bos et al., 1994) we used a "stroboscope technique" to construct an analytical solution for O_2 diffusion into the tissue. With this technique, the erythrocytes were considered only at specific time intervals when each erythrocyte had moved to the exact position where its predecessor was in the former instance. Leaving out the myoglobin, the solution for tissue oxygen pressure p was described by:

$$p = p(r,z) = \frac{\dot{Q}}{4\Theta}\left\{ \Phi(\vec{r}) + \sum_{i=1}^{N} \frac{V}{\pi|\vec{r}-\vec{r_i}|} \right\}$$

(2)

where $\Phi(\vec{r})$ was a function called the "field term", of dimension r^2, and the sources were counted from 1 to N in a cylinder of limited length. The field term $\Phi(\vec{r})$ was a smooth function, dependent on tissue geometry but not on the individual sources whose contribution was accounted for in each of the sum terms. In the geometry here, the sources extend in both directions and their number will be extended to infinite so that after each time step $\Delta z/v$ exactly the same situation occurs as before. This allows a treatment in terms of Fourier analysis (Hoofd, 1992). Rewritten in this form, equation (2) becomes (Hoofd et al., submitted):

$$p(r,z) = \frac{\dot{Q}}{4\Theta}\left\{ \Phi_S(r,z) + R^2 f_S(r,z) - R^2 \ln(r^2/r_c^2) \right\}$$

(3)

where subscript S indicates "stroboscope solution" and:

$$f_S(r,z) = 4 \sum_{n=1}^{\infty} K_0(n\omega r)\cos(n\omega z)$$

(4)

where $\omega=2\pi/\Delta z$. The summation in terms of the modified Bessel function $K_0(\ )$ is finite because this $K_0(\ )$ quickly approaches zero when its argument grows large. The "stroboscopic" Fourier-solution can be extended to a time-dependent one conforming the primary equation (1):

$$p(r,z,t) = \frac{\dot{Q}}{4\Theta}\left\{ \Phi_T(r,z) + R^2 f_T(r,z,t) - R^2 \ln(r^2/r_c^2) \right\}$$

(5)

where now subscript T indicates "time-dependent solution" and:

$$f_T(r,z,t) = 4 \sum_{n=1}^{\infty} \Re e \left\langle K_0(\xi_n r)\exp(-in\omega[z-vt]) \right\rangle$$

(6)

where $\Re e\langle\ \rangle$ means the real part of the respective complex equation, involving $i=\sqrt{-1}$, and the ξ_n can be expressed in terms of a characteristic inverse length $\lambda=v/D$:

$$\xi_n = \sqrt{\tfrac{1}{2}n\omega\{\sqrt{n^2\omega^2+\lambda^2} + n\omega\}} + i\sqrt{\tfrac{1}{2}n\omega\{\sqrt{n^2\omega^2+\lambda^2} - n\omega\}}$$

(7)

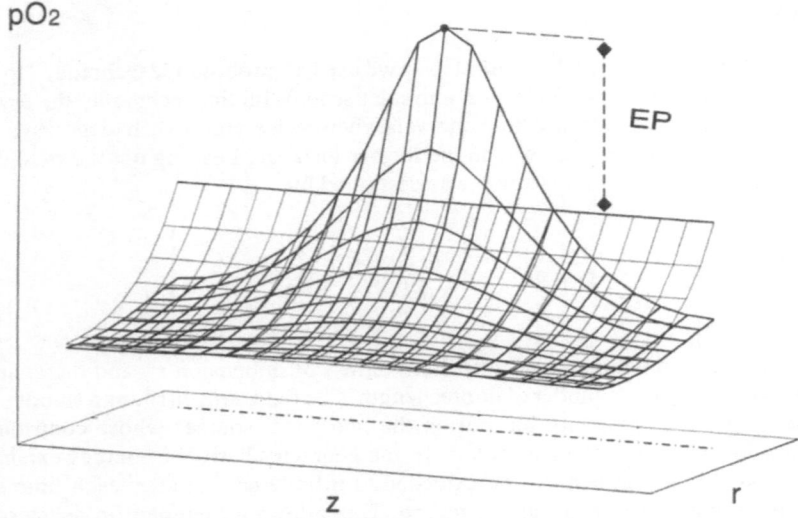

Figure 2. The extraction pressure by example of the condition of v/D = 0. Both the pO_2 profiles for the moving sources (thick lines) and the Krogh-formula (thin lines) are shown. The difference between the boundary condition at the top of moving sources profile and the Krogh profile is the extraction pressure.

Extraction Pressure

According to its definition, EP is obtained from comparison with a solution for homogeneous blood. The situation is depicted in figure 2. The thick grid is the pO_2 field around one of the sources and calculated according to the solution derived above, for a boundary condition with an erythrocyte pO_2 as indicated by the black dot. The thin grid is a solution for homogeneous blood leading to the same (average) pO_2 value at the outer edge of the cylinder. Note, that it starts at a lower capillary level than the erythrocyte pO_2. The difference in level is EP. For a cylinder, the homogeneous solution is the well-known Krogh equation (Krogh, 1919), here written as:

$$p_{(H)}(r,z) = p_c(z) + \frac{\dot{Q}}{4\mathcal{P}}\left\{r^2 - r_c^2 - R^2 \ln(r^2/r_c^2)\right\} \tag{8}$$

where the subscript (H) indicates that the solution is for homogeneous blood and $p_c(z)$ is the corresponding capillary pO_2. For large r, p from equation (5) must be equal to $p_{(H)}$ from equation (8). Note, that $K_0(\xi_n r)$ and thus $f_T(r,z,t)$ disappear for large r, so that:

$$\Phi_T(r,z) \approx \frac{4\mathcal{P}}{\dot{Q}}\, p_c(z) + r^2 - r_c^2 \tag{9}$$

The location (r_e,z_e) where the erythrocyte boundary condition must be taken - the black dot in figure 2 - is not trivial and r_e is not equal to the capillary radius r_c. E.g., for the stroboscope solution, r_e is derived from the erythrocyte volume $V_e = \frac{4}{3}\pi r_e^3$. EP is the difference between $p(r_e,z_e)$ and $p_c(z_e)$, and can be calculated from the combination of equations (5) and (9):

$$EP = \frac{\dot{Q}}{4\bar{Q}}\left\{ r_e^2 - r_c^2 + R^2 f_T(r_e,z_e,0) - R^2 \ln(r_e^2/r_c^2)\right\} \qquad (10)$$

The location (r_e,z_e) can be calculated from the implicit set of equations (Hoofd et al., subm):

$$\begin{cases} (r_e,z_e) = s_e\left(\cos(\alpha),\sin(\alpha)\right) \\ \lambda s_e = \dfrac{2\sin(\alpha)}{1 - \sin(\alpha)} \\ \zeta + \ln(\zeta) = \sin(\alpha) + \ln(\lambda s_e) \\ \lambda^3 V_e = \pi\zeta^2\left(e^{2\zeta} - 1 - \tfrac{2}{3}\zeta\right) \end{cases} \qquad (11)$$

RESULTS AND DISCUSSION

The effect of the movement of the sources on the oxygen release from the capillary can be demonstrated by the EP. To facilitate understanding of the effects of source movement the comparison of pO_2 profiles can be helpful. The data used for the calculation of EP and the pO_2 profiles are valid for rat heart and can be found in the literature. The parameters that we need are:

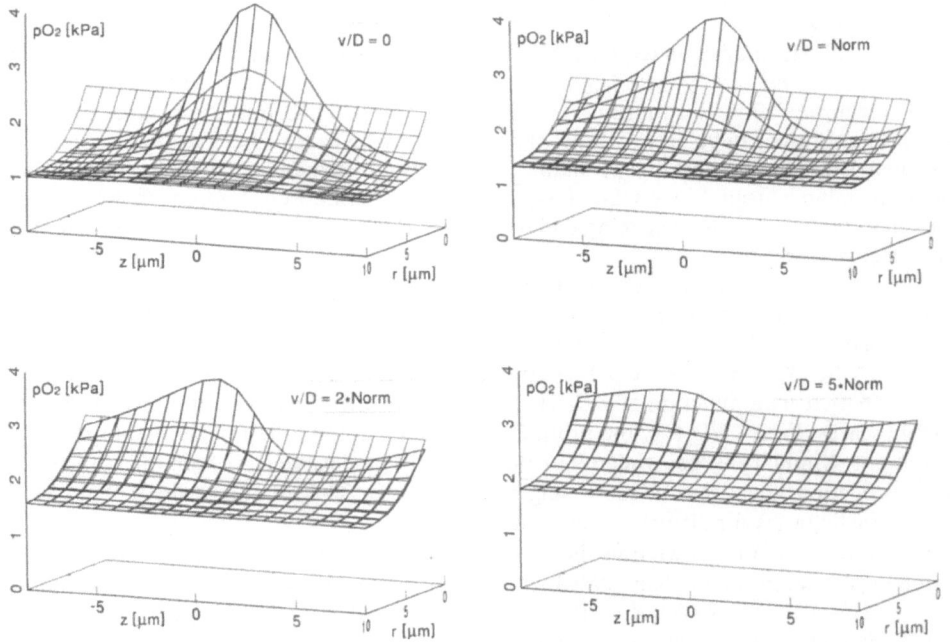

Figure 3. Profiles of pO_2 in the tissue according to the moving sources model (thick lines) and the Krogh model incorporating EP (thin lines), around a single source ($r_c \leq r \leq R$, $-\Delta z/2 \leq z \leq \Delta z/2$). $R = 10$ μm and $\Delta z = 16.9$ μm. Values for v/D are 0 μm^{-1}, normal (0.583 μm^{-1}), 2 times normal (1.167 μm^{-1}), and 5 times normal (2.917 μm^{-1})

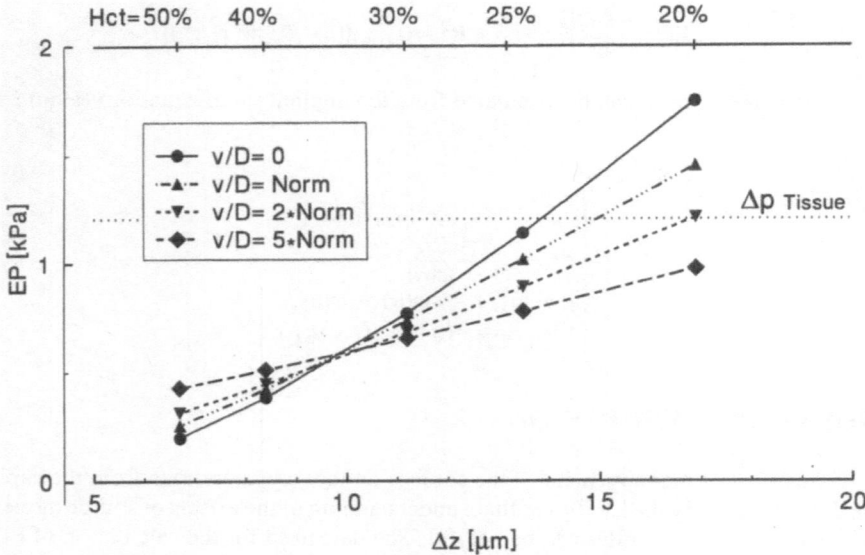

Figure 4. Extraction pressure plotted against source spacing (Δz) for $v/D = 0$ μm^{-1}, normal (0.583 μm^{-1}), 2 times normal (1.167 μm^{-1}), and 5 times normal (2.917 μm^{-1}). Hct values corresponding to Δz are indicated at the upper horizontal axis. For comparison, the pressure drop in pO_2 (Δp) in the second step of the transport, from the capillary wall to the tissue cylinder border, is shown, according to the Krogh formula.

$r_e = 2.44$ μm, calculated from $V_e = 61$ μm^2 (Altman et al., 1958), $r_c = 2.4$ μm and R $= 10$ μm (Turek et al., 1986), and $\dot{Q}/4\wp = 6.33$ $Pa/\mu m^2$ and $\alpha F/\wp = 10.56$ μm (Hoofd et al., 1990) for normal flow, leading to $v/D = (\alpha F/\wp)/(\pi r_c^2) = 0.583$ μm^{-1}. For various species, the coronary flow can be up to 5 times the normal flow in case of hard work (Honig, 1981; Lochner, 1971; Van Citters and Franklin, 1969). As an intermediate value, a v/D of 2 times the basic value is used. Since $v/D = 0$ is identical to the stroboscope approach, this value is considered too. The source spacing and hematocrit (Hct) can be transformed to each other, since Hct = $V_e/\pi r_c^2 \Delta z$.

Figure 3 shows pO_2 profiles in the tissue around a single source for different source velocities. For comparison pO_2 profiles are shown of a calculation with the Krogh formula with substraction of EP that result in the same pO_2 at the tissue cylinder border. In all four panels, an identical erythrocyte pO_2 is used. It is evident that the pO_2 at the tissue border (r $= R$) is an increasing function of v/D. The top of the profiles, however, diminishes when the v/D increases. When $v/D = 0$, the profile is symmetric around the top and the top is at z $= z_e$, but this symmetry is lost when $v/D > 0$ and the distance between the top location and the source location increases for increasing values of v/D. This is due to the differences in characteristic times for diffusion and source velocity.

In figure 2 it is shown how the EP can be calculated in case $v/D = 0$. It is close to the difference between the pO_2 calculated with the current model and that of the Krogh model at ($r = r_c$, $z = 0$). For $v/D > 0$ the EP cannot be calculated from the pO_2 profiles but should be calculated from equation (10). In figure 4 results of EPs calculated with this formula are shown as a function of the source spacing. At low hematocrit the transport of oxygen into the tissue is enhanced by source movement, since the EP is a decreasing function of increasing Δz. For high hematocrit the EP increases with increasing v/D. This negative effect of source movement occurs when the source enters the pO_2 top of the preceding source. The

pO_2 drop further afield, in the tissue, can be calculated with the Krogh formula and is also shown in figure 4. Although the EP depends on the spacing, the EP often equals the tissue pressure drop Δp, since in capillaries the hematocrit is low. It is evident, at least for rat heart, that the particulate nature of blood can be quite important for the oxygen transport into tissue. The importance can be estimated by the EP, and in this way it can also be used to account for the effect of the particulate nature of blood in models that use a homogeneous treatment of blood.

The present model can be used to investigate the effect of erythrocyte spacing and source movement on the oxygen transport to tissue from capillaries and to calculate a single determinant of this effect: the extraction pressure. An increase of sources spacing reduces the average pO_2 in the tissue and increases the EP, but an increase in spacing can be compensated for by an higher source velocity.

REFERENCES

Altman, P.L., Gibson, J. F., and Wang, C. C., 1958, "Handbook of Respiration", Eds. Dittmar, D. S. and Grebe, R. M., Saunders company, Philadelphia and London

Bos, C., Hoofd, L., and Oostendorp, T., 1994, Mathematical model of erythrocytes as point-like sources, *Math.Biosci.* (in press).

Honig, C. R., 1981, Modern cardiovascular physiology, Little, Brown and company, Boston, pp 347

Hoofd, L., 1992, Updating the Krogh model - assumptions and extensions, *In: Oxygen Transport in Biological Systems, Soc. Exper. Biol. Seminar Series 51*, S. Egginton, and H.F. Ross, eds, Cambridge University Press, Cambridge, pp 197-229.

Hoofd, L., Bos, C., and Oostendorp, T., (subm), The effect of blood flow on oxygen extraction pressures calculated in a model of point-like erythrocyte sources for rat heart, *Math.Biosci.* (submitted).

Hoofd, L., Bos, C., and Turek, Z., 1994, Modelling erythrocytes as point-like O_2 sources in a Kroghian cylinder model, *In: Oxygen Transport to Tissue XV, Adv. Exper. Med. Biol. 345*, eds. P. Vaupel, R. Zander, and D.F. Bruley, Plenum Press, New York & London: 893-900.

Hoofd, L., Olders, J., and Turek, Z., 1990, Oxygen pressures calculated in a tissue volume with parallel capillaries, *In: Oxygen Transport to Tissue XII, Adv. Exper. Med. Biol. 277*, eds. J. Piiper, T.K. Goldstick, and M. Meyer, Plenum Press, New York & London: 21-29.

Krogh, A., 1919, The number and distribution of capillaries in muscles with calculations of the oxygen pressure head necessary for supplying the tissue, *J. Physiol. 52*: 409-415.

Lochner, W., 1971, Herz, in: "Physiologie de Kreislaufs I.", Ed. E. Bauereisen, Springer-Verlag, Berlin-Heidelberg-New York

Turek, Z., L. Hoofd, and K. Rakušan, Myocardial capillaries and tissue oxygenation, *Can. J. Cardiol.* 2: 98-103 (1986).

Van Citters, R. L., and Franklin, D. L., 1969, Cardiovascular performance of Alaska sled dogs during exercise, *Circ. Res.*, 24:33-42

22

MYOCARDIAL OXYGENATION AND CONTRACTILE FUNCTION DURING GRADED REDUCTION OF CORONARY FLOW

Eiichiro Imamura,[1,2] Akira Kitabatake,[2] and Mamoru Tamura[1]

[1] Research Institute for Electronic Science
[2] Department of Cardiovascular Medicine
Hokkaido University
Sapporo 060, Japan

INTRODUCTION

In the isolated perfused heart, several studies have shown that there is a linear relationship between coronary flow and myocardial contractile function (1-7). The severe reductions of coronary flow decreases high energy phosphate compounds and increases lactate production. In contrast, the mild reductions cause only reduction in contractile function but not high energy compounds (2,4,5,7,8). Under such conditions it is supposed that reduced energy consumption (oxygen demand) balances the reduced oxygen supply (4,6).

The purpose of the present study is to determine the relationships between cardiac function and myocardial oxygenation during graded reductions of coronary flow and to compare the results with those of previous hypoxic perfusion studies (9-11).

METHOD

Male Wistar rats (200-230 g) were anesthetized with pentobarbital. The hearts were excised rapidly and perfused retrogradely through aortic cannula. The perfusate was a modified Krebs–Henselite buffer, equilibrated with 95% O_2 + 5% CO_2 at pH 7.4, and its temperature was maintained at 27°C. Flow was held constant by a peristaltic pump. A thin latex balloon, attached via water-filled polyethylene tubing to a pressure transducer, was inserted into the left ventricle and was kept isovolumic. The left ventricular developed pressure and heart rate were continuously monitored.

Light was guided on to the surface of the left ventricle of the perfused heart with a flexible light guide and dual-wavelength reflectance spectrophotometry was used to measure the oxygenation of myoglobin (580-617 nm) and the redox state of cytochrome-aa3 (605-617 nm) (9,12).

Oxygen Transport to Tissue XVII, Edited by Ince et al.
Plenum Press, New York, 1996

185

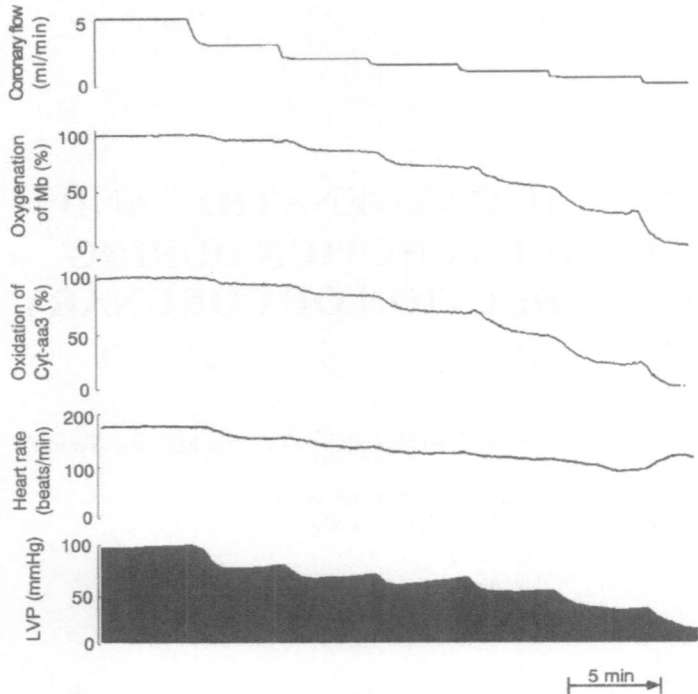

Figure 1. Changes in various cardiac performances of perfused heart. Coronary flow was reduced at 5, 3, 2, 1.5, 1, 0.5 and 0 ml/min by a peristaltic pump and each flow was kept constant for 5 min.

The flow rate was reduced stepwisely by a peristaltic pump. Each flow rate was kept constant for 5 min.

RESULT

Figure 1 shows a typical recording of the experiments. Coronary flow rate was reduced at 5, 3, 2, 1.5, 1 and 0.5 ml/min by a peristaltic pump and then flow was stopped. As the flow rate was reduced, left ventricular peak systolic pressure and heart rate were decreased. The degrees of the oxygenation of myoglobin and the oxidation of cytochrome-aa3 decreased in parallel, both of which reached 0% within 2 minutes when flow was stopped.

Figure 2 shows the effect of coronary flow rate on the rate-pressure product (left ventricular developed pressure x heart rate) and the oxygenation of myoglobin. Data were obtained from 6 hearts. Although the rate-pressure product decreased linearly with flow rate, the oxygenation state of myoglobin was higher than 96% until the flow rate was reduced to 2 ml/min, which started to decrease below 2 ml/min. Figure 2 indicates that the myocardium becomes oxygen deficient at flow rates below 2 ml/min and, therefore, indicates that the decrease in the rate-pressure product at flow rate above 2 ml/min is not due to oxygen deficiency.

The changes of the oxygenation of myoglobin and of the redox state of cytochrome-aa3 also were measured simultaneously in a similar manner to the flow reduction of Figure

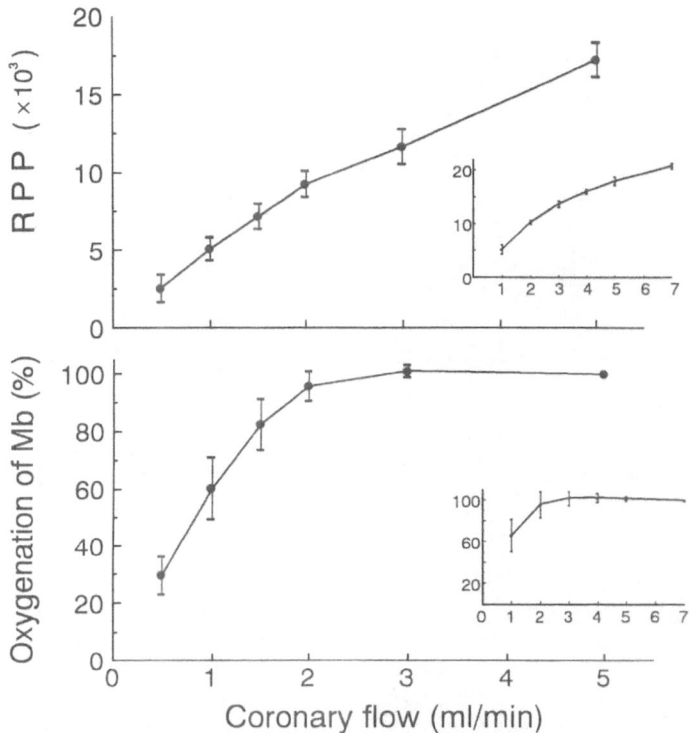

Figure 2. Rate-pressure product and the oxygenation of myoglobin with variable coronary flow. Insets show the results of another group of 5 hearts. The oxygenation of myoglobin was over 96% until the flow rate was reduced below 2 ml/min.

2. Figure 3 shows the summary of the results of 6 hearts. The deoxygenation of myoglobin paralleled the reduction of cytochrome-aa3, showing a straight-line in the plot.

DISCUSSION

As demonstrated in Figure 2, with moderate reductions in coronary flow, the rate-pressure product decreased though myocardial oxygenation state remained at almost normal aerobic level. This means that the decrease in cardiac contractility is not due to the shortage of oxidative energy production. With severe flow reduction below 2 ml/min, instead, the decrease in the rate-pressure product was accompanied by deoxygenation of myoglobin and reduction of cytochrome-aa3, showing the severe hypoxia and energy-shortage.

When the flow rate was reduced below 2 ml/min, the deoxygenation of myoglobin paralleled the reduction of cytochrome-aa3 as shown in Figure 3, in spite of the difference in their oxygen dependence. This indicates that there is a steep oxygen tension gradient in the tissue and this gradient does not alter with the flow reduction. The intracellular oxygen gradient has been shown in hypoxic experiments in the isolated heart cell (13) and in the perfused heart (12), although there are some arguments against this idea (14).

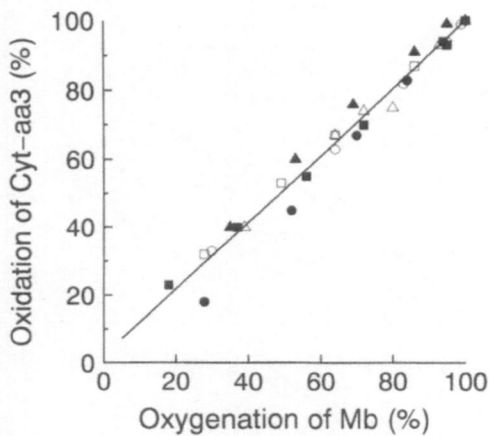

Figure 3. Relationship between the oxygenation of myoglobin and the oxidation of cytochrome-aa3 during graded reductions of coronary flow. Different marks represent different hearts.

From the results of Figure 2, the rate-pressure product during coronary flow reduction is replotted to the intracellular oxygen tension in Figure 4, with the use of an oxygen dissociation curve of myoglobin by Tamura et al. (12), and the results are compared to previous studies of hypoxic perfusion in the isolated heart (9). In both cases the oxygen delivery to tissue decreases, but the response of the heart is quite different. In the case of hypoxia, cardiac contraction and the rate of oxygen consumption remain at control normoxic level until the intracellular oxygen tension decreases below 10 μM (9). In contrast, when coronary flow is moderately reduced (above 2 ml/min in our experiment), the heart decreases its contractility by some mechanism to balance the oxygen consumption with reduced oxygen delivery to maintain the myocardium in high oxygen tension. The mechanism of how the heart adjusts its contractility in response to the reduction of coronary flow or coronary pressure is not fully understood. However, non-energy related mechanism must be

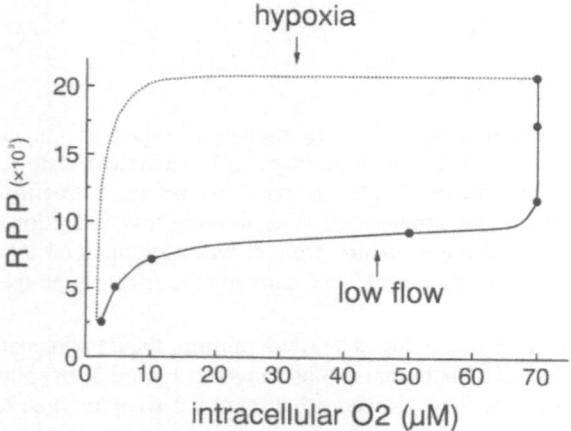

Figure 4. Relationship between contractile function and intracellular oxygen tension during hypoxic perfusion at constant pressure and coronary flow reduction. The dotted line for hypoxia is replotted from Araki et al. (9).

considered. One candidate is alterations in intracellular calcium transient which was shown in isolated ferret heart by Kitakaze et al. (5). This needs further investigation.

REFERENCES

1. Opie, L.H., 1965, Coronary flow rate and perfusion pressure as determinants of mechanical function and oxidative metabolism, J.Physiol. 180:529-541.
2. Clarke, K., and Willis, R.J., 1987, Energy metabolism and contractile function in rat heart during graded, isovolumic perfusion using ^{31}P nuclear magnetic resonance spectroscopy, J.Mol.Cell.Cardiol. 19:1153-1160
3. Marshall, R.C., Nash, W.W., Bersohn, M.M., and Wong, G.A., 1987, Myocardial energy production and consumption remain balanced during positive inotropic stimulation when coronary flow is restricted to basal rates in rabbit heart, J.Clin.Invest. 80:1165-1171
4. Marshall, R.C., and Zhang, D.Y., 1988, Correlation of contractile dysfunction with oxidative energy production and tissue high energy phosphate stores during partial coronary flow disruption in rabbit heart, J.Clin.Invest. 82:86-95
5. Kitakaze, M., and Marban, E., 1989, Cellular mechanism of the modulation of contractile function by coronary perfusion pressure in ferret hearts, J.Physiol. 414:455-472
6. Downing, S.E., and Chen, V., 1990, Myocardial hibernation in the ischemic neonatal heart, Cir.Res. 66:763-772
7. Keller, A.M., and Cannon, P.J., 1991, Effect of graded reductions of coronary pressure and flow on myocardial metabolism and performance: A model of "hibernating" myocardium, J.Am.Coll.Cardiol. 17:1661-1670
8. Kreutzer, U., and Jue, T., 1991, ^{1}H-nuclear magnetic resonance deoxymyoglobin signal as indicator of intracellular oxygenation in myocardium, Am.J.Physiol. 261:H2091-H2097
9. Araki, R., Tamura, M., and Yamazaki, I., 1983, The effect of intracellular oxygen concentration on lactate release, pyridine nucleotide reduction, and respiration rate in the rat cardiac tissue, Cir.Res. 53:448-455
10. Fukuda, H., Yasuda, H., Shimokawa, S., and Tamura, M., 1988, The oxygen dependence of the energy state of cardiac tissue: ^{31}P-NMR and optical measurement of myoglobin in perfused rat heart, Adv.Exp.Med.Biol. 248:567-573
11. Ito, K., Nioka, S., and Chance, B., 1990, Oxygen dependence of energy state and cardiac work in the perfused rat heart, Adv.Exp.Med.Biol. 277:449-457
12. Tamura, M., Oshino, N., Chance, B., and Silver, I.A., 1978, Optical measurements of intracellular oxygen concentration of rat heart in vitro, Arch.Bioch.Biophy. 191:8-22.
13. Kennedy, F.G., and Jones, D.P., 1986, Oxygen dependence of mitochondrial function in isolated rat cardiac myocytes, Am.J.Physiol. 250:C374-383
14.. Wittenberg, B.A., and Wittenberg, J.B., 1989, Transport of oxygen in muscle, Annu.Rev.Physiol. 51:857-878

23

BASAL Q_{10} FOR CEREBRAL METABOLIC RATE FOR OXYGEN ($CMRO_2$) IN RATS

R. Klementavicius, E. M. Nemoto, and H. Yonas

Departments of Anesthesiology/CCM and Neurological Surgery
University of Pittsburgh, A-1017 Scaife Hall
Pittsburgh, Pennsylvania 15261

INTRODUCTION

The effect of hypothermia on cerebral metabolic rate for oxygen ($CMRO_2$) is well established and is effectively expressed by the Q_{10}, i.e. the ratio of total $CMRO_2$ ($TCMRO_2$) over a temperature range of $10°C$. Reported values for Q_{10} vary from 2.3 (Michenfelder and Theye, 1968) in dogs to 4.4 in man (Cohen et al, 1964), reflecting the controversy on the "true" value of Q_{10}. Reports over the last decade investigating the metabolic effects of mild and moderate hypothermia on the brain and their protective effects during global and focal ischemia relative to those of barbiturates, have questioned the notion as to whether the predominant mechanism of brain protection by hypothermia or barbiturates is simply attributable to a reduction of $CMRO_2$. We previously reported a Q_{10} of 2.5 for $TCMRO_2$ in rats but also showed that hypothermia depressed active $CMRO_2$ ($ACMRO_2$) i.e. thiopental suppressible or EEG associated $CMRO_2$ less than basal $CMRO_2$ which is utilized for maintaining cell integrity and viability (Nemoto et al, 1993). On the basis of these results, we hypothesized that Q_{10} for $BCMRO_2$ differs from that for $ACMRO_2$ and determined the Q_{10} for BCMRO2 in rats

METHODS

Certified virus free, male, Wistar rats weighing between 350 and 550 g had free access to food and water up to the time of the experiments. The protocol was approved by the Animal Care and Use Committee of the University of Pittsburgh School of Medicine.

Anesthesia and Surgical Preparations

The rats were anesthetized with 5% isoflurane (AErrane, Anaquest Caribe Inc. Guayama, PR) in oxygen in a plastic jar, intubated with the 14 ga intravenous catheters (Surflo, Terumo Medical Corp. Elkton, MD) and mechanically ventilated (Harvard Appara-

Oxygen Transport to Tissue XVII, Edited by Ince et al.
Plenum Press, New York, 1996

tus Rodent Respirator, South Natick, MA) on 1.5-2% isoflurane 70%N_2)/30%O_2. Pancuronium bromide (Pavulon, Elkins-Sinn, Inc. Cherry Hill, NJ) was used for muscle paralysis. Femoral artery and vein catheters were inserted. Arterial blood pressure as well as scalp EEG and hydrogen clearance were monitored and recorded with polygraph (Grass Model 79E EEG & Polygraph Data Recording System, Grass Instruments Co. Quincy, MA). A catheter with its tip at the exit of the left transverse sinus via a branch of the internal maxillary vein as described by Gjedde et al. (1975) was used to sample cerebral venous blood. A 25 μm tip diameter platinum microelectrode was inserted into the confluence of the sinuses via a burr hole to monitor brain H_2 clearance following equilibration with inspired H_2 of 5%. All surgical wounds were treated with 2% lidocaine jelly (Astra Pharmaceuticals Products, Inc. Westborough, MA). Brain temperature was monitored with a thermistor inserted epidurally over the left parietal cortex and connected to a monitoring thermometer (Thermalert TH-5, Physitemp Instruments, Inc. Clifton, NJ). Continuous EEG was recorded with scalp electrodes placed on left and right parietal bones. Rectal temperature was monitored with a thermistor and telethermometer (YSI Instruments, Inc. Yellow Springs, OH). After the surgical procedures the rats were mechanically ventilated on 0.4% isoflurane/70%N_2O/30% O_2 for 30 min before the experiment protocol was conducted. Arterial blood gases were verified to be within normal limits.

Experimental Protocol

After 30 min of ventilation with 0.4% isoflurane 70% N_2O/30% O_2 measurements of whole brain cerebral blood flow (CBF) and TCMRO$_2$ were made. Rats were equilibrated on 0.5% H_2 in the inspired gas mixture. Arterial and cerebral venous blood samples (0.5 ml each) were withdrawn for analysis of pH and blood gases (Model 238 pH/Blood Gas Analyzer, Ciba Corning, Medfield, MA) and oxygen content (Model 482 CO-Oximeter, Instrumentation Laboratory, Lexington, MA) and equal amounts of donor blood were replaced. Blood gases, pH and oxygen content were measured at 37°C and recorded uncorrected for temperature. Immediately after blood samples were obtained, inspired H_2 was discontinued and desaturation begun during continuous recording of H_2 clearance. After baseline measurements, titrated infusion of thiopental sodium (Pentothal, Abbott Labs, North Chicago, IL) was begun until an isoelectric EEG was obtained(recorded at 50 μV/1cm) and maintained throughout the study. Arterial blood pressure was monitored and mean arterial pressure (MAP) maintained in the range of 80-130 mmHg with titrated infusion of donor blood obtained from another rat. Donor blood infusion was continued throughout the study (total volume 10-18 ml). After obtaining an isoelectric EEG with thiopental infusion, isoflurane was discontinued, the inspired gas mixture changed to 70%N_2/30% O_2 and CBF and TCMRO$_2$ measured. During baseline measurements, temperature was maintained with a heating lamp. Later, hypothermia was induced with ice packs placed on the rat's back. Subsequent measurements of CBF and CMRO$_2$ were performed at 34°, 30°, and 28°C. Statistical analyses were done by ANOVA for repeated measures with post hoc analyses by Dunnet's test with a maximum p value of 0.05 for statistically significant differences.

CBF and CMRO$_2$ Measurements

Whole brain CBF was calculated from the H_2 clearance curve of the brain as monitored in the confluence of sinuses (Singh et al.,1992) The equation was CBF (ml/100g/min) = λ • 0.693 • 60 • 100/T 1/2(S), where λ = 1.0 for H_2. TCMRO$_2$ was calculated from the equation TCMRO$_2$ (ml/100g/min) = CBF •(A-V)D • 0.01

Table 1. Physiologic and cerebral variables during progressive hypothermia and thiopental induced isoelectric EEG

Variable		BSLN38	TP38	TP34	TP30	TP28
T b (^{O}C)	AVG	37.83	37.90	33.80	29.99	28.01
	STD	0.23	0.12	0.12	0.11	0.10
CBF*	AVG	169.88	88.97	61.98	38.75	27.70
	STD	37.61	17.76	12.19	5.23	5.33
CMRO$_2$*	AVG	6.46	4.62	2.39	1.42	1.01
	STD	2.13	1.32	0.67	0.78	0.38
MAP (mmHg)	AVG	122.86	103.57	95.71	88.57	90.00
	STD	15.08	11.56	10.15	10.93	15.12
PCO$_2$ (mmHg)	AVG	36.86	36.43	35.57	34.43	32.86
	STD	4.79	6.30	7.31	7.31	7.16
pH	AVG	7.40	7.42	7.40	7.42	7.47
	STD	0.05	0.06	0.06	0.06	0.05
PO$_2$ (mmHg)	AVG	136.14	122.43	163.86	193.29	201.43
	STD	24.12	39.00	23.16	25.49	30.66

* = (ml/100gm/min)
BSLN = Baseline
TP = Thiopental

RESULTS

Physiologic variables were within ranges that would not be expected to influence cerebral variables. Induction of isoelectric EEG at 38°C with TP decreased CBF by 50% and, correcting for the 20% reduction in CMRO$_2$ by the anesthesia at baseline, reduced TCMRO$_2$ by 50%. With an isoelectric EEG, reduction of brain temperature to 34°C decreased CBF by 30% and BCMRO2 by 48%. Reduction of temperature to 28°C decreased CBF by 66% and basal CMRO$_2$ by 78%.

The Q$_{10}$ for basal CMRO$_2$ was 4.9 ± 1.2 (X ± SD). The decrease in basal CMRO$_2$ with respect to temperature was nonlinear and likely exponential with the largest changes occurring in the first 4^{0}C i.e. 38° to 34°C.

DISCUSSION

Our experiments revealed two interesting findings. First, the Q$_{10}$ for basal CMRO$_2$ of 4.9 was much higher than that for TCMRO$_2$ of 2.0 to 2.5. Over the wide range of Q$_{10}$ values reported, one sees that the lowest numbers were obtained when the experiments were designed to measure total CMRO$_2$. Michenfelder and Theye (1968) reported a value of 2.23 in dog and Hagerdal et al. (1975) 2.0 in rat without barbiturate anesthesia. Higher values were reported in studies using barbiturates for anesthesia: Rosomoff and Holaday (1954)-3.3 in dog; Bering (1961)-3.5 in monkeys; and Mutch et al. (1994)-3.8 in dog. In this aspect the report of Stone et al. (1956) seems relevant though exact calculations of Q$_{10}$ are impossible because of unknown level of narcosis used. However in one patient "deeply anesthetized" with pentobarbital and pentothal hypothermia to rectal temperature 28.3°C (83°F) decreased CMRO$_2$ from 2.8 to 0.9, a O$_{10}$ of 3.11, and in "lightly anesthetized" with pentothal, only patient hypothermia to 29.3°C resulted in decrease from 1.6 to 1.1, a Q$_{10}$ of 1.5. A Q$_{10}$ of 3.9

in dog for basal $CMRO_2$ can be calculated from the work of Lafferty et al. (1978) where they showed that hypothermia and barbiturates depress $CMRO_2$ more than either of them alone. In that study, under barbiturate induced isoelectric EEG, $CMRO_2$ decreased by 64% when dogs were cooled from 37° to 30°C. Steen et al. (1983) reported a Q_{10} of 4.5 between temperatures of 27° and 17°C in dog which was confirmed by Michenfelder et al. (1991) between temperatures of 27°-14°C. In both studies the explanation was related to a progressive functional depression of the brain and disappearance of the EEG. From the data of Woodcock et al. (1987) who studied the effects of hypothermia on CBF and $CMRO_2$ in human patients anesthetized with fentanyl while undergoing cardiopulmonary bypass surgery, the overall Q_{10} of 4.4 can be calculated in the group with thiopental suppressed EEG. Lazenby et al. (1992) reported Q_{10} for dog brain 4.1 anesthetized with sodium pentobarbital and 1.0-1.5% isoflurane. In the same study they also reported a Q_{10} of 2.6 for whole body in dogs. Michenfelder and Theye (1968) also reported Q_{10} for whole body to be in the range of 1.7-2.4. However Q_{10} for brain reported by them was 2.23 and one of the reasons to explain this could be again, different level of anesthesia and anesthetics used. The relationship of the high Q_{10} values and the use of barbiturates was noticed by Siesjo in 1978. However, the explanation given at that time was only about "uncertain effects of barbiturates on Q_{10}" (Siesjo, 1978).

The Q_{10} of 3.65 reported in neonates, infants and children during cardiopulmonary bypass by Greeley et al. (1991) is also relevant to the data described above, because of less developed synaptic activity (and subsequently prevalence of $BCMRO_2$).

The second interesting finding was that effect of hypothermia on basal $CMRO_2$ was not linear when plotted against temperature and the most acute changes could be seen during mild hypothermia. We found that decrease in temperature by 4°C from 38° to 34°C suppressed $BCMRO_2$ by almost 50%. A big decrease in $CMRO_2$, between 37°-32°C compared to other temperatures was also seen in the work of Bering (1961) in monkeys under pentobarbital anesthesia. Schiozaki et al. (1993) also showed a decrease in $CMRO_2$ of more than 50% while applying mild hypothermia (37° to 34°C) to control elevated intracranial pressure in head injured patients treated with high dose barbiturates and hyperventilation.

Our study shows that mild and moderate hypothermia selectively suppresses basal $CMRO_2$, the component attributed to the maintenance of cellular integrity and viability whereas barbiturates selectively suppress, by definition, active $CMRO_2$ associated with brain function. These differential effects of hypothermia and barbiturates on active and basal $CMRO_2$ may be the basis for the difference in efficacy in attenuating ischemic brain damage.

REFERENCES

Bering EA. Effect of body temperature change on cerebral oxygen consumption of the intact monkey. Amer J Physiol 1961;200:417-9

Cohen PJ, Wollman H, Alexander SC, Chase PE, Behar MG. Cerebral carbohydrate metabolism in man during halothane anesthesia. Anesthesiology 1964;25:185-91

Gjedde A, Caronna JJ, Hindfelt B, Plum F. Whole-brain blood flow and oxygen metabolism in the rat during nitrous oxide anesthesia. Am J Physiol 1975;229:113-118

Greeley WJ, Ungerleider RM, Quill T, et al. The effects of hypothermic cardiopulmonary bypass and total circulatory arrest on cerebral metabolism in neonates, infants and children. J Thorac Cardiovasc Surg 1991;101:783-94

Hagerdal M, Harp J, Nillson L, Siesjo BK. The effect of induced hypothermia upon oxygen consumption in the rat brain. J Neurochem 1975;24:311-16

Lafferty JJ, Keykhah MM, Shapiro HM, Van Horn K, Behar MG Cerebral hypometabolism obtained with deep pentobarbital anesthesia and hypothermia (30 C). Anesthesiology 1978;49:159-161

Lazenby WD, Ko W, Zelano JA, Lebowitz N, Shin YT, Ison OW, Krieger KH: Effects of temperature and flow rate on regional blood flow and metabolism during cardiopulmonary bypass. Ann Thorac Surg 1992;53:957-964

Michenfelder JD, Theye RA. Hypothermia: Effect on canine brain and whole-body metabolism. Anesthesiology 1968;29:1107-12

Michenfelder JD, Milde JH. The relationship among canine brain temperature, metabolism and function during hypothermia. Anesthesiology 1991;75:130-136

Mutch WAC, Sutton IR, Teskey JM, Cheang MS, Thomson IR. Cerebral pressure-flow relationship during cardiopulmonary bypass in the dog at normothermia and moderate hypothermia. J Cereb Bl Flow Metab 1994;14:510-18

Nemoto EM, Mellick J, Klementavicius R. Effect of mild hypothermia on compartmentation of cerebral blood flow and metabolism. International Society of Oxygen Transport to Tissue 21st annual meeting, San Diego, CA 1993.

Rosomoff HL, Holaday DA. Cerebral blood flow and cerebral oxygen consumption during hypothermia. Amer J Physiol 1954;179:85-88

Siesjo BK.Brain Energy Metabolism.1978. New York, John Wiley.

Singh NC, Kochanek PM, Schiding JK, Melick JA, Nemoto EM. Uncoupled cerebral blood flow and metabolism after severe global ischemia in rats. J Cereb Bl Flow Metab 1992;12:802-808

Shiozaki T, Sugimoto H, Taneda M, et al. Effect of mild hypothermia on uncontrollable intracranial hypertension after severe head injury. J Neurosurg 1993;79:363-8

Stone HH, Donnelly C, Frobese AS. The effect of lowered body temperature on the cerebral hemodynamics and metabolism of man. Surg Gynecol Obstet 1956;103:313-7

Steen PA, Newberg L, Milde JH, Michenfelder JD. Hypothermia and barbiturates: individual and combined effects on canine cerebral oxygen consumption. Anesthesiology 1983;58:527-532.

Woodcock TE, Murkin JM, Farrar JK, Tweed WA, Guiraudon GM, McKenzie FN. Pharmacological EEG Suppression during cardiopulmonary bypass: Cerebral hemodynamic and metabolic effects of thiopental or isoflurane during hypothermia and normothermia. Anesthesiology 1987;67:218-224.

24

COMPARISON OF *IN VIVO* MICROVASCULAR PERFUSION PATTERN BETWEEN SKELETAL MUSCLE AND HEART

Colloidal Gold Particles as *in vivo* Plasma Marker

Martin F. König, Ewald R. Weibel, and Sanjay Batra

Institute of Anatomy, University of Berne
Bühlstrasse 26, CH-3012 Berne
Switzerland

INTRODUCTION

Direct observation of capillary perfusion is only possible on superficial capillaries. For the examination of capillary perfusion in deeper tissue layers, histological sections are necessary. On light microscopic tissue sections, the presence of a marker injected into the blood stream indicates a capillary perfusion under certain *in vivo* conditions (7,20).

Commonly, macromolecule labeled fluorescent dyes such as dextran-FITC (4) or globulin-FITC (5) are used as plasma tracers. Such tracers, however, are only identifiable on light microscopic sections. Since capillaries are not unambiguously identifiable at these low magnifications, a marker specific to the capillary wall must be employed in addition to the fluorescent dye. A common method for capillary identification is the alkaline phosphatase technique (7,15,18,21). However, it has been shown that the alkaline phosphatase reaction might not stain venular capillary portions (2,12,15). Therefore, alkaline phosphatase (AP) in combination with dipeptidyl peptidase IV (DPP IV) has been used to stain as well as distinguish arteriolar and venular capillary regions (1,3).

Since both endothelial markers, AP and DPP IV, are not specific to only microvessel endothelium, difficulties in establishing valid standards for identifying capillaries from larger vessels in histological sections arise (14). In addition, this method does not work in all organs and species (6,12). So far, various blood vessel markers such as toluidine blue, silver methenamine, fibronectin, laminin and lectins have also been used (9,16,17); but they showed similar disadvantages as AP and DPP IV reactions, without the ability to distinguish the proximal (i.e. arteriolar) from the distal (i.e. venular) portions of the microvascular bed.

Another disadvantage in capillary perfusion studies using fluorescent dyes as a perfusion marker is the fact that a possible gradient in capillary perfusion cannot be detected, since capillaries filled by blood cells or any variation in tissue fixation reduce fluorescence

Oxygen Transport to Tissue XVII, Edited by Ince et al.
Plenum Press, New York, 1996

197

intensity. Further, it should be mentioned that low concentrations of plasma tracer remain below the threshold of recognition (7).

Since both difficulties, capillary and tracer identification, are based on the low resolution of light microscopy, studies of capillary perfusion pattern using electron microscopy (EM) is preferable. Electron microscopy allows unambiguous capillary identification as well as detection of even small amounts of an appropriate plasma tracer. Where the high resolution of EM is ideal for identification of plasma markers and capillaries, light microscopic (LM) sections allow for the identification of many more capillaries on the same section, which is convenient for studies of regional perfusion patterns. Therefore, an optimal plasma tracer should be identifiable by both light and electron microscopy.

To demonstrate pulmonary capillary perfusion under *in vivo* conditions, we have recently developed a new plasma tracer visible at both EM and LM levels (11). A highly concentrated suspension of albumin labeled colloidal gold particles were injected into the blood stream of anaesthetized rabbits. After dispersion of the gold nanospheres in the blood plasma of the whole animal, their amount was high enough for direct identification on ultrathin electron microscopic sections. After silver enhancement they are unambiguously identified at light microscopic level. In the present study, we examined whether this plasma tracer is suitable to demonstrate capillary perfusion in skeletal and heart muscle, both at EM and LM levels.

MATERIAL AND METHODS

Preparation of RSA-Gold Complexes

An initial volume of 3 liter of 8 nm diameter gold particles was prepared (11): solution A [2.370 ml H_2O + 30 ml 1% (wt/vol) $HAuCl_4$ (no. 50800 Fluka) in water] was heated to 60°C and rapidly mixed with solution B [450 ml H_2O, 120 ml 1% trisodium citrate, 6 ml 1% tannic acid (Mallinckrodt), heated to 60°C]. By this reaction ~$4 \cdot 10^{16}$ gold particles were produced (10). To concentrate the gold colloid, the solution was gently boiled down to approximately 50% of its original volume. pH was adjusted by adding 75 ml of 0.2 M sodium phosphate buffer pH 6.1. To stabilize the gold particles against electrolyte induced aggregation, 0.0391 mg RSA (suspended in H_2O added with 10 mM sodium phosphate buffer pH 6.1) per ml gold suspension was rapidly mixed with 525 ml portions of gold solution.

For further concentration, the RSA-gold suspension was centrifuged for 90 minutes at 35'000 g in a Beckman 45 Ti fixed angle rotor to a final volume of 4-5 ml gold concentrate. This concentrate was dialysed against Ringer solution. For storage, the suspension was frozen in liquid nitrogen and stored at -70°C. Before infusion, the gold suspension was Millipore filtered (0.25 μm).

Experimental Method

For this experiment, 3 rabbits with a body weight of ~4 kg were premedicated (atropine 50 mg, heparin 2500 ui, ketamine-hydrochloride 50 mg) and anaesthetized with pentobarbital injected into the ear vein. The trachea was cannulated and oxygen was added to environmental air during the surgical procedures. Oxygen supplementation was stopped 1 minute before gold injection. A thin catheter (OD = 1.27 mm) was advanced to the right atrium through the femoral vein. The chest was opened and the lung was ventilated by means of a small animal respirator (tidal volume 15 ml, frequency 40/min, end expiratory pressure 2 cm H_2O). A snare was placed inside the pericardium and around the base of the heart.

An aliquot of 4-5 ml RSA-gold concentrate warmed to body temperature was injected within 15 seconds through the femoral catheter into the right atrium. After two minutes circulation time, the blood flow was stopped by rapidly closing the snare placed around the base of the heart. From heart, a tissue sample comprising the apical third of the left ventricle was dissected perpendicular by the base-apex axis. From M. soleus a sample from the mid portion was taken. The tissue blocks were immediately covered with half concentrated Karnovsky fixative, cut into smaller tissue blocks (~2 mm side length for EM, ~50 mm for LM) and stored in the same fixative for at least 24 hours before further processing for EM and LM.

For embedding standard techniques were followed (8,13). The 2 mm tissue blocks were cut (thickness ~60 nm) and mounted on uncoated 200 mesh grids and lightly stained with lead citrate and uranyl acetate (each 5 minutes at 20°C).

For light microscopy, tissue was embedded in paraffin and cut to sections of varying thickness (6-30 μm). After removing the paraffin, the sections were placed into citrate buffer (pH 3.8) containing 0.25% (wt/vol) hydrochinone. Then the sections were transferred for 45 minutes to a solution of 0.1% silver acetate dissolved into 0.25% hydrochinone citrate buffer. The sections were fixed with diluted (1:9) Ilford fixative and counterstained with nuclear fast red.

Analyses of the Tissue

From each animal, randomly sampled tissue blocks were examined by EM at a final magnification between x6000 and x200'000. The optimal final magnification for gold particle quantification on the screen of a projection device used for stereological analysis has been found to be ~x50'000.

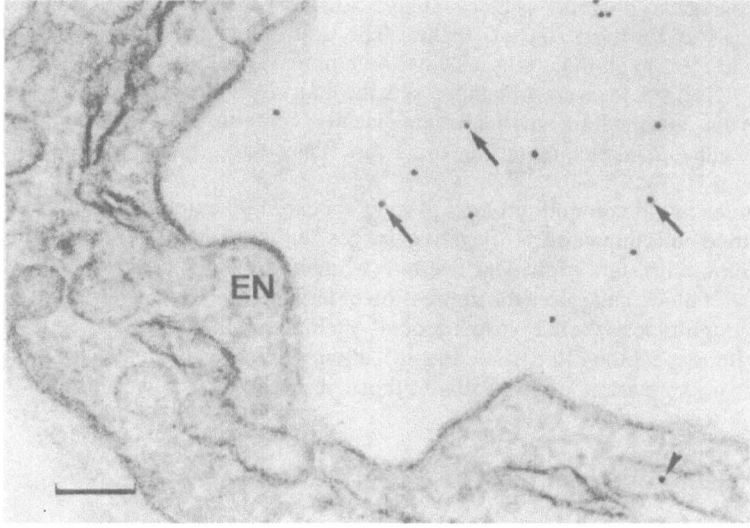

Figure 1. Capillary profile from soleus muscle at high magnification showing colloidal gold particles (arrows) appearing as black nanospheres of uniform size homogeneously dispersed in the plasma space. One single gold particle is recognized in the endothelial cell (EN) in an endosome (arrowhead; scale marker: 0.1 μm).

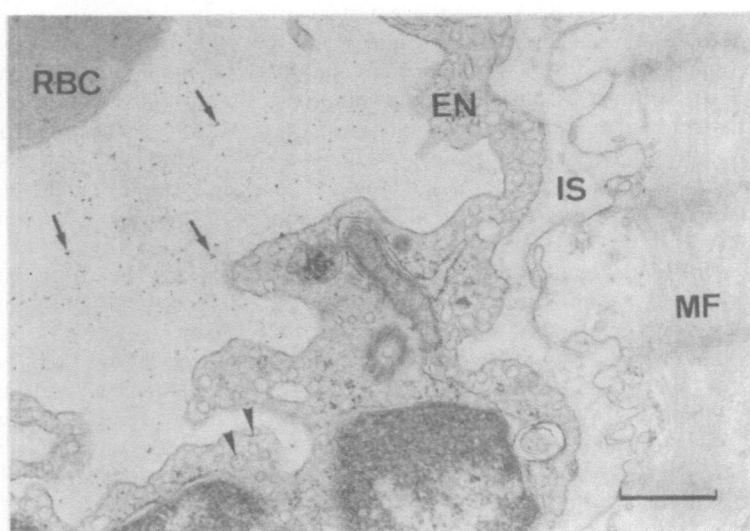

Figure 2. Highly concentrated colloidal gold particles (arrows) are clearly identified at lower magnification. They fill the plasma space between red blood cell (RBC) and capillary endothelium (EN). Single gold particles are recognized in endothelial endovesicles (arrowheads), but no particles are observed in the interstitial space (IS) between capillary and muscle fiber (MF) in this area (heart; scale marker: 0.5 μm)

RESULTS

Gold particles were identified in the plasma space of capillary profiles in both, skeletal muscle and heart. Electron micrographs at higher magnification showed electron dense gold particles appearing as black 'spheres' dispersed in the plasma space (Fig.1). Gold particles were of uniform size of ~8 nm. The gold concentration was high enough to recognize particles even in capillaries where the lumen was collapsed or partly filled with blood cells. The particles were well dispersed, the majority appearing as single particles with less than 10% grouped in small clusters usually containing 2-6 particles (Fig. 2). In endothelial cells of skeletal muscle and heart, vesicular profiles often contained gold particles (Figs. 1, 2).

Whereas gold containing blood plasma appeared black on silver enhanced paraffin sections, blood cells appeared as "negative images" either unstained (red blood cells; RBC) or showing the faint stain of nuclear fast red (white blood cells; Fig. 3). On cross sections, the presence of blood cells gave the impression of less intensely labeled capillaries, but there was still enough black labeled plasma around the blood cells to indicate capillary labeling (Fig. 4). With respect to gold particle identification within the plasma of the capillaries, the results in heart were remarkably similar to those observed in skeletal muscle.

DISCUSSION

In the present study we developed a plasma tracer to study *in vivo* capillary perfusion pattern of heart and skeletal muscle on tissue sections by both, electron and light microscopy. Highly concentrated colloidal gold particles enabled us to visualize plasma perfusion. Even at the level of individual capillaries, plasma labeling was clearly observed. Based on the direct detectability of colloidal gold particles on electron micrographs, no uncertainties about tracer or capillary identification occurred as it may be the case if only light micrographs are considered (11). Therefore, colloidal gold particles are an excellent plasma marker for quantitative perfusion determination at the EM level and at least a semi-quantitative marker for studies of the perfusion pattern of tissue by light microscopy. Electron microscopic sections revealed, that colloidal gold particles are also useful for the demonstration of endocytosis in various tissues.

Colloidal Gold Particles as Plasma Marker

Given the high resolution of EM, the presence of even one gold particle per capillary profile is sufficient for indication plasma perfusion. This method enables us to detect gold particles even in capillaries which are collapsed or partly filled with blood cells since they contain at least some gold particles distributed in the surrounding plasma. These capillaries might be missed by using a plasma tracer which is visible only by light microscopy. The gold concentration (~4·10^{16} particles) used in our experiments was high enough to recognize particles in capillaries containing only a small amount of plasma. In addition EM sections enable us to identify capillaries without arbitrary definitions such as vessel size which is necessary at LM level (4).

Colloidal Gold Particles as Endocytosis Marker

Gold particles were also found to be a useful tool for the description of endocytosis in endothelial endosomes. We recently have shown a steady decrease of the gold particle concentration with a half life of about 30 minutes (11). This decrease is not only explained by endocytosis of gold nanospheres by endothelial cells, but also by the uptake by cells of the reticulo endothelial system (RES) such as liver or spleen. In liver, we found a high

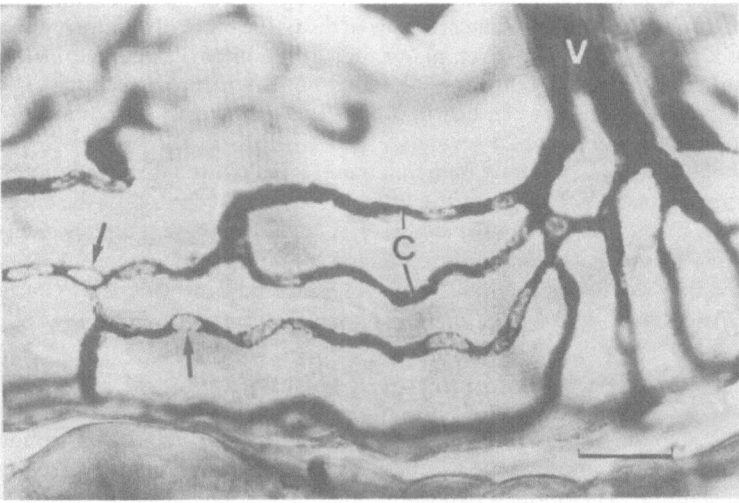

Figure 3. Due to the black labeled plasma, thick silver enhanced paraffin sections show the three dimensional capillary network (C) connected to a larger vessel, presumably a venule (V). Red blood cells appear as negative images surrounded by black plasma (arrows). (M. soleus 30 μm; scale marker: 20 μm).

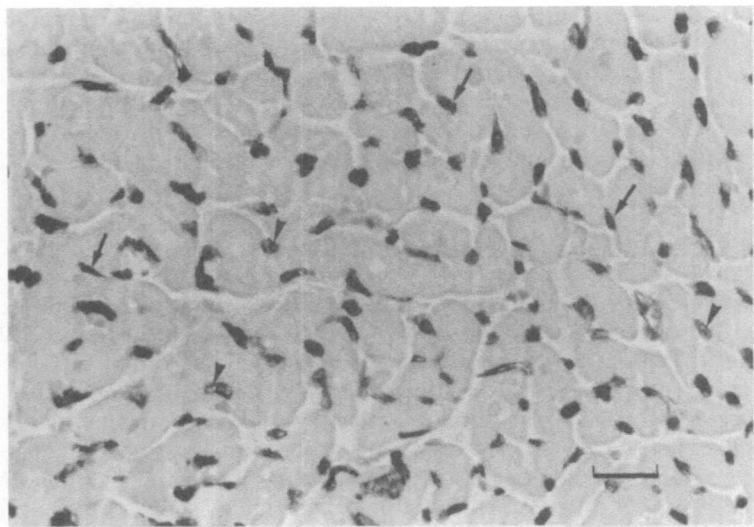

Figure 4. Even on cross sections, marked capillaries are clearly identified by their black labeled plasma (arrows). Red blood cells (arrowheads) filling the capillary lumen falsely might lead to the impression of less intensive labeled capillaries (heart 8 μm; scale marker 20 μm).

concentration of gold particles in endosomes of endothelium, hepatocytes and macrophages (Kupffer cells; (11))

Fixation

In the present study immersion fixation of tissue was performed. This enabled us to preserve the blood in the vasculature. The disadvantage of this method is the time dependent gradient of fixation from outer to inner regions of the tissue block. Therefore, minimal plasma fluxes cannot be excluded. However, major redistribution of plasma with delayed fixation appears unlikely. Weiss and Conway (21) found no differences in capillary labeling after fluorescent dye injection between heart portions instantaneously frozen after removal and samples frozen with a delay of 25 seconds.

Although tissue fixation by immersion preserves plasma distribution, the process of endocytosis may continue to occur in still unfixed regions. Accordingly, the observed degree of endocytosis may overestimate the extent of endocytosis present at the time of the arrest of blood flow. In order to maintain the precise degree of endocytosis at a defined time, perfusion fixation would be preferable. In the present study, our aim was to introduce a new plasma perfusion marker, and therefore immersion fixation was required.

Factors Affecting Silver Enhancement Intensity

In our experiments, silver enhancement was carried out for 45 minutes. Longer exposure times did not improve the capillary labeling intensity, but increased non-specific background staining. The optimal exposure time to silver enhancement may vary between different organs, fixatives and objectives of the study. To study RBC distances in the capillary network of the lung, we reduced the exposure time to 23 minutes which reduced the intensity of the black stain of plasma and therefore improved the visibility of RBC profiles (19).

CONCLUSION

In the present study we demonstrated *in vivo* microvascular perfusion and endocytosis in skeletal muscle and heart by using small colloidal gold particles complexed to albumin as a new tracer which allows both, electron and light microscopic examination. From this study we conclude that colloidal gold particles are an excellent plasma tracer to demonstrate *in vivo* capillary perfusion on light and electron microscopic sections in skeletal muscle and heart.

ACKNOWLEDGMENTS

The authors would like to thank Mr. M. Linder for his excellent technical assistance. Supported by SNSF grant # 31-30946.91

REFERENCES

1. Batra, S., Koyama, T., Gao, M., Horimoto, M., and Rakusan, K., 1992, Microvascular geometry of the rat heart: arteriolar and venular capillary regions, *Jpn. Heart J.* 33:817-828.
2. Batra, S., Kuo, C., and Rakusan, K., 1989, Spatial distribution of coronary capillaries: A-V segment staggering, *Adv. Exp. Med. Biol.* 248:241-247.
3. Batra, S. and Rakusan, K., 1991, Geometry of capillary networks in volume overloaded rat heart, *Microvasc. Res.* 42:39-50.
4. Gobel, U., Theilen, H., and Kuschinsky, W., 1990, Congruence of total and perfused capillary network in rat brains, *Circ. Res.* 66:271-281.
5. Gobel, U., Theilen, H., Schrock, H., and Kuschinsky, W., 1991, Dynamics of capillary perfusion in the brain, *Blood Vessels* 28:190-196.
6. Grim, M. and Carlson, B. M., 1990, Alkaline phosphatase and dipeptidylpeptidase IV staining of tissue components of skeletal muscle: a comparative study, *J. Histochem. Cytochem.* 38:1907-1912.
7. Hargreaves, D., Egginton, S., and Hudlicka, O., 1990, Changes in capillary perfusion induced by different patterns of activity in rat skeletal muscle, *Microvasc. Res.* 40:14-28.
8. Hoppeler, H., Mathieu, O., Weibel, E. R., Krauer, R., Lindstedt, S. L., and Taylor, C. R., 1981, Design of the mammalian respiratory system. VIII. Capillaries in skeletal muscles, *Respir. Physiol.* 44:129-150.
9. Kayar, S. R. and Weiss, H. R., 1991, Capillary recruitment and heterogeneity of perfused capillary distribution in dog myocardium, *Microcirc. Endoth. Lymphatics* 7:77-107.
10. Kehle, T. and Herzog, V., 1987, Interactions between protein-gold complexes and cell surfaces: a method for precise quantitation, *Eur. J. Cell Biol.* 45:80-87.
11. König, M. F., Lucocq, J. M., and Weibel, E. R., 1993, Demonstration of pulmonary vascular perfusion by electron and light microscopy, *J. Appl. Physiol.* 75:1-7.
12. Lojda, Z., 1979, Studies on dipeptidyl (amino) peptidase IV (glycyl-proline naphthylamidase). II. Blood vessels, *Histochem.* 59:153-166.
13. Mermod, L., Hoppeler, H., Kayar, S. R., Straub, R., and Weibel, E. R., 1988, Variability of fiber size, capillary density and capillary length related to horse muscle fixation procedures, *Acta Anat.* 133:89-95.
14. Renkin, E. M., Gray, S. D., and Dodd, L. R., 1981, Filling of microcirculation in skeletal muscles during timed india ink perfusion, *Am. J. Physiol.* 241:H174-H186.
15. Romanul, F. C. A. and Bannister, R. G., 1962, Localized areas of high alkaline phosphatase activity in the terminal arterial tree, *J. Cell Biol.* 15:73-84.
16. Schelper, R. L., Olson, S. P., Carroll, T. J., Hart, M. N., and Witters, E., 1986, Studies of the endothelial origin of cells in systemic angioendotheliomatosis and other vascular lesions of the brain and meninges using ulex europaeus lectin strains, *Clin. Neuropathol.* 5:231-237.
17. Tomanek, R. J. and Aydelotte, M. R., 1992, Late onset renal hypertension in old rats alters left ventricular structure and function, *Am. J. Physiol.* 262:H531-H538.
18. Vetterlein, F., Keitel, U., and Schmidt, G., 1993, Capillary filling kinetics in the rabbit heart during normoxemia and hypoxemia, *Am. J. Physiol.* 264:H287-H293.

19. Weibel, E. R., Federsiel, W. J., Fryder-Doffey, F., Hsia, C. C. W., König, M., Stalder-Navarro, V., and Vock, R., 1993, Morphometric model for pulmonary diffusion capacity, *Resp. Physiol.* 93:125-149.
20. Weiss, H. R., 1988, Measurement of cerebral capillary perfusion with a fluorescent label, *Microvasc. Res.* 36:172-180.
21. Weiss, H. R. and Conway, R. S., 1985, Morphometric study of the total and perfused arteriolar and capillary network of the rabbit left ventricle, *Cardiovasc. Res.* 19:343-354.

25

CONTROL OF OXIDATIVE METABOLISM IN VOLUME-OVERLOADED RAT HEARTS

Effect of Pretreatment with Propionyl-L-Carnitine

J. Moravec , Z. El Alaoui–Talibi , M. Moravec and A. Guendouz

Laboratoire d'Energétique et de Cardiologie Cellulaire
Faculté de Pharmacie
7 Bvd Jeanne d'Arc, 21 000 Dijon, France

Chronic mechanic overloading of the heart has been shown to lead to a significant depletion of tissue carnitine (1, 2) that is possibly related to impaired carnitine transport to the myocardium (3, 4). At the same time, the ability of chronically overloaded hearts to oxidize exogenous palmitate is diminished (5, 6). This decrease of long-chain fatty acid utilization is accompanied by reduced myocardial oxygen consumption (MVO2) and gives rise to an impaired mechanical activity during *in vitro* perfusions (6, 7). Most of the above quoted alterations disappear when exogenous palmitate is replaced by octanoate, a short-chain fatty acid that has free access to mitochondrial matrix (8). This suggests that the respiratory chain of volume-overloaded rat hearts perfused in presence of long-chain fatty acids may be actually substrate limited (7). In this work, we tried to improve NADH delivery to respiratory chain by a prolonged treatment of volume-overloaded rats with millimolar concentrations of propionyl-L-carnitine (9). It has been shown that the administration of this compound significantly increases both blood plasma concentrations and myocardial tissue levels of L-carnitine (10, 11). This, in turn, may improve long chain fatty acid utilization (5) and glucose oxidation (12) via a decreased acetyl-CoA/CoA ratio (13, 14). The control and volume-overloaded hearts were perfused with 11 mM glucose and 1.2 mM palmitate (2.4 mM octanoate) over a range of left ventricular work loads, leading to a progressive increase in the myocardial VO2 (7). The respective relationships between the rates of oxidative phosphorylation and different intracellular energy parameters ((cytosolic phosphorylation potential (ATP/ADPf.Pi), ADPf, and mitochondrial NAD+/NADH ratio)) as obtained in control and volume-overloaded hearts were compared for each metabolic condition examined. The effects of the pretreatment with propionyl-L-carnitine on the kinetics of oxidative phosphorylation were tested under conditions of a high work load (heart ejecting against an increased aortic resistance related to the clamp of the aortic outflow line) as described previously (6, 7). The respective contributions of glucose and fatty acid oxidation to ATP productions were assessed by isotopic techniques (15) under conditions of both moderate and high work loads.

METHODS

Animals. A chronic volume overload was induced in 2-mo-old Wistar rats (IFFA CREDO) by a surgical opening of the aortocaval fistula. Sham operated animals from the same litters were used as controls. Three months after the surgery, the surviving animals were sacrificed using a light ether anesthesia and the hearts of rats presenting a significant increase in heart weight but no signs of congestive heart failure were used for the *in vitro* perfusions.

Heart perfusions. After excision, the hearts were rapidly fastened to the aortic cannula and perfused for 10 min with bicarbonate buffer containing 11 mM glucose. The hearts were then recirculated via the left atrium with the same buffer containing 11 mM glucose and 1.2 mM palmitate bound to 3% bovine serum albumin (Sigma, fraction V). The perfusions were continued for 20 min under conditions of moderate work load (10 Torr preload, 70 Torr afterload). In some cases, cardiac work was increased by clamping the aortic flow line so that the hearts pumped for further 20 min against a high resistance (high work load). Separate groups of control and volume-overloaded hearts were perfused according to Langendorff considered as a low work condition. The perfusion system was derived from that of Neely et al (16).

Monitoring of left ventricular performance. Mechanical activity was assessed in terms of pressure work (product of left ventricular peak systolic pressure and heart rate) and, where possible, in terms of cardiac output (ejecting hearts).

Myocardial VO2 measurements. The rate of myocardial oxygen consumption (umoles O2 per min per g dry weight) was determined during steady state perfusion conditions by polarographic methods (Instech 125/02 Clark electrode and Instech Oxygen Intake package). The coronary flow was measured at 5 min intervals by means of a calibrated cylinder (7).

Biochemical assays. At the end of each perfusion period, the hearts were freeze-clamped and the frozen samples were freeze-dried. They were then deproteinized in 0.6 N perchloric acid and the supernatant, brought to pH 5.8, was used for biochemical assays (17). Tissue contents of selected metabolites were converted to concentrations assuming that 1g of dry tissue contained 2.25 ml of intracellular water (18).

The concentrations of unbound ADP (ADPf) and cytosolic phosphorylation potential (ATP/ADPf.Pi) were calculated according to Starnes et al (18) from creatine phosphokinase equilibrium constant ($K_{eq}CPK$) which, at pH = 7 and 1.0 mM free Mg++, is 1.66×10^{9} M^{-1} (19):

$$ATP / ADPf \times Pi = CP \times H+ / Cr \times Pi \times K_{eq}CPK$$

The redox state of mitochondrial NAD was estimated from the glutamate dehydrogenase reaction, supposed to reach equilibrium in all conditions studied (18, 20) and considering the changes in the intracellular distribution of glutamate and alpha -ketoglutarate as negligible:

$$KGDH = NADH \times KG \times NH_3 / NAD^+ \times glutamate = 3.87 \times 10^{-3} M^{-1}$$

Statistical analysis. Student's t test was employed to determine differences between control and volume-overloaded hearts perfused with the same substrate. The effects of the pretreatment with propionyl-L-carnitine on hearts of the same group were assessed by analysis of variance and the pairs of means that were not equal were compared by Fisher's LSDP test.

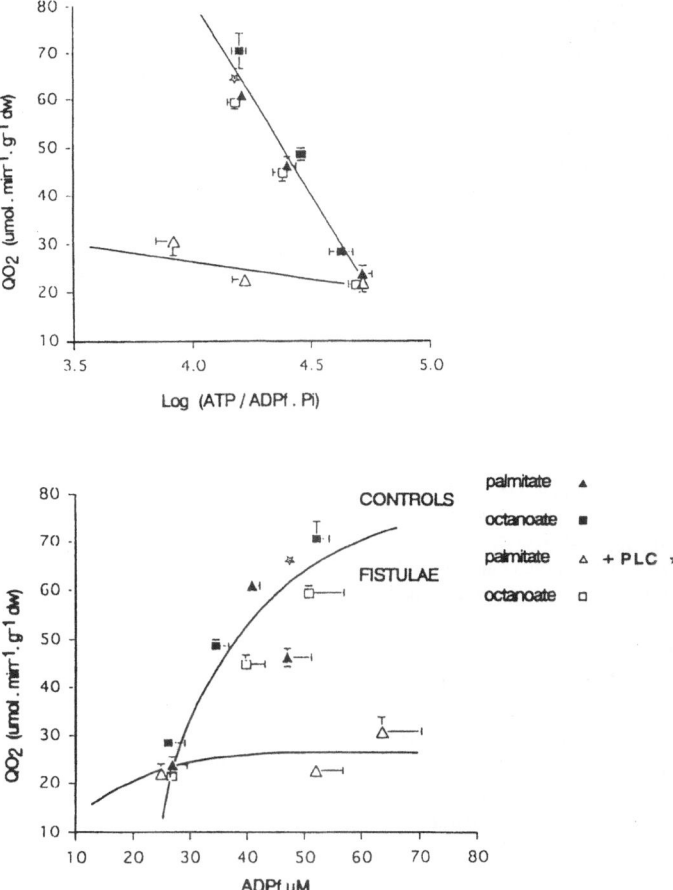

Figure 1. Relationships between MVO2 and cytosolic phosphorylation potential (upper panel) and between MVO2 and unbound ADP (lower panel) as observed in control (closed symbols) and volume-overloaded (open symbols) hearts perfused with different exogenous substrates. The data obtained on palmitate perfused hearts from rats receiving propionyl-L-carnitine are also indicated (asterisks). Note loss of both thermodynamic control and kinetic regulation of oxidative phosphorylation by cytosolic ADP in volume-overloaded hearts perfused with palmitate. The substitution of this substrate by octanoate or a stimulation of an alternative metabolic pathway by propionyl-L-carnitine renders both cytosolic phosphorylation potential and ADP regulatory again.

RESULTS

Figure 1 shows the relationship between MVO2 and cytosolic phosphorylation potential (upper panel) as well as that between MVO2 and unbound ADP (lower panel) as seen in control and volume-overloaded hearts perfused with either glucose + palmitate (triangles) or 2.4 mM octanoate (squares). It appears that, in presence of this latter substrate, the kinetics of oxidative phosphorylation of volume-overloaded hearts is similar to that of controls : the unbound ADP remains regulatory over the range of physiological concentrations (20-60uM). At the same time, a unique linear relationship between MVO2 and Log

Table 1. Steady state of glycolysis and oxidative metabolism in conrol and hypertrophied

	Glycolysis	Glucose Oxidation	Palmitate Oxidation
	nmol / min . g dry wt		
Moderate Load			
Control	5288 ± 466.0	802.38 ± 60.72	1287 ± 40.7
Fistulae	6515 ± 563.5	620.94 ± 76.58	709.7 ± 28.1***
High Load			
Control	7493 ± 855[++]	2359 ± 112 [+++]	1705 ± 75[++]
Fistulae	10981 ± 432* [+++]	2143 ± 88 [+++]	964.7 ± 48.6*** [++]
	nmol / min . g dry . HR.PSP		
Moderate Load			
Control	247 ± 23	39.22 ± 3.31	49.72± 2.79
Fistulae	382 ± 67*	35.37 ± 8.13	33.16 ± 1.74**
High Load			
Control	253 ± 28	69.84 ± 8.21 [+]	56.03 ± 1.16
Fistulae	390 ± 85*	76.85 ± 13.88 [+]	40.48 ± 2.77**

Values are the mean ± SEM, n = 4-7, *p < 0.05; **p < 0.01; ***p < 0.001.
*Significantly different than control hearts perfused at same workload.
[+] Significantly different than comparable hearts perfused at moderate Workload.

ATP/ADPf.Pi indicates that, in presence of this particular lipid substrate, the equilibrium between cytosolic adenylate system and respiratory chain is being maintained. On the other hand, in palmitate perfused volume-overloaded hearts, unbound ADP is not regulatory any longer and significant drops in ATP/ADPf . Pi ratio is not accompanied by any acceleration of oxidative phosphorylation. As we showed previously, this kinetic limitation of the respiration is essentially related to an enhanced pyridine nucleotide oxidation resulting from impaired exogenous palmitate utilization (6, 7).

Table 2.

	Steady -State acetyl-CoA Production		
Perfusion condition	From Glucose oxidation	From Palmitate oxidation	Total
	μmol / min · g dry wt		
Moderate Load			
Control	1.60 ± 0.12	10.29 ± 0.33	11.89
Fistulae	1.24 ± 0.15	5.68 ± 0.23***	6.92
High Load			
Control	4.72 ± 0.22[++]	13.64 ± 0.6[++]	18.36
Fistulae	4.29 ± 0.18[++]	7.72 ± 0.39*** [++]	12.01

Values are the mean ± SEM, n = 4-7, *p < 0.05; **p < 0.01; ***p < 0.001.
*Significantly different than control hearts perfused at same workload.
[+] Significantly different than comparable hearts perfused at moderate workload.

Table 3. Energy parameters in control and volume-overloaded
hearts: Effects of propionyl-L-carnitine

	ATP	CP $umol.g^{-1}\,dw$	ADPf	ATP/ADPf.Pi 10^4M^{-1}	NAD+/NADH	QO2 $umol.min^{-1}.g^{-1}dw$
Controls untreated	16.72 ±1.17	25.02 ±2.88	0.100 ±0.01	1.56 ± 0.09	5.38 ±0.36	68.85 ± 4.30
Fistulae untreated	12.40 ±0.58 *	16.34 ±3.00 **	0.128 ±0.033 *	0.88 ±0.14 ***	8.72 ±0.29 **	43.50 ±3.05 **
Controls PLC	16.23 ±1.31	24.44 ±2.28	0.087 ±0.01	1.50 ±0.06	4.85 ±0.14	68.98 ±3.05
Fistulae PLC	16.02 ±0.99	24.41 ±2.38	0.106 ±0.012	1.26 ±0.07 *	5.59 ±0.32 *	66.75 ±7.09

The hearts were perfused for 15 min at high work load conditions, values are m ± SEM for n = 4

Table 1 shows that the above decrease of palmitate utilization is accompanied by a depression of glucose oxidation and by a slightly increased glycolytic flux. This latter change becomes significant at high work loads. However, under both conditions examined, the total steady state acetyl-CoA production is 30-50% lower in volume-overloaded hearts perfused with glucose and palmitate (Table 2). Consequently, the mitochondrial NAD becomes more oxidized than in the control hearts exposed to the same perfusion conditions. The difference between the two groups of hearts is the most significant at high work loads (Table 3).

This table also shows that the pretreatment with propionyl-L-carnitine (250 mg/kg per day for two weeks) avoids the above oxidation of pyridine nucleotides. This effect does not seem to be related to improved palmitate oxidation, since the cumulative curves of [14]CO2 production from U- [14]C -palmitate are not significantly affected (not shown). The principal reason for the anaplerotic effect of the propionyl-L-carnitine administration seems to be an improved utilization of exogenous glucose. In fact, the pretreatment with propionyl-L-carnitine increases slightly the glycolytic rate (Fig 2) and, more substantially, the rate of glucose oxidation (Fig 3). This amplifies the shift from fatty acid to glucose contribution to ATP production as predicted by earlier studies on mechanically overloaded hearts (21) and the existence of which we confirm experimentally in our work. The propionyl-L-carnitine-related improvement in mitochondrial NADH availability seems to generate conditions in which unbound ADP (and/or ATP/ADPf . Pi ratio) become regulatory. In this way, relatively high respiratory rates can be obtained without any major accumulation of cytosolic ADP or decreased phosphorylation potential. This means that mechanical work-MVO2 coupling becomes more efficient. The resulting preservation of the cytosolic GATP, even at relatively high energy turnover rates, should improve in turn the mechanical activity of volume-overloaded hearts. In this respect, it is worthy to note that the pretreatment with propionyl-L-carnitine proved to be followed by a positive inotropic effect in control rabbit hearts (9, 10). In our hands, long term administration of propionyl-L-carnitine presented also a cardiostimu-

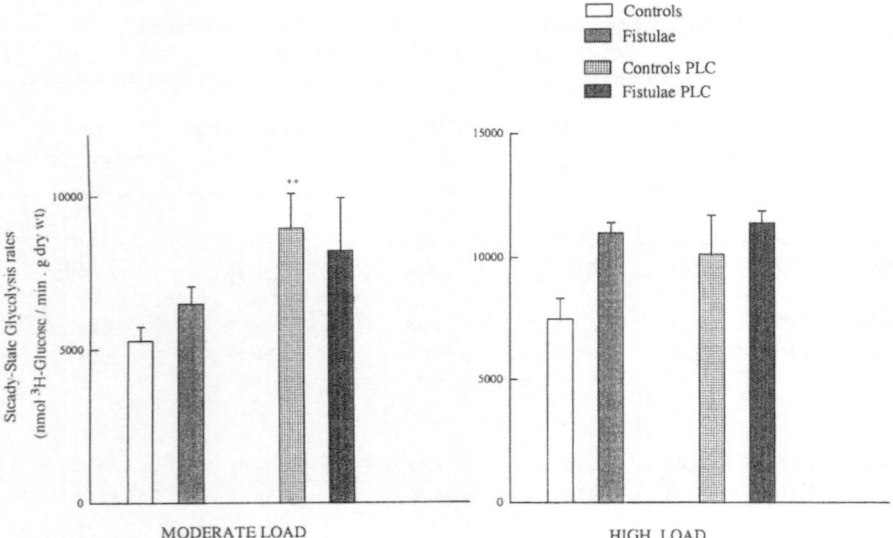

Figure 2. Respective glycolysis rates as observed in control and volume-overloaded hearts perfused under conditions of a moderate and high work loads. Note a slight increase of glycolytic flux in volume-overloaded hearts as compared to controls perfused under same conditions and stimulatory effect of PLC on glycolytic flux of control hearts perfused at moderate work loads. In other groups studied, the effects of PLC are not

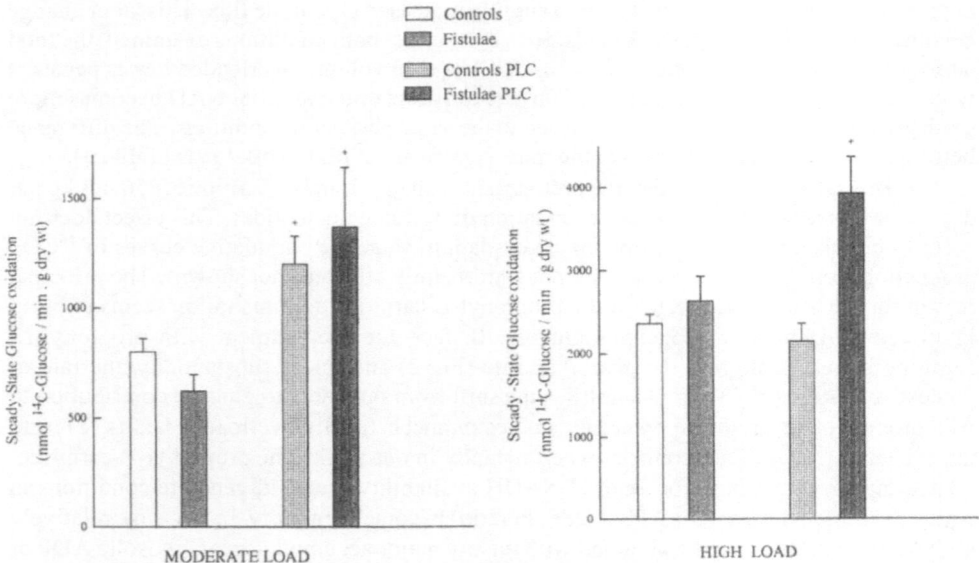

Figure 3. Effect of propionyl-L-carnitine on glucose oxidation in control and volume-overloaded hearts perfused with 1.2 mM palmitate, 11 mM glucose and 10 IU of insulin. At moderate work load, the pretreatment with propionyl-L-carnitine stimulates significantly glucose oxidation in both control and volume-overloaded hearts (left hand panel). Under conditions of high work load (right hand panel), the stimulation still persists in volume-overloaded hearts. In controls, the maximal rates of glucose oxidation are reached already prior the administration of the drug.

Table 4. Mechanical performance of isolated working hearts from control and volume-overloaded rats: Effect of PLC

Parameters measured	Untreated		PLC-treated	
	Controls	Fistulae	Controls	Fistulae
Moderate load				
Aortic flow	186 ± 2.3	84 ± 7.0*	227 ± 12	182 ± 6.2*
Heart rate	251 ± 1.0	210 ± 1.2*	268 ± 6.0	270 ± 6.3
Aortic pressure	106 ± 6.0	98.0 ± 7.0	105 ± 7.0	103 ± 9.0
Pressure work	24.8 ± 1.0	19.5 ± 0.4*	27.6 ± 2.0	27.2 ± 1.5
Oxygen consumption	46.0 ± 1.7	27.0 ± 3.0*	53.0 ± 4.2	45.0 ± 1.8
High load				
Aortic flow	-	-	-	-
Heart rate	245 ± 5.0	208 ± 10*	256 ± 9.0	265 ± 12
Aortic pressure	143 ± 12	121 ± 7.0	151 ± 12	144 ± 10
Pressure work	35.0 ± 1.3	26.0 ± 1.7*	38.1 ± 2.1	38.6 ± 2.5
Oxygen consumption	68.0 ± 3.4	43.0 ± 5.1*	56.0 ± 4.6	66.7 ± 5.3

m ± SEM, n= 5-10

lant action on isolated hearts from rats with volume overload (Table 4). This effect was evident at moderate load and persisted also under conditions of high work load.

DISCUSSION

The results presented in our work are consistent with the previous demonstrations of the anaplerotic effect of propionyl-L-carnitine (13, 14). When control hearts are exposed to millimolar concentrations of this compound, both tissue carnitine level and tissue content of dicarboxylic acids tend to increase (11, 14). These two alterations were shown to lower acetylCoA /CoA ratio and, in this way, stimulate both long chain fatty acid (22) and glucose (12) oxidation. However, a part of the metabolic effects of propionyl-L-carnitine persists after relatively long periods from the last injection of propionyl-L-carnitine (9). At the same time, chronic administration of propionyl- L-carnitine causes a long-lasting prolongation of the plateau phase of action potential. This suggests that at least a part of the cardiostimulant action of the propionyl-L-carnitine may rely on its effects on cellular membranes and, particularly, on intracellular Ca++ movements. If this should be the case, part of the metabolic alterations that we described in our volume-overloaded hearts might be related to Ca++ stimulation of different intracellular dehydrogenases (23). These include rate-limiting steps in glycolysis and, mainly, mitochondrial pyruvate decarboxylase the stimulation of which may promote the shift from fatty acid to glucose oxidation as observed in the present work. The failure of propionyl-L-carnitine to improve the utilization of exogenous palmitate, as observed in this and our earlier work (11), may be related to an alteration of the functional association between palmitoylCoA synthetase and carnitine palmitoyl transferase (6).

REFERENCES

1. Reibel DK, Uboh CE and Kent RL. (1983) Altered CoA and carnitine metabolism in pressure overloaded hypertrophied hearts. Am. J. Physiol. 244 : H839-H843
2. Bowé C, Nzonzi J, Corsin A, MoravecJ, and Feuvray D. (1984) Lipid intermediates in chronically volume-overloaded rat hearts. Pflügers Arch. 402 : 317-320
3. Reibel DK, O'Rourke V and Forster KA. (1987) Mechanism for altered carnitine content in hypertrophied rat heart. Am. J. Physiol. 252 : H561-H565
4. El Alaoui-Talibi Z and Moravec J. (1989) Carnitine transport and exogenous palmitate oxidation in chronically volume-overloaded rat hearts. BBA 1003 : 109-114
5. Wittels B and Spann JF. (1968) Defective lipid metabolism in the failing heart. J. Clin. Invest. 47 : 1787-1794
6. El Alaoui-Talibi Z, Landormy S, Loireau A and Moravec J. (1992) Fatty acid oxidation and mechanical performance of volume-overloaded rat hearts. Am. J. Physiol. 262 : H1068-1074
7. Ben Cheikh R, Guendouz A and Moravec J. (1994) Control of oxidative metabolism in volume overloaded rat hearts : effect of different lipid substrates. Am. J. Physiol. 266 : H2090-H2097
8. Fritz IB, Kaplan E and Yue KTN. (1962) Specificity of carnitine action in fatty acid oxidation by heart muscle. Am. J. Physiol. 202 : 117-121
9. Ferrari R, Di Lisa F, de Jong JW, Ceconi C, Pasini E, Barbato R, Menabo R, Barbieri M, Carbai E and Mugelli A. (1992) Prolonged PLC pretreatment of rabbits : biocemical and electro- physiological effects on myocardium. J. Mol. Cell. Cardiol. 24 : 219-232
10. Ferrari R, Pasini E, Condorelli E, Boraso A, Lisciani R, Marzo A, and Vision O.(1991) Effect of propionyl-L-carnitine on mechanical function of isolated rabbit heart. Cardiovasc. Drugs Ther. 5 : 17-24
11. El Alaoui-Talibi Z and Moravec J.(1993) Assessment of the cardiostimulant action of propionyl-L-carnitine on chronically volume-overloaded rat hearts. Cardiovasc. Drugs Ther. 7 : 357-363
12. Broderick TL, Quinney HA and Lopaschik GD. (1992) Carnitine stimulation of glucose oxidation in the fatty acids perfused isolated working hearts. J. Biol. Chem. 267 : 3758-3763
13. Siliprandi N, Di Lisa F, Menabo R. (1991) Propionyl-L-carnitine : biochemical significance and possible role in cardiac metabolism. Cardiovasc. Drugs Ther. 5 (suppl 1) : 11-16
14. Hülsmann WC. (1991) Biochemical profile of propionyl-L-carnitine. Cardiovasc. Drugs Ther. 5 (suppl 1) : 7-10
15. Lopaschuk GD and Spafford GD. (1991) Glucose and palmitate oxidation during reperfusion of hearts from diabetic rats. In : Nagano M and Dhalla NS (eds) The diabetic Heart, Raven, NY, pp 451-464
16. Neely JR, Whitmer KM and Mochizuki S. (1976) Effect of mechanical activity and hormones on myocardial glucose and fatty acids utilization. Circ. Res. 38 (suppl I) : I22-I30
17. Williamson JR and Corkey B. (1969) Assays of intermediates of citric acid cycle and related compounds. In : Colowick SP and Kaplan NO (eds) Methods in Enzymol., Academic, NY, vol. 13, pp 439-513
18. Starnes JW, Wilson DF and Erecinska M. (1985) Substrate dependance of metabolic state and coronary flow in perfused rat hearts. Am. J. Physiol. 249 : H799-H806
19. Lawson JW and Veech RL. (1979) Effects of pH and free Mg++ on Keq of the creatine kinase and other phosphate hydrolases and phosphate transfer reactions. J. Biol. Chem. 254 : 6528-6537
20. Nuuntinen H, Hiltunen JK and Hassinen IE. (1981) The glutamate deshydrogense system and redox state of mitochondrial free NAD in myocardium. FEBS Letter 128 : 356-360
21. Bishop SP and Altschuld RA. (1970) Increased glycolytic metabolism in cardiac hypertrophy and congestive heart failure. Am. J. Physiol. 218 : 153-159
22. Shug A, Paulson D, Subramanian R and Regitz V. (1991) Protective effects of propionyl-L-carnitine during ischemia and reperfusion. Cardiovasc. Drugs Ther. 5 (suppl 1) : 77-84
23. Mc Cormack JG, Halestrap AP and Denton RM. (1990) Role of calcium ions in regulation of mammalian intramitochondrial metabolism. Physiol. Rev. 70 : 391-425.

26

CYTOCHROME OXIDASE IS THE PRIMARY OXYGEN SENSOR IN THE CAT CAROTID BODY

Sukhamay Lahiri,[1] Deepak K. Chugh,[1] Anil Mokashi,[1]
Sergei Vinogradov,[2] Shinobu Osanai,[1] and David F. Wilson[2]

[1] Departments of Physiology
[2] Biophysics and Biochemistry
   University of Pennsylvania School of Medicine
   Philadelphia, Pennsylvania 19104

INTRODUCTION

The carotid body (CB) senses hypoxia of arterial blood and signals it by increasing the activity in the sensory nerve. Since glomus cells are the only elements that are innervated, by monitoring the activity of this nerve with PO_2 changes it is possible to gauze the oxygen sensing in this tissue. The consensus model is that upon O_2 sensing the cell is depolarized which mobilizes Ca^{2+}. This increase in intracellular Ca^{2+} increases the release of neurotransmitters which then act on the nerve endings to generate the signals. This neural signal then traveling to the brain-stem elicits various respiratory and cardiovascular responses. In fact, this tiny organ initiates the major respiratory reflex effects, because in it's absence hypoxia decreases ventilation, and the animal succumbs to hypoxia.

Previously we showed that inhibitors of mitochondrial oxidative phosphorylation specifically blocked the sensory response to hypoxia while it enhances that to hypercapnia (Mulligan et. al., 1981; Buerk et al. 1993). CO in high concentration specifically prevents the reaction of cytochrome oxidase with O_2 (Keilin, 1970), and causes sensory stimulation which is reversed by white light (Joels and Neil, 1962; Lahiri et al., 1993). In the present study we have used the different wavelengths of light to examine the reversal of activity induced by CO to identify whether it corresponds to those of cytochrome oxidase. Warburg (1926) demonstrated that light reversed the inhibition of respiration by CO, and Warburg and Negelein (1928) showed that the dependence of light wavelengths effect is responsible for most of the respiration by the cells. This technique when it works can be used for identification of biological oxidases. It became ideal for the *in vitro* carotid body preparation showing instantaneous responses to CO and its reversal by light.

Oxygen Transport to Tissue XVII, Edited by Ince et al.
Plenum Press, New York, 1996

213

METHODS

For perfusion and superfusion *in vitro,* the carotid body was vascularly isolated and dissected out with the carotid sinus nerve (CSN) and placed in the perfusion chamber with perfusion starting immediately (Iturriaga et al., 1991). The perfusate and superfusate were modified Tyrode solution containing CO_2-HCO_3^- (pH = 7.38-7.40 at 37 $^{\circ}$C) and which was equilibrated with either 21% O_2 and 5% CO_2 with balance N_2 or replacing N_2 with CO. Paraffin oil was layered over the superfusate to a depth of less than 3 mm, and carotid sinus nerve was lifted above it and the nerve activity was recorded.

Without CO, light had no influence on the activity. With CO the CSN activity was stimulated in the dark and white light silenced the stimulated activity. For photochemical action spectrum, the illuminating light was obtained by passing the light of a tungsten-I_2 bulb through a monochromator into a light guide and was focused on an approximately 1 mm area of the carotid body. The intensity of light as a function of wavelength was determined using a photodiode of known spectral response, allowing correction of the measured response to equal light intensities at each wavelength. The measured light response can be converted to the absorption spectrum of the CO-compound.

RESULTS

The afferent activity increased significantly when the carotid body was perfused with CO containing medium in the dark and was completely reversed by white light. The responses were rapid and the light of different wavelengths separated by dark periods allowed the chemosensory activity to return to steady state. Experimental records illustrating the behavior of the carotid body to wavelengths of lights from 620 nm to 420 nm are shown in

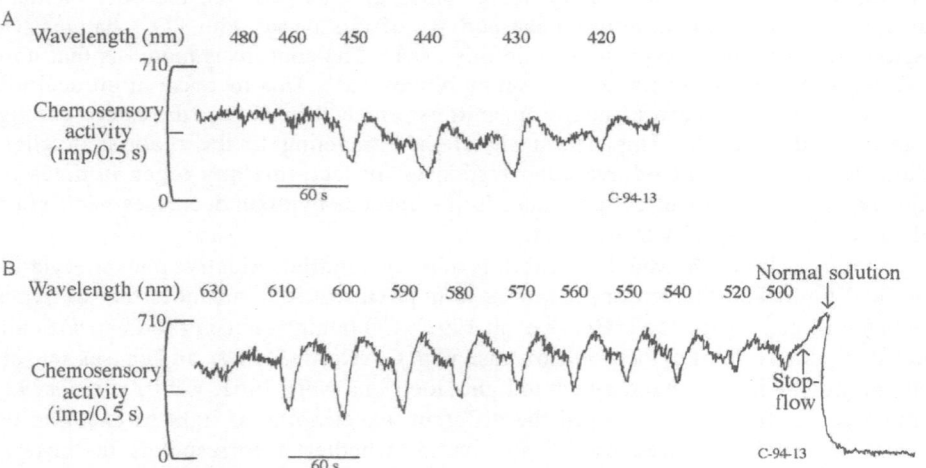

Figure 1. The carotid body's photosensitivity as expressed in chemosensory discharges when perfused with CO. At PCO of 560 Torr the chemosensory response was high in the dark (upper trace). On exposure to monochromatic light (6-7 nm width at half height) of the indicated wavelengths for 6-7 periods the photoresponse is seen. These responses at wavelengths from 480 nm to 420 nm are shown in the upper trace and those at 630 nm to 500 nm in the lower trace. Finally the flow was interrupted in the dark and the perfusion with normal CO-free solution was restored.

Fig. 1. The CO/O_2 ratio was such that it provided about 80% of the response produced by interruption of perfusate flow, and is sufficient to provide energy levels required to maintain stable cellular chemosensory function.

Fig. 2 (upper panel) shows the photosensitive response. The activity of the afferents (arbitrary units) is plotted against the wavelengths of light. There are two peaks: one at 590 nm and another at 432 nm. The response to illuminating light of each wavelength was converted to equal intensity and is shown in Fig. 2, lower panel. A given response at 432 nm required 6-7 times less light than that at 590 nm, indicating the absorption of the CO compound was 6-7 times greater at 432 nm than at 590 nm. There is a small response at 550 nm, although it is too small for clear resolution.

Measurements were made for eight independent carotid body preparations, and the results are highly reproducible. However, each carotid body preparation has a different response to each given intensity of light. Nerve fibers coupled to sensory cells deep in the carotid body would require more light intensity. Thus the absolute intensity of light required might be different but not the spectrum.

DISCUSSION

CO binds with the oxygen reaction site of heme-enzyme displacing O_2, and thereby acts as a specific inhibitor. The cytochrome a_3^{2+} was identified as the primary oxidase in cells, and its combination with CO will give rise to absorption spectra. However, it is impossible to measure it in the carotid body due to its small size, light scattering by the tissue and the presence of several pigments with overlapping absorption spectrum. Even if accurate absorption could be measured, this would not identify the particular component responsible for O_2 sensing. In contrast, the photochemical action spectrum is highly specific for the O_2

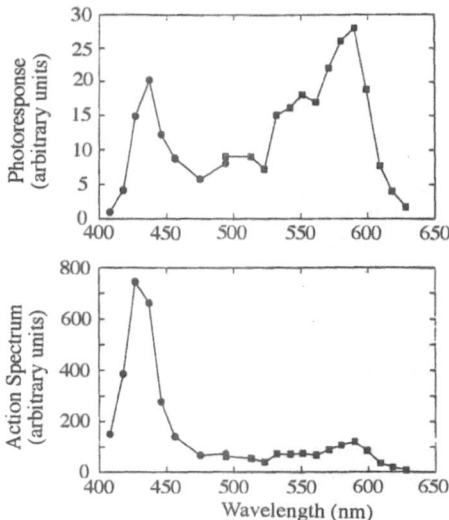

Figure 2. The photochemical action spectrum of the O_2 sensing by the carotid body. The uncorrected photoresponse is shown in upper panel in arbitrary units. The extent of the decrease of sensory discharge by light was then corrected to the same quantum intensity of illuminating light, and plotted against the light wavelengths (lower panel, in arbitrary units). The maximal responses are seen at 432 nm and 590 nm.

sensor. The sensory response of the carotid body is coupled to photochemical action spectrum.

The reversal of CO effect on O_2 sensing by the carotid body has well defined maxima at near 432 nm and 590 nm. The ratio of the peak is between 6 and 7, and corresponds to cytochrome oxidase in yeast as originally determined (Warburg and Negelein, 1928; Kubovitz and Haas, 1932; Castor and Chance, 1955). A similar value has been found for submitochondrial particles from heart muscle (Melnick, 1942). There is no evidence for contribution by other heme-CO compounds, such as CO compound of cytochrome P_{450} or of a b-type cytochrome. We have not seen a distinctive maximum at 450 nm or 550 - 580 nm.

CO induced nearly maximal activity similar to one induced by perfusate flow interruption, and CO induced activity was fully reversed by light. Thus mitochondrial cytochrome a_3^{2+} is responsible for most of the hypoxic response. This effect of CO is accompanied by 30–40% reduction in O_2 consumption which was nearly fully reversed by bright white light (Lahiri et al., 1995), and presumably by 590 nm and 430 nm wavelengths of light. This is consistent with the conclusion that mitochondrial cytochrome a_3^{2+} is the primary O_2 sensor in the carotid body (Wilson et al., 1994).

ACKNOWLEDGMENTS

Supported in part by HL-43413.

REFERENCES

Buerk, D. G., Iturriaga, R., and Lahiri, S., 1994, Testing the metabolic hypothesis of O_2 chemoreception in the cat carotid body *in vitro*, *J. Appl. Physiol.* 76:1317-1323.

Caster, L. N., and Chance, B., 1955, Photochemical action spectra of carbon monoxide-inhibited respiration, *J. Biol. Chem.* 217: 453-464.

Iturriaga, R., Rumsey, W. L., Mokashi, A., Spergel, D., Wilson, D. F., and Lahiri, S., 1991, *In vitro* perfused-superfused cat carotid body for physiological and pharmacological studies, *J. Appl. Physiol.* 70: 1393-1400.

Kubovitz, F., and Haas, E., 1932, Ausbau die chemische Zusammensetzung der Kartoffeloxydase, *Biochem. Z.* 292: 221-229.

Joels, N., and Niel, E., 1962, The action of high tensions of carbon monoxide on the carotid chemoreceptors, *Arch. Intl. Pharmacodyn.* 130: 528-534.

Keilin, D., 1970, The History of Cell Respiration and Cytochromes, London: Cambridge University Press, pp. 221-327.

Lahiri, S., Buerk, D. G., Chugh, D., Osanai, S., and Mokashi, A., 1995, Evidence for cytochrome a_3 as the primary O_2 sensor in the cat carotid body from photoreversible effects of CO on O_2 consumption and chemoreceptor excitation, *Brain Res.* (in press).

Lahiri, S., and Delaney, R.G., 1975, Stimulus interaction in the responses of carotid body chemoreceptor single afferent fibers, *Resp. Physiol.* 24: 249-266.

Lahiri, S., Iturriaga, R., Mokashi, A., Ray, D. K., Chugh, D., 1993, CO reveals dual mechanism of O_2 chemoreception in the cat carotid body, *Resp. Physiol.* 94: 227-240.

Melnick, J. L., 1942, The photochemical spectrum of cytochrome oxidase, *J. Biol. Chem.* 146: 385-390.

Mulligan, E., Lahiri, S., Storey, B. T., 1981, Carotid body O_2 chemoreception and mitochondrial oxidative phosphorylation, *J. Appl. Physiol.* 51: 438-446.

Warburg, O., 1926, Über die Wirkung des Kohlenoxyds auf den Stoffwechsel der Hefe, *Biochem. Z.* 177: 471-486.

Warburg, O., and Negelein, E., 1928, Über die photochemische Dissoziation bei intermittierender Belichtung und das absolute Absorptionsspektrum des Atmungsferments, *Biochem. Z.* 202: 202-228.

Wilson, D. F., Mokashi, A., Chugh, D., Vinogradov, S., Osanai, S., and Lahiri, S., 1994, The primary oxygen sensor of the cat carotid body is cytochrome a_3 of the mitochondrial respiratory chain, *FEBS Lett.* 351: 370-374.

27

CEREBRAL OXYGENATION CHANGES DURING MOTOR AND SOMATOSENSORY STIMULATION IN HUMANS, AS MEASURED BY NEAR-INFRARED SPECTROSCOPY

Hellmuth Obrig, Tilo Wolf, Claudia Döge, Jan Junge Hülsing, Ulrich Dirnagl, and Arno Villringer

Department of Neurology
Humboldt-University
Charité Berlin, Germany

INTRODUCTION

Functional cortical activation is coupled to changes of cerebral hemodynamics and metabolism. These changes have been used in positron emission tomography and recently developed functional magnetic resonance techniques to localize task-related cortical activation. Near-infrared spectroscopy has the potential to monitor changes of intracerebral blood oxygenation during functional activation (Villringer 1993, Hoshi 1993). We proceeded from a rather global, cognitive stimulus to experimental paradigms which entail a more localized cerebral activation and which have been examined by other functional techniques. In establishing a simple motor and a vibratory stimulus we sought a means to further examine the issue of coupling of neuronal function and cerbral hemodynamics. In recent experiments we were able to demonstrate that these stimulation models are useful to retrieve additional information about oxygenation dependent signal changes in functional MRI (Kleinschmidt 1994, Obrig 1994). As near-infrared spectroscopy has a good temporal resolution, we were also interested in the time course of local oxygenation changes.

METHODS

We used a NIRO-500 monitor (Hamamatsu) to detect changes of cerebral blood oxygenation during functional cortical stimulation. The method is explained elsewhere in detail (Cope 1988b, Jöbsis FF 1992). In brief, changes in optical densities are measured at four different wavelengths (775, 825, 850 and 904 nm) and concentration changes of oxy- and deoxy-hemoglobin as well as changes in redox state of cytochrome oxidase are calculated according to an algorithm described by Cope and Delpy (Cope 1988a). The light

Oxygen Transport to Tissue XVII, Edited by Ince et al.
Plenum Press, New York, 1996

produced by a pulsed laser source is guided to the subject's head by a fiber-optic bundle, the so-called optode. It is strongly scattered by brain tissue and thus a second optode fixed to the subject's head some centimeters apart is able to detect some of the light reflected. The changes in optical densities stem from concentration changes of oxy- and deoxy-hemoglobin in a banana-like shaped volume connecting the light emitting and the light collecting optode.

The interoptode distance for both experiments was 3.5 cm; data were collected with a 1.9 kHz sampling rate and recorded averaged over 1s periods.

Experiment 1

Twelve subjects performed a sequential finger opposition task, opposing the thumb to each of the other fingers in one hand either contra- or ipsilateral to the optode position over the left primary motor cortex. A 10 s stimulation period was followed by 50 s of rest. Data over 15 cycles of each condition were acquired and averaged time-locked to movement onset. Optodes were positioned according to a modified 10-20 system (Steinmetz 1989). If no signal change was elicited during the first few cycles, optodes were repositioned until a stimulus associated concentration change was detected. This implied changes of optode position of less than 1 cm in most experiments.

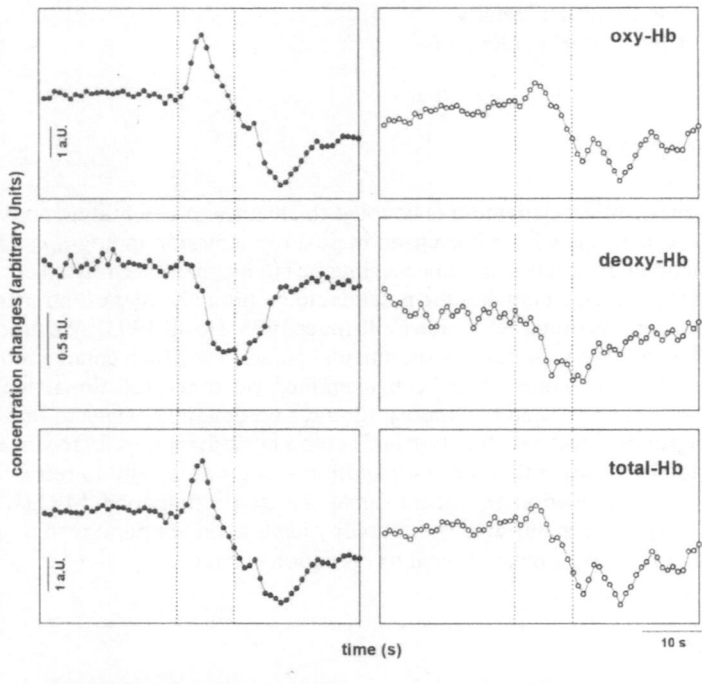

contralateral motor stimulation   ipsilateral motor stimulation

Figure 1. Changes of oxygenated hemoglobin (oxy-Hb), deoxygenated hemoglobin and total hemoglobin (total-Hb=oxy-Hb+deoxy-Hb) during functional motor stimulation. Subjects performed a sequential motor task in either one hand, contra- or ipsilateral to optode positioning over left motor cortex. 10s stimulation period is denoted by dotted lines. Average over 12 subjects performing 10-15 stimulation cycles in each hand. Ipsilateral stimulation led to smaller changes of oxygenation parameters. Also note the different time courses for oxy-Hb and deoxy-Hb concentration changes.

Experiment 2

In 12 subjects the changes in NIRS parameters during somatosensory stimulation were assessed. Temporal pattern and localization procedure were similar to those applied in Experiment 1. In this experiment two different stimulus modalities were tested. Responses to an electrical stimulation of median nerve contralateral to optode positioning was compared to a vibratory stimulus applied to the same hand. Again, 15 cycles of each condition were averaged time-locked to stimulus onset.

For both experiments subjects were lying comfortably in a dimmed silent room so as to minimize distractions from the experiment resulting in movement artifacts. Subjects who showed strong movement artifacts or in whom major baseline drifts were seen were excluded from analysis.

RESULTS

As for motor stimulation, a typical response pattern was found (Figure 1). Oxy-hemoglobin concentration increased during functional stimulation. A decrease in concentration was found for deoxy-hemoglobin during motor stimulation. Concentration returned to baseline values some seconds after cessation of the stimulus. Changes in oxy- and deoxy-hemoglobin followed a different temporal pattern. While oxy-hemoglobin concentration began to decrease during stimulation and showed a pronounced post stimulus undershoot, deoxy-hemoglobin concentration began to increase only a few seconds after cessation of the stimulation period, a post stimulus overshoot was not seen. Concentration changes of total hemoglobin were calculated by adding oxy- and deoxy-hemoglobin changes. This parameter also increased during stimulation period. Comparing the response to contra- and ipsilateral stimulation, concentration changes were clearly more pronounced for the contralateral condition.

A different pattern was elicited in response to somatosensory stimulation (Figure 2). Besides a minor decrease in deoxy-hemoglobin concentration oxy-hemoglobin concentration also decreased during stimulation. There was only a slight difference between the different stimulus modalities, which proved significant some seconds after cessation of the stimulus. Post-stimulus undershoot or overshoot were not seen.

DISCUSSION

Previous studies have demonstrated, that near-infrared spectroscopy has the potential to monitor changes of cerebral oxygenation during functional cortical activation (Villringer 1993, Hoshi 1993, Kato 1993). Cognitive, auditory and visual stimuli were used to demonstrate changes over long stimulation periods. The problem of correct probe positioning was solved by either stimuli, resulting in a rather broad cortical activation, or by the use of multiple NIRS monitors. This study is the first to systematically examine two stimuli resulting in an activation of primary cortical sensory-motor areas. According to MRI studies (Steimetz 1989) central sulcus has a comparatively small spatial variance, as in reference to 10-20 system, relying on external bony landmarks of the skull. Hence two stimuli were chosen, which have been established to activate sensorymotor areas. Signal to noise problems arising from small activated cortical areas, were overcome by repetitive stimulations combined with an averaging procedure.

The results of Experiment 1 demonstrate that a localized hemodynamic response to a motor stimulus is detectable by near-infrared spectroscopy. Apart from the expected

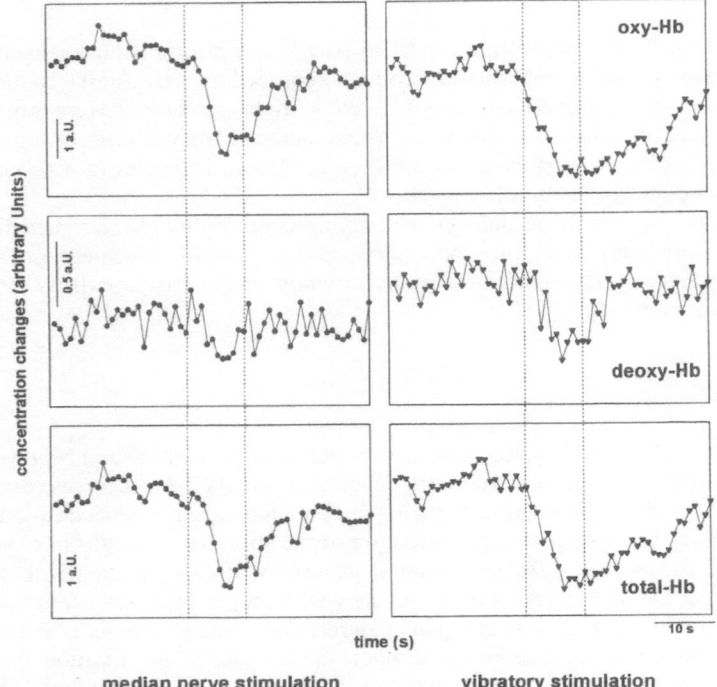

Figure 2. Average of cerebral oxygenation changes associated with electrical median nerve stimulation (left) and vibratory stimulation (right) in 12 subjects (always contralateral to NIRS probe positioning over left sensori-motor cortex). Changes in oxygenated hemoglobin (oxy-Hb), deoxygenated hemoglobin (deoxy-Hb) and total hemoglobin (total-Hb, the sum of the latter two) are shown. Data are averaged time-locked to the onset of the 10 second stimulation period, denoted by dotted lines.
Note, that there is no substantial difference between the different stimulus modalities. Compared to motor stimulation (Fig. 1) the most striking difference is a decrease in oxy-Hb concentration during the stimulation period.

increase in blood oxygenation over contralateral motor cortex, there is also a substantial, though smaller response to ipsilateral stimulation. PET studies (Roland 1980; Kawashima 1993, Grafton 1991) have shown that even rather simple motor tasks lead to an increase of blood flow in premotor areas. Some parts of the premotor area are activated bilaterally. Functional MRI studies foster these views (Rao 1993); a study using the same stimulus as in our study (Kim 1993) demonstrates a cortical asymmetry resulting in an activation of ipsilateral primary motor cortex of left, dominant hemisphere during left finger opposition task. Apart from the activation of ipsilateral motor areas, contamination of the NIRS signal by extracerebral hemodynamic changes cannot be excluded. As depth penetration of near-infrared spectroscopy depends on interoptode spacing (McCormick 1992) the recording of cortical oxygenation critically depends on the interoptode distance. According to pathlength estimations (Delpy 1988) an interoptode spacing of 3.5 cm will result in a depth penetration of about 2 cm thus measuring cerebral oxygenation only of a rather restricted cortical area.

Somatotosensory stimulation led to a decrease in both oxy- and deoxygenated hemoglobin concentration. During the post-stimulus period a slight difference between the different stimulus modalities was seen. Local changes of hematocrit may account for this strking finding (Mchedlishvili 1986). Vibratory stimulation may predominantly lead to an increase in flow velocity as opposed to the increase in blood volume accounting in the main for oxygenation changes during motor stimulation. PET techniques primarily using plasma volume as an indicator of local blood flow changes ($H_2^{15}O$-technique) have demonstrated an increase in blood flow during vibrotactile stimulation (Fox 1986). Glucose metabolism and oxygen uptake, however, show only little increase over somatosensory cortical areas during vibratory tasks as opposed to complex vibrotactile or motor tasks (Ginsberg 1987 &1988). A recent Transcranial Doppler study monitoring flow velocity of the middle cerebral artery during vibratory and a more complex vibrotactile task did not find a significant contralateral increase of flow velocity during mere vibratory stimulation (Kelley 1993).

We conclude from our data , that voluntary movement tasks result in hemodynamic changes, which are qualitatively different from those seen during a somatosensory stimulation. The most striking difference between the two tasks was an increase of oxy-hemoglobin concentration during the sequential motor task as opposed to a decrease in concentration associated with a vibratory stimulus.

REFERENCES

Cope M, Delpy DT, Reynolds EOR, Wray S, Wyatt JS, van der Zee P. Methods of quantitating cerebral near infrared data. Adv Exp Med Biol 222:191-197 (1988a)

Cope M, Delpy DT. A system for long term measurement of cerebral blood and tissue oxygenation in newborn infants by near infrared transillumination. Med Biol Engl Comp 32:1457-67 (1988b)

Delpy DT, Cope M, van der Zee P, et al: Estimation of optical pathlength through tissue from direct time of flight measurements. Phys Med Biol 33:1433-42 (1988)

Fox PT, Raichle ME. Focal uncoupling of cerebral blood flow and oxydative metabolism during somatosensory stimulation in human subjects. Proc Natl Acad Sci 83: 1140-44 (1986)

Ginsberg MD, Chang JY, Kelley RE, Yoshii F, Barker WW, Ingenito G, Boothe TE. Increases in both cerebral glucose utilization and blood flow during execution of a somatosensory task. Ann Neurol 23:152-60 (1988)

Ginsberg MD, Yoshii F, Vibulsresth S, Chang JY, Duara R, Barker WW, Boothe TE. Human somatosensory activation. Neurology 37: 1301-08 (1987)

Grafton ST, Woods RP, Mazziotta JC, Phlebs ME. Somatotopic Mapping of the Primary Motor Cortex in Humans: Activation Studies With Cerebral Blood Flow and Positron Emission Tompgraphy. J Neurophysiol; 66(3):735-43 (1991)

Hoshi Y, Tamura M. Dynamic multichannel near-infrared optical imaging of human brain activity; J Appl Physiol 75(4):1842-46 (1993)

Kato T, Kamei A, Takashima S, Ozaki T. Human Visual Cortical Function During Photic Stimulation monitoring by Means of Near-Infrared Spectroscopy. J Cereb Blood Fl Metab 13:516-20 (1993)

Kawashima R, Yamada K, Kinomura S, Yamaguchi T, Matsui H, Yoshioka, Fukuda H. Regional cerebral blood flow changes of cortical motor areas and prefrontal areas in humans related to ipsilateral and contralateral hand movement. Brain Research, 623:33-40 (1993)

Kelley RE, Chang JY, Suzuki S, Levin BE, Reyes-Iglesias Y. Selective increase in the right hemisphere transcranial doppler velocity during a spatial task. Cortex 29: 45-52 (1993)

Kim SG, Ashe J, Hendrich K, Ellerman JM, Merkle H, Ugurbil K, Georgopoulos AP. Functional Magnetic Resonance Imaging of motor cortex: Hemispheric Asymmetry and Handedness. Science 261:615-617 (1993)

Kleinschmidt A, Obrig H, Requardt M, Merboldt KD, Dirnagl U, Villringer A, Frahm J. Simultaneous Recording of Cerebral Blood Oxygenation Changes During Human Brain Activation by Magnetic Resonance Imaging and Near-Infrared Spectroscopy. J Cereb Blood Fl Metab (1995) (submitted).

McCormick P, Stewart M, Lewis G, Dujovny M, Ausman JI. Intracerebral penetration of near infrared light. J Neurosurg 76: 315-18 (1992)

Mchedlishvili GI; Arterial Behaviour and Blood Circulation in the Brain. Chapter 6.4. pp. 274-291. Plenum Press New York (1986)

Obrig H, Kleinschmidt A, Merboldt KD, Dirnagl U, Frahm J, Villringer A. Monitoring of cerebral blood
    oxygenation during human brain activation by simultaneous high-resolution MRI and near-infrared
    spectroscopy. Society of Magnetic Resonance, 2nd Meeting, Book of Abstracts 67.
Rao SM, Binder JR, Bandettini PA, Hammeke TA, Yetkin FZ, Jesmanowicz A, Lisk LM, Morris GL, Mueller
    WM, Estkowski LD, Wong EC, Haughton VM, Hyde JS. Functional magnetic resonance imaging of
    complex human movement. Neurology 43:2311-18;(1993)
Roland PE, Larsen B, Lassen NA, Skinhoj E. Supplementary motor area and other cortical areas in organization
    of voluntary movements in man. J Neurophysiol 43:118-136 (1980)
Steinmetz H, Fürst G, Meyer BU. Craniocerebral topography within the international 10-20 system. Electroenc
    Clin Neurophysiol 72:499 (1989)
Villringer A, Planck J, Hock C, Schleinkofer L, Dirnagl U. Near infrared spectroscopy (NIRS): a new tool to
    study hemodynamic changes during activation of brain function in human adults; Neurosci Lett
    154(1-2):101-4 (1993)

28

PRESERVATION OF MITOCHONDRIAL MEMBRANE POTENTIAL DURING ANOXIA

Y. Nomura, T. Miyao, and M. Tamura

Biophysics group
Research Institute for Electronic Science
Hokkaido University
Sapporo 060, Japan

INTRODUCTION

Mitochondria (Mt) in living cells can be intensely and selectively stained against a dark cytoplasmic background by several fluorescent cationic dyes. For example, the incorporation of Rhodamine123 (RH123) into Mt depends on the membrane potential, accompanied by red shifts of absorption spectra and quenching of fluorescence intensity (Emaus *et al.*, 1986). Though almost all the fluorescence dyes work in the ultraviolet and visible regions, our major goal is to develop the optically active contrast agents that can enhance and contrast the changes of energization of living tissues in near-infrared region. This is due to the fact that the light in the near-infrared region is more translucent to tissues than visible light. In this paper, we show the possible extrinsic optical probes, RH123 as energy-dependent contrast reagent, though it works in visible region. Expanding to near-infrared region is also discussed.

MATERIALS AND METHODS

Liver Mt and hepatocytes (Ht) were prepared according to the previous report (Andersson *et al.*, 1987). Cell viability of Ht was > 95% as judged by trypan blue exclusion. Unless otherwise stated, respiratory control ratio of Mt used was > 5. Absorption changes of RH123 and heme aa_3 of cytochrome oxidase were measured simultaneously and continuously with a two-channel dual wavelength spectrophotometer (Unisoku, Japan). The wavelength pairs used were 515-488 nm for changes of RH123 and 605-622 nm for redox changes in heme aa_3. Fluorescence changes of RH123 were measured by excitation at 500 nm and emission at 525 nm. The reaction medium for Mt contained 250 mM sucrose/ 2 mM $MgCl_2$/ 5 mM Pi/ 1 mM RH123/ 10 mM Hepes pH 7.4. In the case of Ht, the medium used was the modified Krebs-Ringer bicarbonate buffer for Ht containing 120 mM NaCl/ 4.8 mM KCl/ 1.3 mM $CaCl_2$/ 1.2 mM KH_2PO_4/ 1.2mM $MgSO_4$/ 2.4mM $NaHCO_3$.

Oxygen Transport to Tissue XVII, Edited by Ince et al.
Plenum Press, New York, 1996

225

RESULTS

Responses of Mitochondrial Membrane Potential in Mitochondrial Suspension

Figure 1 (panel 1) and 1 (panel 2) showed changes in mitochondrial energization as $\Delta_{A_{515-488\,nm}}$ and redox state of heme aa$_3$ as $\Delta_{A_{605-622\,nm}}$ from aerobic to anaerobic conditions. Fluorescence intensity of RH123 also responded in a similar fashion to $\Delta_{A_{515-488\,nm}}$ although changes in fluorescence intensity inverted that of $\Delta_{A_{515-488\,nm}}$. As ATP synthesis of mitochondria occurred after additions of ADP under aerobic conditions, $\Delta_{A_{515-488\,nm}}$ decreased and fluorescence intensity increased. Upon complete conversions of added ADP to ATP, $\Delta_{A_{515-488\,nm}}$ and fluorescence intensity returned close to their previous state 4 levels though weakly transient and reversible reduction of heme aa$_3$ was not observed in this sensitivity. When aerobic conditions on state 3 were changed to anoxia, $\Delta_{A_{515-488\,nm}}$ rapidly reduced to 80% and fluorescence intensity increased, concomitantly with the full reduction of heme aa$_3$ (Fig.1 (panel 1)). Addition of oxygen saturated buffer caused the return of $\Delta_{A_{515-488\,nm}}$ and fluorescence intensity to aerobic levels just before anoxia. The $\Delta_{A_{515-488\,nm}}$ during anoxia stepwise decreased with the repetition of reoxygenation. As mitochondrial suspension was conditioned to anaerobiosis on state 4, $\Delta_{A_{515-488\,nm}}$ and fluorescence intensity remained unchanged above 30 min in spite of full reduction of heme aa$_3$ (Fig.1 (panel 2)).

Mitochondrial Membrane Potential of Hepatocytes during Anoxia

Measurement of changes in redox state of heme aa$_3$ in a suspension of Ht by dual wavelength spectrophotometry showed that maximal reduction occurred within a few minutes after anoxia judged by oxygen electrode. No changes of $\Delta_{A_{515-488\,nm}}$ were observed in this suspension under anaerobic conditions where heme aa$_3$ was fully reduced (Fig.1 (panel 3)). Anoxia had also no effect on fluorescence intensity of RH123 in Ht suspension. RH123, however, was taken up mitochondria because an addition of the uncoupler caused a rapid large decrease of $\Delta_{A_{515-488\,nm}}$ and an increase of fluorescence intensity. Changes in cell viability were negligible for 30 min under anoxia. However, even by 30 min, morphological changes of extensive formation of blebs on the cell surface sometimes occurred. Thus mitochondrial membrane potential remained almost unchanged by the 30-min anoxic exposure that caused full reduction of heme aa$_3$ of mitochondria and slightly morphological changes on cell surface.

Differentiation of Critical Stage of Hepatocytes Preceding Cell Death

Ht having normal shape in the aerobic conditions shows intense RH123 fluorescence (Fig.2-A and B). The cell indicated by an arrow had died from the beginning, judged by trypan blue exclusion. As shown in Fig.2-C, clear blebs formed on the surface in almost all cells after 40 min under anoxia. Figure 2-D shows a fluorescence microscope picture of Ht under the conditions. There were Ht with the different mitochondrial membrane potential although they had blebs and excluded trypan blue. In Ht, cell-surface bleb formation generally occurs under anoxia. Such blebbing is a characteristic response of hepatocytes exposed to anoxia *in situ* preparations (Lemasters *et al.*, 1983). If cell does not swell rapidly, plasma membrane blebs resorve completely within 6 min of reoxygenation without a loss of viability (Herman *et al.*, 1988). Though all the cells had the blebs, fluorescence staining by RH123 clearly differentiate the critical stages among these cells. When oxygen was readmitted, the blebs of "fluorescent" cell disappeared. Thus we could discriminate the critical

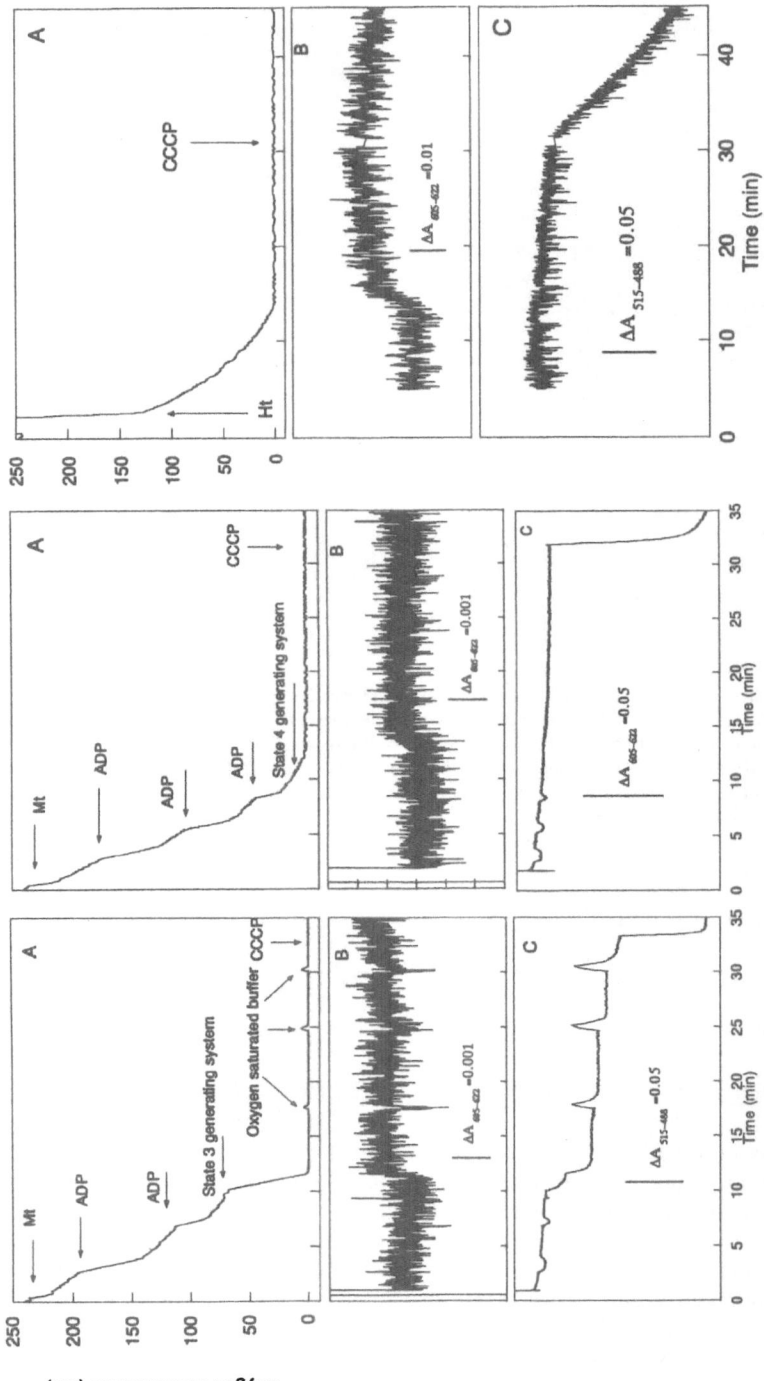

Figure 1. Responses of mitochondrial membrane potential as $\Delta A_{515\text{-}488\ nm}$ to anoxia concomitant with reduction of heme aa3 in cytochrome oxidase as $\Delta A_{515\text{-}488\ nm}$. Left panel (1); mitochondria in state 3. Mitochondria were incubated in the reaction medium containing 1 μM RH123. Hexokinase (15 U/ml), 10 mM glucose, 0.5 mM ADP, 0.5 mM magnesium chloride were used as a state 3 generating system. Three times of oxygen saturated buffer (60 μl) were added into a reaction medium under anoxia and then 1 μM CCCP. Middle panel (2); mitochondria in state 4. Pyruvate kinase (28 U/ml), 3 mM phosphoenolpyruvate and 0.5 mM ATP were used as a state 4 generating system. Right panel (3); hepatocytes. Hepatocytes were incubated in modified Krebs-Ringer buffer containing 20 μM RH123 for 15 min.

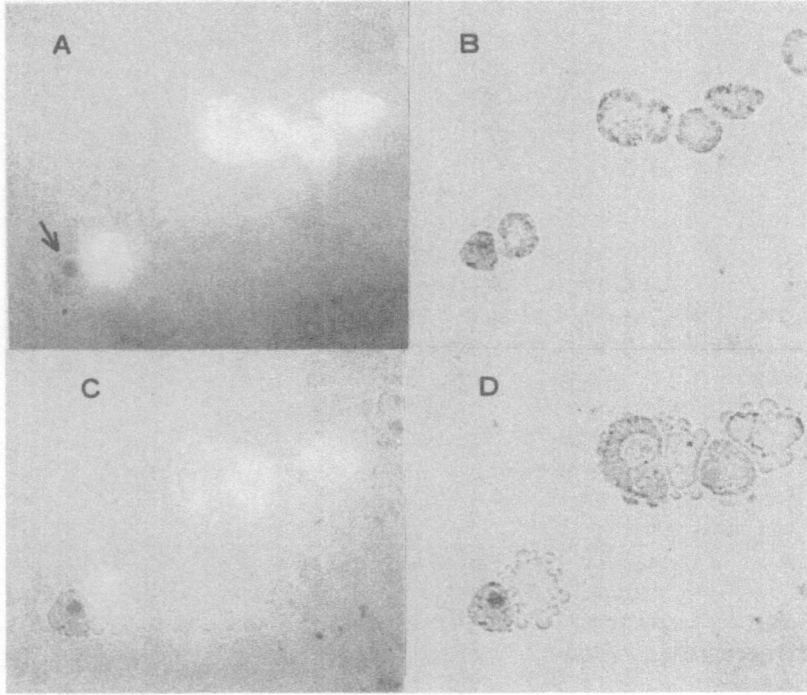

Figure 2. Mitochondrial membrane potential dependent uptake of RH123 and bleb formation in aerobic and anaerobic hepatocytes. A; aerobic cells observed by fluorescence microscope. B; a phase contrast image of aerobic cells. C; same cells after 40 min-anoxia by fluorescence microscope. D; a phase contrast image of C. Hepatocytes were isolated from a 24-h fasted rat and incubated in modified Krebs-Ringer buffer containing 20 μM RH123 for 15 min. The hepatocytes transferred to the slide glass after the addition of trypan blue. The size of each cell was 10-20 μm in diameter.

conditions of cells following to death by means of measurements of RH123 fluorescence under anoxia.

DISCUSSION

Present study demonstrated that the mitochondrial membrane potentials in both Mt and Ht suspension were highly preserved during anoxia. Under anaerobic conditions from state 4, ATP was enzymatically supplied. The motive force to maintain mitochondrial membrane potential would be derived from the reversal reaction of ATP synthase with continuously supplied ATP in the medium according to the chemiosmotic coupling hypothesis of oxidative phosphorylation (Mitchell, 1979). On the other hand, under anaerobic conditions from state 3, the incubation medium did not contain any ATP because ATP in the medium was continuously and enzymatically consumed. If passive transport down an electrochemical gradient or ATP production occurs under anoxia, mitochondrial membrane potential will decrease because of an arrest in proton translocation out of mitochondria. Therefore it appears that passive transport or ATP synthesis without electron transport in respiratory chain caused the rapid decrease of mitochondrial membrane potential to 80%.

And then mitochondria itself rather than an interaction between cytosol and mitochondria would preserve mitochondrial membrane potential of 80% to aerobic state during anoxia due to drastic decreases of ion permeability, i.e., inhibiting the ion transport systems of inner membrane.

Mt take up one of kryptocyanine analogues that has absorption maximum at 650 nm and fluorescent maximum at 720 nm although it becomes toxic in the light (McDonald *et al.*, 1990). By the use of an energy-dependent contrast reagent that acts in near-infrared region, the energization of blood perfused tissues would be monitored.

SUMMARY

Advantages of RH123 as an energy dependent contrast reagent were summarized below. (1) We could differentiate Ht with the different mitochondrial membrane potential when they formed blebs but excluded trypan blue in the critical conditions. (2) Optical responses of RH123 differed from redox state of heme aa_3. (3) Mitochondrial membrane potential was highly preserved during anoxia.

ACKNOWLEDGMENTS

This study was supported in part by grants from the Kowa Life Science Foundation.

REFERENCES

Andersson, B. S., AW, T. Y., and Jones, D. P., 1987, Mitochondrial transmembrane potential and pH gradient during anoxia, *Am. J. Physiol.* 252(*Cell Physiol.* 21): C349-C355.

Emaus, R. K., Grunwald, R., and Lemasters, J. J., 1986, RH123 as a probe of transmembrane potential in isolated rat-liver mitochondria: spectral and metabolic properties, *Biochim. Biophys. Acta.* 850: 436-448.

Herman, B., Nieminen, A. L., Gores, G. J., and Lemasters, J. J., 1988, Irreversible injury in anoxic hepatocytes precipitated by an abrupt increase in plasma membrane permeability, *FASEB J.* 2: 146-151.

Lemasters, J. J., Stemkowski, C. J., Ji, S., and Thurman, R. G., 1983, Cell surface changes and enzyme release during hypoxia and reoxygenation in the isolated perfused rat liver, *J. Cell Biol.* 97: 778-786.

McDonald, J. W., Brash, D. E., Oseroff, A. R., and Harris, C. C., 1990, Photosensitizing dyes for the selection of nontumorigenic revertants from human lung cancer cell lines, *Cancer Res.* 50: 5369-5373.

Mitchell, P., 1979, Keilin's respiratory chain concept and its chemiosmotic consequences, *Science Wash. DC.* 206: 1148-1159.

29

INFLUENCE OF THE TREADMILL SPEED/SLOPE ON QUADRICEPS OXYGENATION DURING DYNAMIC EXERCISE

V. Quaresima,[1] A. Pizzi,[2] R. A. De Blasi,[3] A. Ferrari,[4] and M. Ferrari[1,5]

[1] Department of Biomedical Sciences and Technology, University of
  L'Aquila
  67100 L'Aquila, Italy
[2] Centro di Recupero Funzionale Fondazione Pro Juventute Don C. Gnocchi
  Pozzolatico, Firenze, Italy
[3] Institute of Anesthesiology
  I University of Rome, Italy
[4] Rehabilitation Section, Scandiano Hospital, Reggio E
[5] Istituto Superiore Sanità
  Rome, Italy

INTRODUCTION

One of the main obstacles that has restricted "*in vivo*" studies of skeletal muscle metabolism during exercise has been the invasive nature of tissue biopsy and the high cost of phosphorus nuclear magnetic resonance equipment. Big research efforts have been recently focused on the development of non-invasive optical techniques to monitor tissue functions. Changes of cerebral and muscle oxy- and deoxy-hemoglobin ($[HbO_2]$, $[Hb]$) can be quantitated from near infrared spectroscopy (NIRS) measurements combining absorption and optical pathlength (Ferrari et al. 1992). NIRS has been extensively used in neuroscience to measure cerebral oxygenation, blood flow, and blood volume (Edwards et al. 1988; Wyatt et al. 1990). Some efforts have been focused also on skeletal muscle. In the last four years the aim of our group has been to explore NIRS capabilities for muscle functional monitoring. Pathlength changes in muscle have been measured on volunteers in different experimental conditions by time resolved spectroscopy (Ferrari et al. 1993) and phase modulation (Duncan et al. 1993). Muscle oxygen consumption (VO_2) can be measured by NIRS at rest and during isometric exercise by the application of arterial occlusion (Cheatle et al. 1991; De Blasi et al. 1993). We recently proposed a new method for the simultaneous measurement of forearm blood flow and VO_2 by inducing a venous occlusion (De Blasi et al. 1994).

Oxygen Transport to Tissue XVII, Edited by Ince et al.
Plenum Press, New York, 1996

In this study we investigated quadriceps femoris oxygenation and blood volume changes during dynamic exercise.

METHODS

Measurements were performed on 10 untrained volunteers. Each subject gave informed consent before starting the study. NIRS measurements were obtained by the NIRO500 (Hamamatsu Photonics, Japan). The light generated by four laser diodes operating at 775, 825, 850 and 904 nm is coupled to a fibre optic bundle with an emitting area of 2.5 mm diameter. The transmitted light is collected by a second fibre optic of the same size and detected by a photomultiplier (sampling time 1 s). A special support allowed both the distance (3-4 cm) and the angle between the ends of the optical fibres (the optodes) to be varied. The optode was fixed by double face tape and maintained by a soft elastic wrapping. To quantify changes in [Hb] and [HbO$_2$], the absorption of light was analyzed according to a modified Beer law, using NIRO500 software:

$$\log(I_o/I) = OD = a \cdot c \cdot L \cdot B + G$$

where OD is the optical density, a is the absorption coefficient (μM/cm) of the chromophore, c is the chromophore concentration (μM), L is the physical separation of the optodes on the skin surface (cm), B the differential pathlength factor (DPF) and G is a tissue and geometry dependent factor. The DPF value was measured separately by time resolved spectroscopy (Ferrari et al. 1992). As a, L, and B are known and G remains constant, changes in chromophore concentration can be calculated relative to a starting point considered as zero. Data collected by NIRS were transferred on-line to a computer for storage and subsequent analysis. The results are expressed as micromoles per liter of tissue (μmol/l). Optodes were firmly positioned on the vastus lateralis of the quadriceps muscle by a special support. Measurements were performed on both quadriceps with a 20 min interval. The sum of [HbO$_2$] and [Hb] was considered an index of blood volume (Hbvol) change; the difference [HbO$_2$]-[Hb] was considered an index of hemoglobin saturation change (Hbdiff) only when no variation of Hbvol occurs. These data can be used to monitor muscle hemodynamic changes. Heart rate and arterial saturation were monitored by pulse oximetry. A treadmill ergometer (Runner, Bianchini & Draghetti, Italy) was used to standardize the walk. The subjects were asked to rest for 5 min in the upright position on the treadmill before exercise. Two protocols were performed. In protocol 1 the exercise consisted of 12 min walking on the treadmill at a speed of with 1.6 km/h. The slope was gradually increased from 0 to 16° with a step of 4° and 2 min duration. In the protocol 2 (incremental exhausting run test) the treadmill speed was gradually increased from 1.6 to 11 km/h in 6-10 min.

RESULTS

In the subjects submitted to protocol 1, two different patterns of [Hb] and [HbO$_2$] changes were found. A typical example of the first pattern, found on 4 out of 10 subjects, is shown in Fig.1.

[HbO$_2$] and [Hb] increased since the beginning of the walk. During the elevation of the treadmill slope, [HbO$_2$] and [Hb] increased proportionally up to 8° and more slowly at 12 and 16°. [HbO$_2$] raise was attributable to the increase of the [HbO$_2$] in the venous circulation due to blood flow increase. [Hb] instead reflects the O$_2$ extraction raise. From this point of view the [Hb] increase could be considered an index of the O$_2$ demand in

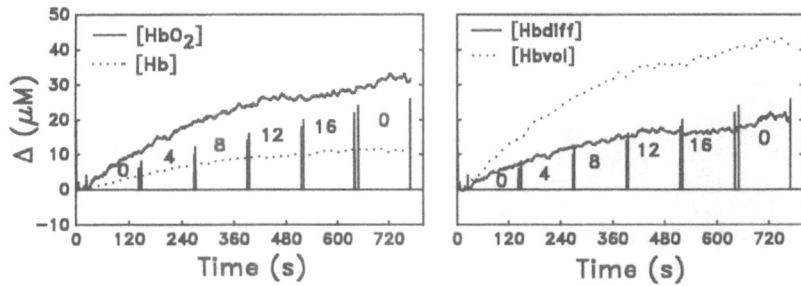

Figure 1. [Hb], [HbO$_2$], [Hbdiff], and [Hbvol] changes on the right quadriceps of a volunteer (M.L. 34 years old) during a walking test (speed = 1.6 km/h). Numbers indicate the slope of the treadmill in degrees.

exercising muscle. A typical example of the second pattern, found on 6 out of 10 subjects, is shown in Fig.2.

[HbO$_2$] and [Hb] increased at the beginning of the walk. During the elevation of the treadmill slope, [HbO$_2$] increased consistently till 4° slope and very slowly in the following steps. Hb was unchanged during the treadmill elevation. This suggests that in these subjects a rapid balance between O$_2$ supply and O$_2$ demand was reached. Heart rate and arterial saturation had no significant variations in both patterns. A right/left difference in [HbO$_2$] higher than 25% was found in 6 out of the 10 subjects analysing the corresponding values at the end of 16° slope. A right/left difference in [Hbvol] higher than 25% was found in 8 out of the 10 subjects analysing the corresponding values at the end of 16° slope. An example of these differences is shown in Fig. 3. No difference was found in the subject of Fig. 2.

More heterogeneous was the response to the incremental exhausting run test. Figure 4 reports the oxygenation changes during different treadmill speed. The greatest [Hb] increase occurred in correspondence to the maximal velocity (11 km/h) when [HbO$_2$] was

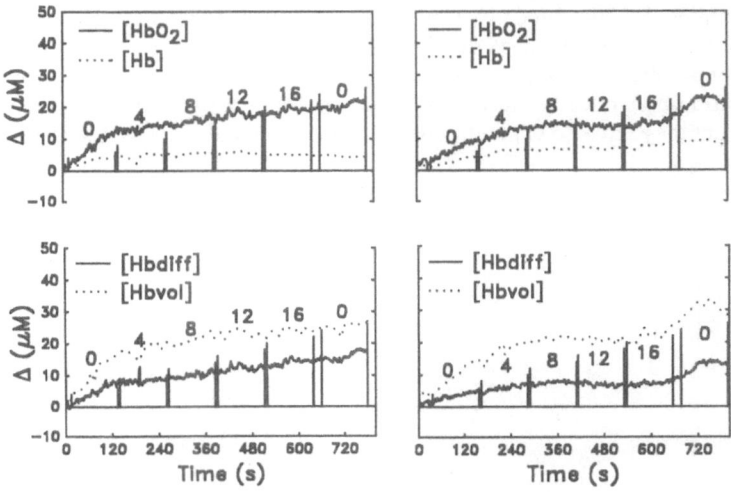

Figure 2. [Hb], [HbO2], [Hbdiff], and [Hbvol] changes in the right (left panels) and left (right panels) quadriceps of a volenteer (16 years old) during a walking test (speed = 1.6 km/h). Numbers indicate the slope of the treadmill in degrees.

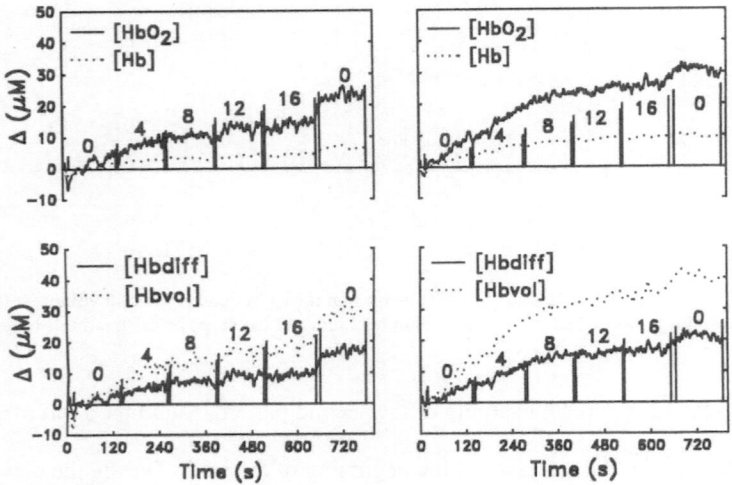

Figure 3. [Hb], [HbO$_2$], [Hbdiff], and [Hbvol] changes in the right (left panels) and left (right panels) quadriceps of a volunteer (S.S. 31 years old) during a walking test (speed = 1.6 km/h). Numbers indicate the slope of the treadmill in degrees.

almost constant. This suggests a limitation of muscle O$_2$ supply. When the run speed declined, [HbO$_2$] and [Hb] rapidly rose and decreased, respectively.

CONCLUSION

NIRS has been extensively applied to investigate muscle pathophysiology (Chance et al. 1988; Chance et al. 1992; De Blasi et al. 1992; Hampson et al. 1988; Mancini et al 1991; Mancini et al. 1994; McCully et al. 1993; McCully et al. 1994; Wilson et al. 1989).

Although the precision of the data reported in this study was limited because no correction for pathlength changes was performed, the results indicate the capability of NIRS for investigating dynamically muscle oxygenation. Many NIRS clinical instruments are on the market, but they have no capability to measure pathlength. A new generation of clinical

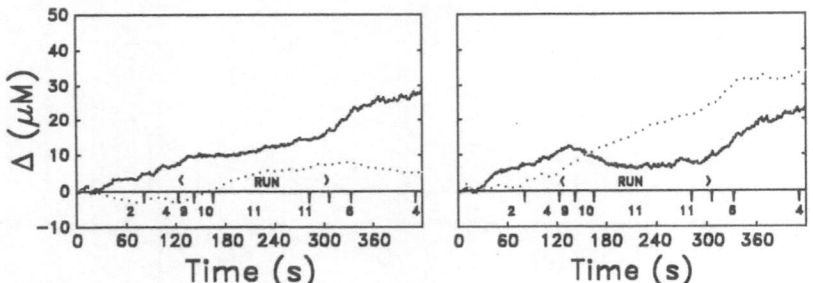

Figure 4. Oxygenation changes on the right quadriceps of a volunteer (C.P. 29 years old) during a running test (slope = 0°). Numbers indicate the treadmill speed. Left panel: [Hb] (dotted line), [HbO$_2$] (solid line) and right panel: [Hbdiff] (solid line), [Hbvol] (dotted line).

instruments capable of combining high sensitivity spectral measurement with pathlength should be available in the near future. The frequency-domain spectrometers under development (Duncan et al. 1993) could be used to investigate accurately muscle oxygenation. In this prospective NIRS could represent a useful non invasive optical tool to investigate dynamically skeletal muscle pathophysiology. We wish to thank Ing. C. Robino of Hamamatsu Photonics Italy who loaned the NIRO500. The financial support of Telethon-Italy (Grant no. 501) and CNR 93.00282/94.02408.CT04 is gratefully acknowledged. This research was supported in part by: Tecnobiomedica/Officine Ortopediche Rizzoli Bologna.

REFERENCES

Chance, B., Nioka, S., Kent, J., McCully, K., Fountain, M., Greenfeld, R., and Holtom, G., 1988, Time-resolved spectroscopy of hemoglobin and myoglobin in resting and ischemic muscle, *Anal. Biochem.* 174, 698-707.

Chance, B., Dait, M.T., Zhang, C., Hamaoka, T. and Hagerman, K., 1992, Recovery from exercise-induced desaturation in the quadriceps muscle of elite competitive rowers, *Am. J. Physiol.* 262:C766-C775.

Cheatle, T.R., Potter, L.A., Cope, M., Delpy, T., Coleridge Smith, P.D., and Scurr J.H., 1991, Near-infrared spectroscopy in peripheral vascular disease, *Brit. J. Surg.* 78:405-408.

De Blasi, R.A., Conti, G., Mattia, C., Mega, A.M., Ferrari, M., and Gasparetto, A., 1992, Non-invasive evaluation of respiratory muscle oxygenation in mechanically ventilated patients by near infrared spectroscopy, *Int. Care Med.* 18:P5.

De Blasi, R.A., Cope, M., Elwell, C., Safoue, F., and Ferrari M., 1993, Non invasive measurement of human forearm oxygen consumption by near infrared spectroscopy, *Eur. J. Appl. Physiol.* 67:20-25.

De Blasi, R.A., Ferrari, M., Natali, A., Conti, G., Mega, A., and Gasparetto A., 1994, Non-invasive measurement of forearm blood flow and oxygen consumption by near infrared spectroscopy, *J. Appl. Physiol.* 76:1388-1393.

Duncan, A., Whitlock, T.L., Cope, M., and Delpy, D.T., 1993, A multiwavelength, wideband, intensity modulated optical spectrometer for near infrared spectroscopy and imaging, *Proc. SPIE* 1888 248-257.

Edwards, A.D., Wyatt, J.S., Richardson, C.E., Delpy, D.T., Cope, M., and Reynolds, E.O.R. 1988, Cotside measurement of cerebral blood flow in ill newborn infants by near infrared spectroscopy, *Lancet* 2:770-771.

Ferrari, M., Wei, Q., Carraresi, L., De Blasi, R.A., and Zaccanti, G., 1992, Time-resolved spectroscopy of human forearm, *J. Photochem. Photobiol.* 16:141-153.

Ferrari, M., Wei, Q., De Blasi, R.A., Quaresima, V., and Zaccanti, G., 1993, Variability of human brain and muscle optical pathlength in different experimental conditions, *Proc. SPIE* 1888:466-472.

Hampson, N.B., and Piantadosi, C.A., 1988, Near infrared monitoring of human skeletal muscle oxygenation during forearm ischemia, *J. Appl. Physiol.* 64:2449-2457.

Mancini, D.M., Ferraro, N., Nazzaro, D., Chance, B., and Wilson, J.R., 1991, Respiratory muscle deoxygenation during exercise in patients with heart failure demonstrated with near infrared spectroscopy, *J. Am. Coll. Cardiol.* 18:492-498.

Mancini, D.M., Wilson, J.R., Bolinger, L., Li, H., Kendrick, K., Chance, B., and Leigh, J.S., 1994, *In vivo* magnetic resonance spectroscopy measurement of deoxymyoglobin during exercise in patients with heart failure. Demonstration of abnormal muscle metabolism despite adequate oxygenation, *Circulation* 90:500-508.

McCully, K.K., Halber, C., and Posner, J.D., 1994, Exercise-induced changes in oxygen saturation in the calf muscles of elderly subjects with peripheral vascular disease, *J. Gerontol.* 49:B128-B134.

McCully, K.K., Iotti, S., Kendrick, K., Wang, Z., Posner, J.D., Leigh, J., and Chance B., 1994, Simultaneous *in vivo* measurement of HbO_2 saturation and PCr kinetics after exercise in normal humans, *J. Appl. Physiol.* 77:5-10.

Wilson, J.R., Mancini, D.M., McCully, K., Ferraro, N., Lanoce, V., and Chance, B., 1989, Noninvasive detection of skeletal muscle underperfusion with near-infrared spectroscopy in patients with heart failure, *Circulation* 80:1668-1674.

Wyatt, J.S., Cope, M., Delpy, D.T., Richardson, C.E., Edwards, A.D., Wray, S., and Reynolds, E.O.R.,1990, Quantitation of cerebral blood volume in human infants by near-infrared spectroscopy, *J. Appl. Physiol.* 68:1086-1091.

30

DISCREPANCY BETWEEN EVIDENCE AND THEORY OF MYOGLOBIN SO$_2$ IN WORK

John W. Severinghaus[*]

Professor of Anesthesiology Emeritus
University of California Medical School
San Francisco California 94143

BACKGROUND

Gayeski, Federspiel and Honig [11] showed that myoglobin PO$_2$ during maximum electrically stimulated exercise in dog gracilis muscle fell as low as 1 torr, but usually was limited to 3-5 torr, and was remarkably uniform, suggesting that much of the diffusion gradient for O$_2$ was between red cells and myocytes. Myoglobin is 50% saturated at about 4 torr, pH = 7, and has no Bohr effect. Its dissociation curve is a single exponential. In recent experiments Richardson et al. [2], using ^{1}H MRS quantification of the quadriceps myoglobin.signal with one leg exercise in human volunteers, indicated that S$_{Mb}$O$_2$ fell with mild exercise (25% $\dot{V}O_{2max}$) to about 50%, but remained at that level at four higher work levels (approaching $\dot{V}O_{2max}$).

I recently modelled exercise O$_2$ transport employing a new hypothesis that at $\dot{V}O_{2max}$ PO$_2$ reaches zero at the muscle cytochrome aa$_3$ oxidase site, which is to say that all O$_2$ diffusing to that site is consumed, leaving no free O$_2$ to diffuse away from the site [4]. I assumed a constant tissue O$_2$ diffusing capacity, uniformly distributed blood flow directly proportional to work and $\dot{V}O_{2max}$ and employed data of others for the relationship of muscle venous PO$_2$ and pH to work [3,5]. At $\dot{V}O_{2max}$ end-capillary (or venous) PO$_2$ has been found to plateau at or below 20 Torr, being sustained as work rises by the lactic acidification through the Bohr effect [51]. The model demonstrated that, in order to reduce myoglobin O$_2$ saturation (S$_{Mb}$O$_2$) at $\dot{V}O_{2max}$ to 20-25%, P$_{mb}$O$_2$ $\simeq$ 2 Torr, about 90% of the O$_2$ diffusion gradient from capillary to cytochrome must lie between blood and myoglobin. The model predicted S$_{Mb}$O$_2$ to. remain above 90% at 25% of $\dot{V}O_{2max}$ and to fall steeply from about 80% at 50% $\dot{V}O_{2max}$ to 25% at $\dot{V}O_{2max}$ Figure 1 illustrates this model with blood flow at rest of 10% of flow at $\dot{V}O_{2max}$ P$_a$CO$_2$ = 40 torr, initial pH$_a$ = 7.4 and BE = 0, and final S$_v$O$_2$ = 10%, final pH, = 7.0. Myoglobin saturation falls non-linearly to a minimum of about 25%, a plateau appearing near $\dot{V}O_{2max}$ due to the effect of lactic acid on P$_v$O$_2$.

---

[*] Mail addr: PO Box 974, Ross, CA 94957.

Oxygen Transport to Tissue XVII, Edited by Ince et al.
Plenum Press, New York, 1996

237

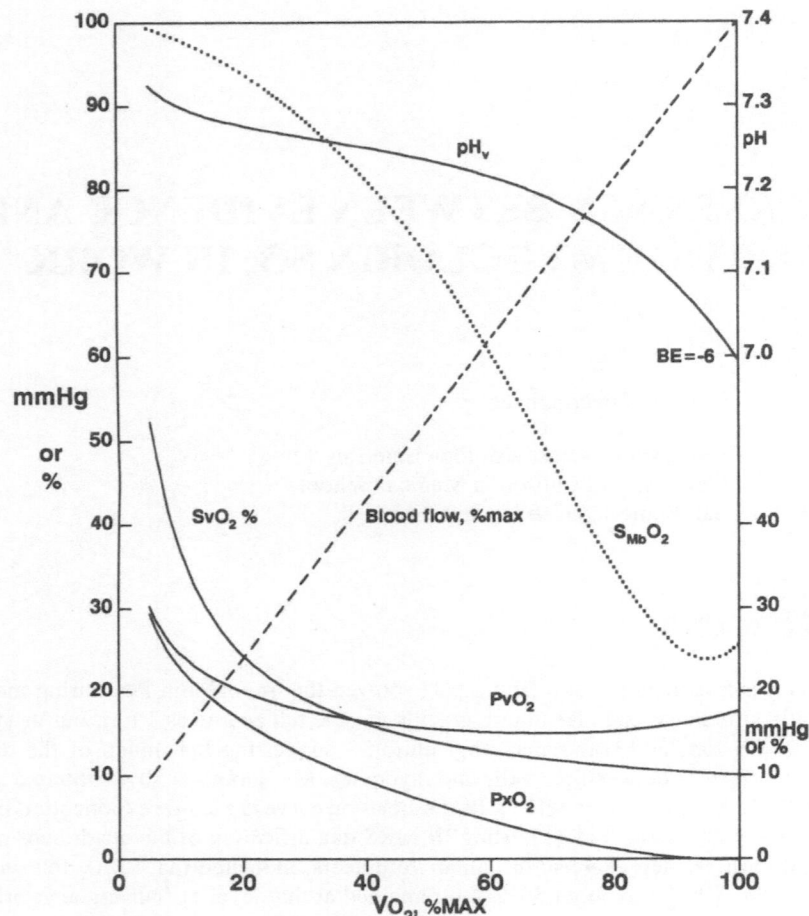

Figure 1. Model of O_2 delivery as a function of O_2 consumption in working muscle, assuming uniform distribution of blood flow and no shunt. Myoglobin saturation falls slowly from near 100% at low work rates, and rapidly at high rates reaching a minimum of about 25% at VO_{2max}. Model is based on data in man [3,51].

THE DISCREPANCY

This was strikingly at variance with the relatively work-independent $S_{Mb}O_2$ data of Richardson et al. I have now attempted to introduce assumptions of capwary recruitment i.e. heterogeneous flow, into the model to determine whether these new data of Richardson et al. could be rationalized. In order to obtain constant $S_{Mb}O_2$ and $P_{Mb}O_2$ as work increases, capillary recruitment would permit the effective diffusing capacity to increase in proportion to O_2 flux (VO_{2max}).

Two-Compartment Model

Muscle venous blood as collected might consist of a mix from non-working and working muscle, with the proportion from non-working (or shunt) areas falling with increased work. In Figure 2, resting blood flow was taken to be 15% of the flow at $\dot{V}O_{2max}$, of which 4% was assumed to come from the muscle, 11% from shunt, skin and other low metabolic rate tissue. Shunt flow was assumed to fall linearly to zero at $\dot{V}O_{2max}$ as capillary recruitment occurred. From 25% to 100% of $\dot{V}O_{2max}$, working muscle capillary PO_2 was assumed to be held constant by recruitment at about 20 torr, thus maintaining myoglobin PO_2 at 4 torr, myoglobin saturation at 50%, with cytochrome PO_2 at or near zero, using all available O_2 at all levels of exercise. In this model, the end-capillary or venous saturation leaving active muscle falls due to the falling blood pH with lactic acidosis (since PO_2 is defined as constant). The diffusing capacity of muscle as estimated from the venous blood

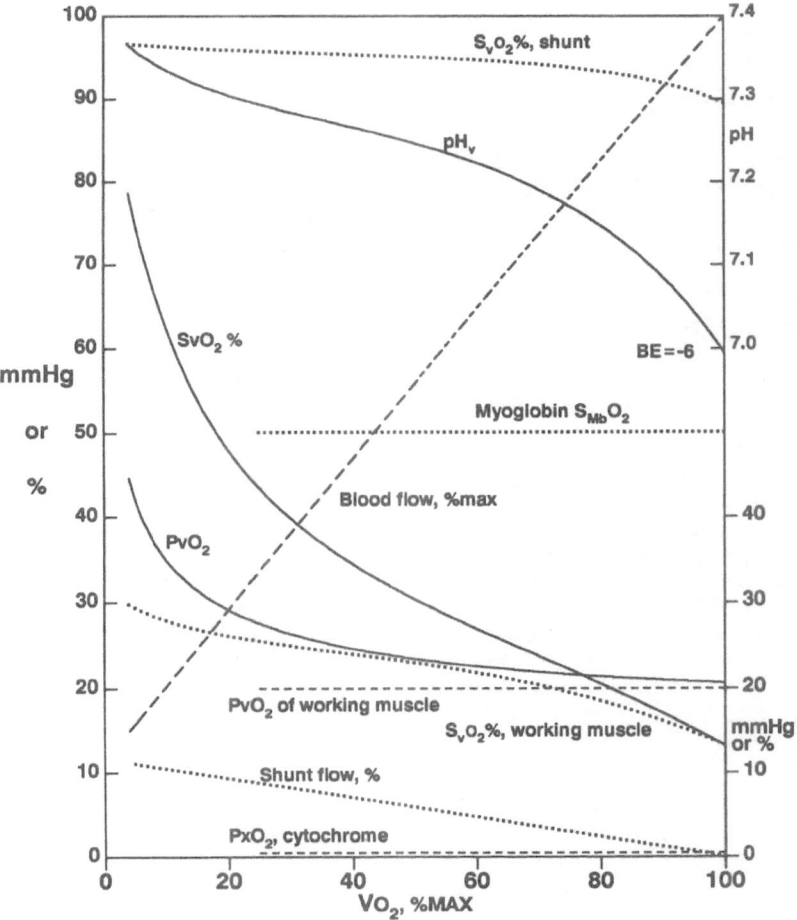

Figure 2. A 2-compartment model with capillary recruitment attempting to account for published data [21] suggesting that myoglobin saturation was approximately constant at work levels varying from 25 to nearly 100 of $\dot{V}O_{2max}$. This model would require a constant effective capillary PO_2 and cytochrome PO_2, independent of work.

would appear to increase with work (the PO_2 gradient from venous blood to myoglobin or to cytgchrome increases with work). The model shunt flow could be modified, e.g. to fall to zero at 25% of VO_{2max} in some curvilinear way, to account for changes at low work rates. This hypothesis is compatible with evidence obtained from muscle venous blood but introduces several problems which I consider sufficient to reject it: 1) Myoglobin must be diflusionally closer to capillary than can be justified if its saturation falls to 25% (lying 80% rather than 90% down the O_2 gradient). 2) The constant cytochrome aa_3 PO_2 at all work levels fails to explain why lactic acid evolution begins only at 50-70% of VO_{2max}.

Several other model ideas were considered briefly but not found reasonable. For example, assume that muscle blood flow were not linearly related to work at the transition from rest, such that, only above some "dog-leg" onset, flow would become proportional to work, and thus capillary SO_2 be work-independent. This fails to account for the gradually falling S_vO_2 and P_vO_2 as work increases. Or assume that each muscle fascicule either rests or contracts maximally, using all its myoglobin O_2 then rests while O_2 is restored. This also appears incompatible with the known relation of lactate production and venous denaturation to increasing work.

CONCLUSION

I was unable to conceive a model which would reconcile the observed lactic acid threshhold, the gradual fall of S_vO_2 and P_vO_2, and the presumed facilitation of O_2 delivery by the Bohr effect with the work-rate-independent myoglobin saturation observations of Richardson et al.

REFERENCES

1. Gayeski, T. E. J., W. J. Federspiel, and C. R. Honig. A graphical analysis of the influence of red cell transit time, carrier-free layer thickness, and intracellular PO_2 on blood-tissue O_2 transport. *Adv. Exp. Me£L Biol.* 222: 25-35, 1988.
2. Richardson R. S., E. A. Noyszewski, K F. Kendrick J. S. Leigh, P. D. Wagner. Evidence for a substantial po_2 gradient from blood to myoglobin during maximal quadriceps exercise in normal man. *Clin Res.* 42:2:254AApr 1994
3. Roca, J., A. G. Agusti, A. Alonso, D. C. Poole, C. Viegas, J. A. Barbera, R. Rodriguez-Rouiin, A. Ferrer, P. D. Wagner. Effects of training on muscle O_2 transport at VO_{2max}. *J. Appl Physiol* 73:1067-1076,1992.
4. Severinghaus, J.W. Exercise O_2 transport model assuming zero cytochrome PO_2 at VO_{2max}. *J Appl Physiol* 77:671-678, 1994.
5. Wasserman, K Coupling of external to cellular respiration during exercise: the wisdom of the body revisited. Am *J. Physiol* 266 (Endrocrinol. Metab 29): E519-E539, 1994.

31

IS THERE A REDUCTION OF ISCHAEMIA/REPERFUSION DAMAGE IN THE ISOLATED RAT HEART BY ALTERNATING OXYGENATED AND DEOXYGENATED REPERFUSION SOLUTIONS?

D. R. Rutgers,[1] J. H. G. M. van Beek,[1] and M. D. Osbakken[2]

[1] Laboratory for Physiology
Institute for Cardiovascular Research (ICaR-VU)
Vrije Universiteit
van der Boechorststraat 7, 1081 BT Amsterdam, The Netherlands
[2] Bristol–Myers Squibb
Princeton, New Jersey and
University of Pennsylvania
Philadelphia, Pennsylvania

INTRODUCTION

Recent pilot experiments (Osbakken *et al*, 1994) suggested that alternate perfusion with normoxic and anoxic solutions ('N_2/O_2-cycling') might have beneficial haemodynamic effects in rat hearts that had developed high diastolic and low developed left ventricular pressures for which the cause was not precisely known. In these pilot experiments, in which the perfusate was recirculated, isolated rat hearts were perfused with Krebs-Henseleit solution according to Langendorff. Even though the hearts developed rigor or ventricular fibrillation, coronary flow was maintained throughout (up to 5 ml/min) with solution gassed with 95% O_2–5% CO_2. Subsequently, arrested hearts recovered when perfused with solution gassed with 95% N_2–5% CO_2, alternating with solution gassed with 95% O_2–5% CO_2, whereas no recovery occurred when hearts were perfused with solution gassed with 95% O_2–5% CO_2 alone. The mechanism of the putative beneficial action of N_2/O_2-cycling is unknown, but it was hypothesized that the alternate gassing regimen might protect the heart from oxygen free radical damage.

In order to further explore the potential of N_2/O_2-cycling as a resuscitation method for the heart, we decided to apply a similar reperfusion regimen to hearts made globally ischaemic. In the present study, globally ischaemic hearts were reperfused with solution

Oxygen Transport to Tissue XVII, Edited by Ince et al.
Plenum Press, New York, 1996

241

gassed with 95% N_2–5% CO_2 alternating with solution gassed with 95% O_2–5% CO_2 and compared to globally ischaemic hearts reperfused with solution gassed only with 95% O_2–5% CO_2. Results from studies such as this might significantly alter regimens of oxygen administration in patients who require resuscitation, either from a sudden death syndrome or when tapered from heart-lung bypass following cardiothoracic surgery.

MATERIALS AND METHODS

Preparation

Adult male Sprague-Dawley rats weighing 260–340 g were anaesthetized with an intraperitoneal injection of 60 mg/kg ketamine (Ketalar®, Parke–Davis S.A., Barcelona, Spain) and 8 mg/kg xylazine (Rompun®, Bayer, Leverkusen, Germany) (Wixson et al., 1987). After an intraperitoneal injection of 750 IU heparin (Leo Pharmaceutical Products, Weesp, The Netherlands), the chest was opened, and the heart was excised and placed in ice-cold saline. The aorta was cannulated and the heart was perfused according to Langendorff with modified Krebs–Henseleit solution at 37°C and pH 7.4. The Krebs-Henseleit solution contained (in mM) NaCl 118.0, KCl 4.7, $CaCl_2$ 1.47, $MgSO_4$ 1.2, KH_2PO_4 1.2, $NaHCO_3$ 25.0 and glucose 10.0. The solution was filtered through a 1.2-μm filter (Millipore S.A., Molsheim, France) before use in perfusing the heart. The flow rate was controlled by a roller pump (Gilson Minipuls 3, Villiers le Bel, France). Two thermostatically controlled vessels were gassed with 95%N_2-5%CO_2 and 95% O_2–5%CO_2 respectively. The heart could be alternately perfused with solutions from both vessels using a glass 2-way stopcock, which was placed proximal to the heart. A pressure transducer (Gould P23ID, Oxnard, California, USA) was located at the proximal side of the cannula to monitor perfusion pressure. A thin latex balloon, connected via a polyethylene tube to a pressure transducer (Gould P23ID), was introduced in the left ventricular chamber through the mitral valve to monitor left ventricular pressure. The pulmonary artery was cannulated to collect coronary venous effluent. The effluent was run through a small glass vessel, in which an O_2 electrode (Radiometer E 5047-0, Copenhagen, Denmark) was placed to monitor venous PO_2 in approximately 2 ml effluent. O_2 consumption, expressed in $\mu mol \cdot min^{-1} \cdot g^{-1}$ dry weight, was calculated according to the equation:

$$O_2 \text{ consumption} = F \cdot \alpha \cdot (PO_{2,a} - PO_{2,v}) / W$$

where F is flow (ml $\cdot$ min^{-1}), α is O_2 solubility ($\mu mol\ O_2 \cdot ml^{-1} \cdot mmHg^{-1}$), ($PO_{2,a}$ - $PO_{2,v}$) is the difference between arterial PO_2 and venous PO_2 (mmHg) and W is dry weight (g).

Finally the heart was placed in a thermostatted double-walled beaker, which was filled with the coronary venous effluent. Perfusion pressure, left ventricular pressure, venous PO_2 and temperature were recorded on an eight-channel recorder (Thermal Arraycorder WR 3600, Graphtec Corporation, Tokyo, Japan).

Measurement of Creatine Phosphokinase

Samples of 1 ml of the coronary venous effluent were collected at different time points from each heart to determine creatine phosphokinase (CPK) activity. Each sample was mixed with 0.5 ml of an albumin solution (75 g/L) to protect the CPK activity during freezing and storage at -80°C (Dunphy & Ely, 1990). CPK activity was determined within six weeks of sample collection using the method of Oliver et al. (1966)

Experimental Protocols

This protocol was approved by the committee for experiments on animals of the Vrije Universiteit, Amsterdam. Hearts were randomly assigned to the experimental groups.

Equilibration Period. All hearts were equilibrated to the perfusion condition for 20 min. During the equilibration period the perfusion pressure was kept at 60 mm Hg by changing the flow rate. From the end of the equilibration period onward the flow rate was kept constant, except when flow was completely stopped to produce global ischaemia. The latex LV balloon was filled with water at the beginning of the equilibration period until the measured left ventricular diastolic pressure was 5 mm Hg. After the experiment, corrections were made for the pressure difference across the wall of the balloon caused by the tension in the balloon itself, which resulted, on average, in an 11 mm Hg increase of left ventricular diastolic and left ventricular developed pressure. If the basal heart rate was $\geq$ 240 beats/min the heart was not paced. Hearts beating at a rate < 240 beats/min were paced at 4 Hz (Grass SD9 stimulator, Quincy, Mass., USA). During the 20th min of equilibration a sample was taken from the coronary venous effluent for CPK determination.

Ischaemia - Reperfusion Period. Group 1: In group 1 (n = 8, N_2/O_2 group), the equilibration period was followed by 60 min no-flow ischaemia. The hearts were immersed in Krebs-Henseleit solution kept at 37°C. This hindered oxygen diffusion from the air into tissue. After this global ischaemic period, hearts were reperfused for 32 min using N_2/O_2-cycling, *i.e.*, hearts were reperfused starting with a solution gassed with 95% N_2–5% CO_2 which was changed after 2 min to a solution gassed with 95% O_2–5% CO_2. Every 2 min, the solutions were switched between N_2 and O_2 gassed solutions until the hearts had been reperfused for 32 minutes. This 32 min period of N_2/O_2-cycling was followed by 28 min with continuous perfusion at 95% O_2–5% CO_2 to monitor the recovery of the heart function. Samples from the coronary venous effluent were taken 1, 10, 21 and 30 min after start of the reperfusion for determination of CPK. The remainder of the reperfusion effluent was collected over 60 min in a vessel placed on ice. At the end of the experiment the effluent was thoroughly mixed and a sample was taken to determine cumulative CPK release.

Group 2: In group 2 (n = 8, O_2/O_2 group), which acted as control, the protocol and setup were exactly the same as for group 1, except that the solutions in the two vessels, which were used to alternately reperfuse the globally ischaemic heart, were both gassed with 95% O_2–5% CO_2. Coronary venous effluent was collected for analysis of CPK using the same timing sequence as used in group 1.

Group 3: In group 3 (n = 4, no-ischaemia group), the stability of isolated perfused heart was documented; in this case the 20 min equilibration period was followed by 120 min of continuous perfusion at constant flow rate using solution gassed with 95% O_2–5% CO_2. Coronary venous effluent was collected for analysis of CPK using the same timing sequence as used in group 1.

Statistical Methods

Data are reported as mean $\pm$ SE. Because in many cases the Bartlett test showed deviation from homogeneity of variance (Snedecor & Cochran, 1980), nonparametric Kruskal-Wallis one-way analysis of variance was used for intergroup comparisons (Snedecor & Cochran, 1980). When comparing two groups, the Mann–Whitney test was used. Repeated measures analysis of variance was used to test for changes over time during reperfusion (Snedecor & Cochran, 1980). Differences were considered statistically significant for p-values < 0.05.

Table 1. Baseline physiological parameters

	Group 1	Group 2	Group 3
	(N_2/O_2 group)	(O_2/O_2 group)	(No ischaemia)
LV diastolic pressure (mm Hg)	14.5 ± 1.7	13.3 ± 2.3	12.5 ± 1.7
LV developed pressure (mm Hg)	40.8 ± 2.4	42.6 ± 2.5	41.3 ± 2.5
O_2 consumption (μmol·min^{-1}·g^{-1} dry wt)	43.9 ± 3.6	41.9 ± 2.2	41.8 ± 6.0
CPK release (mU·min^{-1}·g^{-1} dry wt)	0.09 ± 0.03	0.13 ± 0.03	0.24 ± 0.09
Coronary flow (ml·min^{-1}·g^{-1} dry wt)	71.7 ± 11.8	72.3 ± 2.9	71.0 ± 5.0

Left ventricular (LV) diastolic pressure was corrected for tension in the wall of the LV balloon; CPK, creatine phosphokinase; dry wt, dry weight.

RESULTS

Baseline Physiological Parameters

Cardiac function and CPK release in the 20th minute of the equilibration period before ischaemia are given in Table 1. There were no significant differences in initial left ventricular diastolic pressure, left ventricular developed pressure (*i.e.*, systolic pressure minus diastolic pressure), O_2 consumption, CPK release or coronary flow among the groups.

Post-ischaemic Physiological Parameters

Left ventricular diastolic pressure, shown in Fig. 1, did not differ significantly between the N_2/O_2 group and the O_2/O_2 group at 1, 30 and 60 min after the start of reperfusion, but diastolic pressure was significantly higher in both post-ischaemic groups than in the no-ischaemia group. The time course of the changes in left ventricular diastolic

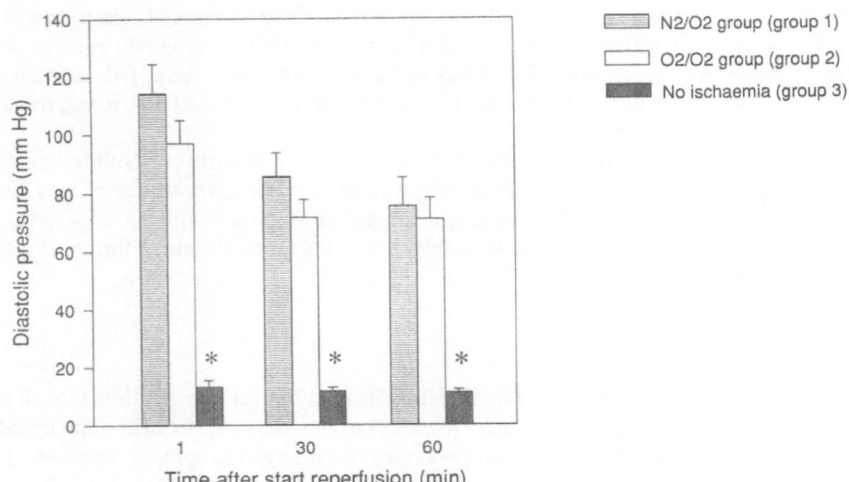

Figure 1. Left ventricular diastolic pressure after the start of reperfusion in groups 1-3. * indicates a significant difference (p<0.05) between the no-ischaemia group and both the N_2/O_2 group and the O_2/O_2 group. Bars give mean ± SE.

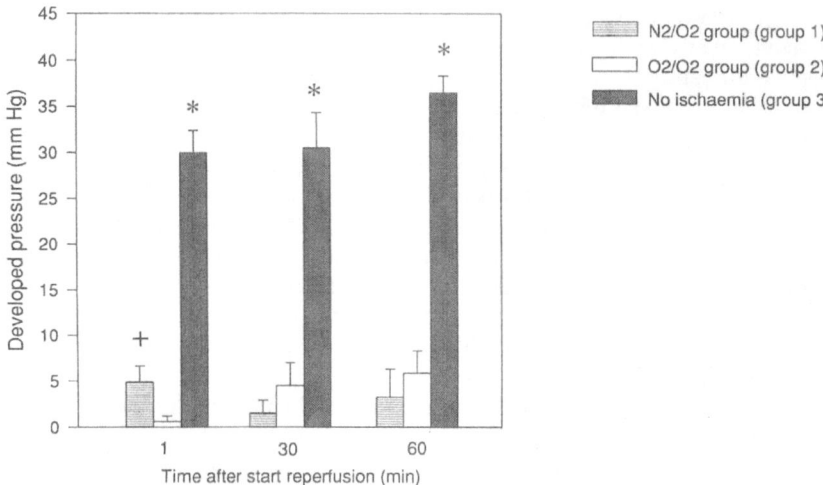

Figure 2. Left ventricular developed pressure, *i.e.* systolic pressure minus diastolic pressure, after the start of reperfusion in groups 1-3. + indicates a significant difference (p<0.05) between the N_2/O_2 group and the O_2/O_2 group. * indicates a significant difference (p<0.05) between the no-ischaemia group and both the N_2/O_2 group and the O_2/O_2 group. Bars give mean ± SE.

pressure during reperfusion, did not differ significantly between the N_2/O_2 group and the O_2/O_2 group.

Left ventricular developed pressure, shown in Fig. 2, was significantly higher in the N_2/O_2 group than in the O_2/O_2 group at 1 min after the start of reperfusion (*i.e.*, developed pressure was higher if reperfusion was started with N_2 than if it was started with O_2), but did

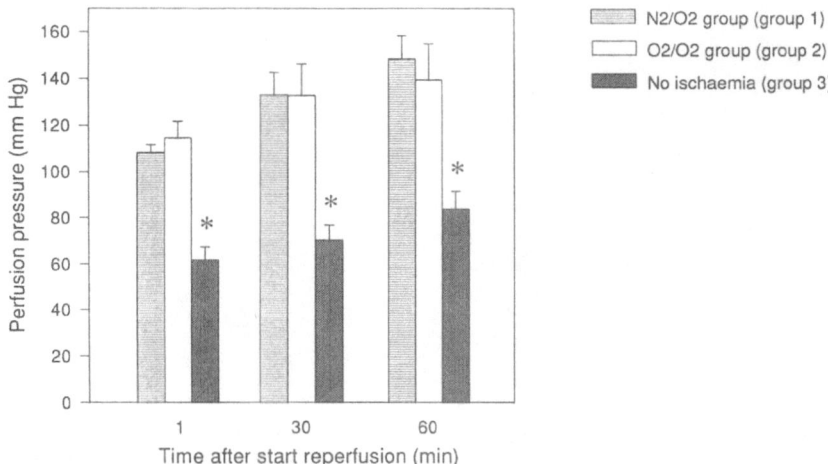

Figure 3. Perfusion pressure after the start of reperfusion in groups 1-3. Flow was kept constant and changes in perfusion pressure therefore indicate trends in coronary vascular resistance. * indicates a significant difference (p<0.05) between the no-ischaemia group and both the N_2/O_2 group and the O_2/O_2 group. Bars give mean ± SE.

not differ significantly between both post-ischaemic groups at 30 and 60 min after the start of reperfusion. The time course of change in left ventricular developed pressure in the N_2/O_2 group was significantly different from the gradual increase observed in the O_2/O_2 group (repeated measures analysis of variance). Developed pressure was significantly lower in both post-ischaemic groups than in the no-ischaemia group. At 60 min after the start of reperfusion, 6 of the 8 hearts in the O_2/O_2 group showed discernible beating on the pressure recording, compared to 1 of the 8 hearts in the N_2/O_2 group.

Perfusion pressure reflects coronary vascular resistance because perfusion flow was kept constant over time and did not differ significantly between the N_2/O_2 group and the O_2/O_2 group at 1, 30 and 60 min after the start of reperfusion (see Fig. 3). For both post-ischaemic groups the change over time in perfusion pressure was not significant. Perfusion pressure was significantly higher in both post-ischaemic groups compared to the no-ischaemia group.

There were no differences in O_2 consumption at 60 min after the start of reperfusion between the N_2/O_2 group and the O_2/O_2 group (22.5 ± 6.6 μmol · min^{-1} · g^{-1} dry wt vs. 22.8 ± 2.4 μmol · min^{-1} · g^{-1} dry wt, respectively). Both post-ischaemic groups showed significantly lower O_2 consumption than the no-ischaemia group (43.4 ± 5.3 μmol · min^{-1} · g^{-1} dry wt).

Creatine Phosphokinase Release during Reperfusion

Fig. 4 shows the time course of CPK release after the start of reperfusion. At 10 and 21 min after the start of reperfusion, CPK release was significantly higher in the N_2/O_2 group compared to the O_2/O_2 group. Differences between both post-ischaemic groups and the non-ischaemic group were significant at all time points after the start of reperfusion. The

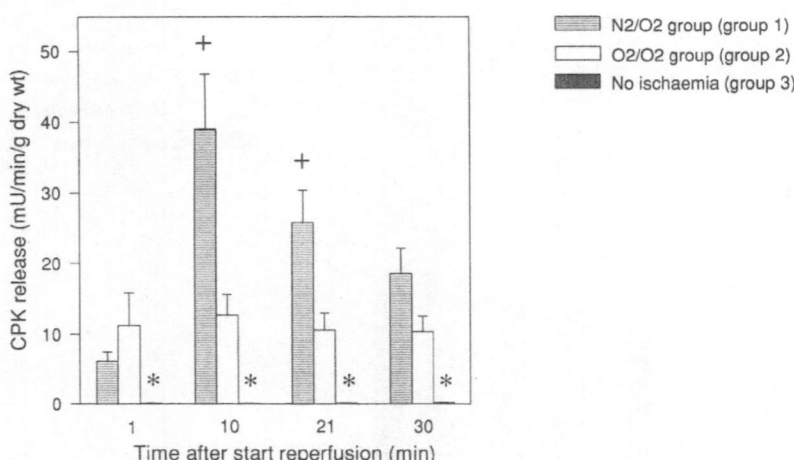

Figure 4. CPK release after the start of reperfusion in groups 1-3. Coronary effluent samples in group 3 (no-ischaemia group) were taken at the same time as those of the two post-ischaemic groups and had very low levels of CPK release: values at 1, 10, 21 and 30 minutes were 0.14 ± 0.05, 0.10 ± 0.03, 0.11 ± 0.06 and 0.22 ± 0.06 mU · min^{-1} · g^{-1} dry wt, respectively and can thus be barely discerned in the figure. + indicates a significant difference ($p<0.05$) between the N_2/O_2 group and the O_2/O_2 group. * indicates a significant difference ($p<0.05$) between the no-ischaemia group and both the N_2/O_2 group and the O_2/O_2 group. Bars give mean ± SE.

difference in time course between the two post-ischaemic groups was significant (see Fig. 4).

Cumulative CPK release during reperfusion was significantly higher in the N_2/O_2 group compared to the O_2/O_2 group (813 ± 72 mU $\cdot$ g^{-1} dry wt vs. 549 ± 87 mU $\cdot$ g^{-1} dry wt, respectively). Cumulative CPK release in both post-ischaemic groups was significantly higher than in the no-ischaemia group (10.6 ± 2.6 mU$\cdot$g^{-1} dry wt).

DISCUSSION

In a recent pilot study, alternate perfusion with normoxic and anoxic solutions ('N_2/O_2-cycling') showed beneficial effects in rat hearts which had developed high diastolic and low developed pressures (Osbakken *et al*, 1994). This suggested that an 'afterconditioning' phenomenon might occur after development of myocardial functional damage. If this were reconfirmed, N_2/O_2-cycling might have clinical applications in resuscitation after cardiac arrest. In the present study we further investigated the role of N_2/O_2-cycling after prolonged ischaemia to determine whether N_2/O_2-cycling might protect the heart from some of the adverse effects of reperfusion after global ischaemia (reperfusion injury). The duration of global ischaemia, 60 min, used in this study was sufficient to cause severe myocardial depression; however, pharmacological intervention with ruthenium red has proved beneficial in such a model (Benzi & Lerch, 1992). During the period of global ischaemia, when the isolated rat heart is immersed in Krebs-Henseleit solution, the diffusion of oxygen across the surface of the rat heart is limited to the outer parts of the heart (van Beek *et al*, 1992). Although in the present study myoglobin oxygenation has not been measured, we found previously that myoglobin oxygenation reached its lowest level after one minute of perfusion with solution gassed with 95% N_2–5% CO_2 and remained at this level in the second minute (van Beek *et al*, this volume).

It has been reported in several studies that oxygen derived free radicals most likely contribute to myocardial stunning after short periods of ischaemia, but whether they play an important role in causing irreversible myocardial damage after a prolonged period of global ischaemia remains unclear (Kloner *et al*, 1989; Reimer *et al*, 1990; Goldhaber & Weiss, 1992; Hearse & Bolli, 1992). Given the possibility of the contribution of oxygen derived free radicals to reperfusion injury, our hypothesis was that N_2/O_2-cycling might be a beneficial reperfusion regimen, because N_2/O_2-cycling reduces mean tissue O_2 concentration during reperfusion (and thus substrate for free radical production), while O_2 is still intermittently available for ATP synthesis. However, in hearts which were exposed to 60 min of global ischaemia, we found that N_2/O_2-cycling was not beneficial, with respect to preservation of mechanical function or prevention of creatine phosphokinase leakage. In contrast, creatine phosphokinase leakage was even increased during N_2/O_2-cycling.

The data from pilot experiments had suggested that N_2/O_2-cycling may be beneficial during resuscitation in hearts that had become asystolic for a variety of etiologies, but were not globally ischaemic. The earlier preliminary results reported by Osbakken *et al* (Osbakken *et al*, 1994), suggesting possible beneficial effects of N_2/O_2-cycling, were obtained in hearts which had been hypoxic and had low perfusion flow but where never no-flow global ischaemia had been applied. In these experiments where N_2/O_2-cycling appeared to be beneficial, the perfusate was recirculated and factors such as vasoactive mediators (Lefer, 1987) may have caused cardiodepressant effects. We did not obtain similar protective effects in the rat heart model of global ischaemia with subsequent reperfusion. This indicates that N_2/O_2-cycling during the early reperfusion period does not protect against myocardial damage caused by prolonged global ischaemia, but does not exclude the possibility that it may be useful in other conditions.

SUMMARY

Recent pilot experiments suggested a beneficial effect of perfusing isolated rat hearts, which had developed high diastolic pressures and low developed pressures due to hypoxia and low flow ischaemia, alternately with two solutions: one gassed with 95% N_2–5% CO_2 and one gassed with 95% O_2–5% CO_2. The aim of the present study was to determine whether this reperfusion method could be generalized to be protective to globally ischaemic hearts. After rats were anaesthetized with ketamine and xylazine, hearts were isolated and perfused for 20 min according to Langendorff. Hearts were then exposed to 60 min of no-flow ischaemia. Reperfusion was reinitiated via two regimens: in group 1 (N_2/O_2 group, n = 8), two solutions were used alternately every 2 min for 32 min; one solution was gassed with 95% N_2–5% CO_2, and the second gassed with 95% O_2–5% CO_2. In group 2 (O_2/O_2 group, n = 8), which served as control, both solutions were gassed with 95% O_2–5% CO_2. After 32 min of alternating the solutions, each group was perfused with 95% O_2–5 %CO_2 for 28 min to assess recovery. Left ventricular (LV) pressure was recorded and creatine phosphokinase (CPK) activity was determined in the coronary effluent. After 60 min reperfusion, neither LV diastolic pressure nor LV developed pressure differed significantly between the N_2/O_2 group and the O_2/O_2 group. In the N_2/O_2 group, CPK release was significantly lower at 1 min, but significantly higher at 10 and 21 min after the start of reperfusion compared to the O_2/O_2 group. These findings suggest no beneficial effect of the alternating N_2/O_2 exposure during reperfusion, with respect to mechanical function or CPK leakage from cardiac myocytes exposed to global ischaemia.

REFERENCES

van Beek J.H.G.M., Loiselle D.S. and Westerhof N., 1992, Calculation of oxygen diffusion across the surface of isolated perfused hearts, *Am. J. Physiol.* 263;H1003-H1010.

van Beek J.H.G.M., Osbakken M.D. and Chance B., This volume, Measurement of the oxygenation status of the isolated perfused rat heart using near infrared detection.

Benzi R.H. and Lerch R., 1992, Dissociation between contractile function and oxidative metabolism in postischemic myocardium, *Circ. Res.* 71:567-576.

Dunphy G. and Ely D., 1990, Decreased storage stability of creatine kinase in a cardiac reperfusion solution, *Clin. Chem.* 36:778-780.

Goldhaber J.I. and Weiss J.N., 1992, Oxygen free radicals and cardiac reperfusion abnormalities, *Hypertension* 20:118-127.

Hearse D.J. and Bolli R., 1992, Reperfusion induced injury: manifestations, mechanisms, and clinical relevance, *Cardiovasc. Res.* 26:101-108.

Kloner R.A., Przyklenk K. and Whittaker P., 1989, Deleterious effects of oxygen radicals in ischemia/reperfusion. Resolved and unresolved issues, *Circulation* 80:1115-1127.

Lefer A.M., 1987, Interaction between myocardial depressant factor and vasoactive mediators with ischemia and shock, *Am. J. Physiol.* 252:R193-R205.

Oliver I.T., 1966, A spectrophotometric method for the determination of creatine phosphokinase and myokinase, *Biochem. J.* 61:116-122.

Osbakken M.D., van Beek J.H.G.M. and Zhang D., 1994, Cardiac resuscitation: Role of N_2 and O_2, *J. Am. Coll. Cardiol.*, Abstract Volume, 43rd Annual Scientific Session. 152A.

Reimer K.A., Tanaka M., Murry C.E., Richard V.J. and Jennings R.B., 1990, Evaluation of free radical injury in myocardium, *Toxicol. Pathol.* 18:470-80.

Snedecor G.W. and Cochran W.G., 1980. "Statistical Methods", The Iowa State University Press, Ames, IA, USA.

Wixson S.K., White W.J., Hughes H.C., Lang C.M. and Marshall W.K., 1987, A comparison of pentobarbital, fentanyl-droperidol, ketamine-xylazine and ketamine-diazepam anesthesia in adult male rats, *Lab. Anim. Sci.* 37:726-749.

32

MODEL ANALYSIS OF OXYGEN ISOTOPE FRACTIONATION IN HUMANS DUE TO DISTURBANCES OF PULMONARY GAS EXCHANGE

K. D. Schuster and H. Heller

Physiologisches Instiut I
Universität Bonn
Nussallee 11, D-53115 Bonn
Germany

INTRODUCTION

The lighter isotopic oxygen molecule $^{16}O_2$ passes through the human oxygen transport system with an approximately 0.8% higher rate than its heavier isotopic species $^{16}O^{18}O$, leading to an overall fractionation effect which has recently been quantified (Schuster and Pflug, 1989). This effect has been measured in healthy persons at rest and during ergometer work (Schuster et al., 1994), situations of hyper-and hypoventilation have been studied (Heller et al.,1993), and patients suffering from anemia (Heller et al., 1994) or lung fibrosis have been investigated (Heller et al., this volume). All these situations affected the overall fractionation effect due to changes of processes of the oxygen transport system. Therefore it is supposed that this parameter and its change contains characteristic information with respect to the pattern of oxygen transport pathways and its variations. To understand the sources of the fractionation effects and their nature, model investigations are necessary. This study models changes of pulmonary gas exchange disturbances with respect to their influence on the overall fractionation effect so as to interpret experimental data.

METHODS

Modelling the Oxygen Transport System

The various steps of the oxygen transport system can be considered as consisting of resistances (or of their reciprocals which are conductancies). The principles which are connected with such a consideration have been already stated by others (Piiper et al., 1971, Dejours, 1975, Weibel, 1984) and have more recently been reviewed by Otis (1987). A model

Oxygen Transport to Tissue XVII, Edited by Ince et al.
Plenum Press, New York, 1996

249

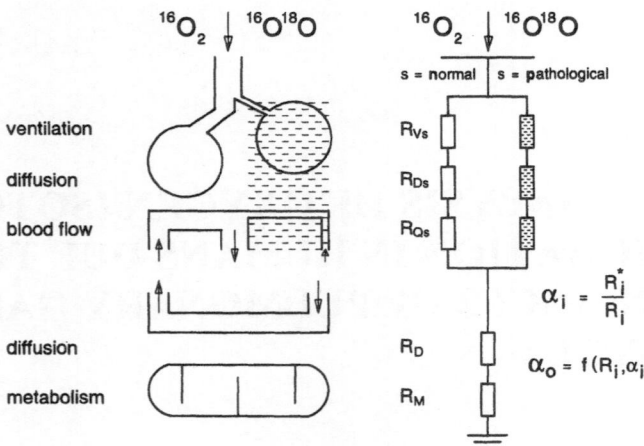

Figure 1. Assuming steady state conditions, the oxygen transport system is modelled as two identical networks of resistances R_i for $^{16}O_2$ and R_i^* for $^{16}O^{18}O$ (not explicitly shown). The lung consists of a healthy area (index s=normal) and a pathological (index s = p) branch. The overall fractionation factor α_o depends on the single fractionation power $\alpha_i = R_i^*/R_i$ of every single transporting step and on the network characteristics.

which is basically similar to that described by Otis has been developed and is given in fig. 1. This consists of two identical networks of resistances R_i for $^{16}O_2$ and R_i^* for $^{16}O^{18}O$, the latter is not explicity shown. The most important postulation for such a model is to assume steady state conditions. The lung is subdivided into two sections of equal size, a normal, healthy area and a pathological branch. Both areas are represented by 3 serially connected resistances, R_{Vs} for ventilation, R_{Ds} for pulmonary diffusion and R_{Qs} for pulmonary blood flow. The index s can admit the two states, n for normal and p for pathological, according to both areas assumed.

Mathematical Description

The model of fig. 1 has been mathematically described with respect to isotopic fractionation of both oxygen species. The fractionation power α_i of a single process i is defined in fig. 1 as ratio

$$\alpha_i = \frac{R_i^*}{R_i} \tag{1}$$

where R_i^* is the resistance for $^{16}O^{18}O$ and R_i that for $^{16}O_2$. After applying Kirchhoffs laws and some mathematical transformations, the following equations were obtained for the fractionation factors α_{ser} and α_{par} of a serial and a parallel network of i resistances:

$$\alpha_{ser} = \sum \alpha_i \cdot \frac{R_i}{R_{ser}} \tag{2}$$

$$\alpha_{par} = \sum \alpha_i \cdot \frac{r_{par}}{R_i} \tag{3}$$

R_{ser} and R_{par} are the overall resistances of the serial and the parallel networks, respectively. These equations have been applied to the model of fig. 1 in order to calculate an overall fractionation factor α_o for the entire network. α_o can be compared with measured overall fractionation effect which is usually given as percentage value δ_{iu} by using the relation

$$\delta_{iu} = (\alpha_o - 1) \cdot 100[\%]. \tag{4}$$

Given in mmHg $\cdot$ ml $O_2^{-1}\cdot$ min, the following resistances have been taken as normal for the healthy area of the lung: $R_{Vn} = 0.33$, $R_{Dn} = 0.067$, $R_{Qn} = 0.23$. These values are twice as high as normal values of an entire healthy lung, since they represent only half a section. Depending on the situation to be simulated, the resistances of the pathological branch have been varied between normal and tenfold higher values. The α_i-values 1.013 for metabolism (Feldman et al., 1959), 1.030 for diffusion, and 1.000 for both convective processes, ventilation and blood flow (Schuster and Pflug, 1989) have been applied.

RESULTS

The resistances of diffusion, blood flow and ventilation respectively have been continuously increased tenfold from their normal values in the pathological branch of the model and the resulting overall fractionations have been calculated. The results of this simulation are shown in fig. 2. Changes of overall fractionation are given as percentage difference $\Delta\alpha_o = (\alpha_{on} - \alpha_{op}) \cdot 100\ [\%]$ where α_{op} is determined for increasing resistances in the pathological branch and α_{on} is the fractionation factor of the basic state of all resistances which are normal (zero-point of the ordinate). As can be seen when singly increasing the diffusion resistance, $\Delta\alpha_o$ increases also to a maximal value, whereas an increase of the resistances of blood flow or ventilation leads to decreasing fractionation factors.

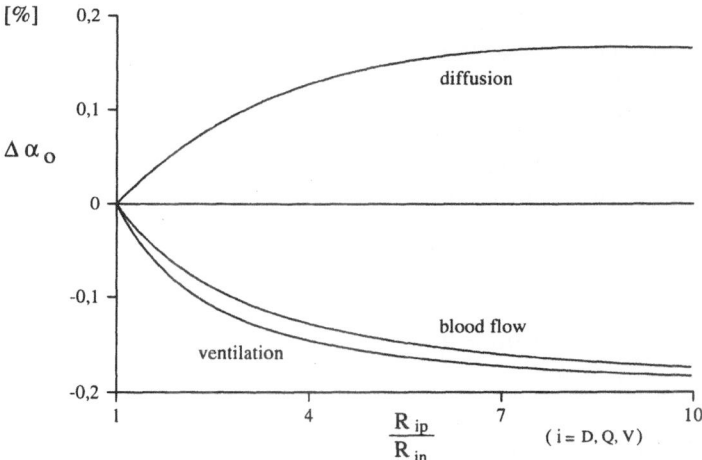

Figure 2. Change of overall fractionation factor as function of increasing resistances R_{ip} in the pathological branch of fig.1. The resistances of diffusion, blood flow or ventilation (indicated with i = D,Q,V) respectively have been continuously increased tenfold from their normal values (R_{in}), given on the abscissa as ratio R_{ip}/R_{in}. On the ordinate the resulting fractionations are shown as the difference $\Delta\alpha_o$ between the overall fractionation factor for the pathological and for the normal state.

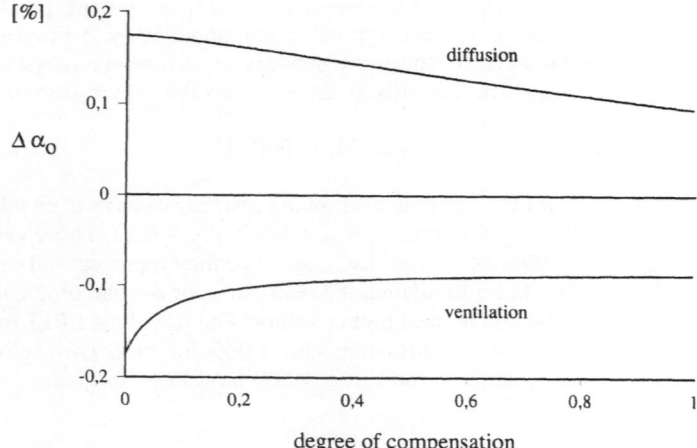

Figure 3. Dependency of the difference between pathological and normal overall fractionation $\Delta\alpha_o$ on degree of compensation by the Euler-Liljestrand effect. In the pathological branch of the lung, resistances for diffusion R_{Dp} or ventilation R_{Vp} have been increased tenfold (last points of fig. 2). Starting with normal values, blood flow resistance in the pathological branch R_{Qp} has also been increased but the resistance has been decreased in the healthy area of the lung, so that pulmonary overall perfusion remains constant. Degree = 1 means R_{QP} is increased proportionally to $R_{Dp} + R_{Vp}$.

Fig. 3 exhibits the influence of compensation on overall fractionation brought about by secondary changes of the perfusion resistances, referred to as Euler-Liljestrand compensation (Euler and Liljestrand, 1947, Sheehan and Farhi, 1993).

Primarily, the resistances of diffusion are increased tenfold within the pathological branch. Starting with normal values, the perfusion resistance has been increased proportionally to the sum of diffusion and ventilation resistance in the pathological branch, and the resistance in the healthy area has been decreased so that pulmonary overall perfusion remains constant. Due to these compensations $\Delta\alpha_o$ returns in zero direction but it stays negative for the graph of increased ventilation resistance (labeled with "ventilation") in fig. 3., and positive for the diffusion graph.

Fig. 4 shows results of a simulation of combined ventilation-diffusion disturbances. This has been realized by changing both the diffusion resistance R_{Dp} and the ventilation resistance R_{Vp} of the pathological branch according to the equations

$$R_{Dp} = R_{Dn} + 10 \cdot R_{Dn} \cdot deg \qquad (5)$$

$$R_{Vp} = R_{Vn} + 10 \cdot R_{Vn} \cdot (1 - deg) \qquad (6)$$

deg increases from 0 to 1 and is the degree of diffusion limitation whereas (1-deg) corresponds to the degree of ventilation limitation in fig. 4. Positive as well as negative values for $\Delta\alpha_o$ can be calculated, depending on the degree of combination but with the ventilation resistance thereby dominating over the diffusion resistance with respect to the influence on the overall fractionation factor.

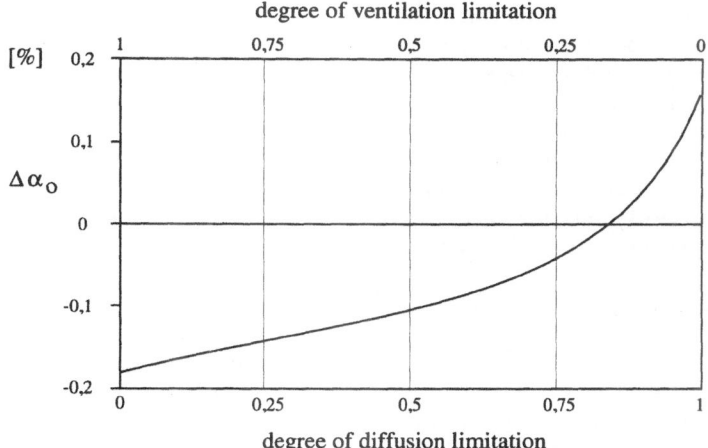

Figure 4. Dependence of the difference between pathological and normal overall fractionation $\Delta\alpha_o$ on combined ventilation-diffusion limitation. From the origin of the lower abscissa, which represents ventilation limitation ($R_{Vp} = 10 \cdot R_{Vn}$), ventilation resistance decreases and diffusion resistance increases to a state of diffusion limitation ($R_{Dp} = 10 \cdot R_{Dn}$).

DISCUSSION

Critique of the Model

For estimating the changes of isotopic fractionation brought about by pulmonary gas exchange disturbances, a very simple resistance model similar to that described by Otis (1987) was used. In reality, the oxygen transport system is much more complicated, containing oxygen storages of changing size, including the non-linear characteristics of oxygen-hemoglobin binding and being further complicated by inhomogeneous distributions of all parameters influencing oxygen transport.

Storages of oxygen can be introduced in a model such as that of fig.1 as capacitors leading to the structure of a compartment model with a time dependent characteristic, and this may represent a better but also a more complicated approach to the oxygen transport system. In order to become independent of varying contents of capacitors, the assumption of steady state conditons has been made which means that the contents of compartments as well as the size of the resistors are constant. In this case, the capacitors can be omitted from the model wihout undergoing any error. In reality, steady state conditions may not be reached at all, but they can be approached by choosing long periods of measurement, for which the parameters needed are taken as mean values.

For convenience, the pattern of oxygen-hemoglobin binding has not been explicitly included in the model. Its influence on overall fractionation may be twofold. First, oxygen binding to hemoglobin exerts its own fractionation power which has been determined to $\alpha = 1.0035$ (Pflug and Schuster, 1989). Second, it represents a resistance which could influence the characteristic of the resistance network of fig.1. Since the fractionation power of oxygen-hemoglobin binding is small and close to 1, which is the fractionation power of a convective process, the resistance brought about by oxygen-hemoglobin reaction can be considered as an additional component of blood flow resistance and it is therefore implicitly included in the model. Nevertheless, oxygen-hemoglobin binding should be taken into account in a more extended model.

The assumption that a lung associated with pulmonary failure consists of only one healthy and one pathological region certainly represents an extreme simplification of the reality in many patients, since pulmonary failure may often be better characterized by a continuous distribution of disturbances rather than by single well-defined compartments. However, using simplified models has the advantage of giving a better insight into the linkage between basic parameter change and resulting overall effect, which is the purpose of this study. Besides this, the main behaviour of highly sophisticated multicompartment or continuous systems can be simulated by two-compartment models.

Interpretation of the Results

When increasing the resistances of ventilation or blood flow, the overall fractionation factor α_o decreases whereas it increases with increasing diffusion resistance (fig. 2). These poor effects are expected in patients suffering only from disturbances of ventilation or perfusion or diffusion respectively. Indeed, reduced overall fractionations have been measured on healthy subjects during alveolar hypoventilation (Heller et al., 1993) and on patients suffering from anemia (Heller et al., 1994). In these cases oxygen transport resistances of ventilation or perfusion are increased, respectively. In the papers cited above the measured changes of fractionation are explained well on the basis of an even more simplified model consisting of only one line of serially connected resistances (referred to as 1-branch model). Such a model can be described completely by eq.(2) under "methods".From this equation it can be easily seen that α_o converges against α_i when R_i / R_{ser} converges against 1. This means that α_o is ruled by the most limiting step of the system. For example, if a diffusion resistance increases, α_o will be more strongly influenced by the fractionation power of diffusion which is the greatest of all the processes, hence α_o will increase also. This is true of the 1-branch as well as of the 2-branch model. However, when increasing the diffusion resistance tenfold, the overall fractionation factor increases by 2% in the 1-branch model but by only 0.2% in the 2-branch model. This latter much lower effect is due to the existence of a healthy area which is parallel to the pathological branch. It has been assumed in the 2-branch model that both sections are of equal size. This means, when increasing the resistances of the pathological branch to infinity, the pulmonary overall resistance only doubles, which explains the lower extent of overall fractionation change of the 2-branch model. When assuming an increasingly greater disturbed area and a decreasingly smaller healthy one, the extent of overall fractionation change increases, converging finally against the value of the 1-branch model.

In fig. 2 the response of fractionation change to variations of ventilation or blood flow is very similar since both are convective processes with a fractionation power 1. The small difference between both processes is due to slightly lower initial resistances assumed for perfusion according to Otis (1987). Due to the similarity of fractionation response to changes of both convective processes, primary variations of only ventilation and diffusion have been taken under further investigation.

In fig. 3 ventilation or diffusion resistances have been increased tenfold above normal within the pathological branch. Starting with normal perfusion, blood flow has been increasingly distributed from the pathological to the healthy area (Euler-Liljestrand compensation). Due to compensation both graphs move toward normal overall fractionation without reaching it. Separate analysis of both components of Euler-Liljestrand compensation showed that the negative slope of the diffusion graph is due to a decrease of perfusion within the pathological section whereas the increasing tendency of the ventilation graph is caused by increasing blood flow within the healthy branch.

In comparison to normal state, the model predicts both higher and lower fractionations due to combined convective-diffusive disturbances, depending on the degree of combination, but the balance is shifted to lower values (fig. 4). Heller et al. (this volume) have investigated patients suffering from pulmonary fibrosis. They measured considerably reduced diffusing capacities. When applying the results of fig.2 and fig. 3 to this data, fractionations higher than normal are expected. However, lower fractionation values have been measured. This can be explained on the basis of fig. 4 when assuming additional limitation brought about by convective processes. Indeed, in patients suffering from lung fibrosis, ventilation-perfusion ratios are known to be also affected (Wagner et al., 1976). However, it is not easy to attribute a measured deficit of gas exchange to various possible origins. The relative role of diffusion limitation versus ventilation-perfusion mismatching as causes of the oxygen exchange deficit in patients with interstitial lung disease has been in dispute for many years and remains an unresolved question (Hughes et al., 1993, Hempelman and Hughes, 1991). With regard to this problem, measuring fractionations can provide additional information.

Due to pulmonary disturbances, the overall fractionation can be increased, normal or decreased, depending on type, magnitude, distribution and combination of various effected processes. Therefore, the fractionation factor cannot be applied alone for diagnosing the real source of gas exchange deficit. But together with other parameters such as diffusing capacity or alveolar-arterial partial pressure difference, it can help partitioning gas exchange disturbances into diffusive and convective components so as to assess certain types of single as well as combined disturbances of pulmonary function.

SUMMARY

A model has been developed describing oxygen isotope fractionation due to pulmonary gas exchange disturbances. Assuming steady state conditions, the various pathways of oxygen transport were modelled as two identical networks of resistances for both isotopic species $^{16}O_2$ and $^{16}O^{18}O$. Within the lung, this network consists of two resistance branches, representing a healthy and a pathological area. For different types of disturbed pulmonary oxygen transfer, oxygen isotope fractionation quantified by an overall fractionation factor α_o was determined, when increasing the resistances of ventilation or blood flow, a decrease of α_o was obtained. However an increase of the diffusion resistance was followed by an increase of α_o. Combinations of these changes as well as compensations due to the Euler-Liljestrand mechanism led to increasing or decreasing fractionations, the extent of which depended on the degrees of combination as well as compensation, respectively. From this investigation it can be concluded: i) measuring isotopic fractionation can help partitioning gas exchange disturbances into diffusive and convective components, ii), decreased isotopic fractionations which have been measured in patients suffering from lung fibrosis is brought about by higher resistances of convective processes which may be combined with the considerably reduced diffusing capacities also measured in these patients.

ACKNOWLEDGMENTS

We gratefully acknowledge the technical assistance of Bernd Eixmann, Christa Pusch, and Barbara Schreiber.

REFERENCES

Dejours, P., Principles of Comparative Respiratory Physiology, Am. Elsv. Publ. Comp., INC, New York, 1975, p.89-112.

Euler, U. S. v., and Liljestrand, G., 1947, Observations on the pulmonary arterial blood pressure in the cat, Acta Physiol. Scand. 12:301-320.

Feldman, D. E., Yost, H. T., jr., and Benson, B. B., 1959, Oxygen Isotope Fractionation in Reactions Catalyzed by Enzymes, Science 129:146-147.

Heller, H., Könen, M., and Schuster, K.-D., 1993, Dependence of overall fractionation effect of respiration on ventilation at rest, Isotopenpraxis 28:133-141.

Heller, H., Schuster, K.-D., and Göbel, B. O., 1994, Dependency of overall fractionation effect of respiration on hemoglobin concentration within blood at rest, Adv. Exp. Med. Biol. 345:755-761.

Heller, H., Könen, M., Overlack, A., and Schuster, K.-D., Fractionation effects of oxygen isotopes within interstitial lung disease, Adv. Exp. Med. Biol., this volume.

Hempleman, S. C., and Hughes, J. M. B., 1991, Estimating exercise and diffusion limitation in patients with interstitial fibrosis, Respir. Physiol. 83:167-178.

Hughes, J. M. B., Lockwood, D. N. A., Jones, H. A., and Clark, R. J., 1991, D_{LCO} / Q and diffusion limitation at rest and on exercise in patients with interstitial fibrosis, Respir. Physiol. 83:155-166.

Otis, A. B., 1987, An overview of gas exchange, in Handbook of Physiology Eds. Farhi, L. E., and Tenney, S. M.) Sect. 3, Vol. 4:1-11, Am. Physiol. Soc., Bethesda, USA.

Piiper, J., Dejours, P., Haab, P., and Rahn, H., 1971, Concepts and basic quantities in gas exchange physiology, Respir. Physiol.13:292-304.

Pflug, K. P., and Schuster, K.-D., 1989, Fractionation of oxygen isotopes due to equilibration of oxygen with hemoglobin, Adv. Exp. Med. Biol. 248:407-411.

Schuster, K.-D., and Pflug, K. P., 1989, The overall fractionation effect of isotopic oxygen molecules during oxygen transport and utilization in humans, Adv. Exp. Med. Biol. 248:151-156.

Schuster, K.-D., Heller, H., and Könen, M., 1994, Investigation of the human oxygen transport system during conditions of rest and increased oxygen consumption by means of fractionation effects of oxygen isotopes Adv. Exp. Med. Biol. 345:747-753

Sheehan, D. W., and Farhi, L. E., 1993, Local pulmonary blood flow: control and gas exchange, Respir. Physiol. 94:91-107.

Wagner, P. D., Dantzker, D. R., Dueck, R., Depolo, J. L., Wasserman, K. and West, J. B., 1976, Distribution of ventilation-perfusion ratios in patients with interstitial lung disease, Chest 69:256-257.

Weibel, E. R., The Pathway of Oxygen. Structure and Function in the Mammalian Respiratory System, Harvard Univ. Press, Cambridge, 1984, p. 49-79

33

SIMULATION OF GAS EXCHANGE AND PULMONARY BLOOD FLOW USING A WATER-DISPLACEMENT MODEL LUNG

E. M. Williams

Nuffield Department of Anaesthetics
University of Oxford
Radcliffe Infirmary
Woodstock Road, Oxford, OX2 6HE, United Kingdom

INTRODUCTION

The transport of oxygen by blood to the tissues begins at the lungs. Here fresh inspired oxygen diffuses into the blood while metabolically produced carbon dioxide diffuses out of the blood. To perform this simple function the lungs require a large surface area. This is achieved by the lungs having a complex sponge-like structure, consisting of millions of microscopic gas-exchanging units each with their own blood supply. In the healthy lung these myriads of units can be treated as one large unit. This simplification has allowed the development of numerous mathematical models describing lung ventilation, gas exchange and perfusion. The simpler mathematical models consist of just three compartments: a dead space, V_D (the ventilated but non gas-exchanging airways of the lungs), an alveolar volume, V_A (the combined volume of the gas exchange units) and a gas-exchanging blood volume, $\dot{Q}_P$ (pulmonary blood flow). Many of these mathematical models make further simplifications and assume that the ventilation of the lungs is continuous rather than tidal.

The original concentric water-displacement mechanical model lung allows the manipulation of V_D, V_A and tidal volume, V_T (Williams *et al.* 1993). This two-compartment water-displacement lung model has been used to test new mathematical ventilation models developed by us to measure V_D and V_A (cf Hahn *et al.* 1993, Williams *et al.* 1994a & b; Sainsbury *et al.* 1995). I have modified the water-displacement lung model so that oxygen and carbon dioxide exchange can be simulated along with pulmonary perfusion. The addition of this third compartment allows the evaluation of more complex three-compartment mathematical models.

Oxygen Transport to Tissue XVII, Edited by Ince et al.
Plenum Press, New York, 1996

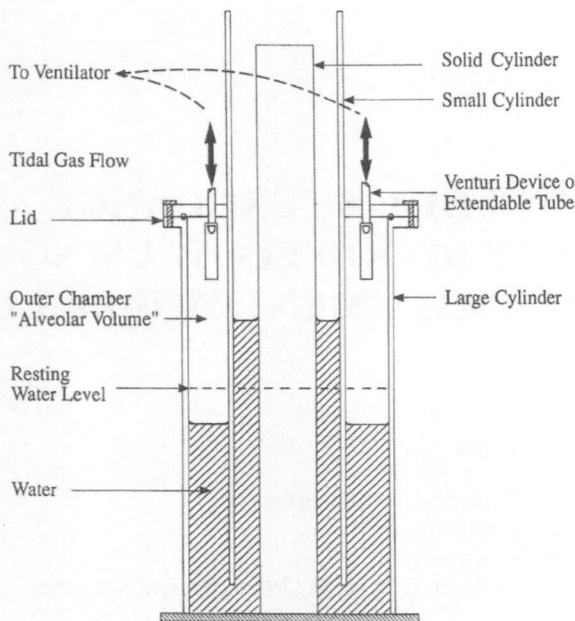

To Ventilator

Tidal Gas Flow

Lid

Outer Chamber
"Alveolar Volume"

Resting
Water Level

Water

Solid Cylinder

Small Cylinder

Venturi Device on
Extendable Tubes

Large Cylinder

Figure 1. The original concentric water-displacement model lung (from J. Biomed. Eng. **15** 420-424).

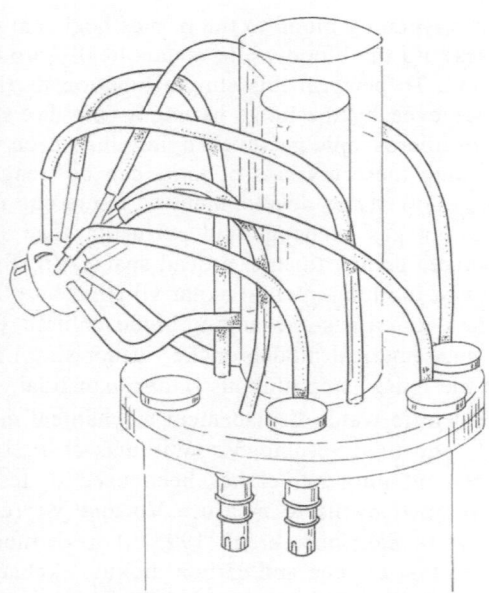

Figure 2. A sketch of the dead space tubing mounted on the model lid (from J. Biomed. Eng. **15** 420-424).

MATERIALS AND METHODS

The original concentric water-displacement model lung (Figure 1) consisted of two ventilated compartments and a non-ventilated compartment (Williams *et al.* 1993). The first ventilated compartment was the dead space and was represented by 12 tubes distributed around the model lid (Figure 2). This volume could be changed by adding or removing an additional length of tube between the ventilator and model. As in the lungs, all ventilated gas passes through this dead space. The second ventilated compartment represented the alveolar volume or alveolar space, V_A. The concentric design of the model dictate that the alveolar volume be concentric in shape and represented by the outer chamber shown in Figure 1. To ensure complete gas mixing throughout the alveolar space the inspired gas provided by a mechanical ventilator, entered the alveolar space through Venturi devices each mounted on the end of one of the 12 dead space tubes (Figure 2). The entrainment produced by the gas entering the alveolar space through these Venturi devices ensures complete mixing. V_A was set by changing the water level in the model. On inspiration the tidal volume entering the alveolar space from the dead space displaced the water in the alveolar space downwards and up into a central non-ventilated chamber (Figure 1, between the small cylinder and the large cylinder). The height to which the water rose in this central column was measured and was proportional to tidal volume (Williams *et al.* 1993).

Simulation of Gas Exchange

Gas exchange was simulated by using two of the 12 dead-space tube inlets on the model lid (Figures 2 and 3). One inlet was used to supply a gas mixture and is analogous to the pulmonary artery. The fractional oxygen concentration in this inlet gas mixture was

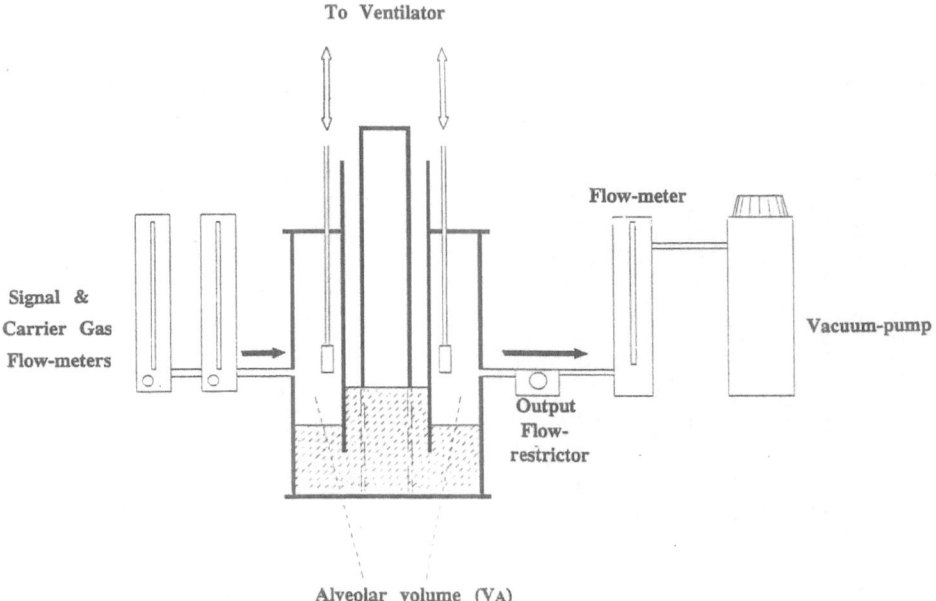

Figure 3. Additional apparatus used to mix, deliver and remove the extra gas flow that is required to simulate gas exchange and pulmonary blood flow.

manipulated so that it was lower than the inspired Fo_2 entering through the 10 remaining dead space tubes. Conversely the Fco_2 was higher in this inlet gas mixture than in the inspired gas entering through the dead-space tubing. The inlet gas was mixed using rotameters and the gas concentrations measured using a respiratory mass spectrometer (Figure 3).

To ensure that the expired tidal volume was the same as the inspired tidal volume and not increased by the additional inlet gas flow, a second inlet on the opposite side of the chamber to the input inlet was used to continuously draw off an equal volume of gas equal to the inlet gas flow (Figure 3). This drawn-off homogeneous gas mixture had the same oxygen and carbon dioxide concentrations as the end-expired gas. This outlet represents the pulmonary vein and the flow of the inlet and outlet gas is analogous to pulmonary blood flow. The outlet gas was drawn from the outer chamber using a vacuum pump (Figure 3). The rate of gas exchange was manipulated by altering the inlet Fo_2 and Fco_2 while keeping the inspired gas concentrations the same. A lower inlet oxygen concentration increases the inspired end-expired difference simulating an increase in oxygen gas exchange while a raised oxygen concentration has the opposite effect.

Simulation of Pulmonary Blood Flow

The modifications to the water-displacement lung model which allowed the simulation of gas exchange also allow pulmonary blood flow, to be simulated. In this case the inspired tidal volume contained a marker gas while the inlet gas flow through the model lid consisted of a similar gas mixture to the inspired gas but without the marker gas. When the water-displacement model was being ventilated, the thoroughly mixed alveolar gas was removed via the outlet tube at an equal rate to the inlet flow ensuring that the inspired and expired tidal volumes were equal. Unlike the inlet flow the outlet flow now contained the

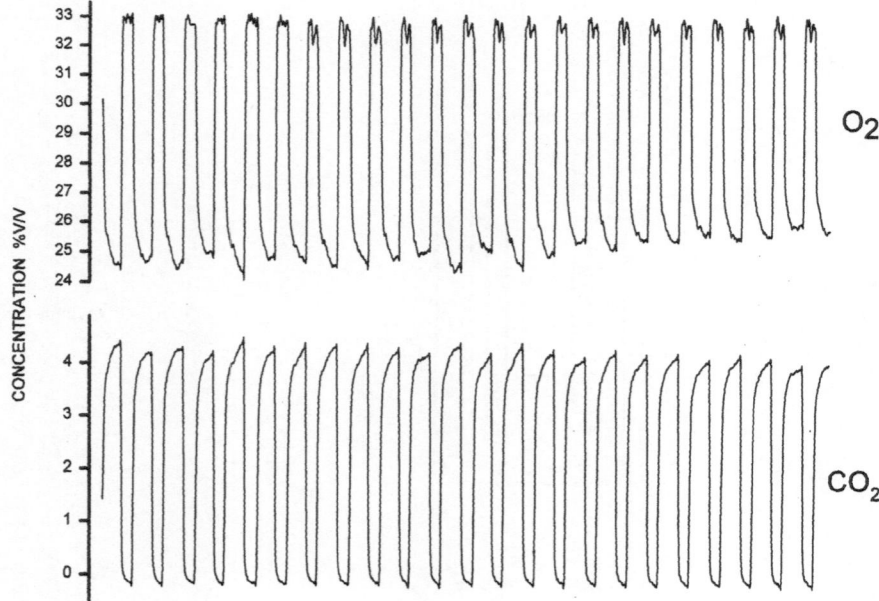

Figure 4. A tracing of inspired and expired oxygen and carbon dioxide concentration in the normally mechanically ventilated human lung.

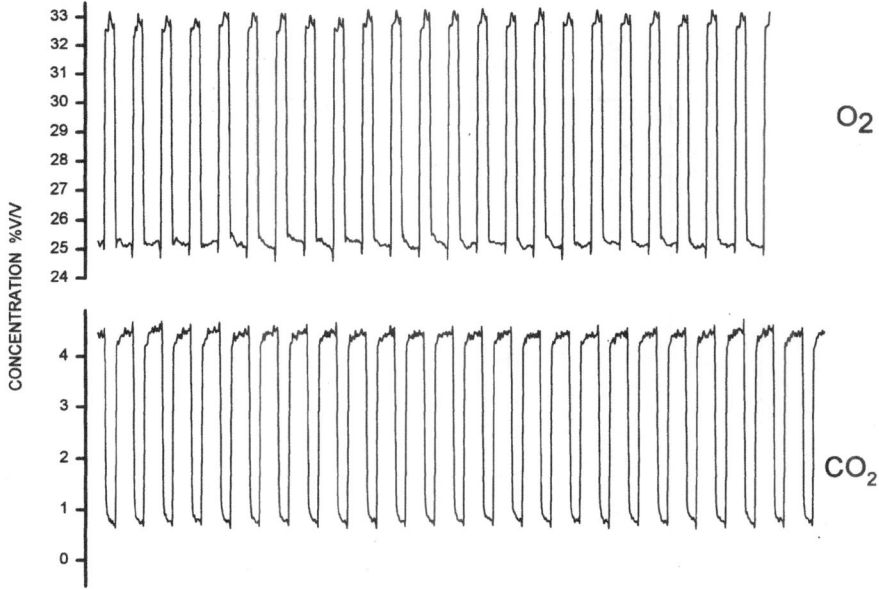

Figure 5. A tracing of simulated oxygen and carbon dioxide exchange using the water-displacement model lung.

marker gas. Once the alveolar gas was in equilibrium the reduction in marker gas end-expired concentration was due only to the rate of inlet-outlet gas flow through the model. The inlet flow therefore simulates pulmonary blood flow or cardiac output.

RESULTS

Simulation of Gas Exchange

The ability of the water-displacement model to simulate normal gas exchange was tested, figure 4 illustrates the changes in oxygen and carbon dioxide concentrations seen in healthy lungs when mechanically ventilated with a mixture of air and oxygen. The difference between the inspired and end-expired oxygen concentration was 8% v/v while the carbon dioxide concentration difference was 4.3% v/v. A similar level of gas exchange was simulated on the water-displacement model lung using a comparable inspired air oxygen mixture. Figure 5 shows an 8% difference between the inspired and end-expired oxygen concentration and a 4.3% end-expired carbon dioxide concentration.

Simulation of Pulmonary Blood Flow

Recent studies have shown that it is possible to use soluble inert gas concentration oscillating forcing signals to non-invasively measure pulmonary blood flow (Hahn *et al.* 1993, Williams *et al.* 1994 b). The above mathematical model predicts that as blood flow and the length of forcing period change, the ratio between the end-expired and inspired inert

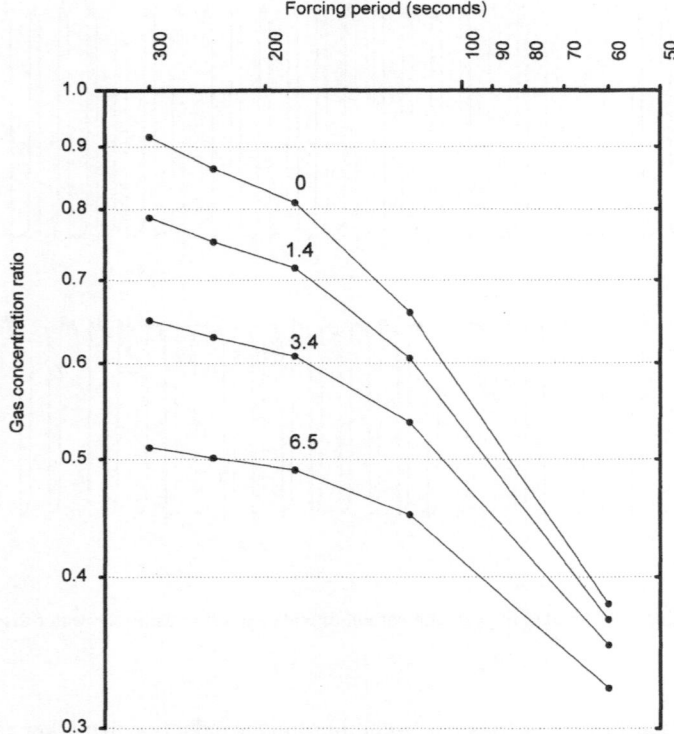

Figure 6. A theoretical log-log plot of changes in pulmonary blood flow as predicted by the sine wave inspiratory forcing technique.

gas concentration amplitude changes in a predictable way. Figure 6 shows the predicted log-log plot of changing pulmonary blood flow on the end-expired/inspired amplitude ratio of a nitrous oxide concentration forcing signal, when applying a range of forcing periods (Williams *et al*. 1994b). In order to demonstrate the water-displacement model's ability to simulate pulmonary blood flow a nitrous oxide concentration forcing signal with a similar range of forcing periods used to create Figure 6 was generated. The forcing gas signal was applied to the modified water-displacement model and the resulting end-expired/inspired amplitude ratios measured (Figure 7). In each case the same respiratory parameters were used.

DISCUSSION

The examples show that the modified water-displacement model lung can usefully simulate gas exchange and pulmonary blood flow. In the unmodified water-displacement model lung the inspired gas concentration was only altered by changes in volume as the gas passed through the dead space and alveolar space of the model. The ability to add and remove gas from the alveolar space of the modified water-displacement model allows gas exchange and pulmonary blood flow to be modelled. This extra gas flow creates a third compartment where inspired gas once in the alveolar space can be mixed with other gases. Thus, the

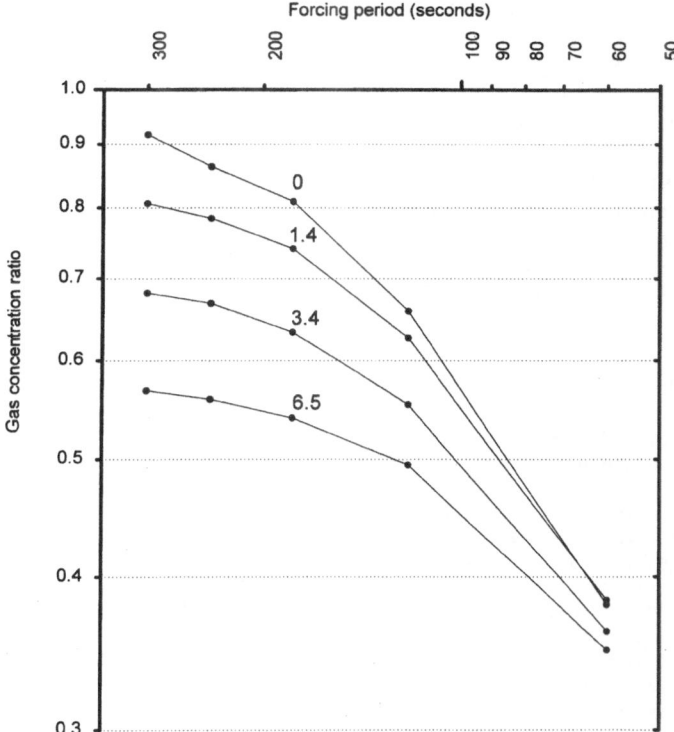

Figure 7. The log-log plot of changes in pulmonary blood flow simulated by the concentric water-displacement model lung.

ventilated gas concentration can be altered independently of volume-induced changes. In the lungs the blood and gas phases are separated by a thin membrane allowing the gas concentrations in these two phases to come into equilibrium during a short period. If this membrane is assumed to have minimal diffusion limitation to gas exchange then the water-displacement model can be used to represent this situation. Instead of a liquid-gas interface, a simple two gas phase interface is produced. These gas phases can be rapidly mixed into a single homogeneous volume due to the entrainment vortices created by the Venturi devices during inspiration. This procedure avoids the need to create the large surface areas needed for gas/liquid membrane gas exchange.

When simulating pulmonary blood flow, different marker gas solubilities can be simulated simply by dividing the required pulmonary blood flow by the solubility coefficient. The open system of gas input and output removes the inherent marker gas recirculation problem seen in closed physiological systems allowing inert gas dilution techniques to be tested under simpler conditions. Further adjustments to the oxygen and carbon dioxide concentrations in the inlet gas can be made separately enabling different respiratory quotients, RQ, to be modelled.

These modifications allow the concentric water-displacement model lung to be used to simulate gas exchange and pulmonary blood flow. The addition of a third perfused but non-ventilated compartment allows the testing of more complex mathematical models of normal lung ventilation and perfusion. With this model lung the conditions can be carefully controlled and results speedily obtained.

ACKNOWLEDGMENTS

The author of this work is supported by the Medical Research Council. He is especially grateful to P.A. Oakley and M.C. Sainsbury for their significant contributions and ideas concerning the simulation of pulmonary blood flow in the water-displacement model lung. He would also like to thank all his colleagues at the Nuffield Department of Anaesthetics for their various contributions to this work.

REFERENCES

Hahn, C.E.W., A.M.S. Black, S.A. Barton and I Scott (1993). Gas exchange in a three-compartment lung model analysed by forcing sinusoids of N_2O. J. Appl. Physiol.**75**, 1863-1876.

Sainsbury, M.C., A. Lorenzi, E.M. Williams, S.A. Barton and C.E.W. Hahn. The reconciliation of continuous and tidal ventilation gas exchange models. J. Appl. Physiol., submitted 1994.

Williams, E.M., L.B. Gale, P.A. Oakley, M.C. Sainsbury and C.E.W. Hahn (1993). Development of a concentric water-displacement lung. J. Biomed. Eng. **15**, 420-424.

Williams, E.M., D.J. Gavaghan, P.A. Oakley, M.C. Sainsbury, L. Xiong, A.M.S. Black and C.E.W. Hahn (1994a). Measurement of airways dead-space in a model lung using an oscillating argon concentration signal. Acta Anaesthesiol. Scand. **38**, 126-129.

Williams E.M., J.B. Aspel, S.M.L. Burroughs, W.A. Ryder, M.C. Sainsbury, L. Sutton, L. Xiong, A.M.S. Black and C.E.W. Hahn (1994b). Assessment of cardio-respiratory function using oscillating inert gas forcing signals. J. Appl. Physiol. **76**, 2130-2139.

34

WHAT DETERMINES CARDIAC OXYGEN CONSUMPTION AND HOW IS IT REGULATED?

Johannes H. G. M. van Beek[1*] and Xinqiang Tian[1,2]

[1] Laboratory for Physiology, Institute for Cardiovascular Research
Free University
1081 BT Amsterdam, The Netherlands
[2] Department of Pathophysiology
Datong Medical College
Datong, People's Republic of China

INTRODUCTION

Tight regulation of cardiac oxidative phosphorylation to ensure fast and accurate adaptation to myocardial metabolic demand is of vital importance. There has been much discussion during the last decade on the nature of the signals that stimulate the mitochondria in the heart muscle to increase mitochondrial ATP synthesis when the workload to the heart is increased. The classic view is that ADP, the direct product of ATP hydrolysis, stimulates the mitochondria to increase ATP synthesis when muscle work increases. This is very likely correct in skeletal muscle but was challenged for cardiac muscle tissue (Balaban et al., 1986). Here we will review recent data that throw new light on this issue. We will particularly consider what happens in the first minute during the dynamic adaptation of ATP synthesis to a sudden change in workload to the heart.

The discussion on which factors are determinants of the rate of oxygen consumption has been quite intense. The determinants of oxygen consumption have often been sought in the past by correlating oxygen consumption with the mechanical performance and inotropic state of the heart. However, because the mechanical performance may be considered to be the output of the heart, and because it would seem awkward to regard the mechanical output as the determinant of the chemical energy input of the heart we will develop a molecular point of view on what determines cardiac oxygen consumption in this paper.

* Address correspondence to: J. H. G. M. van Beek, Laboratory for Physiology, Free University, Van der Boechorststraat 7, 1081 BT Amsterdam, The Netherlands. phone: 31 20 4448110; FAX: 31 20 4448255.

Oxygen Transport to Tissue XVII, Edited by Ince et al.
Plenum Press, New York, 1996

265

HOW IS CARDIAC OXYGEN CONSUMPTION REGULATED?

Regulation via ADP and Inorganic Phosphate

When skeletal muscle it stimulated to work harder and as a consequence its ATP turnover is increased, the phosphocreatine concentration in the muscle in situ decreases. Because of the fast creatine kinase reaction this indicates an increase in the concentration of adenosine diphosphate (Meyer, 1988). An increase in the concentration of inorganic phosphate (Pi) is also found. When sufficient carbon substrate and oxygen is present, isolated mitochondria react to increases of ADP and Pi with an increase in oxygen consumption and ATP synthesis. Thus during increases in workload in skeletal muscle cells, ADP and inorganic phosphate may form an important stimulus to the mitochondria to increase ATP production (Chance and Williams, 1956). Consequently, ADP and inorganic phosphate (Pi) take part in a negative feedback loop that matches mitochondrial ATP synthesis to ATP hydrolysis, see Fig. 1.

In the heart, several nuclear magnetic resonance (NMR) spectroscopy groups did not detect changes in Pi, in phosphocreatine (PCr) and indirectly in ADP when the cardiac workload was increased (Balaban et al., 1986; Balaban and Heineman, 1989). Consequently, it was often maintained in recent years that Dr. Britton Chance's classic view (Chance and Williams, 1956) that mitochondrial oxygen consumption is stimulated by ADP and Pi upon increases in workload (feedback regulation) would not apply to the heart.

Oxygen Transients

Our own approach to study regulation of cardiac oxidative phosphorylation in the intact heart was quite different from NMR spectroscopy. We developed a method to determine the time course of cardiac mitochondrial oxygen consumption during steps in heart rate (Van Beek and Westerhof, 1991). To accomplish this we recorded the time course of oxygen uptake in the heart as a whole by recording the arterial and venous oxygen tensions with fast responding oxygen electrodes. We used the mean response time to characterize the time course of oxygen consumption. This mean response time is similar to the time constant, but is also applicable when the time course is not strictly monoexponential. The response time was corrected for the delay incurred due to diffusion between mitochondria and capillaries and due to vascular transport. To accomplish this we used a mathematical model for tissue oxygen transport whose key parameters were determined using experimental data on the mean transit time of oxygen from artery to vein obtained in the same heart. We found that the mean response time for the increase in oxygen consumption after a fast increase in ATP hydrolysis was about 8 s in the heart at 37°C (Van Beek and Westerhof, 1991). At 28°C

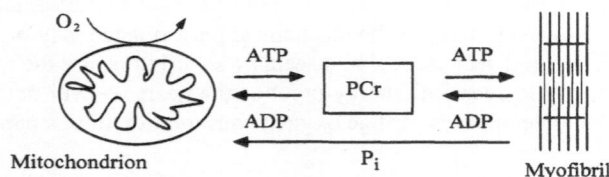

Figure 1. A scheme of feedback regulation of oxidative phosphorylation by ADP and inorganic phosphate (Pi), which are the products of ATP hydrolysis in the muscle during contraction. The myofibrils are an important site of ATP hydrolysis. A lower level of phosphocreatine (PCr), which reacts with ADP to resynthesize ATP, corresponds with an increased concentration of ADP and Pi in the cytosol. ADP and Pi both stimulate the mitochondria to synthesize ATP.

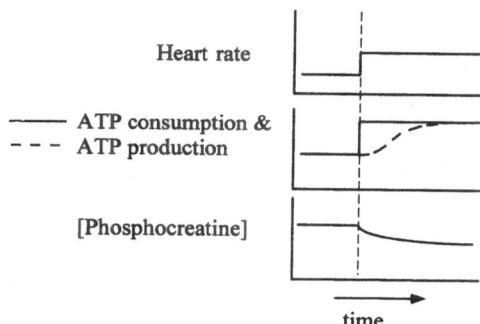

Figure 2. Hypothetical scheme: prediction of change in phosphocreatine (PCr) from the time courses of ATP consumption and production. At the vertical dashed line heart rate is increased stepwise by pacing. The response time of the change in ATP consumption was found to be >5 s. The experimentally obtained response of oxygen consumption and PCr was similar to the predicted time course.

the response time varied from 5-13 s depending on experimental conditions (Hak et al., 1992, 1993 a-d; Eijgelshoven et al., 1993, 1994). These response times indicate a considerable lag in the adaptation of mitochondrial ATP synthesis after changes in ATP hydrolysis. The negative balance between ATP synthesis and ATP hydrolysis thus persists for several seconds, leading to a decrease of the phosphocreatine energy buffer, see Fig. 2. ADP, formed by hydrolysis of ATP, is resynthesized by the reaction between ADP and PCr yielding ATP and creatine, catalyzed by the enzyme creatine kinase. Consequently, the level of PCr should decrease after a step in heart rate and the increases in [ADP] and [Pi] that accompany a decrease in PCr can stimulate the cardiac mitochondria.

We used a new NMR protocol, where we gated NMR signal acquisition to repeated heart rate steps and acculumated 64 NMR scans at 2 sec time resolution relative to the heart rate steps. We found that Pi increased substantially and that PCr decreased significantly by 8-11% after doubling the heart rate (Eijgelshoven et al., 1994). We showed that there very likely was no limitation of cardiac energy metabolism by the oxygen supply. The time constant of the increase in Pi after doubling the heart rate was about 5 s at 28°C, and was about 2.5 s at 37°C. The mean response time of oxygen consumption, determined simultaneously, was significantly larger: 11 and 14 s respectively. Thus the change in the phosphate metabolites precedes the change in oxygen consumption, and may thus contribute to the stimulation of oxidative phosphorylation. However, it has been found that the reaction of the mitochondria to an external stimulus is very fast (<100 ms; Chance, 1965). Therefore, the significant delay between the time constant of the phosphate metabolites in the cytosol and the change in mitochondrial oxygen consumption might be explained by other regulating factors which change more slowly.

Regulation via ADP and Inorganic Phosphate Revisited

The relatively slow upregulation of mitochondrial ATP synthesis after an increase in ATP hydrolysis leads to an increase in concentration of the hydrolysis products of ATP. This raised the question whether the increase in ADP and Pi levels was sufficient in magnitude to stimulate ATP synthesis to match ATP hydrolysis. In order to investigate this we modeled the regulation of oxidative phosphorylation in the cardiac myocyte by using a far-from-equilibrium thermodynamic model (Van Beek et al., 1992). The model calculation showed that the mitochondrial density in the heart muscle cells is so high that even the small decrease in high energy phosphate concentrations that results from the measured delay may stimulate cardiac oxidative phosphorylation significantly. Thus the regulation of the mitochondria by ADP and Pi has a very high feedback loop gain and plays an important role in the regulation of cardiac oxidative phosphorylation during increases in cardiac workload. Analysis with an enzyme kinetic model for stimulation of oxidative phosphorylation by ADP and Pi (Heineman and Balaban, 1990), also indicates that ADP and Pi may take care of 20-100% of the increase in cardiac oxygen consumption in our experiments (Eijgelshoven et al., 1994), depending on the conditions.

Other Factors Participate in the Regulation Of Oxidative Phosphorylation

Several pieces of evidence suggest that, although the phosphate metabolites play an important role in the regulation of cardiac oxidative phosphorylation, other factors also must take part in the stimulation of mitochondrial ATP synthesis. In several NMR spectroscopy studies it was found that when the cardiac workload was increased, PCr and by inference also the concentration of free ADP in the cytosol did not change significantly (Balaban et al., 1986; Balaban and Heineman, 1989). Although some studies did show that PCr may change even in the heart in situ, especially when the cardiac oxygen consumption gets very high (Massie et al., 1994), the idea that PCr does not change with cardiac workload has received much attention (Balaban, 1990). Indeed, our own calculations based on far-from-equilibrium thermodynamic (Van Beek et al., 1992) and kinetic (Heineman and Balaban, 1990) models for the regulation of oxidative phosphorylation indicate that possible changes in ADP and Pi concentration in the cytosol below the detection limit of the NMR experiment of Balaban et al. (1986) are too small to explain the increases in cardiac oxygen consumption. In our experiments (Eijgelshoven et al., 1994) on isolated rabbit hearts the increase in oxygen consumption at 28°C can be fully explained by the increase in ADP and Pi, but less than half of the increase in oxygen consumption at 37°C can be explained by ADP and Pi.

These results on the changes in oxygen consumption and phosphate metabolites in the steady state show that ADP and Pi may play an important role in the regulation of cardiac oxidative phosphorylation, but are presumably not the only regulators of oxidative phosphorylation. It is also apparent from the discordant time courses of mitochondrial oxygen consumption and phosphate metabolites during the dynamic adaptation to an immediate increase in ATP consumption, already discussed above, that this is also true for the first tens of seconds of the response.

The results for the steady state as well as for the dynamic adaptation suggest that, besides the feedback loop via the phosphate metabolites, other mechanisms also appear to play a role in the regulation of cardiac oxidative phosphorylation.

The overall reaction equation for oxidative phosphorylation is:

$$2 \text{ NADH} + 2 \text{ H}^+ + 6 \text{ ADP} + 6 \text{ Pi} + O_2 \underset{\leftarrow}{\overset{\rightarrow}{}} 2 \text{ NAD} + 6 \text{ ATP} + 2 \text{ H}_2O \tag{1}$$

From this equation it can be seen that, besides ADP and Pi, low levels of NADH and oxygen may limit ATP synthesis, and therefore that changes in the levels of these substrates may regulate ATP synthesis. Balaban (1990) has suggested that with increases in cardiac workload calcium increases in the mitochondria and stimulates pyruvate dehydrogenase and two other NADH producing enzymes in the Krebs cycle. The increase in NADH concentration would then stimulate oxidative phoshorylation. However, in the isolated working rabbit heart NADH fluorescence did not change with cardiac workload (Heineman and Balaban,

1993). Recent data even indicate that NADH fluorescence decreases with increased heart rate in the isolated rat heart (Ashruf et al., 1995). These experiments do not show an increase in the mitochondrial NADH concentration with workload and indicate that NADH may not be of preponderant importance in the regulation of oxidative phosphorylation. The results do, however, not preclude that regulation of NADH production, for instance by cytosolic calcium which has entered the mitochondria, does play a role in the regulation of oxidative phosphorylation, because stimulation by calcium might alleviate a decrease in mitochondrial NADH concentration that would otherwise be even bigger. Indeed, blocking the mitochondrial calcium entry channels causes a significant slowing of the adaptation of oxygen consumption after a heart rate step (Hak et al., 1993b).

Regulation of Oxidative Phosphorylation at the Enzyme Level

We have seen that regulation of oxidative phosphorylation may not be completely explained by changes in the concentrations of its substrates ADP, Pi and NADH. One should realize that oxidative phosphorylation is a reaction catalyzed by a sequence of enzymes embedded in the mitochondrial membrane. Regulation of oxidative phoshorylation can thus be accomplished by regulating the activity of one or more enzymes, at constant concentrations of the reactants of the enzyme chain. There is some evidence that ATP synthase (also called the F_0F_1-ATPase) may be regulated via two regulatory proteins (Harris and Das, 1991). The first protein is called IF1, and is sensitive to the electrical potential difference across the mitochondrial inner membrane. The second protein is called CaBI (calcium binding inhibitor protein) and is sensitive to the concentration of Ca^{2+}. The F_0F_1-ATP-ase activity in isolated cardiac myocytes seems to vary with the physiological condition: activity is increased during electrical pacing of the cells, but it is decreased during hypoxia. The entry of calcium into the mitochondrial matrix may be involved in the upregulation of the F_0F_1-ATPase activity (Harris and Das, 1991). Recently the regulation of F_0F_1-ATPase activity has also been shown in dog hearts in situ (Scholz and Balaban, 1994). In the experiments on isolated cardiac myocytes and on in situ hearts the F_0F_1-ATPase activity was determined by measuring the formation of ADP after ATP had been added to the cells after quickly disrupting the cell membranes. It remains to be investigated whether the ATP synthetic activity of the F_0F_1-ATPase changes in parallel with its ATP hydrolyzing activity (Harris and Das, 1991; Scholz and Balaban, 1994). Thus the ATP synthetic activity is perhaps regulated on the enzyme level, independent of the levels of the substrates and products of the ATP synthase, but direct evidence for this direct regulation of the enzyme is still lacking.

Summary: The Regulation of Oxidative Phosphorylation Is Multifactorial

As we have seen the regulation of cardiac oxidative phosphorylation may be partly explained by changes in ADP and Pi. Although these changes are small they may play an important role in the regulation of oxidative phosphorylation because the feedback loop gain is large. About 30-40% of the cardiac myocyte in rabbits and rats consists of mitochondria, emphasizing the importance of aerobic ATP production for cardiac function. However, stimulation of oxidative phosphorylation by ADP and Pi is not the only way to regulate the mitochondria. Regulation of the mitochondrial NADH production or direct regulation of enzyme activity of the ATP synthase complex are possibly playing a role. It is not excluded that the electron transport chain is also regulated at the enzyme level. Thus many factors play a role in the regulation of cardiac oxidative phosphorylation and it is a distinct possibility that the importance of these factors is dependent on the conditions and may change dramatically with species, with age, state of training or with heart disease. This strong dependence on condition may explain the apparently contradictory findings on the extent of changes in PCr and Pi reported in the literature (Balaban et al., 1986; Massie et al., 1994; Eijgelshoven et al., 1994).

WHAT DETERMINES CARDIAC OXYGEN CONSUMPTION?

The Classic Answer

Many investigators have tried to correlate different indices of the mechanical performance of the heart with cardiac oxygen consumption (e.g., Suga, 1979; Rooke and Feigl, 1982). It turned out that such relations were never perfect and that the relation between performance and oxygen consumption tended to vary with the conditions. Some approaches were relatively successful: in a practical way the product of heart rate and systolic left ventricular pressure often shows a high correlation with oxygen consumption. The pressure-volume area (PVA) defined by Suga (1979) is a more fundamental measure which correlates well with oxygen consumption. The PVA is the mechanical work performed by the heart added to the potential energy still stored in the heart muscle fibers at the end of systole. However, all these classical approaches share one drawback: they explain cardiac oxygen consumption, which reflects the chemical energy input of the heart, in terms of the mechanical output of the heart. These approaches have a lot of virtue in that they define the efficiency of cardiac contraction in a physiological way. On the other hand, these approaches are not helpful to answer which factors determines cardiac oxygen consumption, because the answer that the chemical energy input is determined by the mechanical output is not satisfactory.

External and Internal Determinants

Cardiac contraction may be stimulated by a positive inotropic agent, such as adrenaline or dobutamine. This is a clear example of an external determinant of cardiac performance. However, when there are defects in the proteins that make up the cellular machinery in the muscle cells the heart may contract weakly despite stimulation with inotropic agents. The myosin and actin molecules that form the contractile apparatus, or the proteins that take care of the transport of calcium to the cytosol are examples of internal determinants of cardiac performance.

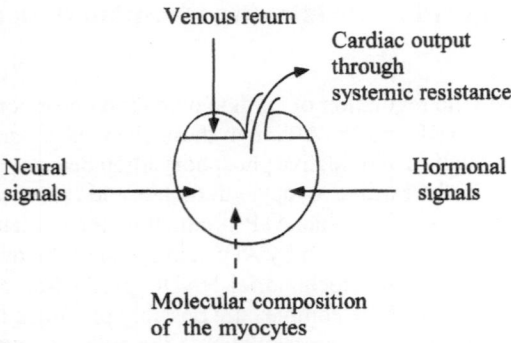

Figure 3. External (solid arrows) and internal (dashed arrows) determinants of cardiac oxygen consumption and mechanical performance.

The isoforms and the amount of protein expressed from the genome may thus be regarded as internal determinants of oxygen consumption. It is widely believed that the amount and type of myosin and actin molecules present and the extent of activation of these molecules by the calcium ions in the cytosol fully determine the amount of ATP hydrolyzed in any given work state. Then the mitochondria would just meet the demand of the myofibrils for ATP, because the mitochondrial capacity for ATP synthesis would be present in excess. Thus, according to this view the sole internal determinant of ATP turnover in the myocyte would be the state of activation of the myofibrillar ATPase.

Our finding above, that Pi may have to increase under some conditions to stimulate oxidative phosphorylation seems to contradict this idea, because an increase in [Pi] does inhibit force development of the myofibrils (Kentish, 1986). To analyze the contribution of various parts of the muscle cell, such as the myofibrillar ATPase and oxidative phosphory-lation, to rate limitation of ATP turnover we make use of Metabolic Control Theory.

Distribution of Rate Limitation

Modern theory on the control of biochemical systems, known as Metabolic Control Theory (Kacser and Burns, 1979; Heinrich and Rapoport, 1985; Westerhoff and Van Dam, 1987), has taught us that there is usually not one single rate limiting step in a biochemical system with many enzymes, but that control of metabolic flux is distributed over several of the enzymes present in the system. The contribution of individual enzymes to the rate limitation is expressed in the Flux Control Coefficient, also called the control strength. The control strength can be determined by influencing the activity of an enzyme, either by partially inhibiting the enzyme, or by altering the expression of the enzyme by genetic techniques. Suppose now that the effect of such an intervention on the activity of the single enzyme can be quantified and that the effect of the change in the single enzyme on the metabolic flux in the system as a whole, e.g., the ATP turnover in the muscle cell, can be determined separately. The control coefficient of the enzyme over the metabolic flux is defined as the percent change in metabolic flux in the system, divided by the percent change in the individual enzyme's activity, in the limit of a small change in enzyme activity. It has been found experimentally that the control strength in a system is usually distributed over several enzymes, for instance in isolated mitochondria control over oxygen consumption is distributed over several mitochondrial enzymes and transport proteins, and in glycolysis control over glycolytic flux is also distributed over several glycolytic enzymes.

Oxidative Phosphorylation Exerts Rate Limitation over Cardiac Oxygen Consumption

We investigated whether the mitochondrial capacity for ATP synthesis would con-tribute to the rate limitation of ATP turnover, and consequently on the mechanical perform-ance of the heart (Hak et al., 1993c). The strategy employed was to partially inhibit the mitochondrial ATP synthase in the intact heart, and then to measure the decline in oxygen consumption and left ventricular pressure that resulted from the inhibition of ATP synthase. Finally, the decrease in mitochondrial capacity for ATP synthesis was measured inde-pendently *in vivo*. This strategy is in line with Metabolic Control Theory, explained above.

Experiments were performed on isolated rabbit hearts which were perfused with a constant flow of Tyrode's solution at 28°C. Pressure was measured in a thin-walled balloon that almost completely filled the left ventricle. After equilibration of the heart to the artificial perfusion conditions the oxygen consumption and systolic and diastolic left ventricular pressures were measured. Then a preset amount of oligomycin was infused and oxygen

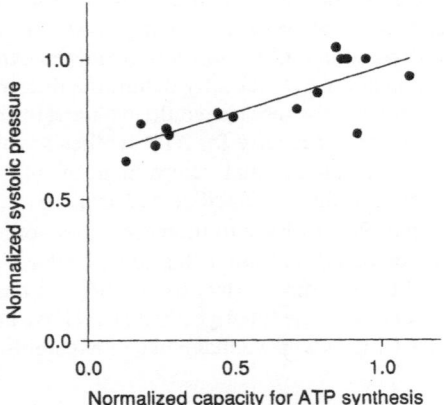

Figure 4. The dependence of cardiac systolic pressure on mitochondrial capacity for ATP synthesis. The normalized capacity for ATP synthesis is the maximal ADP-stimulated oxygen consumption divided by the maximal oxygen consumption during uncoupling with FCCP in mitochondria isolated from the isolated heart after the perfusion experiment. The normalized systolic pressure is the systolic pressure after infusion of oligomycin in the isolated heart divided by the systolic pressure before oligomycin infusion.

consumption and pressure were determined after these had again stabilized. This was done at several electrically paced heart rates: 60, 70, 80, 100 and 120 beats/min. After the measurements on the isolated heart were completed, the heart was homogenized and its mitochondria were isolated. The state 3 oxygen consumption of the mitochondria, i.e., the maximal oxygen consumption obtainable with stimulation by ADP, was determined. Further, the mitochondria were also stimulated with the uncoupler carbonyl cyanide p-(tri-fluoromethoxy)phenylhydrazone (FCCP), which bypasses the inhibition by oligomycin.

Control experiments indeed showed that the FCCP-stimulated oxygen consumption was a good measure of state 3 oxygen consumption before the introduction of oligomycin. Thus the ratio of ADP-stimulated oxygen consumption to FCCP-stimulate oxygen consumption in the isolated mitochondria reflects the fraction of mitochondrial ATP synthetic capacity remaining after inhibition with oligomycin. In Fig. 4 this fraction of the original capacity for ATP synthesis that is preserved after perfusion of the heart with oligomycin is plotted on the ordinate and the fraction of left ventricular pressure that was preserved in the heart is plotted on the ordinate. It may be observed that there is a virtually linear decrease of systolic pressure with the reduction in mitochondrial capacity for ATP synthesis. The oxygen consumption decreased in parallel to the systolic pressure. From the slope of the line in Fig. 4 it can be deduced that the mitochondrial capacity for ATP synthesis exerts considerable control over the ATP turnover and mechanical performance of the heart.

Per 10% decrease in mitochondrial capacity for ATP synthesis cardiac oxygen consumption decreased linearly by 2.6%. This translates into a Flux Control Coefficient, giving the control exerted by the mitochondrial capacity for ATP synthesis over cardiac oxygen consumption, of 0.26. Because Metabolic Control Theory teaches us that the sum of all Control Coefficients of all individual enzymes in the cell is 1, a Control Coefficient of 0.26 suggests that 26% of all rate limitation for ATP turnover exerted by enzymes and transport molecules in the heart is attributable to the mitochondrial ATP synthetic machinery. It is important to realize that this is found at rates of oxygen consumption which are considerably below the maximal values. The Flux Control Coefficients were virtually the same at 60-120 beats/min. The remaining 74% of rate limitation of ATP turnover is located outside the mitochondria, and its site is very likely the myofibrillar ATPase whose activity is regulated by the cytosolic concentration of free calcium ions, previously thought to be the one and only site of rate limitation.

REGULATION AND CONTROL: TWO SIDES OF ONE COIN

The example of rate control exerted by the mitochondria over cardiac ATP turnover, which is closely linked to cardiac mechanic performance invites to draw the following picture. The hormonal and neural control mechanisms of the heart, where adrenaline and the sympathetic cardiac innervation are classic examples, and the resistance of the systemic circulation and venous return request a certain work output from the heart. The myofibrillar ATPase is set to work by the second messenger calcium. However, this ATPase needs ATP to fuel muscle contraction and may be inhibited by ATP's breakdown product inorganic phosphate. Because the inorganic phosphate concentration must increase together with ADP to stimulate the mitochondria it cannot avoid to inhibit ATP hydrolysis. The higher the capacity of mitochondrial ATP synthesis, the less increase in inorganic phosphate is needed to increase ATP synthesis, and the less ATP hydrolysis can rise. This example shows that regulation of metabolic pathways determines the distribution of rate limitation in the system. The interplay of external determinants such as inotropic agents, mechanical load and supply of carbon substrates influence the muscle cells which have a certain repertoire of macro-molecules expressed from the genome. The isoforms and the quantity of macromolecules together with the external signals determine the mechanical performance and energy turnover of the heart. The example of inorganic phosphate as a link between ATP hydrolysis and ATP synthesis is certainly overly simple. Actually, the metabolic system of the heart has multiple control sites and besides inorganic phosphate and ADP other molecules such as NADH are expected to play key roles. Last but not least regulation at the enzyme level, for instance of ATP synthase by regulatory proteins, may play an important role in heart muscle. The levels of metabolites which regulate the mitochondria may also influence the performance of the myofibrils. Thus increases in the levels of metabolites that stimulate cardiac oxygen consumption might also limit ATP hydrolysis and thus cardiac performance. Metabolites form the link between the enzymes and regulate enzyme fluxes. The sensitivity of the enzymes for the metabolites determines what part of the rate limitation is exerted by various enzymes.

ACKNOWLEDGMENTS

This work was partially supported by grants from the Netherlands Organization for Scientific Research NWO and by a travel grant of Datong Medical College to X. Tian.

REFERENCES

Ashruf, J.F., Coremans, J.M.C.C., Bruining, H.A., and Ince, C., 1995 Increase of cardiac work is associated with decrease of mitochondrial NADH in the Langendorff rat heart, *Am.J.Physiol.* 269:H856-H862.

Balaban, R.S., 1990, Regulation of oxidative phosphorylation in the mammalian cell, *Am.J.Physiol.* 258:C377-C389.

Balaban, R.S., and Heineman, F.W., 1989, Interaction of oxidative phosphorylation and work in the heart, *in vivo*, *News in Physiological Sciences* 4:215-218.

Balaban, R.S., Kantor, H.L., Katz, L.A., and Briggs, R.W., 1986, Relation between work and phosphate metabolites and oxygen consumption in the *in vivo* paced mammalian heart, *Science* 232:1121-1123.

Beek, J.H.G.M. van, and Westerhof, N., 1991, Response time of cardiac mitochondrial oxygen consumption to heart rate steps, *Am.J.Physiol.* 260: H613-H625.

Beek, J.H.G.M. van, Hak, J.B., and Eijgelshoven, M.H.J., 1992, Oxidative phosphorylation in the myocardium may be strongly stimulated by undetectably small changes in high-energy phosphates, Pflügers Arch. 420: R105.

Chance, B., 1965, The energy-linked reaction of calcium with mitochondria, *J.Biol.Chem.* 240: 2729-2748.

Chance, B., and Williams, G.R., 1956, The respiratory chain and oxidative phosphorylation, *Adv. Enzymol.* 17:65-134.

Eijgelshoven, M.H.J., Hak, J.B., Beek, J.H.G.M. van , and Westerhof, N., 1993, Adaptation speed of cardiac mitochondrial oxygen consumption decreases with higher heart rate, *Am. J. Physiol.* 265:H1893-H1898.

Eijgelshoven, M.H.J., Beek, J.H.G.M. van, Mottet, I., Nederhoff, M.G.J., Echteld, C.J.A. van, and Westerhof, N., 1994, Cardiac high-energy phosphates adapt faster than oxygen consumption to changes in heart rate, *Circ. Res.* 75: 751-759.

Hak, J.B., Beek, J.H.G.M. van ,Wijhe, M.H. van , and Westerhof, N., 1992, Influence of temperature on the response time of mitochondrial oxygen consumption in isolated rabbit heart, *J. Physiol.* (Lond) 447:17-31

Hak, J.B., Beek, J.H.G.M. van, and Westerhof, N., 1993a, Acidosis slows the response of oxidative phosphorylation to metabolic demand in isolated rabbit heart, *Pflügers Arch.* 423:324-329

Hak, J.B., Beek, J.H.G.M. van, Eijgelshoven, M.H.J., and Westerhof, N., 1993b, Mitochondrial dehydrogenase activity affects adaptation of cardiac oxygen consumption to demand, *Am. J. Physiol.* 264:H448-H453

Hak, J.B., Beek, J.H.G.M. van, Wijhe, M.H. van, Eijgelshoven, M.H.J., and Westerhof, N., 1993c, Reduced cardiac ATP-synthetic capacity slows metabolic regulation and reduces contractility, In: J.B. Hak, Dynamics of Myocardial Oxidative Phosphorylation, Thesis, Vrije Universiteit, Amsterdam, the Netherlands. ISBN: 90-9005935-0

Hak, J.B., Beek, J.H.G.M. van, Wijhe, M.H. van, and Westerhof, N., 1993d, Dynamics of myocardial lactate efflux after a step in heart rate in isolated rabbit heart, *Am. J. Physiol.* 265:H2081-H2085.

Harris, D.A, and Das, A.M., 1991, Control of mitochondrial ATP synthesis in the heart, *Biochem.J.* 280: 561-573.

Heineman,F.W., and Balaban, R.S., 1990, Phosphorus-31 nuclear magnetic resonance analysis of transient changes of canine myocardial metabolism *in vivo, J.Clin.Invest.* 85:843-852

Heineman, F.W., and Balaban, R.S., 1993, Effects of afterload and heart rate on NAD(P)H redox state in the isolated rabbit heart., *Am. J. Physiol.* 264:H433-H440

Heinrich, R., and Rapoport, T.A., 1975, Mathematical analysis of multienzyme systems: II. Steady state and transient control, *Biosystems* 7:130-136

Kacser, H., and Burns, J.A., 1979, Molecular democracy: who shares the controls? *Biochem. Soc. Trans.* 7:1149-1160

Kentish, J.C., 1986, The effects of inorganic phosphate and creatine phosphate on force production in skinned muscles from rat ventricle. *J Physiol* (Lond) 370:585-604

Massie, B.M., Schwartz, G.G., Garcia, J., Wisnewki, J.A., Weiner, M.W., and Owens, T., 1994, Myocardial metabolism during increased work states in the porcine left ventricle *in vivo, Circ.Res.* 74:64-73.

Meyer, R.A., 1988, A linear model of muscle respiration explains monoexponential phosphocreatine changes, *Am.J.Physiol.* 254:C548-C553

Rooke, G.A., and Feigl, E.O., 1982, Work as a correlate of canine left ventricular oxygen consumption, and the problem of catecholamine oxygen wasting, *Circ.Res.* 50:273-286.

Scholz, T.D., and Balaban, R.S., 1994, Mitochondrial F1-ATPase activity of canine myocardium: effects of hypoxia and stimulation, *Am.J.Physiol.* 266: H2396-H2403

Suga, H., 1979, Total mechanical energy of a ventricle model and cardiac oxygen consumption, *Am.J.Physiol.* 236: H498-H505.

Westerhoff, H.V., and Dam K. van, 1987, *Thermodynamics and control of biological free-energy transduction,* Elsevier, Amsterdam, the Netherlands, pp 162-186 and pp 436-494

35

MITOCHONDRIAL NADH IN THE LANGENDORFF RAT HEART DECREASES IN RESPONSE TO INCREASES IN WORK

Increase of Cardiac Work Is Associated with Decrease of Mitochondrial NADH

J. F. Ashruf,[1] J. M. C. C. Coremans,[1] H. A. Bruining,[1] and C. Ince[2]

[1] Department of General Surgery
Erasmus University of Rotterdam, The Netherlands
[2] Department of Anaesthesiology
University of Amsterdam, The Netherlands

INTRODUCTION

The mechanism that tunes the rate of respiration to a changing ATP demand during alterations in work of the intact myocardium has not been elucidated (3, 4, 5, 12). The phosphate potential (ATP/ADP ratio) and the substrate oxidation potential (e.g. redox potential difference of the $NADH/NAD^+$ couple and O_2/H_2O couple) have been proposed as potential regulatory factors of oxidative phosphorylation. Although studies of isolated mitochondria have demonstrated that the rate of respiration can be controlled by changes in the concentration of ADP and P_i (7), in intact myocardium no significant change in the phosphorylation potential has been found (3, 4). Whereas elevation of the NADH concentration can increase the maximum rate of oxidative phosphorylation and stimulate ATP synthesis in isolated mitochondria (18), conflicting data on the NADH redox state during increase of work in intact myocardium have been found. In isolated rabbit heart no change was found in the NAD(P)H redox state during changes in afterload and heart rate (12). An increase of the mitochondrial $NADH/NAD^+$ ratio was found in response to increased heart rate in the rat heart (16). In contrast, a decrease of the intramitochondrial $NADH/NAD^+$ ratio was reported for isolated papillary rabbit muscle in response to twitches (9). Methodological differences could explain these contradictory results. Measurement of the intracellular autofluorescence of NADH (365 nm excitation, 470 nm emission), allows qualitative evaluation of the mitochondrial $NADH/NAD^+$ redox state since NADH fluoresces when excited with ultraviolet light, and NAD^+ does not (6). The NADH fluorescence measurement can be disturbed by ischemia, oximetric effect, endogenous substrate depletion, movement and photo-bleaching (1, 2, 11, 12, 14, 15, 16). Ischemia may develop when oxygen demand

Oxygen Transport to Tissue XVII, Edited by Ince et al.
Plenum Press, New York, 1996

275

exceeds oxygen supply during periods of increased work. During ischemia, lack of O_2 supply will reduce the transfer of reducing equivalents from the respiratory chain and will result in an increase of the mitochondrial NADH concentration. This ischemia is distributed heterogeneously over the myocardium (1, 2, 14) and will be a serious distorting effect in NADH fluorescence measurements which do not take this spatial heterogeneity into account. Changes in the spectral absorption of mitochondrial cytochromes and myoglobin brought about by changes in cardiac metabolism are known as the oximetric effect, and may influence the NADH fluorescence intensity (11). Depletion of endogenous substrates may also affect the NADH fluorescence measurement (1, 12, 16). If NADH fluorescence intensity of a glucose perfused heart decreases during increases in work, one could argue that this decrease may be caused by the inability of glycolysis to increase its reducing equivalents delivery for the Krebs cycle as much as the utilization of reducing equivalents by the Krebs cycle is increased (17). A further complication of the NADH fluorescence measurement is movement of the organ surface with respect to the illumination detection system (1, 12, 16). Finally, prolonged illumination of tissue with UV light, necessary to induce NADH fluorescence, eventually results in decrease of NADH fluorescence intensity due to photo-bleaching (1, 12, 16).

The present study was conducted to investigate whether changes in work of the myocardium are associated with changes in the mitochondrial NADH/NAD$^+$ ratio, taking into account the above mentioned factors influencing NADH fluorescence measurements. A vasodilator was added to the perfusate to avoid the development of ischemia by increasing coronary flow (1, 14). To detect a possible oximetric effect in this study, diffuse reflectance spectra were acquired during low and high cardiac work. Endogenous substrate depletion was prevented by perfusing the heart with 10 mM pyruvate (1, 17). NADH fluorescence intensity was measured in images of the heart surface in identical geometrical configurations, enabling the recognition of heterogeneously distributed ischemic areas and the reduction of motion artifacts. Photo-bleaching was diminished by using discontinuous UV illumination.

MATERIALS AND METHODS

Experimental Set-up

Male Wistar rats weighing 275-375 g were anesthesized by ether inhalation. Before sterneotomy, the rats were given 200 IU/kg heparin intravenously (Thromboliquine, Organon Teknika Oss, the Netherlands). The hearts were perfused according to Langendorff and were paced via aortic and right ventricular leads. Mean flow rates (in ml/min/g of ventricle wet weight) were measured with an electromagnetic flow probe (Skalar-Medical BV, Delft, the Netherlands) placed just before the aortic cannula. Left ventricular pressure was measured as described elsewhere (19), using a 22 gauge cannula inserted through the apex in the left ventricle and connected to a pressure transducer (Hewlett Packard 8805B Corner Amplifier, USA). In case the cannula communicated with the atmosphere, no pressure was developed by the left ventricle and the work output of the heart was low. In order to increase the amount of work performed by the heart, the apex cannula was closed and simultaneously the heartbeat frequency was increased from 5 to 7 Hz. To decrease the work output, the apex cannula was opened and the heartbeat frequency was decreased from 7 to 5 Hz according to the method of Kobayashi and Neely (17). One pulmonary artery was cannulated and its effluent was diverted with a constant flow to an oxygen probe (YSI Biological Oxygen Monitor, model 5300, Yellow Springs Instruments Co., Inc., USA). Oxygen tensions in both the aortic influent and pulmonary effluent were measured with an YSI Biologic Oxygen Monitor. From these measurements oxygen consumption was calculated (expressed as μmol/min/g of ventricle wet weight). The perfusate was a modified Tyrode solution (128 mM NaCl, 4.7 mM KCl, 1 mM $MgCl_2$, 0.4 mM NaH_2PO_4, 1.2 mM Na_2SO_4, 20.2 mM $NaHCO_3$, 1.3 mM $CaCl_2$) containing pyruvate (10 mM). The perfusate was thermostatted at 37°C and equilibrated with either 95% O_2 and 5% CO_2, or 95% N_2 and 5% CO_2. In some experiments 10 μM sodium-nitroprusside, an endothelium independent vasodilator, was added to the perfusate. Aortic pressure was kept at 80 mm Hg. Chemicals were obtained from Merck (Darmstadt, Germany).

NADH Fluorescence Measurements

The videofluorimeter used has been described in Ref. (14). The 365 nm line from a 100 W mercury arc lamp (Olympus, Tokyo) was selected by means of an UG-1 barrier filter to provide the UV light needed for NADH excitation. An image intensified video camera (MXRi second generation CCD video camera, HCS vision, Eindhoven, the Netherlands) with a Micro-Nikkor 105 mm macro-lens was used to record NADH fluorescence images of the left ventricle of the heart. The NADH fluorescence signal was selected by means of a band pass filter centred around 470 ± 20 nm which was placed in front of the camera. To enable correction of images for changes in the sensitivity of the videofluorimeter and fluctuations in the intensity of the light source, a small piece of uranyl calibration glass was placed next to the heart within the excitation field. Images were recorded on a video recorder and computer analyzed off-line. In order to minimize motion artifacts, pacemaker signals and fluorescence images were recorded simultaneously on video tape, thus enabling selection of fluorescence images recorded in the same geometrical configuration (14). In a metabolically stable preparation, these precautions restricted variations in NADH fluorescence intensity due to motion to the 5% margin. NADH fluorescence intensities are expressed in arbitrary units (A.U.), which are defined as the ratio of the NADH fluorescence intensity of myocardium and the fluorescence intensity of the uranyl calibration glass. Changes in intensity of NADH fluorescence are expressed in %, taking initial NADH fluorescence intensity (that is, just prior to the change) as 100 %.

Diffuse Reflectance Spectroscopy

Diffuse reflectance spectra of the heart surface were measured with an OMA Vision System (EG&G Princeton Applied Research), essentially as described in Ref. (11). The heart was illuminated by a Xe(Hg) lamp (100 W, Oriel). Remitted light from a small area of the left ventricle was transmitted to a spectrograph (model 1228, focal length 0.32 m, 25 μm entrance slit, grating 147.5 g/mm blazed at 300nm) by a quartz fibre (KO1655) which was positioned at 45° with respect to the heart surface at a working distance of 1.5 mm. In this configuration the field under observation sized approximately 3 mm^2. The diffuse reflectance spectra (450 - 650 nm, 240 ms scans) were measured with a 1024 element Silicon Photodiode Array (model 1453A, thermoelectrically cooled to 10°C).

Experimental Protocols

After arranging the hearts for Langendorff perfusion, they were allowed to stabilize for ± 15 min before the following experimental protocols were performed:

1. Cardiac Work at Limited O_2 Supply. Perfusion with oxygen-carrier free medium renders the Langendorff heart borderline aerobic (10, 13). Under these conditions ischemia easily develops during increase in oxygen demand (14). To study the effect of increased work on the mitochondrial NADH concentration under conditions which tend to limit O_2 supply and either limit or maximize substrate supply, the heartbeat rate of the Langendorff hearts was increased and pressure development by the left ventricle was induced. The work was changed according to a fixed protocol of 30 min duration. During the first 10 min the hearts were paced at 5 Hz with an open apex cannula (no pressure development by the left ventricle, low cardiac work). Then the apex cannula was closed and the frequency was increased to 7 Hz for 10 min (pressure development by the left ventricle, high cardiac work), during which period flow and oxygen consumption were stabilized. Then the apex cannula was re-opened and the frequency was decreased to 5 Hz (again low cardiac work) for another period of 10 min to stabilize flow and oxygen consumption.

    2. Cardiac Work at Ample O_2 Supply. To study the effect of increased work without development of hypoxic areas, 10 μM nitroprusside was added to the perfusate and protocol 3 was performed. Some hearts underwent this protocol twice, the first time during registration of NADH fluorescence and the second time during registration of diffuse reflectance spectra. During this protocol 2 the hearts were illuminated with UV light 5 s every min except during switches of pace frequency or substrate when the hearts were illuminated continuously for ± 10 s.

    Group means were tested for difference using paired t tests. The criterion for significance was taken to be P < 0.01 for all comparisons. Descriptive data are presented as means ± SE.

RESULTS

    To investigate whether redox changes of the mitochondrial respiratory chain and oxygenation changes of myoglobin influence the absorptive and reflective properties of cardiac tissue during the low and high cardiac work period, diffuse reflectance spectra were measured (protocol 2). Diffuse reflectance spectra obtained during the low and high cardiac work period did not differ from each other.

    To study the effect of changes of work upon epicardial NADH fluorescence during O_2-limited perfusions, 6 hearts were subjected to periods of low and high cardiac work (Fig. 1). These work changes were applied to each heart in the absence of a vasodilator (protocol

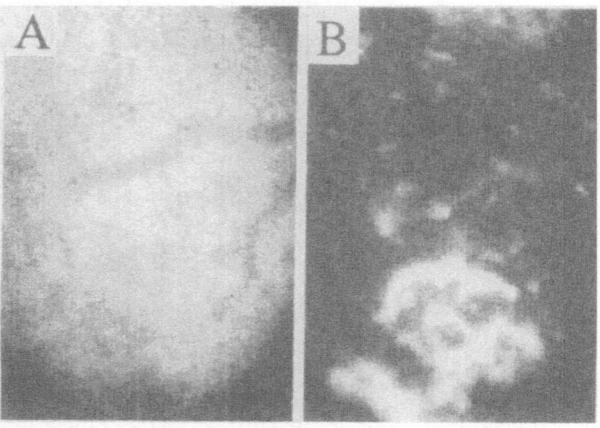

Figure 1. NADH fluorescence images of a heart perfused in the absence of nitroprusside. A: low work. B: high work.

Table 1. Changes of NADH fluorescence associated with changes in myocardial oxygen consumption

	10 mM pyruvate, 10 μM nitroprusside	
LVP cannula	open→closed	closed→open
Frequency change	5→7 Hz	7→5 Hz
ΔNADH fluorescence, %	-20±2.6	17±1.9
ΔMVO$_2$, μmol/min/g	3.6±0.7	-3.7±0.7

Presented values are steady state values expressed in means ± SE (n=6). ΔNADH fluorescence, change of NADH fluorescence relative to the NADH fluorescence just prior to the change in frequency from 5 to 7 Hz; ΔMVO$_2$, difference of oxygen consumption before and after the change in cardiac work.

1). Fig. 1A shows the epicardial NADH fluorescence at low cardiac work. Upon increase in work the NADH fluorescence images revealed the development of very prominent highly fluorescent ischemic areas (Fig. 1B). The ischemic areas were very different in size and distribution pattern from heart to heart. Oxygen consumption increased upon increase of work. After a work change coronary flow did not alter (data not shown).

To avoid the development of ischemic areas during increase in cardiac work, the experiment was repeated with 10 μM nitroprusside present in the perfusate (protocol 2). Due to nitroprusside induced vasodilation, coronary flows were larger than those during protocol 1. Other hemodynamic data were not affected by nitroprusside (data not shown). Oxygen consumption increased in response to increased cardiac work. The NADH fluorescence images of a heart recorded in periods of low and high cardiac work (Fig. 2A and B, respectively), all showed an even distribution of the fluorescence intensity. The corresponding time-trace of NADH fluorescence intensity changes during protocol 2 is shown in Fig. 2C where fluorescence images taken at the moments indicated by A and B are depicted in Fig. 2A and B. Grouped data of the changes in NADH fluorescence intensity and the associated changes in oxygen consumption are summarized in Table 1.

DISCUSSION

This study was performed to investigate whether mitochondrial NADH forms a regulatory factor of oxidative phosphorylation, tuning the rate of respiration to a changing ATP demand during alterations in cardiac work. Following the pioneering work of Chance et al. (6, 8) mitochondrial NADH was evaluated using its fluorescent property. Several studies monitoring NADH fluorescence in isolated working heart preparations under conditions of limited substrate supply (glucose alone), showed a decrease in baseline NADH fluorescence intensity in time, thereby complicating interpretation of work induced NADH fluorescence changes (1, 12, 14, 16) (data not shown). This decrease could possibly be subscribed to the gradual decline in the availability of intracellular substrate as the heart becomes dependent solely on the perfusate glucose because of the depletion of endogenous substrates (such as fatty acids and glycogen) (17). In hemodynamically stable Langendorff hearts illuminated continuously with UV light, we observed a decrease of NADH fluorescence during perfusion with pyruvate, which was not different from the decrease observed during perfusion with glucose (data not shown). Since the presence of pyruvate prevents use

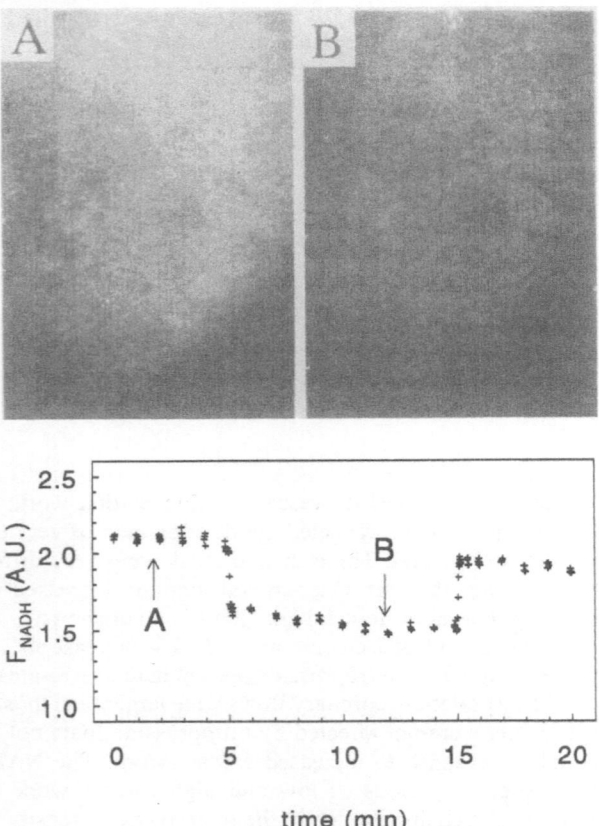

Figure 2. NADH fluorescence images of a heart perfused the presence of nitroprusside. A: low cardiac work. B: high cardiac work. C: the time-trace of NADH fluorescence intensity changes during protocol 2. From 0 to 5 min and 15 to 20 min: low work. From 5 to 15 min: high work. Video images at moments indicated by A and B are shown in Fig. 2A and B.

of endogenous substrates (17), we conclude that the observed decrease in NADH fluorescence intensity is not caused by depletion of endogenous substrates. When hearts were exposed to UV illumination discontinuously, the decrease in NADH fluorescence intensity greatly reduced in both glucose and pyruvate perfused hearts (the trend again did not differ between glucose and pyruvate perfused hearts, data not shown), indicating that photo-bleaching by UV illumination is probably the principal factor responsible for the time dependent decline of NADH fluorescence (20).

We did not observe changes in the reflectance spectra of isolated heart during changes in cardiac work (data not shown). This invariability of the reflectance spectra of vasodilated myocardium indicates that the NADH fluorescence measurement was not affected by changes in tissue absorbance during alterations in cardiac work.

Simultaneous registration of NADH fluorescence images and pacemaker signals allowed acquisition of the NADH fluorescence intensities of a well-defined area on the epicardial surface in exactly the same geometrical configuration. As described before (14), this method provided adequate compensation for motion artifacts. The importance of image formation is also stressed by the fact that isolated rat hearts all developed ischemic tissue areas when subjected to increase in pace-frequency (from 5 to 10 Hz) (14) or to increase in cardiac work (this study) in case no vasodilator was added to the perfusate. These ischemic areas could be alleviated by administration of a vasodilator.

In order to be able to asses the effect of changes in cardiac work on mitochondrial NADH without inducing ischemic spots, nitroprusside was added to the perfusate in protocol 2. In vasodilated rat hearts increase of work was associated with decrease in NADH fluorescence intensity (Fig. 2).

The finding that, apart from the expected increase in coronary flow, nitroprusside addition did not change the hemodynamic parameters of hearts during increase of cardiac work illustrates the superior sensitivity of epicardial NADH-fluorometry for identification of tissue hypoxia. In accordance with our study, twitches of isolated papillary muscle of the rabbit resulted in reduction of the mitochondrial NADH fluorescence intensity (9).

In summary, the present findings suggest that increase (decrease) of respiratory activity induced by increase (decrease) of cardiac work, are associated with decrease (increase) of NADH fluorescence. Under the conditions of this study, an increase of myocardial oxidative phosphorylation is not related to an increase of mitochondrial NADH, making it unlikely that an increase of mitochondrial NADH is a primary stimulus for oxidative phosphorylation. Rather, it seems that changes in mitochondrial NADH concentrations were probably (partially) secondary to changes in respiratory chain activity.

ACKNOWLEDGMENTS

This investigation was supported in part by the Netherlands Heart Foundation (grant number 90298). The authors are thankful for support by Friends of the Rotterdam Blood Bank.

REFERENCES

1. Ashruf, J.F., J.M.C.C. Coremans, H.A. Bruining and C. Ince. Increase of cardiac work is associated with decrease of mitochondrial NADH. Am. J. Physiol. 269 (Heart Circ. Physiol. 38) H856-H862 1995.
2. Avontuur, J.A.M., H.A. Bruining and C. Ince. Inhibition of nitric oxide synthesis causes myocardial ischemia in endotoxemic rats. Circ. Res. 76 (3): 418-425, 1995.
3. Balaban, R.S. Regulation of oxidative phosphorylation in the mammalian cell. Am. J. Physiol. 258 (Cell Physiol. 27): C377-C389, 1990.
4. Balaban, R.S., and F.W. Heineman. Interaction of oxidative phosphorylation and work in the heart, in vivo. NIPS. 4: 215-218, 1989.
5. Balaban, R.S., H.L. Kantor, L.A. Katz, and R.W. Briggs. Relation between work and phosphate metabolites in the in vivo paced mammalian heart. Science Wash. DC 232: 1121-1123, 1986.
6. Chance, B., P. Cohen, F. Jöbsis, and B. Schoener. Localized fluorometry of oxidation-reduction states of intracellular pyridine nucleotide in brain and kidney cortex of the anesthetized rat. Science Wash. DC 136: 325, 1962.
7. Chance, B. and C.M. Williams. The respiratory chain and oxidative phosphorylation. Adv. Enzymol. 17: 65-134, 1956.
8. Chance, B., J.R. Williamson, D. Jamieson, and B. Schoener. Properties and kinetics of reduced pyridine nucleotide fluorescence of the isolated and in vivo rat heart. Biochem. Z. 341: 357-377, 1965.
9. Chapman, J.B. Fluorometric studies of oxidative metabolism in isolated papillary muscle of the rabbit. J. Gen. Physiol. 59: 135-154, 1972.
10. Chemnitius, J.M., W. Burger, and R.J. Bing. Crystalloid and perfluorochemical perfusates in an isolated working rabbit heart preparation. Am. J. Physiol. 249 (Heart Circ. Physiol. 18): H285-H292, 1985.

11. Coremans, J.M.C.C., C. Ince, and H.A. Bruining. NADH fluorimetry and diffuse reflectance spectroscopy on rat heart. In: Medical Optical Tomography: Functional Imaging and Monitoring. G.J. Müller et al. eds., Vol. IS 11: 589-617, SPIE, Washington, USA.

12. Heineman, F.W., and R.S. Balaban. Effects of afterload and heart rate on NAD(P)H redox state in the isolated rabbit heart. Am. J. Physiol. 264 (Heart Circ. Physiol. 33): H433-H440, 1993.

13. Hülsmann, W.C., J.F. Ashruf, H.A. Bruining, and C. Ince. Imminent ischemia in normal and hypertrophic Langendorff rat hearts; effects of fatty acids and superoxide dismutase monitored by NADH surface fluorescence. Biochim. Biophys. Acta. 1181: 273-278, 1993.

14. Ince, C., J.F. Ashruf, J.A.M. Avontuur, P.A. Wieringa, J.A.E. Spaan, and H.A. Bruining. Heterogeneity of the hypoxic state in rat heart is determined at capillary level. Am. J. Physiol. 264 (Heart Circ. Physiol. 33): H294-H301, 1993.

15. Ince, C., J.M.C.C. Coremans, and H.A. Bruining. In vivo NADH fluorescence. Adv. Exp. Med. Biol.: Oxygen Transport to Tissue XIV. 317: 277-296, 1992.

16. Katz, L.A., A.P. Koretsky, and R.S. Balaban. Respiratory control in the glucose perfused heart. A ^{31}P NMR and NADH fluorescence study. FEBS Lett. 221: 270-276, 1987.

17. Kobayashi, K., and J.R. Neely. Control of maximum rates of glycolysis in rat cardiac muscle. Circ. Res. 44: 166-175, 1979.

18. Koretsky, A.P., and R.S. Balaban. Changes in pyridine nucleotide levels alter oxygen consumption and extramitochondrial phosphates in isolated mitochondria: a ^{31}P NMR and fluorescence study. Biochim. Biophys. Acta 893: 398-408, 1987.

19. Nishiki, K., M. Erecinska, and D.F. Wilson. Energy relationships between cytosolic metabolism and mitochondrial respiration in rat heart. Am. J. Physiol. 234 (Cell Physiol. 3): C73-C81, 1978.

20. Udenfriend, S. Fluorescence assay in Biology and Medicine Vol 1. New York, Academic Press. p104, 1962.

36

HYPOXIA/ISCHEMIA AND THE pH PARADOX

Joseph C. LaManna

Department of Neurology
School of Medicine
Case Western Reserve University
Cleveland, Ohio 44106-4938

INTRODUCTION

Hydrogen ions play an important role in cellular processes. There are intimate links between energy metabolism and the control of cell and tissue acid/base balance. Control of this balance is threatened or lost during severe hypoxia or ischemia. Re-establishment of pH balance must occur before the tissue can be considered to have returned to normal operating conditions. Because hydrogen ions influence so many reactions, the timing of re-normalization can be crucial to the entire recovery process. Indeed, in many active tissues, too fast reversal of acidosis during recovery from severe hypoxia or ischemia appears to be detrimental to the overall recovery of homeostasis. That a tissue could restore function more rapidly if mild acidosis were maintained during the immediate post-stress recovery time has been referred to as the "pH paradox" (Currin et al., 1991), in analogy with the so-called "calcium paradox" that has been discussed primarily in the cardiovascular literature. In this paper we will review the changes that occur in pHi during hypoxia and ischemia in rat brain. We will explore the interrelationships of protons with metabolism, and we will propose a scheme for the interaction of protons in brain function. Finally, we will attempt to reach a conclusion concerning the applicability of the concept of pH paradox in brain.

INTRACELLULAR ACIDOSIS DURING ISCHEMIA

Anoxic depolarization occurs within a few minutes of total cerebral ischemia, resulting in a large shift in extracellular ions and water (Hansen, 1985). It is unlikely that permanent cell damage can occur without anoxic depolarization, and residual tissue damage increases with increasing duration of ischemia in a progressive, graded fashion rather than all or none (Plum, 1983). During the anoxic depolarization accompanying ischemia, a significant event is increased H+ (acidosis) of as much or more than 1 pH unit (Harris and Symon, 1984; Blomqvist et al., 1982; Javaheri et al., 1983; Nemoto and Frinak, 1981; Crowell and Kaufmann, 1961; Kraig et al., 1983; Mutch and Hansen, 1984; Mabe et al.,

Oxygen Transport to Tissue XVII, Edited by Ince et al.
Plenum Press, New York, 1996

283

1983; von Hanwehr et al., 1986), followed by an intracellular alkalosis in the recovery period (Blomqvist et al., 1982; Mabe et al., 1983).

Brain intracellular acidosis accompanies ischemia in proportion to lactate accumulation which in turn is proportional to preischemic availability of glucosyl units from glucose and glycogen (LaManna et al., 1992a; Gyulai et al., 1987; Boris-Möller et al., 1988; Paschen et al., 1987; Combs et al., 1990; Chopp et al., 1988; Katsura et al., 1992). Brain tissue lactacidosis has been considered one of the primary mechanisms causing ischenic brain damage (Siesjö, 1988). Acidosis of sufficient magnitude injures neurons by direct denaturation of proteins and nucleic acids, astrocyte necrosis secondary to failure of membrane transport systems, and promotion of iron-dependent free radical formation (Petito et al., 1987). Nevertheless, it is not at all clear that the acidosis during ischemia, even under hyperglycemic preconditions, reaches sufficient intensity to initiate these cell lytic effects.

The pHi of neurons has been measured by microelectrode to fall as low as 6.2 in hyperglycemic ischemia (Kraig and Chesler, 1990), essentially equilibrating with extracellular pH. Glial pHi may be even more acidic under these condition (Kraig and Chesler, 1990). But, neurons and glia can apparently survive exposure to acid below pH 5.2 for significant periods of time (Goldman et al., 1989). Rat brains in vivo have been exposed to hypercarbic induced pHi of 6.5 without any signs of metabolic instability (Cohen et al., 1990), and brain slices have been kept at even lower pHi, down to 6.2 without effect (Espanol et al., 1992). It would appear from these data that it is unlikely that acidosis has a direct role in ischemic tissue damage. But, it is possible that there are heterogeneous pH compartments where pHi might drop to very low levels, and where a direct toxic effect could occur. Attempts to observe compartmentation of pHi by NMR have so far failed (Boris-Möller et al., 1988), but our own data suggest there is a distribution of pHi during ischemia that includes some very acidic sites (Griffith et al., 1992).

METABOLIC ROLE OF PROTONS

A simplified scheme outlining the major interactions of protons with cell energy metabolism is shown in Figure 1. The left hand portion of the figure indicates that glycolysis

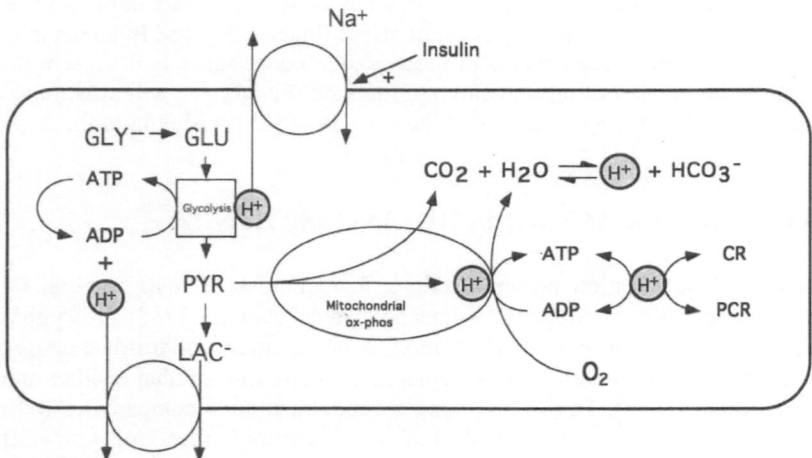

Figure 1. Metabolic role of protons.

is sensitive to hydrogen ions. The rate limiting enzyme of glycolysis, phosphofructokinase (PFK) has a narrow pH control range, being nearly completely activated at pH 7.5 and nearly completely inhibited at pH 6.5 (Trivedi and Danforth, 1966). Acidosis accompanying glycolytic activity is shown as emanating from the breakdown of glycolytically produced ATP rather than from lactic acid accumulation. The effect of insulin (in insulin-sensitive tissues) is shown as activation of the Na^+/H^+ transporter resulting in cellular alkalosis and consequent activation of glycolysis (Busa and Nuccitelli, 1984).

Glycolytically produced pyruvate is metabolized to carbon dioxide which participates in the bicarbonate buffer system. The substrate hydrogen is oxidized by oxygen in coupled oxidative-phosphorylation through chemiosmotic mechanisms. Finally, creatine kinase, catalyzing the phosphocreatine/creatine:ATP/ADP reaction, is in equilibrium with the hydrogen ion concentration.

During severe hypoxia and ischemia tissue intracellular pH becomes acidified primarily due to the turnover of glycolytically produced ATP, retention of CO_2 from residual oxidative -phosphorylation, and net breakdown of ATP (Hochachka and Mommsen, 1983; Dennis et al., 1991).

How are protons removed from the cytoplasm?

Protons are primarily buffered by histidine residues on intracellular proteins and by the bicarbonate buffer system. Buffer strength varies somewhat among tissues, but is usually in the range of 20 - 50 mM in muscle and brain. Depending on the intracellular pH, hydrogen ions are removed from the cytoplasm through the activation of the Na^+/H^+ transporter, through hydrogen/lactate cotransport, through NaH_2PO_4 leak, and through oxidative re-synthesis of ATP.

CARDIAC ARREST AND RESUSCITATION - REVERSIBLE TOTAL CEREBRAL ISCHEMIA

Upon resuscitation and reperfusion after cardiac arrest, brain pHi returns fairly quickly. Although measurements of the rate of realkalinization vary with the method used (LaManna et al., 1992a; Hoffman et al., 1995; Nishijima et al., 1989; Widmer et al., 1992; Silver and Erecinska, 1992), ischemic acidosis is completely reversed within 20 minutes of recovery. We found that pHi was fully restored by 2 minutes after 5 minutes of occlusion, and almost restored by 2 minutes after 10 minutes of bilateral carotid occlusion in both normo- and hyperglycemic gerbils. Intermediate pHi values were recorded at earlier time points. Interestingly, we found that ATP and Phosphocreatine concentrations went through an initial rapid partial recovery at 15 - 30 seconds before falling secondarily. This suggests that there was an initial rapid metabolic recovery that provided the energy source for ionic pumps which began to operate, thereby using ATP (Hoffman et al., 1995). Because the pHi after reperfusion returns so quickly, ischemic acidosis can only be considered a trigger of other events, and may act indirectly to cause tissue damage; or else acidosis may be simply an epiphenomenon not causally linked to ischemic damage. This might be the case, for example, if it is the osmolar increase that accompanies ischemia that is the crucial factor. In tissue culture, lactic acid accumulation by itself, independent of pH, has additional specific cell swelling inducing properties through osmotic mechanisms, yet cells maintain viability very well down to pH as low as 5.6 (Staub et al., 1990). The anaerobic glycolytic production of lactate from glucose produces an osmotic load because two lactates are produced from each glucose and glucosyl unit (from glycogen). Increased osmoles come also from inorganic phosphate accumulation. The osmotic pressure of the tissue thus rises during ischemia by 50-80 mOsm/kg which leads directly to cell swelling with the onset of reperfusion (Hossmann, 1982). This is amplified by acidosis since recovery from acidosis triggers mechanisms which may lead to further cell swelling. The combination is thus additive. The hyperosmosis may occur in general body tissues and be responsible for hemoconcentration which increases viscosity and this together with perivascular and endothelial cell swelling could be responsible for decreased blood flow. On the other hand, the delayed hypoperfusion does not seem to be a primary limiting factor as shown by our own data and that of others (LaManna et al., 1988; Michenfelder and Milde, 1990). The decreased blood flow is thus linked to decreased metabolic demand. The functional capillary density appears to be lower in ischemia (Ennis et al., 1990; Gjedde et al., 1990), and there is no mismatch between flow and metabolism post-ischemia (Mies et al., 1990).

Although acidosis may not be responsible for post-ischemic tissue damage, the subsequent *alkalosis* following reperfusion may be correlated with damage since the duration of its time course (days) seems to be correlated with duration of ischemia (Chopp et al., 1990). Also, significantly, the alkalosis occurs before late edema formation.

We have previously reported the observation that intracellular pH in the hippocampal slice preparation, in vitro, was significantly more alkaline than the buffer pH (Sick et al., 1987). The alkalinization in these brain slices was reversed by amiloride (LaManna et al., 1987) which led us to consider Na^+/H^+ exchange as the most probable mechanism. Na^+/H^+ exchange through an amiloride-sensitive channel has been described in primary astrocyte cell cultures (Kimelberg et al., 1979), C6 glioma cells (Jakubovicz et al., 1987; Benos and Sapirstein, 1983) and neuroblastoma cells (Moolenaar, 1986; Benos and Sapirstein, 1983). This Na^+/H^+ exchange transporter is activated by intracellular acidification, exchanging intracellular protons for extracellular sodium. Under normal conditions, at neutral pH, the transporter is inactive. During tissue acidification, intracellular protons interact with a regulatory site on the cytoplasmic side of the cell membrane to activate the exchanger so that the sodium gradient can be used to exchange protons for sodium ions to return the cell to pH neutrality. Water influx accompanies the sodium influx; and, thus, this transporter has been linked to cell volume regulation in many tissues. Activation of this transporter has been observed to be a main contributor to cell edema and swelling in neuronal and glial cell cultures and brain slice preparations (Kimelberg and Frangakis, 1985; Bourke et al., 1983; Jakubovicz et al., 1987; Kimelberg et al., 1979). It has been suggested (Jakubovicz et al., 1987; Kimelberg et al., 1979) that this same mechanism might be responsible for the initial swelling of astrocytes that occurs soon after reperfusion begins following a reversible ischemic event. The finding of an intracellular alkalinization in the cerebral cortex for up to 3 hours after reversible forebrain ischemia in the rat (von Hanwehr et al., 1986; Blomqvist et al., 1982; Mabe et al., 1983) is compatible with this suggestion.

An amiloride-sensitive Na^+/H^+ exchange transporter has been identified in rat cerebral microvessels and in hippocampal brain slices (Kalaria et al., 1991). We have shown in two preliminary studies that it is likely that the Na^+/H^+ exchange transporter does contribute to post-ischemic alkalinization.

We showed that in amiloride-pretreated rats, pHi recovery during resuscitation after cardiac arrest was slower than in untreated rats. In 4 rats treated with amiloride, the mean cerebral cortical pHi after 5 minutes of reperfusion was 6.94 ± 0.03 (mean $\pm$ sem), compared to control untreated rats where pHi was 7.15 ± 0.05 (n = 5, p < 0.05, two-tailed unpaired t-test)(LaManna et al., 1992a).

We have subsequently showed that methylisobutyl amiloride (MIA) which was one of the most effective amiloride analogs in vitro, was effective in inhibiting the Na^+/H^+ transporter in rat brain in vivo. The single pass extraction fraction for MIA was about the same as that for leucine, in the range 7-13%. There was no significant effect on resting blood flow in cerebral cortex. Fifteen minutes after resuscitation from cardiac arrest, MIA treated rats had cortical pHi of 6.78 ± 0.18 (n = 5, p < 0.05) compared to vehicle treated controls of

7.11 ± .07 (n = 5). This result demonstrated for the first time that the Na^+/H^+ transporter in brain cortex was activated during resuscitation, and that the time course of pHi recovery after cardiac arrest and resuscitation could be altered by inhibition of the transporter (Ferimer et al., 1995).

THE pH PARADOX

There has begun a shift towards the view that moderate acidosis during recovery from stroke may be beneficial (Tombaugh and Sapolsky, 1993). This "pH paradox" (Currin et al., 1991) has been described in other tissues such as isolated perfused rat liver (Currin et al., 1991), heart muscle (Bing et al., 1973) and cardiac myocytes (Bond et al., 1993). In cardiac tissues, inhibition of Na^+/H^+ exchange has been shown to be protective in ischemic recovery (Sack et al., 1994; Meng and Pierce, 1990; Scholz and Albus, 1993; Moffat and Karmazyn, 1993; Meng et al., 1993). No one has reported any brain effects of these inhibitors, however, acidosis as a treatment strategy for focal cerebral ischemia has been tried, primarily based on the effects of pH on glutamate receptors (Tang et al., 1990), and protective effects were found down to a pH of 6.5 (Simon et al., 1993). On the other hand, prolonged post-ischemic acidosis has been correlated with poorer functional recovery in at least one model (Hurn et al., 1991; Nishijima et al., 1989; Maruki et al., 1993).

PHYSIOLOGICAL RATIONALE FOR THE pH PARADOX

The potential benefits of mild acidosis have been ascribed to three main mechanisms. It has been shown that acidosis inactivates the glutamate NMDA receptor, and that this inactivation is neuroprotective after ischemia (Giffard et al., 1990; Kaku et al., 1993). Phospholipase A_2, which generates free radicals from the production of arachidonic acid, is inhibited by acidosis (Harrison et al., 1991). Finally, the K^+-ATP channel is sensitive to protons because they compete with ATP at the ATP-binding site (Davies et al., 1992; Standen et al., 1992). Ordinarily, ATP binds to this channel with high affinity preventing hyperpolarization. Acidosis promotes hyperpolarization, and thus decreases excitability and energy demand. Indeed, at normally encountered ATP concentrations, this channel may be more appropriately termed the proton-sensitive K^+ channel. Lactate may also activate this channel (Keung and Li, 1991).

Two other mechanisms may also be relevant. Blocking Na^+/H^+ exchange may help prevent Na^+ loading which would lead to cell swelling and calcium entry. Also, acidification would inhibit glycolysis.

HYPOXIA AND CELLULAR ACID-BASE BALANCE

The effect of hypoxia on intracellular acid base balance is somewhat different than that of ischemia. Although there appears to be a pH paradox effect in isolated cells and tissues exposed to hypoxia, in the intact animal there are other considerations. During hypoxia in intact animals, vascular perfusion is not only maintained, but is usually elevated (LaManna et al., 1992b). Hypoxia-induced hyperventilation results in decreased blood and tissue CO_2, which results in alkalosis. Systemically this alkalosis is compensated for by excretion of bicarbonate. In the brain, adaptation to prolonged mild hypoxia produces near neutral pHi. The decrease in tissue CO_2 is balanced by an increase in glycolysis, and an increased turnover of glycolytically produced ATP

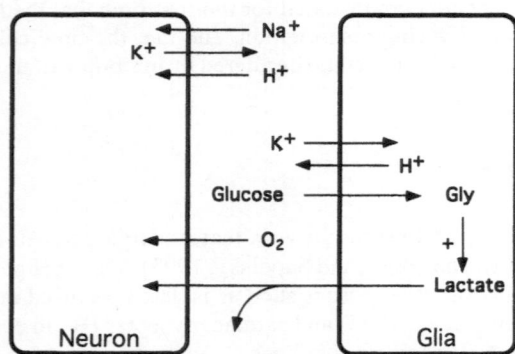

Figure 2. Metabolic interactions between neurons and glia.

(Lauro and LaManna, 1994). ATP and PCr return to control levels. This implies that glycolytic increases in mild hypoxia are driven by acid-base considerations rather than in response to inadequate energy production (Siesjö, 1973).

METABOLIC INTERACTIONS BETWEEN NEURONS AND GLIA

We have previously speculated that the mechanism underlying postischemic alkalinization reflects the activation of normal processes of metabolic coupling, exaggerated by ischemic conditions. Figure 2 presents the proposed scheme. Neuronal activation results in sodium ion influx and potassium ion efflux. This is accompanied by an influx of protons into neurons, initially alkalinizing the extracellular space (Urbanics et al., 1978). Potassium uptake occurs into glial astrocytes, along with a proton efflux from astrocytes which acidifies the extracellular space, but results in alkalinization of the astrocyte (Kraig et al., 1983; Chesler, 1990). Alkalinization of the astrocyte results in activation of glycolysis (Hochachka and Mommsen, 1983) with glycogen breakdown (Swanson, 1992; Swanson et al., 1992), and production of both pyruvate and lactate in excess of glial metabolic utilization needs, and despite normal tissue oxygenation.. The lactate can then leave the astrocyte and be taken up by neurons through membrane monocarboxylic acid transport mechanisms (Assaf et al., 1990), and be oxidatively metabolized by neurons (Dringen et al., 1993). A heterogeneous metabolic anatomy has also been suggested based on cerebral vascular and metabolic distributions compared to enzyme histochemistry maps (Borowsky and Collins, 1989). This mechanism might be responsible for the observation that neuronal activity initially stimulates glucose metabolism to a much greater extent than oxygen consumption (Fox et al., 1988).

CONCLUSION

In conclusion, it is likely that the pH paradox is as significant for brain tissue as it is for other active tissues. Furthermore, it is possible that Na^+/H^+ transport inhibition would be a useful postischemic treatment strategy in brain.

ACKNOWLEDGMENTS

These studies were supported by PHS grants HL42215, HL25830, NS22077 and NS22571.

REFERENCES

Assaf, H.M., Ricci, A.J., Whittingham, T.S., LaManna, J.C., Ratcheson, R.A., and Lust, W.D., 1990, Lactate compartmentation in hippocampal slices: Evidence for a transporter, *Met. Br. Dis.*, 5: 143-154.

Benos, D.J. and Sapirstein, V.S., 1983, Characteristics of an amiloride-sensitive sodium entry pathway in cultured rodent glial and neuroblastoma cells, *J. Cell. Physiol.*, 116: 213-220.

Bing, O.H.L., Brooks, W.W., and Messer, J.V., 1973, Heart muscle viability following hypoxia: Protective effect of acidosis, *Science*, 180: 1297-1298.

Blomqvist, P., Mabe, H., and Siesjö, B.K., 1982, Transient ischemia leads to intracellular alkalosis in the brain, *Acta Physiol. Scand.*, 116: 103-104.

Bond, J.M., Chacon, E., Herman, B., and Lemasters, J.J., 1993, Intracellular pH and Ca^{2+} homeostasis in the pH paradox of reperfusion injury to neonatal rat cardiac myocytes, *Am. J. Physiol.*, 265: C129-C137.

Boris-Möller, F., Drakenberg, T., Elmdén, K., Forsén, S., and Siesjö, B.K., 1988, Evidence against major compartmentalization of H^+ in ischemic rat brain tissue, *Neurosci. Lett.*, 85: 113-118.

Borowsky, I.W. and Collins, R.C., 1989, Metabolic anatomy of brain: A comparison of regional capillary density, glucose metabolism, and enzyme activities, *J. Comp. Neurol.*, 288: 401-413.

Bourke, R.S., Kimelberg, H.K., Dazé, M., and Church, G., 1983, Swelling and ion uptake in cat cerebrocortical slices: Control by neurotransmitters and ion transport mechanisms, *Neurochem. Res.*, 8: 5-24.

Busa, W.B. and Nuccitelli, R., 1984, Metabolic regulation via intracellular pH, *Am. J. Physiol.*, 246: R409-R438.

Chesler, M., 1990, The regulation and modulation of pH in the nervous system, *Prog. Neurobiol.*, 34: 401-427.

Chopp, M., Welch, K.M.A., Tidwell, C.D., and Helpern, J.A., 1988, Global cerebral ischemia and intracellular pH during hyperglycemia and hypoglycemia in cats, *Stroke*, 19: 1383-1387.

Chopp, M., Chen, H., Vande Linde, A.M.Q., Brown, E., and Welch, K.M.A., 1990, Time course of postischemic intracellular alkalosis reflects the duration of ischemia, *J. Cereb. Blood Flow Metab.*, 10: 860-865.

Cohen, Y., Chang, L.-H., Litt, L., Kim, F., Severinghaus, J.W., Weinstein, P.R., Davis, R.L., Germano, I., and James, T.L., 1990, Stability of brain intracellular lactate and ^{31}P-metabolite levels at reduced intracellular pH during prolonged hypercapnia in rats, *J. Cereb. Blood Flow Metab.*, 10: 277-284.

Combs, D.J., Dempsey, R.J., Maley, M., Donaldson, D., and Smith, C., 1990, Relationship between plasma glucose, brain lactate, and intracellular pH during cerebral ischemia in gerbils, *Stroke*, 21: 936-942.

Crowell, J.W. and Kaufmann, B.N., 1961, Changes in tissue pH after circulatory arrest, *Am. J. Physiol.*, 200: 743-745.

Currin, R.T., Gores, G.J., Thurman, R.G., and Lemasters, J.J., 1991, Protection by acidotic pH against anoxic cell killing in perfused rat liver: evidence for a pH paradox, *FASEB J.*, 5: 207-210.

Davies, N.W., Standen, N.B., and Stanfield, P.R., 1992, The effect of intracellular pH on ATP-dependent potassium channels of frog skeletal muscles, *J. Physiol. (Lond.)*, 445: 549-568.

Dennis, S.C., Gevers, W., and Opie, L.H., 1991, Protons in ischemia: Where do they come from; where do they go to?, *J. Mol. Cell. Cardiol.*, 23: 1077-1086.

Dringen, R., Gebhardt, R., and Hamprecht, B., 1993, Glycogen in astrocytes: possible function as lactate supply for neighboring cells, *Br. Res.*, 623: 208-214.

Ennis, S.R., Keep, R.F., Schielke, G.P., and Betz, A.L., 1990, Decrease in perfusion of cerebral capillaries during incomplete ischemia and reperfusion, *J. Cereb. Blood Flow Metab.*, 10: 213-220.

Espanol, M.T., Litt, L., Yang, G.-Y., Chang, L.-H., Chan, P.H., James, T.L., and Weinstein, P.R., 1992, Tolerance of low intracellular pH during hypercapnia by rat cortical brain slices: A $^{31}P/^1H$ NMR study, *J. Neurochem.*, 59: 1820-1828.

Ferimer, H.N., Kutina, K.L., and LaManna, J.C., 1995, Delayed normalization of brain intracellular pH by methyl isobutyl amiloride after cardiac arrest in rats, *Crit. Care Med.*, (in press):

Fox, P.T., Raichle, M.E., Mintun, M.A., and Dence, C., 1988, Nonoxidative glucose consumption during focal physiologic neural activity, *Science*, 241: 462-464.

Giffard, R.G., Monyer, H., Christine, C.W., and Choi, D.W., 1990, Acidosis reduces NMDA receptor activation, glutamate neurotoxicity, and oxygen-glucose deprivation neuronal injury in cortical cultures, *Br. Res.*, 506: 339-342.

Gjedde, A., Kuwabara, H., and Hakim, A.M., 1990, Reduction of functional capillary density in human brain after stroke, *J. Cereb. Blood Flow Metab.*, 10: 317-326.

Goldman, S.A., Pulsinelli, W.A., Clarke, W.Y., Kraig, R.P., and Plum, F., 1989, The effects of extracellular acidosis on neurons and glia in vitro, *J. Cereb. Blood Flow Metab.*, 9: 471-477.

Griffith, J.K., Cordisco, B.R., Lin, C.-W., and LaManna, J.C., 1992, Distribution of intracellular pH in the rat brain cortex after global ischemia as measured by color film histophotometry of neutral red, *Br. Res.*, 573: 1-7.

Gyulai, L., Schnall, M., McLaughlin, A.C., Leigh, J.S.J., and Chance, B., 1987, Simultaneous [31P]- and [1H]-nuclear magnetic resonance studies of hypoxia and ischemia in the cat brain, *J. Cereb. Blood Flow Metab.*, 7: 543-551.

Hansen, A.J., 1985, Effect of anoxia on ion distribution in the brain, *Physiol. Rev.*, 65: 101-148.

Harris, R.J. and Symon, L., 1984, Extracellular pH, potassium, and calcium activities in progressive ischemia of rat cortex, *J. Cereb. Blood Flow Metab.*, 4: 178-186.

Harrison, D.C., Lemasters, J.J., and Herman, B., 1991, A pH-dependent phospholipase A_2 contributes to loss of plasma membrane integrity during chemical hypoxia in rat hepatocytes, *Biochem. Biophys. Res. Comm.*, 174: 654-659.

Hochachka, P.W. and Mommsen, T.P., 1983, Protons and anaerobiosis, *Science*, 219: 1391-1397.

Hoffman, T.L., LaManna, J.C., Pundik, S., Selman, W.R., Whittingham, T.S., Ratcheson, R.A., and Lust, W.D., 1995, Early reversal of acidosis is a first step to metabolic recovery following ischemia, *J. Neurosurg.*, (in press):

Hossmann, K.-A., 1982, Treatment of experimental cerebral ischemia, *J. Cereb. Blood Flow Metab.*, 2: 275-297.

Hurn, P.D., Koehler, R.C., Norris, S.E., Blizzard, K.K., and Traystman, R.J., 1991, Dependence of cerebral energy phosphate and evoked potential recovery on end-ischemic pH, *Am. J. Physiol.*, 260: H532-H541.

Jakubovicz, D.E., Grinstein, S., and Klip, A., 1987, Cell swelling following recovery from acidification in C6 glioma cells: an in vitro model of postischemic brain edema, *Br. Res.*, 435: 138-146.

Javaheri, S., Weyne, J., and Demeester, G., 1983, Changes in the brain surface pH and cisternal cerebrospinal fluid acid-base variables in respiratory arrest, *Resp. Physiol.*, 51: 31-43.

Kaku, D.A., Giffard, R.G., and Choi, D.W., 1993, Neuroprotective effects of glutamate antagonists and extracellular acidity, *Science*, 260: 1516-1518.

Kalaria, R.N., Kroon, S.N., and LaManna, J.C., 1991, Identification and characterization of the Na^+/H^+ antiporter of cerebral microvessels and the choroid plexus, *J. Cereb. Blood Flow Metab.*, 11(suppl): S865(Abstract)

Katsura, K., Asplund, B., Ekholm, A., and Siesjö, B.K., 1992, Extra- and intracellular pH in the brain during ischaemia, related to tissue lactate content in normo- and hypercapnic rats, *Eur. J. Neurosci.*, 4: 166-176.

Keung, E.C. and Li, Q., 1991, Lactate activates ATP-sensitive potassium channels in guinea pig ventricular myocytes, *J. Clin. Invest.*, 88: 1772-1777.

Kimelberg, H.K., Biddlecome, S., and Bourke, R.S., 1979, SITS-inhibitable Cl⁻ transport and Na^+-dependent H^+ production in primary astroglial cultures, *Br. Res.*, 173: 111-124.

Kimelberg, H.K. and Frangakis, M.V., 1985, Furosemide- and bumetanide-sensitive ion transport and volume control in primary astrocyte cultures from rat brain, *Br. Res.*, 361: 125-134.

Kraig, R.P., Ferreira-Filho, C.S., and Nicholson, C., 1983, Alkaline and acid transients in cerebellar microenvironment, *J. Neurophysiol.*, 49: 831-850.

Kraig, R.P. and Chesler, M., 1990, Astrocytic acidosis in hyperglycemic and complete ischemia, *J. Cereb. Blood Flow Metab.*, 10: 104-114.

LaManna, J.C., Assaf, H., Sick, T.J., and Whittingham, T.S., 1987, Amiloride reversal of alkaline intracellular pH in hippocampal slices, *Soc. Neurosci. Abstr.*, 13: 126(Abstract)

LaManna, J.C., Crumrine, R.C., and Jackson, D.L., 1988, No correlation between cerebral blood flow and neurologic recovery after reversible total cerebral ischemia in the dog, *Exptl. Neurol.*, 101: 234-247.

LaManna, J.C., Griffith, J.K., Cordisco, B.R., Lin, C.-W., and Lust, W.D., 1992a, Intracellular pH in rat brain in vivo and in brain slices, *Can. J. Physiol. Pharmacol.*, 70: S269-S277.

LaManna, J.C., Vendel, L.M., and Farrell, R.M., 1992b, Brain adaptation to chronic hypobaric hypoxia in rats, *J. Appl. Physiol.*, 72: 2238-2243.

Lauro, K. and LaManna, J.C., 1994, Cerebral oxygen and metabolic consumption model of the compensatory adaptations in chronic hypobaric hypoxia in the rat, *FASEB J.*, 8: A1047(Abstract)

Mabe, H., Blomqvist, P., and Siesjö, B.K., 1983, Intracellular pH in the brain following transient ischemia, *J. Cereb. Blood Flow Metab.*, 3: 109-114.

Maruki, Y., Koehler, R.C., Eleff, S.M., and Traystman, R.J., 1993, Intracellular pH during reperfusion influences evoked potential recovery after complete cerebral ischemia, *Stroke*, 24: 697-704.

Meng, H.-P., Maddaford, T.G., and Pierce, G.N., 1993, Effect of amiloride and selected analogues on postischemic recovery of cardiac contractile function, *Am. J. Physiol.*, 264: H1831-H1835.

Meng, H.-P. and Pierce, G.N., 1990, Protective effects of 5-(*N,N*-dimethyl)amiloride on ischemia-reperfusion injury in hearts, *Am. J. Physiol.*, 258: H1615-H1619.

Michenfelder, J.D. and Milde, J.H., 1990, Postischemic canine cerebral blood flow appears to be determined by cerebral metabolic needs, *J. Cereb. Blood Flow Metab.*, 10: 71-76.

Mies, G., Paschen, W., and Hossmann, K.-A., 1990, Cerebral blood flow, glucose utilization, regional glucose, and ATP content during the maturation period of delayed ischemic injury in gerbil brain, *J. Cereb. Blood Flow Metab.*, 10: 638-645.

Moffat, M.P. and Karmazyn, M., 1993, Protective effects of the potent Na/H exchange inhibitor methylisobutyl amiloride against post-ischemic contractile dysfunction in rat and guinea-pig hearts, *J. Mol. Cell. Cardiol.*, 25: 959-971.

Moolenaar, W.H., 1986, Effects of growth factors on intracellular pH regulation, *Ann. Rev. Physiol.*, 48: 363-376.

Mutch, W.A.C. and Hansen, A.J., 1984, Extracellular pH changes during spreading depression and cerebral ischemia: Mechanisms of brain pH regulation, *J. Cereb. Blood Flow Metab.*, 4: 17-27.

Nemoto, E.M. and Frinak, S., 1981, Brain tissue pH after global brain ischemia and barbiturate loading in rats, *Stroke*, 12: 77-82.

Nishijima, M.K., Koehler, R.C., Hurn, P.D., Eleff, S.M., Norris, S., Jacobus, W.E., and @REFAUSTY = Traystman, R.J., 1989, Postischemic recovery rate of cerebral ATP, phosphocreatine, pH and evoked potentials, *Am. J. Physiol.*, 257: H1860-H1870.

Paschen, W., Djuricic, B., Mies, G., Schmidt-Kastner, R., and Linn, F., 1987, Lactate and pH in the brain: Association and dissociation in different pathophysiological states, *J. Neurochem.*, 48: 154-159.

Petito, C.K., Kraig, R.P., and Pulsinelli, W.A., 1987, Light and electron microscopic evaluation of hydrogen ion-induced brain necrosis, *J. Cereb. Blood Flow Metab.*, 7: 625-632.

Plum, F., 1983, What causes infarction in ischemic brain?, *Neurol.*, 33: 222-233.

Sack, S., Mohiri, M., Schwarz, E.R., Arras, M., Schaper, J., Ballagi-Pordány, G., Scholz, @REFAUSTY = W., Lang, H.J., Schölkens, B.A., and Schaper, W., 1994, Effects of a new Na$^+$/H$^+$ antiporter inhibitor on postischemic reperfusion in pig heart, *J. Cardiovasc. Pharmacol.*, 23: 72-78.

Scholz, W. and Albus, U., 1993, Na$^+$/H$^+$ exchange and its inhibition in cardiac ischemia and reperfusion, *Basic Res. Cardiol.*, 88: 443-455.

Sick, T.J., Whittingham, T.S., and LaManna, J.C., 1987, Evidence for multiple H$^+$ pools and their significance for electrical function during anoxia in hippocampal slices, *J. Cereb. Blood Flow Metab.*, 7 (suppl. 1): S113(Abstract).

Siesjö, B.K., 1973, Metabolic control of intracellular pH, *Scand. J. Clin. Lab. Invest.*, 32: 97-104.

Siesjö, B.K., 1988, Acidosis and ischemic brain damage, *Neurochem. Pathol.*, 9: 31-88.

Silver, I.A. and Erecinska, M., 1992, Ion homeostasis in rat brain in vivo: intra- and extracellular [Ca^{2+}] and [H$^+$] in the hippocampus during recovery from short-term, transient ischemia, *J. Cereb. Blood Flow Metab.*, 12: 759-772.

Simon, R.P., Niiro, M., and Gwinn, R., 1993, Brain acidosis induced by hypercarbic ventilation attenuates focal ischemic injury, *J. Pharmacol. Exp. Ther.*, 267: 1428-1431.

Standen, N.B., Pettit, A.I., Davies, N.W., and Stanfield, P.R., 1992, Activation of ATP-dependent K$^+$ currents in intact skeletal muscle fibres by reduced intracellular pH, *Proc. Roy. Soc. Lond. B*, 247: 195-198.

Staub, F., Baethmann, A., Peters, J., Weigt, H., and Kempski, O., 1990, Effects of lactacidosis on glial cell volume and viability, *J. Cereb. Blood Flow Metab.*, 10: 866-876.

Swanson, R.A., 1992, Physiologic coupling of glial glycogen metabolism to neuronal activity in brain, *Can. J. Physiol. Pharmacol.*, 70: S138-S144.

Swanson, R.A., Morton, M.M., Sagar, S.M., and Sharp, F.R., 1992, Sensory stimulation induces local cerebral glycogenolysis: demonstration by autoradiography, *Neurosci.*, 51: 451-461.

Tang, C.-M., Dichter, M., and Morad, M., 1990, Modulation of the N-methyl-D-aspartate channel by extracellular H$^+$, *Proc. Natl. Acad. Sci. USA*, 87: 6445-6449.

Tombaugh, G.C. and Sapolsky, R.M., 1993, Evolving concepts about the role of acidosis in hypoxic/ischemic injury, *J. Neurochem.*, 61: 793-803.

Trivedi, B. and Danforth, W.H., 1966, Effect of pH on the kinetics of frog muscle phosphofructokinase, *J. Biol. Chem.*, 241: 4110-4112.

Urbanics, R., Leniger-Follert, E., and Lübbers, D.W., 1978, Time course of changes of extracellular H$^+$ and K$^+$ activities during and after direct electrical stimulation of the brain cortex, *Pflüg. Arch.*, 378: 47-53.

von Hanwehr, R., Smith, M.-L., and Siesjö, B.K., 1986, Extra- and intracellular pH during near-complete forebrain ischemia in the rat, *J. Neurochem.*, 46: 331-339.

Widmer, H., Abiko, H., Faden, A.I., James, T.L., and Weinstein, P.R., 1992, Effects of hyperglycemia on the time course of changes in energy metabolism and pH during global cerebral ischemia and reperfusion in rats: Correlation of ^{1}H and ^{31}P NMR spectroscopy with fatty acid and excitatory amino acid levels, *J. Cereb. Blood Flow Metab.*, 12: 456-468.

37

NON INVASIVE MEASUREMENT OF BRACHIORADIAL MUSCLE VO$_2$-BLOOD FLOW RELATIONSHIP DURING GRADED ISOMETRIC EXERCISE

R.A. De Blasi,[1] R. Sfareni,[2] B. Pietranico,[1] A.M. Mega,[1] and M. Ferrari[2,3]

[1] Institute of Anesthesiology
I University of Rome, Italy
[2] Istituto Superiore Sanità
Rome, Italy
[3] Department of Biomedical Sciences and Technology
University of L'Aquila
67100 L'Aquila, Italy

INTRODUCTION

It is generally agreed that the rate of blood flow in exercising skeletal muscle has a relevant influence on the biochemical processes proceeding in that tissue and on the cause of fatigue. The investigation of the relationship between blood flow and oxygen consumption (VO$_2$) of the limbs is thus of great relevance in exercise physiology of healthy subjects. In patients with cardiovascular disease the reduction of blood supply can limit the efficiency of skeletal muscle influencing functions like movements and respiration (Weber, 1982). As regards blood supply during exercise, although some experimental evidences suggest that in strongly stimulated muscle the vessels are dilated after 15 sec of contraction beginning, the mechanical compression occurring on the vascular bed has been widely reported to induce a blood flow limitation (Bonde–Petersen, 1975, Moller 1979). Different methods have been developed for the assessment of muscle blood flow and VO$_2$, but whereas the measurement of flow can be performed non invasively, the oxygen uptake of the limbs is usually calculated by multiplying blood flow with arterio-venous oxygen content difference (Mottram, 1955, Hartling, 1989).

Near infrared spectroscopy (NIRS) has been used by our group and to measure the VO$_2$ in the human forearm during arterial occlusion at rest and maximum isometric exercise (De Blasi, 1993). The hemoglobin (Hb) desaturation occurring during vascular occlusion has been related to the oxygen extraction for metabolic needs. The Hb desaturation occurring during isometric maximum exercise also without preceding vascular occlusion, has been interpreted as an indirect evidence that during maximal isometric contractions (MVC) the

Oxygen Transport to Tissue XVII, Edited by Ince et al.
Plenum Press, New York, 1996

293

blood flow supply is arrested or strongly restricted (De Blasi, 1993). We recently applied NIRS for the simultaneous measurement of forearm blood flow (FBF) and oxygen consumption (VO$_2$) in healthy volunteers at rest and after maximum exercise during ischemia (De Blasi, 1994).

The aim of the present study was to verify on the forearm of healthy and untrained volunteers if the Hb desaturation rate occurring during incremental levels of isometric exercise without vascular occlusion is proportional to the degree of strength. The relationship between FBF and VO$_2$ after contractions was investigated in order to establish how blood supply counteracts metabolic needs.

METHODS

Subjects

Seven healthy subjects (1 male and 6 female) were studied. Their age ranged between 21 and 41 yrs (26.8 ± 5.9), their mean weight and height were 55.8±9.4 Kg and 164.1±6.2 cm, respectively. All the subjects were untrained, not overweight and provided written informed consent. The study was performed on the brachioradial muscle of the dominant forearm.

Exercise Protocol

On each subject a MVC of 30 s length was followed by three isometric contractions at different strengths with a rest period of 15 min between contractions. After MVC strength recording, each subject was requested to perform 75%, 50% and 25% of their MVC, maintaining it for 30 s. The Hb desaturation rate was taken into account in order to calculate the VO$_2$ at different strengths (De Blasi, 1993). Measurements of FBF and VO$_2$ were obtained at rest and soon after each contraction by a rapid venous occlusion (De Blasi, 1994). This method was used to obtain contemporary and independent measurements of the two variables.

Instruments and Procedures

NIRS measurements were obtained using the NIRO500 (Hamamatsu Photonics, Japan) recently described in detail (Cope, 1988). The DPF value used was 4.3 (Ferrari, 1992). The isometric contractions were performed with a hand grip (MIE Medical Research, England) recording on-line into a computer the achieved strength expressed in Newton. The Hb deoxygenation was evaluated as HbO$_2$-Hb (Hb diff) and the Hb change was measured as HbO$_2$+Hb (Hb sum). The linear regression of the maximum Hb desaturation rate occurring during isometric contractions was considered to calculate VO$_2$ (De Blasi, 1993). VO$_2$ was expressed as μM·min^{-1}·100 g of tissue taking into account 1.33 g·ml as muscle density. The sum of the Hb and HbO$_2$ increment and the increment of Hb due to venous occlusion were considered for calculation of FBF and VO$_2$ respectively as recently described (De Blasi, 1994). The FBF was expressed as ml·100ml^{-1}·min^{-1}.

Statistical Analyses

The values of FBF and VO$_2$ at rest were expressed as the mean of 3 repetitive measurements. For the FBF measurement following the exercise, only the first measurement soon after each contraction was considered. Because of the wide range of force generated

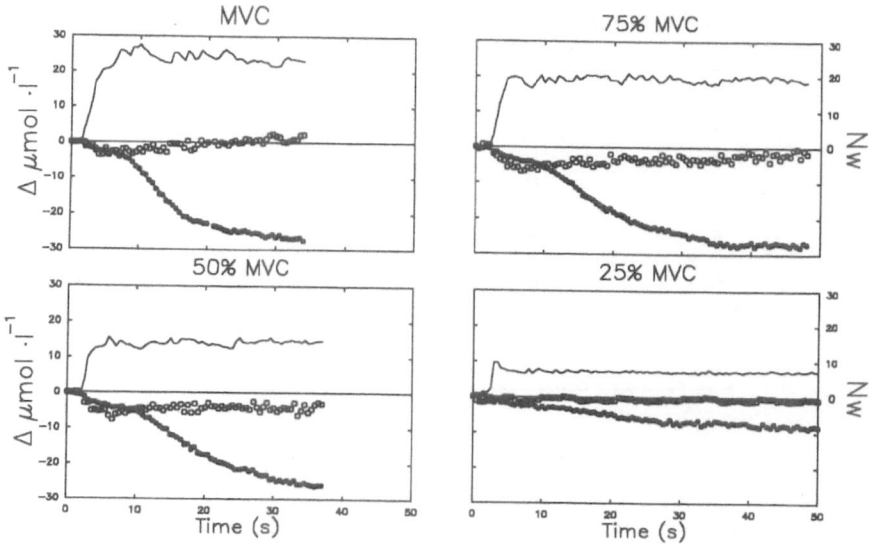

Figure 1. Time course of Hb desaturation (Hb diff), total Hb (Hb sum) and strength during sustained exercise at MVC and 75%, 50%, 25% of MVC.

by the different subjects due to their varied levels of muscular mass and conditions the force was expressed as percentage of maximum for each contraction (0-100%). Individual linear regression analyses of brachioradial VO_2 during and after contraction and FBF with force generated were performed for each subject. The mean coefficient of all the measurements (r^2) was calculated. Paired t tests were used to delineate significant differences among VO_2 during and after contractions and FBF r^2 values. P<0.05 was considered significant. Regression analysis was performed to evaluate the correlation between different variables.

RESULTS

Representative Hb diff and Hb sum changes subsequent to muscle contraction at MVC, 75%, 50% and 25% MVC are shown in Fig. 1.

The strength reached the target value in 3 and 5 s in all subjects and was maintained stable during the whole contraction. In all subjects Hb sum decreased of 2-10% soon after contraction and remained constant until the end of the exercise. The maximum rate of Hb deoxygenation started few seconds after muscle contraction when the Hb sum was stable. Subjects having Hb sum changes during the contraction were excluded from the study. VO_2 values measured during graded isometric exercise are reported in Table 1.

A statistically significant relationship between VO_2 and corresponding force values was found (mean $r^2 = 0.9\pm0.1$). No relationship was found between VO_2 measured after exercise and the corresponding force value (mean $r^2 = 0.4\pm0.3$). Flow and VO_2 were measured at rest and just after each isometric contraction. Flow and VO_2 at rest were 2.37 and 5.37, respectively. Fig. 2 shows representative increments of oxy and deoxy-Hb at rest and after MVC.

Linear regression was executed on the values corresponding to the maximum oxy and deoxy-Hb increment rate. No relationship was found between FBF measured after

Table 1. VO₂ values measured during isometric exercise
at MVC and 75%, 50%, 25% of MVC

	Percentage of MVC			
	25%	50%	75%	100%
1	26.85	4.68	46.81	51.74
2	11.57	11.29	25.01	26.92
3	12.21	29.64	33.56	52.60
4	9.14	13.03	18.35	23.13
5	2.94	3.52	3.85	9.50
6	16.35	22.80	44.68	46.22
7	2.26	8.08	11.60	18.53
MEAN	11.62	19.01	26.30	32.66
SD	7.78	13.31	15.10	16.10

exercise and the corresponding force value (mean r^2 = 0.4±0.3; range = 0.017-0.98). The FBF and the corresponding VO₂ measured in all subjects at rest and soon after each exercise at different strengths were regressed each other. The relationship was statistically significant ($P<0.01$) and the r-square value was 0.76 (Fig. 3).

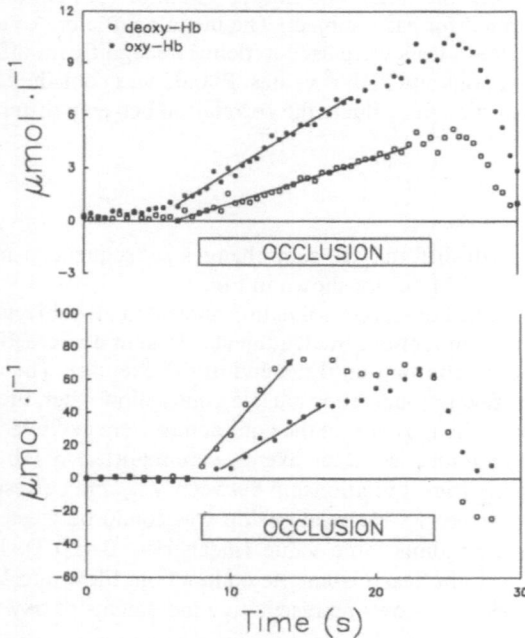

Figure 2. Typical tracing of oxy and deoxy-Hb changes during forearm venous occlusion at rest (upper panel) and after isometric contraction (lower panel).

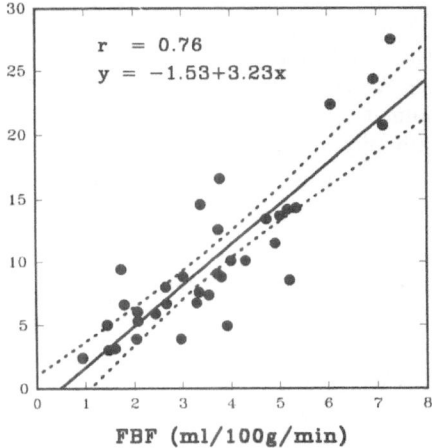

FBF (ml/100g/min)

Figure 3. Relationship between FBF and VO_2 at rest and after incremental isometric contractions measured in seven subjects. Solid line: first order regression that fits all data points. Dashed lines: 95% confidence interval.

DISCUSSION

In this study the evaluation of VO_2 has been performed by two different methods. The calculation of VO_2 by measuring the Hb desaturation occurring during isometric contractions has been performed to verify if at different strengths corresponded different oxygen extraction rates. No validation of this method has been performed; however the results of this study indicate that during isometric contractions the blood flow supply is functionally arrested or it is not sufficient for the enhanced metabolic requests. In our subjects the Hb desaturation occurring during muscle contractions at different forces was found proportional to the effort increments (Table 1).

There is a great deal of evidence that intramuscular pressure correlates linearly with the contraction strength during isometric exercise over a wide range of force (Parker, 1984, Sadamoto, 1983; Sejersted, 1984). At force corresponding to MVC intramuscular pressure of the muscle biceps brachii was reported to reach 153-326 mmHg during contraction (Sadamoto, 1983). In the same study at about 50% of MVC the interruption of blood flow, measured by ^{133}Xe clearance, was shown to be complete in the muscle biceps brachii and in other muscles of the limbs (Sadamoto, 1983). Even though the measurement of muscle blood flow during exercise by ^{133}Xe clearance technique is unreliable and makes the flow estimation useless, nevertheless the adaptation of flow to the increment of energetic requests could be delayed or abolished even at efforts lower than 50%. The fact that in our study Hb sum did not change during contractions support the above mentioned literature.

Another method of calculation of VO_2, based on the measurement of Hb increments after venous occlusion, was used. The same method can provide a FBF measurement (De Blasi, 1994). The independence of VO_2 from FBF makes these methods appropriate for evaluating the relationship between these two variables. The FBF values that we measured at rest in the brachioradial muscle are comparable to the most recent values obtained by the ^{133}Xe clearance technique (Elia, 1993). After contractions the measured VO_2 showed a poor correlation with the effort degree. This suggests that cellular metabolism and subsequent

VO_2 is strictly related to exercise varying soon after the exercise. The relationship between FBF and VO_2 demostrates a dependence of tissue metabolism from blood flow.

Although the precision of the data was limited because pathlength information was missing, the results indicate the capability of NIRS for investigating dynamically muscle oxygenation. Frequency-domain spectrometers under development could be used to investigate accurately muscle pathophysiology during exercise.

ACKNOWLEDGMENTS

The financial support of Telethon-Italy (Grant no. 501) and CNR 93.00282/94.02408.CT04 is gratefully acknowledged.

REFERENCES

Bonde-Petersen, F., Heriksson, J., Lundin, B., 1975, Blood flow in thigh muscle during bicycling exercise at varying work rates, *Eur. J. Appl. Physiol.* 34:191-197.

Cheatle, T.R., Potter, L.A., Cope, M., Delpy, D.T., Coleridge, S. and Scurr, J.H., 1991, Near-infrared spectroscopy in peripheral vascular disease, *Br. J. Sur.* 78:405-408.

Cope, M., Delpy, D.T., Reynolds, E.O.R., Wray, S., Wyatt, J., van der Zee, P., 1988, Methods of quantitating cerebral near infrared spectroscopy data, *Adv. Exp. Med. Biol.* 222:183-189.

De Blasi, R.A., Cope, M., Elwell, C., Safoue, F., and Ferrari, M., 1993, Non invasive measurement of human forearm oxygen consumption by near infrared spectroscopy, *Eur. J. Appl. Physiol.* 67:20-25.

De Blasi, R.A., Cope, M., and Ferrari, M., 1992, Oxygen consumption of human skeletal muscle by near infrared spectroscopy during tourniquet-induced ischemia in maximal voluntary contraction, *Adv. Exp. Med. Biol.* 317:771-777.

De Blasi, R.A., Ferrari, M., Natali, A., Conti, G., Mega, A., and Gasparetto, A., 1994, Non-invasive measurement of forearm blood flow and oxygen consumption by near infrared spectroscopy. *J. Appl. Physiol.* 76:1388-1393.

Elia, M., Kurpad, A., 1993, What is the blood flow to resting human muscle?, *Clin. Sci.* 84:559-563.

Ferrari, M., Wei, Q., Carraresi, L., De Blasi, R.A., Zaccanti, G., 1992, Time-resolved spectroscopy of human forearm, *J. Photochem. Photobiol.* 16:141-153.

Hartling, O.J., Kelbæk, H., Gjørup, B., Schibye, B., Klausen, K., Trap-Jensen K., 1989, Forearm oxygen uptake during maximal forearm dynamic exercise, *Eur. J. Appl. Physiol. Occup. Physiol.* 128:268-276.

Moller, E., Rasmussen, O.C., Bonde-Petersen, F., 1979, Mechanism of ischemic pain in human muscles of mastication: intramuscular pressure, EMG, force and blood flow of the temporal and masseter muscles during biting, *Adv. in Pain Research Therapy.* 3:271-281.

Mottram, R.F., 1955, The oxygen consumption of human skeletal muscle in vivo, *J. Physiol. Lond.* 128:268-276.

Parker, P., Körner, L., and Kadefors, R., 1984, Estimation of muscle force from intramuscular total pressure, *Med. Biol. Eng. Comput.* 22:453-457.

Sadamoto, T., Bonde-Petersen F., and Suzuki, Y., 1983, Skeletal muscle tension, flow, pressure, and EMG during sustained isometric contractions in humans, *Eur. J. Appl. Physiol. Occup. Physiol.* 51:395-408.

Sejersted, O.M., Hargens, A.R., Kardel, K.R., Blom, P., Jensen, O., and Hermansen, L., 1984, Intramuscular fluid pressure during isometric contraction of human skeletal muscle, *J. Appl. Physiol.* 56:287-295.

Weber, K.T., Kinasewitz, G.T., Janicki, J.S., Fishman, A.P., 1982, Oxygen utilization and ventilation during exercise in patients with chronic cardiac failure, *Circulation.* 65:1213-1223.

38

EXTRA- AND INTRACELLULAR OXYGEN SUPPLY DURING CORTICAL SPREADING DEPRESSION IN THE RAT

T. Wolf, U. Lindauer, H. Obrig, A. Villringer, and U. Dirnagl

Department of Neurology
Humboldt University Berlin
Germany

INTRODUCTION

Cortical Spreading Depression was discovered fifty years ago [15], but the complexity of its mechanisms is still far from being understood. It was then the high time of Wolff's vascular theory of migraine [25] when Lashley first calculated the velocity of a supposed electrophysiological phenomenon in the primary visual cortex that could cause the optical hallucinations of the migraine aura as the same velocity that characterizes CSD [12]. A role for CSD in the pathophysiological scenario of migraine has become more and more attractive ever since, mainly by virtue of accumulating evidence for neuronal mechanisms as found in the pursuit of Moskowitz' theory of neurogenic inflammation in migraine [14,17].

Meanwhile an abundancy of studies has been performed that characterized CSD electrophysiologically as a depression of EEG and as a negative shift in cortical DC potential [15] that is caused by simultaneous depolarization of neuronal and glial cells [21], and is associated by large increases in extra cellular potassium and intracellular sodium and calcium [14] and neurotransmitter release [6,19,22]. It is accompanied by blood flow changes, the most reproducible feature of which is an immediate increase in rCBF [16]. Its functional correlates have been subject to study [18] as well as its perceptive value [11]. But all this literature is on all but humans. CSD so far has never been proven in human cerebral cortex in vivo, and for ethical reasons all research depends on scarce occasion if it can't circumvent the obvious obstacles of invasive investigation.

Considering the sturdy data on typical blood flow changes in CSD, to us the optical method of Near Infrared Spectroscopy (NIRS) seemed a promising tool to detect CSD by means of its haemodynamic sequelae.

NIRS is a spectrophotometric technique that up to now has been used clinically in the monitoring of neonates [5]. It operates in a light range in the infrared near the visible end of the spectrum, that penetrates even dense tissue as the bone easily and is mainly absorbed by three chromophores: oxyhemoglobin, deoxyhemoglobin and the mitochondrial cytochrome oxidase, cytochrome aa3. A laser is pulsed successively in four wavelengths. Light

Oxygen Transport to Tissue XVII, Edited by Ince et al.
Plenum Press, New York, 1996

is transmitted to the skull by fibre optics. The photons that have passed through the tissue and leave the skull again, carrying the spectrophotometric information, are collected by another fibre bundle. The signal is amplified in a photomultiplier tube, and the optical densities of the tissue for the different wavelengths are measured continuously. From changes in these optical densities the changes of concentration of the chromophores are determinated[2,26]. If we were able to monitor CSD by NIRS non invasively in the animal model, we could hope to detect it non invasively in humans too.

METHODS

Seven male Wistar rats of 250 to 300 g were anaesthetized with i.p. Pentobarbital 100 mg/kg body weight. After tracheotomy they were ventilated and endexpiratory CO_2 was monitored throughout the experiment. Body temperature was maintained at 38 ± 0.5 centigrade by a heating pad. The femoral artery was cannulated for continuous blood pressure monitoring and periodical arterial blood gas samples. So was the femoral vein for continuous infusion of 1 ml/h saline.

The animals were fixed in a stereotactic frame and two open cranial windows were installed in the frontal bones, the dura was removed to expose the frontal cortex of both hemispheres and artificial cerebrospinal fluid was superfused. In this site the elicitation of CSD was possible by brief superfusion of 150 mM potassium chloride solution.

Five and ten millimetres occipitally from this triggering site, burr holes were drilled with a saline cooled drill. At both trephinations calomel electrodes were placed for registration of DC potential deflection. DC potential was measured differentially between those two electrodes by a differential voltage meter (WPI, FD223), so that from the distance of the two peaks of the resulting biphasic curve the velocity of CSD could be calculated. In the burr hole at 10 mm distance from triggering site additionally a polarographic Clark-Type oxygen probe (Licox-pO2 monitoring system) was slit through the dura and placed on the cortical surface by a micromanipulator.

In the same area the transmitting and the receiving optodes (ø about 4mm) of the four wavelength Near Infrared Spectrometer (Hamamatsu NIRO 500) were positioned over the

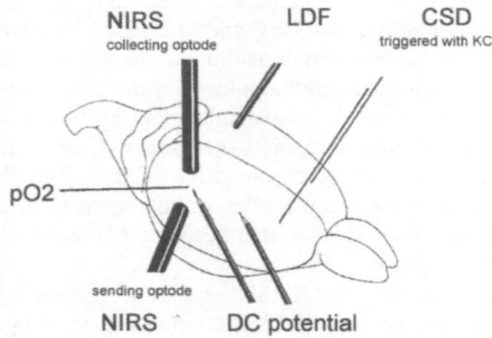

Figure 1. Schematic view of the experimental set-up; the bone is intact apart from an open cranial window over the frontal poles and burr holes for calomel electrodes and polarographic oxygen probe; rCBF is measured by LDF through the thinned bone; the sending and collecting NIRS optodes are placed over the preserved skull. CSD is triggered in the frontal region of both hemispheres by simultaneous superfusion of high potassium aCSF, the propagating DC-deflection wave is followed by calomel electrodes. NIRS, LDF and polarographic pO2 probe are placed 10 mm from triggering site.

intact skull. In order to reduce sampling volume to mainly cortical structures they were arranged at an angle of about 70 degrees and about 6 mm apart from each other in a right angle to the direction of propagation of CSD[23]. The changes of rCBF due to CSD were measured by Laser Doppler Flowmetry (LDF) through the thinned bone[3]. Since the laser signals of both LDF and NIRS interfered if employed in the same site, we chose to measure rCBF symmetrically in the opposite hemisphere at 10 mm distance to the symmetrically situated frontal open cranial window, supposing that CSD behave in an analogue fashion in both hemispheres when triggered at the same time. The setup is illustrated in Fig. 1.

Systemic arterial pressure, cortical DC potential, rCBF, pO2 and the concentration changes of HbO2, Hb and CytO2 were measured continuously. Data were acquired by a PC using the Macmillan ASYST software. A baseline was recorded before CSD was triggered

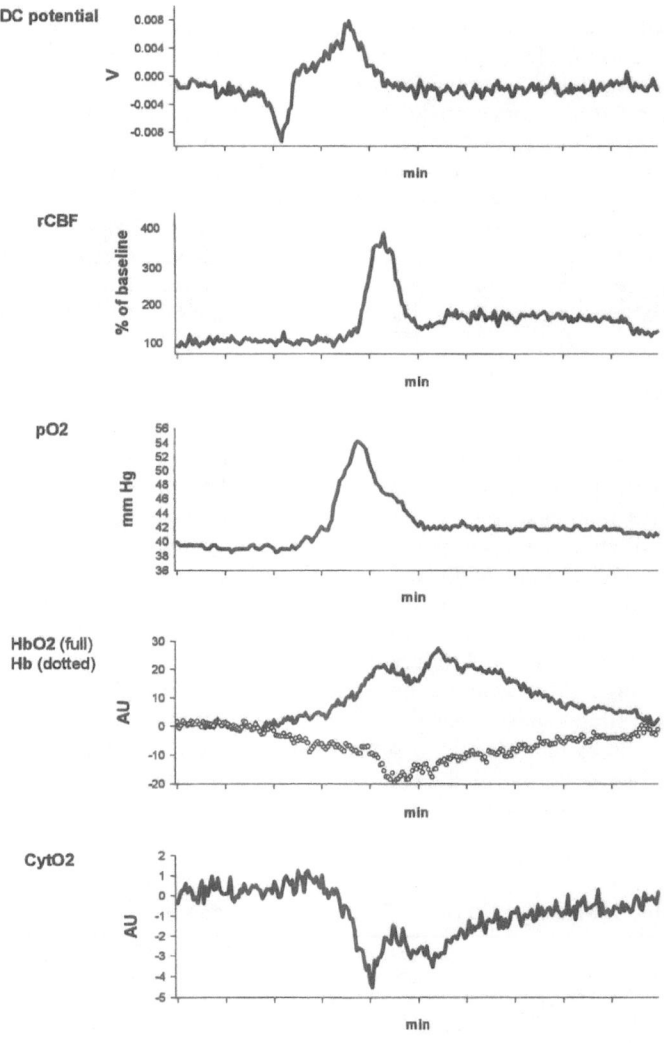

Figure 2. Characteristic changes during a single episode of CSD; CSD was triggered at 30 sec.

in the frontal poles of both hemispheres by superfusion of 150 mM potassium chloride solution for about 30 seconds.

RESULTS

Table 1 summarizes our results and Fig. 2 demonstrates characteristic curve patterns of the different parameters for a single period of CSD. The physiological parameters of the animals were kept in normal range throughout the experiment.

When CSD had been elicited in the frontal open cranial windows, cortical DC potential showed a marked negative shift first after about 90 seconds at the electrode 5 mm from triggering site, returning to normal after a minute or so. Further 90 seconds later it shifted at the more distant electrode, indicating a CSD propagation velocity of 3.4 ± 0.1 mm/min. At the same time LDF showed the expected increase in rCBF, which raised to 360 ± 196 percent of baseline, to nearly normalize after 1 to 2 minutes. Extracellular oxygen partial pressure raised by 15.3 ± 3.3 mm mercury. In some CSD this increase was preceded by a brief drop. Normally it returned to baseline after 2 to 3 minutes.

Parallel to the above mentioned changes, the NIRS signal showed a distinct increase of oxyhemoglobin by 26.1 ± 6.3 Arbitrary Units of concentration (AU) in the measuring site whilst deoxyhemoglobin decreased by 15.4 ± 4.0 AU. The cytochrome aa3 signal showed a decrease of oxydized cytochrome oxydase by 4.6 ± 0.3 AU. All NIRS signals were clearly above the noise range. The NIRS parameters were the first to change and took between 6 and 8 minutes to return to baseline, indicating that the NIRS has the largest sampling volume of all employed probes, thus already detecting CSD-associated changes when they have not reached yet, still detecting them when they have passed the other probes already.

DISCUSSION

Our NIRS data correlate well with the hyperperfusion phase in CSD. The measured increase in oxyhemoglobin concentration correlates well to the established phenomenon of a brief hyperperfusion phase in CSD. This hyperperfusion is known to be due to arterial [24] and arteriolar [20] dilatation. (A prolonged hypoperfusion as described by other investigators [4,13] was not seen, which might be due to the use of barbiturate anaesthesia in our study.) The increased rCBF leads to a faster clearance of deoxyhemoglobin from the tissue, resulting in lowered Hb concentration. The increase of tissue oxygen partial pressure we found in CSD, supports data of Back et al. [1] and reflects the increased perfusion. Our data on cytochrome aa3 though, showing a shift towards a more reduced state inspite of the clearly raised oxygen offer of the extra cellular space, seem paradox and are in discordance with previous findings of other workers on cytochrome aa3 spectrophotometry in the visible light range [9]. Much successful work has been done on the validation of the NIRS cytochrome aa3 signal[7,10]. Still the question remains, as to how much of the cytochrome aa3 calculation is based not on

Table 1. Measurements of CSD-associated changes (n = 7; mean ± SD)

velocity of CSD	increase of rCBF	increase of pO2	increase of HbO2	decrease of Hb	decrease of CytO2
mm/min	% of baseline	mm Hg	AU	AU	AU
3.4 ± 0.1	360 ± 196	15.3 ± 3.3	26.1 ± 6.3	15.4 ± 4.0	4.6 ± 0.3

absorption changes but on alterations in optical pathlength, as it might be caused to a substantial extent by cell swelling in CSD. Although a four wavelength approach has been shown to produce sensible data even in extreme conditions as severe hypoxia, where the optical properties of the brain are much changed[8], simultaneous measurement of optical path length remains to be desired, and we are currently conducting such experiments. Knowing the pathlength would provide the opportunity of exact calibration of Arbitrary Units of concentration to μM. Better separation of cytochrome aa3 spectra from haemoglobin spectra might be achieved by measuring over the entire spectral range (e.g. 600-1000 nm), thus enabling us to obtain reliable information about intracellular and vascular oxygenation simultaneously.

We showed in this study for the first time that NIRS is capable of detecting CSD in the rat noninvasively. Currently we investigate patients in the aura phase of migraine with the same method.

ACKNOWLEDGMENT

This study was supported by Deutsche Forschungsgemeinschaft DFG Vi 93 / 7-1 and Di 454 / 4 - 2.

REFERENCES

1. Back, T., Kohno, K. and Hossmann, K.A., Cortical negative DC deflections following middle cerebral artery occlusion and KCL-induced spreading depression: Effect on blood flow, tissue oxygenation, and encephalogram, *J Cereb Blood Flow Metab*, 14 (1994) 12-19.
2. Cope, M., Delpy, D.T., Reynolds, E.O.R., Wray, S., Wyatt, J. and van der Zee, P., Methods of quantitating cerebral near infrared spectroscopy data, *Adv Exp Biol Med*, 33 (1988) 183-189.
3. Dirnagl, U., Kaplan, B., Jacewicz, M. and Pulsinelli, W., Continuous measurement of cerebral cortical blood flow by laser-Doppler flowmetry in a rat stroke model, *J Cereb Blood Flow Metab*, 9 (1989) 589-596.
4. Duckrow, R.B., Regional cerebral blood flow during spreading cortical depression in conscious rats, *J. Cereb. Blood Flow Metab.*, 11 (1991) 150-154.
5. Edwards, A.D., Brown, G.C., Cope, M., Wyatt, J.S., McCormick, D.C., Roth, S.C., Delpy, D.T. and Reynolds, O.R., Quantification of concentration changes in neonatal human cerebral cytochrome oxidase, *J Appl. Physiol*, 71 (1991) 1907-1913.
6. Fabricius, M., Jensen, L.H. and Lauritzen, M., Microdialysis of interstitial amino acids during spreading depression and anoxic depolarization in rat neocortex, *Brain Res.*, 612 (1993) 61-69.
7. Ferrari, M., Hanley, D.F., Wilson, D.A. and Traystman, R.J., Redox changes in cat brain cytochrome-c oxidase after blood-fluorocarbon exchange, *Am J Physiol*, 258 (1990) H1706-H1713.
8. Hotez, L., Dailey, J.W., Geelhoed, G.W. and Gainer, J.L., The role of oxygen diffusivity in biochemical reactions., *Experientia*, 33/11 (1977) 1424-1425.
9. Jöbsis, F.F., Keizer, J.H., LaManna, J.C. and Rosenthal, M., Reflectance spectrophotometry of cytochrome aa3 in vivo., *J Appl Physiol*, 43 (5) (1977) 858-872.
10. Kariman, K. and Burkhart, D.S., Non-invasive in vivo spectrophotometric monitoring of brain cytochrome aa3 revisited., *Brain Res.*, 360 (1985) 203-213.
11. Koroleva, V.I. and Bures, J., Rats do not experience cortical or hippocampal spreading depression as aversive, *Neurosci. Lett.*, 149 (1993) 153-156.
12. Lashley, K.S., Patterns of cerebral integration indicated by the scotomas of migraine., *Arch Neurol Psychiatry*, 46 (1941) 331-339.
13. Lauritzen, M., Long-lasting reduction of cortical blood flow of the brain after spreading depression with preserved autoregulation and impaired CO2 response, *J. Cereb. Blood Flow Metab.*, 4 (1984) 546-554.
14. Lauritzen, M., Pathophysiology of the migraine aura. The spreading depression theory, *Brain*, 117 (1994) 199-210.
15. Leao, A.P., Spreading depression of activity in the cerebral cortex, *J Neurophys*, 7 (1944) 359-390.

16. Mies, G. and Paschen, W., Regional changes of blood flow, glucose, and ATP content determined on brain sections during a single passage of spreading depression in rat brain cortex, *Exp. Neurol.*, 84 (1984) 249-258.

17. Moskowitz, M.A. and MacFarlane, R., Neurovascular and molecual mechanisms in migraine headaches, *Cerebrovasc Brain Metab Rev*, 5 (1993) 159-177.

18. Oitzl, M.S. and Huston, J.P., Electroencephalographic Spreading Depression and concomitant behavioural changes induced by intrahippocampal injections of ACTH 1-24 and d-Ala2-Met-Enkephalinamide in the rat., *Brain Res*, 308 (1984) 33-42.

19. Pavlasek, J., Haburcak, M., Masanova, C. and Orlicky, J., Increase of catecholamine content in the extracellular space of the rat's brain cortex during spreading depression wave as determined by voltammetry, *Brain Res.*, 628 (1993) 145-148.

20. Shibata, M., Leffler, C.W. and Busija, D.W., Pial arteriolar constriction following cortical spreading depression is mediated by prostanoids, *Brain Res.*, 572 (1992) 190-197.

21. Somjen, G.G., Aitken, P.G., Czeh, G.L., Herreras, O., Jing, J. and Young, J.N., Mechanism of spreading depression: a review of recent findings and a hypothesis, *Can. J. Physiol. Pharmacol.*, 70 Suppl (1992) S248-S254.

22. Szerb, J.C., Glutamate release and spreading depression in the fascia dentata in response to microdialysis with high K+: role of glia, *Brain Res.*, 542 (1991) 259-265.

23. van der Zee, P., Arridge, S.R., Cope, M. and Delpy, D.T., The effect of optode positioning on optical pathlength in near infrared spectroscopy of brain, *in:. Oxygen. transport. to. tissue XII.*, J (1990) 79-84.

24. Wahl, M., Schilling, L., Parsons, A.A. and Kaumann, A., Involvement of calcitonin gene-related peptide (CGRP) and nitric oxide (NO) in the pial artery dilatation elicited by spreading depression, *Brain Res*, 637 (1994) 204-210.

25. Wolff, H.G., *Headache and other head pain.*, Oxford University Press, New York, 1963,

26. Wray, S., Cope, M., Delpy, D.T., Wyatt, J.S. and Reynolds, E.O.R., Characterization of the near infrared absorption spectra of cytochrome aa3 and haemoglobine for the non-invasive monitoring of cerebral oxygenation, *Biochim. Biophys. Acta*, 933 (1988) 184-192.

39

THE ENERGY DEPENDENT REDOX RESPONSES OF HEME AND COPPER IN CYTOCHROME OXIDASE IN RAT BRAIN *IN SITU*

A. Matsunaga,[1,2] Y. Nomura,[1] M. Tamura,[1] and N. Yoshimura[2]

[1] Research Institute for Electronic Science
Hokkaido University
Sapporo 060, Japan
[2] Department of Anesthesiology
Kagoshima University School of Medicine
Kagoshima 890, Japan

INTRODUCTION

With the *in vitro* calibration of isolated mitochondria, it has been shown that the redox state of heme $a+a_3$ in cytochrome oxidase depends on both mitochondrial energy state and respiratory rate, but that of copper does not (1, 2, 3). This finding, however, has not been confirmed *in vivo* living tissues, since hemoglobin absorption masks the redox change of heme $a+a_3$ in the visible region.

In the present study, therefore, to expand the *in vitro* observations to *in vivo*, we employed the hemoglobin-free perfusion technique of living tissues, especially cerebral tissue of rat head. Using this preparation, we investigated the redox behaviors of both heme $a+a_3$ and copper in cerebral tissue, as related to mitochondrial energy state. The data obtained with the present experiment bridged the *in vivo* and *in vitro* observations.

MATERIALS AND METHODS

Materials

The perfusion medium was an emulsion of perfluorotributyl-amine (FC-43; Green Cross Corp., Osaka, Japan) (4) containing 10 mM glucose, equilibrated with a 95% O_2-5% CO_2 gas mixture to maintain PCO_2 40 ± 5 mmHg, PO_2 > 700 mmHg and PH 7.4 ± 0.05. In order to induce the energy depletion, two perfusates were prepared. One was the perfusate containing uncoupler of 400 nM carbonyl cyanide m-chlorophenylhydrazone, CCCP, (un-

Oxygen Transport to Tissue XVII, Edited by Ince et al.
Plenum Press, New York, 1996

305

coupled FC-43) and another 40 mM 2-deoxy-D-glucose (5), 2-DG, (hypoglycemic FC-43). Both perfusates were pre-equilibrated with a 95% O_2-5% CO_2 gas mixture.

Surgical Preparation

Brain perfusion, first reported by Andjus et al. (6), was carried out according to the procedure of Inagaki et al. (7). Both common carotid arteries were exposed and the external carotid arteries were ligated. After injection of heparin the bypass-catheters were implanted in each of the common carotid arteries without interrupting brain circulation. The rat brain was perfused with the FC-43 via the bypass catheters in a non-recirculating system. The flow rate was maintained at 6 ml/min and the perfusate temperature at 30°C. Bipolar electroencephalograms (EEG) were recorded from the parietal region of each hemisphere with electrodes placed in direct contact with the bone.

Spectrophotometric Measurement

The measuring lights from a halogen lamp filtered through four interference filters (wave length; 605, 622, 780, and 830 nm) were illuminated onto the center of the parietal region of a rat skull through a light guide with a 5 mm diameter. The lights transmitted through the cranial bone and cerebral tissue were guided again through another light guide with a 4 mm diameter to the photodetection system. Redox changes in heme $a+a_3$ and copper were measured simultaneously and continuously with a two-channel dual-wavelength spectrophotometer (Unisoku, Japan). The wavelength pairs used were 605-620 nm for heme $a+a_3$ and 830-780 nm for copper, respectively. The redox level of heme $a+a_3$ and copper in cytochrome oxidase were expressed as a percentage of the total absorbance change. The full scale of absorbance change was obtained with perfusion with the oxygenated FC-43 (equilibrated with 95% O_2-5%CO_2) and anoxia with the interruption of perfusion.

RESULTS

Redox Responses of Heme $a+a_3$ and Copper in Normal Energy Conditions

After the rat brain reached steady state conditions by perfusion with the FC-43, brain anoxic insults were induced by transient interruptions of perfusion. Fig. 1 shows the redox responses of heme $a+a_3$ and copper to the transient anoxic insult induced by the interruption. After stopping the perfusion, both heme $a+a_3$ and copper started immediately to reduce. However, the time-courses of the reduction of these chromophores were different. heme $a+a_3$ was triphasic, rapid, slow, and second rapid phases. About 70% of heme $a+a_3$ was reduced at the first rapid phase, whereas little at the slow phase. After 8 min of cerebral ischemia, heme $a+a_3$ was fully reduced. In contrast to heme $a+a_3$, the time-course of the reduction of copper was monophasic. Copper was fully reduced about 2 min later than heme $a+a_3$. In contrast, after perfusion was re instituted following ischemia, the time-course of the oxidation of heme $a+a_3$ was biphasic, and the rapid oxidation phase accounted for 70% of the total absorbance change, which was the same magnitude of the rapid reduction phase. The oxidation of copper was monophasic, the same pattern as observed in the reduction. The spontaneous EEG disappeared within 60 sec after cessation of perfusion. When the reduction of heme $a+a_3$ shifted from the rapid phase to the slow phase, EEG disappeared. At this critical

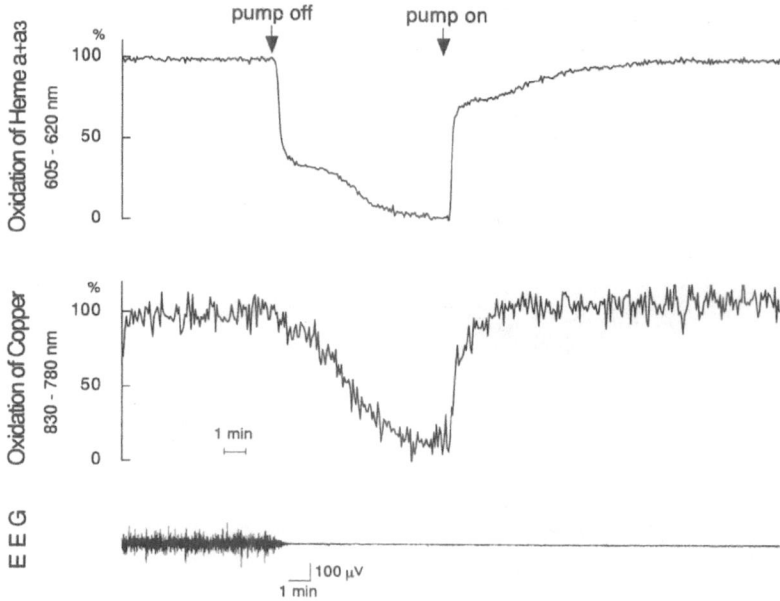

Figure 1. The redox behaviors of heme a+a$_3$ and copper to the anoxic insult induced by the interruption of perfusion under the normal condition. The redox level of heme a+a$_3$ and copper were expressed as a percentage of the total absorbance change. 100% indicated the redox level obtained with perfusion with the oxygenated FC-43 (equilibrated with 95% O$_2$-5%CO$_2$) and 0% anoxia with the interruption of perfusion.

stage, more than 70% of heme a+a$_3$ was reduced, but about 80% of copper was still in the oxidized state.

Redox responses of heme a+a$_3$ and Copper in the Energy Depleted Conditions

Fig. 2 shows the redox responses of heme a+a$_3$ and copper to the transient anoxic insult induced by the interruption of perfusion with the uncoupled preparation. After switching from the normal FC-43 to the uncoupled FC-43, heme a+a$_3$ was reduced transiently (about 70% reduced) and kept reduced state partially (about 40% reduction). While copper was kept in the fully oxidized state, after the transient reduction (about 30%). The spontaneous EEG disappeared, when heme a+a$_3$ and copper were reduced transiently. In the hypoglycemic preparation of Fig. 3, heme a+a$_3$ kept in the full oxidized state after the fluctuation in the redox state, whereas copper became hyper oxidized (about 120% oxidized). The spontaneous EEG disappeared within 20 min after switching to the hypoglycemic FC-43. As seen in these figures, the slow reduction phase of heme a+a$_3$ diminished, which appeared in the normal preparation, and the rapid phase was only observed. The reduction of heme a+a$_3$ became monophasic with both preparations. On the other hand, after perfusion was reinstituted following ischemia, the oxidation of heme a+a$_3$ was different between the two preparations. The oxidation of heme a+a$_3$ with the uncoupled preparation was monophasic, while biphasic with the hypoglycemic preparation. The copper showed monophasic in all the preparations.

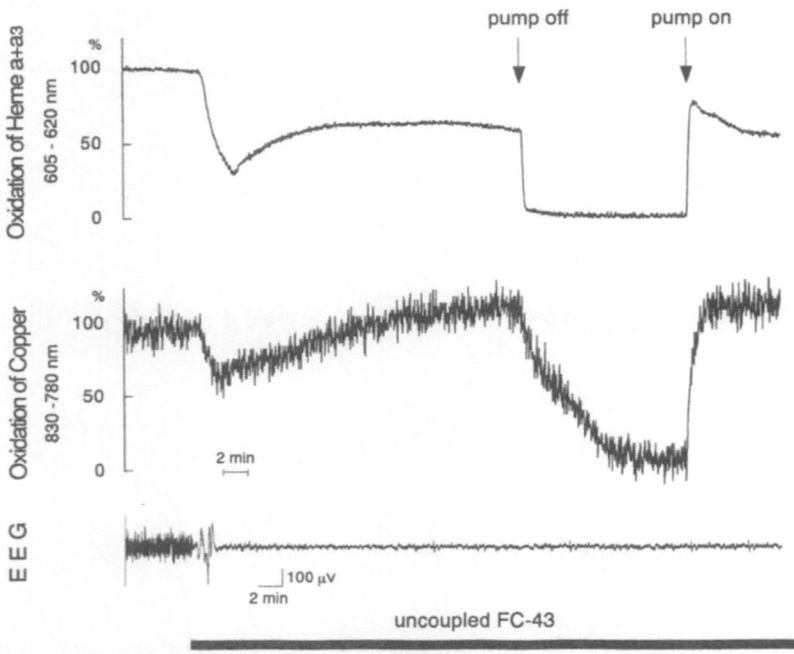

Figure 2. The redox behaviors of heme a+a₃ and copper to the anoxic insult induced by the interruption of perfusion under the uncoupled condition. The brain was perfused with the perfusate containing uncoupler of 400 nM carbonyl cyanide m-chlorophenylhydrazone, CCCP, (uncoupled FC-43).

DISCUSSION AND CONCLUSION

With isolated mitochondria, the reduction of heme a+a₃ at anoxia is biphasic in the presence of ATP, while monophasic in the absence of ATP. In contrast to heme a+a₃, that of copper has been reported to be monophasic independent of the energy state (1,3,8). In this study, the reduction of heme a+a₃ in the perfused brain was triphasic under the normal energy conditions and monophasic under the energy depleted conditions. The precise calibration of isolated mitochondria showed that 72-89% of the total absorbance of heme a+a₃ at 605 nm is attributed to heme a and the remainder to heme a₃ (9). In addition, heme a₃ is energy dependent but not heme a (9,10). Taking these mitochondrial observations, the observed triphase of the reduction of heme a+a₃ in normal cerebral tissue can account for this complex behavior of cytochrome oxidase. The first rapid phase of the reduction, about 70% of total absorbance change, can be attributed to the reduction of heme a and the following second slow and third phases are due to the reduction of heme a₃ accompanied by decrease in the tissue energy state. This was confirmed by addition of the uncoupler and 2-DG, which diminished the second and third phases. The oxidation process after anoxia can also explain the energy dependent redox behavior of heme a+a₃. Thus, heme a+a₃ depends on the energy state in the perfused brain.

The redox behaviors of copper were all monophasic, and were independent of the energy state. Therefore, the degree of oxidation-reduction state of copper depends only on tissue oxygen concentrations. Kariman *et al.* (11), in the perfused rat brain with FC-43, have

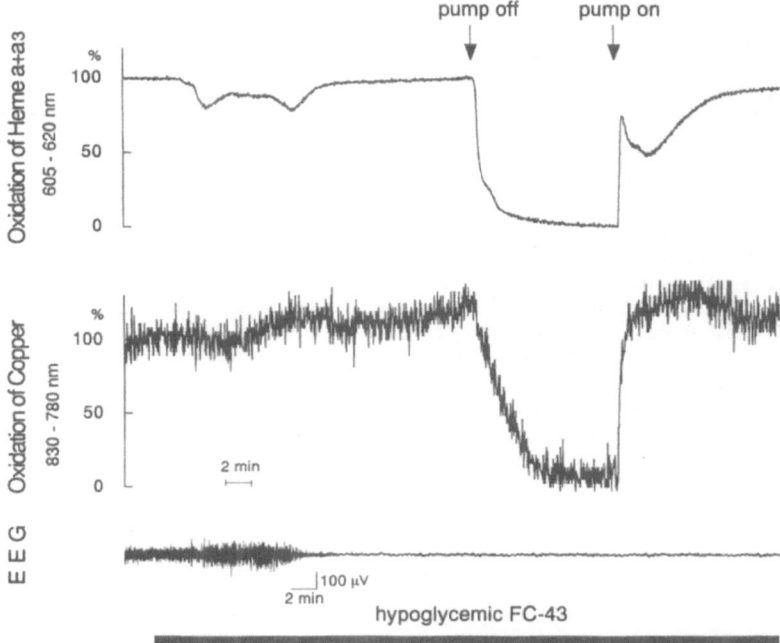

Figure 3. The redox behaviors of heme a+a$_3$ and copper to the anoxic insult induced by the interruption of perfusion under the hypoglycemic condition. The brain was perfused with the perfusate containing 40 mM 2-deoxy-D-glucose, 2-DG, (hypoglycemic FC-43).

reported that the copper shifted toward oxidation despite continuation of the anoxic insult, when heme a+a$_3$ shifted from the rapid phase to the slow phase. But we did not observe this.

In the present study, we confirmed that the redox state of heme a+a$_3$ depends on the energy state and the oxygen concentrations, whereas copper depends only on the oxygen concentrations but not the energy state. Therefore, the redox behavior of heme a+a$_3$ can be used as an indicator of tissue energy conditions and that of copper tissue oxygenation, when both are measured simultaneously. Present data emphasizes the critical importance of measurement of the redox behaviors of heme a+a$_3$ in addition to copper for assessment of tissue energy state. The latter has already been measured in near-infrared oxygen monitoring (12,13). Thus, in spite of the difficulty of heme a+a$_3$ measurement in the visible region under normal blood supply conditions, an effort must be made for bedside energy-monitoring as a replacement of MRS.

REFERENCES

1. Hoshi Y., Hazeki O., and Tamura M., 1989, The oxygen dependency of the redox state of heme and copper in cytochrome oxidase *in vitro*, Adv. Exp. Med. Biol. 248: 71-76.
2. Oshino N., Sugano T., and Chance B.,1974, Mitochondrial function under hypoxic conditions: the steady states of cytochrome a+a$_3$ and their relation to mitochondrial energy states, Biochim. Biophys. Acta. 368: 298-310.
3. Hoshi Y., Hazeki O., and Tamura M., 1993, Oxygen dependence of redox state of copper in cytochrome oxidase *in vitro*, J. Appl. Physiol. 74: 1622-1627.
4. Green Cross Crop., 1976, FC-43 Emulsion, Technical information Ser. No.3 September 4, Osaka.

5. Hlinak Z., and Madlafousek J., 1987, Transition from precopulatory to copulatory behavior in male rats with lesions in medial preoptic area, Act. Nerv. Super. 29: 257-262.
6. Andjus R.K., and Suhara K., 1967, An isolated, perfused rat brain preparation, its spontaneous and stimulated activity, J. Appl. Physiol. 22:1033-1039.
7. Inagaki M., and Tamura M., 1993, Preparation and optical characteristics of hemoglobin-free isolated perfused rat head in situ. J. Biochem. 113: 650-657.
8. Rich P.R., and West I.C., 1988, The location of CuA in mammalian cytochrome c oxidase, FEBS Lett. 233: 25-30.
9. Lindsay J.G., and Wilson D.F., 1972, Apparent adenosine triphosphate induced ligand changes in cytochrome a_3 of pigeon heart mitochondria, Biochemistry 11: 4613-4621.
10. Wilson D.F., and Dutton P.L., 1970, The oxidation-reduction potentials of cytochromes a and a_3 in intact rat liver mitochondria, Arch. Biochem. Biophys. 163: 583-584.
11. Kariman K., and Burkhart D.S., 1985, Heme-copper relationship of cytochrome oxidase in rat brain in situ, Biochem. Biophys. Res. Commun. 126: 1022-1028.
12. Hazeki O., and Tamura M., 1989, Near infrared quadruple wavelength spectrophotometry of the rat brain, Adv. Exp. Med. Biol. 247: 63-69.
13. Hoshi Y., and Tamura M., 1993, Dynamic changes in cerebral oxygenation in chemically induced seizures in rats: study by near-infrared spectrophotometry, Brain Res. 603: 215-221.

40

CRITICAL OXYGEN EXTRACTION IN PIGLET HINDLIMB IS IMPAIRED AFTER INHIBITION OF ATP-SENSITIVE POTASSIUM CHANNELS

Benoit Vallet,[1] Benoit Guery,[1] Jacques Mangalaboyi,[1] Patrick Menager,[1] Scott E. Curtis,[2] Stephen M. Cain,[2] Claude Chopin,[1] and Bernard A. Dupuis[1]

[1] Departments of Pharmacology and Intensive Care and INSERM U279
University of Lille
59045 Lille, France
[2] Departments of Physiology and Biophysics, and Pediatrics
University of Alabama at Birmingham
Birmingham, Alabama 35294

INTRODUCTION

Hypoxic vasodilation is a well-recognized and often studied response. In arterial smooth muscle cells, decreased PO_2 between 80 and 30 Torr caused a strong hyperpolarization (Grote et al., 1988). In vascular smooth muscle, activation of ATP-sensitive K^+ channels by specific channel openers induces vasodilation through hyperpolarization of the cell membranes (Standen et al., 1989). In isolated perfused guinea pig hearts, glibenclamide, a potent inhibitor of ATP-sensitive K^+ channels, prevents coronary hypoxic vasodilation and cromakalim, an ATP-sensitive K^+ channel opener, mimics hypoxic vasodilation (Daut et al., 1990), which suggests that opening of ATP-sensitive K^+ channels may mediate hypoxic vasodilation in guinea pig hearts. Likewise, in rat aortic rings we demonstrated that glibenclamide attenuated the endothelium-independent relaxation induced by a 30 Torr bath oxygen tension (PO_2) (Vallet et al., 1994a).

When oxygen supply to the tissue is lowered, the concept of intrinsic metabolic regulation proposes that precise adjustment to the demand is accomplished by a feedback mechanism which in some way changes local distribution of blood flow (Granger et al., 1976). Arteriolar dilation, modulated by decreased PO_2, determines increased blood flow and capillary exchange capacity and could promote oxygen transport to meet oxygen demand (Segal, 1991). Dilation, which was demonstrated to be retro and anterograde among the highly branched microcirculatory network, must be coordinated by some form of communication to provide blood flow in accord with the local requirements of parenchymal cells

Oxygen Transport to Tissue XVII, Edited by Ince et al.
Plenum Press, New York, 1996

311

(Segal, 1991). Hyperpolarization can be conducted for millimeters along the arteriolar wall (Hirst & Neild, 1978) and it has been shown that these distances are similar to those reported for conducted vasodilation of arterioles (Segal, 1991). Thus, conducted vasodilation can induce red blood cells perfusion of terminal arterioles and capillaries (i.e., microvascular units) in striated muscle. This mechanism is intrinsic to the arteriolar network and can direct oxygen delivery among muscle fibers to satisfy oxygen demand.

To investigate the possibility that ATP-sensitive K^+ channels of vascular smooth muscles may contribute importantly to control of regional blood flow during hypoxia, we proposed to use an isolated hindlimb preparation which provides a model of local hypoxia without confounding baroreceptor and chemoreceptor influence (Cain & Chapler, 1979). We gave glibenclamide, a potent inhibitor of ATP-sensitive K^+ channels, and studied effects of reduced oxygen supply (DO_2) during ischemic hypoxia on limb vascular resistance, oxygen extraction (ERO_2) and consumption (VO_2).

METHODS

Fourteen pigs (2.5 month-old), weighing 22.1 ± 0.4 (SE) kg, were anesthetized and mechanically ventilated with room air to keep arterial PCO_2 between 35 and 40 Torr. A catheter was inserted into the left carotid artery where mean arterial pressure was continuously monitored. A Swan-Ganz catheter was inserted into the pulmonary artery and provided cardiac output measurement and core temperature. Heat lamps suspended above the operating table were used to maintain core temperature near 37ºC. Standard limb leads were used to obtain heart rate continuously by means of a cardiotachometer.

Arterial inflow and venous outflow from the right hindlimb were isolated as previously described (Cain & Chapler, 1979). The proximal 10 cm of the iliac nerve, artery, and vein were dissected free and all vascular branches were tied off. Venous outflow from the limb was restricted to the iliac vein by a tourniquet technique. The isolated iliac vessels and nerve were excluded from this tourniquet. Arterial isolation and reactive hyperemia were documented in all animals at the beginning of each experiment by occluding the iliac artery for 30 sec. Heparin was given intravenously at a dose of 1,000 U/kg before cross perfusion was initiated. Blood flow from the right iliac vein was returned through a flowmeter to a reservoir positioned above and connected to the right jugular vein. A roller occlusive pump (Masterflex) directed blood flow from the left carotid artery to the iliac artery of the vascularly isolated right hindlimb in an alternate perfusion system. A sampling port and pressure transducer were placed in this circuit proximal to the limb.

Hindlimb VO_2 was calculated as the product of its blood flow and the arteriovenous difference in O_2 content. Limb vascular resistance was calculated as the mean perfusion pressure divided by the limb flow. Regional DO_2 was calculated as the product of the appropriate values for blood flow and arterial O_2 content. O_2 extraction ratio (ERO_2) was calculated as the arteriovenous O_2 content difference divided by the arterial O_2 content. Leg blood flow, O_2 delivery (DO_2), and O_2 consumption (VO_2) were reported per kg of muscle mass (mean $\pm$ SE was 0.73 ± 0.05 kg).

After all pressures and flows were stable for at least 30 min, the experiment began with a 20-min control period during which measurements were obtained every 10 min. After this period, half of the pigs (n = 7) received a 10 $\mu g.min^{-1}.kg^{-1}$ infusion of glibenclamide (GLIB group) into the right femoral artery. Glibenclamide was dissolved in 5% glucose solution containing 0.01 NaOH. Other pigs (n = 7) received 5% glucose infusion and were used as controls (CTRL group). After glibenclamide or 5% glucose infusions were begun and when all pressures and flows were again stable, measurements were obtained every 10 min during another 20-min control period. Pigs with or without glibenclamide were then exposed to a progressive regional ischemic hypoxia. The isolated hindlimb flow, initially set at approximately 50 ml.min^{-1}.kg^{-1}, was decreased every 10 min with 10 decreased steps of approximately 8 ml.min^{-1}.kg^{-1}. Measurements were obtained at the end of each period of 10 min.

Following each study, regression lines were fitted to the delivery independent and delivery dependent portions of the O_2 delivery-uptake curve using a dual-line, least squares method. The intercept of the two lines defined the critical DO_2. The critical ERO_2 was taken as the ratio of VO_2 to DO_2 at critical DO_2.

Data were analyzed within and between groups using repeated measures analysis of variance and unpaired t-tests as appropriate. Statistical significance was accepted at $p<0.05$ for all comparisons.

RESULTS

Whole body arterial blood gas tensions, lactate, core temperature, plasma levels of glucose and K^+, and cardiovascular parameters at baseline were all within normal ranges and did not differ between groups before glibenclamide infusion. These values remained essentially unchanged between groups and until the end of experiments and were within expected limits for paralyzed-anesthetized animals.

The stepwise decrements in hindlimb oxygen delivery were well matched in the two groups, from initial values of 6.7 ± 0.4 and 6.8 ± 0.7 ml.min^{-1}.kg^{-1} to final values of 1.1 ± 0.3 and 0.6 ± 0.5 ml.min^{-1}.kg^{-1}, in CTRL group and GLIB group, respectively. Baseline VO_2 (prior to treatment) did not differ between groups and averaged 2.6 ± 0.2 ml.min^{-1}.kg^{-1}. Following glibenclamide or solvent infusion, hindlimb VO_2 did not change and average hindlimb VO_2 for O_2 deliveries above critical was not different between groups (2.5 ± 0.2 ml.min^{-1}.kg^{-1} in GLIB group vs 2.4 ± 0.2 ml.min^{-1}.kg^{-1} in CTRL group). With further decreases in hindlimb flow, all groups showed typical biphasic DO_2-VO_2 relations (Fig 1).

O_2 supply dependency became apparent at significantly higher DO_2 in GLIB group (4.9 ± 0.5 ml.min^{-1}.kg^{-1}) than in CTRL group (3.4 ± 0.3 ml.min^{-1}.kg^{-1}). The oxygen extraction

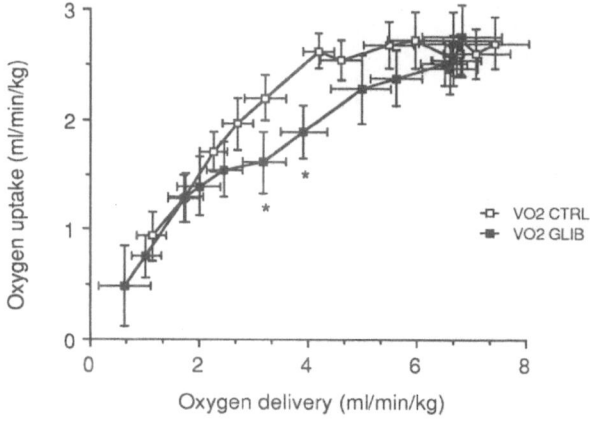

Figure 1. Hindlimb O_2 consumption as a function of O_2 delivery in controls and glibenclamide-treated animals. Data (mean ± SE) were binned by measurement time. * p <0.05 between controls (CTRL) and glibenclamide-treated animals (GLIB) at same measurement time.

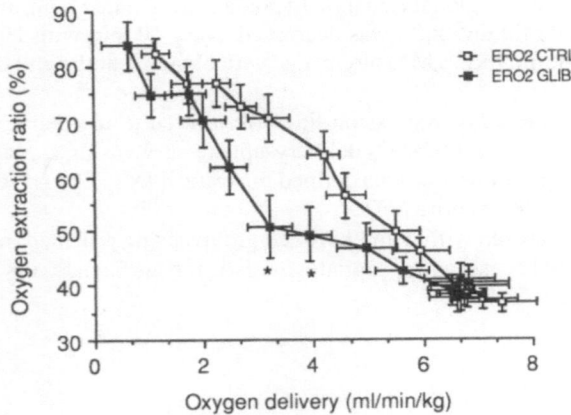

Figure 2. Hindlimb O_2 extraction as a function of O_2 delivery in controls and glibenclamide-treated animals. Data (mean ± SE) were binned by measurement time. * p <0.05 between controls (CTRL) and glibenclamide-treated animals (GLIB) at same measurement time.

ratio (ERO_2) at critical DO_2 was significantly lower in GLIB group (52.6 ± 3.6%) than in CTRL group (72.5 ± 4.7%)(Fig 2).

With decreases in hindlimb flow, there was a significant increase in hindlimb resistance in the glibenclamide-treated group (Fig 3).

DISCUSSION

It has been demonstrated that local hypoxic and ischemic vasodilation is mediated by opening of ATP-sensitive K^+ channels (Daut et al., 1990). Our question for this study was whether ATP-sensitive K^+ channels-induced hypoxic vasodilation was essential to adjust oxygen transport to meet oxygen demand and optimize oxygen extraction. We restricted

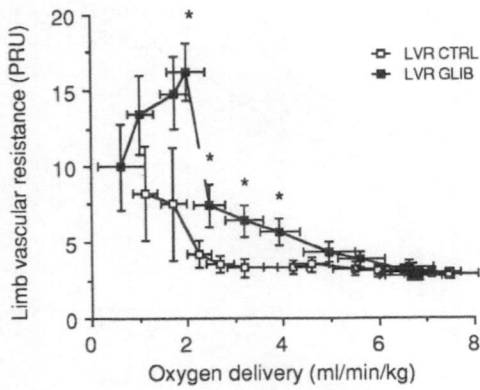

Figure 3. Limb vascular resistance as a function of O_2 delivery in controls and glibenclamide-treated animals. Data (mean ± SE) were binned by measurement time. * p <0.05 between controls (CTRL) and glibenclamide-treated animals (GLIB) at same measurement time.

hypoxia to the right hindlimb skeletal muscle by controlling blood flow to the area in the presence and absence of glibenclamide, a potent blocker of ATP-sensitive K^+ channels.

We used a glibenclamide dose (10 μg.min^{-1}.kg^{-1}; total dose 1.25 mg/kg) and local infusion to avoid any systemic effect and to observe selective effects only at the level of the isolated hindlimb. Systemic effects that were observed by Moreau et al. (Moreau et al., 1994) with higher doses (20 mg/kg) were only mild with a 17% decrease in cardiac output and a 31% increase in systemic vascular resistance. Plasma glucose concentrations were not significantly affected by glibenclamide infusion. Similarly, Imamura et al. (Imamura et al., 1992) did not notice any change in plasma levels of glucose, K^+ and Na^+ before and after glibenclamide infusion (50 μg.min^{-1}.kg^{-1}). In our study, the low total dose of glibenclamide we used did not induce any change in systemic hemodynamic or metabolic parameters.

Locally, when flow was controlled and stable, glibenclamide infusion induced an 11% increase in vascular resistance. This was not statistically significant. Buckingham et al. (1989), and Quast and Cook (1989) demonstrated that glibenclamide *per se* has no effect on vascular tone *in vitro* or *in vivo* at doses that block the effect of cromakalim administration. In contrast, Imamura et al. (1992) noticed that glibenclamide at a dose of 5 μg.min^{-1}.kg^{-1} significantly decreased coronary blood flow, suggesting to the authors that ATP-sensitive K^+ channels play an important role in the maintenance of basal coronary blood flow under physiological conditions. Also, Jackson (1994) assessed the role of ATP-sensitive K^+ channels in determining resting arteriolar tone and vasodilator reactivity in superfused, hamster microcirculatory beds studied via intravital microscopy. Under resting conditions, glibenclamide produced concentration-dependent vasoconstriction in microcirculatory beds, suggesting that the activity of ATP-sensitive K^+ channels determines, in part, resting arteriolar tone.

Increases in limb vascular resistance became significant in glibenclamide-treated animals when DO_2 was reduced in stepwise decrements, suggesting that ATP-sensitive K^+ channel mediated hypoxic vasodilation was impaired with glibenclamide infusion. Also, during decreased limb DO_2, we observed that hindlimb VO_2 became supply dependent at significantly higher DO_2 in glibenclamide-treated pigs (4.9 $\pm$ 0.5 ml.min^{-1}.kg^{-1}) than in controls (3.4 $\pm$ 0.3 ml.min^{-1}.kg^{-1}). The oxygen extraction ratio (ERO_2) at critical DO_2 was significantly lower in glibenclamide-treated pigs (52.6 $\pm$ 3.6%) than in controls (72.5 $\pm$ 4.7%).

The concept of intrinsic metabolic regulation states that the control of microvascular tone is locally coupled to the metabolic demand of the parenchyma in order to maintain intracellular PO_2 above the critical level at which oxygen availability limits the aerobic production of ATP (Granger et al., 1976). The mechanisms that mediate this response, however, remain uncertain. Recent efforts have focused on ideas that, for example, hypoxia promotes the release of a vasodilator metabolite or it causes the release of endothelial cell-dependent vasodilators (Coburn et al., 1986; Pohl & Busse, 1989). Prostaglandin I_2 (PGI_2), and endothelium-derived relaxing factor (EDRF), which is believed to be nitric oxide (NO) or a related compound, are usually released simultaneously, and hypoxia stimulates PGI_2 production and release (Michiels et al., 1993). However, the release of EDRF/NO by hypoxia has been uncertain (Rengasamy & Johns, 1991; Vallet et al., 1994a). We and others (Vallet et al., 1994a; Vallet et al., 1994b; Pohl & Busse, 1989) have addressed the role of EDRF/NO in vasodilation during hypoxia by *in vitro* and *in vivo* experiments, and showed that EDRF/NO was not necessary for vasodilation during hypoxia (at PO_2 30 Torr). Furthermore, the ability of the hindlimb muscle to increase O_2 extraction was not diminished by EDRF/NO synthase inhibition. In contrast, we provided evidence that cyclooxygenase blockade and PGI_2 production inhibition significantly altered oxygen extraction ability during progressive ischemia in canine hindlimb (Winn et al., 1994). Interestingly, gliben-clamide was shown to inhibit prostacyclin-induced vasodilation of the rabbit coronary circulation and rabbit and bovine coronary arteries (Jackson, 1994). However, glibenclamide had no significant effect on cAMP accumulation in bovine coronary artery smooth muscle induced by iloprost, a stable analog of prostacyclin. Therefore, PGI_2 is released during hypoxia and at least part of its effects are mediated by ATP-sensitive K^+ channels.

Muscle ability to extract oxygen during an ischemic challenge may be considered as an integrative test of microcirculatory coordination. There is a mechanism, intrinsic to the arteriolar network, whereby localized vasomotor stimuli can increase O_2 delivery to specific regions of the muscle. In this manner, when O_2 delivery is lowered O_2 shunting is minimized through areas with a high rate of perfusion relative to their O_2 uptake. It has been proposed that an O_2 sensor is situated downstream from the arteriole in the supplied tissue, and that there is a retrograde propagation of the local metabolic signal from the capillary to the supplying arteriole (Segal, 1991). The existence of conduction of signals along microvessels has been demontrated in arterioles of the hamster cheek pouch, and it has been hypothesized (Segal, 1991) that the propagation was initiated by a receptor-mediated change in membrane potential, involving an electronic spread through gap junctions between smooth mucle cells and/or endothelial cells. Hyperpolarization in response to hypoxia occurs through activation of ATP-sensitive K^+ channels which are physiologically closed by millimolar concentrations of ATP. The vasodilation triggered locally by hyperpolarization on a constricted terminal arteriole can be conducted proximally, against the direction of blood flow, and converge onto the feeding arteriole, increasing flow into the stimulated vessel. In presence of glibenclamide, when the limb muscles were made progressively ischemic in the absence of ATP-sensitive K^+ channels-induced hyperpolarization, local hypoxia was insufficient to direct blood flow according to regional O_2 demand.

In summary, the ATP-sensitive K^+ channel inhibitor glibenclamide impaired oxygen extraction ability in hindlimb muscle during progressive ischemia and lowered the critical level at which oxygen availability limits aerobic production of ATP. These results strongly support the hypothesis that ATP-sensitive K^+ channel induced vasodilation in hypoxia is essential to adjust oxygen transport to meet oxygen demand and optimize oxygen extraction and tissue oxygenation. We propose that ATP-sensitive K^+ channels play a critical role in the local metabolic control of microcirculation.

ACKNOWLEDGMENTS

Funds for these studies were provided by Institut National de la Santé et de la Recheche Médicale (contrat MGEN), Centre Hospitalier Régional et Universitaire de Lille (contrat CIVIS), and Lilly Company (France).

REFERENCES

Buckingham, R.E., T.C. Hamilton, D.R. Howlett, S. Mootoo, and C. Wilson. (1989) Inhibition of glibenclamide of the vasorelaxant action of cromakalim in the rat. *Br. J. Pharmacol.* 97: 57-64

Cain, S.M., and C.K. Chapler. (1979) Oxygen extraction by canine hindlimb during hypoxic hypoxia. *J. Appl. Physiol.* 46: 1023-1028

Coburn, R.F., R. Eppinger, and D.P. Scott. (1986) Oxygen-dependent tension in vascular smooth muscle. Does the endothelium play a role? *Circ. Res.* 58: 341-347

Daut, J., W.M. Rudolph, N. von Beckerath, G. Meherke, K. Gunther, and L.G. Meinen. (1990) Hypoxic dilation of coronary arteries is mediated by ATP-sensitive potassium channels. *Science Wash. DC* 247: 1341-1344

Granger, H.J., A.H. Goodman, and D.N. Granger. (1976) Role of resistance and exchange vessels in local microvascular control of skeletal muscle oxygenation in the dog. *Circ. Res.* 38: 379-385

Grote, J., G. Siegel, K. Zimmer, and A. Adler. (1988) The influence of oxygen tension on membrane potential and tone of canine artery smooth muscle. *Adv. Exp. Med. Biol.* 222: 481-487

Hirst, G.D.S., and T.O. Neild. (1978) An analysis of excitatory junctional potentials recorded from arterioles. J. Physiol. Lond. 280: 87-104

Imamura, Y., H. Tomoike, T. Narishige, T. Takahashi, H. Kasuya, and A. Takeshita. (1992) Glibenclamide decreases basal coronary blood flow in anesthetized dogs. *Am. J. Physiol.* 263: H399-H404

Jackson, W.F. (1994) Arteriolar tone is determined by activity of ATP-sensitive potassium channels. *Am. J. Physiol.* 265: H1797-H1803

Michiels, C., T. Arnould, M. Dieu, and J. Remacle. (1993) Stimulation of prostaglandin synthesis by human endothelial cells exposed to hypoxia. *Am. J. Physiol.* 264: C866-C874

Moreau, R., H. Komeichi, P. Kirstetter, S. Yang, B. Aupetit-Faisant, S. Cailmail, and D. Lebrec. (1994) Effects of glibenclamide on systemic and splanchnic haemodynamics in conscious rats. *Br. J. Pharmacol.* 112: 649-653

Pohl, U., and R. Busse. (1989) Hypoxia stimulates release of endothelium-derived relaxing factor. *Am. J. Physiol.* 256: H1595-H1600

Quast, U., and N.S. Cook. (1989) In vitro and in vivo comparison of two K channels openers, diazoxide and cromakalim, and their inhibition by glibenclamide. *J Pharmacol. Exp. Ther.* 250: 261-271

Rengasamy, A., and R.A. Johns. (1991) Characterization of endothelium-derived relaxing factor/nitric oxide synthase from bovine cerebellum and mechanism of modulation by high and low oxygen tensions. *J. Pharmacol. Exp. Therap.* 259: 310-316

Segal, S.S. (1991) Microvascular recruitment in hamster striated muscle: role for conducted vasodilation. *Am. J. Physiol.* 261: H180-H189

Standen, N.D., J.M. Quayle, N.W. Davies, J.E. Brayden, Y. Huang, and M.T. Nelson. (1989) Hyperpolarizing vasodilators activate ATP-sensitive K^+ channels in arterial smooth muscle. *Science Wash. DC* 245: 177-180

Vallet, B., N.K. Asante, M.J. Winn, and S.M. Cain. (1994a) Separation of the effects of EDRF/NO from vascular relaxation by the availability of oxygen. *J. Cardiovasc. Pharmacol.*, in press

Vallet, B., S.E. Curtis, M.J. Winn, C.E. King, C.K. Chapler and S.M. Cain. (1994b) Hypoxic vasodilation does not require nitric oxide (EDRF/NO) synthesis. *J. Appl. Physiol.* 76: 1256-1261

Winn, M.J., B. Vallet, S.E. Curtis, C.K. Chapler, C.E. King, and S.M. Cain. (1994) Critical oxygen extraction in dog hindlimb after inhibition of nitric oxide synthase and cyclooxygenase systems. *Adv. Biol. Exp. Med.*, in press

41

RESPONSES OF ELECTRICAL ACTIVITY AND REDOX STATE OF CYTOCHROME OXIDASE TO OXYGEN INSUFFICIENCY IN PERFUSED RAT BRAIN *IN SITU*

Y. Nomura, A. Matsunaga, and M. Tamura

Biophysics group
Research Institute for Electronic Science
Hokkaido University
Sapporo 060, Japan

INTRODUCTION

During the transition from normoxia to hypoxia in brain tissue, the energy related metabolites remain unchanged until the cerebral oxygen consumption decreases. For example, ATP levels are kept normal at arterial oxygen tensions higher than around 25 mmHg. Below this, ATP begins to decrease together with the increases of ADP and AMP, where oxygen consumption decreases. The electrical activity is also attenuated. The activation of glycolytic flux prevents, in part, the decline of ATP (Sylvia and Piantadosi, 1988, Erecinska and Silver, 1989). In addition, it has been reported that neuronal activity is suppressed prior to decreases in ATP. Thus, the energy failure is not solely responsible for suppression of brain function in hypoxic conditions. Previously, we have calibrated the oxygen dependence of the redox centers of cytochrome oxidase, heme aa_3 and copper, in isolated mitochondria (Hoshi et al., 1993). The redox state of heme aa_3 depends on both the energy state and the respiratory rate, whereas the redox state of copper is independent of both the energy state and the respiratory rate. Thus, simultaneous measurements of these chromophores in cytochrome oxidase can give both oxygen concentration and energy state at mitochondria. On the basis of the in vitro data of isolated mitochondria, we measured the redox state of cytochrome oxidase and the electrical activity in perfused rat brain *in situ*. The relationship between functional failure and oxygen insufficiency became much clearer from the present study, which is presented here.

Oxygen Transport to Tissue XVII, Edited by Ince et al.
Plenum Press, New York, 1996

319

MATERIALS AND METHODS

According to the method of our previous report (Inagaki and Tamura, 1993), blood-free perfused rat brain was prepared except for perfusion *in situ*. The perfusate through the brain was collected from the vena cava to avoid difficulties of a jugular venous cannulation. Holes in the skull over the parietal and frontal cortex were bored before vascular surgery. A bipolar electroencephalogram (EEG) was recorded from silver ball electrodes placed in these holes and fixed using EEG electrode paste. The EEG was amplified and displayed using a pen recorder. Amplified EEG signals were digitized at a sampling rate of 250 Hz. Digitized data were collected at an interval of 65 sec throughout the perfusion period and analyzed employing the Fast Fourier Transform algorithm (Wavemetrics). Redox changes in cytochrome oxidase were measured with a two-channel dual wavelength spectrophotometer (Unisoku, Japan). The wavelength pairs used were 605-620 nm for heme aa_3 and 780-830 nm for copper.

RESULTS

Relationship between Cerebral Oxidative Metabolism and Function under Hypoxic Condition

When the oxygen concentration of the inflowing perfusate was decreased acutely with a constant flow perfusion by exchanging the oxygen saturated buffer to the nitrogen saturated one, the amplitude of the spontaneous EEG decreased concomitantly with the reduction of heme aa_3 and copper (Fig. 1-A). There was clear dissociation of the time courses of the reduction of heme aa_3 and copper, though both are components of the cytochrome oxidase molecule. The reduction of heme aa_3 preceded the reduction of copper (Fig. 1-B). When 10% of heme aa_3 was reduced where copper remained fully oxidized, a slow wave pattern appeared on EEG (c in Fig. 1-C). When reduction of the heme aa_3 reached 30%, the spontaneous EEG almost disappeared (d in Fig. 1-C). At this stage, more than 90% of the copper was still in the oxidized state. Under anoxia, both heme aa_3 and copper were fully reduced when the EEG was flat. Fourier transformed data (i.e., amplitude vs. frequency) was averaged within the frequency bandwidths used by others (Harvey et al., 1991). As shown in Fig. 1-D, decreases in activities of a- and b-bands by hypoxia were larger than those of d- and q-bands. When oxygen concentration in perfusate was decreased, the perfused brain showed a slow wave pattern of EEG and then a flat EEG.

When oxygen concentration in the inflowing perfusate was decreased gradually with a constant flow perfusion, the slowing of EEG was also accompanied by a reduction of heme aa_3 and copper. In a steady state, digitized EEG data was sampled for 2-3 min and then the perfusate was changed to new conditions with the decreased oxygen concentration. The relationship between the redox state of heme aa_3 and mean frequency of EEG obtained with gradual hypoxia was similar to that in acute hypoxia. Figure 2 summarizes changes in mean frequency of EEG as a function of redox states of heme aa_3 and copper when oxygen concentrations were varied. The mean frequency of EEG declined to half of the normal aerobic state at the stage of 90% oxidation of heme aa_3 although copper remained fully oxidized. The EEG became flat when 40-50% of heme aa_3 was reduced and 10-30% of copper. We frequently found burst suppression also under the condition. The slowing of EEG was more sensitive than changes in redox state of cytochrome oxidase against oxygen insufficiency.

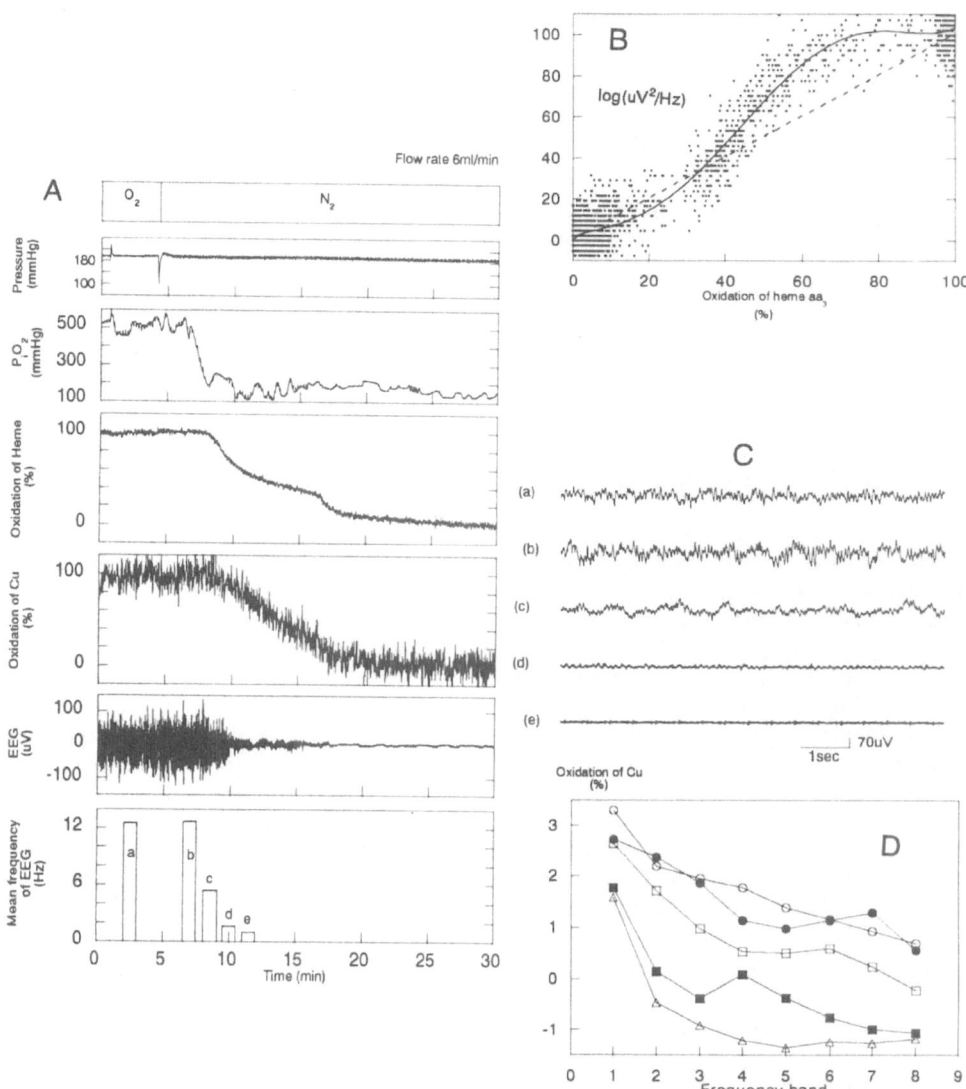

Figure 1. A: Changes in electrical activity and redox state of heme aa$_3$ and copper caused by acute hypoxia in the perfused brain using a constant flow perfusion. B: Relative oxidation-reduction state of copper with respect to that of heme aa$_3$ replotted from Fig.1-A. Solid line represented a polynomial curve with five terms for best fit of all data points. Broken line was the straight line expected for the case of oxidation of heme aa$_3$ equal to that of copper. C: Changes in the EEG caused by acute hypoxia. EEG samples (a)-(e) were recorded at the mark on the bottom of Fig.1-A. (a) and (b), before; (c), slow wave; (d), low amplitude; (e), flat. D: EEG power spectra (a, ●; c, □; d, ■; e, △; whole rat, ○). Frequency bands are defined in Ref.(Harvey *et al.*, 1991): 1 (0.1-1.50 Hz) and 2 (1.51-3.60) are δ bands, 3 (3.61-5.56) and 4 (5.57-7.52 Hz) are θ bands, 5 (7.53-9.47 Hz) and 6 (9.48-12.50 Hz) are α bands, 7 (12.51-17.48 Hz) and 8 (17.49-25.0 Hz) are β bands.

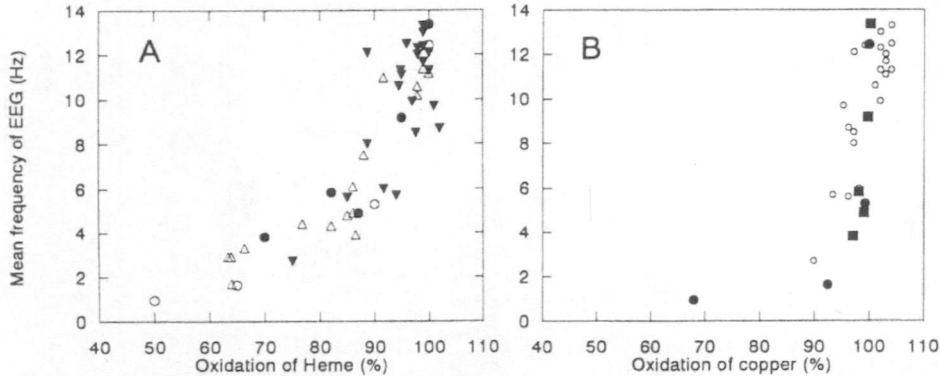

Figure 2. Relationship between mean frequencies of EEG and redox states of heme aa$_3$ (A) and copper (B). Each symbol shows the result for an individual rat.

Effect of Reoxygenation on Cerebral Oxidative Metabolism and Function

Reoxygenation returned both redox states of cytochrome oxidase and mean frequency of EEG close to their previous aerobic states. Figure 3 shows reversible responses of redox state of cytochrome oxidase and mean frequency of EEG from the various oxygen insufficiencies. A small change of redox state of heme aa$_3$ caused large responses of mean frequency of EEG, particularly in the case of 80% oxygenated perfusate. At this stage, copper did not change. The more severe the hypoxia, the less recovery of mean frequency was observed. Figure 4 summarizes restorations of mean frequency of EEG from various stages of oxygen insufficiencies with different redox states of heme aa$_3$ and copper. When oxidation of heme aa$_3$ was above 80% in hypoxia, mean frequency of EEG returned to that of the aerobic state within 15 min by reoxgenation. When oxidation of heme aa$_3$ was in the the range 60-80%, the recovery of EEG was incomplete. After the severe hypoxic conditions where oxidation of heme aa$_3$ was below 40%, flat EEG continued over 30min in the aerobic condition. From the hypoxic conditions where a small reduction of copper occurred, the incomplete EEG lasted for 15 min after reoxygenation.

Chemically Induced Seizure during Hypoxia

As shown in the previous section, EEG became flat when 40% of heme aa$_3$ was reduced. From the conditions, however, a significant recovery of brain function could be observed by the treatment of reoxygenation. It is important whether the energy required for the neuronal activity is left in cerebral tissues with the flat EEG under hypoxia or not. Therefore, we gave pentylenetetrazol (PTZ) to the perfused brain under hypoxia, which activated the neuronal activity. Figure 5-A shows changes in redox state of heme aa$_3$ and copper and mean frequency of EEG caused by PTZ administration under hypoxia. Figure 5-B shows the EEG. In the case of initial aerobic state, mean frequency was almost 12 Hz. When the perfusate was changed to 60% oxygenated FC43, 13-15% of heme aa$_3$ was reduced and 5-15% of copper was gradually reduced. In the condition of perfusion with 40% oxygenated FC43, 36% of heme aa$_3$ was reduced and 28% of copper. At the same time, the EEG became almost flat (c in Fig. 5-B). As shown in Fig. 5-A, PTZ administration caused a rapidly transient reduction of heme aa$_3$ and a small reduction of copper. The epileptic discharges almost simultaneously occurred on the EEG within 1 sec from the start of transient

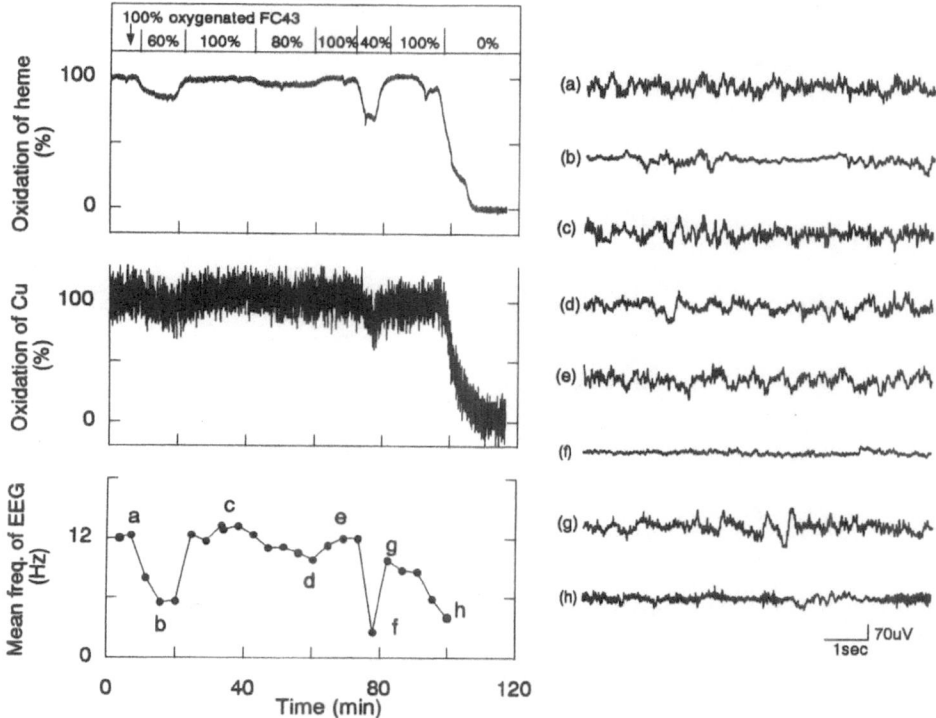

Figure 3. Reversible responses of redox state of cytochrome oxidase and EEG to oxygen insufficiencies. EEG samples were recorded at the points marked on the time course of mean frequency.

reduction of heme aa$_3$ (d-1 in Fig. 5-B). And then bursts of spikes appeared on the EEG (d-2 in Fig. 5-B). In this period, heme aa$_3$ started to be reoxidized to 60% though copper was reduced. Spike activity lasted about 2 min and then ceased (e in Fig. 5-B). The injection of physiological saline caused no changes in either EEG or the redox state of heme aa$_3$ and copper. Furthermore, second administration of pentylenetetrazol (same dose as the first) did not cause bursts of spikes. Finally, when the perfusate was changed to 100% oxygenated FC43, heme aa$_3$ was reoxidized somewhat slower than copper in contrast to Fig.3-A but EEG remained flat. The electrical activity including after-discharges caused by PTZ administration lasted for more than 15 min under aerobic conditions.

DISCUSSION

There are few reports that show correlation between the redox state of cytochrome oxidase and quantitative EEG. In a blood-perfused brain, light absorption by hemoglobin interferes with the measurement of redox state of cytochrome oxidase, especially heme aa$_3$. The perfused brain employed in this study had many advantages as follows. 1) Since blood was completely removed from the brain, optical artifacts due to hemoglobins were negligible. 2) The oxygen supply to cerebral tissue could be precisely controlled. Present study demonstrates that the cerebral tissue responded

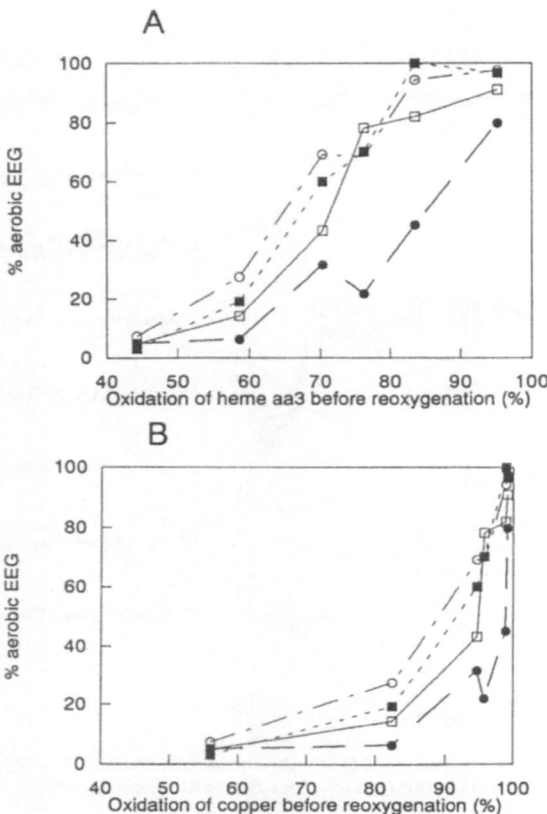

Figure 4. Restoration of electrical activity from different hypoxic conditions due to reoxygenation. Electrical activity was shown as a ratio of observed mean frequencies to those under initial aerobic conditions. Each symbol shows the time after reoxygenation (●, 0 min; □, 5; ■, 10; ○,15).

to various oxygen insufficiencies with the redox state of cytochrome oxidase and mean frequency of EEG. When copper, whose redox state is expected to give information about oxygen concentrations, was reduced, EEG became very weak. Furthermore, the mean frequency of EEG decreased in hypoxia though copper remained fully oxidized. Falls of the ratio of creatine phosphate to inorganic phosphate paralleled the reduction of copper in cytochrome oxidase (Tamura et al., 1988). Although ATP as well as creatine phosphate remained unchanged as judged by the full oxidation of copper, EEG became slow wave. These results suggest that mechanisms independent of ATP detected oxygen insufficiencies and suppressed EEG in order to recover EEG by the following oxygen supply before mitochondrial oxidative phosphorylation significantly declined. They may include in part the decline of various oxidase systems having higher apparent Km value than that of cytochrome oxidase, especially the modulation by neurotransmitters of catechol and indolamine that do not require ATP but need molecular oxygen as a substrate for their continued synthesis (Sylvia and Piantadosi, 1988). These mechanisms, however, collapsed due to the compulsory activation of neuron with PTZ under hypoxia. In the hypoxic condition that the EEG became flat, the typical epileptic discharge was caused by PTZ administration. At the time, reduction

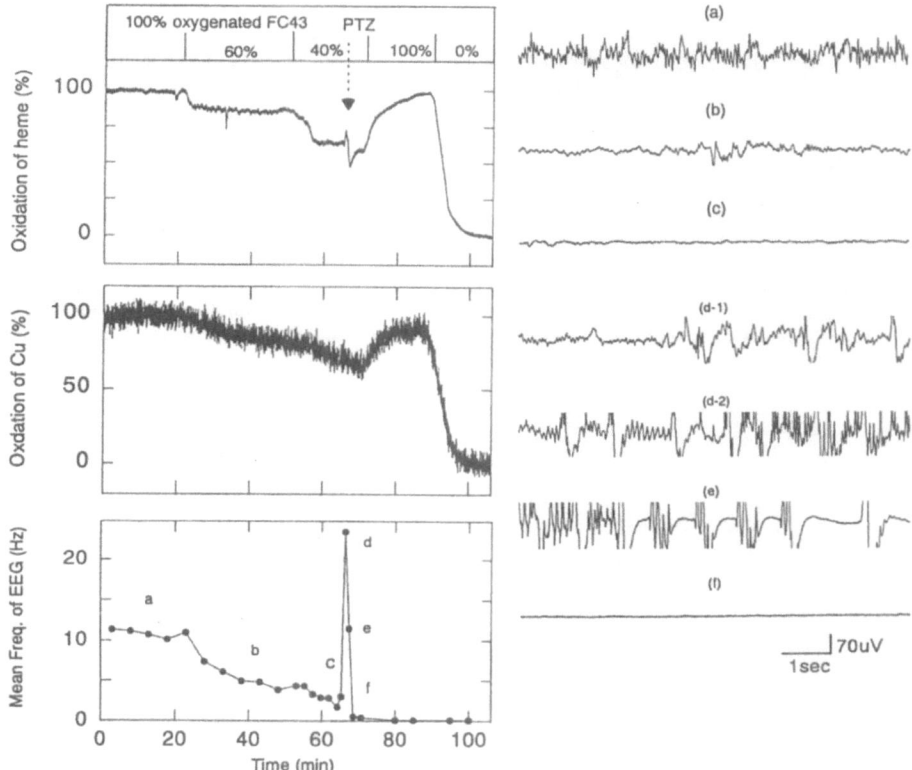

Figure 5. Changes in the redox state of cytochrome oxidase and electrical activity caused by PTZ administration under hypoxia. PTZ (5 mg/100 g b.w.) was injected at the indicated point. EEG samples were recorded at the points marked on the time course of mean frequency.

of heme aa_3 and copper were 40 and 30%, respectively. The EEG remained flat although it appeared that EEG was recovered by the readmission of oxygen from Fig.3 and 4. The redox behavior of cytochrome oxidase gave an adequate reason for the failure of mechanisms. Before the addition of PTZ, reduction of heme aa_3 preceded that of copper. After this, however, heme aa_3 was reduced in a similar fashion to copper. In the isolated mitochondria, the deviation from a straight line, which is expected if both components had had the same oxygen affinity, was larger in state 3 than state 4 and was the smallest in the uncoupled state (Hoshi et al., 1993). Therefore, it was thought that mitochondria in the brain were uncoupled by neuronal activation under hypoxia.

ACKNOWLEDGMENTS

This study was supported in part by a Grant-in Aid for Scientific Research (06858087) from the Ministry of Education, Science and Culture of Japan and by grants from Tateishi foundation (942006).

REFERENCES

Erecinska, M., and Silver, I. A., 1989, ATP and brain function, J. Cereb. Blood Flow Metab. 9: 2-19.

Harvey, S. A. K., Trankina, M. L., Olson, M. S., and Clark, J. B., 1991, Fluorocarbon perfusion of the isolated rat brain: measurement of tissue spaces, EEG and oxygen uptake, Biochim. Biophys. Acta. 1073: 486-492.

Hoshi, Y., Hazeki, O., and Tamura, M., 1993, Oxygen dependence of redox state of copper in cytochrome oxidase in vitro, J. Appl. Physiol. 74: 1622-1627.

Inagaki, M., and Tamura, M., 1993, Preparation and optical characteristics of hemoglobin-free isolated perfused rat head *in situ*, J. Biochem. 113: 650-657.

Sylvia, A. L., and Piantadosi, C. A., 1988, O2 dependence of in vivo brain cytochrome redox responses and energy metabolism in bloodless rats, J. Cereb. Blood Flow Metab. 8: 163-172.

Tamura, M., Hazeki, O., Nioka, S., Chance, B. and Smith, D. S., 1988, The simultaneous measurements of tissue oxygen concentration and energy state by near-infrared and nuclear magnetic resonance spectroscopy, Adv. Exp. Med. Biol. 222: 359-363.

42

DIFFUSION COEFFICIENT OF OXYGEN IN VARIOUS MODEL LAYERS AS DETERMINED BY ANALYSIS OF TIME-DEPENDENT DIFFUSION

Astrid M. C. van Dijk, Louis J. C. Hoofd, and Berend Oeseburg

Department of Physiology
Faculty of Medical Sciences
University of Nijmegen
P.O.Box 9101, 6500 HB Nijmegen, The Netherlands

INTRODUCTION

To obtain an accurate oxygen diffusion coefficient for muscle tissue, the tissue is divided into several fractions. These fractions can be represented by myoglobin solutions, protein solutions, endothelium, lipid layers, etc. For each of these fractions an oxygen diffusion coefficient can be measured in vitro in a diffusion chamber. In general, the methods for measuring oxygen diffusion across these various layers or membranes can be divided into two types; steady state and nonsteady state measurements. The steady state methods yield the oxygen permeability of a layer ($\wp O_2$), which is equal to the product of the oxygen diffusion coefficient (DO_2) and the oxygen solubility (αO_2). This implies that the determination of the diffusion coefficient from steady state measurements requires accurate solubility data. These may not be available for every type of model layer. From the nonsteady state measurements, however, the diffusion coefficient can be obtained directly [1, 2]. Therefore, in this study the nonsteady state method was applied to determine the DO_2. The model layers that were used in the experiments were plain water layers, either confined to Millipore® filters or applied on top of a porous PTFE membrane.

MATERIALS

Two types of Millipore® filters (Millipore Corp.), composed of cellulose acetate & nitrate esters, were used. Type GS (pore size 0.22 μm, thickness 150 μm), and type RA (pore size 1.2 μm, thickness 165 μm). Layer thickness was determined with a micrometer, and this did not change when the filters were soaked in water. The porous PTFE membrane was

Oxygen Transport to Tissue XVII, Edited by Ince et al.
Plenum Press, New York, 1996

327

supported by a polyester membrane and had hydrophobic properties. In this study type TE 36 (pore size 0.45 µm) from Schleicher & Schuel was used.

METHOD

For each experiment a model layer was placed in the diffusion chamber which was based on the ones described previously [3, 4], dividing the inner chamber into a top and a bottom part. Throughout the experiment oxygen partial pressures (pO_2) in the top and bottom chamber were recorded by polarographic oxygen electrodes [5]. At first, the gas phases above and below the layer were flushed with identical gas mixtures until the pO_2 reached a constant level and a steady state could be assumed in the layer. Then, at time t = 0, the oxygen pressure in the bottom chamber was changed instantly. After a delay time (t_d) this resulted in a gradual change in pO_2 in the top chamber. This change was compared with the resulting concentration changes as predicted by the mathematical analysis of an instant concentration change at one side of a flat layer (see *Appendix*). Combination of the experimental model and the mathematical analysis led to the calculation of the oxygen diffusion coefficient as:

$$DO_2 = 0.15 \, L^2 \, / \, t_d$$

To obtain the delay time for just the water layer t_d was calculated as $t_1 - t_2$, where t_1 was the delay time for the supporting layer plus water, and t_2 was the delay time for the supporting layer alone. This way the delay time was corrected for possible diffusion distances in the chamber or diffusion resistances of the supporting layers. The delay times t_1 and t_2 were derived from the recordings of pO_2 during the time course of the experiment. They were obtained by taking the intercept of the initially linear increase of pO_2 with the constant pO_2 level at the beginning of the experiment.

RESULTS

Figure 1 shows the recording of a typical experiment with a Millipore® filter type GS, thickness 150 µm. Delay time t_1= 10.3 sec and t_2= 2.5 sec. From the resulting t_d= 7.8 sec we calculated DO_2= 0.432 10^{-9} m^2/sec.

For type RA with thickness 165 µm Figure 2 shows that t_1= 8.5 sec and t_2= 2.25 sec, resulting in t_d= 6.25 sec and thus DO_2= 0.653 10^{-9} m^2/sec. Both experiments were carried out at 20°C.

The diffusion measurements with plain water layers on top of the TE 36 membrane were performed at 25°C, as were the nonsteady state measurements of Goldstick and Fatt. The values for t_1 and t_2 were 51 sec and 6 sec, respectively (Figure 3). This corresponds with t_d= 45 sec and therefore DO_2= 2.13 10^{-9} m^2/sec.

DISCUSSION AND CONCLUSIONS

The value for the DO_2 in the 800 µm water layers at 25°C compared favourably with the one reported by Goldstick and Fatt at this temperature (2.13 10^{-9} m^2/sec), thus validating the nonsteady state measuring method. The DO_2 (20°C) for the water-soaked Millipore® filters, however, were 0.432 10^{-9} and 0.653 10^{-9} m^2/sec for type GS and type RA, respectively. This was 77% and 65% lower than the reported value of 1.86 10^{-9} m^2/sec at this temperature. Part of the explanation for the values in the water-soaked Millipore® filters could be found

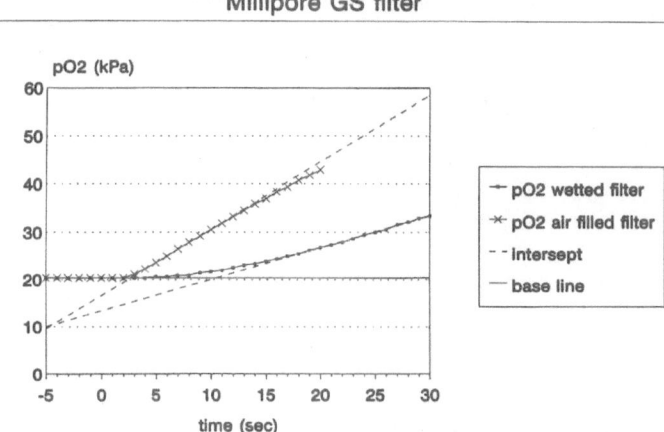

Figure 1. pO_2 as a function of time after concentration change for wetted and air filled GS filter.

in the structure of the filters. Because of the porosity of the filter the diffusion path for the oxygen would be longer than the given thickness, thus increasing L. It was estimated [6] that the lengthened pathway (L^*) could be up to 157% of L, which would explain a decrease in DO_2 between 36% and 59%. Since this was not the complete explanation for the drastic decrease we found there must be other properties of the Millipore® filter which interfere with the oxygen diffusion. Possibly the filter material itself may retain some oxygen, thereby lengthening t_d and thus decreasing DO_2.

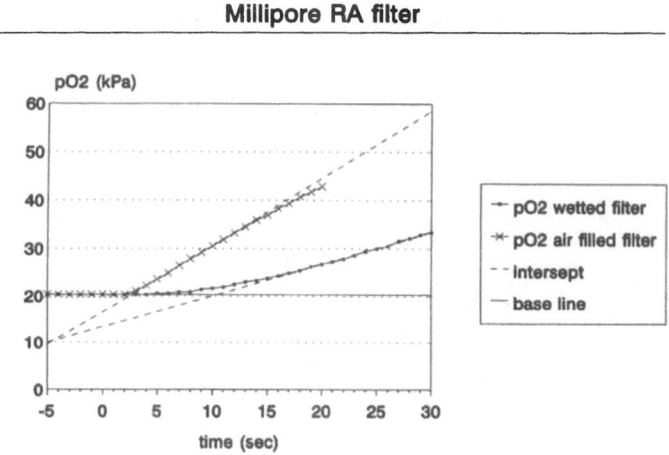

Figure 2. pO_2 as a function of time after concentration change for wetted and air filled RA filter.

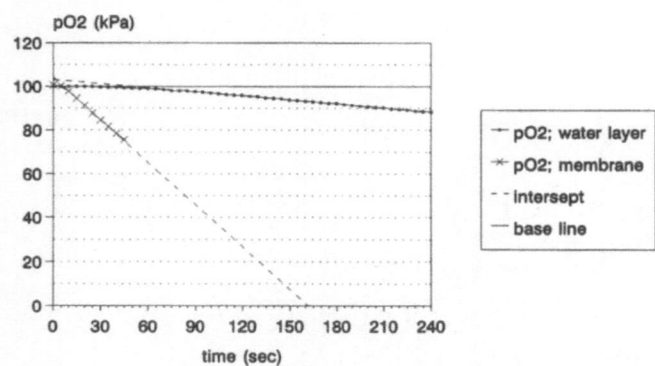

Figure 3. pO_2 as a function of time after concentration change for PTFE membrane with or without water layer.

APPENDIX

When a concentration of a species is suddenly changed from c_0 to c_1 at the border of a semi-semi-infinite flat layer [$z : 0 \rightarrow \infty$] the resulting concentration c is classically described as:

$$c = c_0 + (c_1 - c_0) \operatorname{erfc}\left(\frac{z}{2\sqrt{Dt}}\right)$$

where D is the species' diffusion coefficient and t is time. The *complementary error function* erfc is an integrated gaussian function with limiting values 1 for the argument being zero

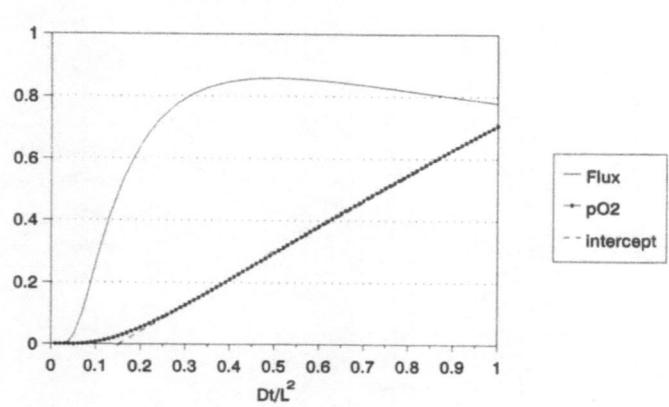

Figure 4. pO_2 as a function of $Dt/L2$.

and 0 when the argument approaches infinity. The species of interest here is oxygen. For a finite layer, with thickness L, the solution has to be expanded to:

$$c = c_0 + (c_1 - c_0) \operatorname{erfc}\left(\frac{z}{2\sqrt{Dt}}\right)$$

$$+ (c_1 - c_0) \sum_{k=1}^{\infty} [\operatorname{erfc}\left(\frac{2kL + z}{2\sqrt{Dt}}\right) - \operatorname{erfc}\left(\frac{2kL - z}{2\sqrt{Dt}}\right)]$$

From this formula, the flux at $z=L$ can be derived, out of the layer:

$$J_L = -D\frac{\partial c}{\partial z}\Big|_{z=L} = D(c_1 - c_0)2\sqrt{\frac{\pi}{Dt}}\, e^{-L^2/4Dt} + \dots$$

where J_L is this flux, D is the oxygen diffusion coefficient of the layer and the dots represent the higher-order terms. As a consequence, the oxygen pressure p in the adjacent chamber will change:

$$\frac{dpV}{dt} = J_L A R T$$

where V is chamber volume, A is layer surface area, R is the gas constant and T is the absolute temperature. From integration of dpV/dt against time the time course of p can be solved.

Both J_L and p as functions of the dimensionless time variable Dt/L^2 are shown in Figure 4 (arbitrary ordinate units) for not-too-long times ($Dt/L^2 < 1.0$).

It is easily seen that the time course for p asymptotically resembles a straight line starting at offset $Dt/L^2 = 0.15$, which was also found by Kreuzer (1953), so that analysis of the measurements can be done in this simplified way.

REFERENCES

1. Goldstick, T.K. & Fatt, I. (1970). *Chem. Eng. Prog. Symp. Ser.*, **66**, 101-113.
2. Kreuzer, F. (1953). *Habilitationsschrift*, Fribourg University
3. Hoofd, L. & Lamboo, A. (1985). *in*: "Oxygen Transport to Tissue - VII," Plenum Press, N.Y., *Adv. Exp. Med. Biol.*, **191**, 565-570
4. de Koning, J., Hoofd, L.J.C. & Kreuzer, F. (1981). *Pflügers Arch.*, **389**, 211-217.
5. Kimmich, H.P., Kreuzer, F. (1969). *"Oxygen Pressure Recording in Gases, Fluids, and Tissues"*, *Progr. Resp. Res.*, **Vol. 3**, 22-41.
6. Kreuzer, F., Hoofd, L.J.C. (1972). *Respir. Physiol.*, **15**, 104-124

43

REALISTIC MODELLING OF CAPILLARY SPACING IN DOG GRACILIS MUSCLE GREATLY INFLUENCES THE HETEROGENEITY OF CALCULATED TISSUE OXYGEN PRESSURES

Louis Hoofd and Zdenek Turek

Department of Physiology
University of Nijmegen
P.O. Box 9101, 6500 HB Nijmegen
The Netherlands

INTRODUCTION

Model calculations of tissue oxygen partial pressure (pO_2) in rat heart have shown that the heterogeneity of capillary spacing is a very important factor. It causes quite a scatter in the pO_2, resulting in broad histograms (Turek et al., 1991; Hoofd, 1992; Hoofd and Turek, 1992). Such histograms conform quite well with oxygen electrode measurements (Turek et al., 1991).

On the other hand, it was found that myoglobin (Mb) saturation in tetanic dog gracilis muscle was overall low, quite below 100% (Honig et al., 1984; Gayeski and Honig, 1986). The pO_2 calculated from these saturations was always low, leading to narrow histograms at low pO_2. These findings seemed supported by mathematical modelling (Federspiel, 1986; Gayeski and Honig, 1986; Groebe, 1990). However, in such modelling the heterogeneity of capillary spacing was disregarded.

Here, we present calculations for dog gracilis muscle incorporating a heterogeneous capillary spacing. Resulting histograms again are broad, in disagreement with the pO_2 inferred from Mb measurements. As found earlier in the rat heart situation (Hoofd and Turek, 1992; Hoofd and Turek, 1994b), there was only little influence of other factors, in particular blood flow distribution, on the shape of these histograms.

Oxygen Transport to Tissue XVII, Edited by Ince et al.
Plenum Press, New York, 1996

MATERIALS AND METHODS

System Layout

Two different capillary configurations were chosen, both with capillaries running parallel to four sides of a block of tissue. The first was according to Groebe (1990), 7 capillary segments either starting from arterioles or draining into venules. Block length was 500 μm where the dimensions of the plane perpendicular to the capillaries (the cross-section) was a square with sides of 80 μm. It is represented in the left panel of figure 1. The second was a cross section according to (a part of) figure 4 of Honig et al. (1984), with dimensions 250 μm × 145 μm. It is shown in the right panel of figure 1. Block length was chosen such that capillaries either started at an arteriolar end or ended into a venular end.

The first layout is an example of literature modelling without taking into account heterogeneity of capillary spacing and will be denoted by REG (regular). The second layout is an example of true capillarity and will be denoted by HET (heterogeneous capillary locations).

Theoretical Treatment

Tissue pO_2 was calculated according to the Multicapillary model (Hoofd and Turek, 1992; Hoofd, 1992). It incorporates facilitation of O_2 diffusion by Mb. With this theoretical treatment, tissue pO_2 in each cross-sectional plane is obtained from the equation:

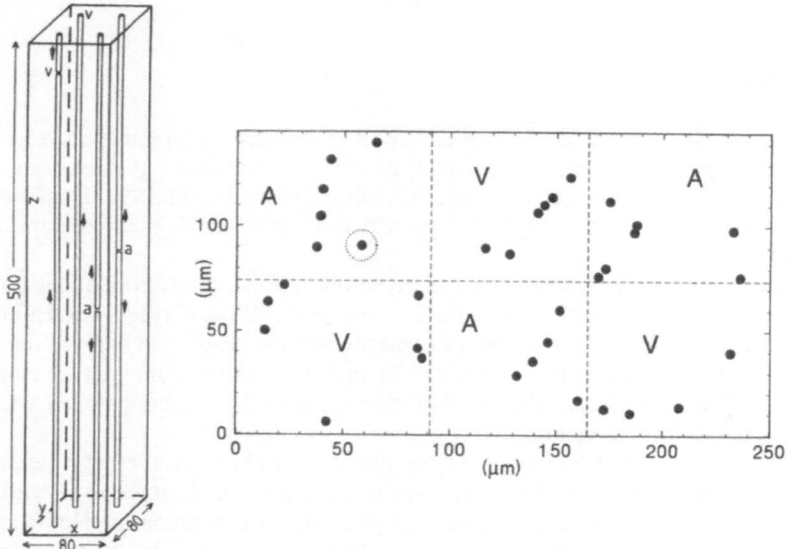

Figure 1. Layout of capillary course as taken from Groebe (1990) (left panel) and heterogeneous capillary locations from Honig et al. (1984) (right panel). For explanation see text. In the left panel, arrows indicate flow direction. In the right panel, around one of the capillaries (black dots) the 5 μm-zone is indicated which is left out from the pO_2 histogram calculations.

$$p + p_F s = \frac{\dot{Q}}{4\wp} [\Phi(\vec{r}) - \sum_{i=1}^{N} \frac{A_i}{\pi} \ln \left(\frac{|\vec{r} - \vec{r_i}|^2}{r_{ci}^2} \right)]$$

(1)

where p is oxygen partial pressure, p_F is facilitation pressure (Hoofd, 1992), s is myoglobin (Mb) oxygen saturation, $\dot{Q}$ is tissue oxygen consumption, $\wp$ is tissue oxygen permeability (Krogh's diffusion coefficient), $\Phi(\ )$ is the so-called background function, of coordinates $\vec{r} =$ (x,y) in the plane, N is the number of capillaries and A_i, $\vec{r_i}$ and r_{ci} are oxygen supply area, location and radius of the ith capillary respectively. The background function is that for a rectangular field (Hoofd, 1994a). $\Phi(\vec{0})$ and A_i are calculated from the boundary conditions, stating that the total O_2 supply goes into the total area is A and that for each capillary:

$$p_{ti} = p_{ci} - EP \frac{A_i N}{A}$$

(2)

where p_{ti} and p_{ci} are tissue and erythrocyte pO_2 for the ith capillary respectively and EP is the (mean) Extraction Pressure (Hoofd, 1992). Each capillary supplies oxygen to the plane; for a tissue slab of thickness Δz this means that, for the next plane (Hoofd, 1994b):

$$c_{i,z+\Delta z} = c_{i,z} - \frac{\dot{Q}}{F_i} (A_i - \pi r_{ci}^2)\Delta z$$

(3)

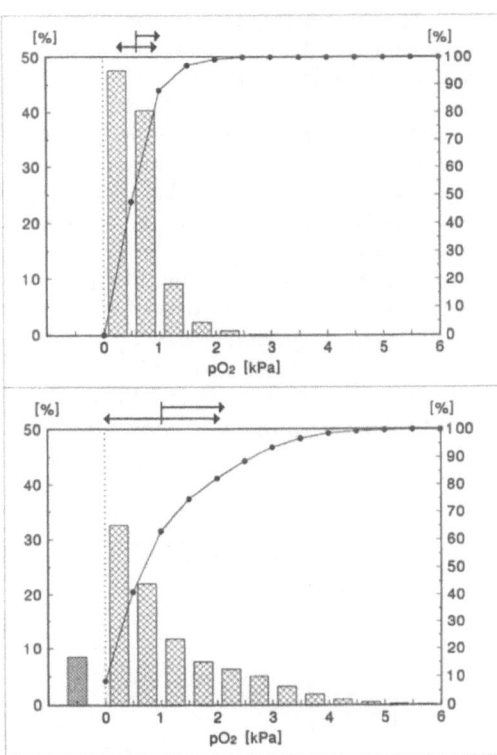

Figure 2. pO_2 histogram (bars; left axis), in classes of 0.5 kPa (3.75 mm Hg), and cumulative histogram (symbols connected by lines; right axis). Top panel is for the REG case, bottom panel for the HET case. Indicated at the top of each panel are mean (vertical line), standard deviation (double arrow) and skewness (single arrow) of the pO_2 distribution.

Table 1. Data values common for both layouts

Hb			Capillary		O_2		Mb	
c_{Hm}/α kPa	p_{50Hb} kPa	n	p_a kPa	r_c μm	$\dot{Q}/\wp$ kPa·μm^{-2}	$F\alpha/\wp$ μm	p_F kPa	p_{50Mb} kPa
1336	3.52	2.65	8.07	2.3	0.01212	28.65	5.28	0.707

where c is total oxygen content in the capillary (solved plus bound to hemoglobin (Hb)), z is the vertical coordinate of the plane and F_i is the blood flow in the ith capillary. Note, that F_i can be negative, for countercurrent flow; then, the capillary looses oxygen in the reverse direction.

Saturation of the Hb in the blood was calculated according to the Hill equation, with half-saturation pressure p_{50Hb} and Hill coefficient n, which does very well in the saturation range considered (O'Riordan et al., 1985). Saturation of the Mb in the tissue was calculated from a hyperbolic function with half-saturation pressure p_{50Mb}.

Input Data Values

The data for the REG case were obtained from Groebe (1990), combined with a capillary blood velocity of 2 mm/sec resulting in a hematocrit of 46% which seems to conform to his figure 3. The venular ends where at (x,y,z) = (20,20,460) and (20,60,500); the arteriolar ends at (60,20,220) and (60,60,260). EP was estimated as 1.4 kPa (10.5 mm Hg) according to Bos et al. (1994). Capillary length was 912 μm, in order to reach an end-capillary Hb saturation equal to the venular saturation of 42%.

The same basic data were used for the HET case. Vertical block size was equal to the capillary length of 912 μm for non-staggered capillaries or half this value, 456 μm, for two-zone staggering. Staggering was such that capillaries either started with arteriolar pO_2 (zones A in figure 1) or with a pO_2 according to the pO_2 distribution that the other capillaries reached half-way (zones V in figure 1). The standard case, of figure 2 below, is staggered, with a V-zone initial pO_2 distribution of 5.30±0.38(SD) kPa (39.8±2.8 mm Hg). Because the capillary density (CD) was higher in this case, EP was 0.9 kPa here (6.8 mm Hg).

The data values used in the calculations are in tables 1 and 2; some data are only needed in combined form, such as c_{Hm}/α where c_{Hm} is the blood oxygen binding capacity and α is the blood oxygen solubility; p_a is arteriolar pO_2 and F is the mean absolute value of the blood flow; in the standard case, blood flow is equal for each capillary. Heterogeneity of capillary spacing (Het) - table 2 - was determined as logarithmic Standard Deviation of the domain areas (Hoofd et al., 1985). The number of layers (#layers) was chosen so that the resulting Δz of 4.56 μm was close to the spacing of the grid where the pO_2 for the histogram was calculated, resulting in a cubic grid. Grid points closer than 5 μm to a capillary were

Table 2. Data values different for both layouts

	Dimensions			Capillaries			Histogram	
	x	y	z	CD				Border
Case	μm	μm	μm	μm^{-2}	Het	#layers		μm
REG	80	80	500	625	0	110		0
HET	250	145	456	966	0.16	100		10
			912			200		

omitted from the histogram since Mb saturation measurements also were this distance away. Also, in the HET case a border zone of 10 μm was omitted from the histogram calculations (Hoofd and Turek, 1994a).

Results

In figure 2, the resulting pO_2 histograms are shown. The top panel is for REG, the bottom panel is for HET (with staggering). Obviously, the histograms are quite different for both cases. This can also be seen from the histogram characteristics, indicated on top of each panel. Mean pO_2 is indicated by a vertical bar, standard deviation as a double arrow, and skewness as a single arrow - here, to the right. These values are 0.61, 0.36 and 0.43 kPa for the REG case and 1.02, 1.03 and 1.12 kPa for the HET case, respectively (1 kPa = 7.5 mm Hg). For the REG case, no pO_2 values below zero were calculated whereas this amounted to 8.4% for the HET case. Calculated pO_2 below zero indicates the presence of anoxic areas, but the model does not handle these areas correctly since it still assumes that oxygen is consumed there. Instead, the oxygen is flowing to other tissue parts in oxygen debt. So, only part of the column represents truly anoxic tissue and the rest must be added to the low-pO_2 column(s).

The effect of heterogeneity of capillary spacing on the oxygen histogram is so predominant, that it masks other effects. This was found earlier for rat heart (Hoofd and Turek, 1992; Hoofd and Turek, 1994b) and confirmed here for the dog gracilis muscle, as can be seen from figure 3. In this figure, pO_2 histograms for different configurations of the HET case are compared. Case A is without staggering, all capillaries starting from an arteriolar pO_2 at the same level (p_a at z=0). Although this seems an unlikely configuration, the resulting histogram hardly differs from the standard HET case. Case C is staggered again, but the capillaries in the middle two zones (90 μm < x < 165 μm) had countercurrent flow

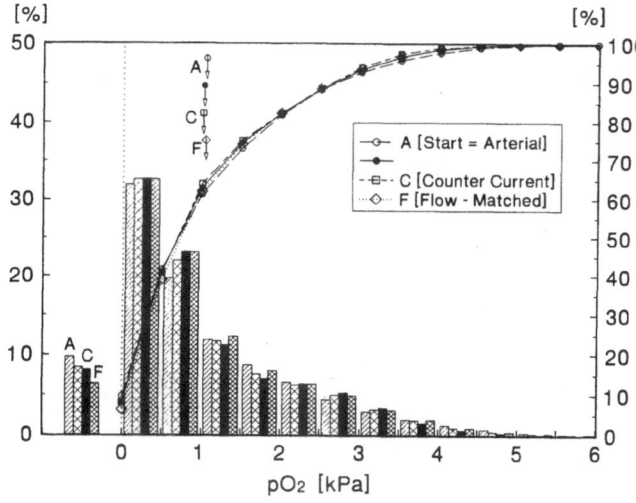

Figure 3. pO_2 histograms (bars; left axis) and cumulative histograms (symbols connected by lines; right axis) for heterogeneous capillarization as in figure 2 (no letter), and for modified conditions (indicated by a letter). A, parallel capillaries all starting with arteriolar pO_2; C, countercurrent flow in the middle two sections of figure 1, right panel; F, flow in each capillary matched to O_2 supply. Arrows indicate mean pO_2.

($F_i < 0$). For case F, the blood flow F_i in each capillary was chosen individually so that it was matched to the capillary oxygen supply, leading to identical end-capillary pO_2s within each zone type (A or V of figure 1). The latter two cases could be expected to lead to a better tissue oxygenation but the histograms again were hardly different from the former two cases. This can be judged also from table 3, where the histogram characteristics are compared. In this table, also the number of locations (#locations) is indicated where pO_2 is calculated, as well as the percentage of tissue where calculated pO_2 either was subzero (< 0) or "high" (> 1.5 kPa = 11.3 mm Hg). Also from these data, it can be seen that the REG case is much narrower than the HET cases and that the latter are hardly different from each other. In this table 3, also some other results are shown, to be discussed below.

DISCUSSION

It was the aim of this paper to show, that for a realistic modelling of tissue oxygenation the heterogeneity of capillary spacing should be taken into account, also for skeletal (dog gracilis) muscle. The two histograms of figure 2 are very different indeed and cannot be brought to conformity by adjusting blood flow patterns.

One might raise arguments against the data used for the models, but that is not relevant for the conclusion drawn above. For instance, CD of the HET case was larger than for the REG case, resulting in higher end-capillary pO_2s (4.14 ± 0.35 kPa) than the venular pO_2 of the REG case (3.12 kPa). Adapting by, e.g., lowering the blood flow left-shifts the whole histogram resulting in a large portion of tissue with subzero pO_2, but the histogram remains quite heterogeneous, even more different from the REG case.

The histograms of figure 2 are almost identical to those of Hoofd and Turek (1994a), where only a 2-μm zone around the capillaries was left out. Further exclusion is not to be expected to have significant influence on the histograms. When analyzing the calculated pO_2s in the tissue, it is found indeed that there are steep gradients around the capillaries and shallow pO_2 levels away from it (Federspiel, 1986; Groebe, 1990; Hoofd, 1992). However, with heterogeneous capillary spacing, these shallow levels are different for each cell. There

Table 3. Comparison of calculated and literature results - see text

Case	HET			REG	KROGH		From Mb	
	C	F	A		22.6	18.2	Tbl1	
#locations	116251		231351	28527	6432	3015	—	
pO_2 [kPa]								
% <0	8.4	8.2	6.4	9.7	0	0	0	
% >1.5	25.3	24.7	25.6	26.8	3.0	9.7	80.1	
Minimum	—	—	—	—	0.08	0.00	1.11	
Mean	1.02	1.01	1.04	1.05	0.61	0.62	2.33	0.87	0.48
SD	1.03	1.01	1.02	1.08	0.36	0.56	0.89	0.37	0.34
Skewness	1.12	1.08	1.11	1.19	0.43	0.63	0.79		

are cells where the level is just above zero, but also subzero and quite above zero. That makes the resulting histograms broad.

For comparison, also calculations were done for a Krogh cylinder layout (one single capillary surrounded by a cylinder of tissue), incorporating EP and facilitation by Mb (Hoofd, 1992). The CDs of 625 and 966 μm^{-2} correspond with cylinder radii of 22.6 and 18.2 μm, respectively. Resulting pO_2 histogram characteristics are represented in table 3 in the columns indicated with these radii. The 22.6 results are quite close to the REG case, the 18.2 results quite different from the HET case. This is not surprising because the Krogh cylinder is utmostly regular. Indeed, it reinforces the above conclusion, that the heterogeneity is a major determinant for tissue pO_2 calculations. Homogeneous capillarity leads to much narrower histograms. Note, that the SD of the 18.2 μm Krogh case is somewhat broader because Mb no longer facilitates oxygen diffusion - overall high pO_2 leads to overall saturated Mb.

Also for comparison, the data from table 1 and table 3 of Gayeski and Honig (1986) were used to calculate mean and standard deviation of Mb saturation and, from these, mean and standard deviation in corresponding tissue pO_2. These data are represented in the rightmost columns of table 3 above. Note, that these data conform to the characteristics of the REG histogram quite well, suggesting agreement between theory and measurements as mentioned in the introduction.

CONCLUSIONS

From the extensive treatment presented here, concerning the overall low pO_2s derived from Mb measurements it must be concluded that these are not in agreement with theoretical calculations. The seemingly concordant results of the REG case - see table 3 - are not for a realistic situation. In sorting out the disagreement, two different arguments can be discerned:

- theory is not correct, in particular, oxygen is not distributed according to diffusion (free and bound to Mb). This would raise fundamental questions about oxygenation of tissue;
- myoglobin saturation does not reflect tissue pO_2, i.e., there is no equilibrium between s and p in hard-working muscle tissue.

In this context, it is interesting to note, that it was reported that myoglobin plays a role in the oxidative phosphorylation process, indicated as Myoglobin-Mediated Oxidative Phosphorylation (Wittenberg and Wittenberg, 1987; Doeller and Wittenberg, 1991; White and Wittenberg, 1993). It is quite conceivable, that such a process would lead to Mb desaturation in disequilibrium with the ambient pO_2. Even small desaturations can have a large effect on the calculated equilibrium pO_2. For example, for 95% saturation this equilibrium pO_2 is 13.4 kPa (100 mm Hg) but it is more than halved (6.4 kPa - 48 mm Hg) for 90% and as low as 2.8 kPa (21 mm Hg) for 80%. This makes determination of pO_2 values from disequilibrium Mb saturations completely unreliable and offers a likely explanation for the discrepancy found here.

The two conclusions can be summarized as:

- data of tissue pO_2 derived from myoglobin saturation measurements do not agree with modelling of tissue oxygenation when incorporating realistic capillary spacing;
- the most likely explanation for this discrepancy is, that there is no equilibrium between local Mb saturation and local O_2.

REFERENCES

Bos, C., Hoofd, L., and Oostendorp T., 1994, Mathematical model of erythrocytes as point-like sources, *Math. Biosci.*, in press.

Doeller, J.E., and Wittenberg, B.A., 1991, Myoglobin function and energy metabolism of isolated cardiac myocytes: effect of sodium nitrite, *Am. J. Physiol.* 261: H53-H62.

Federspiel, W.J., 1986, A model study of intracellular oxygen gradients in a myoglobin-containing skeletal muscle fiber, *Biophys. J.* 49: 857-868.

Gayeski, T.E.J., and Honig, C.R., 1986, O_2 gradients from sarcolemma to cell interior in red muscle at maximal VO_2, *Am. J. Physiol.* 251: H789-H799.

Groebe, K., 1990, A versatile model of steady state O_2 supply to tissue; application to skeletal muscle, *Biophys. J.* 57: 485-498.

Honig, C.R., Gayeski, T.E.J., Federspiel, W., Clark, A., Jr., and Clark, P., 1984, Muscle O_2 gradients from hemoglobin to cytochrome: new concepts, new complexities, *In: Oxygen Transport to Tissue V, Adv. Exper. Med. Biol.* 169: 23-38.

Hoofd, L., 1992, Updating the Krogh model - assumptions and extensions. *In: Oxygen Transport in Biological Systems. Modelling of pathways from environment to cell, Soc. Exper. Biol. Seminar Ser.* 51, eds. S. Egginton, and H.F. Ross, Cambridge University Press, Cambridge, U.K.: 197-229.

Hoofd, L., 1994a, Calculation of oxygen pressures in tissue with anisotropic capillary orientation. I: Two-dimensional analytical solution for arbitrary capillary characteristics, *Math. Biosci.*, in press.

Hoofd, L., 1994b, Calculation of oxygen pressures in tissue with anisotropic capillary orientation. II: Coupling of two-dimensional planes, *Math. Biosci.*, in press.

Hoofd, L., and Turek, Z., 1992, Oxygen pressure histograms calculated in a block of rat heart tissue. *In: Oxygen Transport to Tissue XIV, Adv. Exper. Med. Biol.* 317: 561-566.

Hoofd, L., and Turek, Z., 1994a, Effect of realistic capillary spacing on pO_2 calculations in dog gracilis muscle, *Med. Biol. Eng. Comput.* 32: 436-439.

Hoofd, L., and Turek, Z., 1994b, The influence of flow redistributions on the calculated pO_2 in rat heart tissue, *In: Oxygen Transport to Tissue XV, Adv. Exper. Med. Biol.* 345: 275-282.

Hoofd, L., Turek, Z., Kubat, K., Ringnalda, B.E.M., and Kazda, S., 1985, Variability of intercapillary distance estimated on histological sections of rat heart, *In: Oxygen Transport to Tissue VII, Adv. Exper. Med. Biol.* 191: 239-247.

O'Riordan, J.F., Goldstick, T.K., Vida, L.N., Honig, C.R., and Ernest, J.T., 1985, Modelling whole blood oxygen equilibrium: comparison of nine different models fitted to normal human data, *In: Oxygen Transport to Tissue VII, Adv. Exper. Med. Biol.* 191: 505-522.

Turek, Z., Rakusan, K., Olders, J., Hoofd, L. and Kreuzer, F., 1991, Computed myocardial Po_2 histograms: effects of various geometrical and functional conditions, *J. Appl. Physiol.* 70: 1845-1853.

White, R.L., and Wittenberg, B.A., 1993, NADH fluorescence of isolated ventricular myocytes: effects of pacing, myoglobin, and oxygen supply, *Biophys. J.* 65: 196-204.

Wittenberg, B.A., and Wittenberg, J.B., 1987, Myoglobin-mediated oxygen delivery to mitochondria of isolated cardiac myocytes, *Proc. Natl. Acad. Sci. USA* 84: 7503-7507.

44

OXYGEN TRANSPORT IN TUMORS

Characteristics and Clinical Implications

Peter Vaupel

Institute of Physiology and Pathophysiology
University of Mainz
D-55099 Mainz, Germany

INTRODUCTION

Experimental evidence suggests that the hypoxic fraction in solid tumors may influence its growth, may increase its malignant potential, and may reduce its sensitivity towards non-surgical treatment modalities (e.g., standard irradiation, certain anticancer drugs). The role of the tumor O_2 status in radio-/chemotherapy and its impact on relevant tumor biological characteristics of tumors are summarized in Table 1.

The clinical importance of tumor hypoxia remains uncertain since valid methods for the routine measurement of intratumoral O_2 tensions in patients have so far been lacking (for reviews see [1 - 7]). During the last five years a clinically applicable standardized procedure has been established which enables the determination of intratumoral O_2 tensions in primary tumors and metastatic lesions of patients by use of a computerized polarographic needle electrode system (pO_2 histography, Eppendorf, Hamburg, Germany). It is the purpose of this review to summarize presently available pO_2 data of human tumors obtained with this reliable system which is well-tolerated by patients and has no effect on metastatic potential and proliferation characteristics [8].

OXYGENATION STATUS OF PRIMARY HUMAN TUMORS

Extensive studies on the tissue oxygenation of *breast cancers* have been performed by several groups [9 - 11]. As a result of a compromised and anisotropic microcirculation (see Table 2), many breast cancers reveal hypoxic tissue areas which are heterogeneously distributed within the tumor mass. Mean and median O_2 tensions (pO_2) obtained from different pathological stages and histological grades are, on average, distinctly lower than in the normal tissue [9, 11]. Oxygen tensions measured in the normal breast revealed a mean (and median) pO_2 of 65 mmHg, whereas in cancers of the breast of stages pT1-4, the median pO_2 was 28 mmHg (Fig. 1). Thus far, one third of the breast cancers investigated exhibited pO_2 values between 0 and 2.5 mmHg, i.e., tissue areas with less than half maximum

Oxygen Transport to Tissue XVII, Edited by Ince et al.
Plenum Press, New York, 1996

341

Table 1. Role of tumor O_2 status in radio-/chemotherapy and impact on tumor biological characteristics [12-22]

Tissue oxygenation can alter

    efficacy of standard radiotherapy (direct effect)

    cell cycle position (G_0 cells!), cell proliferation

    pharmacodynamics of anticancer agents

    sensitivity of thermo-radiotherapy

    ability to repair sublethal and potentially lethal damage

    invasive and metastatic potential

    expression of genes responsible for drug resistance

    pH (glycolytic rate, ATP hydrolysis, glutaminolysis)

Tissue oxygenation is an

    independent predictor of radiation response [23]

---

radiosensitivity [9]. In contrast, in the normal breast pO_2 values ≤ 12.5 mmHg could not be detected [9]. In all systematic studies on breast cancers, bimodal pO_2 distribution curves have been obtained [9, 11], either indicating the co-existence of normoxic and hypoxic tumor areas or a relevant contribution of pO_2 readings in the stromal compartment of breast cancers [9].

When pooled data for stages pT1 & 2, and pT3 & 4 breast cancers are compared, there is no evidence of statistically significant differences between the two groups (median pO_2 in pT1 & 2 tumors: 26 mmHg; pT3 & 4 tumors: 30 mmHg, Fig. 1). This implies that the oxygenation in breast cancers and the occurrence of hypoxia and/or anoxia does not correlate with the clinical stage [9 - 11]. Similarly, there was no association between tumor size and blood flow [24, 25]. The proportion of pO_2 values between 0 and 2.5 mmHg was

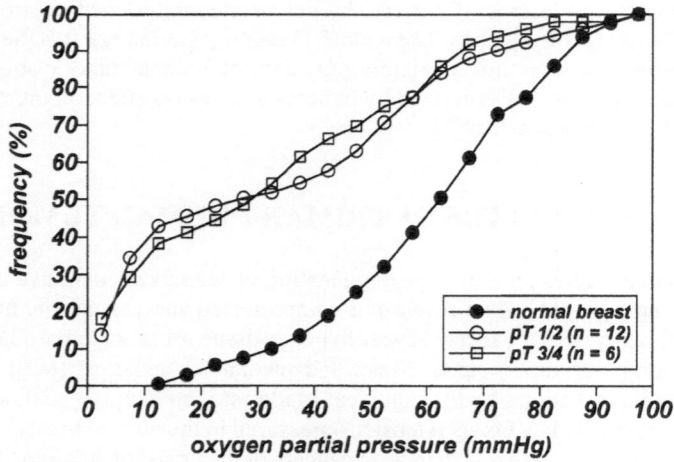

Figure 1. Cumulative frequency distributions of measured tissue pO_2 values in normal breast and in breast cancers of stages pT1 & 2 and pT3 & 4.

Table 2. Relevant structural and functional abnormalities of tumor neovasculature/microcirculation and their (possible) consequences (examples)

Pathological/pathophysiological condition(s)	Lead(s) to	Consequence(s)
Incomplete or missing endothelial lining interrupted or absent basement membrane blood channels lined by tumor cell cords	Increased vascular permeability RBC extravasation	Hemoconcentration increase in viscous resistance to flow perfusion-limited supply reduced Fahraeus-Lindqvist effect significant interstitial fluid flow interstitial hypertension
Lack of contractile wall elements lack of pharmacological/physiological receptors irregular occurrence of pericytes	Lack of responsiveness to pharmacological/physiological stimuli	Absence of vasomotion absence of flow regulation
Existence of arterio-venous shunt vessels	Arterio-venous shunt perfusion	Global flow > nutritive flow
Contour irregularities, tortuosity, blind endings, elongation of vessels, loss of hierarchy	Increase in geometric resistance to flow	Perfusion-limited supply & drainage
Lack of formation of lymphatic vessels	Interstitial fluid flow & hypertension	Impaired extravasation of immune cells, facilitation of intravasation of cancer cells
Expansion of intercapillary space decrease of vessel density	Increase of diffusion distances	Diffusion-limited supply & drainage (e.g., diffusion-limited hypoxia)
RBC sludging, leukocyte sticking, platelet aggregation, micro-/macrothrombosis	Ischemia, intermittent flow, regurgitation, unstable speed and direction of flow	Intermittent ischemia (intermittent hypoxia, re-perfusion injury ?)
Sinusoidal vessels arising from and ending in veins ("veno-venous rete")	Perfusion with hypoxemic, nutrient deprived blood	Tissue hypoxia, substrate depletion

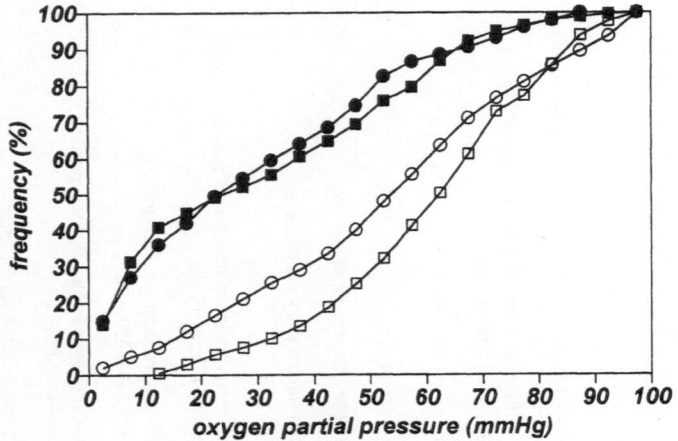

Figure 2. Comparison of cumulative frequency distributions of measured tissue pO$_2$ values in normal breast (open symbols) and breast cancers (filled symbols) obtained in two different institutions. Squares indicate data obtained by Vaupel et al. [9], and circles data obtained by Runkel et al. [11].

~6% in pT1 & 2, and ~7% in pT3 & 4 tumors. In addition, there is substantial evidence that the oxygenation patterns do not correlate with the histological grade [9 - 11], the menopausal status, the tumor histology (ductal vs. lobular), the extent of necrosis or fibrosis, and with a series of other clinically relevant parameters (e.g., hormone receptor status, hemoglobin level, smoking habits [9, 10]).

When comparing pO$_2$ data of breast cancers from different institutions, there is good agreement in the oxygenation status observed (Fig. 2). The tumor data are nearly identical whereas the pO$_2$ data for the normal breast exhibit some differences [9, 11]. Most probably, the lower pO$_2$ values in the study by Runkel et al. [11] are due to a higher number of postmenopausal breasts being included (94% vs. 67% in the study by Vaupel et al. [9]).

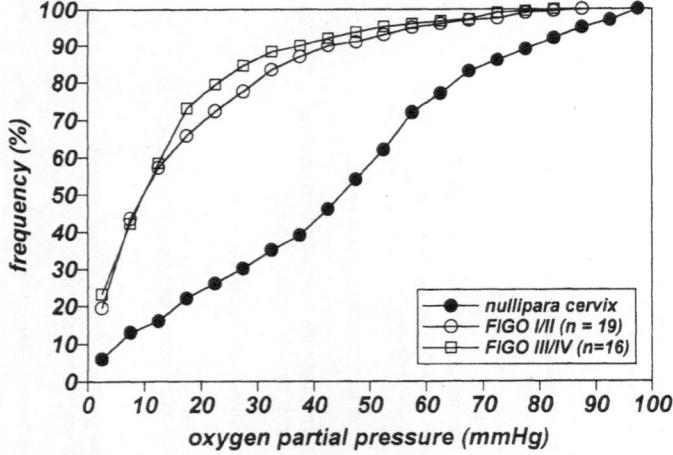

Figure 3. Cumulative frequency distributions of measured tissue pO$_2$ values in normal cervix of nullipara and in cervical cancers of stages FIGO I & II and FIGO III & IV.

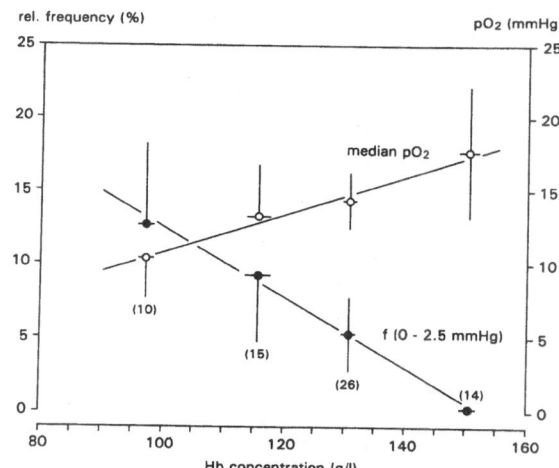

Figure 4. Tumor median pO_2 and fraction of pO_2 values ≤ 2.5 mmHg as a function of hemoglobin concentration in cervical cancer patients. Values are means ± SEM, with the number of tumors investigated given in brackets.

Analysis of O_2 tensions measured in the normal cervix of nulliparous women resulted in oxygenation patterns characteristic for normal, adequately supplied tissues (median pO_2: 48 mmHg, Fig. 3). As a rule, the mean (and median) pO_2 values were distinctly lower in the normal cervix of parous women (most probably due to tissue changes following vaginal delivery). Here, the median pO_2 was 13 mmHg (with ~14% of the pO_2 readings in the 0 - 2.5 mmHg class). In *cervical cancers* (stages FIGO I-IV), the median pO_2 is 11 mmHg (Fig. 3). To date, one third of the cervical cancers investigated exhibited pO_2 values between 0 and 2.5 mmHg. The relative number of pO_2 readings between 0 and 2.5 mmHg ranged from 1% (in a FIGO IV cancer) to 82% (in a FIGO III tumor)[26].

In FIGO I & II tumors, the median pO_2 is 11 mmHg, and in FIGO III & IV tumors it is 10 mmHg, with 6% of the readings, and 18%, respectively, in the lowest pO_2 class (0 - 2.5 mmHg). As was the case with the breast cancers, the oxygenation pattern in cervical

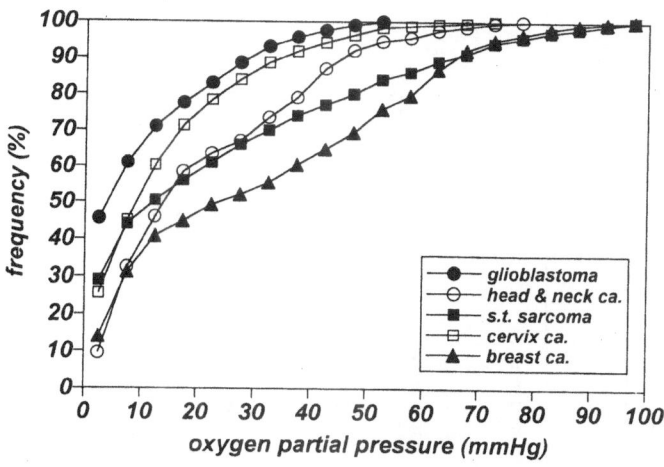

Figure 5. Compilation of cumulative frequency distributions of measured pO_2 values in different primary tumors. Data have been adapted from published values [9, 26, 31-33].

Table 3. Pretherapeutic oxygenation status of human malignancies (assessed by computerized pO2 histography). n = number of tumors investigated, f = fraction

Tumor	n	f (0 - 2.5 mmHg) [%]	f (0 - 5 mmHg) [%]	f (< 10 mmHg) [%]	Mean pO$_2$ [mmHg]	Median pO$_2$ [mmHg]	Size dependency	Reference
Breast cancer	18	5.9	15.4	32.3	32	28	Ø	[9]
	5	4.5		13.0		24	Ø	[10]
	18	7.5	15.5	26.0		23	Ø	[11]
Cervical cancer	37	11.1	25.9	45.6	18	10	Ø	[23]
	6	15.0	17.0	21.0	29	21		[27]
Rectal cancer	5		35.0	10.0		30		[29]
	5					25		[30]
Lung cancer	6	12.7		36.3		14		[10]
Soft tissue sarcoma	4	19.0	17.0	44.0	21	20	+ (hypoxic fr.)	[30]
	9		29.0		24	21		[32]
	18		10.0			23		[39]
Glioblastoma	10	38.0	45.5	61.0		7	Ø	[31]
Head & neck cancer	7	22.5	9.8	33.0	22	19	+	[33]
head & neck mets.	35		32.0	50.0	21	10		[35, 36]
	31		22.8			12		[39]
Melanoma (mets.)	13	16.0	28.5	49.5	25	10		[37, 38]

cancers and the occurrence of hypoxia and/or anoxia did not correlate with the above-mentioned clinically relevant parameters with the exception of the hemoglobin concentration.

From a recent evaluation of oxygenation data of 65 cervical cancers there is an indication that severely anemic patients ([Hb] ≤ 100 g/l) tend to have lower pO_2 values in cervical cancers than patients with normal Hb concentrations ([Hb] > 140 g/l; p = 0.05). The fraction of "hypoxic" pO_2 values (0 - 2.5 mmHg) is significantly higher in the anemic group than in patients with normal Hb concentrations (p = 0.004; Fig. 4). Blood transfusion in severely anemic patients ([Hb] ≤ 100 g/l) resulted in an increase in the median pO_2 values and, at the same time, a drastic reduction in the number of pO_2 readings in the 0 - 2.5 mmHg range occurred.

From our clinical studies on breast and cervical cancers there is clear indication that the oxygenation status of individual tumors before therapy cannot be predicted on the basis of tumor staging and/or grading. The lack of predictability is predominantly due to pronounced tumor-to-tumor variabilities even if tumors of the same pathological stage and histological grade are compared. Tumor-to-tumor variability in the oxygenation pattern is more pronounced than intratumor heterogeneity.

Oxygen tension measurements have also been performed in patients with untreated squamous cell carcinomas of the cervix by Lartigau et al. [27]. These authors observed slightly higher median pO_2 values of 21 mmHg, together with a higher proportion of pO_2 readings in the lowest class (0 - 2.5 mmHg). At the time of brachytherapy after external radiotherapy, the median pO_2 was ~15 mmHg in cervical cancers [28].

Relevant parameters of the pretherapeutic oxygenation status of various human malignancies are listed in Table 3. In 6 out of 7 tumor entities, no significant correlation with tumor size was found. Only in soft tissue sarcomas was the hypoxic fraction found to increase with enlarging tumor volume [32].

In Fig. 5, all available oxygenation data obtained from primary tumors using the pO_2 histography system are plotted as cumulative pO_2 distribution curves [9, 26, 31-33]. From this compilation of pO_2 data there is evidence that different oxygenation patterns and different fractions of hypoxic tissue volumes have to be expected when considering different

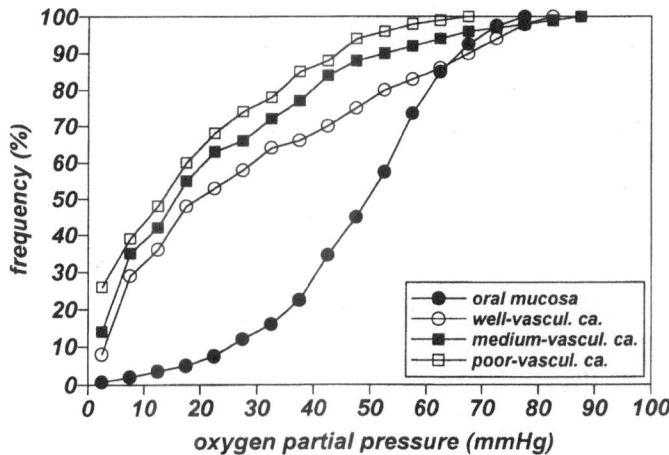

Figure 6. Cumulative frequency distributions of pO_2 values in the normal oral mucosa and in squamous cell carcinomas of the oral cavity as a function of vascular density. Data have been calculated from oxyhemoglobin saturation values of individual red blood cells in tumor microvessels measured in cryobiopsies under *ex vivo* conditions [34].

tumor entities. So far, breast cancers exhibit a significantly better oxygenation status than glioblastomas, with the other primary malignancies grouped between these tumors.

Vascularity of tumors seems to have a significant impact on the oxygenation status of many human malignancies. This statement is supported by the data presented in Fig. 6: In squamous cell carcinomas of the oral cavity, oxyhemoglobin saturation (HbO_2) of individual red blood cells in tumor microvessels has been measured in cryobiopsies under *ex vivo* conditions [34]. From these intravascular HbO_2 values, tissue oxygen tensions have been calculated and plotted together with the respective data of the normal oral mucosa. There is a clear correlation between the oxygenation status and vascular density. Well-vascularized tumors showed a better oxygenation status and less hypoxia than poorly vascularized lesions. Tumors with medium-quality vascularization exhibited an oxygenation status between these two extreme patterns.

OXYGENATION STATUS OF METASTATIC LESIONS

As was the case with the primary tumors, the oxygenation of metastatic lesions is generally anisotropic and compromised as compared to normal tissues at the site of metastatic growth. The median pO_2 values of the secondary tumors are distinctly lower than those recorded in the tumor surroundings (Fig. 7). This holds true for both metastatic lesions of squamous cell carcinomas [35, 36] and metastatic melanomas [37].

When comparing the oxygenation status of primary head and neck carcinomas [33] with metastatic lesions of this entity [35, 36], the fraction of pO_2 readings ≤ 5 mmHg is distinctly higher in the latter group (Fig. 8). Whether this pattern is a characteristic of metastatic lymph nodes of the head and neck region or a general biological phenomenon has to be elucidated in ongoing studies.

In contrast to primary tumors investigated so far, the oxygenation status of metastatic lesions of the head and neck appears to be linked to tumor size with the lower median pO_2 values preferentially occurring in larger nodal sizes [36].

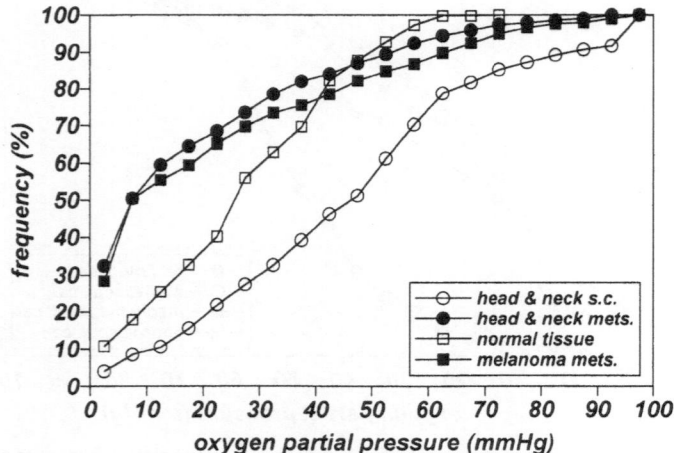

Figure 7. Cumulative frequency distributions of pO_2 values in metastatic lesions of squamous cell carcinomas of the head and neck region and of metastatic melanomas together with values obtained in the normal tissue adjacent to the lesions. Data have been adapted from published values [35 - 37].

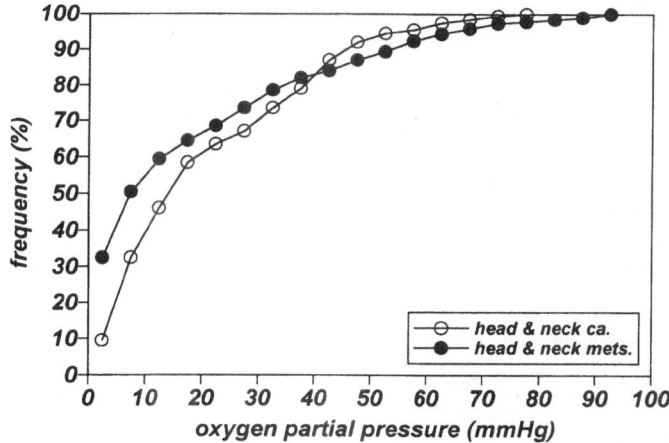

Figure 8. Cumulative frequency distributions of pO_2 values in primary head and neck tumors [33] compared to metastatic lesions of this tumor entity [35, 36].

CONCLUSIONS

Using the pO_2 histography system for assessment of tumor oxygenation status, several investigations have clearly shown that:

a. tumor oxygenation, as a rule, is anisotropic and compromised as compared to normal tissues,

b. on average, the median pO_2 values in primary and metastasic tumors are lower than in the normal tissue at the site of growth,

c. many solid tumors (~30-35%) contain hypoxic tissue areas ($pO_2 \leq 2.5$ mmHg),

d. tumor-to-tumor variability in oxygenation is significantly greater than intratumor variability,

e. tumor oxygenation is unpredictable considering staging and grading,

f. tumor oxygenation is independent of other known oncologic parameters,

g. tumor oxygenation status is dependent on the vascularity,

h. head and neck metastases of squamous cell carcinomas seem to have a higher hypoxic fraction than the primary tumors, and

i. tumor oxygenation appears to be a new, independent prognostic factor influencing overall survival and local control upon radiotherapy (± chemotherapy) in locally advanced cervical cancer.

As an optimistic outlook, tumor oxygenation status evaluated by computerized pO_2 histography could become an important new oncologic parameter for locally advanced cervical cancers (and probably also for other solid tumors). Determination of the oxygenation status may enable pretherapeutic selection of hypoxic tumors as candidates for modified treatment approaches. Considering tumor-to-tumor variability, the consequences for future clinical studies are as follows: Do not randomize, do individualize!

REFERENCES

1. P. Vaupel, F. Kallinowski & P. Okunieff. Blood flow, oxygen and nutrient supply, and metabolic microenvironment of human tumors: a review. Cancer Res. *49*, 6449 - 6465 (1989).
2. P. Vaupel. Oxygenation of human tumors. Strahlenther. Onkol. *166*, 377 - 386 (1990).
3. P. Vaupel, K. Schlenger & M. Höckel. Blood flow and tissue oxygenation of human tumors. Funktionsanal. Biol. Syst. *20*, 165 - 185 (1991).
4. P. Vaupel, K. Schlenger & M. Höckel . Blood flow and tissue oxygenation of human tumors: an update. Adv. Exp. Med. Biol. *317*, 139 - 151 (1992).
5. P. Vaupel. Physiological properties of malignant tumours. NMR Biomed. *5*, 220 - 225 (1992).
6. P.W. Vaupel. Oxygenation of solid tumors. In: Drug Resistance in Oncology. Ed. Teicher, B.A.. Marcel Dekker, New York, pp. 53 - 85 (1993).
7. P. Vaupel. Blood flow, oxygenation, tissue pH distribution and bioenergetic status of tumors. Ernst Schering Research Foundation, Lecture 23, Berlin (1994).
8. E. Lartigau, F. Lespinasse, L. Vitu & M. Guichard. Does the direct measurement of oxygen tension in tumors have any adverse effects? Int. J. Radiat. Oncol. Biol. Phys. *22*, 949 - 951 (1992).
9. P. Vaupel, K. Schlenger, C. Knoop & M. Höckel. Oxygenation of human tumors: evaluation of tissue oxygen distribution in breast cancers by computerized pO_2 tension measurements. Cancer Res. *51*, 3316-3322 (1991).
10. S.J. Falk, R. Ward & N.M. Bleehen. The influence of carbogen breathing on tumour tissue oxygenation in man evaluated by computerised pO_2 histography. Br. J. Cancer *66*, 919 - 924 (1992).
11. S. Runkel, A. Wischnik, J. Teubner, E. Kaven, J. Gaa & F. Melchert. Oxygenation of mammary tumors as evaluated by ultrasound-guided computerized-pO_2-histography. Adv. Exp. Med. Biol. *345*, 451 - 458 (1994).
12. J.M. Brown. Tumor hypoxia, drug resistance, and metastases. J. Natl. Cancer Inst. *82*, 338 - 339 (1990).
13. C.N. Coleman. Hypoxia in tumors: A paradigm for the approach to biochemical and physiologic heterogeneity. J. Natl. Cancer Inst. *80*, 310 - 317 (1988).
14. R.E. Durand. Keynote address: The influence of microenvironmental factors on the activity of radiation and drugs. Int. J. Radiat. Oncol. Biol. Phys. *20*, 253 - 258 (1991).
15. G.C. Rice, C. Hoy & R.T. Schimke. Transient hypoxia enhances the frequency of dihydrofolate reductase gene amplification in Chinese hamster ovary cells. Proc. Natl. Acad. Sci. USA, *83*, 5978 - 5982 (1986).
16. S.D. Young, R.S. Marshall & R.P. Hill. Hypoxia induces DNA overreplication and enhances metastatic potential of murine tumor cells. Proc. Natl. Acad. Sci. USA, *85*, 9533 - 9537 (1988).
17. L.H. Gray, A.D. Conger, M. Ebert, T.S. Hornsey & O.C.A. Scott. The concentration of oxygen dissolved in tissues at the time of irradiation as a factor in radiotherapy. Br. J. Radiol. *26*, 638 - 648 (1953).
18. B.A. Teicher, S.A. Holden, A. Al-Achi & T.S. Herman. Classification of antineoplastic treatments by their differential toxicity toward putative oxygenated and hypoxic tumor subpopulations in vivo in the FSallC murine fibrosarcoma. Cancer Res. *50*, 3339 - 3344 (1990).
19. E.M. Zeman, D.P. Calkins, J.M. Cline, D.E. Thrall & J.A. Raleigh. The relationship between proliferative and oxygenation status in spontaneous canine tumors. Int. J. Radiat. Oncol. Biol. Phys. *27*, 891 - 898 (1993).
20. D.G. Hirst & J. Denekamp. Tumor cell proliferation in relation to the vasculature. Cell Tissue Kinet. *12*, 31 - 42 (1979).
21. F. Monschke, W.-U. Müller, U. Winkler & C. Streffer. Cell proliferation and vascularization in human breast carcinomas. Int. J. Cancer *49*, 812 - 815 (1991).
22. R.P. Hill. Tumor progression: potential role of unstable genomic changes. Cancer Met. Rev. *9*, 137 - 147 (1990).
23. M. Höckel, C. Knoop, K. Schlenger, B. Vorndran, E. Baussmann, M. Mitze, P. Knapstein & P. Vaupel. Intratumoral pO_2 predicts survival in advanced cancer of the uterine cervix. Radiother. Oncol. *26*, 45 - 50 (1993).
24. C.B.J.H. Wilson, A.A. Lammertsma, C.G. McKenzie, K. Sikora, T. Jones. Measurements of blood flow and exchanging water space in breast tumors using positron emission tomography: a rapid and noninvasive dynamic method. Cancer Res. *52*, 1592 - 1597 (1992).
25. E.M. Grischke, M. Kaufmann, M. Eberlein–Gonska, T. Mattfeldt, Ch. Sohn, G. Bastert. Angiogenesis as a diagnostic factor in primary breast cancer: microvessel quantitation by stereological methods and correlation with color Doppler sonography. Onkologie *17*, 35 - 42 (1994).
26. M. Höckel, K. Schlenger, C. Knoop & P. Vaupel. Oxygenation of carcinomas of the uterine cervix: evaluation by computerized O_2 tension measurements. Cancer Res. *51*, 6098 - 6102 (1991).

27. E. Lartigau, L. Martin, P. Lambin, C. Haie-Meder, A. Gerbaulet, F. Eschwege & M. Guichard. Mesure de la pression partielle en oxygène dans des tumeurs du col utérin. Bull. Cancer Radiother. *79*, 199 - 206 (1992).

28. E. Lartigau, E., L. Vitu, C. Haie-Meder, M.F. Cosset, M. Delapierre, A. Gerbaulet, F. Eschwege & M. Guichard. Feasibility of measuring oxygen tension in uterine cervix carcinoma. Eur. J. Cancer *28A*, 1354-1357 (1992).

29. M. Molls, F. Kallinowski & H.J. Feldmann. Radiosensitivity of tumors depending on oxygenation. Proc. IORT Meeting, Munich, Germany (1992).

30. H.J. Feldmann. Optimierungsansätze und Limitationen in der regionalen Thermoradiotherapie von Beckentumoren. Thesis, University of Essen, Germany (1994).

31. R. Rampling, G. Cruickshank, A.D. Lewis, S.A. Fitzsimmons & P. Workman. Direct measurement of pO_2 distribution and bioreductive enzymes in human malignant brain tumours. Int. J. Radiat. Oncol. Biol. Phys., *29*, 427 - 432 (1994).

32. D.M. Brizel, G. Rosner, J. Harrelson, L.R. Prosnitz & M.W. Dewhirst. Pretreatment oxygenation profiles of human soft tissue sarcomas. Int. J. Radiat. Oncol. Biol. Phys., *in press* (1994).

33. W. Fleckenstein, J.R. Jungblut, M. Suckfüll, W. Hoppe & Ch. Weiss. Sauerstoffdruckverteilungen in Zentrum und Peripherie maligner Kopf-Hals-Tumoren. Dtsch. Z. Mund. Kiefer GesichtsChir. *12*, 205 - 211 (1993).

34. W. Mueller-Klieser, P. Vaupel, R. Manz & R. Schmidseder. Intracapillary oxyhemoglobin saturation of malignant tumors in humans. Int. J. Radiat. Oncol. Biol. Phys. *7*, 1397 - 1404 (1981).

35. L. Martin, E. Lartigau, P. Weeger, P. Lambin, A M. Le Ridant, A. Lusinchi, P. Wibault, F. Eschwege, B. Luboinski & M. Guichard. Changes in the oxygenation of head and neck tumours during carbogen breathing. Radiother. Oncol. *27*, 123 - 130 (1993).

36. E. Lartigau, A.M. Le Ridant, P. Lambin, P. Weeger, L. Martin, R. Sigal, A. Lusinchi, B. Luboinski, F. Eschwege & M. Guichard. Oxygenation of head and neck tumors. Cancer *71*, 2319 - 2325 (1993).

37. M. Guichard & E. Lartigau. Personal communication (1994).

38. E. Lartigau, H. Randrianarivelo, L. Martin, S. Stern, C.D. Thomas, M. Guichard, P. Weeger, A.M. Le Ridant, B. Luboinski, T. Nguyen, J.-C. Ortoli, F. Grange, M.-F. Avril, A. Lusinchi, P. Wibault, C. Haie-Meder, A. Gerbaulet & F. Eschwege. Oxygen tension measurements in human tumors: The Institut Gustave-Roussy experience. Radiat. Oncol. Invest., *in press* (1994).

39. M. Nordsmark, S.M. Bentzen & J. Overgaard. Measurement of human tumour oxygenation status by a polarographic needle electrode. An analysis of inter- and intratumour heterogeneity. Acta Oncol. *33*, 383-389 (1994).

45

FLUCTUATIONS IN THE OXYGENATION OF EXPERIMENTAL TUMOURS

Andrew I. Minchinton and Karen H. Fryer

B.C. Cancer Research Centre
601 West 10th Avenue
Vancouver, B.C. V5Z 1L3, Canada

INTRODUCTION

The outcome of therapy for solid tumours is modulated by many factors (Hall, 1978). The most important modulator of response to radiation therapy is the presence of oxygen. Cells that are severely hypoxic are approximately 3 times more resistant to low LET radiation (used in radiotherapy) than well oxygenated cells. Chemotherapeutic efficacy may also be adversely affected by oxygen deprivation (Tannock & Guttman, 1981).

The nature of tumour oxygenation has been studied for several decades. Forty years ago histological observations of lung tumours led to the suggestion that no viable cells existed at a distance greater than 150 μm from the nearest blood vessel (Thomlinson & Gray, 1955). It was postulated that cells at the margin of the viable regions of tissue were hypoxic and therefore likely to represent a population of cells resistant to radiation therapy. Their potentially clonogenicity meant they could act as the focus for new tumour growth after radiation therapy. This so called 'diffusion limited' model of hypoxia is now widely accepted, though its prevalence in human tumours has not been adequately characterized. A second mechanism giving rise to tumour hypoxia was suggested more recently (Brown, 1979). This mechanism involved the temporary dysfunction of some blood vessels within tumours resulting in hypoxia occurring in the cells fed by these dysfunctional vessels. As with 'diffusion limited' hypoxia this model of 'perfusion limited' hypoxia has been characterized in a few experimentally induced tumour systems in mice, but has not been studied widely, or at all in human tumours.

Preliminary studies performed in this laboratory to monitor the oxygenation of SCCVII tumours in control untreated animals revealed unexpected fluctuation in the overall oxygenation with time. Prompted by these results we have performed measurements of tumour oxygenation over several days in four types of experimental mouse tumour to determine if the overall tumour oxygenation is stable.

Oxygen Transport to Tissue XVII, Edited by Ince et al.
Plenum Press, New York, 1996

353

MATERIALS AND METHODS

The SCCVII, RIF-1, EMT6 and KHT tumours were grown in their syngeneic hosts the C3H/km and Balb/C mice. The study of oxygenation was initiated when tumours reached the following sizes (mean ± standard deviation): SCCVII, 265 ± 66 mm³; RIF-1, 435 ± 65 mm³; EMT6 330 ± 39 mm³ and KHT 468 ± 35 mm³. During the oxygen tension measurements, mice were placed in lead irradiation jigs and the tumour gently immobilized with tape. No anaesthetics were used. An Eppendorf pO_2 Histograph (KIMOC 6650, Eppendorf, Hamburg) was used to measure tumor oxygenation. A puncture in the skin covering the tumour was made with a 26g needle to facilitate the insertion of the microelectrode. The reference electrode was attached to the tail of the mouse using ECG gel. The microelectrode was inserted approximately 1.5-2 mm through the puncture in the skin covering the tumour and the oxygen tension measured by the microelectrode was allowed to stabilize before measurements were collected. The microelectrode was advanced through the tumour in a 'pilgrims step' fashion comprising a forward movement of 0.7 mm followed immediately by a backward movement of 0.3 mm. Individual pO2 measurements were collected between each step through the tissue. The tumour was steadied using a finger which also acted to sense the advance of the electrode through the tumour. As the electrode was felt to be approaching the distal edge of the tumour the measurement was halted and the electrode withdrawn to the starting position 1-2 mm from the surface of the tumour mass. The angle of the electrode was then adjusted so that it would penetrate a different area of tumour tissue and the recording of the measurement was re-commenced. This process was repeated until 30-40 measurements were made. Sub-cutaneous pO2 measurements were also made in some animals by inserting the microelectrode through an indwelling catheter.

RESULTS

At any time during the study oxygen tension measurements made within individual SCCVII, RIF-1 and EMT6 tumours showed considerable intra-tumour homogeneity. Even when multiple tracks through the tumour were measured the pO2 varied by only a few mmHg. Figure 1 shows examples of representative pO2 measurements along tracks through

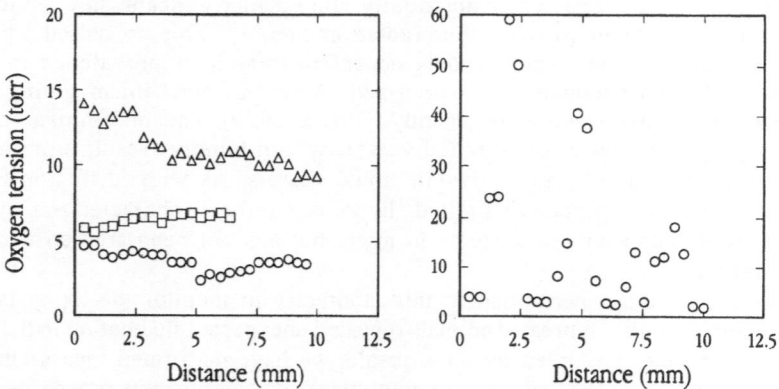

Figure 1. Panel a illustrates typical oxygen tension measurements made along tracks through an SCCVII tumour (squares), a RIF-1 tumour (circles) and an EMT6 tumour (triangles). Panel b illustrates typical oxygen tension measurements made through a KHT sarcoma.

SCCVII, RIF-1 and EMT6 tumours. In contrast to the consistency of intra-tumour pO2 measurements in these tumours the KHT sarcoma exhibited considerably greater heterogeneity.

Because the measurements within individual tumours showed great consistency at any one time-point in the SCCVII, RIF-1 and EMT6 tumours we were able to confidently determine the mean oxygenation of the tumour with relatively few readings. This had the advantage of sparing the tumour excessive damage caused by repeated insertion of the needle probe. Separate histological studies (data not shown) indicated that one track of pO2 measurements performed each day for five days produced insignificant trauma to the tumour.

We then followed the tumour oxygen tension of SCCVII, RIF-1 and EMT6 tumours over several days. Instead of averaging the data from a group of animals we chose to study any natural perturbations that might occur with time within individual tumours. Figure 2 shows the measured oxygen tension of 3 individual SCCVII tumours over a period of 60 hours. Similar patterns of fluctuations in tumour oxygenation were observed in the RIF-1 and EMT6 tumours. The periodicity of the variations observed appear random and different for each of the individual tumours studies.

DISCUSSION

The findings of relatively homogeneous pO2 values throughout the SCCVII, the RIF-1 and the EMT6 tumours support our previous findings in the SCCVII tumour (Minchinton, In press). We believe that the homogeneous histological structure of the SCCVII, the RIF-1 and the EMT6 underlies the homogeneity of the pO2 measurements. All three tumours are anaplastic rapidly growing tumours and in this study none of the tumours exhibited significant areas of necrosis. The cells comprising these tumours appear homogeneous exhibiting numerous mitotic bodies and supported by a dense vascular network. The oxygen microelectrode measures the pO2 in a volume probably comprising more than 40 cells and since the histological structure is relatively consistent the pO2 measured at any location within the tumour reflects this consistency. The KHT is also anaplastic, but has large areas of focal necrosis. These areas of cellular debris would be expected to exhibit very low oxygen levels and this may be the cause of the heterogeneity seen in the pO2 measurements made in this tumour.

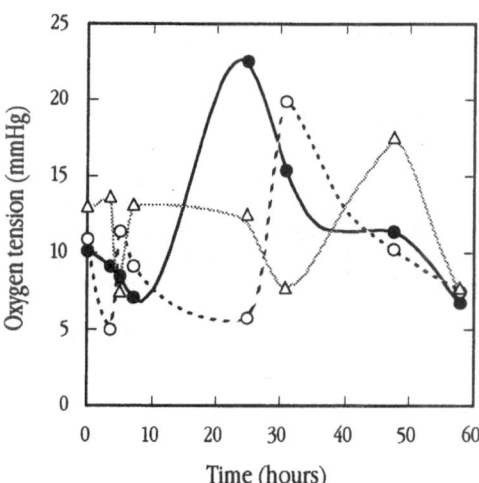

Figure 2. An example of the fluctuations in tumour oxygen tension observed in three individual SCCVII tumours.

One advantage of the tumours which exhibit homogeneous intra-tumour oxygenation is that it is possible to estimate the tumour oxygenation from a limited number of measurements. This allows frequent measurements to be made in the same tumour without producing significant damage to the tumour, damage which could influence subsequent measurements of tumour oxygenation. When the SCCVII, the RIF-1 tumour and the EMT6 tumours were followed over several days we observed a considerable variation in the measured pO2 with apparently random periodicity. This observation suggests that the oxygenation of these anaplastic tumours may be temporally unstable. The mechanism underlying this temporal fluctuation in global tumour oxygenation is not known, but could involve arteriolar vasomotion, changes in intra-tumour pressure, or systemic effects on blood flow to the tumour.

The measurement of pO2 using microelectrodes averages the oxygen tension over a small, but finite volume of tissue. Although the number of cells that comprises the volume sampled by the microelectrode is not known it is likely that it encompasses cells at range of oxygen tensions. It therefore follows that the oxygen tension within these sub-populations exhibit oxygen tensions both higher and lower than the average pO2 measured by the microelectrode. The oxygen tension of these sub-populations must also fluctuate in order to bring about a change in the mean measured oxygen tension observed. It is therefore likely that the population of cells that exist at pO2 which impart radiation resistance may fluctuate in number, and therefore the fraction of cells within the tumour which are radiobiologically hypoxic will change. If this phenomena is reflected in human neoplasms, then it has important implications for the treatment of solid tumours with radiation because it suggests that the radiation resistance of these tumours may vary within short time periods.

CONCLUSION

Measurements of pO2 using microelectrodes in several types of anaplastic murine tumours suggests that the overall oxygen tension fluctuates with time. The periodicity of the changes in pO2 appear to be random and the underlying cause is not known. If these types of changes in tumour pO2 occur in human tumours it could have important implications for therapy.

ACKNOWLEDGMENTS

This work was supported by the National Cancer Institute of Canada.

REFERENCES

Brown, J. M. (1979). Evidence for acutely hypoxic cells in mouse tumours and a possible mechanism of reoxygenation. *Br. J. Radiol.*, **52**, 650-656.

Hall, E. J. (1978). *Radiobiology for radiologists* (2nd ed.). Philadelphia: Harper & Row.

Minchinton, A. I. (In press). Differences in the tumour oxygen tension in mice and man. *Acta Oncologica*,

Tannock, I., & Guttman, P. (1981). Response of Chinese hamster ovary cells to anti-cancer drugs under aerobic and hypoxic conditions. *Br. J. Cancer*, **43**, 245-248.

Thomlinson, R. H., & Gray, L. H. (1955). The histological structure of some human lung cancers and the possible implications for radiotherapy. *Br. J. Cancer*, **9**, 539-549.

46

THE EFFECTS OF CHRONIC HYPEROXIA (80%) ON THE BEHAVIOR OF RAT LITTERS

Secil Binokay and Tuncay Özgünen

University of Çukurova
Medical Faculty
Department of Physiology
Division of Neurophysiology
01330 Balcali/Adana, Turkey

INTRODUCTION

In our previous reports it was shown that oxygen at 40% and 80% inhibited pregnancy and birth rate of rats (4, 5). In this study, pregnant female rats in the third trimester were exposed to hyperoxia and the births were performed in a hyperoxic cabin. The action of hyperoxia is known to be stressful for living organisms. Brain and lungs are two organs which are most sensitive to hyperoxia, and have been intensively investigated (3, 7, 8, 11). Extensive literature exists regarding the acute effects and toxicity of normobaric hyperoxia (3, 12). Few reports have appeared to examine behavioral changes in rats that were exposed chronically to hyperoxia (8-10). On the other hand, the cerebral circulation is sensitive to changes in blood oxygen. Also, hyperoxia reduced cerebral blood flow (3). Low blood flow will especially influence the brain stem, a region that is vulnerable, and develop specific necrosis during hyperoxic oxygen exposure (1). There is also evidence that the central nervous system (CNS) is particularly sensitive to the toxic effects of hyperoxia, even three hours exposure with 95% oxygen produced neural necrosis in the rat (2). The study has been carried out to obtain chronic hyperoxic rat litters and to investigate the effects of hyperoxic conditions on the changes of behavioral patterns of the rat litters (6). This is the reason, we chose these parameters to get a suitable method for testing ontogeny of open field behavior to determine hyperoxic-caused modifications of locomotor activity.

MATERIAL AND METHODS

Subjects

The subjects for this study were 20 Wistar rats. Ten animals were controls and 10 were a hyperoxic group; each group consisted of five males and five females. The average

Oxygen Transport to Tissue XVII, Edited by Ince et al.
Plenum Press, New York, 1996

357

weight of the animals was182 ± 6.10 g (± S.E.M). The stocks of rats originated from the Center of Experimental and Surgery of Çukurova University (DECAM). The animals were handled seven days prior to hyperoxic exposure. The rats were housed in stainless steel colony cages (50 X 35 X 15 cm) and five animals in a temperature controlled room in our department. Food and water were available ad libitum.

Apparatus

Animals were tested for levels of gross motor activity in a square open field apparatus after hyperoxic exposure. The apparatus (100 x 100 x 50 cm) was made of clear varnished plywood. Interior of the apparatus was painted matt black and the floor was divided into 25 squares (20 X 20 cm) by white lines. The open field was illuminated by a 60-W red bulb located one meter above the center.

Procedure

All open field tests were carried out between 9.00 a.m. and 16.00 p.m. The animals were taken into the experiment room 30 minutes before tests. Each rat was tested by the same experimenter during the 9-min session at about the same time of day on three consecutive days. At the beginning of examination the animals were weighed and transferred into a starting box (20 X 15 X 15 cm) and were put in the middle square of the open field, allowed to stay 90 seconds in the starting box, then the front door of the starting box was opened. If the animal does not leave the starting box in 90 seconds, the animal was kindly held and left in the middle square of the open field. The following behavior categories were recorded in each period of three minutes. Latencies to entry into the field (in seconds), duration of first rearing (in seconds), the number of rearing in the first, second and the last periods of three minutes, the total number of rearing, the total number of fecal boli (in both field and starting box), the number of urine drops (in both field and starting box), the duration of first crossing middle line (in second), the number of crossing square in the first, second and last periods of three minutes, the total number of crossing square, the number of middle line in the first, the second and the last period of three minutes, the total number of crossing middle line, the total number of grooming in the first, second and last periods of three minutes, the total number of grooming.

Hyperoxic Cabinet

The exposure cabinet has a capacity of 45 L. We designed this cabinet in which environmental parameters including temperature, pressure, humidity and concentrations of oxygen can be controlled; this cabinet was found to be efficient in a previous study (4). Hyperoxic concentration (80% ± 1) was controlled with a Hudson (5590) mark oxygen meter and control box and solenoid valve (10). To absorb water vapor and CO_2 in the cabinet, calcium chloride and soda lime were used. The cabinet was cleaned a couple of days for a two-hour period. The cabinet temperature, pressure and humidity were recorded; they were 24.00 ± 0.30 °C, 744 ± 1.00 mmHg and 53 ± 3%, respectively.

Statistical Analysis

Statistical analyses were performed with 3-way MANOVA [Groups (male and female) X Pre/post (Control and hyperoxic exposure) X Days] with repeated measures. Whenever significant results were found in MANOVA analysis, post-hoc analysis was accomplished with LSD or Newman-Keul's multiple range test for significance at $p<0.05$.

RESULTS

The Duration of the First Rearing

The only main effects of pre/post treatments were significant ($F(1,8 = 26.96$ p < 0.01). The duration of the first rearing in the control group was 12.76 ± 2.19 seconds, but it was 33.23 ± 9.63 seconds in the hyperoxic groups (Table 1).

Number of Rearing in the First, Second and Third Periods

There were significant differences between pre/post treatments at first and second sessions [$F(1,8 = 5.52$ p < 0.05, $F(1,8 = 7.03$ p < 0.05), respectively]. Also the main effects of days treatments were significant in all sessions [$F(2,16 = 11.0$ p < 0.01, $F(2,16 = 7.64$ p < 0.01 and $F(2,16 = 4.09$ p < 0.05), respectively]. In addition, the pre/post X days interactions at the first period and group X pre/post X days interactions at the third period were significant [$F(2,16 = 20.17$ p < 0.001) $F(2,16 = 5.81$ p < 0.05), respectively.

Table 1. Means of observed parameters in open field of hyperoxic and control groups analyzed with three-way MANOVA ($\pm$S.E.M = standard error of means), *p<0.05 **p<0.01

Parameters in open field	Control group ($\pm$S.E.M)	Hyperoxic group ($\pm$S.E.M)
Latency of entry into field	63.63±11.00	76.00±6.74
Duration of the rearing	12.76± 2.19	33.23±9.63
Number of rearing at 1st period	11.06 ±0.86	8.80±1.38 *
Number of rearing at 2nd period	8.03±1.45	10.56±1.31 *
Number of rearing at 3rd period	7.69±1.13	9.70±1.38
Total number of rearing	26.80±2.00	29.06±2.77
Duration of first crossing middle line	106.6±25.1	65.26±33.42
Number of crossing middle line 1st period	0.70±0.20	0.46±0.23 *
Number of crossing middle line 2nd period	0.36±0.14	0.20±0.12
Number of crossing middle line 3th period	0.20±0.07	0.10±0.07 *
Total number of crossing middle line	1.26±0.27	0.76±0.28 *
Number of crossing square 1st period	35.76±5.21	35.76±3.46
Number of crossing square 2nd period	20.90±4.40	34.10±4.96 **
Number of crossing square 3rd period	19.36±3.29	26.26±3.7
Total number of crossing square	75.70±6.35	101.3±9.02 **
Number of grooming 1st period	1.10±0.23	0.80±0.25
Number of grooming 2nd period	1.20±0.25	1.13±0.23
Number of grooming 3rd period	1.20±0.27	1.03±0.28 *
Total number of grooming	3.56±0.50	2.90±0.41 *

The Total Numbers of Rearing

The only main effects of days were found to be significant (F(2,16 = 14.88 p<0.01).

Number of Crossing Squares in the First, Second and Third Periods

There were significant differences between pre/post treatments at only second session F(1,8 = 16.63 p < 0.01). The mean number of crossings in the control group was 20.90 ± 4.40, it was 34.10 ± 4.96 in the hyperoxic group (Table 1). In the first period pre/post X days interactions were found to be significant F(2,16 = 5.39 p < 0.05). The number of crossings on the first, second and third days in the control group was 45.70 ± 3.30, 28.90 ± 4.04 and 32.70 ± 3.06, respectively; in hyperoxic group, the number of crossings was 33.30 ± 5.20, 37.50 ± 5.88 and 52.20 ± 4.55, respectively.

Total Numbers of Square Crossings

The main effects of the groups were found to be significant F(1,8 = 32.03 p < 0.01) in terms of the total numbers of crossing of middle square. The three-way MANOVA showed main effects of pre/post were significant F(1,8 = 13.98 p<0.01). The values for hyperoxic group (101.36 ± 9.02) were significantly higher than control group (75.70 ± 6.35).

The Number of Crossing Middle Line in the First, Second and Third Periods

The main effects of pre/post treatments and pre/post X days interactions were significant [F(1,8 = 7.53 p < 0.05, F(2,16 = 4.83 p < 0.05)] in the third period. No differences were found at the second period. In the third period, a three-way MANOVA of the performance in three days after the hyperoxic exposure revealed significant overall main effects of groups, pre/post days and group X pre/post, groups x days, pre/post X days and group X pre/post X days interactions [(F(1,8 = 23.14 p < 0.01), F(1,8) = 6.00 p < 0.05), F(2,16 = 14.18 p < 0.001, F(1,8 = 6.00 p < 0.05, F(2,16 = 14.18 p < 0.01, F(2,16 = 7.42 p < 0.01, F(2.16 = 7.428 p < 0.01), respectively].

Total Numbers of Crossing Middle Line

If the rat crossed the middle directly to the opposite wall and crossed the middle square, one crossing of the middle line was recorded. The results of MANOVA showed that the main effects of pre/post and pre/post X days interactions were significant [F(1,8 = 11.25 p < 0.05, F(2,16 = 7.41 p < 0.01), respectively].

Total Numbers of Boli

The analyses of the total number of boli values showed that the only main effects of days were significant F(2,16 = 5.07 p < 0.05).

Total Numbers of Urine

The main effects of groups and pre/post X days interaction were significant [F(1,8 = 67.13 p < 0.001, F(2,16 = 4.90 p < 0.05), respectively].

Number of Grooming in the First, Second and Third Periods

While there were significant differences between the main effect of group and days treatments [$F(1,8 = 10.90$ p < 0.05, $F(2,16 = 3.79$ p < 0.05)] in the first period, respectively, no differences were found at the second period. In the third period, a three-way MANOVA of the performance in three days after the hyperoxic exposure revealed the significant main effects of pre/post treatments $F(1,8 = 5.40$ p < 0.05).

Total Numbers of Grooming

In the grooming periods, continuous grooming duration longer than three seconds was accepted as one grooming. Three-way Manova with repeat measures revealed that differences between the main effects of group and group X pre/post X day interactions were significant [$F(1,8 = 11.57$ p < 0.01, $F(2,16 = 5.03$ p < 0.05, respectively].

DISCUSSION

The purpose of this study was to determine the effect of hyperoxic exposure on behavior of rat litters. In this experiment, we studied the effect of chronic hyperoxia on behavior of rats, and obtained first litters that their fecundations occurred normoxic environment and the births were performed and they survived in hyperoxic cabinet. The major result of this experiment is that the effects of hyperoxic exposure on rat litters were less efficient than expected. In terms of most parameters such as the duration of latency of entry into the field, the duration of the first rearing, the number of rearing at the third period, the total number of rearing, the number of fecal boli and urine, the duration of first crossing middle line, the number of crossing square at the first and third periods, the numbers of grooming at the first period the hyperoxic groups did not differ from the control group. Also, the numbers of rearing and crossing square in the hyperoxic group were higher than control groups. When we compare to our previous study (1). Whereas, we reported in the previous study that, all parameters of their parents decreased in hyperoxic groups, compared to values of litter seen in Table 1 (5).

These results suggest that hyperoxia did not affect the rat litter's behavior. On the other hand, the activity analysis experiment showed mild hyperactive changes in qualitative aspects of locomotor activity: The treated rats were not willing to remain in the corners as the indication of the locomotion activity. In addition, the numbers crossing of square also increased. Since rats normally possess strong motivation to continually explore their environment by changing their place, after hyperoxic exposure high behavioral diversity indicated a changed structure of behavior, and may be interpreted as increased attention or exploratory motivation.

REFERENCES

1. Balantine, J.D., 1982, Pathology of oxygen toxicity. New York, Academic Press.
2. Barmada, A.M., Moossy, J., Nemoto, M.E., Lin, R.M., 1986, Hyperoxia produces neural necrosis in rat. J. of Neuro. Pat. and Exp. Neurology 45:233-246.
3. Bean, W.J. 1945, Effect of oxygen increased pressure. Physiological Reviews 25:1.
4. Binokay, S., 1986, The Effect of different oxygen concentrations on viability and fertility of mice and the controlled cabinet production. Journal of Çukurova University Inst. of Health Sciences 1:1-2-3 60-1.

5. Binokay, S., Özgünen, T., 1994, The Effect of Chronic Hyperoxic (%80) exposure on the Behavior of Rats. XXth Annual Meetings of EUBS on Diving and Hyperbaric Medicine. Istanbul Turkey.
6. Coursin, B.D., Cihla, P.H., Will, A.J., Mccreary, L.J., 1987, Adaptation to Hyperoxia. Am. Rev. Respir. Dis.135:1002-1006.
7. Fremann, B.A., Crapo, J.D., 1982, Biology of disease Lab. Invest. 47:412-26.
8. Gerben, M.J., 1968, Hyperoxia and locomotor activity. Percept. Mot. Skills 26(3):745-746.
9. Greenbaum, M., Gunberg, L.D., 1962, The effect of Neonatal hyperoxia on sexual arousal and emotionality in male rat. Anim. Behav. 10:28-33.
10. Philp, B.R., Fields D.N.G., Johnson, J.P., 1986 The effect of nitrogen and helium at pressure on drug-induced stereotype and locomotion mice. Aviat. Space Environ. Med. 57(8):769-776.
11. Prestwich, N.K., Buss, D.D., Posner, P., 1984, A new method for raising neonatal rabbits in a hypoxic environment. J. Appl. Physiol. Respirat. Environ. Exercise Physiol. 56(6):1913-1916.
12. Roth, M.E., 1964, Space-Cabin Atmospheres Part-1 Oxygen Toxicity. National aeronautics and space administration, Washington, D.C.

47

THE CHANGES OF SKIN AND MESENTERIC MIROVASCULAR PERMEABILITY IN ADAPTATION TO HYPOXIA IN RATS

M. V. Dubina and N. A. Gavrisheva

St. Petersburg Pavlov Medical University
Department of Pathophysiology
6/8 Lev Tolstoy street 197089, St. Petersburg, Russia

INTRODUCTION

Character of changes in permeability of vessels in hypoxia depends on degree of tension of process supplying the lack of oxygen in tissues. The changes of microcirculation in hypoxia depend on many factors, in particular - level and duration of activity, compensate possibilities of organism, endothelial sensitivity to vasoactive substances etc. One of the effective methods to raise the resistance of an organism to oxygen lack is intermittent hypoxia (Vorontsov & Rusanova, 1987). It is known that the endothelium has a barrier function, with vessel permeability as one of its indices (Ogawa, et al., 1990). Barrier characteristics of the endothelium of microvessels are defined in many respects by structure-functional peculiarities of organs. Hence, in our experiments we examined the permeability of microvessels of rat skin and mesentery on different stages of adaptation to intermittent hypobaric hypoxia.

METHODS

Procedure

We used mongrel white female rats (150 - 180 g). The animals were anesthetized by injection of Nembutal (5mg/100g body weight subcutaneously), then we catheterized the femoral vein. The permeability of microvessels was determined by contact fluorescence biomicroscopy (microscope LUMAM I-2 LOMO). The probe of microscope was set by the vessel in the region of good vascularization on prepared part of tissues. Na-fluoresceine was used as an indicator of permeability in a dose of 0.25 mg/100 g body weight (1mg/ml of physiological salt solution) by intravenous injection.   Autofluorescence of tissues was measured prior to fluoresceine injection.

Oxygen Transport to Tissue XVII, Edited by Ince et al.
Plenum Press, New York, 1996

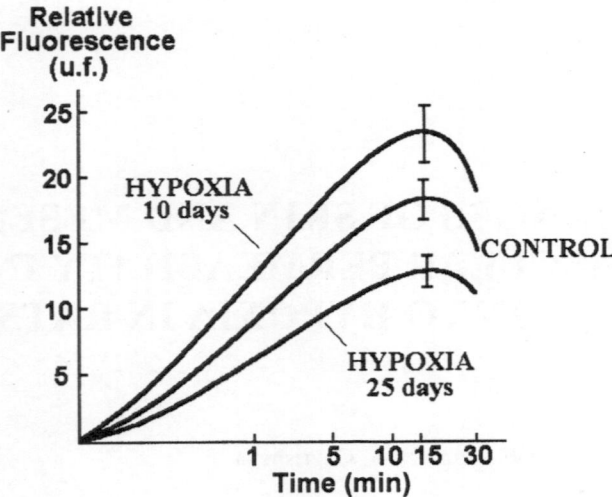

Figure 1. The permeability of microvessels of skin for Na-fluorescein at different stages of intermittent hypobaric hypoxia.

Intermittent Hypobaric Hypoxia

Intermittent hypobaric hypoxia was induced in a hypobaric chamber (volume - 75 litres) by gradual lowering the pressure to 47.17 kPa (equivalent to 6000 meters altitude) over the course of 25 daily treatments.

Permeability Data and Statistical Methods

The permeability of microvessels was investigated on the 10-th and 25-th day of hypoxia. To quantify the changes of vascular permeabilitiy, we measured the time between fluorescein injection and the peak fluorescence. Data were analyzed using the Student's t-test.

RESULTS

Permeability of Skin Microvessels

During the examination of skin microvascular permeability in control and in hypoxia animals the fluorescence increased practically immediately after fluorescein injection. Fluorescence increased and reached its maximum (18.75±0.75 u.f.) after 14-16 minutes. After that it diminished slowly. In hypoxia animals we saw a significant increase of maximum fluorescence on the 10-th day and a decrease on the 25-th day in comparison with control. The time to maximal fluorescence was not effected by hypoxia.

Permeability of Mesentery Microvessels

During examination of mesentery microvascular permeability in control animals, the fluorescence reached its maximum 5-6 minute after injection (6.75±0.46 u.f.). On the 10-th

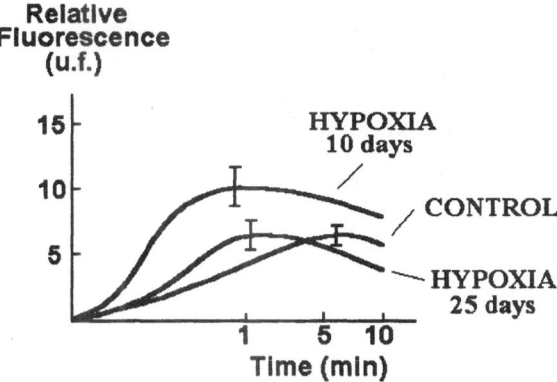

Figure 2. The permeability of microvessels of mesentery for Na-fluorescein at different stages of intermittent hypobaric hypoxia.

day of hypoxia we observed significant increase of permeability of microvessels, while on the 25-th day there was only a non-significant tendence to decrease in comparison with control. Thus, in rats the process of adaptation to intermittent hypobaric hypoxia leads to phasic changes of skin and mesentery microvascular permeability.

DISCUSSION

Comparative analysis of characteristics of permeability, received by investigation of skin and mesentery microvessels, showed the monodirected changes of fluorescence in skin and mesentery at different times of hypoxia, i.e. on the 10-th day there is an increase of permeability, but on the 25-th day - a decrease. As it was pointed, intermittent hypobaric hypoxia is one of the effective ways to increase organism's resistance to oxygen "hunger", in the process of adaptation to it the complex of compensator-adaptative reactions is formed, directed on the recovery of the adequate oxygen maintenance of the organism. In particular, occur phasic changes of acid-base blood condition, physico-chemical features of erythrocytes (Vorontsov & Rusanova, 1987), activity of the sympatho-adrenergic system and vasoactive substances (Ivashkevich & Nagnibeda, 1989), vascular-platelet hemostasis (Petrishchev & Stepanova, 1991). In our examination this regularity was confirmed by the monodirected phasic changes of permeability of skin and mesentery microvessels.

Observed regional differences of permeability indices are obviously connected with better skin vascularisation and also with structure-functional features layer of skin and mesentery vessels. This is accorded to adopted ideas about the organ gradient of permeability in dependence of correlation of large and small holes in vascular wall (Chernukh, et al, 1984). Thus, on the base of our results one can conclude there are the monodirected phasic character of changes in rat skin and mesentery microvascular permeability in the process of adaptation to hypoxia.

REFERENCES

Chernukh, A.M., Aleksandrov, P.N., Alexeyev, O.V., 1984, Microcirculation, Meditsina, 432 p.

Ivashkevich, A.A., and Nagnibeda, N.N.,1989, Changes in neutral peptid-hydrolases of blood and catecho-
lamines of tissues in organism adaptated to alpine hypoxia, Physiol. Journ. 35(5):59-64.
Ogawa, S., et al., 1990, Modulation of endothelial function by hypoxia: perturbation of barrier and anticoagu-
lant function; and induction of a novel factor X-activator, Adv. Exp. Med. Biol. 281:303-312.
Petrishchev, N.N., and Stepanova, M.N., 1991, The thrombocyte-vascular hemostasis in hypobaric hypoxia,
Sechenov Physiol. Journ. USSR. 77(6):42-47.
Vorontsov, V.A., and Rusanova, N.R., 1987, Gas regimen of the organism in the process of adaptation and
deadaptation to the intermittent hypobaric hypoxia, Physiol. Journ. 33(3):33-38.

48

DETECTION OF MICROREGIONAL FLUCTUATIONS IN ERYTHROCYTE FLOW USING LASER DOPPLER MICROPROBES[*]

S. A. Hill[†] and D. J. Chaplin

CRC Gray Laboratory
PO Box 100, Mount Vernon Hospital
Northwood HA6 2JR, Middlesex, United Kingdom

INTRODUCTION

It is well established that experimental tumours develop microregions of nutrient-deprived and oxygen-deficient cells as a consequence of their heterogeneous vascular supply. Spatial heterogeneity can lead to the existence of chronically hypoxic cells at a distance from blood vessels, as originally described by Thomlinson and Gray (1955). However, it has also been suggested that a temporal heterogeneity in perfusion could lead to a transient, potentially reversible, acute hypoxia (Brown, 1979). Evidence for temporary non-perfusion of vessels has been provided by direct observation of 'sandwich' tumour preparations (Reinhold et al., 1977) as well as by histological techniques which involve the injection of two fluorescent markers, separated in time (Trotter et al., 1989). These perfusion markers have provided evidence of intermittent blood flow in a number of different experimental murine tumours, but they do have limitations; in some systems the dyes themselves are vasoactive, the techniques are not clinically applicable and they provide no kinetic information on the duration of vessel non-perfusion.

The recent development of a multichannel laser Doppler system, utilising 300 μm microprobes (the Oxford Array, Oxford Optronix Ltd., Oxford, U.K.), has made feasible the real-time monitoring of microregional blood flow in tumours. Each insertable probe has a nominal sampling volume of 10^{-2} mm^3, making it much more sensitive to changes at the microregional level than conventional needle probes, which sample volumes of 0.5 to 1 mm^3. The aim of the present study was to characterise the temporal changes in microregional perfusion in a transplantable murine tumour using the Oxford Array and to investigate the ability of nicotinamide to modify such changes.

[*] This work was wholly supported by the Cancer Research Campaign.

[†] Address requests for reprints to Dr. Hill.

Oxygen Transport to Tissue XVII, Edited by Ince et al.
Plenum Press, New York, 1996

367

MATERIALS AND METHODS

Tumour Model

The SaF is a rapidly growing, undifferentiated sarcoma which has been maintained for many generations in the murine strain of origin. Tumours were grown subcutaneously on the backs of 12 to 16 week old CBA/Gy f TO mice from an inoculum of 0.05 ml of a crude tumour cell suspension. Animals were selected 10 to 14 days later, when their tumours reached a geometric mean diameter (GMD) of 5.5 to 6.5 mm (150-300 mg).

Laser Doppler Flowmetry

Relative changes in erythrocyte flux were measured using the Oxford Array Laser Doppler system (Oxford Optronix Ltd., Oxford, U.K.). Signals were recorded at up to five separate locations within each tumour, using custom-designed microprobes of approximately 300 μm diameter.

Throughout the experiments, the mice were restrained in Perspex jigs which were taped to a heating pad maintained at 37°C; this in turn was placed on a stone slab, resting on a piece of high density foam to minimise vibrations from the bench top. A Perspex box placed over the jig served to anchor the probes in position, via rubber teeth, while also eliminating visual disturbance to the mouse. Once stable readings were obtained from each probe, erythrocyte flux was monitored for 60 minutes. In a second series of experiments, 500 mg/kg nicotinamide was injected i.p., via an in-dwelling catheter, at the end of this 60 minute period and blood flow was monitored for a further 60 minutes. At the end of the experiment a lethal dose of sodium pentobarbitone was injected intravenously.

Data Analysis

From the twenty readings per second recorded, a single average was calculated for each 2 minute interval for each channel. The 2 minute average calculated after the death of the animal was then subtracted from these values. Apparent changes in erythrocyte flux were compared with the original recorded data so that changes associated with animal movement or probe movement (detected as an abrupt change in the backscatter signal) could be eliminated.

RESULTS

In two separate experiments, performed on different batches of animals, 43 and 39 evaluable records of erythrocyte flux were obtained from different microregions of untreated SaF tumours. A further 50 traces were recorded following the i.p. injection of 500 mg/kg nicotinamide. When averaged, data from a single batch of untreated animals showed no significant fluctuations in RBC flux over the 60 minute observation period, as shown in Figure 1. Following nicotinamide injection, a mean increase in erythrocyte flux of approximately 20% was apparent.

Although the averaged data suggest an apparently constant level of perfusion in untreated tumours, examination of the individual traces measured in different tumour microregions reveals both spatial and temporal heterogeneity of flow. Figure 2 shows the data obtained from four different microprobes inserted into a single tumour and indicates that changes in RBC flux can occur independently in different microregions. Data from each

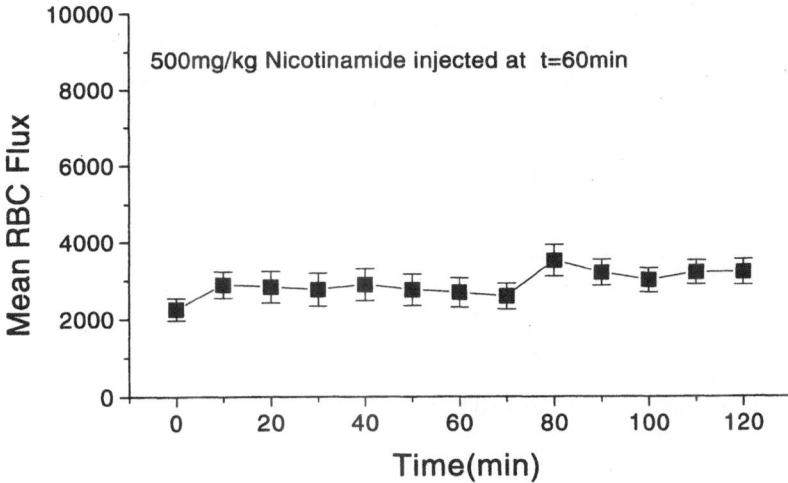

Figure 1. Averaged erythrocyte flux measured in multiple locations in 11 individual tumours. Nicotinamide (500 mg/kg) was injected at the mid-point of the 2 hr measurement period. Points represent means ± 1 sem.

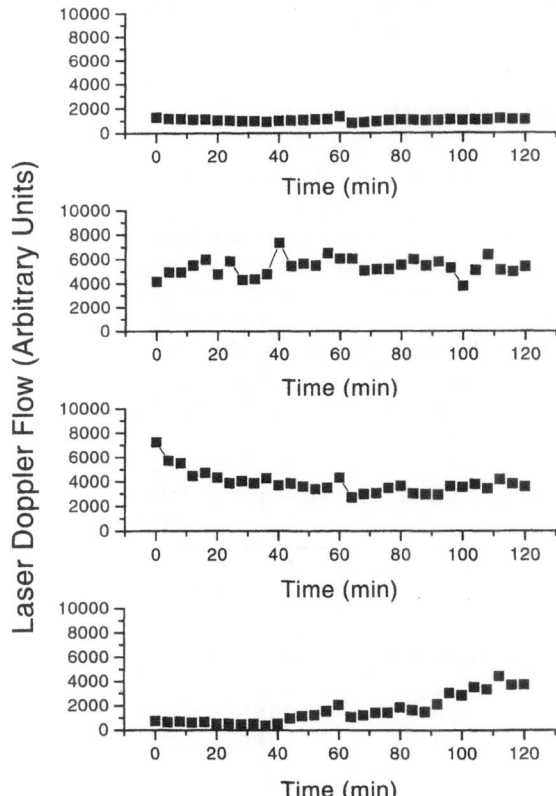

Figure 2. Four individual traces from a single tumour. Each point represents the mean of 2400 readings taken over a 2 minute period.

Table 1. Changes in microregional erythrocyte flux during 60 minute observation period. Treatment: None

Factor of change	Experiment 1		Experiment 2	
	% of traces showing change	Direction of change	% of traces showing change	Direction of change
≥ 2	56	18↑ 25↓	49	13↑ 13↓
≥ 5	19	3↑ 6↓	15	4↑ 4↓
≥ 10	5	0↑ 2↓	10	1↑ 3↓
	n = 43		n = 39	

probe was analysed separately and the percentage of traces showing significant changes in erythrocyte flux was determined before and after nicotinamide injection. Table 1 shows the data for the two groups of untreated animals. The measured changes in erythrocyte flux are remarkably similar; in each experiment approximately 50% of traces showed a change of at least a factor of two, while 15 to 19% showed more than a five-fold change. The direction of change, i.e. whether flow increased or decreased, was also very similar for the two experiments, an approximately equal number of increases and decreases were measured. Although the data are not shown, approximately 20% of the changes seen were reversed during the monitoring period. The incidence of fluctuations measured after nicotinamide injection is shown in Table 2 and for comparison the data for the two groups of untreated animals have been pooled. Nicotinamide did not appear to reduce the frequency of measured changes in erythrocyte flux, but the majority of variations took the form of an increase rather than a decrease in flow.

DISCUSSION

This study has demonstrated that it is possible to measure significant fluctuations in erythrocyte flux in unperturbed tumour tissue using the laser Doppler microprobes of the Oxford Array. Two separate experiments demonstrated the reproducibility of the technique when the same tumour type was examined. Approximately 50% of the regions sampled showed either an increase or a decrease in RBC flux, by at least a factor of two, over the one hour observation period and 5 to 10% of regions showed a greater than ten-fold change. Although it is not possible to determine directly the impact of this temporal heterogeneity of perfusion on cellular oxygen tensions, the data suggest that up to 25% of tumour microregions in the Sarcoma F suffered some reduction in oxygenation during the sampling

Table 2. Changes in microregional erythrocyte flux during 60 minute observation periods pre- and post-nicotinamide

Factor of change	Treatment: none		Treatment: 500 mg/kg nicotinamide i.p.	
	% of traces showing change	Direction of change	% of traces showing change	Direction of change
≥ 2	52	31↑ 38↓	72	43↑ 18↓
≥ 5	15	7↑ 10↓	22	12↑ 1↓
≥ 10	7	1↑ 5↓	12	6↑ 0↓
	n = 82		n = 50	

period. The size of the sampling volume (approx. 10^{-2} mm^3) is such that a single vessel could be responsible for the recorded signal, in which case a halving of erythrocyte flux would correspond to a halving of oxygen delivery. Alternatively, a small number of vessels may show a smaller but equal reduction in flow, thus having a smaller impact on oxygen tension, or the measured changes may reflect a very large or complete flow reduction in some vessels with no change in others. Certainly the existence of dynamic changes in microregional perfusion, consistent with induced changes in oxygenation, is indicated in this tumour system. The observation that 20% of changes were reversed within 60 minutes implies that at least some of the cells are subjected to a transient, acute hypoxia.

The nature of tumour hypoxia, whether dominated by diffusion or perfusion limitations, will have important implications for the choice of therapeutic approach required for its modulation. Increasing tumour blood flow or the oxygen carrying capacity of the blood will have little influence on cells rendered hypoxic via a temporary vascular disruption. However, the tumour-specific radiosensitiser nicotinamide has been reported to act primarily by reducing or preventing transient fluctuations in tumour blood flow (Chaplin *et al.*, 1990). In the current study, we saw no evidence of a decrease in the incidence of fluctuations. However, whereas untreated tumours showed an equal distribution of increases and decreases in flow, 70 to 100% of the changes measured after drug injection were manifest as an increase in the level of perfusion, i.e. the incidence of flow reductions was more than halved. This is consistent with an overall reduction in the level of acute hypoxia, as previously postulated. The preponderance of flow increases translated into a global increase in tumour blood flow of approximately 20%.

In summary, the data presented demonstrate that real-time spatial flow mapping of tumour blood flow is possible at the microregional level using insertable Doppler microprobes. Significant temporal changes have been measured in the Sarcoma F, consistent with the transient induction of hypoxia. Nicotinamide reduced the incidence of flow reductions.

REFERENCES

Brown, J.M. Evidence for acutely hypoxic cells in mouse tumours and a possible mechanism for reoxygenation. *Br. J. Radiol.*, **52**, 650-656, 1979.

Chaplin, D.J., Horstman, M.R. and Trotter, M.J. Effect of nicotinamide on the microregional heterogeneity of oxygen delivery within a murine tumour. *JNCI.*, **82**, 672-676, 1990.

Reinhold, H.S., Blachiwiecz, B. and Blok, A. Oxygenation and reoxygenation in "sandwich" tumours. *Bibl. Anat.*, **15**, 270-272, 1977.

Thomlinson, R.H. and Gray, L.H. The histological structure of some human lung cancers and the possible implications for radiotherapy. *Br. J. Cancer*, 9, 539-549, 1955.

Trotter, M.J., Chaplin, D.J., Durand, R.E. and Olive, P.L. The use of fluorescent probes to identify regions of transient perfusion in murine tumors. *Int. J. Radiat. Oncol. Biol. Phys.*, **16**, 931-934, 1989.

49

FRACTIONATION EFFECTS OF OXYGEN ISOTOPES WITHIN INTERSTITIAL LUNG DISEASE

H. Heller,[1] M. Könen,[1] A. Overlack,[2] and K. -D. Schuster[1]

[1] Physiologisches Institut I
Universität Bonn
Nussallee 11, D-53115 Bonn, Germany
[2] Medizinische Universitäts-Poliklinik Bonn
Wilhelmstr. 35, D-53111 Bonn, Germany

INTRODUCTION

Up to now diffusion limitation versus ventilation-perfusion inhomogeneities are in dispute as the main reason for impairing pulmonary gas exchange within interstitial lung diseases (Finley et al., 1962, Wagner et al., 1976, Jernudd-Wilhelmsson et al., 1986, Agusti et al., 1987, Hempleman and Hughes, 1991, Hughes et al., 1991, Yamaguchi et al., 1991). In order to assess pulmonary gas transport, such methods as the estimation of the arterial partial pressure of oxygen during nitrogen washout (Finley et al., 1962), the multiple inert gas elimination technique (Wagner et al., 1976, Jernudd-Wilhelmsson et al., 1986, Agusti et al., 1987, Hughes et al., 1991, Yamaguchi et al., 1991) and the determination of the single breath diffusing capacity for carbon monoxide (Finley et al., 1962, Wagner et al., 1976, Hughes et al., 1991) have been applied.

In order to introduce a new approach to this discussion, the quantification of the oxygen isotope transport within human respiration is presented.

In this connection the stable isotopic oxygen molecules $^{16}O_2$ and $^{16}O^{18}O$ are used. Due to different molecular weights, $^{16}O_2$ passes through diffusion and the respiratory chain with a 3% or 1.3% higher rate than the heavier isotopic species $^{16}O^{18}O$ (Schuster and Pflug, 1989, Feldman et al., 1959). However, during convective pathways such as ventilation and perfusion, these isotopic molecules are transported at the same rate. Therefore the respective influence of the different pathways can be evaluated measuring the overall change in isotopic composition of oxygen transported from the inspiratory gas mixture (I) to the respiratory chain of mitochondria (R). This change is quantified as overall fractionation effect of respiration Δ_{IR}. The more Δ_{IR} increases from normal up to values of 3% the more the entire transport system is influenced by diffusion and/or the less it is influenced by convection. In other words diffusive pathways become more limiting and/or convective ones less limiting. The opposite constellation occurs when Δ_{IR} is decreased down to 0%.

Oxygen Transport to Tissue XVII, Edited by Ince et al.
Plenum Press, New York, 1996

In the present study Δ_{IR} is estimated at rest in patients suffering from pulmonary fibrosis. The results are compared to normal values in order to assess whether diffusion limitation is the predominant reason disturbing pulmonary gas exchange within pulmonary fibrosis.

METHODS

Measurement and Quantification of the Overall Fractionation Effect of Respiration Δ_{IR}

Δ_{IR} can be estimated comparing the isotopic ratio $^{16}O^{18}O/^{16}O_2$ of inspired oxygen (X_I) to that beyond the respiratory chain of mitochondria (X_R). Since the mitochondria is inaccessible for isotopic analysis, X_R cannot be measured. Therefore a term has been developed to calculate Δ_{IR} from the difference between X_I and the $^{16}O^{18}O/^{16}O_2$-ratio of expired oxygen (X_E), weighted by the ratio of expiratory flow and consumption of $^{16}O_2$ (f):

$$\delta_{IR} = -\left[\frac{X_I - X_E}{X_I}\right] * f$$

(1).

This equation is an analogous term to the above mentioned difference between X_I and X_R (Schuster and Pflug, 1989).

With the aid of a respiratory mass spectrometer (Varian MAT, M3) the $^{16}O^{18}O/^{16}O_2$-ratio of inspired and expired oxygen is determined by a simultaneous detection of the ions $^{16}O^{18}O^+$ and $^{16}O_2^+$ at two ion collectors. For this purpose the mass spectrometer has been especially adapted to direct and continuous measurements of $^{16}O^{18}O$-ratios of stable oxygen isotopes within gas mixtures (Schuster et al., 1979, Heller et al., 1993). This method is clearly distinguished from the usually performed isotopic analysis of $^{16}O^{18}O$-ratios within carbon dioxide, which is to be obtained from the combustion of oxygen.

Gas Sampling Procedure

Investigations are performed on 6 patients and 6 healthy test subjects respectively. All test subjects breathe room air in a sitting position. Since eq. (1) presupposes steady state conditions of oxygen isotope transport, the expiratory gas mixture is sampled after a period of at least 5 minutes of breathing to attain these conditions. During the gas sampling procedure the test subjects expire approx. 35 L to 50 L into two bags in succession, retaining the breathing pattern complied before. The sample gas is taken from this in order to measure X_E. This value is compared to X_I which is obtained from measurements of room air.

Experimental Protocol and Test Subjects

In all patients the diagnosis of pulmonary fibrosis is made on clinical grounds, chest X-ray examination, routine pulmonary function tests such as body plethysmography (Masterlab, E. Jäger) and blood gas analysis taken from the earlobe (AVL 995), bronchoscopy, and bronchoalveolar lavage respectively. Furthermore patients undergo measurements of the diffusing capacity for carbon monoxide (D_{LCO}) estimated by the single breath method (Cotes, 1982). The pulmonary function tests and measurements of D_{LCO} are also carried out on healthy test subjects.

All examinations are performed at hospital, except for the Δ_{IR}-determination.

Table 1. Mean values ± standard deviations of age, hemoglobin concentration within blood ([Hb]), arterial partial pressures of oxygen and carbon dioxide (P_{AO_2}, P_{ACO_2}), and percentage of the forced vital capacity expired in 1 second (FEV1/FVC) respectively, according to patients (A, n = 6) and healthy humans (B, n = 6)

	Age	[Hb]	P_aO_2	P_aCO_2	FEV1/VC
		g / 100 ml	mmHg	mmHg	%
(A)	58.33 ± 4.9	14.07 ± 1.3	84.37 ± 6.0	39.1 ± 3.4	112.7 ± 7.6
(B)	40.83 ± 8.1	13.17 ± 2.1	92.70 ± 8.1	36.7 ± 4.1	102.2 ± 8.5
P (A Vs B)	< 0.0001	not significant	< 0.05	not significant	< 0.025

RESULTS

The most important data is listed as mean values in both tables, including a paired t-test which is applied to comparable data. It should be taken into account that Δ_{IR} and arterial partial pressures are not estimated at the same time. Therefore the alveolar-arterial oxygen gradient could not be calculated. Blood gas analysis, lung function tests, D_{LCO}- and Δ_{IR}-measurements respectively are performed at rest.

DISCUSSION

Interpretation of the Results

As expected, the paired t-test yields significantly smaller respective means of P_{AO_2}-, VC-, and D_{LCO}-values and significantly "supernormal" FEV1/VC-values in patients. These results confirm the diagnosis. In detail these fibrotic mean values are estimated to be larger compared to those of literature (Finley et al., 1962, Wagner et al., 1976, Hughes et al., 1991). This indicates that our patients do not suffer from a high graded pulmonary fibrosis. Typically the fibrotic P_{ACO_2}-mean value of 39 mmHg is inconsistent with a significantly increased resting minute ventilation and consecutive hypocapnia. Since patients are older than healthy humans, VC- and D_{LCO}-data is in particular related to predicted values taken from Quanjer (1982).

Table 2. Mean values ± standard deviations of vital capacity and single breath diffusing capacity for CO related to the values predicted from Quanjer (1982) (VC % pred., D_{LCO} % pred.), and of the overall fractionation effect of respiration Δ_{IR} respectively, according to patients (A, n = 6) and healthy humans (B, n = 6)

	VC	D_{LCO}	Δ_{IR}
	% pred.	% pred.	%
(A)	72.8 ± 8.8	53.7 ± 29.9	0.65 ± 0.03
(B)	120.0 ± 11.0	103.7 ± 11.3	0.711 ± 0.07
P (A Vs B)	< 0.0001	< 0.005	< 0.05

D_{LCO}- and Δ_{IR} -Data

Applying two or more compartment models to experimental data, Piiper and Sikand (1966) and Piiper (1992) could show that within a functionally inhomogeneous lung an unequal distribution of ventilation to pulmonary perfusion is a possible cause of pulmonary gas exchange inefficiency. This leads inter alia to reduced single breath D_{LCO}-results if such inhomogeneities are ignored. Therefore the decreased D_{LCO}-values in patients should not necessarily be interpreted as a diffusion limitation of pulmonary gas exchange. Furthermore pure diffusion limitation which is nevertheless assumed would be contrary to the parallel detected drop of Δ_{IR}-values in patients, as Δ_{IR} should increase in such a limitation.

Ventilation-Perfusion Inhomogeneities versus Diffusion Limitation

Corresponding to clinical experience, the chest X-ray examination yielded in all patients diffusively distributed increased markings, particularly pronounced in the lower regions of the lung. This points out that the fibrotic changes within the parenchyma are clearly inhomogeneously distributed.

In fact Wagner et al. (1976) found in patients with interstitial lung disease that at rest the ventilation-perfusion ratios were distributed differently when compared to healthy lungs, with an increased contribution of low ventilation-perfusion ratios (< 0.1). In these shunt perfused lung regions the influence of convection on pulmonary oxygen transport is more pronounced than that of diffusion. This leads to a stronger weighting of the convective pathways within the lungs as well as within the entire oxygen transport system. Since within convective pathways the isotopic composition of oxygen does not change, these particular ventilation-perfusion inhomogeneities can explain the significant decrease of Δ_{IR}-values in patients.This also applies to high ventilation-perfusion ratios which are due to low perfusion values.

At rest pulmonary gas exchange insufficiency is reported to be brought about completely (Wagner et al., 1976, Jernudd–Wilhelmsson et al., 1986) or mostly (Agusti et al., 1987) by ventilation-perfusion inhomogeneities. In the present study a possible diffusion limitation is covered up by an increased influence of convective resistances, otherwise Δ_{IR} should have been increased in patients.

Application of a Model Analysis to the Data

The above mentioned conclusion is confirmed by a model analysis of the human oxygen transport according to Schuster and Heller (this volume). Within this the entire oxygen transport system is considered as consisting of two identical networks of resistances for $^{16}O_2$ and for $^{16}O^{18}O$. As each pathway is represented by a singular resistance the particular fractionation effect of each pathway (e.g. 3% for diffusion) can be weighted by the contribution of its resistance to the overall resistance which respiration exerts on oxygen transport. Therefore Δ_{IR} depends on the single fractionation effects weighted in such a matter. In order to assess the effect of functionally inhomogeneous lungs on pulmonary gas transport, within lungs the two networks are again subdivided into a pathological and a healthy branch. Each of these branches consists of a ventilatory, diffusive, and perfusive resistance respectively. As the main results of this model, Δ_{IR} decreases when the resistance of ventilation is increased. However, Δ_{IR} increases when the resistance of pulmonary diffusion is increased. Furthermore the extent of a Δ_{IR}-drop caused by increased pulmonary convective resistances depends on the contribution of the remaining diffusive resistance to the entire pulmonary disturbance. But since D_{LCO}-measurements are also influenced by mismatched ventilation-perfusion ratios, and alveolar-arterial oxygen gradients could not be calculated in the present study, the amount of limitation contributed by diffusive or convective pathways cannot be exactly calculated from the presented data.

Critique of Methods

In the case of a direct measurement of $^{16}O^{18}O/^{16}O_2$-ratios as applied here, all sources of error connected to the usually performed combustion procedure are excluded. But side effects and drift errors remain. These effects are excluded by a reference gas technique. Within this the sample gas is repeatedly compared to a reference gas which is composed of the same main gas components (O_2, CO_2, N_2). Consequently, relatively large amounts of sample gas are needed. This leads to extended sampling periods so that a contamination of the sample gas by room air becomes probable. Nevertheless no significant difference could be found between the sampled gas volume taken from patients (43863 ± 3165 L_{BTPS}) and healthy humans (45550 ± 3105 L_{BTPS}). There is also no significant relationship between individual Δ_{IR}-values and sampled gas volumes.

Previous studies (Heller et al., 1993, Heller et al., 1994) confirm the above mentioned dependency of Δ_{IR} on convective pathways. Heller et al. (1993) could show that Δ_{IR} increases with increasing ventilatory rate or alveolar partial pressure of oxygen (P_{AO2}). In the present study P_{AO2}- values are simultaneously determined to Δ_{IR}-values and are estimated to be larger in patients (113 ± 9.8 mmHg) than in healthy test subjects (107 ± 12.5 mmHg). Therefore an application of this Δ_{IR}-P_{AO2}-relationship to the data would raise the significance of the Δ_{IR}-drop in patients. The same result is obtained when [Hb]-values of patients (14.07 ± 1.3 g/100ml) and healthy humans (13.17 ± 2.1 g/100 ml) are considered as Δ_{IR} increases with increasing hemoglobin concentration within blood (Heller et al., 1994).

CONCLUSIONS

The quantification of the oxygen isotope transport is presented as a new approach to assess pulmonary gas exchange within interstitial lung disease. In this connection the stable isotopic oxygen molecules $^{16}O_2$ and $^{16}O^{18}O$ are used. Due to different molecular weights, $^{16}O_2$ passes through diffusion and the respiratory chain with a 3% or 1.3% higher rate than $^{16}O^{18}O$. Since during convective pathways both isotopic molecules are transported at the same rate, the respective influence of these different pathways on the entire transport system can be evaluated. For this purpose the overall change in isotopic composition of oxygen within entire respiration Δ_{IR} is determined. Besides this investigation, 6 patients suffering from pulmonary fibrosis undergo measurements of their single breath diffusing capacity of carbon monoxide (D_{LCO}). As expected, the fibrotic D_{LCO}-values are significantly smaller than those obtained from 6 healthy humans. But in contrast to one's expectation Δ_{IR} is decreased in patients indicating that diffusion limitation cannot be the main reason disturbing pulmonary gas exchange in interstitial lung diseases. With the aid of a resistance model of respiration it is concluded that in pulmonary fibrosis diffusion limitation of oxygen isotope transport is covered up by increased ventilation-perfusion inhomogeneities which lead to a stronger weighting of the convective resistances of respiration.

ACKNOWLEDGMENTS

We gratefully acknowledge the technical assistance of Bernd Eixmann, Christa Pusch, Barbara Schreiber, and L. Schmidt-Schilling.

REFERENCES

Agusti, A.G.N., Roca, J., Rodriguez-Roisin, R., Gea, J., Xaubet, A., and Wagner, P.D., 1987, Role of O_2
  diffusion limitation in idiopathic pulmonary fibrosis, Am. Rev. Respir. Dis. 135:A307.
Cotes, J. E., 1982, The transfer factor (diffusing capacity), in "Standardized lung function testing" (Quanjer
  Ph. H., Editor), Report of the European Community for Coal and Steel, Luxembourg.
Feldman, D.E., Yost jr., H.T., and Benson, B.B., 1959, Oxygen isotope fractionation in reactions catalyzed by
  enzymes, Science, 129: 146-147.
Finley, T.N., Swenson, E.W., and Comroe, J.H., 1962, The cause of arterial hypoxemia at rest in patients with
  'alveolar-capillary block syndrome', J. Clin. Invest., 41: 618-622.
Heller, H., Könen, M., and Schuster, K.-D., 1993, Dependence of overall fractionation effect of respiration on
  ventilation at rest, Isotopenpraxis 28:133-141.
Heller, H., Schuster, K.-D., and Göbel, B. O., 1994, Dependency of overall fractionation effect of respiration
  on hemoglobin concentration within blood at rest, Adv. Exp. Med. Biol. 345:755-761.
Hempleman, S.C., and Hughes, J.M.B., 1991, Estimating exercise D_{LO2} and diffusion limitation in patients
  with interstitial fibrosis, Respir. Physiol., 83: 167-178.
Hughes, J.M.B., Lockwood, D.N.A., Jones, H.A., and Clark, R.J., 1991, D_{LCO}/Q and diffusion limitation at
  rest and on exercise in patients with interstitial fibrosis, Respir. Physiol., 83: 155-166.
Jernudd-Wilhelmsson, Y., Hornblad, Y., and Hedenstierna, G., 1986, Ventilation-perfusion relationships in
  interstitial lung disease, Eur. J. Respir. Dis. 68:39-40.
Piiper, J., and Sikand, R.S., 1966, Determination of D_{CO} by the single breath method in inhomogeneous lungs:
  theory, Respir. Physiol., 1: 75-87.
Piiper, J., 1992, Diffusion-perfusion inhomogeneity and alveolar-arterial O_2 diffusion limitation: theory,
  Respir.Physiol. 87:349-356.
Quanjer, Ph.H., 1982, Standardized lung function testing, in: Working party "Standardization of lung function
  tests" of the European community for Coal and Steel, Ph.H. Quanjer, ed., Luxembourg.
Schuster, K.-D., Pflug, K.P., Förstel, H., and Pichotka, J.P., 1979, Adaptation of respiratory mass spectrometer
  to continuous recording of abundance ratios of stable oxygen isotopes, in: "Recent developments in
  mass spectrometry in biochemistry and medicine", 2:451-462, A. Frigerio, ed., Plenum Publishing
  Corp., New York.
Schuster, K.-D., and Pflug, K.P., 1989, The overall fractionation effect of isotopic oxygen molecules during
  oxygen transport and utilization in humans, Adv.Exp.Med.Biol., 248: 151-156.
Schuster, K.-D. and Heller, H., 1994, Model analysis of oxygen isotope fractionation in humans due to
  disturbances of pulmonary gas exchange, Adv. Exp. Med. Biol. this volume.
Wagner, P.D., Dantzker, D.R., Dueck, R., dePolo, J.L., Wasserman, K., and West, J.B., 1976, Distribution of
  ventilation-perfusion ratios in patients with interstitial lung disease, Chest 69: 256-257.
Yamaguchi, K., Kawai, A., Mori, M., Asano, K., Takasugi, T., Umeda, A., Kawashiro, T., and Yokoyama, T.,
  1991, Distribution of ventilation and diffusing capacity to perfusion in the lung, Respir. Physiol.
  86:171-187.

50

CHANGES IN BLOOD FLOW IN THE COMMON FEMORAL ARTERY RELATED TO INACTIVITY AND MUSCLE ATROPHY IN INDIVIDUALS WITH LONG-STANDING PARAPLEGIA

M. T. E. Hopman,[1*] W. N. J. C. van Asten,[2] and B. Oeseburg[1]

[1] Department of Physiology
[2] Clinical Vascular Laboratory
University of Nijmegen
The Netherlands

INTRODUCTION

A spinal cord lesion refers to a partial or total disruption of the structural and functional integrity of the spinal cord, which is often caused by some sort of trauma, and results in impairments such as paralysis, loss of sensation and autonomic dysfunctioning.

Paralysis of the leg muscles leads to muscle atrophy and loss of muscle pump activity in the legs. In addition, the autonomic dysfunction, such as the lack of sympathetic innervation of the vascular system in the legs, may affect blood flow in the legs.

Previous studies (Davis 1993, Hopman 1994) have shown that both muscle pump inactivity and sympathetic dysfunction result in a disturbed blood redistribution during arm exercise in spinal cord injured individuals (SCI). Consequently, preload and stroke volume will not increase in SCI as usually seen in able-bodied subjects (ABS) during arm exercise.

To date, only a few studies have been directed towards peripheral circulation in SCI at rest and during exercise (Bidart and Maury 1973, Kinzer and Convertino 1989, Hopman et al. 1993). In these studies plethysmography was used; however, this method is vulnerable to movement artifacts and, therefore, is more suitable for the measurement of blood volume rather than blood flow changes.

Hardly any studies have investigated blood flow to the inactive and atrophied muscles in SCI individuals. Blood flow measurements at rest may provide information regarding long-term adaptation to inactivity and atrophy in SCI individuals. Similarly, analysis of the

* Address for correspondence: Maria T.E. Hopman, PO Box 9101, 6500 HB Nijmegen, The Netherlands. Fax: 080-540535; Phone: 080-614200; Email: M.Hopman@fysio.kun.nl

Oxygen Transport to Tissue XVII, Edited by Ince et al.
Plenum Press, New York, 1996

379

changes in blood flow during arm exercise may yield further information related to the redistribution of blood during exercise.

The purpose of this study was, therefore, to examine the blood flow in the common femoral artery in individuals with long-standing paraplegia in comparison to able-bodied subjects and, in addition, to examine changes that occur in the blood flow in the common femoral artery during submaximal arm exercise.

MATERIAL AND METHODS

Subjects

10 male individuals with paraplegia (P) and 10 non-lesioned male control subjects (C) participated in this study. P had complete spinal cord lesions between thoracic 4 and thoracic 12. None of the subjects had cardiovascular diseases or used medication likely to affect the results of the study. The study was approved by the Faculty Ethics Committee and all subjects gave their written informed consent.

Exercise Protocol

Exercise was performed using an electromagnetic arm-crank ergometer. Subjects were seated in a wheelchair with the elbows slightly flexed at the point of maximal extension. The axis of the arm crank was at shoulder level.

Subjects performed arm exercise at 50% of the individual maximal load during 25 minutes. The individuals maximal load was determined in a maximal pretest a few weeks before.

Measurement

Blood flow (BF) in the common femoral artery (CFA) was measured two times at rest and once every 5 minutes during arm exercise, while the subjects continued arm cranking. The mean of the two rest-measurements was used as rest value.

Apparatus

BF in the CFA was determined using an Echo Doppler Ultrasound device (Toshiba, SSA 270A).

BF was calculated from the mean velocity of red blood cell movement (Vmean) and the diameter (D) of the CFA. Pulsatility index (PI), an indication of peripheral resistance, was calculated from the Doppler spectra [(maximal velocity - minimal velocity)/mean velocity].

Statistical Analysis

A Student's t-test was applied to determine differences between P and C in physical characteristics and in blood flow at rest. The blood flow changes during arm exercise were compared to the initial blood flow at rest using a Student's t-test. Statistical significance was accepted at $p < 0.05$.

Table 1. Hemodynamic characteristics of the common femoral artery in paraplegic individuals and able-bodied control subjects at rest. Vmax = maximal velocity of the red blood cells; Vmin = minimal velocity of the rd blood cells; Vmean = mean velocity of the red blood cells; D = diameter of the common femoral artery; Flow = blood flow in the common femoral artery; PI = pulsatility index

	Paraplegics	Controls	p-value
Vmax (m/s)	0.56 ± 0.16	0.76 ± 0.15	0.01
Vmin (m/s)	-0.10 ± 0.08	-0.26 ± 0.08	<0.01
Vmean (m/s)	0.12 ± 0.04	0.14 ± 0.03	0.21
D (mm)	5.5 ± 1.3	10.1 ±1.1	<0.01
Flow (ml/min)	192 ± 114	687 ± 168	<0.01
PI	3.65 ± 0.32	3.35 ± 0.25	0.03

RESULTS

Physical characteristics of the subjects were not significantly different between P and C.

At rest, maximal velocity (Vmax), diameter (D) and BF of the CFA were significantly lower in P than in C, whereas minimal velocity (Vmin) and PI were significantly higher in P compared to C (table 1).

During exercise, blood flow in the CFA in C increased significantly, based on an increase in mean velocity (Vmean). No significant increase in BF or Vmean was observed in P (figure 1).

DISCUSSION

In this study Echo Doppler Ultrasound, which has typically been used usually for diagnostic purpose in relation to cardiovascular abnormalities, was used to assess BF in CFA at rest and during arm exercise in P and C.

The very low BF in the CFA in P is in agreement with the clinically observed cold and blue lower extremities in these individuals. The lower BF and the higher PI-value indicate that P have no excessive vasodilatation in the legs as a result of the sympathetic denervation. This casts doubt on the existence of a "venous blood pooling phenomena" in the lower limbs, as has been suggested by several researchers as an explanation for the disturbance in blood redistribution during arm exercise.

The vascular system in the legs of P-individuals has been adapted to inactivity and muscle atrophy as evidenced by a marked decrease in D and BF in the CFA. In other words, the oxygen delivery appears to be geared to a lower oxygen utilization in the atrophied muscle in P. Therefore, it is likely that the redistribution of blood during arm exercise is disturbed in P due to a limited amount of blood available in the legs. This is in agreement with previous studies; Hopman et al. 1994a found a decrease in venous capacity and an increase in outflow resistance in the legs of SCI using occlusion plethysmography, Walden et al. 1991 reported low blood flows and poor pulsatility in the vascular system of the legs in SCI individuals.

The unchanged BF of the CFA during arm exercise in P is in agreement with result found by Messenger et al. 1988, who reported no alterations in circulation in the legs of SCI subjects when either sitting or standing, measured by transcutaneous blood gas pressure

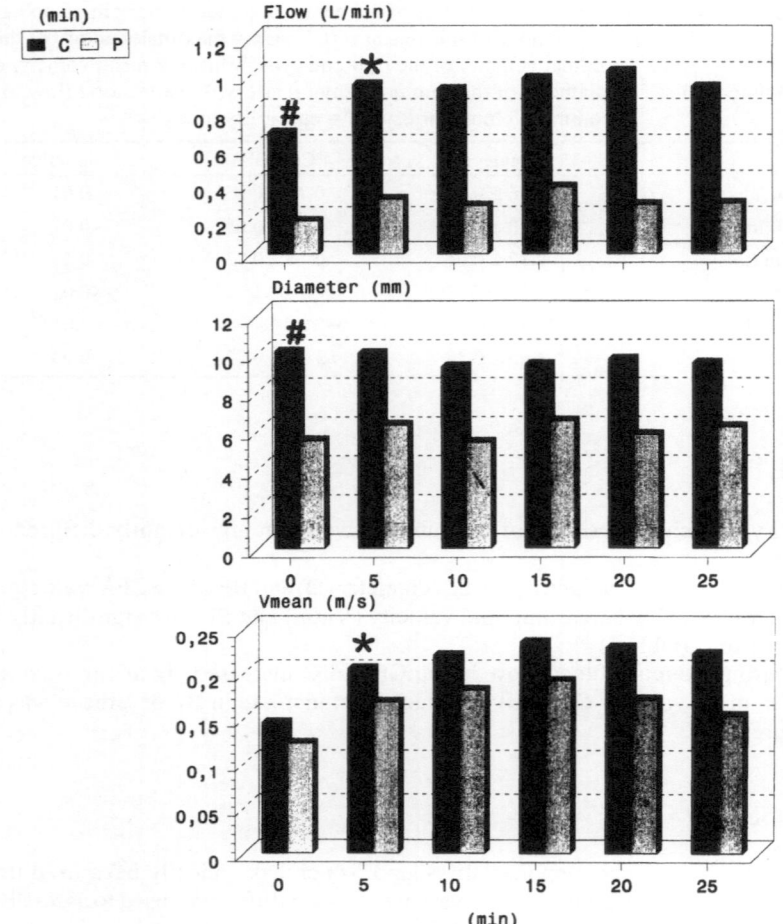

Figure 1. Blood flow (flow), diameter and mean velocity (Vmean) of the common femoral artery at rest and during 25 minutes (min) arm exercise in paraplegic individuals (P) and able-bodied control subjects (C). # = significantly different between P and C. * = significantly different between rest and exercise in C.

changes. The vascular system in the legs of SCI appears to act as a shunting system, which means no active regulation of vasoconstriction or vasodilation. It is known that the skin vascular system acts as a shunting system of which the capacity of shunting is determined by the skin temperature. The results of this study, therefore, suggest that skin blood flow represents the major part of the total leg blood flow in the P-individuals.

IN CONCLUSION

This study demonstrates a decrease in BF in the CFA at rest in P compared to C; this likely represents a vascular adaptation to inactivity an atrophy of the leg-muscles.

REFERENCES

Bidart Y, and Maury M. The circulatory behavior in complete chronic paraplegia. Paraplegia 11:1-24, 1973.

Davis GM. Exercise capacity of individuals with paraplegia. Med Sci Sports Exerc 25:423-432, 1994.

Hopman MTE, Verheijen PHE, Binkhorst RA. Volume changes in the legs of paraplegic subjects during arm exercise. J Appl Physiol 75:2079-2083, 1993.

Hopman MTE Circulatory responses during arm exercise in individuals with paraplegia. Int. J. Sports Med 15:126-131, 1994.

Hopman MTE, Nommensen E, Van Asten WNJC, Oeseburg B, Binkhorst RA. Properties of the venous vascular system in the lower extremities of individuals with paraplegia. Paraplegia, in press, 1994a.

Kinzer, S.M., V.A. Convertino. Role of leg vasculature in the cardiovascular response to arm work in wheelchair-dependent populations. Clin Physiol 9:525-533, 1989.

Messenger N, Rithalia SVS, Bowker P, Ogilvie C. Effects of ambulation on the blood flow in paralysed limbs. J Biomed Eng 11:249-252, 1988.

Walden R, Bass A, Ohry A, Schneiderman J, Adar R. Pulse volume recording disturbances in paraplegic patients. Paraplegia 29:457-462, 1991.

51

TUMOUR OXYGENATION

The Influence of Normobaric and Hyperbaric Oxygen

A. J. van der Kleij,[1] R. Kooijman,[1] J. B. A. Kipp,[2] and H. Obertop[1]

[1] Department of Surgery
[2] Department of Radiobiology
Academic Medical Center
University of Amsterdam
PO Box 22700, 1100 DE Amsterdam, The Netherlands

INTRODUCTION

The presence of hypoxic fractions has been shown in experimental tumour models[3, 12 13, 16,19,] as well as in human tumours[6,19], related to the tumour size and tumour growth rate. In clinical oncology several treatment modalities have been applied to overcome tumour hypoxia e.g., hyperthermia, chemical modifiers of tumour blood flow, hypervolemic blood transfusion, hypoxic cell sensitizers (e.g. misonidasole), hyperbaric oxygen (HBO)[7]. Of all the different radiation enhancement factors, oxygen possesses the highest enhancement ratio , 2.7 to 3.0, provided that molecular oxygen is present during irradiation to act as such[20]. The efficacy of HBO in decreasing radioresistance in tumours has been shown in experimental studies[5,11,] as well as in clinical trials.[9] Furthermore, it has been shown that oxygen tension distributions in human breast tumours[19] and cervix carcinoma[2,8] are sufficient to estimate the radiation response. Controlled trials with HBO have shown a 60% benefit, whereas in trials with misonidasole only a 21% benefit has been established[14]. A recent phase 2 study in our institution[21] showed an increase of the twenty-eight months cumulative survival rate from 12% to 28% in patients treated with radioactive Methyl-[131]Iodine Benzyl Guadinine (M-[131]IBG) combined with HBO, compared with a similar group of patients in a preceded period treated previously with M-[131]IBG without HBO. Two factors can be identified which could have contributed to these results. First, an unsealed source radiation brachytherapy was used during the HBO therapy. Secondly, a neuroblastoma cell is biochemically characterized by reduced endogenous defence mechanisms to oxygen derived free radicals[9,17]. These encouraging results prompted us to investigate the feasibility of sealed source brachytherapy combined with HBO therapy for solid tumours. Therefore we were interested in the time resolution of the oxygen tension distribution in an experimental tumour model with a polarographic needle electrode during normobaric (FiO_2: 21%) and hyperbaric (FiO_2: 21% and FiO_2: 100%) conditions without irradiation.

Oxygen Transport to Tissue XVII, Edited by Ince et al.
Plenum Press, New York, 1996

385

Table 1. Protocol oxygen distribution measurements in a R-1
Rhabdomyosarcoma rat tumour model

	T 1	T 2	T 3	T 4	T 5
ATA	1	3	3	3	1
FiO$_2$	21%	21%	100%	100%	21%
Time scale in minutes	–	20	35	105	158

MATERIAL AND METHOD

A R-1 rhabdomyosarcoma rat tumour model was used as previously described[1, 10]. Small fragments of R-1 rhabdomyosarcoma were subcutaneously implanted in both flanks and upperlegs of six Wag/Rij female rats during aether inhalation anaesthesia. After 2-3 weeks the tumours reached the minimum size (≥ 1 cm^3) for the tumour-oxygenation assessment. Rats (n=10) were placed in a multiplace hyperbaric chamber. Spontaneously breathing and measuring tumour oxygen distribution during adequate analgesia was obtained by a 0.25 ml aescoket (15 mg ketamine, 0.5 mg xylasine, 0.1 mg atropine) intramuscular injection. Ambient temperature was kept constant between 28 and 29°C. After one normobaric control oxygen distribution measurement the hyperbaric chamber was pressurized from one ATA to three ATA followed by a second oxygen distribution measurement. Thereafter oxygen was delivered to the rat by a mask as previously described [15] at a constant flow of 4 L. min^{-1}. Fifteen and eighty-five minutes later a third and a fourth oxygen distribution measurement was performed (table 1). The last oxygen distribution measurement was performed after decompression during a FiO2 of 21%

Tumour Oxygen Distribution Measurement

Tumour oxygen distribution measurements were determined with a polarographic pO$_2$ needle electrode (outside diameter 300 μm) as previously described[4]. The method of pO$_2$ measurement[4] by insertion of the probe by a motor-driven computerized micromanipulator, and this pO$_2$ electrode has been widely used to assess tissue oxygenation in experiments[22,23] as well as in clinical conditions[8,14]. Before each tumour pO$_2$ distribution measurement a calibration procedure was performed with ambient air and 100% nitrogen during normobaric conditions and hyperbaric conditions. The pO$_2$ measurement steps consisted of a rapid forward movement of 700 μm and an equally rapid backward movement of 300 μm ("pilgrim step"), resulting in an effective forward step of 400 μm and as such reducing as much as possible mechanical pressure on the tip of the pO$_2$ electrode. Because of the relatively small size of the tumours only between 10 and 15 local pO$_2$ values could be obtained in one puncture track. When after a pO$_2$ measurement a bleeding from the puncture channel occurred a new direction of the puncture channel was chosen. Negative values resulting from bending of the pO$_2$ electrode or too much pressure at the tip of the pO$_2$ electrode were excluded from calculation procedures. The results of ten rats were pooled and expressed in a pO$_2$ histogram.

RESULTS

All rats survived the experiment. At T = 1 the natural variation of tumour oxygenation is demonstrated (figure 1). At T = 2 the median pO$_2$ value decreased from 25.0 mm Hg to

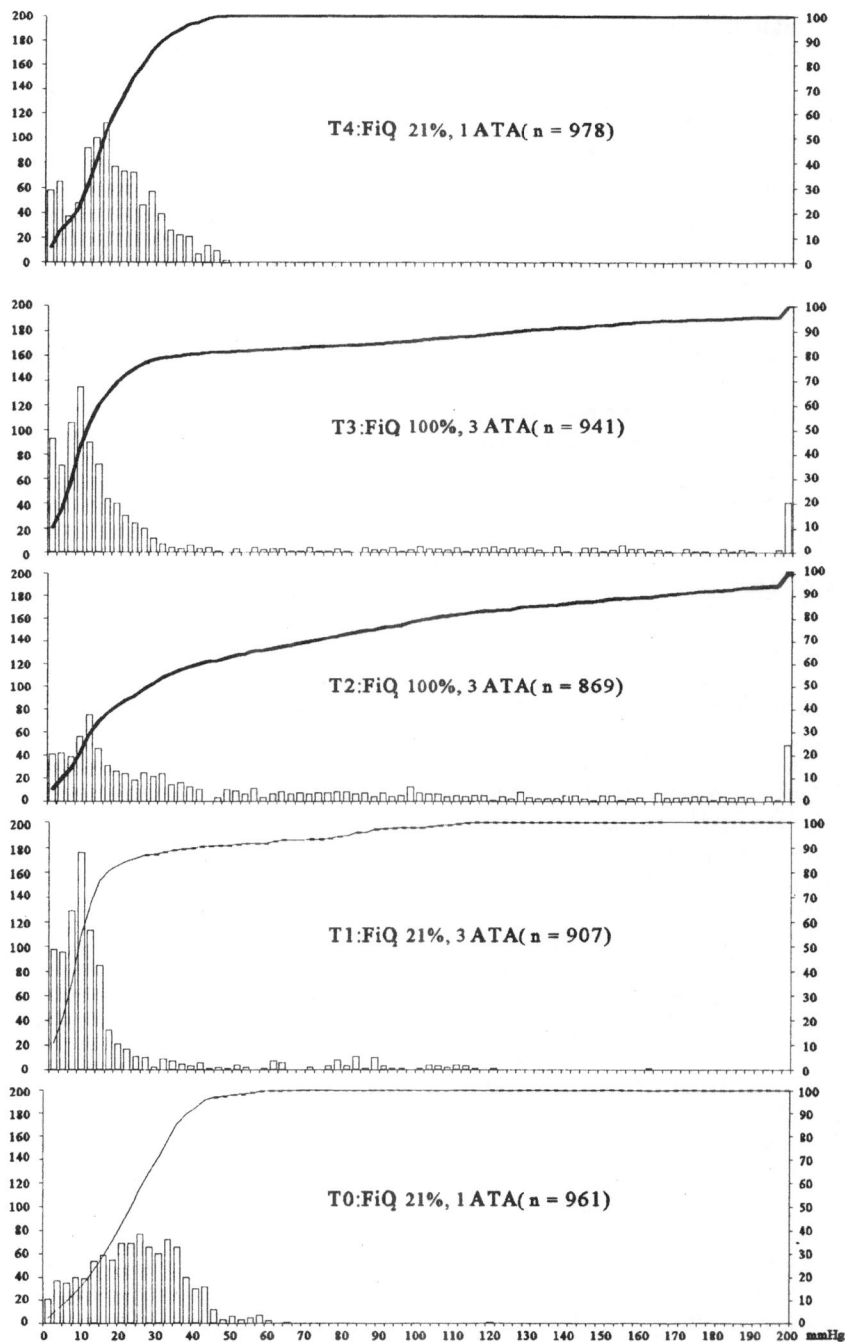

Figure 1. Changes in time of the tumour oxygen distribution in the rhabdomyosarcoma of the rat. A shift to the left induced by increasing ambient pressure from 1 ATA to 3 ATA (lower panel, control state and 2nd panel). Secondly, a shift to the right induced by increasing FiO_2 from 21% to 100% at 3 ATA (3rd panel) followed by a spontaneous shift to the left (4th panel) and finally (upper panel) the oxygen distribution at 1 ATA and FiO_2: 21%

Table 2. Tumour oxygenation during normobaric and hyperbaric oxygenation.
Results are derived from pooled data and expressed in mean, 90% , 50%
(=median), and 10% percentile mm Hg

	T1	T2	T3	T4	T5
Mean	24.8	16.7	54.8	36.9	18.2
10 percentile	7.0	2.0	4.0	2.0	3.0
50 percentile	25.0	9.0	26.0	11.0	17.0
90 percentile	40.1	40.0	151.1	128.0	33.0

9.0 mm Hg (table 2). At T = 3 an increase of the median pO_2 value from 9.0 mm Hg to 26.0 mm Hg was observed accompanied by a slight increase of the 10% percentile from 2.0 mm Hg to 4.0 mm Hg suggesting not an optimal reduction of the hypoxic values. After a longer period of exposure to hyperbaric oxygenation (T = 4) the median pO_2 value returned to 11.0 mm Hg, whereas the 10% percentile value returned to 2.0 mm Hg. At T = 5 the oxygen distribution is still shifted to the left (hypoxia) compared with the control values (T = 1).

DISCUSSION

This study was designed to investigate the changes in time of the oxygen distribution in a rhabdomyosarcoma rat tumour model. To obtain changes in tumour oxygenation, first the rats were subjected to pressurization (from 1 ATA to 3 ATA) and subsequently the FiO_2 was increased from 21% to 100% at 3 ATA. The control (normobaric) oxygenation was nearly Gaussian distributed with a median pO_2 value of 24.8 mm Hg and a 10 percentile value of 7.0 mm Hg. The relative small hypoxic fraction may be associated with the subcutaneous site of the tumour[12] . Using a micro pO_2 electrode to assess tumour oxygenation other investigators[16] found a mean value of ± 13 mm Hg in a rat rhabdomyosarcoma pretreated with perfluorchemicals. These differences may be caused by heterogeneity of naturally-occurring hypoxic cells. Since no analysis of cell survival was performed in this study, measured hypoxic values may be caused by tumour necrosis or real tumour hypoxia. Pressurization from 1 to 3 ATA with ambient air revealed a shift to more hypoxic values; a median value of 9.0 mm Hg and a 10 percentile value of 2.0 mm Hg. This observation may be associated with the physiologically vasoconstrictive reaction initially caused by increased arterial oxygenation which may occur in the absence of anaesthetics. This event may indicate that the distribution of oxygen in this tumour model is subject to alterations in ambient air. A similar phenomenon was found during increasing FiO_2 from 21% to 100% at 3 ATA intending to reduce the number of hypoxic fractions. However, considerable less pronounced reduced hypoxic values were registered: the 10 percentile value increased from 2.0 mm Hg to 4.0 mm Hg. Perhaps the timing of increasing FiO_2 before or after pressurization may influence the final result. Using the cryophotometric micro-method to quantify indirectly tumour oxygenation of a subcutaneously implanted DS-carcinosarcoma[18], pressurization from 1 to 3 ATA with pure oxygen revealed a considerably enhanced reduction of hypoxic values compared with our results. In order to induce a maximum reduction of the hypoxic values this finding suggests that in this rhabdomyosarcoma tumour model first the FiO_2 has to be raised and subsequently the ambient air. The duration of exposure to an elevated FiO_2 and an increased ambient pressure seem to be other factors influencing the oxygen distribution in this tumour model. After seventy minutes pressurization at 3 ATA with a FiO_2 of 100% a "spontaneous" shift to more hypoxic values of the tumour oxygenation was observed and providing evidence that tumour oxygenation during hyperbaric conditions is subjected to changes in time.

It is concluded from this experiment that the oxygen tension distribution in a rhabdomyosarcoma rat tumour model changes in time during hyperbaric oxygenation. These observations must be taken into account and may be therapeutically important for the application of radiation enhancement by hyperbaric oxygen for solid tumours.

REFERENCES

1. Barendsen, G.W., Broerse, J.J. 1969, Experimental radiotherapy of a rat rhabdomyosarcoma with 15 MeV neutrons and 300 kV x-rays. I. Effects of single exposures. *Europ. J. Cancer 5*: 373-391.

2 Bergsö, P. Evans, J.C., 1971, Oxygen tension of cervical Carcinoma during the early phase of external irradiation. *Scand. J. Clin Lab Invest.* 27:71-82

3 Cole, M. A., Crawford, D. W., Warner, N. E., Puffer, H. W. 1983,Correlation of regional disease and in vivo pO_2 in rat mammary adenocarcinoma. *Am J Pathol.* 112:61-67.

4 Fleckenstein, W., Heinrich, R., Kersting, TH., Schomerus, H., Weiss, Ch. 1984, A new method for the bedside recording of tissue pO_2 histograms.*Verh. Dtsch. Ges. Inn. Med.* 90: 439.

5 Fujimura, E. 1974, Experimental studies on radiations effects under high oxygen pressure. *J. Osaka Dent. Univ.* 19:100-108.

6 Gathenby, K., Kessler, H.B., Rosenblum, J.S., Coia, L.R., Moldofsky, P.J., Hartz, W. H., Broder, G.J. 1988, Oxygen distribution in squamous cell carcinoma metastases and its relationship to outcome of radiation therapy. *Int. J. Radiat. Oncol. Biol. Phys.* 14: 831-838.

7 Henk, J.H., Smith, C.W. 1977, Radiotherapy and hyperbaric oxygen in head and neck cancer. *Lancet* 2:104-105.

8 Höckel, M., Vorndran, B., Schlenger, K., Buaßmann ,E., Knapstein, P.G. 1993, Tumour oxygenation: A new predictive parameter in locally advanced cander of the uterine cervix. *Gynecology Oncology* 51:141-149.

9 Iancu, T. C., Shiloh, H., Kedar, A. 1988, Neuroblastomas contain iron-rich ferritin. *Cancer (Phila.).* 61:2497-2502.

10 Kipp, J.B.A., Kal, H.B., Gennip van, A.H., Berkel van, A.H. 1993, Treatment of the rat R-1 rhabdomyosarcoma with methotrexate and radiation; effects of timing on cell survival and tumour growth delay.*J. Cancer Res Clin Oncol.* 119:215-220.

11 Milas, L., Hunter, N.M., Ito H., Brock W.A., Peters L.J., 1985. Increase in radiosensitivity of lung micrometastases by hyperbaric oxygen. *Clin. Exp.Metastasis* 3:21-27

12 Moulder, J.E., Rockwell S., 1984. Hypoxic fractions of sloid tumours: Experimental techniques, methods of analysis, and a survey of exsisting data. *Int. J. Radiation Oncology Biol. Phys.* 10:695-712.

13 Meuller-Klieser, W., Vaupel, P., 1983. Tumour oxygenation under normobaric and hyperbaric conditions. *British J. of Radiologie.* 56:559-564.

14. Nias, A.H.W., 1991. The oxygen problem in radiotherapy. In: *New Developments in Fundamental and Applied Radiobiology.* Colin.B. Seymour & Carmel Mothersill (eds). Taylor & Francis Ltd. 4 John St. London WC 1N 2 ET. 318-327.

15 Richey, K. J., Engrav, L. H., Avlin, E. G., Murrat, M. J., Gottlieb, J. R., Walkinishaw, M. D., 1989. Topical growth factors and wound contraction in the rat: Part I. Literature review and definition of the rat model. *Ann. Plast. Surg.* 23:159-165.

16 Sostman, H.D., Rockwell, S., Sylvia, A.L., Madwed, D., Cofer, G., Charles, H.C., Negro–Vilar, R., Moore, D., 1991. Evaluation of BA 1112 rhabdomyosarcoma oxygenation with microelectrodes, optical spectro-photometry, radiosensitivity, and magnetic resonance spectroscopy. *Mag. Res. Med.* 20:253-267.

17 Steinkühler, C., Mavelli, I., Melini, G., Piacentini, M., Rossie, L., Weser, U., Rotilio, G., 1988. Antioxidant enzyme activities in differentiating human neuroblastoma cells. *Ann NY Acad Sci.* 551:137-140.

18 Torres Folko, I.P. Leunig, M., Yuan F., Intagietta M., Jain R.K., 1994. Non-inavsive measurements of Microvascular and Interstitial O_2 profiles in human tumors and in SCID. *Proc. Natl Acad. Sci. USA.* 91: 2081-2085.

19 Vaupel, P., Schlenger, K., Knoop, C., Höckel, M., 1991. Oxygenation of human tumours; Evaluation of tissue oxygen distribution in breast cancers by computerized O_2 tension measurements. *Cancer Res.* 51: 3316-3322.

20. Vaupel P., Jain R.K. (eds), 1991. Blood flow and oxygenation of human tumours. *In: Tumor blood supply and metabolic microenvironment; characterization and implications for therapy.* Gustav Fischer Verlag, Stuttgart-New-York, 165-185.

21 Voûte, P.A., Kraker de J., Hoefnagel C.A., 1992. Tumours of the Sympatethic Nervous System: Neurobalstoma, Ganglioneuroma and Phaeochromocytoma. *In: Cancer in Children. Clinical Management. 3rd edition.* Voûte P.A., Barrett A., Lemerle J. (eds), Springer Verlag, Heidelberg (ISBN 0-387-55186-7)

22 Wiedemann, G., Roszinski, S., Biersack, A., Mentzel M., Weiss C., Wagner T., 1992. Treatment efficacy, intratumoural pO_2 and pH during thermochemotherapy in xenotransplanted human tumours growing in nude mice. *Contrib Oncol. Karger.* 42: 556-565.

23 Zywietz, F., Reeker W., Kochs E., 1994. Studies in tumour oxygenation in a rat rhabdomyosarcoma during fractionated irradiation. *Presented at Isott, 22nd Annual Meeting, Plenum Press (in press).*

52

RISE OF TISSUE TEMPERATURE INDUCED BY REDUCED BLOOD PERFUSION CAUSED BY EXTERNAL PRESSURE

Y. Yamada,[1] H. Ishiguro,[2] M. Yamashita,[2] T. Tanaka,[3] M. Takeuchi,[4] and H. Kawamura[5]

[1] Mechanical Engineering Laboratory
1-2 Namiki Tsukuba, Ibaraki 305, Japan
[2] University of Tsukuba
1-1-1 Ten-Oh-Dai Tsukuba, Ibaraki 305, Japan
[3] Electrotechnical Laboratory
Tsukuba, Ibaraki 305, Japan
[4] Toin University of Yokohama
Midori-ku, Yokohama 227, Japan
[5] Science University of Tokyo
Yamazaki Noda, Chiba 278, Japan

INTRODUCTION

When homeostatic control is suppressed by anesthetizing patients, their body temperatures tend to lower to room temperature. To keep the patients' body temperatures in an appropriate range, they are laid on warming mattresses in which warm water is circulated through flexible tubes embedded in plastic mattresses. Even though the water temperature is accurately controlled within a safe limit to avoid burn injuries, there have been cases where the body portions contacting the mattresses suffered from severe damage, showing symptoms of thermal injuries after a prolonged surgery of more than five hours (Crino & Nagel 1963, Scott 1967). Although the damage seemed light at the surface, it was often severe at deeper tissues. While it is likely that reduced perfusion is the origin of the injury, the process involved in the injury is still unknown.

By close look at the various conditions related to the damage by warming mattresses, it has been suggested that the damage may be caused by some thermal conditions leading to burn injuries. Simple analysis of heat transport (Tanaka et al. 1991) has been reported showing that such burn injury may be possible if metabolic heat generation exists even under the conditions of reduced blood perfusion and thermal insulation to outside.

From the above thermal point of view for the damage to anesthetized patients, we have conducted experiments using anesthetized pigs. Many physical and physiological parameters were measured, and clear temperature rises were observed at the body portions

Oxygen Transport to Tissue XVII, Edited by Ince et al.
Plenum Press, New York, 1996

391

where pressure was loaded. The blood perfusion rate, pH, and oxygenation state of the tissue decreased with the pressure, and the tissue temperatures increased as result. The bioheat transfer equation was employed to analyze the temperature distribution of the tissue. The heat balance between the heat removal by blood perfusion and heat generation by metabolism has been found to be the key phenomenon for the tissue temperature rise. We have obtained a good agreement on the temperature rise between the experimental and theoretical results. It is suspected that the measured temperature rise at the pressurized tissues can be a trigger of the damage.

EXPERIMENTAL METHOD

Anesthetized pigs were laid on a warming mattress to measure various physical and physiological parameters of locally pressurized tissues, such as temperature, heat flux, blood perfusion, pH, and oxygenation. The commercially available mattress was made of plastic measuring 500 mm by 800 mm. A flat flexible tube with a 10 mm longer axis was embedded in the mattress, and warm water flowed inside the tube to keep the body temperature of the pigs (usually of human patients). The pigs' body parts contacting the mattress were shaved. Total of six pigs with average weight of 25 ± 3 kg were used, and the period of experiment lasted four to 8 hours from the start of anesthetization. The water temperature was controlled under 42°C which is believed to be the beginning temperature of protein metamorphosis. To simulate the reduced blood perfusion by external pressure, a sand bag up to 25 kg was loaded on the haunch portion of the pigs.

We measured temperatures, heat flux, blood perfusion rate, pH, and oxygen saturation at the local body parts and monitored blood pressure, heart pulse rate, and arterial blood gas of the pigs throughout the experiments. The temperatures were measured at six locations in the haunch; three at 30 mm depth and three at 60 mm depth from the mattress, as shown in Fig. 1. Sheathed thermocouples with 0.2 mm diameter and 0.1°C resolution were inserted into the haunch. Also measured were the temperatures of the skin contacting the mattress and sand bag, the haunch temperature 30 mm deep from the sand bag, and the rectal temperature which was considered as the core temperature of the whole body. Heat flux meter (Showa Denko, HFM-EL6) was located at the interface between the mattress and haunch, and a pH sensor with 3 mm diameter (Corning, M-220) and a blood flow meter (Biomedical Science, LBF-III) were inserted between the six thermocouples in the haunch. Oxygen saturation of venous blood at the haunch was monitored in real time by a non-inva-

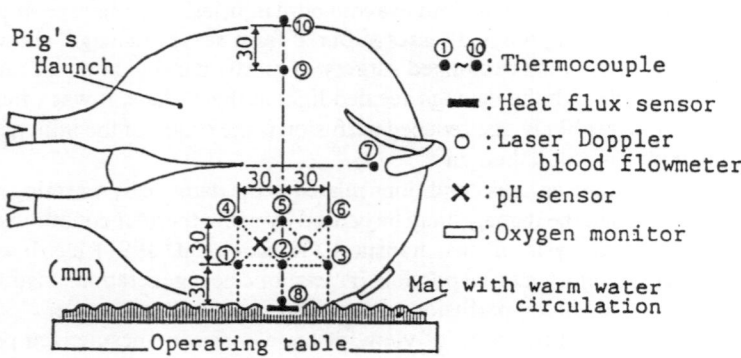

Figure 1. Locations of sensors.

sive oxygen monitor using near-infrared light at 780 nm, 805 nm and 830 nm (Shimadzu Corp., OM-100A) (Tamura et al. 1989). The distance between the emitting and detecting optodes was about 5 cm.

We also measured the pressure distribution at the contact surface of the pressurized haunch. The maximum pressure was 9 g/cm^2 for no load while it was 125 g/cm^2 for 25 kg load. The pressure distribution was localized in a small region for 25 kg load. Cyanosis was observed at the pig's lower extremity showing deoxygenation.

EXPERIMENTAL RESULTS

The results of one of six experiments are shown below as the typical one. The other results showed similar trends as those shown here. Figure 2 (a) (b) (c) show the temporal variation of the temperatures, heat flux, pH, blood perfusion, and oxygenation for the case of the same water temperature as the body core temperature. The abscissa is the time from the start of anesthetization, and in this case the pressure loading began at the time 110 min after confirming the stable state. Pressure loading continued for 130 min until the time 240 min. Water temperature was electronically controlled quite accurately as 39.1°C, but the rectal temperature of the pig showing the core temperature was subject to fluctuation because of the suppressed homeostasis by anesthetization. Therefore, it was manually controlled by covering the body with a blanket when it was lower than the desired temperature and by

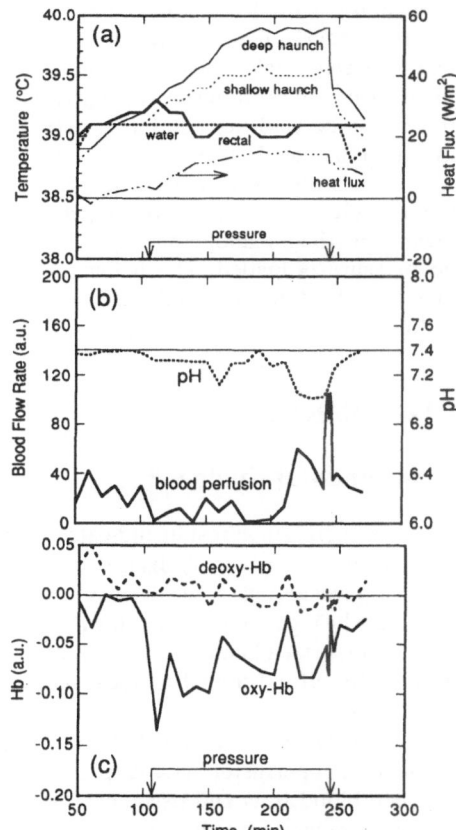

Figure 2. Time course of temperature, heat flux, pH, blood perfusion rate, and oxygenation for the case of same water temperature as the core temperature.

cooling the body with ice bags when it was higher. The rectal and water temperatures were precisely controlled to 39.0 ± 0.2°C during the experiment. Soon after the start of loading the haunch temperatures began to rise and they reached equilibrium temperature at about the time 200 min, 90 minutes after loading started. The equilibrium temperatures at deep (60 mm) and shallow (30 mm) haunch tissues were 39.9°C and 39.5°C, respectively. Only two temperatures out of six thermocouples are shown in Fig. 2(a), but the other temperatures also showed similar behavior. Generally speaking, the deeper thermocouples recorded higher temperatures than the shallow ones. Because the haunch temperatures were higher than the water temperature, the heat flux was positive during the pressure loading, showing the heat is flowing from the haunch to water. [Ed: the f

The blood perfusion shown in Fig. 2(b) decreased during the pressure than the level before the pressure although a higher peak was observed at the time 220 min. Immediately after the unloading, the blood perfusion showed a very sharp rise and resumed to the level before the loading. The record of blood perfusion shows a fluctuation throughout the experiment and this is supposed to be characteristic either to the blood vessel network or to the blood flow meter because the tissue volume being monitored was unknown. Also a small motion of the pig body is attributable to the fluctuation. pH gradually decreased from the normal value of 7.4 to the minimum of 7.0 at the time about 230 min. But immediately after the release it resumed the normal value. Oxygenation during the loading clearly decreased although deoxygenation was kept almost constant as shown in Fig. 2(c).

All of blood flow, pH and oxygen monitors qualitatively suggested the decrease of blood perfusion which agreed with the observed cyanosis. If we assume that the blood perfusion is completely blocked at the pressurized portion, the temperature of the portion must be equal to the water temperature same as the core temperature because no metabolic heat generation takes place without blood perfusion. However, the experimental results of the pressurized portion showed higher temperatures than the water/core temperature. From this it can be said that the blood perfusion was not completely blocked but decreased to maintain the metabolism.

It is observed that the temperatures of the haunch quickly restored their normal temperatures same as the core temperature immediately after the release of the loading. This means that the metabolic heat generation is smoothly transported by blood perfusion when no pressure is loaded. The balance between the heat transport by blood perfusion and metabolic heat generation determines the equilibrium temperature. This is discussed in the following.

DISCUSSION

Experimental Evidence

Experimental results indicate that both the heat transport by blood perfusion and the heat generation by metabolism decreased because of the reduced blood perfusion by the external pressure. Now we discuss and compare the rates of reduction in the heat transport and in the heat generation from the measured temperatures. From the measured temperatures of the pressurized haunch, we can estimate the metabolic heat generation at the part, Q_s, by using a simplified model of local balance of heat transport in which the metabolic heat generation is balanced with the sum of the heat transport to water and to blood perfusion.

Now let us assume that the blood perfusion rate under pressure has decreased to 10% of the normal perfusion without pressure; blood perfusion ratio $R_b = 0.1$. The heat transport between blood flow and tissue can be proportional to the blood perfusion. The measured average temperature at haunch for this case is 39.7°C and these values are used to estimate

Q_s for pressurized tissues. For the case of Fig. 2, Q_s is estimated as $Q_s = 0.33 \times 10^3$ W/m^3 which is 31% of the normal value of 1.5×10^3 W/m^3 (Bendict, 1974); the ratio of heat generation under pressure to that of no pressure $R_Q = 0.31$. This means that the decrease in the heat generation caused by the reduced blood perfusion is 69% although the decrease in heat transport by blood perfusion is 90 %. As a result, the tissue temperature rises to 39.7°C to achieve the equilibrium which is 0.6°C higher than that without pressure.

Because of the uncertainty in the data of blood flow meter, we assumed the reduction of blood perfusion to 5% ($R_b = 0.05$) and 20 % ($R_b = 0.2$). then the estimation of Q_v varied as shown in the column 1-1 in Table 1. Other data were also analyzed to give other columns (2-1, 3-1, 3-2, 3-3) in Table 1. The averaged R_Q values are 0.15, 0.37 and 0.80 for $R_b = 0.05$, 0.10 and 0.20, respectively. The ratios of R_Q to R_b, $R = R_Q/R_b$, are 3.1, 3.7 and 4.0 for $R_b = 0.05$, 0.10 and 0.20, respectively. From these data, it has been shown that tissue heat generation does not decrease so much as heat transport by blood flow does and this leads to a higher equilibrium temperature. It should be noted that the value of 1.60 for R_Q in Table 1 is unrealistic thus indicating that $R_b = 0.2$ is unrealistic too.

Theoretical Consideration

In order to verify the experimental evidence described above, we have tried to formulate the heat balance in tissues. According to Pennes (1948) transient heat balance in tissues is expressed by the following bioheat transfer equation.

$$\rho C \frac{\partial T}{\partial t} = \nabla(k\nabla T) - \rho_b C_b W_b (T - T_a) + Q_m$$

(1)

where T, t, r, C, k, W are temperature, time, density, specific heat, thermal conductivity and perfusion rate, respectively, with subscripts b and a representing blood in general and arterial blood, respectively. The metabolic heat generation can be written in the following form.

$$Q_m = \frac{W_b}{W_{bs}} \frac{\Delta S(\text{pH}, T, P\text{CO}_2, P\text{O}_2)}{\Delta S_S} Q_S$$

(2)

Here ΔS is the drop of oxygen saturation of hemoglobin which is a function of temperature, pH, O_2 partial pressure, CO_2 partial pressure and so on, as shown in Fig. 3. The

Table 1. Estimated ratios of heat generation under pressure to that at normal perfusion

Data number	1-1	2-1	3-1	3-2	3-3
Time(min)	240	460	160	250	450
Water temp., T_w (°C)	39.1	38.2	38.4	38.4	42.4
Core temp., T_b (°C)	39.1	38.2	37.0	36.0	37.5
Press. haunch temp., Th (°C)	39.7	38.7	37.9	36.9	39.7
Ratio of heat generation, R_Q					
$R_b = 0.05$	0.18	0.15	0.14	0.07	0.22
$R_b = 0.10$	0.31	0.26	0.33	0.27	0.68
$R_b = 0.20$	0.57	0.47	0.71	0.66	1.60

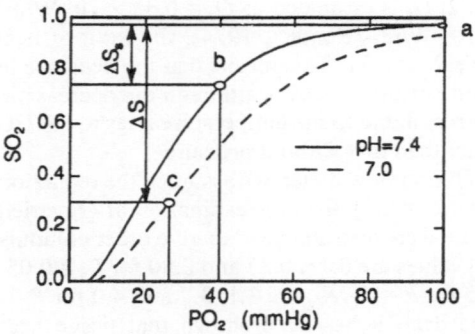

Figure 3. Oxygen dissociation curves and the drop of the oxygen saturation for hemoglobin.

subscript s represents the resting state. We have assumed that the oxygen transport from hemoglobin to tissue is fast enough for oxygen to be distributed to tissue volume homogeneously and for hemoglobin to achieve an equilibrium state with tissue oxygen partial pressure. At normal state with pH = 7.4, hemoglobin in arterial blood exists at the state **a** with $P_aO_2 = 100$ Torr and O_2 saturation of $S_a = 0.975$ in Fig. 3, and hemoglobin in venous blood which is equilibrium with tissue reaches the state **b** with $P_sO_2 = 40$ Torr and $S_S = 0.741$ to release oxygen with the saturation drop $\Delta S_s = 0.234$. When the blood perfusion is blocked by external pressure the tissue oxygen partial pressure and pH decreases with the state **c**. then the oxygen saturation drop ΔS becomes greater than ΔS_s. This greater drop makes the heat generation by metabolism decrease less than that of heat transport by blood perfusion.

For steady state and homogeneous distribution of temperature, eqn. (1) is simplified, and the equilibrium temperature is given by eqn. (3).

$$T = T_a + \frac{\Delta S(\text{pH}, T, P\text{CO}_2, P\text{O}_2)}{\rho_b c_b W_{bs} \Delta S_S} Q_s \qquad (3)$$

Note that the blood perfusion rate W_b does not appear in the above equation and that the equilibrium temperature always exceeds the arterial blood temperature (body core) T_a because the second term in the right hand side is positive. To estimate the equilibrium temperature from eqn. (3), we use the formulation of oxygen dissociation curve given by Kelman (1966). Using Kelman's formula, the equilibrium temperatures shown in Fig. 4 are obtained as a function of tissue oxygen partial pressure $P\text{O}_2$ with pH and arterial blood (body

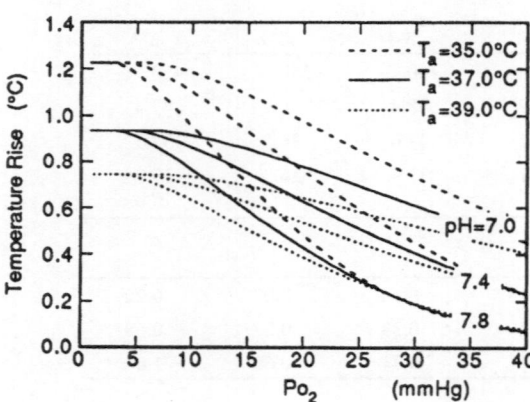

Figure 4. Temperature rises at equilibrium state as a function of oxygen partial pressure with pH and core temperature as parameters.

core) T_a temperature as parameters. PCO_2 is fixed as the standard value of 40 Torr. Other standard values and physical properties used in the calculation are given as appropriate

It is shown in Fig. 4 that the tissue temperature rises about 1°C when the blood perfusion is blocked and tissue oxygen pressure reduces to almost zero. The temperature rise depends on the arterial blood (core) temperature and pH. The lower the core temperature and pH, the higher the tissue temperature rises. This amount of temperature rise agrees very well with experimental results of the maximum temperature rise obtained in Fig. 2.

The observed and predicted temperature rises in the above are not so great as to lead to injury of tissues under normal environment, but it may possible that this small temperature rise can trigger the occurrence of the injury. It is necessary to study the injury more from physiological and pathological points of view in addition to the heat transport point of view in this report.

CONCLUSION

By the measurements of temperature, pH, oxygenation etc., of pressured haunch of anesthetized pigs and the theoretical study of heat transport of tissues, it is concluded that the decrease in the blood perfusion rate caused by external pressure induces temperature rises of the pressurized tissues when the heat loss to the environment is minimal. This is closely related to the linear decrease of heat transport and nonlinear decrease of oxygen transport by blood perfusion in tissues. This temperature rise may potentially trigger the occurrence of burn injuries under normal temperature environment.

REFERENCES

Bendict, D., 1974, *Thermoregulation and Bioengineering*, American Elsevier, 114.

Crino, M. H. and Nagel, E. L., 1963, Thermal burns by warming blankets in the operating room, *Anesthesiology*, 29, 149-150.

Housain, T., 1953, Experimental study of some pressure effects on tissues with reference to bed-sore problem, *J. Path. & Bact.*, 66, 347, 1953.

Kelman, G. R., 1966, Digital computer subroutine of oxygen tension into saturation, *J. Appl. Physiol.*, 21, 1375-1376.

Lindan, O., 1961, Etiology of decubitus ulcers: Experimental study, *Arch. Phys. Med.*, 42, 774-783.

Pennes, H. H., 1948, Analysis of tissue and arterial blood temperatures in the resting forearm, *J. Appl. Physiol.*, 1, 93-122.

Scott, S. M., 1967, Thermal blanket injury in the operating room, *Arch. Surg.*, 94, 181.

Tamura, M., Eda, H., Takada, M., and Kubodera, T., 1989, New instrument for monitoring hemoglobin oxygenation, *Adv. Exp. Med. Biol.*, Vol. 248, *Oxygen Transport to Tissue XI*, 103-107.

Tanaka, T., Yamada, Y., and Ishiguro, H., 1991, A consideration of thermal injuries occurring in anesthetized patients undergoing operations, *Proc. 3rd ASME/JSME Thermal Engineering Joint Conference*, Eds. Lloyd, J. R. and Kurosaki, Y., ASME Book No. 10309A, 299-304.

53

SYSTEM PARAMETER ANALYSIS OF NIR-TRS SPECTRA FROM HOMOGENEOUS MEDIA WITH AND WITHOUT AN ABSORBING BOUNDARY AND FROM HETEROGENEOUS MEDIA WITH A SINGLE ABSORBER

Kyung A. Kang,[1] Duane F. Bruley,[1] T. Kitai,[2] and Britton Chance[2]

[1] Bioengineering Program, College of Engineering
University of Maryland Baltimore County (UMBC)
5401 Wilkens Ave., ECS 202, Baltimore, Maryland 21228
[2] Johnson Research Foundation
Department of Biochemistry and Biophysics
University of Pennsylvania
Philadelphia, Pennsylvania 19104

INTRODUCTION

Traditional near infrared-time resolved spectroscopy (NIR-TRS) analysis includes fitting a spectrum with a known analytical solution to obtain the media absorption and scattering coefficients or computing mean time-of-flight for heterogeneity localization. The benefit of applying frequency response analysis to NIR-TRS data reduction is that magnitude ratio (MR) and phase shift (ϕ) information in a wide frequency range can be obtained from a single TRS spectrum. The sharper the input pulse is the wider the frequency range values one can obtain. This analysis has been applied to NIR-TRS spectra to obtain optical parameters from homogeneous systems (Kang et al., in press) and for localizing absorbers in heterogeneous systems (Kang et al., 1994). MR and ϕ can also be obtained from phase modulation spectroscopy (PMS). PMS, however, provides these values at a fixed modulation frequency unless the instrument is sophisticated enough to change the modulation frequency in a wide range. Another benefit is that multiple system parameters can be obtained from a single TRS spectrum, i.e., MR, ϕ, steady state gain (K), break frequency (f_b)/time constant (τ), and system order (n). Steady state gain (K) is a valuable parameter that can be used to check system linearity. Other new parameters (f_b/τ, n) are still to be studied in an optical sense.

Oxygen Transport to Tissue XVII, Edited by Ince et al.
Plenum Press, New York, 1996

The objective of this study is to better understand highly scattering media, such as biological systems, through multiple parameter analysis of TRS spectra via frequency response analysis. Changes in multiple parameters with changes in optical parameters were studied for homogeneous and heterogeneous media. Two major investigations were performed; (1) Experimental: the change in system parameters with respect to the distance between the boundary and the source-detector (S-D) unit. (2) Computer simulation: the sensitivity of system parameters with respect to frequency in homogeneous and heterogeneous systems with an absorber. In this paper, the sensitivity is defined as the degree of the change in parameters relative to the optical property changes in the system.

THEORY

Transforming time domain NIR-TRS information to the frequency domain has been undertaken by many researchers (Duncan et al., 1993). However, frequency response analysis adds more information to the simple transformation between two domains. The basic theory of frequency response analysis resulting from NIR-TRS pulse testing is described by Kang and Bruley (Kang et al., 1994; Kang et al., in press; Bruley, this volume). A brief overview of the technique is presented here.

This method analyzes the system transfer function, [G(s)], which is defined to be the Laplace transformed output, Y(s), divided by Laplace transformed input, X(s). When the Laplace independent variable, s, is substituted by $j\omega$ the transformation becomes a Fourier transformation:

$$G(\omega) = \frac{Y(\omega)}{X(\omega)} = \frac{\int_0^{T_y} y(t)e^{-j\omega t}dt}{\int_0^{T_x} x(t)e^{-j\omega t}dt} \tag{1}$$

where j is an imaginary number, $x(t)$ and $y(t)$, respectively, are input and output function in the time domain and $X(\omega)$ and $Y(\omega)$, respectively, are input and output function transformed to the frequency domain.

Equation (1) was integrated by a computer code (Bruley, 1974) using Filon's numerical integration method (Filon, 1928). This quadrature method provides a more stable solution than other numerical integration methods.

The results can be expressed in terms of magnitude ratio (MR) and phase shift (ϕ):

$$MR = |G(\omega)| = \sqrt{Re^2(\omega) + Im^2(\omega)} \tag{2}$$

$$\phi = \phi|_{G(\omega)} = \tan^{-1}\left[\frac{Im(\omega)}{Re(\omega)}\right] = \phi|_{Y(\omega)} - \phi|_{X(\omega)} \tag{3}$$

It should be noted that ϕ, as presented here, is the negative of *phase* typically used among researchers in the NIR spectroscopy field.

Normalized frequency content (NFC) is a good indicator for the reliability of the reduced data.

$$NFC = \frac{\int_0^{T_x} x(t)e^{-jwt}dt}{\int_0^{T_x} x(t)dt}$$

(4)

Conventionally, the MR and ϕ values at a frequency where NFC values are less than 0.3 are considered to be spurious. The break frequency, f_b, is the minimum frequency at which the system becomes incapable of transferring the signal from the high frequency input to the output of the process and the signal starts to attenuate. The time constant, τ, is the inverse of break frequency and represents the time for the response function to reach to thirds of the total response.

Detailed descriptions of these parameters are well illustrated in many engineering control books and articles (Bruley and Prados, 1964; Clements and Schnelle, 1963; Coughanowr, 1991; Lewis et al., 1967; Leuben, 1991).

INSTRUMENTS AND MATERIALS

The following instruments and materials were used to carry out the experiments and computer simulation.

1. Near infrared - time resolved spectroscope (NIR-TRS): The TRS system was manufactured by Hamamatsu Photonics, KK (Hamamatsu, Japan). The wavelength used was 780 nm. This TRS is a single photon counting system with a pulse repetition rate of 5 MHz. The detector system was a multi-microchannel plate photomultiplier tube (MCP-PMT) (Fig. 1).

2. Optical fibers for the source and the detector: Glass fiber with a diameter of 200 μm was used for the source (OZ Optics Corp., ONT, Canada) and fiber with a diameter of 3 mm was used for the detector (Twardy, Daren, CT).

3. Intralipid™ (Kabi Pharmacia, Clayton, NC): 20% Intralipid™ was generously donated by Kabi Pharmacia and used as the scatterer in the continuum.

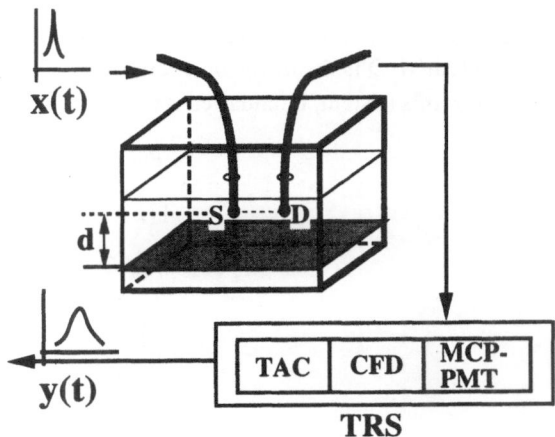

Figure 1. Schematic diagram of experimental system. A source and a detector were located in the middle of a container filled with Intralipid™ solution and a black plate simulating an absorbing boundary was placed at the bottom of the container. The distance (d) between the plate and the S-D unit was varied from 0.5 to 3 cm.

4. Neutral density filter (Kodak Co., NY, NY): Neutral density gelatin filters of various N.D. numbers were used between the source and the detector for measuring the input TRS spectrum.

5. Computer used for simulation: Cray Y-MP at Illinois Super Computing Center and Cray Y-MP EL, College of Engineering, UMBC.

METHOD

Experimental System Description

A large container (20 cm x 40 cm x 26 cm) filled with 5% Intralipid™ solution was used as a continuum (Fig. 1). The source (S) and the detector (D) were immersed in the solution. A black plate, which simulates an absorbing boundary, was placed at the bottom of the container. The distance (d) between the black plate and the S-D unit was controlled by lowering or raising the plate, while the S-D unit remains in a fixed position. The distances between the S-D unit and the boundary were 0.5, 1.0, 1.5, 2.0, 2.5, and 3.0 cm. For each d, TRS spectra were obtained at various distances between the source and the detector (S-D distance). For each TRS spectrum, frequency response analysis was performed and multiple system parameters were obtained. TRS spectra were also obtained from the system without the boundary. The MR and ϕ values from the system without the boundary were used as a reference and compared with values obtained from the system with the boundary.

Description of the System used for Computer Simulation

Computer simulation was used to understand the change in system parameters (1) with change in absorption and scattering of homogeneous systems and (2) with the distance from the absorber. The simulated system was very similar to the experimental system. The dimensions of the container were 30 cm x 30 cm x 30 cm. The governing equation was a photon diffusion equation with a first order consumption (Ishimaru, 1978; Zhu et al., 1991).

$$\frac{\partial P}{\partial t} = C_n D \frac{\partial^2 P}{\partial \rho^2} - C_n \mu_a P$$

$$(5)$$

where ρ is the S-D distance (cm), t is the time (ps), and C_n is the speed of the light in the medium (cm/s). D is the diffusion coefficient (cm), which is $1/3(\mu_a + \mu'_s)$.

For this system, boundaries were assumed to be absorbing.

$$P(x = 0.0) = 0.0 \tag{6}$$

$$P(x = 30.0) = 0.0 \tag{7}$$

$$P(y = 0.0) = 0.0 \tag{8}$$

$$P(y = 30.0) = 0.0 \tag{9}$$

$$P(z = 0.0) = 0.0 \tag{10}$$

$$P(z = 30.0) = 0.0 \tag{11}$$

For homogeneous systems, the photon fluence rates (TRS spectra) in media of various absorption and scattering were computed and frequency response analysis was performed using the information for the input and the computed output values in time. For the study of system parameters with respect to change in medium absorption, the scattering coefficient of the media was set to be constant, 10 cm^{-1}; for the study of change in medium scattering, the absorption coefficient of the media was 0.02 cm^{-1}. MR and ϕ values over the entire modulation frequency were compared with those of the reference spectra.

For heterogeneous systems, a black absorber (1 x 1 x 1 cm) was placed in the middle of the same container. The absorption and scattering coefficients of the continuum were 0.02 and 10.0 cm^{-1}, respectively. A source-detector unit was placed on the plane 1 cm above the absorber in the z direction (from the center of the absorber, 1.5 cm) and TRS spectra were computed at 1.0 cm S-D distance and at 1 cm intervals (Fig. 2) on the plane, x = 12 - 18 cm and y = 12 - 18 cm. The photon fluence rates (TRS spectra) at various S-D unit positions on the plane of study were computed by the probabilistic numerical method, BWK technique (Bruley, 1994). With the computed spectra frequency response analysis performed. Five system parameters were computed and the values were compared with those of the reference system.

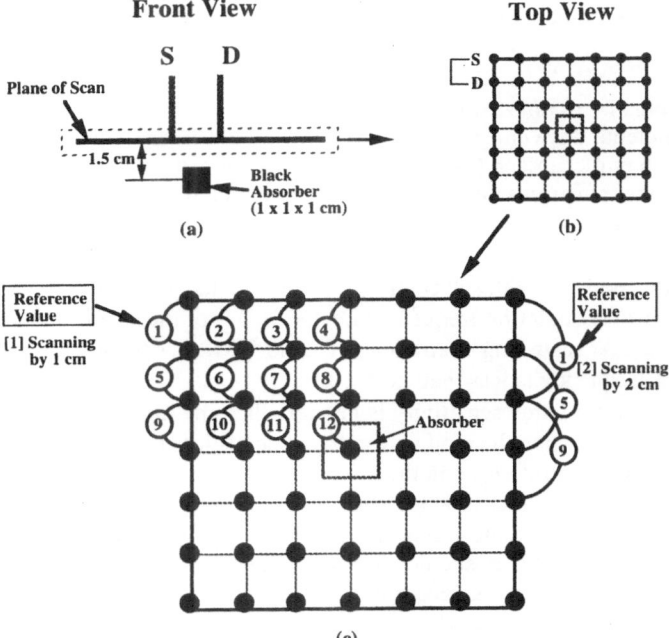

Figure 2. The plane of TRS spectra computation for imaging. The absorber (1 x 1 x 1 cm) was placed in the middle of the container and a plane (6 x 6 cm) located 1.0 cm above the absorber (1.5 cm above the center of the absorber) was scanned at S-D distance of 1 cm at every 1 cm interval.

RESULTS AND DISCUSSION

I. Homogeneous Systems (Computer Simulation Study)

A method for obtaining optical values (absorption and scattering coefficients) using MR and φ from the frequency response analysis of TRS spectra was presented by Kang et al. (in press). MR and φ in a wide range of frequencies, τ, and n with respect to the change in medium optical properties has been also studied by Kang et al. (1994). To obtain system optical properties from NIR-TRS spectra using frequency response analysis, MR and φ values at any frequency may be used. The parameter values at the frequency that yields the highest sensitivity may provide more accurate optical values. In this computer simulation study, the optimal modulation frequency range that provides the highest sensitivity was studied with respect to the changes in optical properties of homogeneous media. For this analysis, computer simulation results were used to avoid the possible ambiguity resulting from experimental noise. Experimental data were compared with the simulation results.

To study the most sensitive frequency region with change in optical properties, MR and φ values in the entire range of frequency were (1) divided by (the ratio), and (2) subtracted with (the difference), the reference values at the same frequencies. The S-D distance for this simulation remained constant (4 cm). When MR or φ values for a system are relatively large, the ratio alone may represent the sensitivity. When both absolute MR and φ values of a system of interest and the reference system are small and within the noise range, the ratio values are not always meaningful, although the value may be large. Hence, for the sensitivity test, the ratio and the difference need to be studied together. As the difference is farther away from 0.0 and the ratio is farther from 1.0 (both positive and negative direction), the sensitivity is greater.

A. Change in Media Absorption at a Constant Scattering. Four different media absorption coefficients were chosen; 0.04, 0.06, 0.08, and 0.1 cm^{-1} with a constant scattering coefficient, 10.0 cm^{-1}. Fig's. 3 (a), (b), (c), and (d) demonstrate the ratio and the difference in MR and φ values at various media absorption, compared with those of the reference system ($\mu_a = 0.02$ cm^{-1}, $\mu'_s = 10.0$ cm^{-1}). When the system optical properties have a larger deviation from those of the reference system both the difference and the ratio become larger.

For MR, both the difference [Fig. 3 (a)] and the ratio [Fig. 3 (b)] show better sensitivity at lower modulation frequency; i.e. the values of the steady state gain, K, has the largest sensitivity. At scattering coefficient, 10.0 cm^{-1}, MR has the greatest sensitivity when the modulation frequency is less than 20 MHz.

For φ, the ratio value sensitivity is greater when the modulation frequency is lower [Fig. 3 (c)]. However, when the modulation frequency is less than 10 MHz the difference in φ between the system of interest and the reference is very small [Fig. 3 (d)], indicating that the difference may not be measurable. The optimum modulation frequency values for the most sensitive phase shift range when the scattering coefficient is 10 cm^{-1}, approximately, is between 40 - 400 MHz. It should be noted that the frequency where the maximum difference occurs increases as the absorption increases.

The order, n, of the system is between 1 and 2, and n increases as the absorption of the system increases [Fig. 4 (a)]. The break frequency, f_b, increases rapidly and linearly as the absorption increases. This means that the signal attenuation due to the fast input modulation frequency declines as the absorption increases [Fig. 4 (b)]. The time constant, τ, decreases as the absorption increases (not shown).

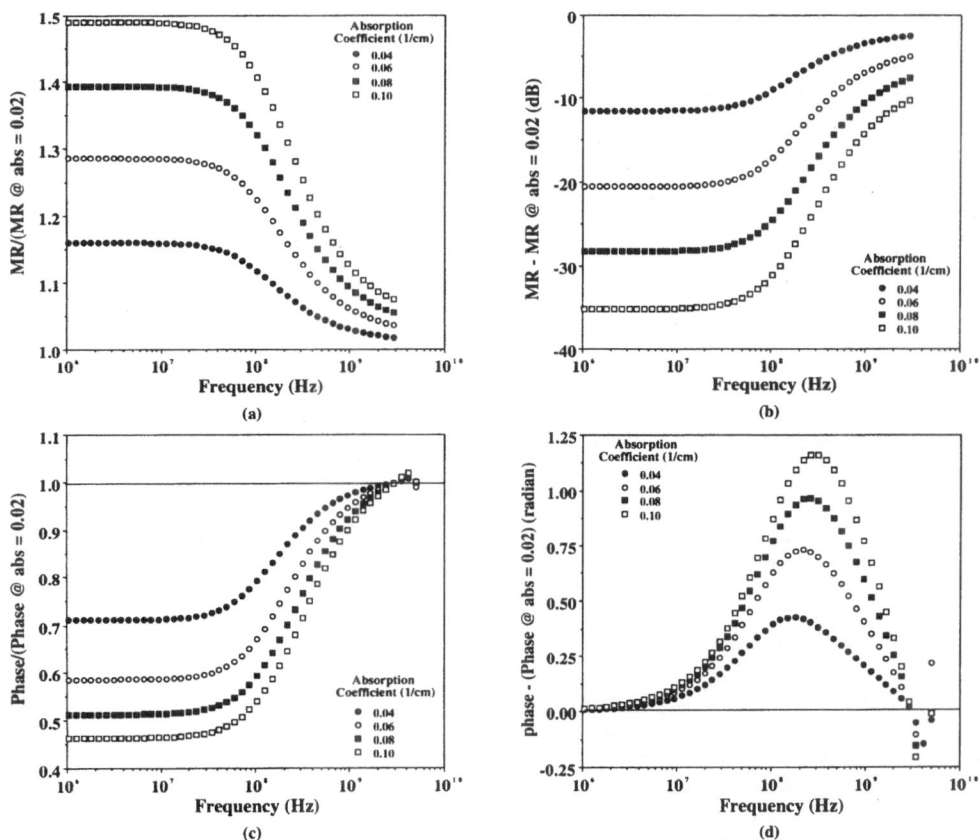

Figure 3. (a) Ratio and (b) difference in MR, and (c) ratio and (d) difference in ϕ, of systems with various absorption compared with the values of the reference system with $\mu_a = 0.02$ cm^{-1}.

B. Changes in Media Scattering at a Constant Absorption. Four different media scattering coefficients, 10.0, 15.0, 20.0, and 25.0 cm^{-1} were compared with the reference system ($\mu_a = 0.02$ cm^{-1}, $\mu'_s = 5.0$ cm^{-1}).

For MR, unlike the case of absorption change (Fig. 3), there is a frequency range which has the highest sensitivity, for both the difference and the ratio. At absorption coefficient, 0.02 cm^{-1}, when the scattering is in the range of 5 and 25 cm^{-1} the most sensitive frequency range seems to be between 300 - 3000 MHz. The frequency of the maximum sensitivity in MR, in both (a) the ratio and (b) the difference, decreases as the scattering increases.

The sensitivity in ϕ increases as the modulation frequency increases although the ratio remains constant until the frequency value is close to 500 MHz. If system scattering is to be analyzed using both MR and ϕ the most sensitive frequency range seems to be between 800 - 2000 MHz.

As the scattering of a system increases, n increases, as in the absorption case. Break frequency, f_b, decreases as the scattering coefficient increases [Fig. 4 (b)].

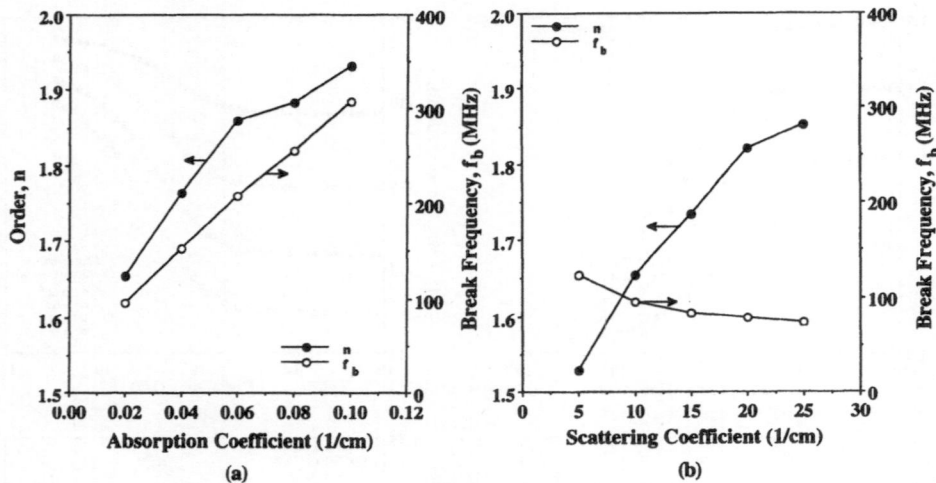

Figure 4. Order, n, and break frequency, f_b, with change in (a) absorption and (b) scattering coefficient of the media.

Systems with Boundaries (Experimental Study)

A. Changes of MR and φ with Change in the Distance between S-D Unit and the Boundary. The purpose of this experiment was to observe the change in system parameters due to the presence of the boundary at various distances from the S-D unit. It is important to study systems with boundaries since many physiological systems can not be assumed to be infinite media. The experiment is also to study the penetration depth of near infrared (Cui, 1991), which is directly related to absorber detectability in depth.

Fig. 6 shows the difference and the ratio of MR and φ from NIR-TRS spectra compared to those from the system without a boundary. The experiments were performed at various S-D distances. The results shown in Fig. 6 are S-D distance at 3 cm. The ratio and difference in MR and φ compared to the system without a boundary have the greatest deviation when the absorbing boundary is the closest to the S-D unit. Spectra at other distances were also studied; however, all have similar trends. Both MR (a) ratio and (b) difference have higher sensitivity at lower frequencies. The changes in sensitivity with respect to frequency may be used for three-dimensional localization (depth of the heterogeneity). For example, in Fig. 6 (a), when the boundary is 2.0 cm away from the S-D unit, at frequency 10 MHz, the difference in MR is 5 dB and change in the ratio is approximately 5%. At 1 GHz, both difference and ratio indicate that there is little difference in MR between the systems with and without a boundary. In the same figure, when the boundary is 0.5 cm away from the S-D unit, at frequency 10 MHz, the difference in MR is about 17 dB and change in the ratio is 15%. When the frequency is 1 GHz, the difference is 8 dB and ratio change is 7%. These values may be used to find the distance of an absorber from the S-D unit and a systematic study needs to be performed in near future. For φ, the ratio sensitivity [Fig. 6 (c)] decreases and the difference sensitivity [Fig. 6 (d)] increases as the frequency increases. In this experimental data the input was not sharp enough to see the results at frequencies greater than 1 GHz. However, simulation results (not included in this paper), illustrates that there is a frequency that demonstrates the maximum difference value in φ. For the experimental system the frequency range that shows the best sensitivity is achieved when the frequency is greater than 100 MHz.

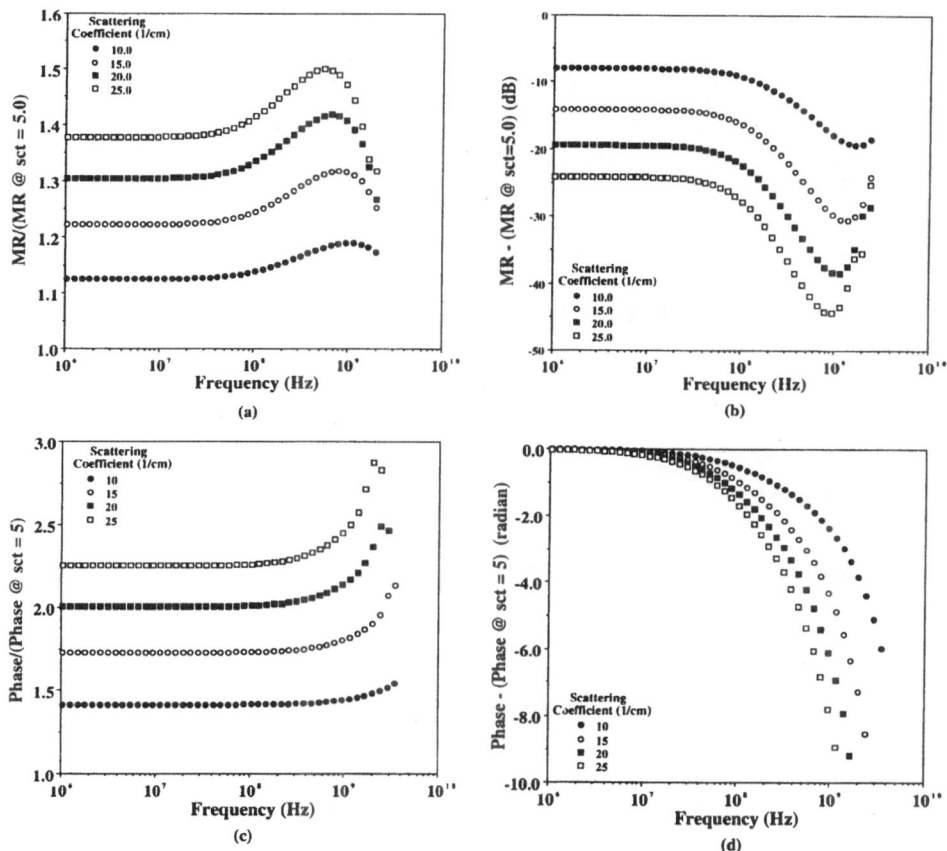

Figure 5. (a) Ratio and (b) difference in MR, and (c) ratio and (d) difference in ϕ, of systems with various scattering compared to the reference system with $M > \mu'_s = 10.0$ cm^{-1}.

B. Changes of Break Frequency (f_b) and Time Constant (τ). Fig. 7 summarizes (a) f_b and (b) τ with changes in the distance between the S-D unit and the absorbing boundary at various S-D distances. The break frequency exponentially decreases [Fig. 7 (a)] with the increase in S-D distance. This implies that as the distance between the source and the detector increases the output signal attenuation due to the input signal modulation occurs at lower modulation frequency. Thus when the S-D distance is large only the lower frequency component can reach the detector. This figure also shows that as the boundary is closer to the S-D unit, break frequency values increase. When the boundary is 3 cm away from the S-D unit the break frequency values deviate only slightly from those of the reference system. This suggests that the NIR wavelength used for this experiment (780 nm), in a system with an absorption and scattering similar to the experimental system at an S-D distance of 3 cm, can be reached probably maximum of 3 cm. If the boundary is more than 3 cm away from the S-D unit, it can be ignored as a boundary and the system can be assumed to be an infinite medium.

Fig. 7 (b) shows the time constants of the system with and without boundaries. Time constant, τ, is the inverse of break frequency, and, appears to have a linear relationship with

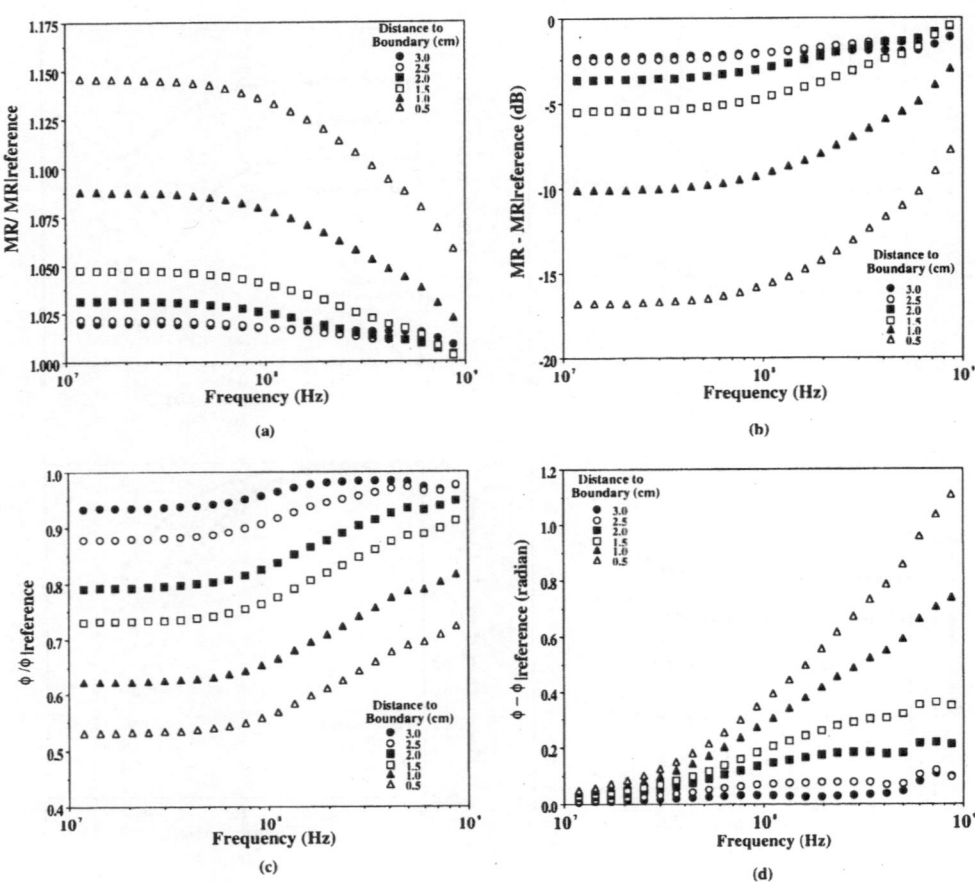

Figure 6. (a) Ratio and (b) difference in MR and (c) ratio and (d) difference in φ, of systems with boundaries at various distances from the S-D unit, compared with values of the system without a boundary.

the distance between the S-D unit and the absorbing boundary. It is also shown that, as the S-D distance increases, the time constant also increases.

Heterogeneous System with a Black Absorber (Simulation Study)

A. Sensitivity of MR and φ in Heterogeneous System. For the heterogeneous system study, as described in Fig. 2, the absorber was located 1 cm below the source and the detector (1.5 cm, from the center of the absorber). The S-D distance was 1 cm.

Fig. 8 (a) and (b) show the difference and the ratio in MR at various S-D unit positions compared to the reference position values. In this particular case, the reference point was chosen to be the one farthest from the absorber. Since the values that we are interested in are the difference and the ratio, any point can be chosen as a reference.

For MR, both the difference and the ratio at lower frequencies have better sensitivity; i.e., a lower modulation frequency provides much better localization capability than a higher frequency. For the system simulated (continuum $\mu_a = 0.02$ cm^{-1}, $\mu'_s = 10.0$ cm^{-1}), the

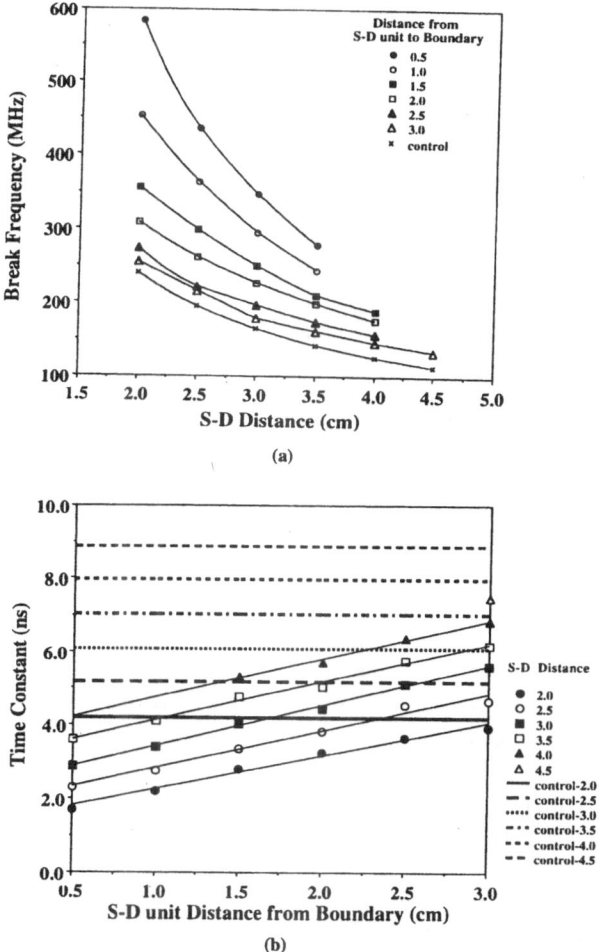

Figure 7. Break frequency (f_b) and time constant (τ) with change in the distance between the boundary and the S-D unit at various S-D distances.

sensitivity starts to decrease when the modulation frequency is greater than 100 MHz. The maximum difference value was greater than 1 dB and the maximum ratio value is 3%.

For ϕ, the maximum difference is approximately 5 degrees [Fig. 8 (d)]. Since the S-D distance is only 1 cm it is difficult to obtain greater phase shift differences among points. The maximum sensitive frequency value increases as the S-D unit moves closer to the absorber; i.e., the S-D position closest to the absorber is position 12. The most sensitive frequency range of phase shift is between 80 - 500 MHz. Fig. 8 demonstrates that it could be very difficult to localize the heterogeneity unless MR and ϕ values were selected at the most sensitive frequency.

Fig. 9 shows three dimensional display of (a) the steady state gains, K, (MR values before the attenuation due to the high modulation frequency), (b) ϕ values at 182 MHz and (c) time constants (τ) of TRS spectra at various points on the simulated plane shown in Fig. 2. Since the highest sensitivity is at lower frequency, the steady state gain values at various

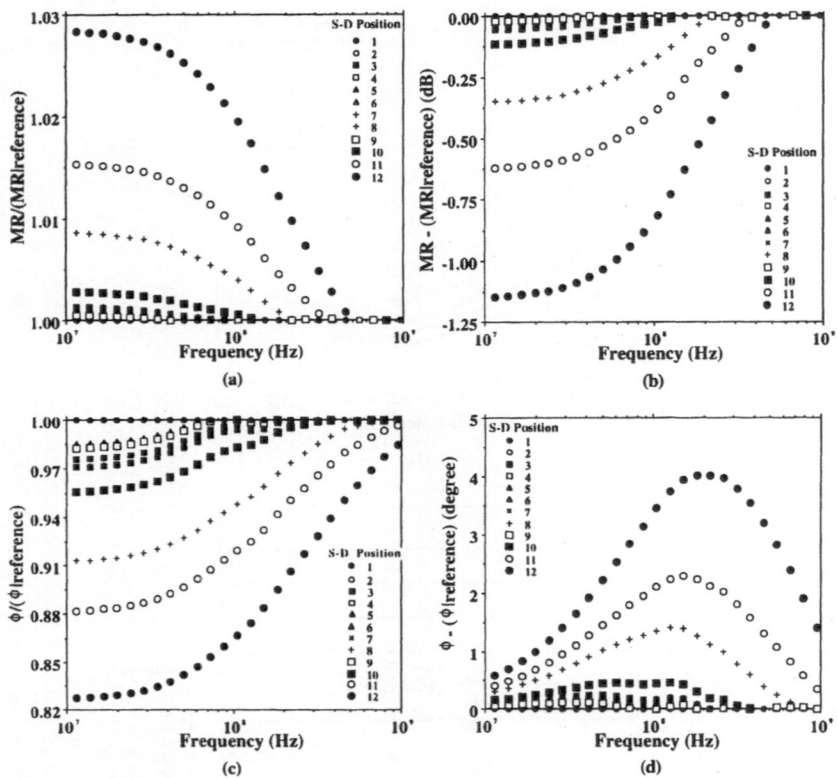

Figure 8. (a) Ratio and (b) Difference in MR, and Ratio and (d) Difference in ϕ, at various positions of the S-D units compared with those at the reference point.

S-D positions are chosen to be plotted for the heterogeneity localization. Steady state gain (K) values presented in Fig. 9 (a) are the difference between MR at the point of interest and those at the reference points. This figure indicates that the photons were absorbed more readily when the S-D unit was closer to the absorber. Phase shift values [ϕ; Fig. 9 (b)] are the differences between the values at various points and those at reference points at frequency 182 MHz. This frequency was chosen because the difference seemed to be greater at this frequency compared to other frequencies. As previously mentioned, ϕ is the negative value of the 'phase' which is used by researchers in the NIR field. The positive ϕ values compared to the reference value indicates that the mean pathway became shorter when the S-D unit is closer to the absorber. Time constants, τ's [Fig. 9 (c)], are plotted by absolute values at various points. As can be seen in this figure, when the S-D unit is closer to the absorber, τ becomes shorter. The order of the system, n, has a definite tendency to increase when the S-D unit is close to the absorber. The change of the value itself was rather small and it is necessary, therefore, to study more systems to better understand the behavior of this parameter.

    This study demonstrates frequency response analysis provides at least three different images to locate an absorber in two dimensions. Further studies are needed to be performed to develop a method to combine these multiple parameter images to improve resolution and sensitivity. Additionally, the location of the absorber in z direction needs to be correlated with the frequency information to complete three-dimensional localization.

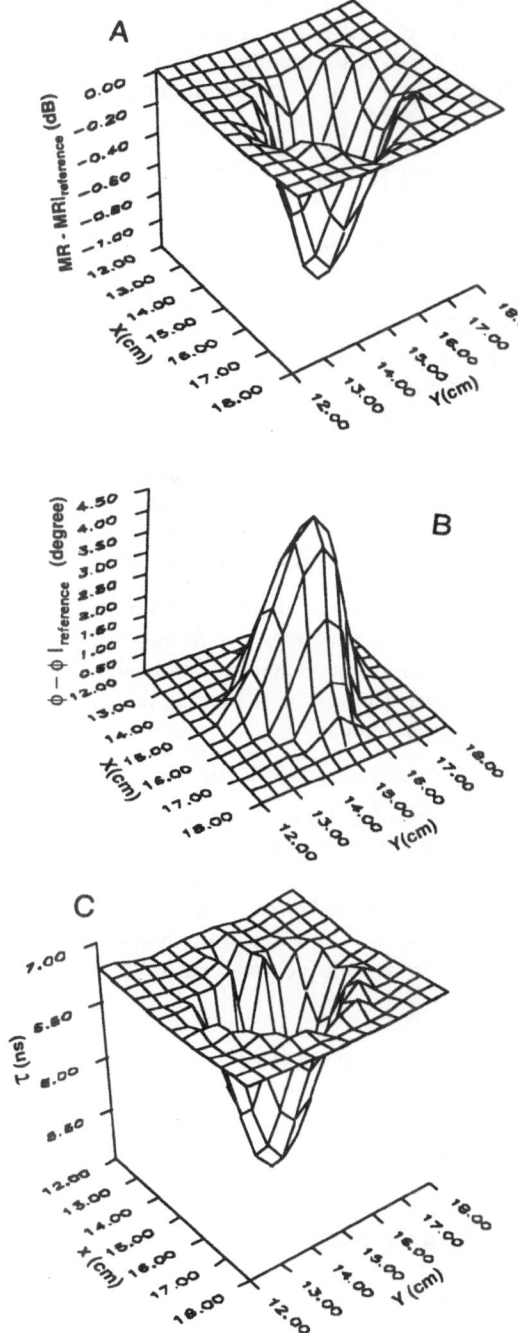

Figure 9. Three dimensional display of (a) the steady state gain, K, (b) φ values at 182 MHz, and (c) τ's on the plane of interest shown in Figure 2.

We have studied values at an S-D distance of 2 cm (measurement configuration shown in Fig. 2) at a 1 cm interval. All values shown in Fig. 8 at S-D distance of 2 cm were greater than those at an S-D distance of 1 cm [except $\phi/(\phi|_{reference})$]. The relationship among the interval, the S-D distance for the measurements, and the resolution of heterogeneity localization also needs to be studied in the future.

CONCLUSIONS

For analyzing systems with various absorption properties or for localizing heterogeneities, using MR, lower modulation frequency values are more sensitive (both ratio and difference). Therefore, if a system is to be analyzed by the intensity information using phase modulation spectroscopy, the system with low frequency will have better sensitivity. For these two cases, ϕ has the maximum sensitivity frequency.

For the analysis of scattering change in a system using MR, there is a maximum sensitivity frequency. The sensitivity of ϕ seems to have the optimum frequency range for the cases in absorption and heterogeneity localization although the sensitivity in ϕ with change in scattering increases as the frequency increases.

If a system is to be analyzed with both MR and ϕ information, the optimal modulation frequency needs to be chosen to obtain the best sensitivity.

Frequency response analysis of NIR-TRS spectra can be utilized for:

1. Improving sensitivity for homogeneous systems by selecting the MR and ϕ values at the most sensitive frequency.
2. Improving heterogeneity localization capability by using the maximum deviation in MR and ϕ values at the most sensitive frequency and by using multiple parameters.

Future studies will focus on:

1. the development of methods for obtaining optical properties for non-infinite media using frequency response analysis,
2. the development of a more concrete method for the localization of heterogeneities in the third dimension (z direction), and
3. the development of methods for intensifying images by combining information from multiple parameters.

ACKNOWLEDGMENT

This work was partially supported by The Whitaker Foundation, Special Opportunity Award, and Cray Research, Inc., and Illinois Super Computing Center.

REFERENCES

Bruley, D. F., Empirical testing to model oxygen transport processes in tissue, *Proc. ISOTT XVIi (this volume)*.
Bruley, D. F. and Prados, J.W., 1964, The frequency response analysis of a wetted wall adiabatic humidifier, *AIChE Journal*, 10(5): 612-616.
Bruley, D. F., 1974, Pulse reduction code written for process identification.
Bruley, D. F., 1994, Modeling Oxygen Transport: Development of Methods and Current State Symposium in Honor of Professor Dr. Gehard Thews, *proc. ISOTT XV*, 345: 807-814.

Clements, W. C., Jr. and Schnelle, K. B., 1963, Pulse testing for dynamic analysis, *I & EC Process Design and Development*., 2(2):94-102.

Coughanowr, D. R., 1991, Process systems analysis and control, McGraw-Hill, New York.

Cui, W., Wang, N., and Chance, B., 1991, Study of photon migration depths with time-resolved spectroscopy, *Optics Letters*, 16(21):1632-1634.

Duncan, A., Whitlock, T. L., Cope, M., and Delpy, D. T., 1993, A multiwavelenth, wideband, intensity modulated optical spectrometer for near infrared spectroscopy and imaging, *Proc. SPIE*, 1888.

Filon, L. N. G., 1928, On a quadrature formula for trigonometric integrals, *Proc. of the Royal Soc. Edinburgh*, 49:38-47.

Ishimaru, A. , 1978, Diffusion of a pulse in densely distributed scatters, *J. Opt. Soc. Am.*. 68(8):1045-1050.

Jacques, S., 1991, Principle of phase-resolved optical measurements, *Proc. SPIE* 1525:143-153.

Kang, K. A., Bruley, D. F., Londono, J. M., and Chance, B., 1994, Highly Scattering Optical System Identification via Frequency Response Analysis of Optical NIR-TRS Spectra, *Annals of Biomedical Engineering* 22:241-253.

Kang, K. A., Bruley, D. F., Londono, J., and Chance, B., (in press), Frequency Response by Pulse Reduction for the Analysis of TRS Spectra, *Proc. ISOTT XVI*.

Koyama, K. and Fatlowitz, D., 1987, Application of MCP-PMTs to time correlated single photon counting and related procedures, Hamamatsu technical information, No. ET-03/OCT.

Lewis, C. I., Jr., Bruley, D. F., and Hunt, D. H., 1967, Evaluation of temperature pulse characteristics and pulse testing for thermal dynamic analysis, *I&EC Process Design and Development*. 6 (3): 281-286.

Luyben, W. L., 1990, Process Modeling Simulation, and Control for Chemical Engineers, 2nd edition, McGraw-Hill, New York.

Zhu, J. X., Pine, D. J., and Weitz, D. A., 1991, Internal reflection of diffusive light in random media, *Phys. Rev. A. 44: 3948-3959.*

54

RESPONSE OF CORTICAL OXYGEN PRESSURE AND STRIATAL EXTRACELLULAR DOPAMINE IN THE BRAIN OF NEWBORN AND ADULT ANIMALS TO HYPOXIA

Anna Pastuszko, Dekun Song, Marta Olano, Chau–Ching Huang, and David F. Wilson

Departments of Pediatrics and of Biochemistry and Biophysics
University of Pennsylvania
Philadelphia, Pennsylvania 19104

INTRODUCTION

The pathophysiologic mechanisms underlying hypoxic-ischemic brain damage at any age are very complex and compared to adults, newborns respond differently to similar hypoxic-ischemic insult (Duffy et al., 1982; Raichle, 1983). These is general agreement that newborns are more resistant to hypoxia/ischemia and can survive hypoxia much longer than adults of the same species (Glass et al., 1944). The greater resistance of the immature animals has been attributed to its lower cerebral metabolic rate; a greater ability to maintain energy reserves through glycolysis; lesser complexity of the synaptic connections and/or differences in the enzyme content (Duffy et al., 1972; Duffy and Vannucci, 1977; Vannucci, 1989). On the other hand, evidence had been presented that exposure to moderate hypoxia during early postnatal life may result in disruption of functional activity at selected synapses even when it does not cause histologically measurable neuronal damage (Hedner and Lundborg, 1979; 1980). Similarly, it has been proposed that exposure of immature brain to moderate hypoxia may cause permanent changes in synaptic function and have significant impact on further neuronal development (Ihle et al., 1985; Lun et al., 1986).

The oxygen dependence of brain metabolism in newborns and adults has been extensively studied, but this work has suffered from the lack of effective methods for measuring oxygen in tissue. An optical method for measuring oxygen pressure, developed in our laboratory, has now made it possible to obtain quantitative measurement of the oxygen distribution in the veins and capillaries of the brain cortex *in vivo*.

The present study was designed to investigate and compare the responses of cortical oxygen pressure and striatal dopamine metabolism in the newborn, as represented by

Oxygen Transport to Tissue XVII, Edited by Ince et al.
Plenum Press, New York, 1996

newborn piglets, and adult animals, as represented by cats. These two model systems have often been used as representative of the respective age groups, and the newborn piglets are at a level of development comparable to term newborn humans.

MATERIALS AND METHODS

The study was carried out in ventilated, anesthetized newborn piglets (3-5 days old) and adult cats. Newborn piglets, of this age, were chosen because the level of development of the piglet brain is comparable to that of a term newborn.

The cats were anesthetized with chloralose (50 mg/kg) and urethane (200 mg/kg) and tracheostomy performed. The animals were then mechanically ventilated and the femoral artery and femoral vein were cannulated.

In piglets, anesthesia was induced with halothane (4% mixed with 96% oxygen) and 1.5% lidocaine was used as a local anesthetic. The halothane was reduced to 0.6-0.8% after tracheotomy and the femoral artery and femoral vein were cannulated. Once the vessels were cannulated, halothane was withdrawn entirely and Fentanyl (30 µg/kg) was injected intravenously at approximately one hour intervals throughout the experiments. The animals were paralyzed with tubocurarine and mechanically ventilated with a mixture of oxygen and nitrous oxide (in control conditions with 21-22% oxygen and 78-79% N_2O).

The head of the animal was then placed in a Kopf stereotaxic holder and an incision was made along the midline of the scalp. The scalp was removed to expose the skull and a hole approximately 8 mm in diameter was made in the skull over one parietal hemisphere for measuring the oxygen pressure in the cortex. The surface of the brain under the hole was flushed with artificial CSF throughout the study. Another hole about 4 mm diameter was drilled in the skull contralaterally to the window and the microdialysis probe was implanted into the striatum

Cerebral oxygenation was measured optically by phosphorescence quenching (Rumsey et al., 1988; Wilson et al., 1991) and extracellular dopamine by *in vivo* microdialysis. Controlled, graded levels of hypoxic insult to the brain of animals were generated by decreasing of the oxygen fraction in the inspired gas (FiO_2) every 18 min over an about 1 hr period (FiO_2 from 21% to 14%, 11% and 9%). At the end of the hypoxic period the animals were given 21% O_2 to breathe for a subsequent 1 hr recovery period.

RESULTS AND DISCUSSION

In the present study the animal models used were newborn piglets and adult cats. Although these are of different species, each represents an important model, representative of its age group, to the study of the response of the brain to periods of hypoxia/ischemia. It is from models such as these that many of the generalizations concerning the differences in response of newborns and adults have been derived. Our goal was to establish the responses of cortical oxygen pressure and extracellular dopamine levels to decreases of FiO_2, such that both the alterations in oxygen pressure and the dependence of dopamine levels on cortical oxygen pressure in the two models could be directly compared.

The effect of graded hypoxia obtained by reducing the fraction of inspired oxygen on the physiological parameters of newborn piglets and cats are shown in Table 1. The values of $PaCO_2$ and PaO_2 were very similar for both types of animals. The arterial blood pressure was not significantly affected by hypoxia in either piglets and cats. However, there was a major difference in effect of hypoxia on blood pH. In cats, during these pathological conditions the blood pH did not change where as in piglets it decreased from a control value

Table 1. Effect of hypoxia on physiological parameters of adult cats and newborn piglets

Conditions	MAP (Torr)	Blood pH	PaCO$_2$ (Torr)	PaO$_2$ (Torr)
Adult cats (n = 4)				
Control	135 ± 9	7.38 ± 0.01	35 ± 6	107 ± 20
Hypoxia (14%)	135 ± 8	7.37 ± 0.02	31 ± 3	54 ± 2
Hypoxia (11%)	128 ± 9	7.40 ± 0.01	28 ± 2	35 ± 1
Hypoxia (9%)	118 ± 10	7.38 ± 0.03	24 ± 1	25 ± 1
Reoxygenation	149 ± 2	7.40 ± 0.01	27 ± 2	129 ± 3
Newborn piglets (n = 8)				
Control	93 ± 12	7.39 ± 0.04	40 ± 3	104 ± 19
Hypoxia (14%)	97 ± 19	7.37 ± 0.05	39 ± 2	47 ± 13
Hypoxia (11%)	94 ± 23	7.35 ± 0.06	37 ± 4	32 ± 4
Hypoxia (9%)	82 ± 28	7.25 ± 0.08	33 ± 7	21 ± 4
Reoxygenation	94 ± 20	7.05 ± 0.02	35 ± 5	108 ± 11

Data are mean ± SEM.

of 7.4 to 7.37, 7.35 and 7.25, respectively, as the FiO$_2$ was decreased to 14%, 11%, and 9%. During the reoxygenation period, in piglets the pH further decreased to 7.05. This acidosis observed in newborn piglets, may be associated with disturbance of cerebral blood flow, periventricular hemorrhage, leucomalacia, increased vascular resistance and decreased myocardial function (see review, Walter, 1992)

The effects of graded decreased of FiO$_2$ on cortical oxygen pressure in cats and piglets are shown in Figure 1.

Stepwise decreases in the FiO$_2$ from 21% to 14%, 11%, and 9%, holding the value for 18 minutes at each level, caused in cats progressive decrease in cortical oxygen pressure from 43 ± 4 Torr to 34 ± 4 , 28 ± 5 and 19 ± 3 Torr (Figure 1A). Return of FiO$_2$ to control values after hypoxia resulted in the oxygen pressure initially rising to above control (to approx. 48 ± 5 Torr) and then decrease to control value. In piglets, the same hypoxic conditions caused a decrease in cortical oxygen pressure from 33 ± 4 Torr to 24 ± 3 , 15 ± 2 and 4 ± 1 Torr (Figure 1B). During first few minutes of reoxygenation, similar as in cats, oxygen pressure rose above control (to approx. 40 ± 3.5 Torr) and then decreased to control value.

The presented data show that in both newborn piglets and cats the cortical oxygen pressure decreases with decrease in FiO$_2$, but decrease is substantially greater in the newborn. The cortical oxygen pressure in newborn piglets decreased almost proportional to decrease in FiO$_2$ and when FiO$_2$ decreases to 9% the brain was severely hypoxic with cortical oxygen pressure decreasing to about 4 Torr. During the same protocol of decreasing FiO$_2$, cortical oxygen pressure in the cortex of cats decreased only to about 19 Torr, values which can be characterized as indicating mild/moderate hypoxia.

These results are consistent with newborns having less well developed vascular regulation by PO$_2$. It had been reported that the immature cerebrovascular system in newborn infants is less capable of invoking autoregulatory mechanisms, presumably due in part to a deficiency of the muscular lining of cerebral arterioles (Hill, 1991). Cerebrovascular autoregulation is a homeostatic regulatory system which ensures preservation of relatively constant cerebral perfusion over a wide range of systemic arterial blood pressure by

A

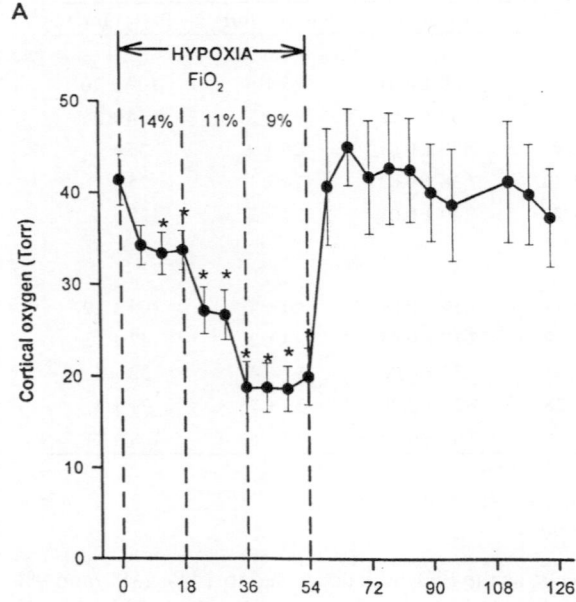

B

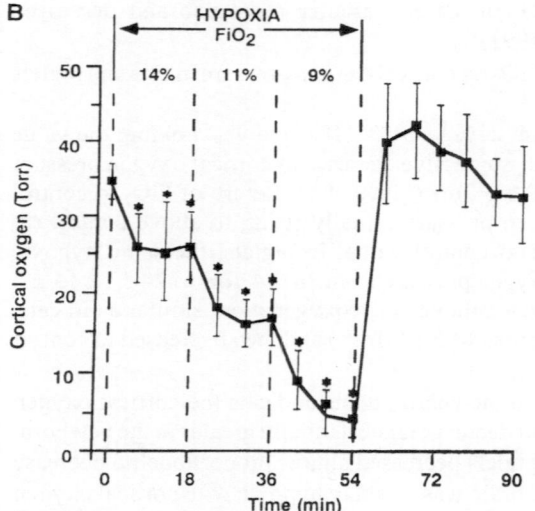

Figure 1. The effect of graded hypoxia on the cortical oxygen pressure in cats (A) and newborn piglets (B). The cortical oxygen concentration was measured using the oxygen dependent quenching of phosphorescence as described in Material and Methods. Graded hypoxia was initiated by decreasing the FiO_2 from 21% (control) to 14%, 11% and 9% holding the FiO_2 constant at each level for 18 min and than returning the FiO_2 to control for a 30-60 min. The results are expressed as the means ± SEM for 4 (A) and 8 (B) experiments. *$p < 0.05$ for significant difference from baseline values as determined by the Student's t test.

regulation of vascular resistance. Impaired cerebrovascular autoregulation has been observed even in the context of relatively mild/moderate perinatal hypoxic/ischemic insult.

The response of extracellular dopamine to the stepwise decrease in FiO_2 in cats and piglets are shown in Figure 2. In cats (Figure 2A), a statistically significant increase of dopamine, to about 800% of control, was observed only in last step of hypoxia. During the period of reoxygenation the dopamine levels in the extracellular medium declined to control values by about 20-30 min of reoxygenation. In piglets, the extracellular level of dopamine

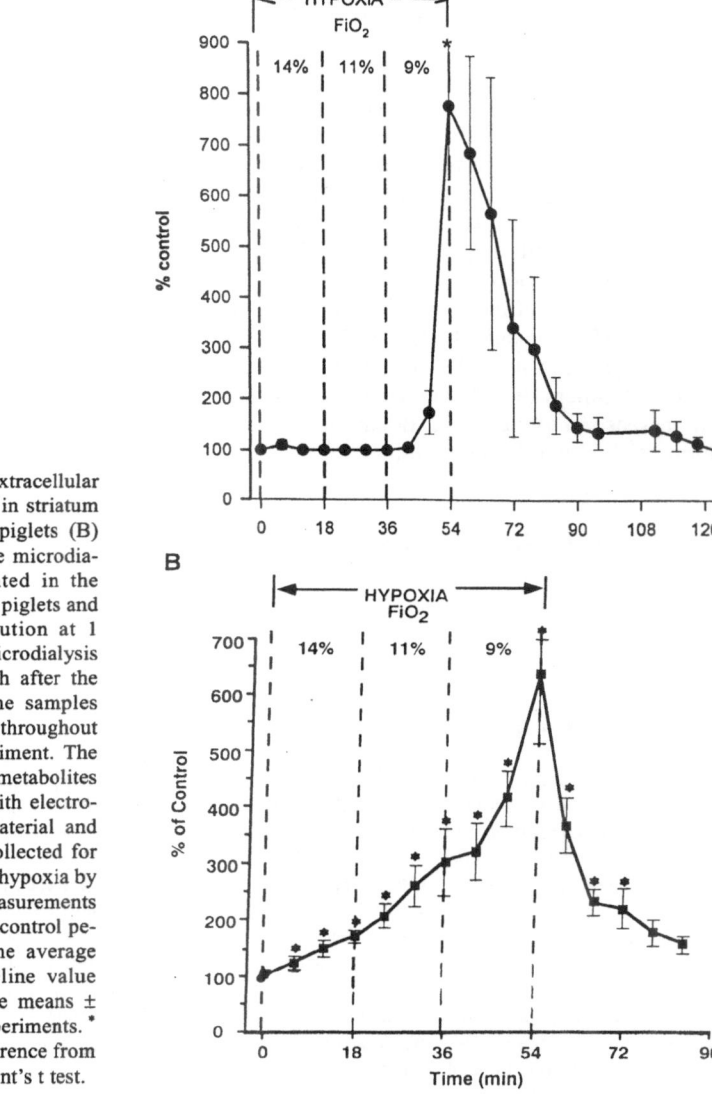

Figure 2. Changes in the extracellular concentration of dopamine in striatum of cats (A) and newborn piglets (B) during graded hypoxia. The microdialysis probes were implanted in the striatum of cats or newborn piglets and perfused with Ringer solution at 1 μl/min. Collection of the microdialysis samples was initiated 1.5 h after the probe was inserted and the samples were collected every 6 min throughout the remainder of the experiment. The levels of dopamine and its metabolites were analyzed by HPLC with electro-chemical detection (see Material and Methods). Samples were collected for 30 min prior to initiation of hypoxia by lowering the FiO_2. The 3 measurements of the dopamine during the control period were averaged and the average value considered the baseline value (100%). The results are the means ± SEM for 4 (A) and 8 (B) experiments. * p < 0.05 for significant difference from baseline values by the Student's t test.

rose stepwise to 180%, 300% and 650% of control, respectively, as FiO_2 decreased. As was the case for cats, the dopamine levels returned to control values during reoxygenation period.

The changes in extracellular dopamine in both groups of animals were dependent on changes in cortical oxygen pressure and extracellular levels of this neurotransmitter increase with decreases in oxygen pressure. In the newborn piglets there is clearly no "oxygen reserve" in brain to protect the cells from decrease in oxygen pressure. The extracellular dopamine levels begin to rise as soon as the oxygen pressure begins to decrease. In contrast, in adult cats the vascular system in the brain acts to compensate for the decrease in oxygen pressure in the arterial blood and the decrease in cortical oxygen pressure is much less than in piglets. The extracellular dopamine levels begin to increase when the cortical oxygen pressures decrease below about 28 Torr, not very different from the oxygen pressures for dopamine increase in newborn piglets. As a result, in cats a statistically significant increase in extracellular dopamine was observed only for the lowest inspired oxygen pressure (FiO_2 of 9%).

The increase in extracellular dopamine may be result from an increase in neurotransmitter efflux due to neuronal depolarization, to inhibition of the reuptake transport system or, more probably, to some combination of increased release and decreased uptake. There is substantial evidence implicating increased extracellular levels of dopamine in mediating ischemic injury to the striatum. Dopamine depletion by lesion of the substantia nigra gives significant protection against ischemic/hypoxic injury to the striatum (Globus et al., 1987). In addition, depletion of catecholamines by pretreatment with α-methyl-ρ-tyrosine exerts a strong protective effect on postischemic damage to nerve terminals in the gerbil (Weinberger et al., 1985). Matsui et al. (1991), suggesting that catabolism of dopamine which is reaccumulated in the cells during reoxygenation after hypoxic insult plays a major role in the development of ischemic necrosis of striatal neurons. Thus, in both newborn and adult, the increase in extracellular dopamine, can be an important determinant of striatal vulnerability to hypoxic/ischemic injury.

ACKNOWLEDGMENTS

This work was supported by a grant NS-31465 from the U.S. National Institutes of Health.

REFERENCES

Duffy T.E., Cavazzuti M., Cruz N.F. and Sokoloff L. 1982. Local cerebral glucose metabolism in newborn dogs: effects of hypoxia and halothane anesthesia. Ann. Neurol. 11: 233-246.

Duffy T.E., Nelson S.R. and Lowry O.H. 1972. Cerebral carbohydrate metabolism during acute hypoxia and recovery. J. Neurochemistry 19: 959-977.

Duffy T.E. and Vannucci R.C. 1977. Metabolic aspects of cerebral anoxia in the fetus and newborn. in Berenberg S.R. ed.: Brain, Fetal and Infant. The Hague, Martinus Nijhoff, pp. 316-323.

Glass H.G., Snyder F.F. and Webster E. 1944. The rate of decline in resistance to anoxia of rabbits, dogs and guinea pigs from the onset of viability to adult life. Am. J. Physiol. 140: 609-615.

Globus, M.Y-T., Ginsberg, M.D., Dietrich, W.D., Busto, R. and Scheinberg, P. 1987. Substantia nigra lesion protects against ischemic damage in the striatum. Neurosci. Letters. 80: 251-256.

Hedner T. and Lundborg P. 1979. Regional changes in monoamine synthesis in the developing rat brain during hypoxia. Acta Physiol. Scand. 106: 139-143.

Hedner T. and Lundborg P. 1980. Catecholamine metabolism in neonatal rat brain during asphyxia and recovery. Acta Physiol. Scand. 109: 169-175.

Hill A. 1991. Current Concepts of Hypoxic-Ischemic Cerebral Injury in the Term Newborn. Pediatr. Neurol. 7: 317-325.

Ihle W., Gross J. and Moller R. 1985. Effect of chronic postnatal hypoxia on dopamine uptake by synaptosomes from striatum of adult rats. Biomed. Biochem. Acta, 44: 433-437.

Lun A., Gross J., Beyer M., Fischer H.D., Wustmann C., Schmidt J. and Hecht K. 1986. The vulnerable period of perinatal hypoxia with regard to dopamine release and behavior in adult rats. Biomed. Biochem. Acta, 45: 619-627.

Matsui Y.and Kumagae Y. 1991. Monoamine oxidase inhibitors prevents striatal neuronal necrosis induced by transient forebrain ischemia. Neuroscience Letters, 126: 175-178.

Raichle M.E. 1983. The pathophysiology of brain ischemia. Ann. Neurol. 13: 2-10.

Rumsey W.L., Vanderkooi J.M. and Wilson D.F. 1988. Imaging of phosphorescence: a novel method for measuring oxygen distribution in perfused tissue Science, 241: 1649-1651.

Walter J.H. 1992. Metabolic acidosis in newborn infants. Arch. Dis. Child, 67: 767-769.

Vannucci R.C. 1989. Acute Perinatal Brain Injury: Hypoxia-Ischemia. in Cohen W.R., Acker D.B., and Friedman E.A. eds. Management of Labor. Aspen Publish. Inc. Rockville pp. 183-244.

Weinberger J., Nieves-Rosa J. and Cohen G. 1985. Nerve terminal damage in cerebral ischemia: protective effect of alpha-methyl-para-tyrosine. Stroke, 16: 864-870.

Wilson D.F., Pastuszko A., DiGiacomo J.E., Pawlowski M., Schneiderman R. and Delivoria-Papadopoulos M. 1991. Effect of hyperventilation on oxygenation of the brain cortex of newborn piglets. J. App. Physiol. 70(6): 2691-2696.

55

THE ROLE OF PAF AND THE EFFECT OF A SPECIFIC PAF ANTAGONIST ON LOCAL TISSUE PO$_2$ AND NEURONAL INTEGRITY DURING AND AFTER PHOTOTHROMBOTIC BRAIN INFARCTION IN UNANESTHETISED RABBITS

Koen van Rossem, Herman Vermariën, Karin Decuyper, and
René Bourgain

Laboratory of Physiology and Pathophysiology
University of Brussels VUB
Laarbeeklaan 103
B-1090 Brussels, Belgium

INTRODUCTION

Platelet-activating factor (PAF) is a phospholipid which acts as a mediator in inflammation and thrombosis (for review, see Braquet et al., 1987). A growing body of evidence supports the hypothesis that PAF may be a key mediator in neuroinjury (Frerichs and Feuerstein, 1990). Brain tissue and blood cells are able to synthesize PAF and this production may be enhanced during stroke and thrombosis. Locally increased levels of PAF during brain ischemia may be detrimental as PAF induces vasoconstriction and endothelial damage with consequent blood brain barrier damage and edema. In addition, high levels of PAF have been shown to be neurotoxic. Specific receptors are involved and have been demonstrated to be present in blood cells and brain tissue.

The present study was conducted in order to evaluate the acute effects of a PAF receptor antagonist (RP 48740) on tissue oygenation, neuronal function and morphologic damage in photothrombotic brain infarction. The applied technology allows simultaneous recording of local tissue PO$_2$ and somesthetic evoked potentials (SEP) during and after the induction of a focal cortical infarction without applying anesthesia. In this way anesthesia related alterations of local hemodynamic responses, neuronal function and resistance to ischemia and anoxia are avoided.

Oxygen Transport to Tissue XVII, Edited by Ince et al.
Plenum Press, New York, 1996

MATERIALS AND METHODS

Electrodes and Measuring Apparatus

PO_2 and SEP-measurements were performed simultaneously with platinum electrodes having a cylindrical measuring tip (length 1 mm, ø 100 μm) covered with a homogeneous cellulose acetate membrane (thickness 20 μm) (van Rossem et al., 1992 a). Applying these electrodes, mean values of tissue PO_2 and biopotential are measured over 1 mm of cortical thickness. They were fixed into polymethylmetacrylate frames (6 x 8 mm) containing a central shaft in which an optic fiber (ø 3 mm) can be mounted above a perfectly delineated transparent area (ø 3.17 mm) (van Rossem et al., 1992 b) and were positioned in the centre of and 0.5 mm and 2.5 mm rostral to the illuminated area.

Measurements were performed with a laboratory made 4-channel device ($FYSPpO_2 2$) allowing simultaneous biopotential and polarographic PO_2 measurement with the same electrode (Vermariën et al., 1992). All electrodes were used for PO_2 recording. The central electrode and the proximal electrode (0.5 mm) were used for simultaneous SEP derivation.

Regarding PO_2 measurement the electrodes are polarized (- 600 mV) with respect to a common circular Ag/AgCl electrode (ø 10 mm) which is fixed to the animal's ear. The error induced in the PO_2 measurement by shifts in brain DC potentials during focal ischemia has been shown to be negligible (van Rossem et al., in press). The electrodes were calibrated for PO_2 in aerated and deoxygenated Ringer solution at 38°C. They permanently show perfectly linear behaviour (van Rossem et al., 1992 a) but as their long-term stability is only moderate we prefer to mention values of in vivo measured current (iO_2) rather than PO_2.

The stimuli for evoking SEPs were rectangular electric pulses (frequency 1 Hz, duration 0.1 ms) generated by a conventional stimulator (Digitimer DSZ) and applied to the median nerve of the right forepaw through a couple of hypodermic needles. Pulses were applied with an amplitude exceeding the motoric threshold value of motor fibers by 1 V (range 7.5 to 9.5 V). Unipolar SEPs were derived with a subcutaneous stainless steel needle electrode in the nose serving as a reference. SEPs were averaged (interval 90 ms, 16 responses) with a conventional 2-channel apparatus (Medelec).

Animal Preparation and Treatment

Male Dutch rabbits (HSD/POC) weighing 1.7 to 2.7 kg were used for the experiments which were approved by the ethical committee of the Belgian Fund for Medical Scientific research. Six rabbits received an i.v. injection of 10 mg/kg of RP 48740 dissolved in saline (40 mg/ml) 10 min before onset of infarction. A control group of 6 rabbits received an equivalent volume of saline. No anesthesia was applied during the infarction experiments until onset of the procedure for perfusion fixation of the brain.

The frame with fixed electrodes was implanted above the cortical projection area of the right forepaw as described earlier (van Rossem et al., 1992 b). Ten days after implantation a cortical infarction was induced photochemically: after i.v. injection of rose bengal (10 mg/kg) an optic fiber was mounted into the shaft of the implanted frame and the brain cortex was illuminated with cold green light (spectral width 450 to 600 nm, intensity 110 mW/cm²) during 20 min. Thirty min before infarction, arterial blood samples were collected for determination of blood gases, pH, haematocrit, platelet count and plasma glucose level. Rectal temperature was monitored during the entire experiment. Fifteen minutes before onset of illumination 2 SEPs were obtained; mean amplitudes were taken as initial values. Subsequent SEPs were recorded 5 min (number 1), 10 min (2), 15 min (3), 30 min (4) and from then each half an hour (numbers 5 to 10) after onset of illumination. PO_2 was monitored continuously from 30 min before until 3.5 h after onset of illumination. At that time the procedure for perfusion fixation of the brain was started.

After induction of general anesthesia (pentobarbital 30 - 60 mg/kg i.v.) the brain was transcardially perfused with Karnovsky's fixative, containing 2% paraformaldehyde and 2.5% glutaraldehyde, delivered at a pressure of 16 kPa for 5 min after a 45 s rinse with Haemacel®. The brain was kept in situ in Karnovsky's fixative during 4 to 24 h. After isolation of the cerebrum coronal Vibratome (T.P.I., St. Louis, USA) sections were cut alternately at 100 μm and 200 μm. All 100 μm sections were stained with azure A - eosin B at pH 4.5. The infarct area was measured in each 100 μm section using an image analysis program (Image 1.41®). The perimeter of the infarct was manually traced and the area was automatically computed. The volume of the infarct was calculated by numerical integration.

Statistical Analysis

The experiments were conducted blind. After collection of all data the composition of the two groups was revealed and statistical analysis was performed. Differences within groups were evaluated applying the randomisation test for paired data. Differences between groups were evaluated applying the randomisation test for independent samples.

RESULTS

Physiological parameters obtained 30 min before infarction did not differ significantly between the treated and control group. Rectal temperature did not show major fluctuations during the experiment.

An example of the simultaneous measurement of iO_2 and SEPs during and after infarction is represented in Figure 1.

Oxygen Measurement

Electrode sensitivity and residual current in vitro respectively were 10.2 ± 3.2 nA/kPa and 15.3 ± 6.7 nA. Before onset of illumination all animals showed a normal pattern of iO_2 characterised by fluctuations (6 to 11/min) around a mean value (IO_2).

The general pattern of changes in iO_2 was comparable in both groups and was the same as that described earlier (van Rossem et al., 1992 b). In the centre of the illuminated area iO_2 started decreasing significantly after a latency period (LP 1; range 1.3 to 9.3 min, n = 12) and reached a low and stable residual current without fluctuations after a decline interval (DI, range 2.0 to 29.3 min). The residual current, assumed to correspond to zero PO_2, was maintained until the end of the experiment in all animals. At both locations in the border zone recurrent slow waves of iO_2 occurred after a latency period (LP 2, range 5.3 to 13.0 min) and were observed during a slow variation period (SVP, range 40 to 100 min). No significant differences in LP 1, DI, LP 2 and SVP were demonstrated between treated and control animals. The number of slow waves was smaller in RP 48740 treated animals (9 ± 3 versus 12 ± 3, mean $\pm$ SD) but this difference was not significant (p < 0.1). The amplitudes of the slow waves were comparable in both groups.

In the control group iO_2 recorded 0.5 mm rostral to the illuminated area significantly decreased following the SVP (Table 1). This was not the case in RP 48740 treated animals where iO_2 remained unaltered. At 2.5 mm rostral to the irradiated zone no significant changes in iO_2 were noticed in any group.

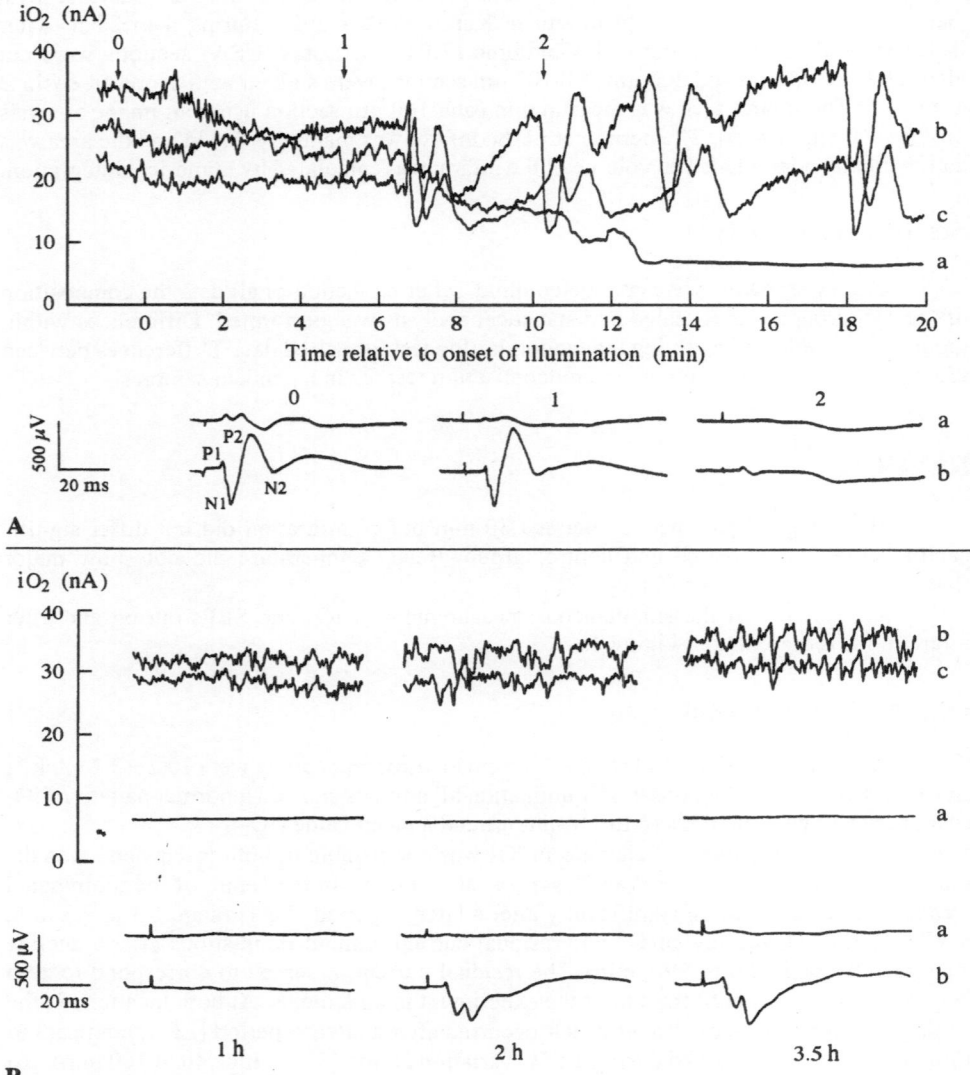

Figure 1. Simultaneous recording of iO_2 and SEP in an RP 48470-treated rabbit during (**A**) and after (**B**) photothrombotic brain infarction in the centre of the illuminated area (a) and respectively 0.5 mm (b) and 2.5 mm (c) rostral to the illuminated area. Infarction at location (a) is characterised by a drop of iO_2 to a residual current, and a complete flattening of the SEP. In the border zone (b and c) iO_2 shows recurrent slow waves during a slow variation period (SVP) and the SEP completely flattens. During the hours following the SVP, iO_2 in this area remains comparable with pre-infarction level and both the P1N1 complex and the P2N2 complex show partial recovery.

Table 1. iO$_2$ (nA) at different time periods after onset of illumination and at different locations relative to the illuminated area

		0 h	1 h	2 h	3 h	3.5 h
centre	controls	27.0 ± 8.4	2.5 ± 2.0**	3.7 ± 2.3**	3.4 ± 2.3**	3.6 ± 2.5**
	RP 48740	26.7 ± 6.3	7.0 ± 4.2**	5.7 ± 4.7**	5.5 ± 4.6**	4.6 ± 3.6**
0.5 mm	controls	25.1 ± 10.9	21.3 ± 5.0	17.6 ± 6.0	16.2 ± 3.7*	15.6 ± 2.1**
	RP 48740	24.7 ± 6.8	23.0 ± 8.2	21.8 ± 5.9	22.1 ± 5.9	20.6 ± 6.4
2.5 mm	controls	28.7 ± 8.1	33.6 ± 11.3	28.1 ± 8.7	26.9 ± 5.7	27.7 ± 5.3
	RP 48740	29.4 ± 8.4	27.3 ± 10.1	27.4 ± 7.9	26.6 ± 6.6	26.9 ± 7.3

Somesthetic Evoked Potentials

Before onset of illumination all animals showed normal SEPs characterised by an initial positive deflection followed by a negative peak (P1N1 complex) and a subsequent positive and negative deflection (P2N2 complex) (Figure 1). It is generally accepted that the first complex represents presynaptic activity of the thalamocortical tract whereas later components reflect cortical activity originating from postsynaptic excitation.

In the centre of the illuminated area the time course of the amplitudes was comparable in treated and control animals. The P2N2 wave rapidly decreased and was reduced to zero in both groups 15 min after onset of illumination. This wave showed no recovery in any animal. Except for one control rabbit the P1N1 complex also completely disappeared after 15 min but partial recovery was observed in both groups. This recovery occurred earlier in the control group but had become comparable in both groups 3 h after onset of illumination.

The time course of the SEPs 0.5 mm rostral to the irradiated area is summarised in Figure 2. At this location the P2N2 complex completely disappeared within 15 min after onset of illumination. Four of the RP 48740 treated animals showed a partial or even complete recuperation of this wave whereas no recovery was noticed in the control group. Considerable recovery of the P1N1 complex at this site was noticed in both groups.

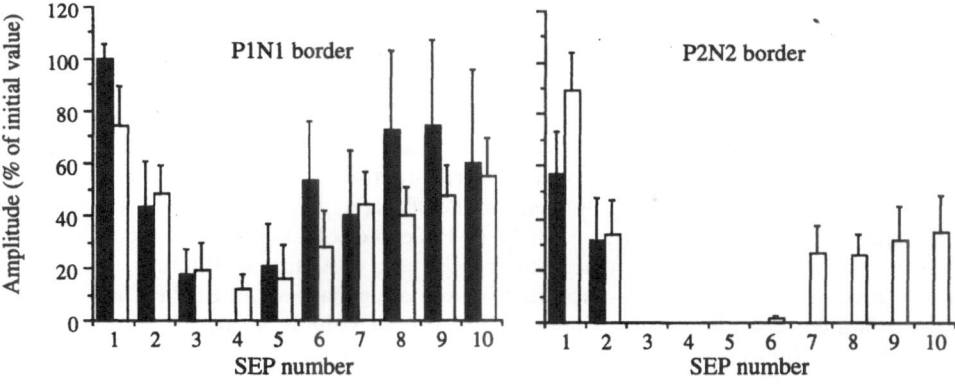

Figure 2. Time course of different wave components of the SEPs 0.5 mm rostral to the illuminated area of the brain cortex. Open bars: RP 48740 treated animals (n = 6); shaded bars: control animals (n = 6). Values are mean ± SE. SEP numbers : see Materials and Methods.

Morphological Analysis

At 4 h after onset of illumination well demarcated lesions were observed in all animals. Due to the slight acidophilia in the infarcted area the lesion could be easily recognised. In all animals the infarct extended from the brain surface to the corpus callosum. Infarct volumes amounted to 18.61 ± 5.08 mm^3 and 15.69 ± 3.32 mm^3 in control animals and treated animals, respectively. Infarct volume was smaller in the RP 48740 treated group but the difference with controls was found not to be significant.

DISCUSSION

PAF antagonists inhibit PAF induced platelet aggregation and thrombosis (review, see Braquet et al., 1987) and exert neuroprotective effects in different types of neuro-injury (review, see Frerichs and Feuerstein, 1990). RP 48740 is a synthetic competitive PAF antagonist which has been shown to inhibit or antagonise a variety of PAF induced events in different species (Lefort et al., 1988). In this study we evaluated the acute effects of RP 48740 on tissue oxygenation, neuronal function and morphologic damage during and after photothrombotically induced brain infarction in unanesthetised rabbits.

In the centre of the infarcted area the time course of PO_2 was comparable in RP 48740 treated and control animals. This suggests that photochemically induced platelet aggregation is not or only minimally PAF dependent or is too intense to be inhibited by PAF receptor blockade only. In the border zone of the illuminated area changes of PO_2 were comparable in treated and control animals shortly after onset of infarction: comparable slow waves of PO_2 (SWO_2s) occurred. The number of SWO_2s was smaller in the RP 48740 treated group but this difference was not significant. As the SWO_2s are related to recurrent waves of cortical spreading depression (CSD) (van Rossem et al., in press) these results suggest that PAF is not a major trigger of ischemia-induced CSD nor critically determines vasomotoric or metabolic changes related to CSD.

RP 48740 clearly abolished the gradual decrease of PO_2 0.5 mm rostral to the illuminated area during the hours following onset of infarction. This decrease thus appears to be PAF related. PAF may be produced due to a number of photothrombosis related events including platelet stimulation and brain ischemia and may cause a decrease of PO_2 via an increased oxygen consumption (Kochanek et al., 1988) or induction of local vasoconstriction and edema (Braquet et al., 1987). According to our results such effects of PAF appear to be involved only in the area closely surrounding the infarct core.

A partial or complete recovery of the P2N2 complex of the SEP 0.5 mm rostral to the illuminated area was noticed in 4 of 6 RP 48740 treated animals but never in the control group. The ability of RP 48740 to prevent the decrease of local tissue PO_2 at this location probably explains the early partial recovery of cortical activity. Indeed, hypoxia significantly reduces or abolishes the cortical components of the SEPs (Haghighi et al, 1992). In addition, in the area where PAF involvement is likely, PAF receptor blockade may protect neuronal integrity, as high levels of PAF are neurotoxic (Kornecki and Ehrlich, 1988).

In contrast with the improvement of tissue oxygenation and SEP recovery close to the infarcted area, the observed reduction of infarct volume in the RP 48740 treated group was not significant. As in our model the penumbral zone - the area in which tissue is believed to be threatened and yet potentially salvagable by therapeutic intervention - is rather small (van Rossem et al., in press), a significant reduction of infarct volume may be hard to accomplish when vascular congestion in the illuminated area cannot be delayed or inhibited.

In conclusion, the present findings indicate the importance of PAF as key mediator of detrimental vascular and/or functional changes related to brain ischemia. Evaluation of long-term effects of PAF receptor blockade may further elucidate the relevance of our results.

ACKNOWLEDGMENTS

This work was supported by FGWO contract 3.0023.91 (Belgian Fund for Medical Scientific Research) and by the OZR VUB.

REFERENCES

Braquet, P., Touqui, L., Shen, T.Y. and Vargaftig, B.B., 1987, Perspectives in platelet-activating factor research, *Pharmacol Rev*, 39:97-145.

Frerichs, K.U. and Feuerstein, G.Z., 1990, Platelet-activating factor - Key mediator in neuroinjury ? *Cerebrovascular and brain metabolism reviews*, 2: 148-160.

Haghighi, S.S., Oro, J.J., Gibbs S.R. and McFadden, M., 1992, Effect of graded hypoxia on cortical and spinal somatosensory evoked potentials, *Surg Neurol*, 37(5), 350-355.

Kochanek, P.M., Nemoto, E.M., Melick, J.A., Evans, R.W. and Burke, D.F., 1988, Cerebrovascular and cerebrometabolic effects of intracarotid infused platelet-activating factor in rats. *J Cereb Blood Flow Metab*, 8, 546-551.

Kornecki, E. and Ehrlich, Y.H., 1988, Neuroregulatory and neuropathological actions of the ether-phospholipid platelet-activating factor. *Science*, 240: 1792-1794.

Lefort, J., Sedivy, P., Desquand, S., Randon, J., Coëffier, E., Maridonneau-Parini, I., Floch, A., Benveniste, J., and Vargaftig, B.B., 1988, Pharmacological profile of 48740 RP, a PAF-acether antagonist, *Eur J Pharmacol*, 150:257-268.

van Rossem, K., Vermariën, H. and Bourgain, R., 1992 a, Construction, calibration and evaluation of pO$_2$ electrodes for chronical implantation in the rabbit brain cortex, *Adv Exp Med Biol*, 316:85-101.

van Rossem, K., Vermariën, H., Decuyper, K., Van Reempts, J., Laureys, M. and Bourgain, R., 1992 b, Local tissue pO$_2$ during and after focal brain cortical infarction in rabbits, *Adv Exp Med Biol*, 317:717-722.

van Rossem, K., Vermariën, H. and Decuyper, K., Slow waves of tissue PO$_2$ in the border zone of photothrombotic brain infarction and their relation to spreading depression-like events, *Adv Exp Med Biol*, in press.

Vermariën, H., van Rossem, K., Altan, R.T., and Decuyper, K., 1992, A method for simultaneous recording of tissue pO$_2$ and EP in the brain cortex of a test animal with a single electrode, *Adv Exp Med Biol*, 317:653-658.

56

MONITORING OF MITOCHONDRIAL NADH LEVELS BY SURFACE FLUORIMETRY AS AN INDICATION OF ISCHAEMIA DURING HEPATIC AND RENAL TRANSPLANTATION

Maureen S. Thorniley,[1] Nick Lane,[1] Sandra Simpkin,[1] Barry Fuller,[2]
Mandana Z. Jenabzadeh,[1] and Colin J. Green[1]

[1] Department of Surgical Research
Northwick Park Institute for Medical Research, Northwick Park Hospital
Harrow, Middlesex, HA1 3UJ, United Kingdom
[2] University Department of Surgery
Royal Free Hospital School of Medicine
Pond Street, London NW3 2QG, United Kingdom

INTRODUCTION

One of the major causes of dysfunction in transplanted organs is ischaemia-reperfusion (IR) injury. Impairment of mitochondrial function is likely to be central to many of the known consequences of ischaemia; these include loss of cellular homeostasis involving a fall in intracellular pH (Fuller et al., 1988), mitochondrial calcium loading and cellular swelling (Calman et al., 1973), accumulation of reduced pyridine nucleotides, inhibition of mitochondrial electron transfer, and a fall in ATP levels (Hardy et al., 1991). In irreversibly damaged cells, respiratory control is lost and is accompanied by oxidation of cytochromes a and a$_3$ and NADH (Taegtmeyer et al., 1985). The latter was attributed originally to substrate deficiency (Chance and Williams, 1955) but more recent studies indicate that an enzymological defect develops resulting in an inability to metabolise NADH-linked substrates (Taegtmeyer et al., 1985 and Hardy et al., 1991). *In vitro* studies of the respiratory chain (RC) complexes have been made in several tissues including cardiac and renal tissue, subjected to ischaemia-reperfusion injury and it was found that complexes I and IV are major defective sites (Hardy et al., 1991 and Veitch et al., 1992). Return of function may, therefore, relate to preservation of inner mitochondrial membrane integrity, and the structure and activities of the RC complexes. The integrity of oxidative metabolic pathways and capacity to resynthesise ATP rather than the immediate post-ischaemic ATP levels appears to determine the return of function (Taegtmeyer et al., 1985).

Clinically it would be of benefit if a non-invasive method of assessing ATP synthesis could be used as a predictive indicator of organ viability pre-transplantation. The value of

Oxygen Transport to Tissue XVII, Edited by Ince et al.
Plenum Press, New York, 1996

non-invasive measurements of the intrinsic mitochondrial fluorophores NADH and oxidised flavoprotein, as indicators of intracellular hypoxia was emphasised by Chance and co-workers (Chance and Williams, 1955, Chance et al., 1979 and Mayevsky, 1984). It was shown that NADH fluorescence is predominantly mitochondrial in origin (O'Conner, 1977). At physiological temperatures, oxidised flavoprotein has only a weak intrinsic fluorescence but this can be enhanced at low temperatures. This enhancement was used to enable oxidised flavoprotein to NADH ratio measurements in freeze trapped biopsy samples from paediatric livers and it was shown that post-transplant the ratio remained lower than that observed in control livers, indicative of ischaemia (Tokuka et al.,1993). A close correlation was found between *in vitro* measurements of energy charge and mitochondrial phosphorylative activity with non-invasive SF NADH measurements in perfused blood-free rat livers (Okamura et al., 1992). However, it is not sufficient to measure NADH oxidation alone as a function of ischaemia and the magnitude and kinetics of the reduction of NAD^+ to NADH is thought to be a far better indicator of RC function. The latter is considered to be a function of the reverse direction of the RC (Tokunga et al., 1987 and Okamura et al., 1992) which is energetically unfavourable and would, therefore, be more critically dependent on integrity of the respiratory chain and, hence, more sensitive to damage.

To study further the possibility of developing a predictive assay we have investigated the changes in *in vivo* surface NADH fluorescence measurements associated with storage and transplantation of rat livers (isografts) and rabbit kidneys (autografts). In each model, we have used two groups comparing organs subjected to minimum storage (control groups 1) with organs stored for longer and damaging periods of cold ischaemia. In the hepatic model, transplantation was carried out after 25 min storage in hypertonic citrate solution (HCA)(1-2^{0}C) whereas in the renal model kidneys were autografted immediately. In both models we expect 100% survival in the recipient animals. In the longer storage groups (2): livers were transplanted after 24 h storage in 1-2^{0}C HCA and this resulted in approximately 10% of the rats surviving with good hepatic function (unpublished observation); in kidneys autografted after a 72 h cold storage period we expect only between 20 to 30% of rabbits to survive with good renal function (Gower et al., 1989). To further identify the site of IR damage, the effect of sodium pentobarbitone inhibition of NADH dehydrogenase (complex 1) on the rate of NADH production was determined (Renault et al., 1987).

METHODS

Liver Transplantation

All animal procedures were carried out according to the Animals (Scientific Procedures) Act, 1986. Male Lewis rats (200-300 g) were used and surgical anaesthesia was maintained using enflurane at 1-2% with O_2 via a face mask at 0.5 L/min. Donor operation: The rat was placed in a supine position and a mid-line incision made. The liver was freed from all adherent ligaments and then all the hepatic vessels prepared ligating any side-vessels present. The portal vein (PV) was cannulated distal to the splenic veins with a 16G Abbocath T catheter. The liver was flushed with 30 ml of 4°C hypertonic citrate solution (HCA), using a pressure of 30 cm H_2O. The liver was stored in HCA at 1-2^{0}C. The total time for the donor operation was 45 min, with a graft cold ischaemic time of 25 min (Group 1, control), or 24 h (Group 2, stored).

Recipient operation: The rat was placed in a supine position and a midline incision made. All peritoneal attachments to the liver were removed. The hepatic vessels were then prepared and clamped, and the liver was removed. The donor liver was placed in 10 ml 4°C HCA and the PV was perfused with 3 ml human albumin (5% in saline) solution at room temperature and the hepatic artery was perfused with 1 ml of the same solution. The donor liver was placed orthotopically and then the suprahepatic venaecavae (SVC) was anastomosed using continuous 8/0 sutures. The PV was anastomosed using a 10/0 continuous suture. Portal flow was then re-established and the SVC unclamped. The inferior vena cava (IVC) was anastomosed using a continuous 10/0 suture and the clamp was removed to establish flow. The arterial anastomoses were then completed using a continuous 10/0 suture. The clamp was removed and arterial flow re-established. The mean time to completion of the SVC, the PV and the IVC anastomoses were approximately 9, 17 and 25 min, respectively. The arterial anastomosis took a further 10 min.

Experimental Groups. Group 1 (n = 11): Control livers were stored for a minimal period of 25 min in HCA surrounded by ice to maintain a temperature of 1-2°C, whilst the recipient was prepared and then isografted as described above. In three Group 1 transplants both NADH and FP, SF measurements were made.

Group 2 (n = 9): Livers were stored in HCA surrounded by ice to maintain a temperature of 1-2°C for 24 h before isografting. In three Group 2 transplants both NADH and FP, SF measurements were also made.

Group 3 (n = 8): Control studies were conducted to determine the quenching effect of deoxygenated haemoglobin on NADH SF measurements. An additional 8 livers were removed, unflushed and subjected to periods of warm ischaemia (25°C) between 50 and 120 min before flushing with 25°C HCA. SF measurements made pre-and post-flushing.

Group 4 (n = 12): The time course of the haemoglobin deoxygenation compared to NADH emission measurements was investigated on untransplanted livers. The median lobe was clamped and SF measurements were made pre- and up to 30 min post clamping and following release of clamping. The effect of portal vein occlusion on fluorescence emission measurements was also performed.

Group 5 (n = 3): Control studies to investigate whether the measured fluorescence emission signal (excitation wavelength 366 nm) is responsive to the systemic pO_2. NADH measurements were continuously made of the *in situ* hepatic surface in response to decreasing the inspired oxygen from 80 to 10%.

Surface Fluorometric (SF) Measurements of NADH. SF measurements were made using a Perkin Elmer LS 50 with a fibre-optic attachment. An SF probe was gently placed on the surface of the *ex vivo* or *in situ* liver or kidney and fixed in position by attachment to a retort stand. Emission spectra of NADH from 366-600 nm were obtained by excitation at a fixed wavelength of 366 nm (slit widths 10/10 nm). In order to investigate any haemodynamic effect, the relationship between the 366 nm and 480 nm (fluorescence emission due to NADH) intensities were compared and if the 366 nm signal differed by more than 10% the results were not used. We used uncorrected fluorescence readings in agreement with Frank et al.,1976. To assess the effect of movement and normal cyclical variations in Hb oxygenation levels on NADH fluorescence emission, each spectrum was performed at least in duplicate, within 1 minute. The values thus obtained agreed within 5 to 10%. The results are presented in terms of relative fluorescence units (arbitrary values). Flavoprotein (FP) was measured between 500 and 600 nm, excitation wavelength 436 nm.

Experimental Protocol. The NADH fluorescence emission (excitation wavelength 366 nm), and spectral properties, were measured after each of the surgical procedures outlined below:

1. Liver exposure
2. Preparation of hepatic vessels prior to ligation

3. Ligation of hepatic artery
4. Hepatic perfusion with 30 ml 4°C HCA via the portal vein at a pressure of 30 cm water
5. Storage at 1-2° in HCA, Groups 1 and 2
6. Flushing with 3 ml human albumin (room temperature) via portal vein and hepatic artery
7. Suprahepatic vena cava, portal vein and descending vena cava anastomoses
8. Hepatic artery anastomosis

Renal Transplantation

Female NZW rabbits (2.5 kg) were anaesthetised with an i.m. injection of ketamine (50 mg/kg) and xylazine (8 mg/kg), trachaeotomized and artificially ventilated with a 50:50 oxygen:nitrous oxide mixture for periods of up to 8 h. Surgical anaesthesia was maintained by continuous i.v. infusion of ketamine and xylazine (50 mg:8 mg/kg/h). The transplantation procedure was identical to that previously described from this laboratory (Gower et al.,1989).

Experimental Groups. Group 1: Freshly nephrectomized left kidneys were flushed with 30 ml of cold (1-2°C) HCA via the renal artery and autografted immediately into the right renal bursa using standard microsurgical techniques.

Group 2: Kidneys were flushed as in Group 1 and then stored in HCA surrounded by ice to maintain a temperature of 1-2°C for 72 hours before autografting.

Experimental Protocol. The intensity of NADH fluorescent emission, and spectral properties, were measured in:

1. *Ex-vivo* Group 1 and Group 2 kidneys
2. *In situ* Group 1 and Group 2 kidneys, prior to transplantation and reperfusion with blood and measured in both Groups up to 6 hours post-transplant at 5, 15, 45, 60 minutes and at 2, 4, 6 hours.
3. In response to alterations in the inspired FiO_2
4. Rapid infusion of sodium pentobarbitone to renal artery: 200 mg/kg (i.v.) (<5 sec) whereafter NADH fluorescence was monitored for a further 20 minutes.

Statistical Analysis

Data are presented as mean ± S.E.M. Statistical significance between groups was tested by unpaired Student's t-test.

RESULTS

A broad fluorescence emission maximum of NADH (excitation wavelength 366 nm) was measured in the ex vivo (1-2°C) and in situ devascularised liver or kidney (Figs. 1,2). An increase in fluorescence emission of NADH (470 nm) from the surface of the perfused rabbit kidney or rat liver, excited at 366 nm, was obtained in response to decreasing the systemic pO_2 from 21 to 5 Kpa.

Emission spectra of NADH (excitation 366 nm) from the hepatic surface made after the following procedures: 1) Liver-exposure, 2) Preparation of vessels prior to ligation, 3) Ligation of hepatic artery and 4) Hepatic perfusion with 30 ml hypertonic citrate (1-2°C HCA) via portal vein.

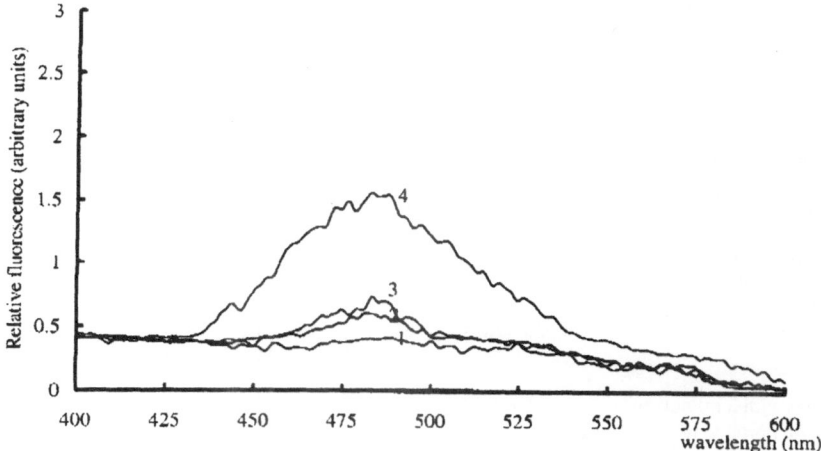

Figure 1. *In vivo* surface fluorimetric measurements of NADH during harvesting of the liver.

Emission spectra of NADH (excitation 366 nm) from the hepatic surface after the following procedures: 1) Storage in 1-2°C HCA (25 min), 2) Flushing with 3 ml human albumin (room temperature) via portal vein and hepatic artery, 3) Suprahepatic venacava, portal vein and descending venacava anastomoses and 4) Hepatic artery anastomosis.

Emission spectra of NADH (excitation 366 nm) from the hepatic surface after the following procedures: 1) Storage in 1-2°C HCA for 24 h, 2) Flushing with 3 ml human albumin (room temperature) via portal vein and hepatic artery, 3) Suprahepatic vena cava, portal vein and descending vena cava anastomoses 4) Hepatic artery anastomosis.

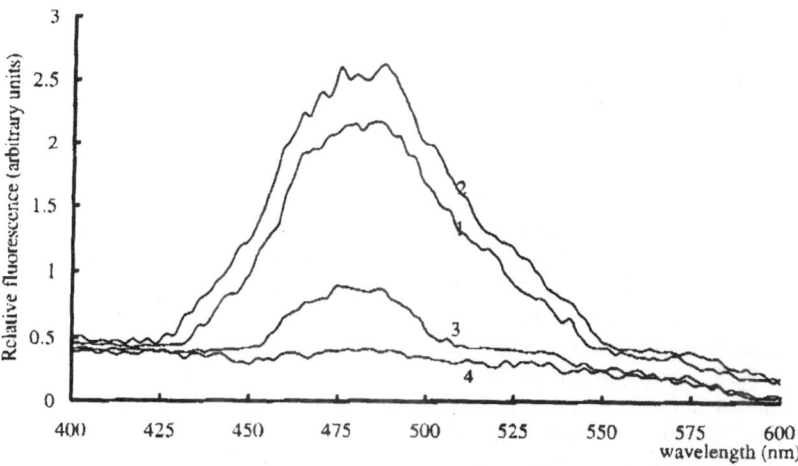

Figure 2. *In vivo* surface fluorimetric measurements of NADH during revascularisation of a control (minimally stored) liver.

Table 1. NADH fluorescence emission measurements during harvesting and transplantation in minimally stored, 25 min and 24 h stored livers

Surgical procedure	Storage time	n	Fl Rel units mean±sem	% NADH* reduced	p-Student's t-test
1. Liver exposure		14	0.43±0.02	11.5	
2. Preparation of vessels prior to harvesting		14	0.52±0.04	17.3	<0.03[1]
3. Ligation of hepatic artery		14	0.59±0.014	21.3	<0.0003[1] ns[2]
4. Complete hepatic ligation and perfusion with 30 ml HCA via portal vein		14	1.22±0.10	57.8	<0.0001[1] <0.0001[3]
5. Storage in 1-2°C HCA	25 min 24 h	86	1.95±0.41 1.63±0.15	100 81.5	<0.002[1] <0.02[4] ns[a,b]
6. Flushing with 3 ml human albumin via portal vein	25 min 24 h	76	1.65±0.17 1.72±0.16	82.4 86.7	<0.004[4] ns[5] ns[a,b]
7. Suprahepatic vena cava, portal vein and descending vena cava anastomoses	25 min 24 h	86	0.56±0.07 0.23±0.04	19.5 0.0	<0.004[a]
8. Hepatic artery anastomosis	25 min 24 h	76	0.42±0.03 0.22±0.02	11.5 0.0	<0.0003[b]

Superscript numbers refer to the group tested against. ns not statistically significant.
[a,b] p value between the 25 min and 24 hr storage groups following: [a] suprahepatic vena cava, portal vein and descending vena cava anastomoses; [b] hepatic artery anastomosis.
* Assuming that the measured fluorescence emission at 480 nm (excitation 366 nm) attributable to NADH is maximal after 25 min storage time and equal to 100% reduction. This also assumes that any differences in relative fluorescence are linearly related to NADH concentration. Similary, the lowest relative fluorescence observed (0.22 ± 0.02) is equal to 100% oxidation (NAD^+).

In vivo Hepatic Surface Fluorescence Measurements of NADH during Harvesting and Transplantation

Table 1 shows the results of the NADH measurements made during harvesting and transplantation of both groups of livers: Group 1 (control-minimum cold ischaemic time, n = 8) or Group 2 livers (24 h, n = 6) which were stored in HCA at 1-2°C.

A mean NADH fluorescence value of 0.43 (±0.02)(n = 14) was measured on the surface of the *in situ* exposed liver. The effect of dissection of the blood vessels prior to any ligation resulted in a statistically significant increase in fluorescence, 0.52±0.04 (n = 14, p<0.03) compared to that obtained on opening the abdomen. It was found that hepatic artery ligation did not result in any further significant increase in the NADH fluorescence. However, complete hepatic ligation and perfusion with 30 ml 4°C HCA resulted in a highly significant increase in NADH fluorescence compared to ligation of the hepatic artery alone (1.22±0.10, n = 14, p<0.0001). The NADH fluorescence increased significantly upon storage at 1-2°C in HCA compared to pre-storage (p<0.02). However, there was no significant difference in the NADH measurements between the 2 storage groups 1.95±0.41, n = 8 (Group 1) and 1.63±0.15, n = 6 (Group 2). It was found that flushing with 3 ml human albumin (25°C) had no further effect on the NADH levels of the 2 groups: 1.65± 0.17 (Group 1, n = 7 livers) and 1.72 ± 0.16 (Group 2, n = 6 livers). In Group 1 restoration of blood supply following suprahepatic vena cava, portal vein and descending vena cava anastomoses resulted in a significant decrease in the NADH fluorescence to the same values as observed following preparation of the vessels prior to ligation. There was a highly significant

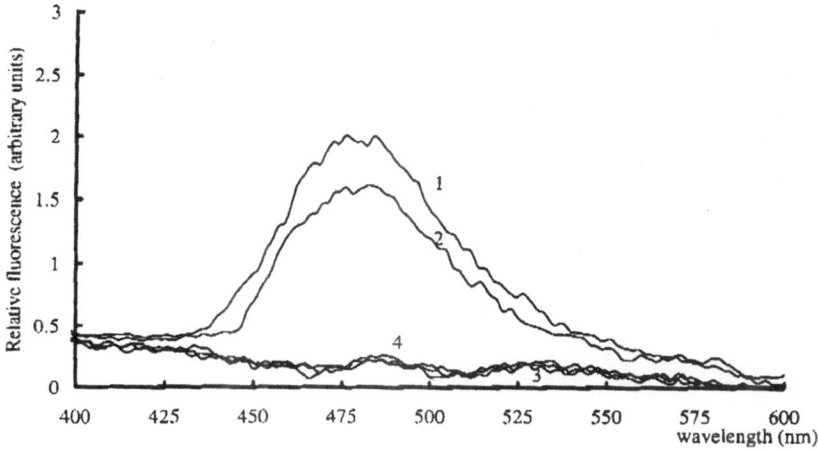

Figure 3. *In vivo* surface fluorimetric measurements of NADH during revascularisation of a 24 hour stored liver.

difference (p<0.004) in the NADH measurements between the 25 min and the 24 h stored groups, 0.56±0.07 versus 0.23±0.04, respectively, after restoration of blood supply.

The final revascularisation step (hepatic artery anastomosis) resulted in the SF NADH measurements returning to the same values as measured on exposure of the liver in Group 1, the minimally stored livers. However, there was a highly significant difference between Groups 1 and 2, 0.43±0.03 versus 0.22±0.022, p<0.0003) respectively (Figs. 2, 3). This significant difference between the NADH levels of the two groups following revascularisation and hence introduction of oxygen, resulted in a maximal level of NADH oxidation in the 24 h stored livers. In studies in which the median lobe vessels were clamped, the quenching effect of deoxygenated haemoglobin upon NADH emission was only apparent after approximately 5 min (n = 12). However, the fluorescence emission due to NADH was unquenched if the portal vein was clamped (unimpeded outflow) even up to 30 min post clamping, and was of the same magnitude as that observed upon harvesting.

In six hepatic isografts (three from Group 1 and three from Group 2) both flavoprotein and NADH measurements were made (excitation 436 and 366 nm, respectively) and the results are shown in Table 2. The emission assumed to be due to flavoprotein decreased initially but then increased, this apparent increase in oxidised flavoprotein may be attributed to low temperature enhancement resulting from the cold HCA flush. The FP/NADH ratio decreased during harvesting and transplatation and did not return to pre-ischaemic values. The FP/NADH ratio was less in Group 2 than in Group 1 in the final stages of the transplants.

In vivo Surface Fluorescence Measurements of NADH in Harvested Kidneys Pre and Post-Transplantation

The fluorescence emission of control kidneys which were removed and flushed with HCA and stored on ice for up to 90 min did not significantly change after storage (3.03 ± 0.09 (mean ± S.E.M., n = 19 estimates), and there was no significant difference between the relative fluorescence intensity of the two groups *in situ* (pre-reperfusion): 2.85± 0.196 (Group 1, n = 9 kidneys) and 2.94 ± 0.26 (Group 2, n = 7 kidneys). Table 3 shows the % change in fluorescence from the pre-reperfusion values during the initial phase (5-20

Table 2. NADH and Flavoprotein surface fluorescence measurements during harvesting and transplantation in minimally stored, 25 min and 24 h stored livers

Surgical procedure	Storage time	FP/NADH* ratio
1. Liver exposure		0.3±0.007
2. Preparation of vessels prior to harvesting		0.15±0.008
3. Ligation of hepatic artery		0.2±0.0156
4. Complete hepatic ligation and perfusion with 30 ml hypertonic citrate (HCA) via portal vein		0.25±0.037
5. Storage in 1-2°C HCA	25 min	0.16
		0.13
		0.18
	24 h	0.12
		0.13
		0.17
6. Flushing with 3 ml human albumin via portal vein	25 min	0.17
		0.22
		0.15
		0.18
	24 h	0.14
		0.16
7. Suprahepatic vena cava, portal vein and descending vena cava anastomoses	25 min	0.11
		0.038[a]
		0.098
	24 h	0.070
		0.055
		0.06
8. Hepatic artery anastomosis	25 min	0.16
		0.019[a]
		0.11
	24 h	0.096
		0.078
		0.06
Ex-vivo recipient HCA flushed livers. Minimal cold ischaemia and immediately measured after storage at -70°C for 3 h	25 min	0.18
		0.19
		0.18

*Denotes Flavoprotein/NADH ratio. NADH and FP excitation wavelength 366 nm, 436 emission 480 nm and 540 nm, respectively.
[a]Poorly perfused liver upon revascularisation.

minutes) post reperfusion of Group 1 and Group 2 kidneys. There was no significant difference between the % oxidation observed in the two groups; [82.85 ± 11.29 (Group 1) compared to 89.85 ± 11.92 (Group 2)]. Figure 4 shows emission spectra obtained by continuous scanning of NADH fluorescence pre- and post-reperfusion in a Group 1 kidney. Immediate oxidation of NADH between 70 and 100% occurred within 1 min (Table 3) and values oscillated within this range. In comparison in a Group 2 kidney, the magnitude and rate of oxidation of NADH was less (Fig. 5). However, 20 min following reperfusion, 100% of the NADH had become oxidised.

1) Pre-reperfusion, 2) 1 min after reperfusion, 3) 12 min after reperfusion, 4) 60 min after reperfusion, 5) 5 min after sodium pentobarbitone infusion (200 mg/kg)

Figure 4 shows emission spectra from continuous measurements of NADH pre and post reperfusion in an unstored kidney.

1) Pre-reperfusion, 2) 1 min after reperfusion, 3) 6 min after reperfusion,

Table 3. NADH fluorescence measurements during reperfusion and on sodium pentobarbitone administration of Group 1 and Group 2 kidneys

Individual Group kidneys	% NADH oxidised upon reperfusion	% of pre-reperfusion NADH formed on pentobarbitone infusion
1	90	95
1	72	76
1	96	69
1	70	70
1	87	70
1	72	70
1	88	100
1	95	80
2	96	0
2	67	82
2	96	0
2	84	8
2	86	0
2	100	0
2	100	37.5

4) 10 min after reperfusion, 5) 20 min after reperfusion and 6) 5 min after sodium pentobarbitone infusion and remained constant until the end of monitoring (20 min).

Figure 5 shows emission spectra from continuous measurements of NADH pre- and post-reperfusion in a stored (72 h) kidney.

The Effect of Sodium Pentobarbitone Infusion on Kidney NADH Measurements

In Group 1 kidneys, infusion of sodium pentobarbitone 200 mg/kg (< 5 sec) resulted in rapid production of NADH (in all 8) kidneys within 1 min, as expected for the inhibition

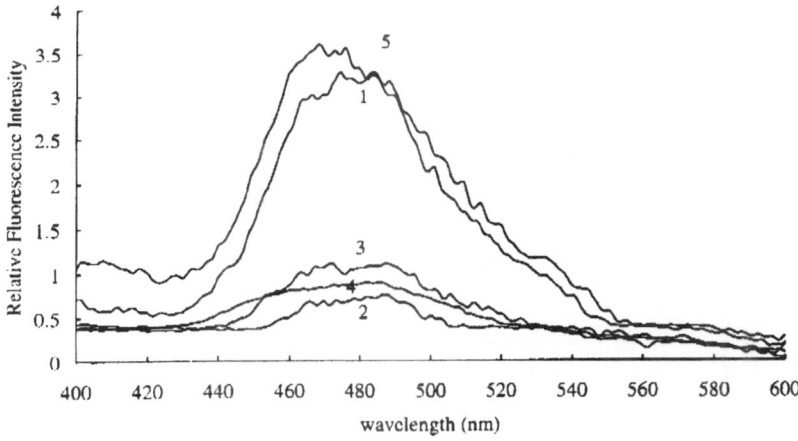

Figure 4. *In vivo* surface fluorimetric measurements of NADH in a transplanted fresh kidney.

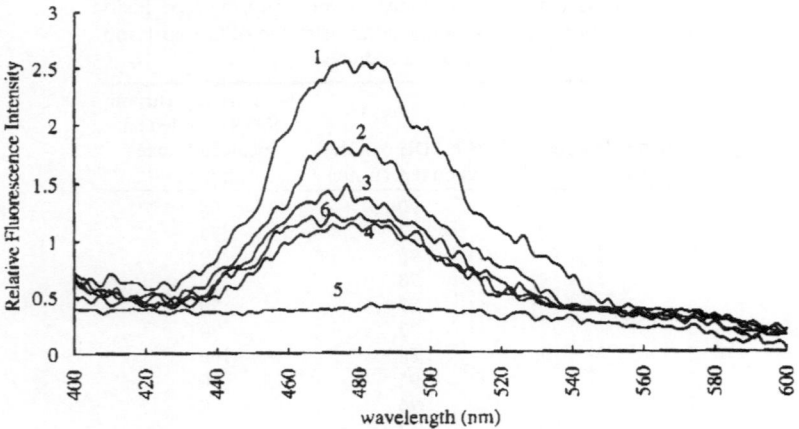

Figure 5. *In vivo* surface fluorimetric measurements of NADH in a transplanted 72 hour-stored kidney.

of NADH dehydrogenase (Table 3). In Group 2 kidneys, sodium pentobarbitone resulted in a decreased rate and magnitude or no NAD$^+$ reduction in 5 out of 7 kidneys within the monitoring period of 20 min (Table 3). In one kidney, sodium pentobarbitone resulted in an NADH emission to a maximum of 37% of the pre-reperfusion value over 5 min and remained constant. In the remaining kidney, the NADH level reached values as observed in Group 1. Figure 6 shows the rate of oxidation and reduction of NADH as measured by SF in response to reperfusion and sodium pentobarbitone. In Group 1 there was a rapid oxidation of NADH immediately on reperfusion and minor oscillations until infusion with sodium pentobarbitone. Approximately 50-70% NADH production occurred within 2 min. In group 2 kidneys the NADH oxidation was slower and with a poor response to sodium pentobarbitone.

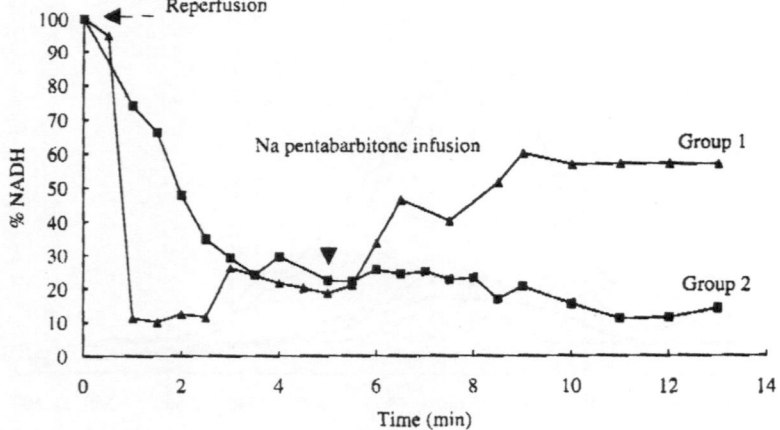

Figure 6. Rate of oxidation/reduction of NADH during the first 13 minutes of reperfusion in Group 1 and Group 2 kidneys.

DISCUSSION

The hepatic surface fluorescence NADH emission spectra were similar to the characteristic spectra excited at 340 nm and 366 nm reported *in vivo* and *in vitro* in rabbit kidney reported by Thorniley et al. (1994) and in rat brain by Mayevsky et al. (1994). The origin of the NADH fluorescence, (i.e. whether cytoplasmic or intra-mitochondrial) has been investigated by several groups and is thought to be predominantly mitochondrial from data in studies utilising various substrates, RC inhibitors and alterations in oxygen supply (Chance et al., 1979; Scholz et al., 1969). Surface measurements of NADH were responsive to fluctuations in the blood flow and the systemic pO_2. These findings are in agreement with others who correlated surface fluorescence monitoring of reduced pyridine nucleotides with physiological function of perfused *ex vivo* rat livers when subjected to various hypoxic insults (Chance et al., 1979; Frank et al., 1976).

Our *ex vivo* pre-transplant NADH measurements showed no statistical difference between the two groups in either livers or kidneys; hence, it is unlikely that fluorescence of NADH was altered as a result of enhancement or quenching during storage, as distinct from changes in the pyridine nucleotide levels. Although we have not measured nucleotide levels in tissue homogenates we believe that there are several factors which support the view that the fluorescent emission can be attributed to NADH: for example, there are few other fluorophores whose emission characteristics are identical to that of NADH (whose emission is oxygen dependent); which are responsive to devascularisations which are responsive to changes in metabolic demands and substrate availability and, finally, which are responsive to sodium pentobarbitone whose mode of action is to inhibit NADH-dehydrogenase resulting in an increase in NADH (Renault et al., 1987). In warm ischaemia studies we found that the NADH emission was only quenched after several minutes; this difference in the rate of change between fluorescence changes due to oxidative effects and those due to haemoglobin deoxygenation has also been reported by others in the rat kidney and brain (Frank et al., 1976 and Mayevsky, 1984).

Our results demonstrate that it is possible to make both discrete and continuous measurements of NADH fluorescence in the *in situ* liver and kidney during harvesting and transplantation. The technique is sufficiently sensitive to detect changes occurring during preparation of vessels prior to ligation (constant FiO_2). In all livers and kidneys there was a large NADH peak before reperfusion in response to ischaemia which significantly increased upon cold storage but then decreased as revascularisation occurred. Following restoration of blood flow there was a statistically significant difference in SF measurements between the two storage groups, the longer ischaemic period being associated with 100% NAD^+ resulting perhaps from a more pronounced IR injury being exacerbated in this group. This is consistent with complete NADH oxidation to a greater than normal physiological variation which can be due to mitochondrial uncoupling and, hence, less effective production of ATP (Chance and Williams, 1955; Taegtmeyer et al., 1985 and Hardy et al., 1991).

In the present study, in both Group 1 and Group 2 livers NADH was rapidly oxidised to NAD^+ and this is consistent with the energetics of the RC; moreover, we observed a significant difference in the reduction state of NADH post revascularisation between the two groups; the 100% NAD^+ observed in the longer preserved grafts is consistent with a diminished rate of NADH production compared to the usual 11% level of reduction in Group 1, control livers. In continuously monitored kidneys the rate of oxidation was slower in Group 2 kidneys than in Group 1 and, moreover, in Group 2 there was no oscillation of NADH levels in response to minor alterations in PO_2 or vascular changes, which also suggests damage to the RC.

The effect of sodium pentobarbitone administration was quite distinct in the two groups. In Group 1 kidneys there was a rapid increase in the relative fluorescence intensity of NADH to pre-reperfusion levels within 1 min, as expected. In Group 2 that there was little or no response to sodium pentobarbitone infusion. We previously found that there were minimal histological changes in Group 1 kidneys; these data strongly suggest that 72 hour storage resulted in mitochondrial dysfunction. The fact that sodium pentobarbitone inhibits NADH dehydrogenase suggests that the damage (lack of response to sodium pentobarbitone) may be at complex I or substrate deficiency leading to complex 1. Similarly, it has been reported that mitochondria isolated from perfused rat hearts subjected to global ischaemia and reperfusion were unable to metabolise NAD^+-linked substrates and the site of damage was shown to be complex 1 (Veitch et al., 1992). It is probably not coincidental that the two main sites of mitochondrial damage during ischaemia, complex I and III, are also the major sources of oxygen derived free radical production during normal electron transfer in mitochondria *in vitro* (Cadenas et al., 1977).

Tokunga et al. (1987) reported a decrease in the rate of NAD^+ reduction in perfused stored rat livers in response to ischaemia (inhibition of flow) on an *ex vivo* circuit following storage. However, Tokunga et al. (1987) and Ozaki et al. (1989) used blood free perfusates and did not make any *in vivo* SF measurements including transplantation to determine the viabilty of their technique in the blood-perfused organ. In our study, in both Group 1 and Group 2 kidneys the NADH was rapidly oxidised to NAD^+ and this is consistent with the energetics of the RC. However, the magnitude and rate of the NAD^+ to NADH reaction was considerably reduced in Group 2 compared to Group 1 kidneys.

Some researchers have used changes in the FP/NADH ratio as an indicator of changes in the intracellular redox state. In theory, during normoxia the FP population will be predominantly in the oxidised (fluorescent) form, whereas NADH fluorescence will be minimal resulting in a high FP/NADH ratio. As the tissue becomes progressively ischaemic during, for example, harvesting of the organ, this ratio will decrease owing to a reduced FP and an enhanced NADH fluorescence. Similarly, following anastomosis the ratio should increase indicating a return to normoxic mitochondrial conditions. However, this presumes that both the NADH and FP signals originate from the mitochondrial space and in many tissues several extramitochondrial flavoproteins have been isolated which can lead to erroneous conclusions (Scholz et al., 1969). In addition, the fluorescent emission attributable to flavoprotein from the *in situ* hepatic surface is not well defined. Thus the determination of the FP fluorescence value is critically dependent upon the choice of emission wavelength and is subject to considerable variability. In spite of this, and as indicated in Table 3, the FP/NADH ratio did fall as expected during ischaemia in agreement with similar findings of Chance et al., (1979) who reported that the FP/NADH ratio in blood-free perfused livers decreased during hypoxia. However, they utilised low temperatures to enhance the fluorescence emission of oxidised flavoprotein which has a negligible fluorescence at room temperatures compared to that of NADH. More recently, Tokuka and colleagues (1993) exploited these early studies and determined the ratios in biopsies obtained from human paediatric freeze-trapped livers and reported that post surgery the ratio was approximately 50% less than control values. In our studies, in both Group 1 and 2 livers the oxidised flavoprotein/reduced NADH ratio was less than control measurements made on exposure of the liver and did not return to pre-harvest levels. Moreover, in Group 2, the longer ischaemic store time, the ratio was even less, suggestive of even greater mitochondrial dysfunction. Moreover, we only measured for a maximum of 5 min post transplant. In order to measure oxidised flavoprotein fluorescence, low temperatures are required and biopsies have to be taken after revascularisation; in the clinical situation in an already compromised tissue this would be undesirable. Since SF measurements of NADH show a reproducible, highly significant difference between both groups following anastomosis we believe that NADH levels alone, rather than FP and NADH, are an adequate index of tissue redox states which may be predictive of subsequent organ function.

We have shown that the degree of ischaemic insult, hence ischaemic damage, can be measured using surface fluorescence measurements of pyridine nucleotides in kidneys and in livers and, indeed, can be applied to many tissues. We have previously reported that changes in the redox level of NADH and cyt aa_3 as measured using near infra-red spectroscopy (NIRS) correlate with changes in morphology and viability. As far as we are aware, this was the first report correlating two non-invasive tests with organ function and morphology following storage and transplantation. Our findings suggest that this mitochondrial defect is extremely important in the failure of reperfused ischaemic tissue. We now perform *in vitro* measurements of respiratory control ratios (RCRS) and the activities of the individual respiratory chain complexes in our storage/transplantation tissues. We hope that by comparing invasive test of mitochondrial function with organ function and histological changes, in addition to non-invasive methods of metabolic/physiological assessment, this may aid in the prevention/limitation of ischaemia-reperfusion injury. Non-invasive measurements of NADH and cyt aa_3 could provide a valuable indication of organ viability and function if measured on an *ex vivo* circuit prior to transplantation. From our studies, we believe that SF measurements could be useful in the clinical transplant situation in which damage may have occurred during storage and, hence, provide a valuable indication of tissue metabolic integrity and function.

ACKNOWLEDGMENTS

I would like to gratefully acknowledge the continuous support, encouragement and expertise given by the late Dr Keith Dalziel FRS, Department of Biochemistry, University of Oxford.

We would like to acknowledge the help, support and encouragement of all the Surgical Research Group.

REFERENCES

Cadenas, E., Boveris, A., Ragan, C.I. and Stoppani, O.K. 1977. Production of superoxide radicals and H_2O_2 by NADH Ubiquinone reductase and Ubiquinone cyt-c reductase from beef heart Mitochondria. *Arch Biochem Biophys* 180:248-257.

Calman K.C, Quinn R.O, and Bell P.R. 1973. In: Organ Preservation, (Ed. Pegg, D.E.), Churchill Press, London, pp: 225-240.

Chance, B. and Williams, G.R: 1955. Respiratory enzymes oxidative phosphosylation. I. Kinetics of Oxygen utilisation. *J Biol Chem* 217:383-393.

Chance, B., Schoener, B., Oshino, R., Itshak, F. and Nakase, Y. 1979. Oxidation-reduction ratio studies of mitchondria in freeze-trapped samples. *J Biol Chem* 254:4764-4771.

Frank, C.H., Barlow, C.H and Chance, B: 1976. Oxygen delivery in perfused rat kidney: NADH fluorescence and renal functional state. *Am J Physiol* 231:1082-1089.

Fuller, B.J., Gower, J.D. and Green, C.J. 1988. Free radical damage and organ preservation, fact or fiction? *Cryobiology* 25:377-393.

Gower, J.D., Healing, G., Fuller, B.J., Simpkin, S. and Green, C.J. 1989. Protection against oxidative damage in cold-stored rabbit kidneys by desferrioxamine and indomethacin. *Cryobiology* 26:309-317.

Hardy, L., Clark, J.B., Darley-Usmar, V.M., Smith, D.R. and D. Stone. 1991. Reoxygenation-dependent decrease in mitochondrial NADH: CoQ reductase (Complex I) activity in the hypoxic/reoxygenated rat heart. *Biochem J* 274:133-137.

Mayevsky, A. 1984. Brain NADH redox state monitored *in vivo* by fiber optic surface fluorometry. *Brain Res Rev* 7:49-68.

O'Conner, M.J. 1977. Origin of labile NADH tissue fluorescence. Oxygen and Physiological Function (Ed. F.F. Jobsis), Professional Information Library, Dallas pp: 90-99.

Okamura, R., Tanaka, A., Uyama, S and Ozawa, K. 1992. Low-temperature fluorometric technique for evaluating the viability of rat liver grafts after simple cold storage. *Transplant Int* 5:165-169.

Ozaki, N., Tokunaga Y., Ikai, I., et al. 1989. Pyridine nucleotide fluorometry in preserved porcine liver with fluorocarbon emulsion. *Transplantation*. 48:198-201.

Renault, G., Muffat-Joly, M., Polianski, J., Hardy, R.I., Boutineau, J-L., Duvent, J-L. and Pocidalo, J-J. 1987. NADH in situ lazer fluorimetry: effect of pentobarbital on continuously monitored myocardial redox state. *Lasers Surg Med* 7:339-346

Scholz, R., Thurman, R.G., Williamson, J.R., Chance, B., and Bucher, T. 1969. Flavin and Pyridine Nucleotide Oxidation-Reduction Changes in perfused rat liver. *J Biol Chem* 244:2317-2324.

Taegtmeyer, H., Roberts, A.F.C. and Raine, A.E.G. 1985. Energy Metabolism in reperfused heart Muscle: Metabolic correlates to return of function. *J Am Coll Cardiol*: 6: 864-870.

Thorniley, M.S., Lane, N.J., Manek, S. and Green, C.J. 1994. Non-invasive measurement of respiratory chain dysfunction following hypothermic renal storage and transplantation. *Kidney Int* 45:1489-1496.

Tokunga, Y., Ozaki, N., Wakashiro, S., Ikai, I., Kimoto, H., Hovimoto, T., Shimahara, Y., Kamiyama, Y., Yamoaka, Y., Ozawa, K., and Nakase, Y: 1987. Fluorometric study for the non-invasive determination of cellular viability in perfused rat liver. *Transplant*. 44:701-706.

Tokuka, A., Tanaka, A., Kitai, T., Yamaoka, Y., Tanaka, K. and Ozawa, K. 1993. Delayed oxidation of intramitochondrial pyridine nucleotide oxido-reduction state as compared with tissue oxygenation in human liver transplantation. [Abstract] 6th Congress Eur Soc for Organ Transplantation, P236.

Veitch, K., Hombroeckx, A., Caucheteux, D., Pouleur, H. and Hue, L. 1992. Global ischaemia induces a biphasic response of the mitochondrial respiratory chain. *Biochem J* 281:709-715.

57

STUDIES ON TUMOR OXYGENATION IN A RAT RHABDOMYO–SARCOMA DURING FRACTIONATED IRRADIATION

F. Zywietz,[1] W. Reeker,[2] E. Kochs[2]

[1] Institute of Biophysics and Radiobiology
University of Hamburg
[2] Institute of Anesthesiology
Technische Universität München
Germany

INTRODUCTION

The oxygenation status of a tumor is certainly one of the major factors which influences the response of a tumor to radiotherapy (7, 9, 16, 22). Due to a chaotic and temporarily fluctuating tumor blood perfusion (5, 24, 25) oxygenation of tumor tissue is inadequate, and consequently, most experimental and human tumors contain varying proportions of hypoxic tumor cells (11, 18, 19, 25) which are radioresistant. It is supposed that these cells limit the curability of some types of tumors by radiotherapy. There is some experimental evidence that after irradiation with either single or fractionated doses of X-rays, a proportion of surviving cells which were initially hypoxic becomes reoxygenated, thereby changing the oxygenation status of the tumor (1, 14, 21, 23). However, most studies on reoxygenation have been carried out in experimental tumors using the paired survival curve assay (18) and doses up to about 30 Gy. Grau and Overgaard (8) reported on the reoxygenation of an irradiated mouse mammary carcinoma using the clamped local tumor control assay. After a priming dose of 20 Gy reoxygenation occurred completely. However, after a priming dose of 40 Gy only a transient reoxygenation was observed, followed by an increase of the hypoxic fraction slightly above the oxygenation level of untreated tumors.

Since there is only limited knowledge about the changes in tumor oxygenation during radiotherapy (2, 4, 6), it was the aim of this study to determine systematically the oxygenation status of a rat tumor *in vivo* in the course of a fractionated irradiation.

MATERIAL AND METHODS

The studies were performed on superficially growing rhabdomyosarcomas R1H transplanted in the flank of 8- to 9-week-old male WAG/Rij rats (200 ± 12 g). The origin

Oxygen Transport to Tissue XVII, Edited by Ince et al.
Plenum Press, New York, 1996

445

and maintenance of this tumor have been described previously (12). Tumors with volumes of 1.8 ± 0.3 cm^3 at day 21-25 after transplantation were selected for the radiation treatment. Tumor volume V_t was determined by measuring three orthogonal diameters a, b, c and using the formula for an ellipsoid $V_t = \pi/6(a x b x c)$.

The studies were approved by the Hamburg Ministry of Health Ethics Committee and were conducted according to the German Law for Animal Protection from 1987.

Irradiation of the tumors was carried out at a ^{60}Co-therapy unit (Gammatron 2, Siemens, FRG) with a dose rate of 1.00 Gy/min. Tumors were irradiated with a total dose of 60 Gy, given in 20 fractions of 3 Gy each, 5 times a week, for 4 weeks. The field size was limited by a special collimator to 3x3 cm^2 allowing the irradiation only of the tumor area. For irradiation the animals (mean body weight$\pm$SD 258$\pm$12 g) were anesthetised by an i.m. injection of 50 mg/kg b.w. ketamine (Ketavet^{TM}, Parke-Davis, FRG) and 6 mg/kg b.w. xylazine (Rompun^{TM}, Bayer, FRG).

Oxygen partial pressure (pO$_2$) of tumors was measured in anesthetised animals at weekly intervals (on Monday) using polarographic needle probes of 350 µm (outer diameter) and the Eppendorf pO$_2$-device (25).

To minimize the effects of anesthesia the pO$_2$ measurements were performed under controlled mechanical ventilation and monitoring of hemodynamic parameters (Fig. 1). Anesthesia was carried out by a continuous infusion of fentanyl (0.05-0.1 mg/kg b.w./h, Fentanyl^{TM}, Janssen, FRG) and midazolam (0.9-1.8 mg/kg b.w./h, Dormicum^{TM}, Hoffmann-LaRoche, FRG) into the left femoral vein. Artificial ventilation was performed at an inspired oxygen fraction (FiO$_2$) of 33% to maintain the arterial pO$_2$ within the range of spontaneous

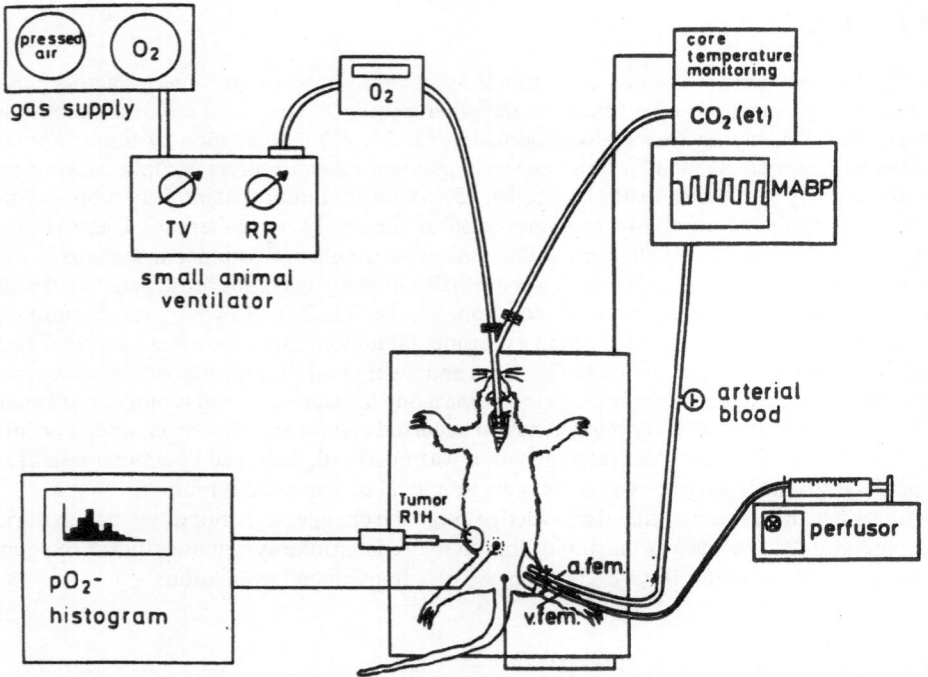

Figure 1. pO$_2$ measurement with the Eppendorf pO$_2$-device in a rhabdomyosarcoma R1H of the rat under artificial ventilation. RR: respiratory rate, TV: tidal volume, O$_2$: oxymeter, CO$_2$(et): endtidal CO$_2$, MABP: mean arterial blood pressure, a.fem.: femoral artery, v.fem.: femoral vein.

breathing rats (98 ± 29 mmHg) (20, 26). By regulating respiration rate (RR) and tidal volume (TV) the endtidal carbon dioxide of the rat could be maintained at a constant level of 35 ± 2 mmHg. Mean arterial blood pressure (MABP) was kept in the range of 115 ± 20 mmHg monitored by a catheter in the left femoral artery and a transducer. The same catheter was used for taking arterial blood for blood gas analysis. Core temperature of 37.0 ± 0.3°C was maintained by temperature-controlled infrared radiation.

The pO_2 measurements were carried out in 8 to 12 tracks per tumor with a step length of 0.7 mm. The pO_2 probe was inserted under 4 defined directions: from above, from two sides under ±45 degree and from the distal side. Recorded pO_2 data were transformed into pO_2 histograms and pooled with a class width of 2.5 mmHg. From the histogram the following parameters were derived: the mean and median pO_2, the relative frequency of pO_2 values falling in the two lowest classes (0-5 mmHg) and the pO_2 range between the 10th and 90th percentile [$\Delta pO_2(10/90)$]. Histograms showing negative pO_2 values were excluded from data analysis. All pO_2 measurements were immediately followed by an analysis of the arterial blood. Statistical analysis was performed using the Mann–Whitney test.

RESULTS

Figure 2 shows the changes in tumor volume in the course of the fractionated radiation with a total dose of 60 Gy. After the first week of irradiation (15 Gy) tumor volume remained approximately unchanged in comparison to untreated controls. During further irradiation tumors shrank in size continuously and decreased to 17% of the starting volume (1.8 cm³) at the end of irradiation.

The changes of the oxygenation status in the R1H tumors recorded after weekly doses of 5x3 Gy during the fractionated irradiation are presented in Figure 3. Untreated controls show a broad distribution of the pO_2 values up to 88 mmHg and a median pO_2 of 23±2 mmHg (±SEM). The histograms after 15 and 30 Gy do not change significantly in the range and in the median pO_2 values. After a dose of 45 Gy, and especially after 60 Gy (4th week), a marked reduction in the range of the histograms, a progressive shift to lower pO_2 values and a steady

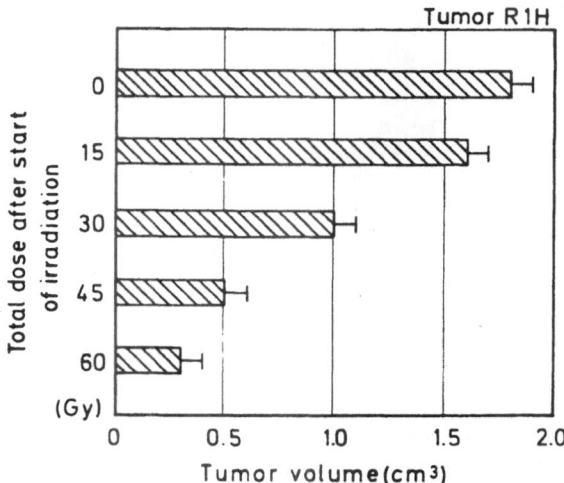

Figure 2. Changes in tumor volume measured at weekly intervals in the course of the fractionated irradiation with a total dose of 60 Gy, given in 20 fractions for 4 weeks. Mean values±SEM.

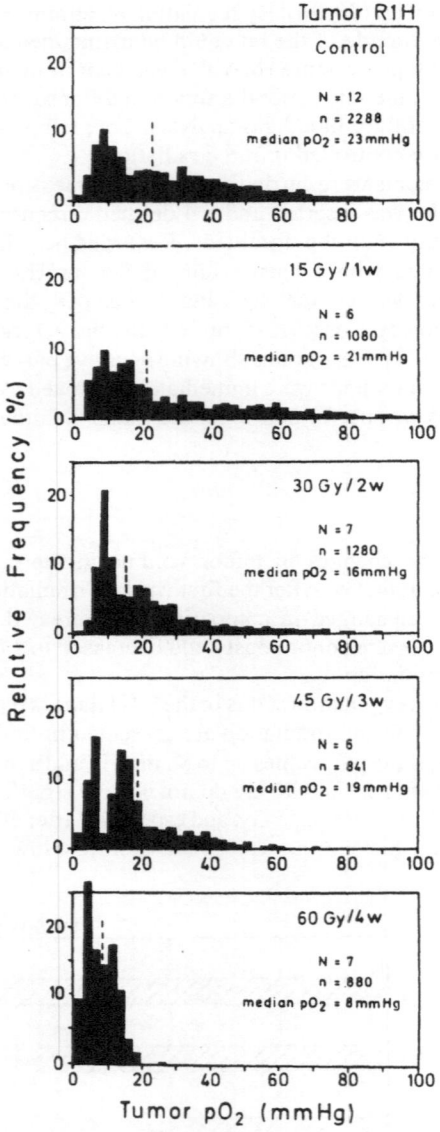

Figure 3. pO_2 histograms of pooled data recorded in R1H tumors at weekly intervals in the course of the fractionated irradiation with 15 Gy per week and a total dose of 60 Gy. N: number of tumors, n: number of pO_2 measurements, broken line: median pO_2.

increase in the frequency of values in the two lowest classes (0-5 mmHg) was observed. After a total dose of 60 Gy nearly all of the recorded pO_2 values were below 20 mmHg. A significant decrease of tumor oxygenation to a median pO_2 of 8 ± 2 mmHg was obtained.

The changes in tumor oxygenation and the corresponding arterial p_aO_2 values as a function of dose during the fractionated irradiation are presented in Figure 4. There is only little variation in the oxygenation up to a total dose of 45 Gy followed by a significant drop

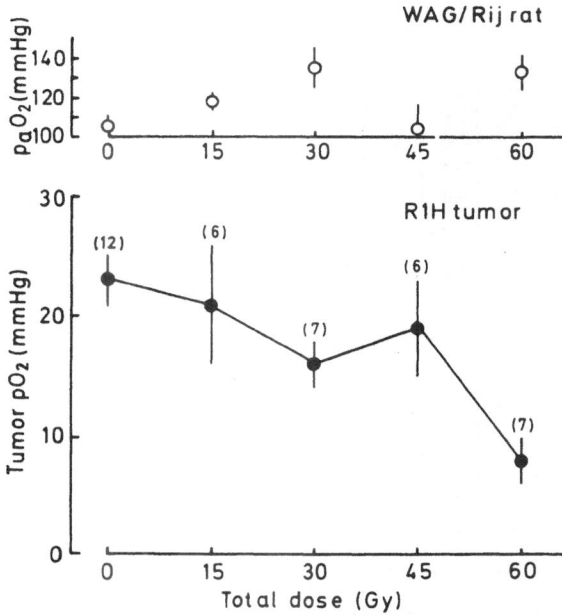

Figure 4. Changes of the median tumor pO_2 and corresponding arterial p_aO_2 values determined at weekly intervals after the start of irradiation. Numbers in brackets indicate number of tumors/animals studied. Mean values±SEM.

after 60 Gy. After a radiation dose of 15 Gy (1st week) a slight increase in the median pO_2 of some tumors was observed which was not significantly enhanced in comparison to untreated controls. With ongoing irradiation a decrease in tumor oxygenation was generally ascertained.

That these changes are not caused by anesthetic effects of the animals is proved by the corresponding values of the analysis of arterial blood, the mean arterial blood pressure, and the respiratory volume during the weeks of treatment and measurement (Figure 5). The values of hemoglobin, arterial pO_2, oxygen saturation and pCO_2 (not shown), the mean arterial blood pressure and respiratory volume demonstrate that the animals were kept in a relatively constant respiratory and metabolic state, which is expressed by the relatively constant pH-value of 7.41±0.03 (±SEM). It must be noted that the values of hemoglobin are not corrected for rat blood. Thus, they are of relative value only.

The analysis of the pO_2 histograms showed further that the proportion of pO_2 values in the two lowest classes (0-5 mmHg) and the range between the 10th and 90th percentile $\Delta p(10/90)$ as function of dose also varied (Figure 6). A decrease in the frequency of pO_2 values below 5 mmHg and a constant or increased pO_2 range should indicate tumor reoxygenation. After a total dose of 30 Gy a reduction in the frequency of low pO_2 values was observed which could be interpreted as an improved oxygenation (reoxygenation). However, the fraction of low pO_2 values increased significantly after a dose of 45 Gy and reached a value of 35±4% after 60 Gy. As far as the range $\Delta p(10/90)$ is concerned a transient increase after a dose of 15 Gy was observed followed by a steady decrease with increasing dose. After 60 Gy $\Delta p(10/90)$ was only 11 mmHg. The decrease in the range and the increase in the frequency of low pO_2 values indicate clearly that the oxygenation status of the R1H

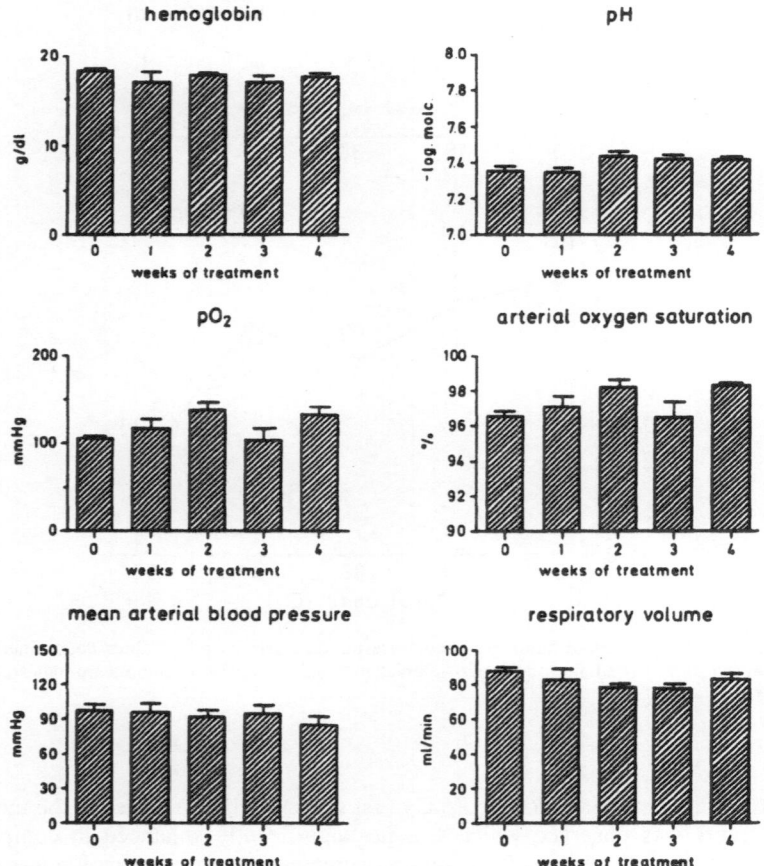

Figure 5. Main parameters obtained from the analysis of the arterial blood, mean arterial blood pressure and respiratory volume of WAG/Rij rats determined during the 4 weeks of irradiation. Mean values±SEM.

tumor has deteriorated during the fractionated irradiation, especially after doses of more than 45 Gy. Table 1 summarizes the data of this study.

DISCUSSION

pO$_2$-histography (Eppendorf-pO$_2$-device) has been used to measure *in vivo* tumor oxygenation in the course of a fractionated irradiation. Reproducible pO$_2$ measurements with this commercially available device could be achieved when the influence of anesthesia on respiratory and hemodynamic parameters was minimized. With the experimental set-up used (Fig. 1) an intratumoral reproducibility of the median pO$_2$ of 9% was achieved. Previous studies on the R1H tumor clearly demonstrated that tumor pO$_2$ correlates strongly with the arterial oxygen pressure and the tumor volume (20). Thus, the median pO$_2$ of 23 ± 2 mmHg of the R1H tumor is related to an arterial p$_a$O$_2$ of 105 ± 14 (±SD) and a volume of 1.8 ± 0.3 cm^3. This value is similar to pO$_2$ values measured with the same device in other rat tumors

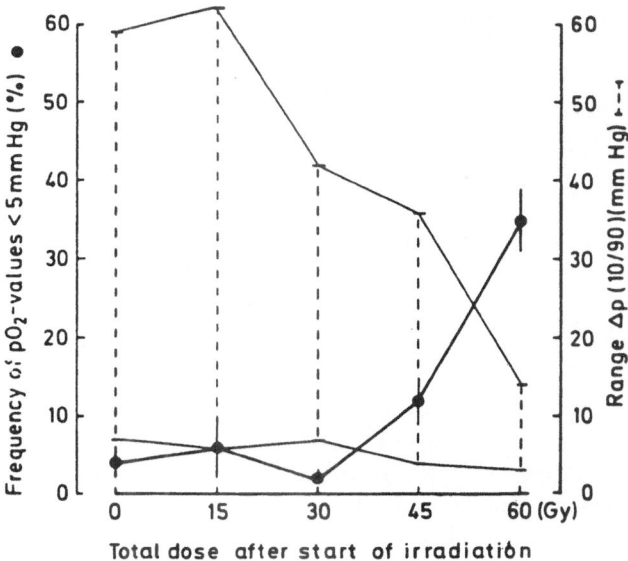

Figure 6. Alterations in the frequency of pO_2-values below 5 mmHg and in the range $\Delta p(10/90)$ after the start of irradiation. Mean values±SEM.

such as in human tumor xenografts (13) and in DS-sarcomas (27), bearing in mind the differences in the tumor systems and in the anesthetics used.

The pO_2 measurements have demonstrated that the oxygenation status of the R1H tumor varied during the fractionated irradiation. In the early phase of irradiation (up to 30 Gy) a slightly improved tumor oxygenation was observed. The nearly unchanged pO_2 distributions (Fig. 3) and the decrease in the proportion of pO_2 values below 5 mmHg (Fig. 6) are indications for this process. This result is in accordance with the general statement for animal tumors that most of the rodent tumors could reoxygenate after the last fraction and presumably between fractions (14, 15, 21). That this might also be true for the R1H tumor was shown in another subline (R2C5) from the original rat rhabdomyosarcoma R-1 (3) by Afzal et al. (1). After a conditioning dose of 15 Gy tumors reoxygenated rapidly during the first 6 hours postirradiation. Since the oxygenation status in the R1H tumor was determined 72 hours after the last fraction (5x3 Gy), tumor oxygenation might have returned meanwhile to the preirradiation level. But it must be mentioned in a limited sense that measurements of tumor oxygenation by electrodes and by the paired survival curve method are not directly comparable (10). Whereas the latter method determines the hypoxic fraction of clonogenic tumor cells, the electrodes measure the degree of oxygen tension in microenvironments within a tumor that may be hypoxic. For a direct *in vivo* determination of tumor oxygenation with radiation doses of more than 30 Gy only pO_2-sensitive electrodes can be used.

In the later phase of the fractionated irradiation the pO_2 measurements clearly demonstrated a decrease in tumor oxygenation. Total doses of 45 Gy and 60 Gy led to a shortening in the range of the pO_2 distribution and a constant increase in the proportion of pO_2 values below 5 mmHg (Figs. 3 and 6) indicating that higher radiation doses induce tumor hypoxia. This is not in agreement with the results obtained after irradiation with single large doses. Vaupel et al. (23) described an increase in pO_2 in a mouse mammary carcinoma 72-74 hours after a single dose of 60 Gy. A detailed analysis of the pO_2 values showed that an

Table 1. Oxygenation status in rhabdomyosarcomas R1H of the rat in the course of a fractionated Co-γ-irradiation with a total dose of 60 Gy, given in 20 fractions for 4 weeks

Treatment dose/time	Number of tumors	Average tumor volume, cm^3	Median pO$_2$ mmHg	pO$_2$-values below 5 mmHg, %	pO$_2$-percentile, mmHg	
					10%	90%
Control	12	1.8±0.1^a	23±2^a	4±2^a	7	59
15 Gy/ 1 wk	6	1.6±0.1	21±5	6±4	6	62
30 Gy/ 2 wks	7	1.0±0.1	16±2	2±1	7	42
45 Gy/3 wks	6	0.5±0.1	19±4^b	12±3^b	4	36
60 Gy/4 wks	7	0.3±0.1	8±2^b	35±4^b	3	14

[a]Values are means ±SEM.
[b]Statistically different ($p < 0.05$).

improvement in tumor oxygenation is particularly evident in the very low pO$_2$ range (0-3 mmHg). Koutcher et al. (17) found a significant increase in the mean pO$_2$ of a mouse mammary carcinoma after a single dose of 65 Gy at day 1, 2 and 4 postirradiation. After a dose of 32 Gy an increase in mean pO$_2$ was noted at the first and 2nd day. A decrease at the 4th day to less than the initial value was measured, although the fraction of pO$_2$-values below 2.5 mmHg remained lower than that of control tumors. The decrease in the hypoxic fraction calculated from differences between normoxic and clamped tumors (TCD$_{50}$-values) after a priming dose of 32 Gy indicated further a better tumor oxygenation. This does not agree with the deterioration of tumor oxygenation in the R1H tumors and indicates that in a multifraction regimen the process of reoxygenation is not constantly repeating and is a function of radiation dose.

The decrease in tumor oxygenation may be based, in part, by morphological changes in the R1H tumor occurring during the fractionated irradiation. In a previous study it could be demonstrated that the shrinkage of the tumor (Fig. 2) led to alterations in the intratumoral structure, including the vasculature (28). Doses of 45 and 60 Gy induced small and moderate reductions of the tumor vasculature. Ultrastructural studies on tumor capillaries have shown that the vascular wall was progressively damaged during the fractionated radiation treatment (Figure 7). After a dose of 30 Gy/10 fractions early signs of radiation damage were seen. The inner surface of the lumen became rough, endothelial cells and pericytes were swollen. Perivascular a heavy edema with collagen fibrils under the basal lamina occurred. With ongoing irradiation the damage to the vascular wall progresses. After a dose of 60 Gy/20 fractions the state of the vascular wall was not very different from that found after 75 Gy/25 fractions. The endothelial cells vary more and more in shape and size, from swollen cells to atrophy. Endothelial cells became elongated, sometimes detached from each other and from the basal lamina. A perivascular edema was still present. These structural changes during the fractionated irradiation might have adverse effects on tumor blood flow and, thus, might reduce tumor oxygenation.

In conclusion, the present study has shown that the oxygenation status of a tumor during a fractionated irradiation is a dynamic and constantly changing process. In the early phase of the fractionated irradiation (up to 30 Gy) an improved tumor oxygenation was observed followed by a distinct decrease in oxygenation in the later phase of irradiation (above 45 Gy). Tumor shrinkage and oxygenation status did not correlate in the R1H tumor. The present findings suggest that the results on tumor reoxygenation with noncurative doses must be considered very critically with respect to doses used in the radiation treatment of human tumors.

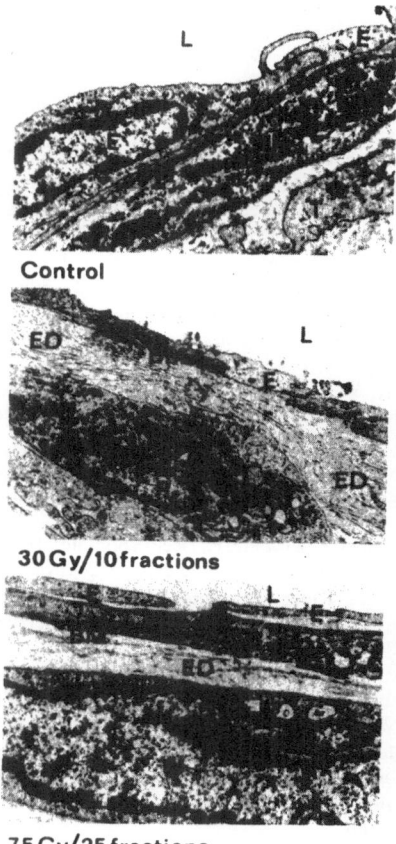

Figure 7. Electron micrographs of the vascular wall of tumor capillaries of rhabdomyosarcomas R1H during fractionated irradiation with Co-γ-iradiation. L: lumen, E: endothelial cell, P: pericyte, T: tumor cell, BM: basal lamina, ED: edema. **a.** untreated control, x 6500. **b.** after 30 Gy/10 fractions, x 4000. **c.** after 75 Gy/25 fractions, x 7100.

ACKNOWLEDGMENT

The authors would like to thank Mrs. S. Scheibner, I. Stark, I. Dwenger and D. Droese for their excellent technical assistance.

REFERENCES

1. Afzal, S.M.J., Tenforde, T.S., Kavanau, K.S., Curtis, S.B., 1991, Reoxygenation in a rat rhabdomyosarcoma tumor following X-irradiation. Int. J. Radiat. Oncol. Biol. Phys. 20: 473-477.
2. Badib, A.O., Webster, J.H., 1969, Changes in tumor oxygen tension during radiotherapy. Acta Radiologica, Therapy, Physics, Biology. Stockholm 8: 247-257.
3. Barendsen, G.W., Broerse, J.J., 1969, Experimental radiotherapy of a rat rhabdomyosarcoma with 15 MeV neutrons and 300 kV X-rays. I. Effects of single exposures. Europ. J. Cancer 5: 373-391.

4. Bergsjö, P., Evans, J.C., 1971, Oxygen tension of cervical carcinoma during the early phase of external irradiation. Scand. J. Clin. Lab. Invest. 27: 71-82.

5. Chaplin, D.J., Olive, P.L., Durand, R.E., 1987, Intermittent blood flow in a murine tumor: radiobiological effects. Cancer Res. 47: 597-601.

6. Fleckenstein, W., Jungblut, J.R., Suckfüll, M., 1990, Distribution of oxygen pressure in the periphery and centre of malignant head and neck tumors. In: Clinical Oxygen Pressure Measurement II (A.M. Ehrly, W. Fleckenstein, J. Hauss, R. Huch, eds.), Blackwell Ueberreuter Wiss., Berlin, pp. 81-90.

7. Gatenby, R.A., Kessler, H.B., Rosenblum, J.S., Coia, L.R., Moldofsky, P.J., Hartz, W.H., Broder, G.J., 1988, Oxygen distribution in squamous cell carcinoma metastases and its relationship to outcome of radiation therapy. Int. J. Radiat. Oncol. Biol. Phys. 14: 831-838.

8. Grau, C., Overgaard, J., 1990, The influence of radiation dose on the magnitude and kinetics of reoxygenation in a C3H mammary carcinoma. Radiat. Res. 122: 309-315.

9. Höckel, M., Knoop C., Schlenger, K., Vorndran, B., Baußmann, E., Mitze, M., Knapstein, P.G., Vaupel, P., 1993, Intratumoral pO_2 predicts survival in advanced cancer of the uterine cervix. Radiother. Oncol. 26: 45-50.

10. Horsman, M.R., Khalil, A.A., Sieman, D.W., Grau, C., Hill, S.A., Lynch, E.M., Chaplin, D.J., Overgaard, J., 1994, Relationship between radiobiological hypoxia in tumors and electrode measurements of tumor oxygenation. Int. J. Radiat. Oncol. Biol. Phys. 29: 439-442.

11. Horsman, M.R., 1993, Hypoxia in tumors: its relevance, identification and modification. in: Current Topics in Clinical Radiobiology of Tumors (H.P. Beck-Bornholdt, ed.), Springer-Verlag, Heidelberg, pp. 99-112.

12. Jung, H., Krüger, H.J., Brammer, I., Zywietz, F., Beck-Bornholdt, H.P., 1990, Cell population kinetics of the rhabdomyosarcoma R1H of the rat after single doses of X-rays. Int. J. Radiat. Biol. 57: 657-589.

13. Kallinowski, F., Schlenger, K.H., Kloes, M., Stohrer, M., Vaupel, P., 1989, Tumor blood flow: the principal modulator of oxidative and glycolytic metabolism, and of the metabolic micromilieu of human tumor xenografts in vivo. Int. J. Cancer 44: 266-272.

14. Kallman,.R.F., Dorie, M.J., 1986, Tumor oxygenation and reoxygenation during radiation therapy: their importance in predicting tumor response. Int. J. Radiat. Oncol. Biol. Phys. 12: 681-685.

15. Kitakabu, Y., Shibamoto, Y., Sasai, K., Ono, K., Abe, M., 1991, Variations of the hypoxic fraction in the SCC VII tumors after single dose and during fractionated radiation therapy: assessment without anesthesia or physical restraint of mice. Int. J. Radiat. Oncol. Biol. Phys. 20: 709-714.

16. Kolstad, P., 1968, Intercapillary distance, oxygen tension and local recurrence in cervix cancer. Scand. J. Clin. Lab. Invest. 106: 145-166.

17. Koutcher, J.A., Alfieri, A.A., Devitt, M.L., Rhee, J.G., Kornblith, A.B., Mahmood, U., Merchant, T.E., Cowburn, D., 1992, Quantitative changes in tumor metabolism, partial pressure of oxygen, and radiobiological oxygenation status postirradiation. Cancer Res. 52: 4620-4627.

18. Moulder, J.E., Rockwell, S., 1984, Hypoxic fractions of solid tumors: experimental techniques, methods of analysis, and a survey of existing data. Int. J. Radiat. Oncol. Biol. Phys. 10: 695-712.

19. Rampling, R., Cruickshank, G., Lewis, A.D., Fitzsimmons, S.A., Workman, P., 1994, Direct measurement of pO_2 distribution and bioreductive enzymes in human malignant brain tumors. Int. J. Radiat. Oncol. Biol. Phys. 29: 427-431.

20. Reeker, W., Zywietz, F., Kochs, E., 1994, Determination of partial oxygen pressure (pO_2) in a rat rhabdomyosarcoma: a methodical study, In: Tumor Oxygenation (P. Vaupel, D.K. Kelleher, M. Günderoth, eds.), Gustav Fischer Verlag, Stuttgart, New York (in press).

21. Rofstad, E.K., 1989, Hypoxia and reoxygenation in human melanoma xenografts. Int. J. Radiat. Oncol. Biol. Phys. 17: 81-89.

22. Stone, H.B., Brown, J.M., Phillips, T.L., Sutherland, R.M., 1993, Oxygen in human tumors: correlations between methods of measurement and response to therapy. Radiat. Res. 136: 422-434.

23. Vaupel, P., Frinak, S., O'Hara, M., 1984, Direct Measurement of reoxygenation in malignant mammary tumors after a single large dose of irradiation. Adv. Exp. Med. Biol. 180: 773-782.

24. Vaupel, P., Kallinowski, F., Okunieff, P., 1989, Blood flow, oxygen and nutrient supply, and metabolic microenvironment of human tumors: a review. Cancer Res. 49: 6449-6465.

25. Vaupel, P., Schlenger, K., Knoop, C., Höckel, M., 1991, Oxygenation of human tumors: evaluation of tissue oxygen distribution in breast cancers by computerized O_2 tension measurements. Cancer Res. 51: 3316-3322.

26. Vaupel, P., Manz, R., Müller-Klieser, W., Grunewald, W.A., 1979, Intercapillary HbO_2 saturation in malignant tumors during normoxia and hyperoxia. Microvasc. Res. 17: 181-191.

27. Vaupel, P., 1993, Effects of physiological parameters on tissue response to hyperthermia: new experimental facts and their relevance to clinical problems, In: Hyperthermic Oncology 1992 (E.W. Gerner and T.C. Cetas, eds.), Arizona Board of Regents, pp. 17-23.

28. Zywietz, F., 1990, Vascular and cellular damage in a murine tumor during fractionated treatment with radiation and hyperthermia. Strahlenther. Onkol. 166: 493-501.

58

ISCHEMIC HEART

Microcirculatory Aspects

Karel Rakusan, Nicholas Cicutti, and Tomas Sladek

Department of Physiology
Faculty of Medicine
University of Ottawa
Ottawa, Ontario, Canada

INTRODUCTION

Tissue becomes ischemic when the blood supply does not match its needs resulting in an inadequate oxygen delivery, i.e., hypoxia. In contrast to other forms of hypoxia, the washout of metabolites and heat are also impeded. Myocardial ischemia is by far the most common cause of cardiac hypoxia. In clinical terms, it may be defined as dysfunction of cardiac myocytes resulting from hypoxia due to limited coronary blood flow. It is usually detected by a shift of an ST-segment on the ECG. Most experimental and virtually all clinical studies of myocardial ischemia concentrate on the pathology of coronary arteries. The coronary microcirculation, however, may play an important role in cardiac ischemia both as its possible cause as well as in the tissue response to this insult.

An example of a clinical syndrome with a possible microvascular component is the so- called syndrome X. This syndrome was first reported by Likoff et al. (1967) in a group of patients with angina, ischemic electrocardiographic responses to exercise and a normal coronary arteriogram. The term syndrome X was coined by Kemp (1973) in an editorial discussing a similar study by Argobast and Bourassa (1973). Kemp (1991) recently reviewed the topic and related literature. The main features of the syndrome are chest pain and evidence of ischemia with reduction of coronary vascular reserve but without depression of left ventricular performance or a negative effect on survival. An abnormality of the coronary microvasculature may be a possible underlying mechanism in syndrome X (Maseri et al. 1991). Microvascular angina is also present in a large proportion of hypertensive patients with left ventricular hypertrophy (Scheler et al. 1992). Architectural or functional abnormalities at the microcirculatory level have been suggested as a possible mechanism in the development of angina in patients with left ventricular hypertrophy but with a normal coronary arteriogram (Kaski 1993). It is of interest to note that patients with the clinical diagnosis of microvascular angina, as a group, are also resistant to insulin-mediated glucose uptake. This is a feature of syndrome X as the term is used in diabetology (hyperinsulinemia, hypertension, obesity and glucose intolerance)

Oxygen Transport to Tissue XVII, Edited by Ince et al.
Plenum Press, New York, 1996

(Reaven 1994). This cluster of metabolic abnormalities, known to increase the risk of coronary heart disease may share with the cardiologic version of syndrome X an involvement of the coronary microvasculature.

Microvascular involvement in the ischemic response of the heart has been studied in animal experiments by several authors. For instance, Polansky and Weiss (1993) reported an increase in heterogeneity of coronary blood flow in response to the reductions of overall blood flow, regardless of whether vasopressin was administered or flow was restricted with a clamp. Such flow heterogeneity may indicate that some areas are at higher risk of ischemia and cell death during flow restriction. Spatial heterogeneity of anoxic zones has been described during the hypoxic state of perfused rat hearts examined by NADH fluorescence (Steenbergen et al. 1977). Based on the size of these discrete areas of anoxic tissue, it was concluded that coronary perfusion appears to be regulated at the level of the arterioles. This issue was recently reexamined by Ince and coworkers (1993) using a combination of NADH and Pd-porphine videofluorometry on perfused rat heart. Recovery from anoxia was accompanied by transient patchy areas of persistent anoxia, patterns of which were always the same for a given heart. Similar patterns were elicited by microsphere embolization of capillaries, but not of arterioles. The authors concluded that these anoxic regions were of anatomic origin and were located at the capillary level.

MATERIAL, METHODS AND RESULTS

Our own studies concentrate on microcirculatory responses to coronary occlusion in experimental animals. We will present two examples, one dealing with an acute reaction to coronary artery occlusion and another with a chronic response. Acute experiments were performed on dogs while chronic experiments were conducted on rats.

In acute experiments on dogs, we found that ligation of the left anterior descending artery extends the perfusion boundary of the neighboring left circumflex artery (Cicutti et al. in press). This was based on methodology which we used previously to detect microvascular anastomoses along the coronary perfusion territory interface (Cicutti et al. 1992). Simultaneous *in vivo* infusion of two distinctly colored microsphere suspensions into the left anterior descending and left circumflex coronary arteries identified a specific region of canine myocardium perfused by both arterial branches. Subsequently, the left anterior descending artery was ligated and a third set of microspheres of a different color was infused into the patent circumflex artery. Morphometric analysis of the microspheres within capillaries in various regions of the heart revealed an expansion of the territory receiving blood from both sources. The coronary perfusion territory of the patent artery extended by approximately 2 mm following coronary artery ligation. The expansion was significantly greater in subepicardial than in subendocardial regions. These microvascular collaterals would play a significant role especially in protecting smaller regions of the myocardium from ischemic damage. Thus, we demonstrated the capability of microvascular anastomoses in providing blood flow to the periphery of an ischemic region, and that the coronary perfusion interface was labile and amenable to manipulation.

In chronic experiments on rats, we examined morphometrically capillary supply in various regions of normal and infarcted heart. The analysis took place three weeks after coronary artery ligation or sham operation. Half of the animals were treated with the AT_1 receptor antagonist, Losartan. Our preliminary results seem to indicate that the therapy improved cardiac contraction mechanics of surviving muscle, diminished the hypertrophic response of cardiac myocytes and increased the overall capillary

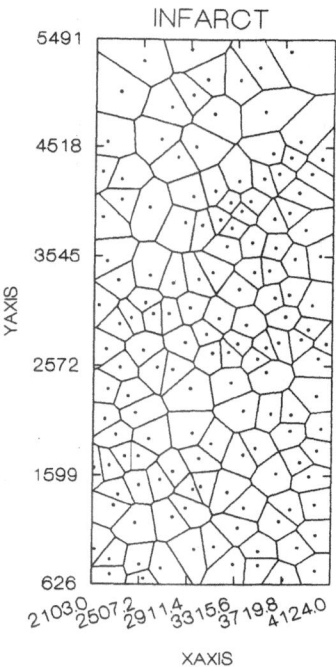

Figure 1. Computer printout of capillaries (represented by dots) and their individual capillary domains at distances increasingly removed from infarcted region of myocardium. The areas subtended by these polygonal regions of tissue are closer to the enclosed capillaries than any other neighboring capillaries.

density of infarcted hearts, albeit to levels which were still lower than in normal hearts.

Our morphometric method of delineating capillary domains is eminently suitable for the evaluation of local capillary supply at the level of individual myocytes. The capillary domain is defined on tissue cross-sections as an area of tissue surrounding a capillary which is nearer to the enclosed capillary than any other neighboring capillary (Turek et al. 1986). Therefore, we decided to evaluate changes in capillary domain size as a function of distance from the infarcted area. In this case, one needs to locate a relatively large area of tissue close to the infarcted zone in which both capillaries and muscle cells are cross-sectioned. A typical computer printout of capillary domain size at various distances from the infarcted region is illustrated in Fig. 1.

We were able to find 7-8 similar regions in groups of rat hearts with experimental infarction either treated or untreated with losartan and in hearts from sham operated animals. The average values of capillary domain areas as a function of a distance from the infarcted region are depicted in Fig. 2.

The graph clearly demonstrates that the domain area is significantly larger in the infarcted hearts. The difference is highest in regions close to the infarction and gradually decreases. The distance is expressed in layers of capillary domains which also correspond to the myocyte layers. The effect of losartan in this example is limited only to the area nearest to the infarction. Capillary domain areas in control hearts do not exhibit any noticeable variations over the same distances.

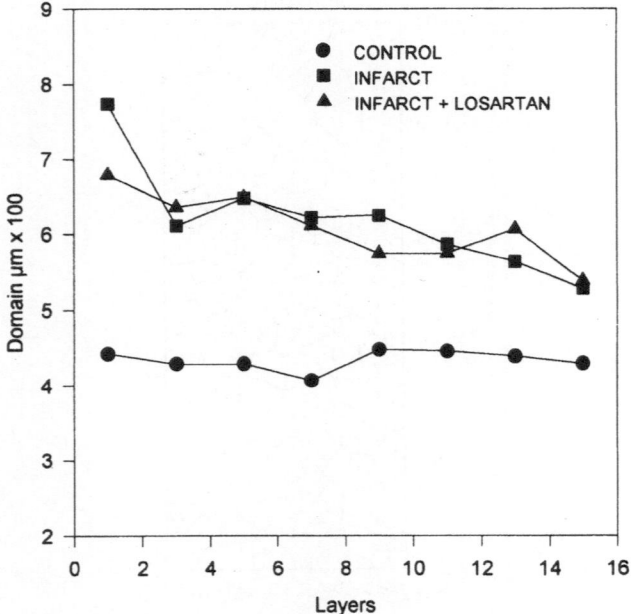

Figure 2. Average values of capillary domain areas as a function of distance (capillary domain layers) from infarcted region in control, infarct untreated, and infarct+losartan groups.

CONCLUSION

Both acute and chronic cardiac ischemia are associated with profound changes at the microcirculatory level. These changes are of a functional as well as of a structural nature and they significantly influence the outcome of the cardiac response to this injury.

ACKNOWLEDGMENT

This study was supported by the Medical Research Council of Canada. The authors wish to thank Mrs. Barbara Hebert and Mrs. Ching Kuo for their expert technical assistance.

REFERENCES

Arbogast, R. and Bourassa, M.G., 1973, Myocardial function during atrial pacing in patients with angina pectoris and normal coronary angiograms, *Am. J. Cardiol.* 32:257-263.

Cicutti, N., Rakusan, K. and Downey, H.F., in press, Coronary artery occlusion extends perfusion territory boundaries through microvascular collaterals, *Bas. Res. Cardiol.* in press.

Cicutti, N., Rakusan, K. and Downey, H.F., 1992, Colored microspheres reveal interarterial microvascular anastomoses in canine myocardium, *Bas. Res. Cardiol.* 87:400-409.

Ince, C., Ashruf, J.F., Avontuur, J.A.M., Wieringa, P.A., Spaan, J.A.E. and Bruining, H.A., 1993, Heterogeneity of the hypoxic state in rat heart is determined at capillary level, *Am. J. Physiol.* 264:H294-301.

Kaski, J.C., 1993, Myocardial ischaemia in the hypertensive patient - the role of coronary microcirculation abnormalities, *Europ. Heart J.* 14(Suppl. J):32-37.

Kemp, H.G. Jr., 1973, Left ventricular function in patients with the anginal syndrome and normal coronary arteriograms, *Am. J. Cardiol.* 32:375-376.

Kemp, H.G. Jr., 1991, Syndrome X revisited, *J. Am. Coll. Cardiol.* 17:507-508.

Likoff, W., Segal, B.L. and Kasparian, H., Parades of normal selective coronary arteriograms in patients considered to have unmistakable coronary heart disease, *N. Engl. J. Med.* 276:1063-1066.

Maseri, A., Crea, F., Kaski, J.C. and Crake, T., 1991, Mechanisms of angina pectoris in syndrome X, *J. Am. Col. Cardiol.* 17:499-506.

Polansky, L. and Weiss, H.R., 1993, Effect of flow reduction on coronary blood flow heterogeneity, *Proc. Soc. Exp. Biol. Med.* 202:97-102.

Reaven, G.M., 1994, Syndrome X: is one enough?, *Am. Heart J.* 127:1439-1442.

Scheler, S., Motz, W., and Strauer, B.E., 1992, Transient myocardial ischaemia in hypertensives: missing link with left ventricular hypertrophy, *Eur. Heart J.* 13 (Suppl D): 62-65.

Steenbergen, C., Deleeuw, G., Barlow, C., Chance, B. and Williamson, J.R., 1977, Heterogeneity of the hypoxia state in perfused rat heart, *Circulation Res.* 41:606-615.

Turek, Z., Hoofd, L. and Rakusan, K., 1986, Myocardial capillaries and tissue oxygenation, *Can. J. Cardiol.* 2:98-103.

59

THE LUNG IN DISTRESS

K. L. So and B. Lachmann

Erasmus University Rotterdam
Department of Anesthesiology
Postbox 1738, 3000 DR Rotterdam, The Netherlands

INTRODUCTION

The lung functions as a gas exchanger by bringing air and blood into as close proximity as possible. For this purpose, the alveolo-capillary membrane is relatively thin (0.2 μm), making it at the same time very vulnerable for stresses of all kind. Especially in the intensive care unit, the magnitude of the stresses and the cumulative impact of many disorders make the lung accident prone.

An impressive variety of insults is capable of damaging the alveolo-capillary membrane, resulting in severe respiratory failure (Acute Lung Injury (ALI) or Acute Respiratory Distress Syndrome (ARDS)) [1]. Although the mechanisms responsible for injury to the alveolo-capillary membrane are complex and still under discussion, the clinical problems are similar regardless of the etiology: severe arterial hypoxemia refractory to oxygen therapy, increased intrapulmonary shunting, decreased lung compliance, decreased lung volumes, and the absence of indicators of left ventricular failure [2,3].

The therapeutic goal in ALI/ARDS is to supply ventilatory support in order to maintain gas exchange until alveolo-capillary membrane integrity is re-established. Therefore, the main critical factors in clinical management of ALI/ARDS are supply of adequate arterial and tissue oxygenation and treatment of the primary disorder, whereas progress of lung injury (e.g. ventilator-induced injury) is to be minimalized or even prevented [4].

A number of possible treatment modalities are now under investigation:

SURFACTANT THERAPY

In 1967 Ashbaugh and colleagues described 12 patients with a similar pattern of acute respiratory distress: acute onset of tachypnea, hypoxemia and loss of compliance after a variety of stimuli; however, none of the patients responded to conventional therapy [5]. These patients exhibited a clinical, physiological, and pathological course of events that was remarkably similar to the infantile respiratory distress syndrome; thus the term adult respiratory distress syndrome (ARDS) was introduced [6]. However, recently the name was

Oxygen Transport to Tissue XVII, Edited by Ince et al.
Plenum Press, New York, 1996

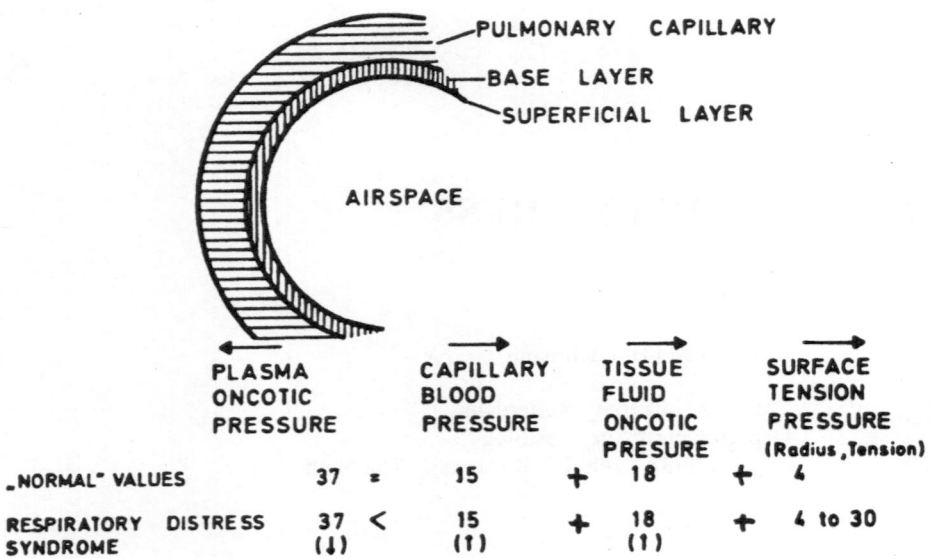

	PLASMA ONCOTIC PRESSURE	CAPILLARY BLOOD PRESSURE	TISSUE FLUID ONCOTIC PRESURE	SURFACE TENSION PRESSURE (Radius, Tension)
"NORMAL" VALUES	37 =	15	+ 18	+ 4
RESPIRATORY DISTRESS SYNDROME	37 < (↓)	15 (↑)	+ 18 (↑)	+ 4 to 30

Figure 1. Schematic water balance in the lung with respect to the surfactant system.

conveyed again to Acute Respiratory Distress Syndrome because ARDS can already occur at two months of age [7].

In their first report, Ashbaugh et al. suggested that surfactant dysfunction might contribute to the pathophysiologic findings in ARDS [5]. In later studies, they demonstrated both qualitative and quantitative surfactant abnormality in patients who died of ARDS, along with reduced compliance of the whole, fresh, excised human lungs, compared with normal control subjects [8,9]. These findings in ARDS patients have been confirmed by others [10,11,12]. Recently, Gregory et al. demonstrated that several of these surfactant alterations already occur in patients at risk of developing ARDS, suggesting that these abnormalities of surfactant occur early in the disease process [13].

The central role of surfactant deficiency can further be illustrated by multiple studies in animal models of ARDS which demonstrated that exogenous surfactant instillation dramatically improves blood gases and lung mechanics [for review: 14].

These results suggest that following a variety of indirect or direct injuries to the alveolo-capillary membrane, alveoli become flooded by plasma-derived proteins. These plasma-derived proteins inhibit the endogenous surfactant [15,16,17], which leads to alveolar instability and, in turn, to an increase in hydrostatic forces favoring pulmonary edema.

Apparently, after the initial insult to the lung, the surfactant system is responsible for further pathophysiological changes and a vicious circle is set in action, finally resulting in the failure of the lung as a gas-exchange organ.

Thus, the rationale for giving surfactant to (at-risk) ARDS patients is to break through this vicious circle by means of recruiting collapsed alveoli and stabilize them with the applied ventilator settings. Before exogenous surfactant therapy is applied, one therefore has to evaluate by lung function tests whether or not sufficient parts of recruitable lung areas are still available. Thus, surfactant should not be given to patients with heavily consolidated and/or fibrotic lungs.

Preliminary work on the administration of surfactant to adults with ARDS was made by Lachmann in 1987; he reported a dramatic improvement in gas exchange and improvement of the chest radiograph after a dose of 300 mg/kg was administered via tracheal instillation to a

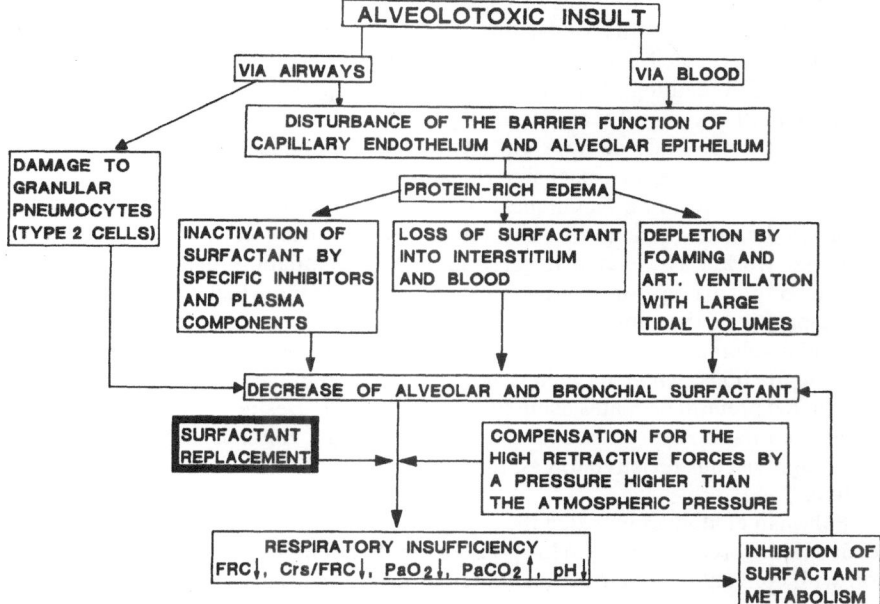

Figure 2. Pathogenesis of ARDS with respect to the surfactant system.

terminal patient with sepsis and severe ARDS [18]. To date, merely case-reports concerning the use of exogenous surfactant in patients with ARDS have been published [19-29]. Although the results of these studies were not consistent, the best results were seen in patients treated with high concentrations, or multiple doses, of surfactant. Evidently, for treatment of ARDS, high concentration of surfactant is required to overcome the inhibitory effect of plasma components. This implies that if after surfactant instillation there is no, or only transient, improvement of blood gases in these patients (fibrotic lungs excluded) this does not mean that surfactant treatment does not work, it only means that the concentration of the exogenous surfactant used is too low in relation to the amount of surfactant inhibitors in the lung.

To build upon and support the findings of recent anecdotal work, clinical trials are necessary to provide information on the preferred dose, route of administration, dose volume, dose frequency, mode of application, time of administration and type of surfactant preparation needed to establish the clinical support role of exogenous surfactant in ARDS.

Recently, Gregory reported in a pilot study on adult patients with ARDS that mortality could be decreased from 43.8 to 17.6 % by instillation of 400 mg surfactant per kg body weight [30]. At current prices, the costs of surfactant treatment for one adult would be about $75,000. This prohibitive price and the non-availability of large amounts of surfactant, make surfactant therapy not yet feasible in adults.

PARTIAL FLUID VENTILATION

In 1929 Von Neergaard demonstrated that lungs that were collapsed would open up more readily if the effect of surface tension was completely nullified by using liquid rather than air as the expanding medium, i.e by eliminating the alveolar gas-liquid interface [31].

In this respect, the findings of Clark et al. in 1966 were of utmost importance, for they demonstrated the ability of healthy small mammals to successfully breathe while submerged in oxygenated perfluorocarbon (PFC); moreover these animals could even be reconverted to air breathing [32]. These findings are explained by the special properties of PFCs: a high ability to dissolve respiratory gases and a low surface tension; furthermore, in the physiologic range PFCs do not show any in vivo metabolisation.

Since this publication, extensive research has been performed to study the efficacy of PFCs in liquid breathing techniques; furthermore, it would be rational to apply these techniques in diseases characterized by high alveolar surface forces.

Initial work was directed towards total fluid ventilation, using PFCs oxygenated outside the body. This type of liquid ventilation is a process in which the gaseous functional residual capacity of the lung is filled with PFCs, and gas exchange is accomplished by pumping tidal volumes of PFCs in and out of the lung, which are guided through a membrane lung outside the body, where oxygen is added and CO_2 removed. Greenspan et al. success-fully ventilated preterm neonates using this principle [33]. Besides its technical complexity, total liquid ventilation causes the movement of liquid tidal volumes through the airway and generates high viscous resistive forces, rendering the work of spontaneous liquid breathing prohibitive.

Fuhrman et al. demonstrated the feasibility of liquid ventilation without the need of a modified liquid breathing system [34]. This technique combined intratracheal PFC admini-stration with conventional ventilation, which brought the use of PFC for liquid ventilation closer to clinical practice. Fuhrman's group named this type of oxygenation: "in vivo bubble oxygenation".

It is established that increased alveolar surface tension plays a central role in the pathophysiology of the respiratory distress syndrome of prematurity; furthermore, it is thought to contribute to lung dysfunction in ARDS [28,35]. Therefore, our group investigated the efficacy of partial fluid ventilation in an animal model of acute respiratory failure.

Thus, Tütüncü et al. were the first to demonstrate in adult animals with acute respiratory failure, using a combination of conventional mechanical ventilation and intratra-cheal PFC administration in increasing doses (yet below functional residual capacity

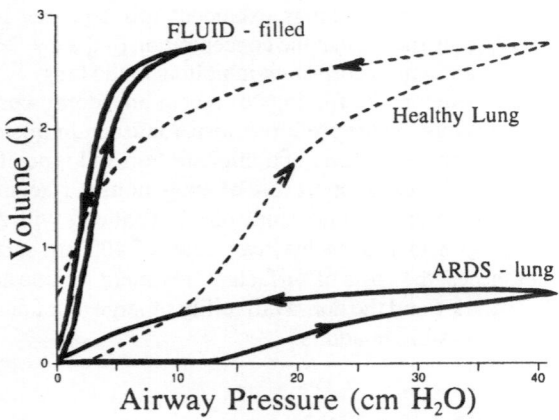

Figure 3. Air-inflated and fluid-distended lungs. When the lungs are inflated with gas (normal surface tension) to a high pressure and then deflated, the dashed line results. If the same lungs are made gas-free, filled with fluid to the same pressure and then allowed to empty, the "fluid-filled" curve (no surface tension) results. The "ARDS-lung" curve (high surface tension) shows that the lungs inflate significantly less when alveolar lining has a high surface tension.

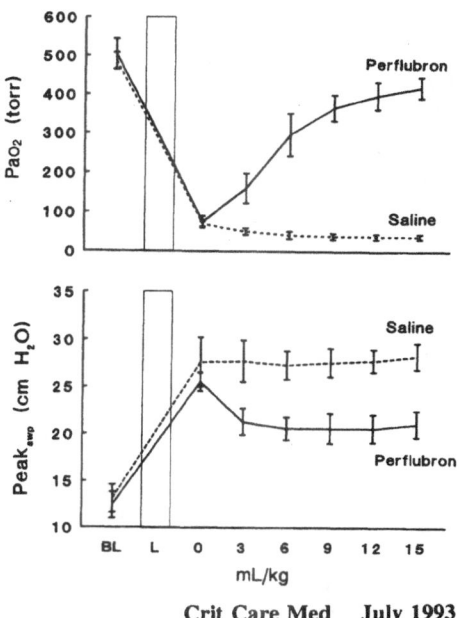

Crit Care Med   July 1993

Figure 4. Top: P_aO_2 and bottom: peak airway pressure (mean ± SD) before lavage (BL), lavage (L), after lavage (0), and after treatment with perfluorocarbon or saline.

volume), that oxygenation can be improved in a dose-dependent manner at reduced airway pressures [36,37].

Subsequently, Tütüncü et al. demonstrated in the same animal model, using the same technique yet with PFC doses approximating functional residual capacity volume, that pulmonary gas exchange was improved and maintained stable throughout the observation period at lower airway pressures; also respiratory lung compliance improved; discernible treatment-related alveolar damage was not seen on histological analysis [38].

From these studies they concluded that the dose-dependent improvement of gas exchange supported that large doses of PFC, approaching normal functional residual capacity of the animal, are required to correct hypoxia as fully as possible. On the other hand, respiratory system compliance and airway pressures can be improved even with a low dose of PFC and further doses do not make significant changes in the lung mechanical properties of lung-lavaged animals.

As for the different mechanisms involved, they proposed that even low doses of PFCs diminish the surface tension forces that oppose lung inflation, but that the space-occupying characteristics of the PFCs play a major role in its restoration of functional residual capacity and gas exchange (e.g. recruitment of previously collapsed alveoli, thus preventing end-expiratory collapse).

In conclusion, these studies demonstrate the feasibility of partial fluid ventilation in improving pulmonary gas exchange and respiratory mechanics in animals with ALI. Moreover this technique, considering the remarkable reductions in airway pressures during partial fluid ventilation, appears to be an alternative modality to minimize or prevent the progress of lung injury (e.g. ventilator-induced injury). Now clinical studies are warranted in order to investigate the clinical efficacy of this new technique.

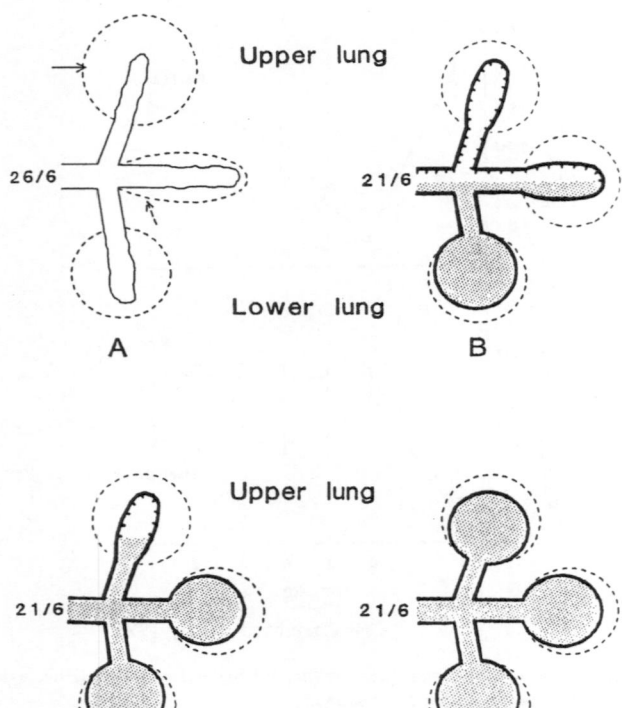

Figure 5. Showing from A to D the effect of increasing dosages of PFC (represented by the shaded areas in B and C) where even at low PFC dose (B) the alveolar mechanical behaviour of the entire lung is improved due to evaporation (represented by the small lines on the inside of the alveoli) of low surface tension PFC (see text). D shows the entirely PFC filled lung where expiratory collapse is physically prevented. All figures are drawn at equal ventilator pressure settings (i.e. 26/6).

NO INHALATION

ARDS is characterized by intrapulmonary shunting that results in arterial hypoxemia, and by acute pulmonal hypertension due to vasoconstriction and widespread occlusion of the pulmonary microvasculature. Pulmonal arterial hypertension contributes to pulmonary edema, therefore, interventions aiming to reduce pulmonal hypertension are warranted.

Intravenously infused vasodilators reduced pulmonary artery pressure; however, the unavoidable dilatation of the systemic circulation causes arterial hypotension, possibly affecting blood flow to various organs. Moreover, since the pulmonary vasodilatation is not selective, increased blood flow to areas of intrapulmonary shunt causes a detrimental effect on oxygenation, thereby minimizing its clinical relevance.

In 1980 Furchgott and Zawadski reported that the vasodilator responses to many agents are mediated by the release of a vasodilator substance from endothelial cells [39], later known as the endothelium-derived relaxing factor (EDRF). In 1987 two independent research groups published results which implied that nitric oxide (NO) accounts for the vasodilatory action of EDRF [40,41]. These results immediately suggested that, since NO is a potent pulmonary vasodilatator, inhalation of NO might be effective as a selective pulmonary vasodilatator in view of its short half-life. Rapid combination of NO with hemoglobin contained in red blood cells would rapidly inactivate any NO reaching the systemic circulation, thereby limiting vasodilatation to pulmonary vessels.

Thus, Falke et al. [42] were the first to treat an ARDS patient successfully with administration of NO inhalation. However, besides a selective pulmonary vasodilatation, they observed an increase in gas exchange. They hypothesized that inhaled NO selectively improved the perfusion of ventilated regions (e.g. no influence on hypoxic pulmonary vasoconstriction reflex), thus reducing intrapulmonary shunting and improving arterial oxygenation.

In an uncontrolled pilot study, they confirmed this concept by demonstrating in nine ARDS patients that inhalation of low doses of NO decreased intrapulmonary shunting and improved arterial oxygenation while reducing the pulmonary-artery pressure; furthermore, inert-gas analyses revealed that this beneficial effect was due to a redistribution of pulmonary blood flow away from nonventilated regions of the lungs and toward ventilated regions, thereby improving the matching of ventilation and perfusion [43]. Subsequently, they investigated dose-response relationships between the concentration of inhaled NO and P_aO_2, as well as between inhaled NO and pulmonary artery pressure (PAP) in patients with ARDS [44]. They showed that in patients with ARDS in order to improve arterial oxygenation NO inhalation should be performed in the parts per billion region or low (1-10) ppm range; however, individual dose-responses are recommended.

An exciting area is the use of other drugs in combination with inhaled NO. To date, combination of inhaled NO with either intravenous L-Arginine or NO synthase inhibitor showed no additional effect on the intrapulmonary shunt in ARDS. Recently, we investigated the possible effect of intratracheal surfactant administration on the effect of inhaled NO on gas exchange. Beforehand, we demonstrated in a pilot study in rats that NO inhalation did not influence endogenous surfactant function, characterized by blood gas analysis and BAL analysis (surface tension, total phospholipids and differentiation) [45]. Subsequently, we demonstrated in a rabbit model of acute respiratory failure that intratracheal surfactant administration has a synergistic effect of NO inhalation on arterial oxygenation [46].

From these results, we hypothesized that surfactant probably increases the ventilated area in the lung by recruiting previously collapsed alveoli; inhaled NO selectively improves the perfusion of the ventilated areas, thus causing the synergic effect of intratracheal surfactant administration on the effect of NO inhalation on arterial oxygenation, without significant changes in blood pressure.

Further studies have to focus on the problems concerning the dose, toxicity, weaning and monitoring aspects of NO inhalation in the clinical setting.

CONCLUSION

All discussed investigational forms of therapy have been shown in experimental and in the clinical setting to be promising therapies to prevent and/or minimize the progress of lung injury in patients suffering from acute respiratory failure.

Prospective, double-blind, controlled, randomized trials are necessary to investigate whether these therapies can make a difference in outcome in patients suffering from a "lung in distress".

ACKNOWLEDGMENTS

Our thanks to Laraine Visser-Isles for language editing.

REFERENCES

1. Spragg, R.G., and Smith, R.M., 1991, Biology of acute lung injury In: Crystal, R.G., and West, J.B., (eds): The Lung: scientific foundations, Raven Press, Ltd., New York, 2003-2017.
2. Rinaldo, J. E., and Christman, J. W., 1990, Mechanisms and mediators of the adult respiratory distress syndrome, Clin.Chest.Med. 11:621-632.
3. Beer, D. J., 1992, ARDS: evolving concepts of a systemic disease, Hosp. Pract. 27(a):57-80.
4. Macnaughton, P. D., and Evans, T. W., 1992, Management of adult respiratory distress syndrome, N.Engl.J.Med. 469-472.
5. Ashbaugh, D. G., Bigelow, D. B., Petty, T. L., and Levine, B. E., 1967, Acute respiratory distress in adults, Lancet 2:319-323.
6. Petty, T. L.,and Ashbaugh, D. G., 1971, The adult respiratory distress syndrome: clinical features, factors influencing prognosis and principles of management, Chest 60:233-239.
7. Bernard, G. R., Artigas, A., Brigham, K. L., Carlet, J., Falke, K., Hudson, L., Lamy, M., LeGall, J. R., Morris, A., Spragg, R., and the Consensus Committee, 1994, The American-European Consensus Conference on ARDS: definitions, mechanisms, relevant outcomes, and clinical trial coordination, Am.J.Respir.Crit.Care.Med. 149:818-824.
8. Petty, T. L., Reiss, O. K., Paul, G. W., and Elkins, N. D., 1977, Characteristics of pulmonary surfactant in adult respiratory distress syndrome associated with trauma and shock, Am.Rev.Resp.Dis. 115:531-536.
9. Petty, T. L., Silvers, W., Paul, G. W., and Stanford, R. E., 1979, Abnormailities in lung elastic properties and surfactant function in adult respiratory distress syndrome, Chest 75:571-574.
10. Lachmann, B., Bergmann, K.-C., Enders, K., Friebel, L., Gehlmannn, B., Grossman, G., Hoffmann, D., Kuckelt, W., Malmquist, E., Robertson, B., Seidel, M., Vogel, J., and Winsel, K., 1977, Konnen pathologische veranderungen im surfactant-system der lunge zu einer akuten rtespiratorischen insuffizienz beim erwachsenen fuhren? In: Danzmann, E. (ed): Anaesthesia 77, Proceedings of the 6th congress of the society of Anaesthesiology and Resuscitation of the G.D.R., Berlin, 337-353.
11. Hallman, M., Spragg, R., Harrell, J. H., Moser, K. M., and Gluck, L., 1982, Evidence of lung surfactant abnromality in respiratory failure, J.Clin.Invest.70:673-683.
12. Pison, U., Seeger, W., Buchhorn, R., Joka, T., Brand, M., Obertacke, U., Neuhof, H., and Schmit-Neurenburg, K. P., 1989, Surfactant abnormalities in patients with respiratory failure after multiple trauma, Am.Rev.Resp.Dis. 140:1033-1039.
13.. Gregory, T. J., Longmore, W. J., Moxley, W. J., Whitsett, J. A., Reed, C. R., Fowler III, A. A., Hudson, L. D., Maunder, R. J., Crim, C., and Hyers, T. M., 1991, Surfactant chemical composition and biophysical activity in acute respiratory distress syndrome, J.Clin.Invest. 88:1976-1981.
14. Lachmann, B, and Van Daal, G. J., 1992, Adult respiratory distress syndrome: animal models, In: Robertson, B., Van Golde, L. M. G., Batenburg, J. J. (eds): Pulmonary surfactant. Amsterdam, Elsevier, 635-663.
15. Tierney, D. F., and Johnson, R. P., 1965, Altered surface tension of lung extracts and lung mechanics, J.Appl.Physiol. 20:1253-1260.
16. Seeger, W., Stohr, G., Wolf, H. R. D., and Neuhof, H., 1985, Alteration of surfactant function due to protein leakage: special interaction with fibrin monomer. J.Appl.Physiol. 58:326-338.
17. Fuchimukai, T., Fujiwara, T., Takahashi, A., and Enhorning, G., 1987, Artificial pulmonary surfactant inhibited by proteins, J.Appl.Physiol. 62:429-437.
18. Lachmann, B., 1987, The role of pulmonary surfactant in the pathogenesis and therapy of ARDS, In: Vincent,J.L.(ed), Update in intensive care and emergency medicine. Berlin Heidelberg New York, Springer-Verlag, 123-134.
19. Richman, P. S., Spragg, R. G., Robertson, B., Merritt, T. A., and Curstedt, T., 1989, The adult respiratory distress syndrome: first trials with surfactant replacement, Eur.Respir.J. 2:109s-111s.
20. Joka, Th., and Obertacke, U., 1989, Neue medikamentöse Behandlung im ARDS: Effekt einer intrabronchialen xenogenen Surfactantapplikation, Z.Herz-,Thorax-,Gefäßchir., 3:21-24.
21. Nosaka, S., Sakai, T., Yonekura, M., and Yoshikawa, K., 1990, Surfactant for adults with respiratory failure, Lancet 1:947-948.
22. Marraro, G., Casiragi, G., and Riva, A., 1991, Effets d'un apport de surfactant chez deux adolescents leucémiques atteints de détresse respiratoire, Cahiers d'Anesthesiologie 39:227-232.
23. Weg, J., Reines, H., Balk, R., Tharratt, R., Kearney, P., Killian, T., Scholten, D., Zaccardelli, D., Horton, J., Pattishall, E., and the Exosurf-ARDS sepsis study group, 1991, Safety and efficacy of an aerosolized surfactant (Exosurf®) in human sepsis-induced ARDS, Chest 100:137S.
24. Wiedemann, H., Baugham, R., deBoisblanc, B., Schuster, D., Caldwell, E., Weg, J., Balk, R., Jenkinson, S., Wiegelt, J., Tharratt, R., Horton, J., Pattishall, E., Long, W., and the Exosurf ARDS sepsis study group,

1992, A Multicenter trial in human sepsis-induced ARDS of an aerosolized synthetic surfactant (Exosurf®), Am.Rev.Respir.Dis. 145:A184.

25. Stubbig, K., Schmidt, H., Bohrer, H., Hulster, Th., Bach, A., Motsch, J., 1992, Surfactantapplikation bei akutem Lungenversagen, Anaesthesist 41:555-558.

26. Spragg, R. G., Gilliard, N., Richman, P., Smith, R. M., Hite, R. D., Pappert, D., Robertson, B., Curstedt, T., and Strayer, D., 1994, Acute effects of a single dose of porcine surfactant on patients with the adult respiratory distress syndrome, Chest 105:195-202.

27. Reines, H.D., Silverman, H., and Hurst, J., 1992, Effects of two concentrations of nebulized surfactant (Exosurf) in sepsis-induced ARDS, Int Care Med 20:S61.

28. Gommers, D., and Lachmann, B., 1993, Does surfactant play a role in adults?, Clinical Intensive Care 4: 284-295.

29. Heikinheimo, M., Hynynen, M., Rautiainen, P., Andersson, S., Hallman, M., and Kukkonen, S., 1994, Successful treatment of ARDS with two doses of synthetic surfactant, Chest 105(4): 1263-1264.

30. Gregory, T. J., Longmore, W. J., Moxley, M. A., Cai, G-Z., Gadek, J. E., Weiland, J. E., Hyers, T. M., Crim, C., Hudson, L. D., Steinberg, K. P., Maunder, R. A., Spragg, R. G., Smith, R. M., Tierney, D. F., and Gipe, B., 1994, Surfactant repletion following Survanta supplementation in patients with acute respiratory failure (ARDS), Am.J.Respir.Crit.Care.Med. 149:A124.

31. Von Neergaard, K., 1929, Neue auffassungen uber einen grundbegriff der atemmechanik. Die retraktionskraft der Lunge, abhangig von der oberflachenspannung in den alveolen. Z.Ges.Exp.Med. 66:373-394.

32. Clark, L. C., and Golan, F., 1966, Survival of mammals breathing organic liquids equilibrated with oxygen at atmosphere pressure, Science 152:1755-1756.

33. Greenspan, J. S., Wolfson, M. R., Rubenstein, S. D., and Shaffer, T. H., 1990, Liquid ventilation of human preterm neonates, J.Pediatr. 117:106-111.

34. Fuhrman, B. P., Paczan, P. R., and DeFrancis, M., 1991, Perfluocarbon-associated gas exchange, Crit.Care.Med. 19:712-722.

35. Lewis, J. F, and Jobe, A. H., 1993, Surfactant and the adult respiratory distress syndrome, Am.Rev.Resp.Dis. 147:218-233.

36. Tütüncü, A. S., Faithfull, N. S., and Lachmann, B., 1993, Intratracheal perfluorocarbon administration combined with mechanical ventilation in experimental respiratory distress syndrome: dose-dependent improvement of gas-exchange, Crit.Care.Med. 21:962-969

37. Tütüncü, A. S., Faithfull, N. S., and Lachmann, B., 1993, Intratracheal perfluorocarbon administration as an aid in the ventilatory management of respiratory distress syndrome, Anesthesiology 79:1083-1093.

38. Tütüncü, A. S., Faithfull, N. S., and Lachmann, B., 1993, Comparison of ventilatory support with intratracheal perfluorocarbon administration and conventional mechanical ventilation in animals with acute respiratory failure, Am.Rev.Resp.Dis. 148:785-792

39. Furchgott, R. F., and Zawadzki, J. V., 1980, The obligatory role of endothelial cells in the relaxation of arterial smooth muscle by acetylcholine, Nature 288:373-376.

40. Ignarro, L. J., Buga, G. M., Wood, K. S., Byrns, R. E., and Chaudhuri, G., 1987, Endothelium-derived relaxing factor produced and released from artery and vein is nitric oxide, Proc.Natl.Acad.Sci.USA. 84:9265-9269.

41. Palmer, R. M., Ferrige, A. G., Moncada, S., 1987, Nitric oxide release accounts for the biological activity of endothelium-derived relaxing factor, Nature 327:524-526.

42. Falke, K. J., Rossaint, R., Pison, U., Slama, K., Lopez, F., Santak, B., and Zapol, W.M., 1991, Inhaled nitric oxide selectively reduces pulmonary hypertension in severe ARDS and improves gas exchange as well as right heart ejection fraction: a case report, Am.Rev.Resp.Dis. 143:A248.

43. Rossaint, R., Falke, K. J., Lopez, F., Slama, K., Pison, U., and Zapol, W. M., 1993, Inhaled nitric oxide in adult respiratory distress syndrome, N.Engl.J.Med. 328:399-405.

44. Gerlach, H., Rossaint, R., Pappert, D., and Falke, K. F., 1993, Time-course and dose-response of nitric oxide inhalation for systemic oxygenation and pulmonary hypertension in patients with adult respiratory distress syndrome, Eur.J.Clin.Invest. 23:499-502.

45. Houmes, R. J., Verbrugge, S., Zimmerman, L., Gommers, D., Lachmann, B., 1994, The influence of nitric oxide on the pulmonary surfactant system, Int.Care.Med. 20:S43

46. Gommers, D., Houmes, R. J. M., Olsson, S. G., So, K. L., Lachmann, B., 1994, Exogenous surfactant and nitric oxide have a synergetic effect in improving respiratory failure, Am.J.Respir.Crit.Care.Med. 149:A568.

60

SIGNIFICANCE OF ENDOTHELIUM-DERIVED RELAXING FACTOR (EDRF) ON PULMONARY VASOCONSTRICTION INDUCED BY HYPOXIA AND HYPERCAPNIA

K. Yamaguchi, T. Takasugi, M. Mori, H. Fujita, Y. Oyamada, K. Suzuki, A. Miyata, T. Aoki, and Y. Suzuki

Department of Medicine
School of Medicine
Keio University
Tokyo 160, Japan

INTRODUCTION

Distribution of pulmonary blood flow is physiologically regulated by alveolar PO_2 and PCO_2 surrounding the microcirculation in the lung (Fishman, 1976; Yamaguchi et al., 1994). Decrease in PO_2 and/or increase in PCO_2 has been considered to evoke a rise in the pulmonary microvascular resistance to blood flow (Sylvester et al., 1986; Rodman and Voelkel, 1991). It has been suggested that hypoxic pulmonary vasoconstriction (HPV) as well as hypercapnic-induced vasoconstriction may divert blood flow away from the poorly ventilated areas (Fishman, 1976). The importance of the endothelial-derived vasoactive agents in modulating blood flow has been increasingly recognized since the discovery of endothelium-derived relaxing factor (EDRF), a noble vasodilator of both the systemic and pulmonary circulation (Palmer et al., 1987; Ignaro et al., 1987; Moncada et al., 1991). Many authors have attempted to clarify a possible role of EDRF in the occurrence of HPV, however definite conclusion has not been attained. Although hypercapnia caused by alveolar hypoventilation is considered as the additionally important factor affecting pulmonary hemodynamics (O'Brodovich et al., 1982; Marshall et al., 1984; Sylvester et al., 1986; Yamaguchi et al., 1994), there have been no systematic studies showing the potential effects of hypercapnic-induced acidosis on EDRF modulation for pulmonary circulation. The first aim of this study is systematically to reinvestigate the effects of EDRF on HPV applying varied inhibitors or inactivators for EDRF acting on different levels of EDRF metabolic pathway. The second aim is to make clear the possible roles of EDRF modulating the pulmonary vascular response to acute hypercapnia.

Oxygen Transport to Tissue XVII, Edited by Ince et al.
Plenum Press, New York, 1996

MATERIALS AND METHODS

The isolated perfused lung preparation was used to evaluate the hypoxic and hypercapnic pulmonary pressor response. Japanese male rabbits were anesthetized and heparinized. Tracheotomy was performed and the lungs were ventilated with room air. A median sternotomy was done and cannulae were inserted into the pulmonary artery and left atrium. Animals were exsanguinated via the left atrial cannula. The isolated lungs were then perfused with the solution without blood at a constant flow rate of 70 ml/min and inspired gas was switched to that containing 21% O_2 and 5% CO_2 in N_2. As the perfusate, the modified Krebs-Henseleit solution was used and 3% bovine serum albumin was added to maintain iso-osmotic pressure. Further, 20 µM indomethacin was also added in order to restrain the production of prostaglandins. Pulmonary arterial (Ppa) and airway pressures were continuously measured by force displacement pressure transducers connected to the proximal pulmonary arterial cannula and the tracheotomy tube, respectively. Since perfusion rate is constant, changes in pulmonary arterial pressure simply reflect changes in pulmonary vascular resistance.

Hypoxic Gas Breathing

After institution of perfusion, stable baseline pulmonary arterial pressures were established over at least 20 min. Thereafter, inspired gas was changed from 21% O_2 (normoxic ventilation) to 3% O_2 (hypoxic ventilation) for 10 min but keeping CO_2 concentration in the inspired gas at 5%. When the plateau of Ppa was attained during hypoxic ventilation, normoxic ventilation was begun again for 20 min. Such a cycle was repeated two times and prior to the second cycle, various agents affecting the activity or production of EDRF in the pulmonary endothelium were administered into the reservoir. The difference of Ppa between normoxic and hypoxic ventilation was used as a measure of HPV. To assess the effects of EDRF on HPV, the following agents were added to the reservoir and the second HPV was measured. Hemoglobin: 0.25 mmol/l, methemoglobin: 0.15 mmol/l, methylene blue: 0.3 mg/ml and L-argininosuccinic acid: 0.03 mg/ml. L-argininosuccinic acid is an irreversible antagonist for L-arginine, the substrate for nitric oxide (NO: main substance of EDRF). Preliminary experiments done by Gold et al. (1989) showed that the ability of L-argininosuccinic acid to inhibit NO synthesis in the endothelium was quantitatively comparable to that of L-NMMA as well as L-NAME.

Hypercapnic Gas Breathing

The lungs were firstly ventilated with the inspired gas mixture containing 21% O_2 and 1% CO_2 (hypocapnic ventilation) for 20 min, allowing us to adjust the perfusate pH at 7.8. Subsequently, the inspired gas was changed to that consisting of 21% O_2 and 10% CO_2 (hypercapnic ventilation) for 10 min, leading to the perfusate pH at 7.1. Finally, inspired gas was returned to the initial hypocapnic gas for 20 min. This constitutes one cycle in a series of experiments. The same procedure was repeated and EDRF inhibitors (hemoglobin, methemoglobin, methylene blue and L-argininosuccinic acid) were added to the reservoir just before the commencement of the second cycle. Effects of hypercapnic-gas breathing on the pulmonary vasculature was estimated from the difference of Ppa observed between hypocapnic and hypercapnic condition.

Determination of NO Metabolites

As a measure of EDRF generation at varied experimental conditions, we examined the concentration of end products of NO metabolism, NO_2^- and NO_3^- in the perfusion media spectrophotometorically with the method described by Green et al. (1982). The color of the products yielded by the diazotization reaction was measured using a spectrophotometer at the absorbance of 540 nm.

Determination of Cyclic GMP in Lung Tissue and Perfusate

In parallel experiments, the effects of hypoxia and hypercapnia on the accumulation of cyclic GMP in the lung tissue as well as the efflux from the lung tissue into the perfusion medium with or without the addition of L-argininosuccinic acid were examined. In this series of experiments, hypoxic or hypercapnic ventilation was continued for 20 min. Thereafter, isolated lung was detached from the perfusion circuit and immediately frozen by immersion in dry ice-cooled acetone and stored at -80°C for subsequent cyclic GMP and protein determination. Cyclic GMP and protein contents were examined with radioimmunoassay as proposed by Johns et al. (1989) and with the method of Lowry et al. (1951), respectively. The extrusion of cyclic GMP from the lung tissue into the extracelluar medium was examined at 0, 5, 10 and 20 min after an exposure to either hypoxic or hypercapnic gas with or without adding L-argininosuccinic acid.

Statistical Analysis

Significant differences in vascular response to hypoxic or hypercapnic conditions including the change of Ppa and of NO synthesis before and after administration of varied drugs were estimated by paired t test or by Wilcoxon test. Cyclic GMP contents in the lung tissue were statistically analyzed by unpaired t test or by Mann–Whitney test, while the time course of cyclic GMP released into the medium was examined by means of multiple comparison analysis.

RESULTS

HPV and EDRF

PO_2, PCO_2 and pH in the perfusate during normoxic gas breathing were 141 ± 6 Torr (mean±SD), 28 ± 2 Torr and 7.38 ± 0.03, respectively. Hypoxic gas ventilation largely

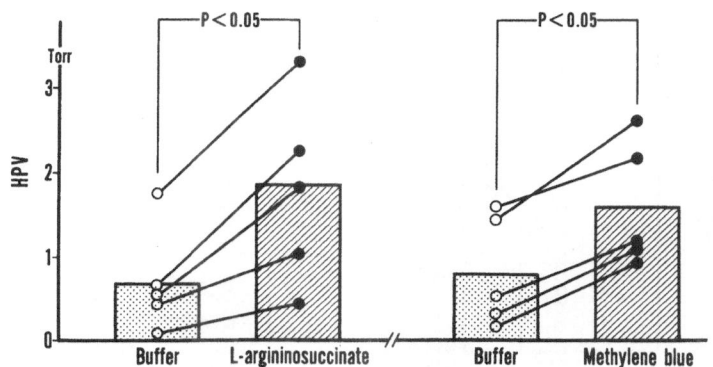

Figure 1. Effects of inhibiting EDRF genesis and of restraining guanylate cyclase on HPV. Left: L-argininosuccinic acid, right: methylene blue.

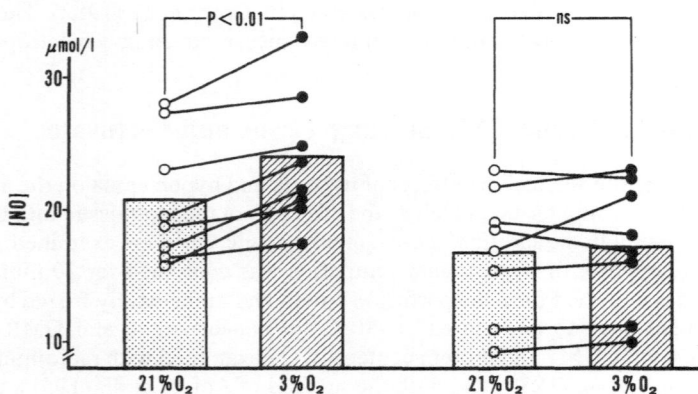

Figure 2. Inhibition of EDRF production by L-argininosuccinic acid. Left: without, right: with L-argininosuc-cinic acid.

reduced PO_2 to 33 ± 3 Torr without any significant alteration in PCO_2 and pH in the perfusate, evoking HPV by about 1 Torr. Addition of hemoglobin solution to the reservoir significantly augmented HPV, while administration of methemoglobin failed to enhance HPV. Both methylene blue and L-argininosuccinic acid considerably increased the degree of HPV (Figure 1). Hypoxic gas breathing for 20 min yielded 2.8 ± 0.3 μM of NO-related metabolites (Figure 2). Addition of L-argininosuccinic acid to the reservoir successfully restrained the augmentation of NO generation during the exposure to hypoxic gas (Figure 2). The total contents of cyclic GMP in the lung tissue averaged 2.3 ± 1.0 pmol/(mg protein) for normoxic ventilation, and 2.0 ± 0.7 pmol/(mg protein) for hypoxic ventilation, respectively. These values were not statistically different. As the time exposed to hypoxic gas was prolonged, the concentration of cyclic GMP in the perfusion medium steadily increased under a condition without addition of any inhibitor for NO production, while administration of L-argininosuccinic acid significantly suppressed the accumulation of cyclic GMP (Figure 3). Statistical difference of cyclic GMP content in the perfusate between the two conditions was observed at 10 and 20 min after initiation of hypoxic ventilation.

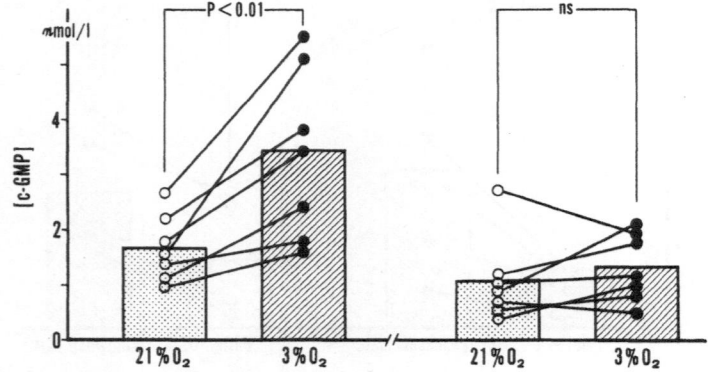

Figure 3. Restraint of cyclic GMP formation by L-argininosuccinic acid. Left: without, right: with L-argini-nosuccinic acid.

Hypercapnic-Induced Vascular Response and EDRF

The change of inspired CO_2 concentration from 1% to 10% but maintaining inspired O_2 at 21% showed a fast increase followed by a gradual decrease of Ppa. Therefore, we estimated the extent of hypercapnia-induced vasoconstriction accompanied with vasodilatation by calculating the difference in Ppa between hypocapnic and hypercapnic ventilation (dP). Vasoconstriction was estimated from the maximum dP (dPmax), while consecutive vasodilatation was appraised based on the relative slope computed by a linear-regression analysis applied between the dPmax and dP at 10 min (%dPslope). PO_2, PCO_2 and pH in the perfusion media during hypocapnic gas breathing were, respectively, 143 ± 2.0 Torr, 10 ± 3 Torr and 7.84 ± 0.02. Hypercapnic ventilation did not change PO_2 values but increased PCO_2 up to 57 ± 3 Torr and decreased pH to 7.12 ± 0.03 in the perfusate. Addition of either hemoglobin, methylene blue or L-argininosuccinic acid to the reservoir showed significant enhancement of dPmax (Figure 4), while methemoglobin did not do so. Although time course of absolute values of dP varied significantly in different series of experiments, the values of %dPslope did not differ statistically irrespective of the experimental conditions (Figure 4). Hypercapnic gas breathing for 20 min increased NO-related metabolites in the perfusion media by 2.6 ± 1.5 μM. This was significantly larger than that obtained during 20-min hypocapnic gas breathing. L-argininosuccinic acid fully inhibited the enhancement of NO synthesis during hypercapnic ventilation. Total cyclic GMP content in the isolated perfused lung tissue after 20-min hypercapnic gas breathing without any inhibitor of NO generation was 1.2 ± 0.4 pmol/(mg protein), while that with L-argininosuccinic acid was 1.3 ± 0.2 pmol/(mg protein). The values were not significantly different. Hypercapnic ventilation for 20 min augmented cyclic GMP concentration in the efflux from the lungs by 1.0 ± 0.7 nM. Addition of L-argininosuccinic acid to the perfusion circuit distinctly restrained the enhanced release of cyclic GMP into the extracellular media, the increment of which was solely 0.2 ± 0.2 nM, being appreciably smaller than that obtained without L-argininosuccinic acid.

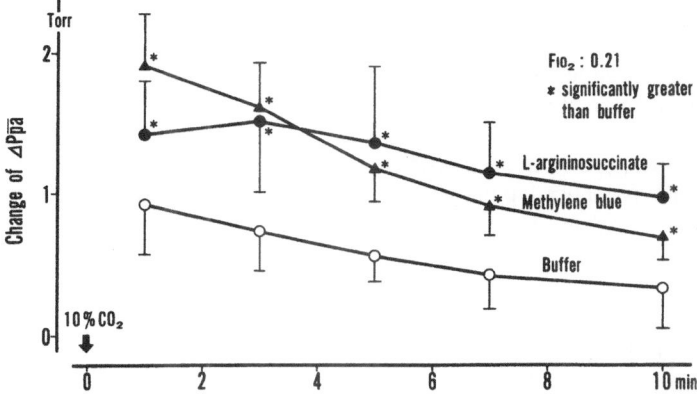

Figure 4. Time course of pulmonary arterial pressure during hypercapnic gas breathing in the presence of L-argininosuccinic acid or methylene blue.

DISCUSSION

Modulation of HPV by EDRF

The relationship between HPV and EDRF has not reached a full agreement. Using isolated rings of rabbit first-branch pulmonary artery, Johns et al. (1989) showed that hypoxia reduced basal cyclic GMP content in the rings and suggested that HPV might be mediated through the inhibition of basal EDRF production in the pulmonary arteries. Qualitatively similar findings were obtained in rat branch pulmonary artery (Rodman et al., 1990). Warren et al. (1989) showed that exposure of cultured bovine pulmonary artery endothelial cells to hypoxic environment distinctly diminished the EDRF production, indicating that reduction of EDRF by hypoxia might be involved in the occurrence of HPV. However, entirely opposing results were obtained in the system of isolated perfused lungs of the rat (Brashers et al., 1988; Archer et al., 1989; Mazmanian et al., 1989; Liu et al., 1991; Hasunuma et al., 1991). These authors showed that alveolar hypoxia enhanced EDRF release in the pulmonary circulation and inhibition of EDRF generation increased the degree of HPV. Exceptionally, Hyman et al. (1991) reported that inhibitor of guanylate cyclase, methylene blue, decreased the extent of HPV in isolated perfused cat lungs. Since there was no report focussed on the possible roles of EDRF for HPV in rabbit lungs, we attempted systematically to clarify the modulating effects of EDRF on HPV in isolated perfused rabbit lungs by using varied agents affecting the pathway mediated by EDRF at different levels. The presence of hemoglobin, a scavenger of EDRF in vascular lumen, significantly augmented HPV, while methemoglobin did not alter the extent of HPV. This is explicable from the fact that hemoglobin can absorb EDRF but methemoglobin containing oxidized iron can not. Similarly, L-argininosuccinic acid, a noble antagonist for NO formation from L-arginine in endothelial cells, as well as methylene blue, an inhibitor of soluble guanylate cyclase in vascular smooth muscle cells, significantly enhanced HPV. Our results were highly consistent with those previously reported in isolated perfused rat lungs, indicating that EDRF generation might be augmented secondary to a hypoxic spell and it might act as a modulator of HPV to avoid an excess of vasoconstriction in the pulmonary circulation. The increasing concentration of NO-related metabolites and of cyclic GMP released in the perfusate and suppression of them by L-argininosuccinic acid during hypoxic gas breathing seems certainly to support the concept as given above. The conflicting views between the experiments including cultured endothelial cells as well as rings of pulmonary artery and those of isolated perfused lungs may be ascribed to the presence or absence of shear stress imposed by active flow. Perfused lung preparation yields an additional shear stress caused by a finite flow in the pulmonary vascular lumen, whereas endothelial-cell culture or vascular ring may be lacking such shear stress. Continuous shear stress induced by the flow in the pulmonary circulation is one of the most important factors stimulating the synthesis of EDRF in a physiological condition. (Moncada et al., 1991). The experimental findings as described above may lead to the conclusion that hypoxia in itself restrains the synthesis of EDRF but increasing shear stress caused by hypoxia-induced pulmonary vasoconstriction, in opposition, enhanced EDRF synthesis in *in vivo* condition.

Effects of EDRF on Pulmonary Vascular Response to Hypercapnic Gas

Although alteration of pulmonary hemodynamics caused by hypercapnia-induced acidosis has been studied since the 1940's (Von Euler and Liljestrand, 1946), no decisive conclusion has been reached concerning the effects of hypercapnia. Most of the studies were designed to test the synergistic effects of extracellular or intracellular acidosis on HPV. The previous reports generally indicate that alkalosis inhibits but hypercapnia and/or acidosis potentiate the hypoxic pressor response (cf. Sylvester et al., 1986). However, there are some studies showing that HPV is attenuated by either increasing or decreasing perfusate pH from normal range (Malik and Kidd, 1973; Marshall et al., 1984). Recently, Raffestin and McMurtry (1987) elicited the opinion that intracellular alkalosis and acidosis, respectively, potentiated and blunted vasoconstrictive response to hypoxia as well as other stimuli in isolated rat lungs. However, there are only a few studies analyzing a direct effect of hypercapnia on basal tone of the pulmonary circulation. Using isolated cat lungs, Viles and Shepherd (1968) showed that hypercapnia under a normoxic condition might be a weak stimulus for vasoconstriction of pulmonary vasculature. Measuring the blood flow in the left lower lobes (LLL) of the cat and dog, Barer and McCurrie (1969) found that the blood flow in LLL decreased along with a moderate increase of PCO_2 in the inspired gas. Farrukh et al. (1989) found that base-line pulmonary arterial pressure was minimum at normal pH but increased at a condition of either extracellular acidosis or alkalosis in isolated rabbit lungs. Direct measurements of vascular diameter in the pulmonary microcirculation was made by Kato and Staub (1966), who showed that normoxic hypercapnia did not change the diameter of cat small muscular arterioles. On the other hand, Koyama and Horimoto (1982) succeeded in showing that microvessels on the surface of the bullfrog lung were significantly constricted when they were exposed to normoxic-hypercapnic gas. They also found that hypercapnia-induced vasoconstriction was a transient phenomenon and was not sustained for more than 6-10 min. The transition of Ppa during 10% CO_2 gas breathing without addition of any agent showed that alveolar hypercapnia evoked a brief vasoconstriction followed by prolonged vasodilatation, the finding being highly consistent with that observed by Koyama et al. (1982). dPmax, an indicator of vasoconstriction induced by hypercapnia, was considerably augmented by addition of either hemoglobin, methylene blue or L-argininosuccinic acid to the perfusate, but not by methemoglobin, suggesting that EDRF was actively released and might function as a modulator coping with an excessive constriction of pulmonary vasculature during hypercapnic gas breathing. This may be reliably supported by the fact that concentrations of both NO-related metabolites and cyclic GMP in the perfusion medium were concomitantly increased after initiation of a gas mixture containing high CO_2. In the presence of L-argininosuccinic acid, the enhanced NO synthesis during hypercapnic ventilation was notably restrained and was closely associated with increasing dPmax, vindicating further the justice of the above findings. %dPslope, the measure of vasodilatation successive to vasoconstriction during hypercapnic ventilation, was not appreciably different among varied experimental conditions. Although addition of L-argininosuccinic acid obviously suppressed the formation of both EDRF and cyclic GMP detected in the perfusate, the transition of %dPslope in the presence of L-argininosuccinic acid was not significantly different from that obtained for control conditions without administration of any agent. The findings are highly compatible with the idea that relaxation of pulmonary vasculature during exposure to hypercapnic gas breathing are nearly independent of EDRF-induced cyclic GMP synthesis, and should be explained from the effects of hypercapnia on smooth muscle contractile machinery. Since CO_2 molecules readily penetrate across the cell membrane, increasing PCO_2 in extracellular environment quickly lowers intracellular pH (pHi). As discussed in detail by Orchard and Kentish (1990), reduction of pHi greatly inhibits the excitation-contraction coupling pathway. Lowering of pHi potentiates Na^+/H^+ exchange (Orchard and Kentish, 1990) which compensatorily regulates pHi, indicating that pHi in the smooth muscle cells is variable during the load of hypercapnic gas. The rate of change in pHi of squid giant axon after exposure to the gas with high CO_2 was systematically studied by Boron and Weer (1976), demonstrating that pHi initially fell sharply followed by a plateau for about 5 min and began to slowly climb at an average rate of 0.1 pH units/hr. This may indicate that pHi of pulmonary vasculature in our experiments (maximum CO_2 loading time: 20 min) are approximately maintained at a constant level during hypercapnic gas breathing.

Significance of Cyclic GMP in the Perfusate

Although cyclic GMP contents in the perfusate released from the lungs were signifi-cantly suppressed by L-argininosuccinic acid, those in the lung tissue were not. This finding may be of much importance to learn the sensitivity for detecting EDRF-mediated cyclic GMP production. Our experimental results may indicate that cyclic GMP content in the perfusate is much more sensitive than that in lung tissue for examining the behavior of cyclic GMP related to EDRF. Using rat isolated aortic preparations, Schini et al. (1989) showed that agonists greatly increased tissue cyclic GMP content in a time-dependent manner. In their study, maximal increases in tissue cyclic GMP content were obtained 1 min after admini-stration of agonists and it rapidly declined thereafter, while that in the incubation medium steadily increased and reached a plateau 30-60 min after addition of agonists. The study of Schini et al. (1989) demonstrated that detection of enhanced cyclic GMP content in the lung tissue might be exceedingly difficult because its maximal increase was observed at a very early time, while measurements on its extracellular accumulation might be more reliable to know the behavior of EDRF-mediated cyclic GMP. Their results appear to be in qualitative agreement with our observations on cyclic GMP in isolated perfused lungs.

REFERENCES

Archer, S.L., Tolins, J.P., Raiji, L., Weir, E.K., 1989, Hypoxic pulmonary vasoconstriction is enhanced by inhibition of the synthesis of an endothelium derived relaxing factor, Biochem. Biophys. Res. Commun. 164: 1198-1205.

Barer, G.R., and McCurrie, J.R., 1969, Pulmonary vasomotor responses in the cat; the effects and interrela-tionships of drugs, hypoxia and hypercapnia, Q. J. Exp. Physiol. 54: 156-172.

Boron, W.F., and Weer, P.D., 1976, Intracellular pH transients in Squid Giant axons caused by CO_2, NH_3, and metabolic inhibitors, J. Gen. Physiol. 67: 91-112.

Brashers, V.L., Peach, M.J., and Rose, C.E., 1988, Augmentation of hypoxic pulmonary vasoconstriction in the isolated perfused rat lung by in vitro antagonists of endothelium-dependent relaxation, J. Clin. Invest. 82: 1495-1502.

Farrukh, I.S., Gurtner, G.H., Terry, P.B., Tohidi, W., Yang, J., Adkinson, F., and Michael, J.R., 1989, Effect of pH on pulmonary vascular tone, reactivity, and arachidonate metabolism, J. Appl. Physiol. 67: 445-452.

Fishman, A.P., 1976, Hypoxia and pulmonary circulation, Cir. Res. 38: 221-231.

Gold, M.E., Wood, K.S., Buga, G.M., Byrns, R.E., and Ignarro, L.J., 1989, L-arginine causes whereas L-argininosuccinic acid inhibits endothelium-dependent vascular smooth muscle relaxation, Bio-chem, Biophys. Res. Commun. 161: 536-543.

Green, L.C., Wagner, D.A., Glogswski, J., Skipper, P.L., Winshnok, J.S., and Tannenbaum, S.R., 1982, Analysis of nitrate, nitrite and (^{15}N) nitrate in biological fluid, Anal. Biochem. 126: 131-138.

Hasunuma, K., Yamaguchi, T., Rodman, D.M., O'Brien, R.F., and McMurtry, I.F., 1991, Effects of inhibitors of EDRF and EDHF on vasoreactivity of perfused rat lungs, Am. J. Physiol. 260: L97-L104.

Hyman, A.L., Lippton, H.L., and Kadowitz, P.J., 1991, Methylene blue prevents hypoxic pulmonary vasocon-striction in cats, Am. J. Physiol. 260: H586-H-592.

Ignaro, L.J., Buga, G.W., Wood, K.S., Byrns, R.E., and Chaudhuri, G., 1987, Endothelium-derived relaxing factor produced and released from artery and vein is nitric oxide, Proc. Natl. Acad. Sci. USA. 84: 9265-9269.

Johns, R.A., Linden, J.M., and Peach, J., 1989, Endothelium-dependent relaxation and cyclic GMP accumula-tion in rabbit pulmonary artery are selectively impaired by moderate hypoxia, Cir. Res. 65: 1508-1515.

Kato, M., Staub, N.C., 1966, Response of small pulmonary arteries to unilobar hypoxia and hypercapnia, Cir. Res. 19: 426-440.

Koyama, T., and Horimoto, M., 1982, Pulmonary microcirculatory response to localized hypercapnia, J. Appl. Physiol. 53: 1556-1564.

Liu, S., Crawley, D.E., Barnes, P.J., and Evans, T.W., 1991, Endothelium-derived relaxing factor inhibits hypoxic pulmonary vasoconstriction in rats, Am. Rev. Respir. Dis. 143: 32-37.

Lowry, O.H., Rosebrough, N.J., Farr, A.L., and Randall, R.J., 1951, Protein measurement with the Folin phenol reagent, J. Biol. 193: 265-275.

Malik, B., and Kidd, B.S.L., 1973, Independent effects of changes in H^+ and CO_2 concentrations on hypoxic pulmonary vasoconstriction, J. Appl. Physiol. 34: 318-323.

Marshall, C., Lindgren, L., and Marshall, B.E., 1984, Metabolic and respiratory hydrogen ion effects on hypoxic pulmonary vasoconstriction, J. Appl. Physiol. 57: 545-550.

Mazmanian, G-M., Baudet, B., Brink, C., Cerrina, J., Kirkiacharian, S., and Weiss, M., 1989, Methylene blue potentiates vascular reactivity in isolated rat lungs, J. Appl. Physiol. 66: 1040-1045.

Moncada, S., Palmer, R.M., and Higgs, E.A., 1991, Nitric oxide: physiology, pathophysiology, and pharmacology, Pharmacol. Rev. 43: 109-142.

O'Brodovich, H.M., Stalcup, S.A., Pang, L.M., and Mellins, R.B., 1982, Hemodynamics and vasoactive mediator response to experimental respiratory failure, J. Appl. Physiol. 52: 1230-1236.

Orchard, C.H., and Kentish, J.C., 1990, Effects of changes of pH on the contractile function of cardiac muscle, Am. J. Physiol. 258: C967-C981.

Palmer, R.M., Ferrige, A.G., and Moncada, S., 1987, Nitric oxide release accounts for the biological activity of endothelium-derived relaxing factor, Nature (London) 327: 524-526.

Raffestin, B., and McMurtry, I.F., 1987, Effects of intracellular pH on hypoxic vasoconstriction in rat lungs, J. Appl. Physiol. 63: 2524-2531.

Rodman, D.M., Yamaguchi, T., Hasunuma, K., O'Brien, R.F., and McMurtry, I.F., 1990, Effects of hypoxia on endothelium-dependent relaxation of rat pulmonary artery, Am. J. Physiol. 258: L207-214.

Rodman, D.M., and Voelkel, N.F., 1991, Regulation of vascular tone, In: The Lung: Scientific Foundations edited by R.G. Crystal, J.B. West, Raven Press., New York, vol. II, p.1105-1119.

Schini, V., Schoeffter, P., and Miller, R.C., 1989, Effect of endothelium on basal and on stimulated accumulation and efflux of cyclic GMP in rat isolated aorta, Br. J. Pharmacol. 97: 853-865.

Sylvester, J.T., Rock, P., Gottlieb, J.E., and Wetzel, R.C., 1986, Acute hypoxic responses, In: Abnormal Pulmonary Circulation, edited by E.H. Bergofsky, Churchill Livingstone, New York, p.127-165.

Viles, P.H., and Shepherd, J.T., 1968, Evidence for a dilator action of carbon dioxide on the pulmonary vessels of the cat, Cir. Res. 22: 325-332.

Von Euler, U., and Lilijestrand, G., 1946, Observations on the pulmonary arterial blood pressure in the cat, Acta. Physiol. Scand. 12: 301-320.

Warren, J.B., Maltby, N.H., MacCormack, D., and Barnes, P.J., 1989, Pulmonary endothelium-derived relaxing factor is impaired in hypoxia, Clin. Sci. 77: 671-676.

Yamaguchi, K., Mori, M., Kawai, A., Takasugi, T., Umeda, A., Kawashiro, T., and Yokoyama, T., 1994, Regulation of blood flow in pulmonary microcirculation by vasoactive arachidonic acid metabolites-analysis in acute lung injury, Adv. Exp. Med. Biol., 345: 113-120.

61

THE ROLE OF REACTIVE OXYGEN SPECIES IN ISCHEMIA-REPERFUSION INJURY OF RAT LUNG

A. S. Tütüncü,[1*] F. A. Genç,[2] A. Telci,[3] and L. Telci[1]

[1] Department of Anesthesiology
[2] Department of General Surgery
[3] Department of Biochemistry
 Medical Faculty of University of Istanbul
 Turkey

INTRODUCTION

Oxygen-derived free radicals have been implicated as important mediators of injury in various disease processes, including postischemic reperfusion injury (1-3). Reactive oxygen intermediates are toxic to cells because of their oxidizing effects on proteins, membrane polyunsaturated fatty acids and DNA (4).

Looking at the local changes at the site of tissue ischemia, studies in skeletal muscles in animals suggest that there is a burst of oxygen free radicals resulting in tissue damage during reperfusion after ischemia (5,6). Reperfusion injury mediated by oxygen free radicals has been reported in heart, liver, kidney, gastric mucosa and intestine. Relevant investigations regarding the systemic effects of ischemia-reperfusion injury occurring at lower extremities are very rare.

This study was performed to determine the systemic effects of local ischemia-reperfusion injury of a lower extremity. For this purpose, taking the lungs as the end-organ, the effects on lung functions in terms of pulmonary gas exchange were evaluated in an ischemia-reperfusion injury of rat hind limb and, the protective effect of the enzyme, superoxide dismutase (SOD), on possible lung injury was detected in this model.

* Address correspondence to: Ahmet S. Tütüncü, MD, Dept. of Anesthesiology, Medical Faculty of University of Istanbul, 34390 Capa, Istanbul, Turkey. Fax: 90-212-5332083; Tel: 90-212-6318767.

Oxygen Transport to Tissue XVII, Edited by Ince et al.
Plenum Press, New York, 1996

483

MATERIALS AND METHODS

Twelve male adult Sprague–Dawley rats (weighing 238 ± 45 g) were anesthetized with Enflurane (1.5%) and intraperitoneal injection (0.3 ml/kg) of Hypnorm (10 mg fluniason and 0.2 mg fentanyl per ml; Janssen, The Netherlands). Additional doses of Hypnorm were given to maintain anesthesia throughout the experiment. The animals were tracheotomized, carotid artery cannulated and mechanically ventilated with Servo 900C (Siemens-Elema, Solna, Sweden) at pressure-controlled mode (inspiratory/expiratory pressure = 12/2 cm H_2O, f = 30/min and FiO_2 = 1.0).

Both hind limbs were prepared leaving only the femoral artery and vein intact in all the animals. Four hours of ischemia was performed in one hind limb by clamping the femoral vessels. Ischemia was diagnosed by observing the changes in the color of the limb. Thereafter, a reperfusion period of 4 h was maintained by releasing the femoral clamp in all the animals.

Animals were studied in two groups (n = 6 each). In the first group, SOD (20.000 U/kg) was administered intravenously via the jugular vein 5 min before reperfusion started. The second group received saline (isovolumic) 5 min before reperfusion to serve as controls. At the end of the 4 h reperfusion period, tissue samples from ischemic limb, non-ischemic limb and lung were collected for measurement of malondialdehyde (MDA) levels, which is a lipid peroxidation product. MDA determination was performed according to the method of Wong et al. (7) which includes reaction of samples with thiobarbituric acid.

Arterial blood gases and pH were determined at 1 h intervals throughout the study (ABL-300, Radiometer, Copenhagen, Denmark). Statistical analysis was performed by ANOVA for repeated measurements and Student's t-test. A P value less than 0.05 was considered statistically significant.

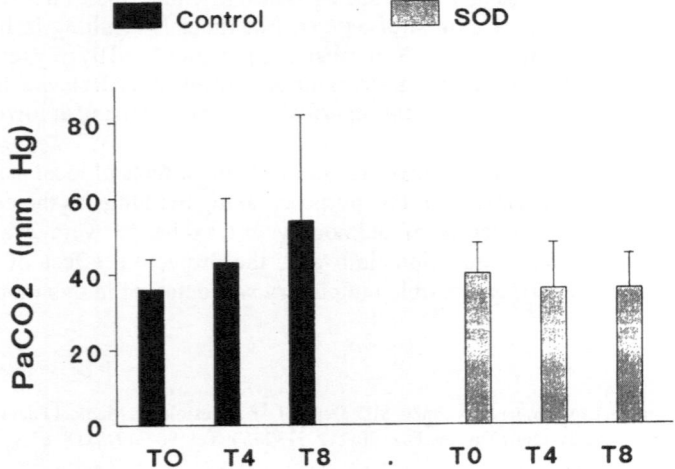

Figure 1. Arterial pO_2 values in the experimental groups at baseline (T0), before reperfusion (T4) and at the end of 4 h of reperfusion period (T8). * = statistically significant when compared to values at T0 and T4 within the group.

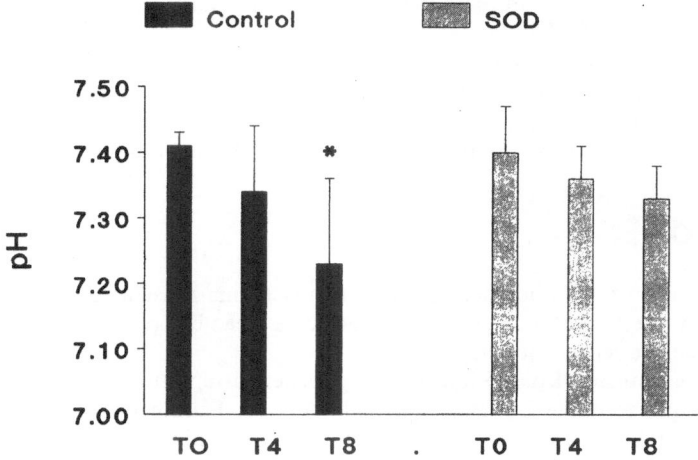

Figure 2. The mean values for $PaCO_2$ in the experimental groups at baseline (T0), before reperfusion (T4) and at the end of reperfusion period (T8).

RESULTS

In both groups, gas exchange and pH values remained unchanged during the 4 h ischemias (Figure 1-3). While PaO_2 levels remained unchanged during reperfusion period in the SOD-treated group, PaO_2 and pH values were significantly less in the control group at 4 h after reperfusion started when compared to the ischemic period. On the other hand, $PaCO_2$ gradually increased during the reperfusion period in the control group and this was

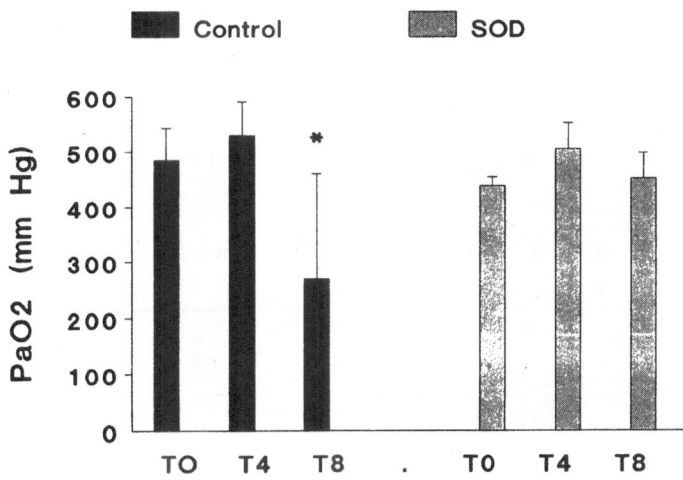

Figure 3. The arterial pH data in the experimental groups at baseline (T0), before reperfusion (T4) and at the end of reperfusion period (T8). * = statistically significant when compared to data at baseline and ischemia period within the group.

accompanied by a decrease in pH. Both $PaCO_2$ and pH data remained unchanged in the SOD-treated group.

There was a significant increase in the MDA level in ischemic hind limb as compared to the non-ischemic hind limb (Table 1). Moreover, the MDA was measured to be significantly high in ischemic hind limb of the control group in comparison to the SOD-treated group. The MDA levels were not different in lung samples in both groups.

DISCUSSION

The results of this study demonstrated that ischemia-reperfusion of the rat hind limb causes oxygen free radical mediated lung injury which can be prevented by the administration of SOD before reperfusion occurs.

The mechanism of tissue injury during ischemia-reperfusion is unclear, but this is currently thought to occur by some molecular events. During ischemia, degradation of cellular stores of ATP to the purines and conversion of xanthine dehydrogenase to xanthine oxidase hold for the critical events. The supply of molecular oxygen to tissue during reperfusion allows xanthine oxidase to produce superoxide radical and hydrogen peroxide. These byproducts further participate in a reaction to produce hydroxyl radicals which are potent oxidants and known to cause tissue injury by initiating lipid peroxidation of cell membranes and oxidative inactivation of cell proteins. During reperfusion, the above reactions can be stopped and injury prevented by inhibiting xanthine oxidase or by neutralizing free radicals with scavenging reagents.

This study demonstrated that above-mentioned local reactions occurring after ischemia-reperfusion may have systemic effects. In contrast to the SOD-treated group, the lung injury as detected by impaired gas exchange in the control group suggested that other mechanisms rather than initial reactions should play a role in this systemic reaction. Free radicals are known to act as both oxidants and reducing agents. When a free radical reacts with a nonradical compound, other free radicals can be formed. This enables free radicals to induce chain reactions that may be thousands of events long. Although the initial free radical produces only local effects, the secondary radicals formed from it and the degradation products produced by reactions involving free radicals can have biologic effects distant from the site where the first free radical was formed.

Although superoxide anion radical is not extremely reactive, it is potentially toxic since it can be transformed into hydrogen peroxide by the superoxide dismutase enzyme and then transformed into highly toxic hydroxyl radical. This radical reacts with molecules, further damaging proteins and initiating lipid peroxidation. Peroxidation of the polyunsatu-

Table 1. Malaondialdehyde (nmol/ 10 mg wet tissue)

Ischemic (limb)	Non-ischemic (limb)	Statistics
0.60 ±0.23	0.25 ± 0.04	$p < 0.05$
control (limb)	SOD (limb)	
0.60 ± 0.23	0.16 ± 0.04	$p < 0.001$
control (lung)	SOD (lung)	
0.12 ± 0.02	0.14 ± 0.01	NS

rated fatty acids in lipid membranes damages the cell membrane on the one hand and, on the other, the end products of such a lipid peroxidation (aldehydes, malondialdehyde) can diffuse away from the site of the chain reaction and can give rise to various pathologic changes (e.g. cell edema, inflammation, increased vascular permeability, chemotaxis). These products may also induce release of arachidonic acid and subsequent formation of prostaglandins (8).

Studies have demonstrated that reactive oxygen intermediates can induce dose and time-dependent effects on pulmonary endothelial cells, and these changes vary from a reversible increase in lung permeability to cell lysis (9-11). Although we have not performed histologic examinations and lung permeability measurements in this study, the present data revealed that the significant impairment of arterial oxygenation occurred at 4 h following reperfusion, suggesting a time dependent alteration in lung function. Thus, we may speculate that reperfusion induced an accumulative pathological process in the lung that was detectable by the fourth hour. In line with the present experimental result, Seekamp et al. documented that ischemia-reperfusion of the rat hind limb resulted in both local injury to skeletal muscle as well as injury to lungs, as measured by increased vascular permeability and hemorrhage (12). The lung injury which developed progressively over a 4-h period in their study proved to be a result of neutrophil-dependent systemic changes during reperfusion of ischemic lower extremity.

This study demonstrated that the systemic effects of local ischemia-reperfusion injury can be prevented by the administration of the superoxide free radical scavenger. The protective effect of exogenous SOD is through acceleration of a reaction which removes superoxide anions, thereby preventing the formation of hydroxyl radical. This extracellular enzyme induces a concentration gradient for superoxide anions to move from the intracellular compartment to the extracellular area. Since the biologic half-life for SOD is too short, this enzyme needs to be administered immediately before the reperfusion starts in order to achieve an adequate tissue concentration for protection against free radical mediated injury.

Because oxygen free radicals have a short life-span, detection of free radicals is difficult by the techniques routinely used, rather than by electron spin resonance spectroscopy. Therefore, detection of products of lipid peroxidation has been a common practice to speak of the free radical reactions as an indirect measure. As a product of lipid peroxidation, we tested the MDA levels in various tissue samples in this study. Despite the increased level of lipid peroxidation product (MDA) in ischemic skeletal muscle after 4-h of reperfusion, the MDA level in the lung samples at 4 h remained unchanged in the experimental groups in this study. This result suggested that measurement of MDA level in the lung tissue is not specific for detection of lung injury resulting from a local ischemia-reperfusion injury induced by reactive oxygen intermediates. In this study, the MDA was measured in the homogenated tissue samples. Increased MDA levels have been detected in the plasma of patients after surgical revascularization operations for kidney transplantation or thromboembolic disease of the limbs (13).

As conclusion, the present data indicated that ischemia-reperfusion injury of rat lower extremity causes oxygen free radical-induced local and systemic changes and, the lung is susceptible to reperfusion injury which can be prevented by the administration of superoxide scavenger immediately before the reperfusion occurs.

REFERENCES

1. Bulkley GB. The role of oxygen free radicals in human disease processes. Surgery 1983; 94:407-411
2. Granger DN, Höllwarth ME, Parks DA. Ischemia-reperfusion injury: role of oxygen-derived free radicals. Acta Physiol Scand 1986; 548 (Suppl):47-63
3. Bulkley GB. Pathophysiology of free radical-mediated reperfusion injury. J Vasc Surg 1987; 5:512-517

4. Jenkinson SG. Free radical effects on lung metabolism. Clin Chest Med 1989; 10:37-47
5. Korthius R, Granger N, Townskey M, Taylor A. The role of oxygen free radicals in ischemia induced increases in canine skeletal muscle vascular permeability. Circ Res 1985; 57:599-609
6. Walter PM, Lindsay TE, Labbe R, Mickle DA, Romaschin AD. Salvage of skeletal muscle with free radical scavengers. J Vasc Surg 1987; 5:68-75
7. Wong SHY, Knight JA, Hopfer SM, Zaharia O, Leach CN, Sunderman FW. Lipoperoxides in plasma as measured by liquid-chromatographic separation of malondialdehyde-thio-barbituric acid adduct. Clin Chem 1987; 33:214-220
8. Del Maestro RF. An approach to free radicals in medicine and biology. Acta Physiol Scand 1980; 492 (Suppl):153-168
9. Johnson KL, Fantone JC, Kaplan J, et al. In vivo damage of rat lungs by oxygen metabolites. J Clin Invest 1981; 67:983-993
10. Harlan JM, Levine JD, Callahan KS, et al. Glutathione redox cycle protects cultured endothelial cells against lysis by extracellularly generated hydrogen peroxide. J Clin Invest 1984; 73:706-713
11. Jolliet P, Polla B, Donath A, Slosman D. Early hydrogen peroxide-induced pulmonary endothelial cell dysfunction: detection and prevention. Crit Care Med 1994; 22:157-162
12. Seekamp A, Mulligan MS, Till GO, Smith CW, Miyasaka M, Tamatani T, Todd RF, Ward PA. Role of B2 integrins and ICAM-1 in lung injury following ischemia-reperfusion of rat hind limbs. Am J Pathol 1993; 143:464-472
13. Rabl H, Khoschsorur G, Colombo T, Tatzber F, Esterbauer H. Human plasma lipid peroxide levels show a strong transient increase after successful revascularization operations. Free Rad Biol Med 1992; 13:281-288

62

POTENTIATION OF THE AGE-DEPENDENT Ca UPTAKE INTO CORONARY ARTERIES OF RATS BY THE RISK FACTOR GENETIC HYPERTENSION[*]

F. Thimm and G. Fleckenstein–Grün

Study Group for Calcium Antagonism
Physiological Institute
University of Freiburg
Hermann-Herder-Str. 7, D-79104 Freiburg

INTRODUCTION

The relationship between arterial calcification and the loss of elasticity and contractility in biological ageing is well established (1-3). Moreover, a progressive mural calcium (Ca) incorporation appears to be an underdying feature of most arteriosclerotic processes known to produce finally mural stiffness and to threaten tissue O_2 supply. Interestingly, in various types of experimental and human arteriosclerosis, an initial increase in the mural Ca uptake may already charac│erize the early lesions (4; 5). It was the aim of the present investigations to study in rats the influences of age and genetic hypertension - an important risk factor of arteriosclerosis - on the Ca incorporation into coronary arteries.

MATERIALS AND METHODS

Experiments were performed during 20 months on spontaneously hypertensive Okamoto rats (SHR; n = 51) and on normotensive Wistar Kyoto rats (WKY; n = 48). Systolic blood pressure was measured weekly according to the tail cuff method. After 6, 15, and 20 months of life, rats of both groups were sacrificed to excise coronary arteries. Arterial segments were dried at 95°C overnight to determine dry weight (188.8 ± 0.06 µg) and dissolved in teflon tubes at 150°C with nitric acid (65%) and perchloric acid (70%) "supra purquality", Merck, Darmstadt, FRG, mixture 2:1) for 6 hours. Ca was analyzed by atomic absorption spectroscopy (AAS, atomic absorption spectrometer, Perkin-Elmer, type 3030),

---

[*] Dedicated to Prof.Dr.Dr.med.h.c.mult. Albrecht Fleckenstein (1917-1992).

Oxygen Transport to Tissue XVII, Edited by Ince et al.
Plenum Press, New York, 1996

489

after addition of lanthanum oxide to a final concentration of 0.5% with a heated graphite tomizer (HGA-600). 20 μl were pipetted into small graphite tubes and heated electrically up to 2,600°C to achieve successively drying, thermic dispersion of the matrix and thermic dissociation into free atoms (atomization). Vapours were removed with inert gas flow that, intermittently, was interrupted to prolong the free atoms contact time with light (wave length for Ca: 422.7 nm) by a factor of 10^3 compared to conventional AAS. Thus large amounts of atoms were excited simultaneously to light absorption to determine Ca in microprobes as picomol/μg. Data are presented as mean ± S.E.M.. Statistical analyses were carried out by Student's t-test.

RESULTS

Blood Pressure

Figure 1 reflects the age-dependent course of blood pressure in SHR and WKY. In SHR, within the first 6 months of life, systolic blood pressure (mmHg) increased from 129.2 ± 1.8 (1st month; n = 51) to 196.3 ± 2.5 (3rd month; n = 51) and 227.9 ± 3.5 (6th month; n = 51) and then remained constantly elevated until txe 20th month (215.5 ± 8.4; n = 23). During the same observation period in WKY normotension (mmHg) was demonstrated: 118.1 ± 1.8 (1st month; n = 48), 121.8 ± 1.9 (3rd month; n = 48), 130.5 ± 2.0 (6th month; n = 48), 131.4 ± 2.7 (20th month; n = 21).

Age-Dependent Calcium Uptake into Coronary Arteries of WKY

In the 6th month of life the Ca content in coronary arteries of WKY amounted to 8.42 ± 0.42 pmol/μg dry weight (n = 14) (Fig.2). Within the next 9 months, the coronary Ca uptake increased by a factor of 1.92, i.e. up to 16.19 ±0.53 pmol/μg (n = 13), p < 0.001 in the 15th month. Finally, the coronary Ca content in coronary arteries of WKY reached 19.3 ± 2.3 pmol/μg (20th month; n = 3).

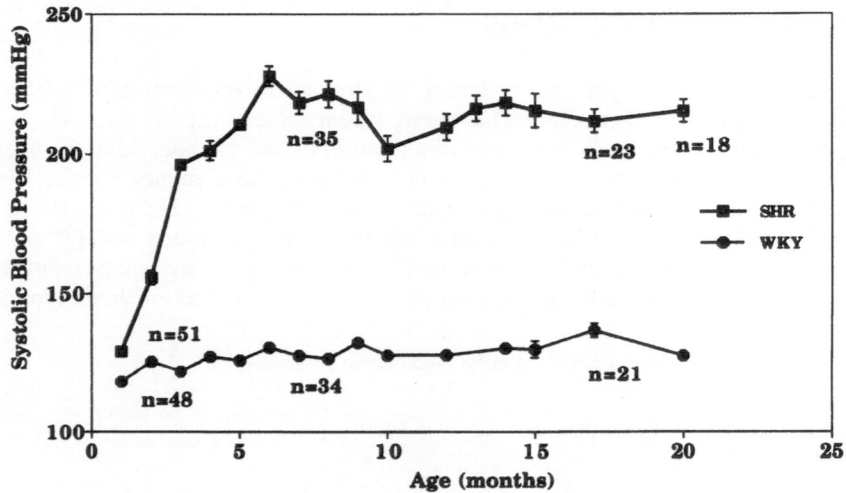

Figure 1. Age-dependent course of systolic blood pressure in SHR and WKY.

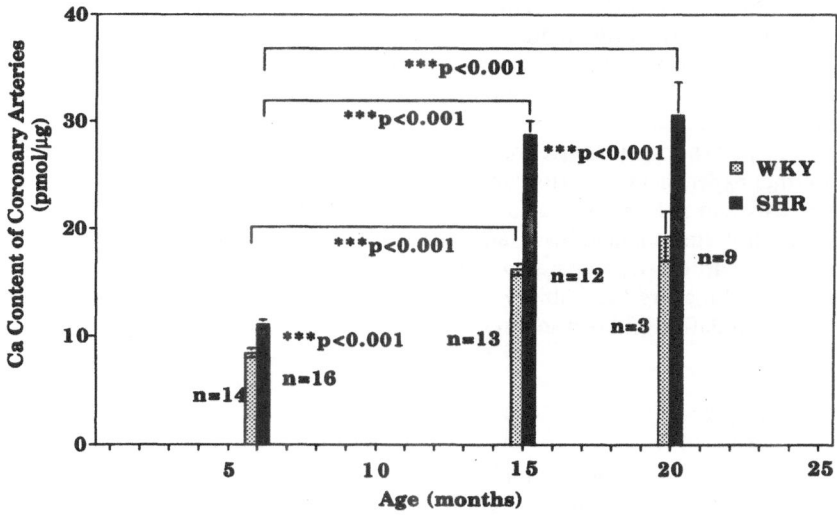

Figure 2. Age-dependent Ca contents (pmol/μg) of coronary arteries of SHR and WKY.

Potentiation of the Coronary Ca Incorporation by Genetic Hypertension

In SHR too, an age-dependent increase in the coronary Ca content (pmol/μg dry weight) was observed: 11.1 ± 0.4 (6th month; n = 16); 28.75 ± 1.3 (15th month; n = 12) p< 0.001; 30.6 ± 3.1 (20th month; n = 9) (Fig.2).

DISCUSSION

According to the cholesterol hypothesis of atherogenesis the Ca accumulation in arteriosclerotic walls is a final "dystrophic" event without any pathogenetic impact (6). However, our group has demonstrated that the development of conventional arteriosclerotic plaques of human coronary arteries (age 40-90 years) is governed by a progressive mural Ca uptake from roughly 55 nmol/mg in healthy segments up to about 4,740 nmol/mg in fully developed stenosing plaques that had led to severe myocardial oxygen deficiency and infarction; in contrast, only a 1.6-fold increase of total cholesterol was found in such plaques compared to normal coronary walls (4).

There does not exist any perfectly fitting experimental model of conventional human arteriosclerosis (7-9). But the present data indicate that the risk factor hypertension in SHR is associated with an age-dependent progressive Ca uptake into coronary arteries exceeding the corresponding Ca contents of WKY by a factor of maximal 1.9 within 20 months. At least 2 components of the arterial wall may be involved in this Ca uptake (3): (i) elastic fibres equipped with specific Ca-binding sites, and (ii) vascular smooth muscle cells (VSMC) known to possess potential- and receptor-regulated Ca channels as well as Ca/Na antiporters of the sarcolemmal membrane, IP_3-sensitive intracellular Ca pools and various ATP-fueled membrane Ca pumps (10;11). In SHR the Ca homeostasis of VSMC is defectively controlled following genetic dysregulation of Ca transport proteins (12). Consequently, Ca is accumulated in the cells' interior with functional and morphological consequences (5;11): Ca-dependent vasoconstriction is increased. Moreover, various Ca-mediated functions of VSMC in atherogenesis are intensified in SHR, i.e. migration, proliferation, and matrix production. Finally necrotization of Ca-overloaded VSMC occurs (13). In the present study, the increased Ca incorporation into coronary arteries of SHR was established with a highly sensitive microchemical atomic absorption technique, long before mural Ca deposits could be visualized in histological cross sections with staining techniques. It obviously represents an "early lesion" of experimental arteriosclerosis in SHR.

Further experiments have to clarify (i) the molecular mechanism underlying coronary physiosclerosis and risk factor-mediated Ca incorporation, (ii) a possible interaction with the vascular cholesterol metabolism, and (iii) vasoprotective potencies of various types of antihypertensive drugs. In comparative animal studies with Ca antagonists, ACE-inhibitors, diuretics, and β-blockers their ability to prevent Ca-mediated arteriosclerotic lesions, involved in dysregulation and restriction of myocardial O_2 supply, should be examined.

CONCLUSION

1. In normotensive WKY the Ca content in coronary artery walls significantly increases between the 6th and the 20th month of life.

2. This age-dependent arterial Ca incorporation is potentiated by genetic hypertension in SHR.

3. WKY and SHR represent experimental models to study the molecular mechanisms underlying coronary Ca uptake in physiosclerosis and its potentiation by the risk factor hypertension, that may initiate the developement of arteriosclerotic lesions, involved in dysregulation and restriction of myocardial O_2 supply.

ACKNOWLEDGMENTS

These experimental studies were financed with a grant (no. 0706320/5) from the German Federal Ministry for Research and Technology .

The authors would like to express their gratitute for the secretariat work of Mrs. S. Jahn and for the continuous technical help of Mrs. L. Dumont.

REFERENCES

1. Fleckenstein A.,1983, Calcium Antagonism in Heart and Smooth Muscle - Experimental Facts and Therapeutic Prospects. Monograph, John Wiley Publishing Company, New York.
2. Karnbaum S.,1961, Elastizität und Morphologie des Aortenwindkessels beim Bluthochdruck. Arch. Kreisl.-Forschung; 27: 721-745.
3. Fleckenstein A, Frey M, Zorn J, Fleckenstein-Grün G.,1990, Calcium, a neglected key factor in hypertension and arteriosclerosis. Experimental vasoprotection with calcium antagonists or ACE inhibitors. In: Hypertension: Pathophysiology, Diagnosis, and Management, J.H.Laragh a. B.M.Brenner eds. Raven Press, New York; 471-509.
4. Fleckenstein A, Frey M, Thimm F, Fleckenstein-Grün G., 1990, Excessive mural calcium overload - A predominant causal factor in the development of stenosing coronary plaques in humans. Cardiovasc. Drugs and Therapy; 4: 1005-1014.
5. Fleckenstein-Grün G.,1994, Cellular Ca disarray: Positive influence of nitrendipine on vascular Ca disorders in animals. Cardiovasc Risk Factors; 4 (Suppl. 1): 11-16.
6. Brown MS, Goldstein JL., 1984, How LDL receptors influence cholesterol and atherosclerosis. Scientific American; 251 (suppl 5): 58-66.
7. Ross R., 1986, The pathogenesis of arteriosclerosis. Atheriosclerosis Reviews; 314: 488-500. .

8. Fleckenstein A, Frey M, Zorn J, Fleckenstein-Grün G.,1987, The role of calcium in the pathogenesis of experimental arteriosclerosis. Trends in Pharmacological Sciences (TIPS); 8: 496-501.

9. Fleckenstein-Grün G, Thimm F, Frey M, Czirfusz A.,1994, Role of calcium in arteriosclerosis. Experimental evaluation of antiarteriosclerotic potencies of Ca antagonists. Basic Res. Cardiol.; 89 (Suppl. 1): 145-159.

10. Fleckenstein-Grün G, Fleckenstein A.,1991, Calcium - A neglected key factor in arteriosclerosis. The pathogenetic role of arterial calcium overload and its prevention by calcium antagonists. Ann. Med.; 23: 589-599.

11. Fleckenstein-Grün G, Thimm F, Czirfusz A, Matyas S, Frey M.,1994, Experimental vasoprotection by calcium antagonists against Ca-mediated arteriosclerotic alterations. J. Cardiovasc. Pharmacol.; 24 (Suppl. 2): S75-S84.

12. Baudouin-Legros M, Meyer P.,1990, Hypertension and atherosclerosis. J. Cardiovasc. Pharmacol.; 15 (Suppl. 1): S1-S6.

13. Fleckenstein-Grün G.,1994, Intracellular calcium overload - a cytotoxic principle. Cellular protection by calcium antagonists. In: Myocardial Protection by Calcium Antagonists, ed. L.Opie, Author's Publishing House, New York: 29-45.

63

OXYGEN TENSION IN ISOTRANSPLANTED MAMMARY CARCINOMAS AND OSTEOSARCOMAS BEFORE AND AFTER IRRADIATION

L. Weissfloch,[1] T. Auberger,[1] H. J. Feldmann,[1] R. Senekowitsch,[2]
K. Tempel,[3] and M. Molls[1]

[1] Clinic and Policlinic for Radiotherapy and
    Radiological Oncology
[2] Clinic and Policlinic for Nuclearmedicine
   Klinikum Rechts der Isar
   TU München
   Ismaninger Str. 22, D-81675 München
[3] Institute for Pharmacology, Toxicology und Pharmacy
   Veterinary Faculty
   LMU Munich
   Königinstr. 16/II, D-80539 München
   Germany

INTRODUCTION

The studies of Chapman (1), Thomlinson and Gray (27,28) infered the presence and significance of hypoxic cells in malignant tumors. Numerous publications indicate, that tumor cells with low oxygen tensions are relatively resistant towards irradiation and some chemotherapy regimes (3,9,10,13,14,24). However, the prognostic significance of hypoxia and reoxygenation remain uncertain (2,3,6,9,10,11,12,19). With the Eppendorf pO_2-histograph[R] a novel technique for oxygen measurement became practicable (5,17,21,22,32,33). We started to use this device under experimental conditions. The objective of this study was to investigate pO_2-distributions in isotransplanted mammary carcinomas and osteosarcomas of various sizes before and after irradiation with different single doses.

Oxygen Transport to Tissue XVII, Edited by Ince et al.
Plenum Press, New York, 1996

METHODS

Animals and Tumors

The studies were performed on two different mouse tumors. AT17-mammary carcinoma was isotransplanted into the right abdominal wall of female C3H-mice. The AT17-tumor is known as a slow-growing adenocarcinoma (15,20) with very rare necroses up to large tumor volumes and the tendency to keratinize like human epithelial tumors. It has been induced by radiation.

The other tumor is an osteosarcoma transplanted into the right hind leg of female balb C-mice. By repeated incorporation of ^{90}Sr the osteosarcoma was induced in balb C-mice. No data exist on proliferation or oxygenation.

Anesthesia and Temperature

For tumor transplantation and pO_2-measurement the mice were narcotized by i.p.-injection of 0.1 g/kg Ketamine [Ketanest^R, Parke-Davis] and 0.016 g/kg Xylazine [Rompun^R, Bayer] (4,8). The body weight of mice varied from 28 to 32 g.

The temperature was measured by a thermosensitive, intratumoral catheter probe [ASEA fiber thermometer 1010] and maintained at 36°C, using an infrared lamp or hot-water plastic bottle.

pO_2-Measurement

To determine the tissue oxygenation an Eppendorf pO_2-histograph^R [Eppendorf-Netheler-Hinz, Hamburg, FRG] with polarographic needle electrodes of 0.3 mm diameter was used. The needle probe was placed into the tumor periphery via a trocar and was automatically moved through towards the center of the tumor. The local pO_2 was measured in steps of 0.7 mm. Depending on tumor size 64 to 600 pO_2-values in mammary carcinomas and 58 to 600 pO_2-values in osteosarcomas were evaluated.

Technical data of the Eppendorf pO_2-histograph, oxygen measurement, needle electrodes and procedures have been described elsewhere (5,6,12,17,33).

Oxygen tensions were measured in 19 unirradiated mammary carcinomas and 9 unirradiated osteosarcomas with different sizes.

Three groups of mammary carcinomas were formed, depending on tumor volume: A_m < 1.000 mm^3, B_m 1.000 - 2.000 mm^3, C_m > 2.000 mm^3.

In osteosarcomas three groups were formed: A_o < 600 mm^3, B_o 800 - 1.300 mm^3, C_o > 1.300 mm^3 [Table 1].

Nine AT17-mammary carcinomas were followed-up over three weeks and tumor pO_2 was measured at weekly intervals [Table 2].

Irradiation

Irradiation was performed using a 100 kV X-ray machine [Siemens RT 100, 1.7mm Al-filter, 8 mA]. Tumors were irradiated while the bodies of mice were shielded with 2 mm of lead. 22 mice bearing AT17-mammary carcinomas were treated with single doses of 2 or 20 Gy, respectively.

The pO_2 was measured in all mice before irradiation as well as 2, 12 and 24 hours, 7, 14 and 21 days afterwards. At every point of time 3 to 16 animals were evaluated.

Four osteosarcomas were irradiated with a single dose of 20 Gy and pO_2 was measured before, 2, 24 hours and 7 days after 20 Gy.

RESULTS

pO_2 as a Function of Tumor Size

Below 1.000 mm^3 tumor volumes (group A_m) the median pO_2 was 11 mmHg in mammary carcinomas. In group B_m the median pO_2 decreased to 7 mmHg and a further decrease of the median pO_2 down to 4 mmHg was observed in very large tumors of group C_m [Table 1]. Also in the nine mammary carcinomas, which were evaluated weekly, a distinct correlation of oxygenation pattern with tumor size was observed. As an example Figure 2 presents measurements in an AT17-mammary carcinoma, which shows a typical decrease of the median pO_2 within three weeks. As a rule the pO_2-histogram shifted to the left with increasing tumor volumes.

Osteosarcomas showed smaller increase of volumes within the same growth period. The osteosarcomas became not as large as mammary carcinomas in these series of tests. The pooled data of nine osteosarcomas revealed a better oxygenation status than AT17-mammary carcinomas of comparable volumes. The median pO_2 of osteosarcomas did not correlate with the tumor volume. In group A_o it was 12 mmHg, in B_o 11 mmHg and in C_o the pO_2 increased to 16 mmHg [Figure 1].

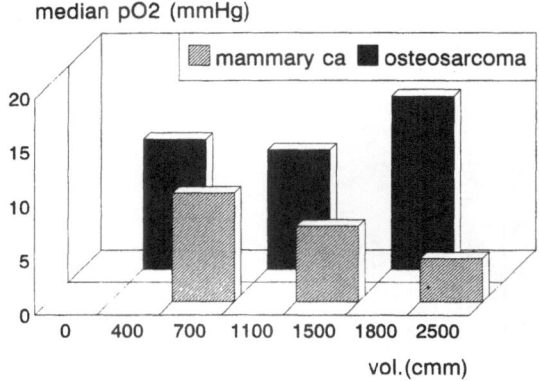

Figure 1. Median pO_2-values of nineteen AT17-mammary carcinomas and nine osteosarcomas with different volumes (pooled data).

Mammary carcinomas:	A_m	<1.000 mm^3	n = 1967
	B_m	1.000 - 2.000 mm^3	n = 2707
	C_m	> 2.000 mm^3	n =1260
Osteosarcomas:	A_o	< 600 mm^3	n = 662
	B_o	600 - 1.300 mm^3	n = 583
	C_o	> 1.300 mm^3	n = 200

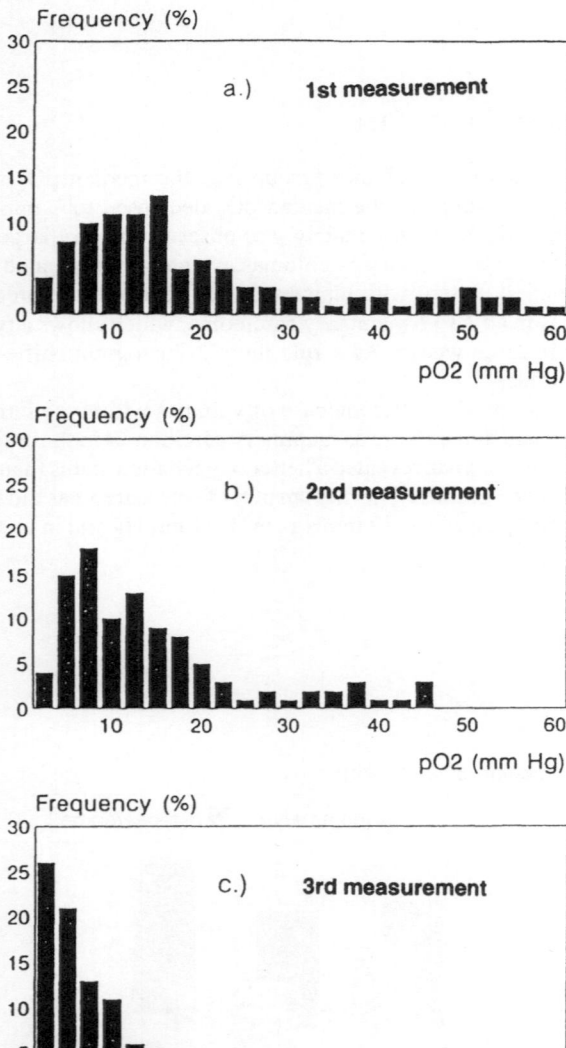

Figure 2. Weekly follow-up of pO$_2$-distribution in an AT17-tumor.

	Volume (mm^3)	med.pO$_2$ (mmHg)	n
1st measurement, day 0	476	13	248
2nd measurement, day 7	1257	10	340
3rd measurement, day 14	3317	4	680

pO₂ after Irradiation

Six AT17-tumors ranging from 100 to 400 mm³ of volumes were irradiated with 2 Gy. The median pO₂ increased from 8 mmHg before irradiation to 27 mmHg 2 hours afterwards. 24 hours after irradiation the median pO₂ decreased to 6 mmHg below the level before irradiation and to a value of 2 mmHg at day 21 [Figure 3].

After irradiation with higher single doses (20 Gy) the increase of the median pO₂ at 2 hours did not occur.

Changes in tumor oxygenation after a single dose of 20 Gy were studied in small (< 1.000 mm³) AT17-tumors under the aspect of volume depending changes in kinetics and magnitude of tumor oxygenation.

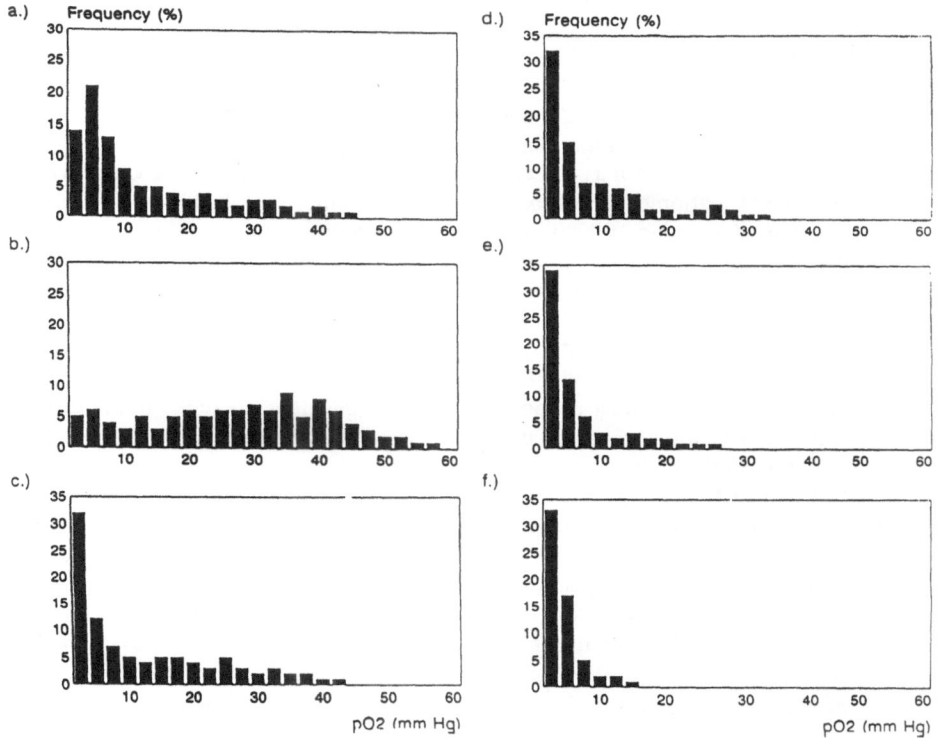

Figure 3. pO₂-histograms of six AT17-mammary carcinomas before and at different points of time after irradiation with 2 Gy.

	Volume (mm³)	med.pO₂ (mmHg)	n
a. before	100–408	8	1144
b. 2 h		27	1047
c. 24 h		6	1308
d. 7 d	262–366	3	942
e. 14 d	346–1.371	2	1354
f. 21 d	880–2.136	2	1554

In small tumors [Figure 4a] a steady decrease started immediately after irradiation. After a single dose of 20 Gy the median pO_2 dropped from 8 to 4 mmHg 2 hours later, to 2 and 1 mmHg 12 and 24 hours after irradiation.

During the further follow-up the median pO_2 continued to decline to nearly 0 mmHg on day 21.

Table 4b shows the kinetics of the median pO_2-values in large tumors. In the large tumors the median pO_2-values declined from 2 to 1 mmHg after 20 Gy 2 hours after irradiation, remained at 1 mmHg at 12 hours and increased to 3 mmHg above the starting-point 24 hours after 20 Gy. Afterwards pO_2 decreased continously to 0 mmHg on day 21.

Osteosarcomas were evaluated after irradiation with 20 Gy. The initial tumor volumes ranged from 440 to 1.400 mm³. The pooled data of the median pO_2 dropped from 20 mmHg before, to 11 and nearly 0 mmHg 2 and 24 hours after irradiation. Lateron, it increased again to 6 mmHg on day 7.

DISCUSSION

It is obvious, that poor and heterogeneous oxygenation of many solid tumors influences the response on radiation therapy or chemotherapy. But whether it has a major impact on treatment result is still a subject of controversy (2,30). Different results have been published on the correlation of tumor oxygenation with tumor size. Kallinowski et al. (18) described a distinct worsening of the tissue oxygenation at larger tumor volumes in pO_2-histograms measured in FSa II mouse fibrosarcoma and in MCa IV mouse mammary adenocarcinoma . Vaupel et al. (34) reported also of a decrease in murine fibrosarcomas as tumor sizes enlarged. However, in contrast the oxygenation of xenografted human gliomas remained uncharged up to large volumes . In several human tumors like breast and cervix cancer the oxygenation status did not correlate with tumor size (7,12,21,23,33).

In our first investigations a correlation of oxygen tension with tumor size was observed in mammary carcinomas, but not in osteosarcomas. Generally, the pO_2-values of osteosarcomas revealed a better oxygenation status than mammary carcinomas of comparable volumes. In AT17-tumors the correlation of median pO_2 with the tumor size was evident, but marked tumor-to-tumor variations of pO_2-values were also noticed in the same volume range.

Kallinowski et al. (16,18) pointed out, that tumors of the same cell line and growth stage can exhibit pronounced variation in their tissue oxygenation at different growth sites. They described remarkable difference in median pO_2 in tumors growing in the thigh and in the dorsal hind leg.

Concerning to the results of the radiation experiments, only a cautious interpretation is possible. In mammary carcinomas, which were measured 2 hours after irradiation with 2 Gy, the increase of oxygen tension occurred much earlier than reoxygenation is described in the literature (19,28,29,31).This pO_2-increase is an interesting observation with regard to clinical radiooncology. It might reflect an increase in tumor blood supply. The combined radiotherapy and chemotherapy is a frequently used concept in clinical oncology. It might be useful, that chemotherapy given as a bolus is applied within a short time period after irradiation. The increase of blood flow and oxygen supply might influence the bioavailability of chemotherapeutical drugs in a positive way. However, whether this fast increase is based on irradiation processes or influenced by anesthetic side effects must be studied further. Howes et al. (15) observed in a mammary carcinoma a reoxygenation to a minimum in the proportion of hypoxic cells, that was lower than the respective values before treatment and occurred not until 3 days after irradiation. This indicates, that the extent and rapidity of

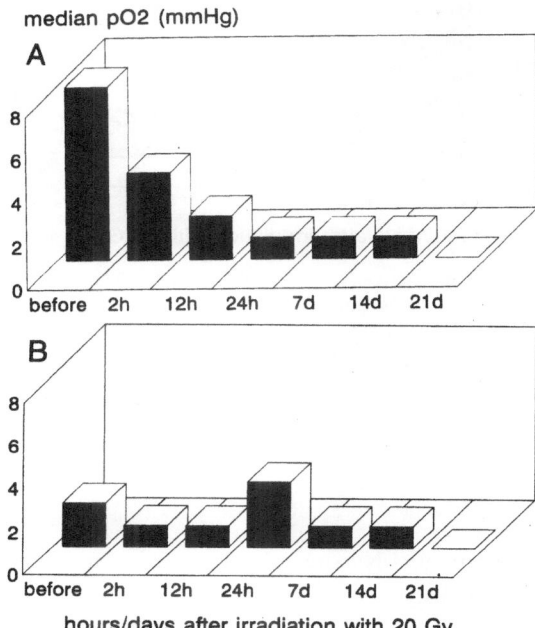

Figure 4. Changes of median pO$_2$ in AT17-mammary carcinomas irradiated with 20 Gy.

		Volume (mm^3)	med. pO$_2$ (mmHg)	n
a. small AT17-tumors (< 1.000 mm^3)				
before	(10 mice)	105–995	8	2122
2 h	(5 mice)		4	1031
12 h	(8 mice)		2	1584
24 h	(10 mice)		1	2085
7 d	(10 mice)		1	1782
14 d	(7 mice)		1	1198
21 d	(6 mice)	105–3.356	0	2010
b. large AT17-tumors (> 1.000 mm^3)				
before	(6 mice)	1.194–1.843	2	2176
2 h	(4 mice)		1	1755
12 h	(4 mice)		1	1756
24 h	(6 mice)		3	2475
7 d	(5 mice)		1	2004
14 d	(3 mice)		1	1118
21 d	(3 mice)	1.513–2.513	0	1600

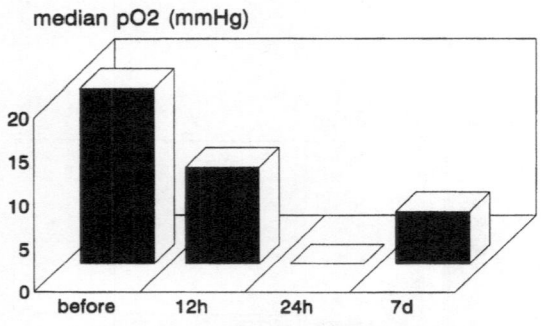

Figure 5. pO$_2$-histogram of four osteosarcomas after irradiation with 20 Gy.

	Volume (mm^3)	med. pO$_2$ (mmHg)	n
before	442-1.414	20	984
2 h		11	792
24 h		0	805
7 d	817–2.990	6	1656

reoxygenation in different tumors is extremely variable and impossible to predict as also described by Rockwell et al. (25).

ACKNOWLEDGMENTS

Financial support from Georg und Inge Weissfloch, Wilhermsdorf, Ilse Knauer, Bamberg and the Bayerisches Staatsministerium für Landesentwicklung und Umweltschutz, München is gratefully acknowledged.

REFERENCES

1. Chapman J.D. (1984): The detection and measurement of hypoxic cells in solid tumors. Cancer 54, 2441-2449

2. Coleman C.N. (1988): Hypoxia in tumors: paradigm for the approach to biochemical and physiologic heterogenity.
J.Natl.Cancer Inst. 80, 310-317

3. Clifton K.H., Briggs R.L., Stone H.B. (1965): Quantitative radiosensitivity studies of solid carcinoma in vivo: methodology and effect of anoxia. J.Natl.Cancer Inst. 36, 965-974

4. Erhardt W. (1989): Anästhesie Tierärztliche Fakultät, LMU München

5. Fleckenstein W., Weiss C., Heinrich R., Schomerus H., Kersting T. (1984): A new method for the bed-side recording of tissue pO$_2$-histograms. Verh.Dtsch.Ges.Inn.Med. 90, 439-443

6. Fleckenstein W., Jungblut J.R., Suckfüll M. (1991): Distribution of oxygen pressure in the periphery and centre of malignant head and neck tumors. In Clinical Oxygen Pressure Measurement Vol.II.
Eds: Ehrly A.M., Blackwell Wissenschaft, Berlin, pp. 81-90

7. Füller J., Feldmann H.J., Molls M., Sack H. (1994): Untersuchungen zum Sauerstoffpartialdruck im Tumorgewebe unter Radio- und Thermoradiotherapie. Strahlenther.Onkol. 170, 453-460

8. Gabrisch K., Zwart P.: Krankheiten der Heimtiere. Schlütersche Verlagsanstalt, Hannover 1987 (2)

9. Gatenby R., Kessler H,B., Rosenblum J.S., Coia L.R., Moldofsky P.J., Hartz W.H., Broder G.J. (1988): Oxygen distribution in squamous cell carcinoma metastases and its relationship to outcome of radiation therapy. Int.J.Radiat.Oncol.Biol.Phys. 14, 831-838

10. Gray L.H., Conger A.D., Ebert M., Hornsey S., Scott O.C.A. (1953): The concentration of oxygen dissolved in tissue at the time of irradiation as a factor in radiotherapy. Br.J.Radiol. 26, 638-642

11. Hall E.J. (1988): The oxygen effect and reoxygenation. In Radiobiology for the Radiologist. Eds: Hall E.J., Lippincott J.B., Philadelphia, pp. 152-159

12. Hoeckel M., Schlenger K., Knoop C., Vaupel P. (1991): Oxygenation of carcinomas of the uterine cervix: evaluation by computerized O_2-tension measurements. Cancer Res. 51, 6098-6102

13. Horsman M.R., Khalil A.A., Nordsmark M., Grau C., Overgaard J. (1993): Relationship between radiobiological hypoxia and direct estimates of tumor oxygenation in a mouse tumour model. Radiother.Oncol. 28, 69-71

14. Howard-Flander P., Wright A. (1955): Effect of oxygen on the radiosensitivity of growing bone and a possible danger in the use of oxygen during radiotherapy. Nature 75, 428-435

15. Howes S.E. (1969): An estimation of changes in the proportion and absolute numbers of hypoxic cells after irradiation of transplanted C3H mouse mammary tumors. Br.J.Radiol. 42, 441-447

16. Kallinowski F., Schlenger K.H., Runkel S., Kloes M., Stohrer M., Okunieff P., Vaupel P. (1989): Blood flow metabolism, cellular microenvironment and growth rate of human tumor xenografts. Cancer Res. 49, 3759-3764

17. Kallinowski F., Zander R., Hoeckel M., Vaupel P. (1990): Tomur tissue oxygenation as evaluated by computerized pO_2 histography. Int.J.Radiat.Oncol.Biol.Phys. 19, 953-961

18. Kallinowski F., Wilkerson, Moore R., Strauss W., Vaupel P. (1991): Vascularity, perfusion rate and local tissue oxygenation of tumors derived from ras-transformed fibroblasts. Int.J.Cancer 48, 121-127

19. Kallmann R.F., Bleehen N.M. (1968): Post-irradiation cyclic radiosensitivity changes in tumors and normal tissues. In Proceedings of the Symposium on dose rate in mammalian radiobiology. Eds: Brown D.G., Cragle R.G., Nooman J.R. Oak Ridge, pp.20.1-20.23

20. Kummermehr J. (1985): Measurement of tumour clonogenes in situ. In Cell Clones: Manual of Mammalian Cell Techniques. Eds: Potten C.S., Hendry J.H. Churchill Livingstone, Edinburgh, pp. 215-222

21. Lartigau E., Vitu L., Haie-Meder C., Cosset M.F., Delapierre M., Gebaulet A., Eschwege F., Guichard M. (1992): Direct oxygen tension measurements in carcinoma of the uterine cervix: preliminary results. In Clinical Oxygen Pressure Measurements III Ed: Ehrly A.M., Blackwell Wisschschaft, Berlin, pp. 117-120

22. Molls M., Feldmann H.J., Füller J. (1994): Oxygenation of locally advanced recurrent rectal cancer, soft tissue sarcoma and breast cancer. In Oxygen Transport to Tissue XV. Ed: Vaupel P., Plenum Publishing Co., New York, pp. 459-463

23. Okunieff P., Hoeckel M., Dunphy E.P., Schlenger K., Knoop C., Vaupel P. (1993): Oxygen tension distributions are sufficient to explain the local response of human breast tumors treated with radiation alone. Int.J.Radiat.Oncol.Biol.Phys. 26, 631-636

24. Powers W.E., Tolmach L.J. (1963): A multicomponent X-ray survival curve for mouse lymphosarcoma cells in irradiated in-vivo. Nature 197, 710-711

25. Rockwell S., Moulder J.E. (1985): Biological factors of importance in split-course radiotherapy. In Optimization of Cancer Radiotherapy. Eds: Paliwal B.R., Herbert D.E., Orton C.G. American Institute of Physics, New York, pp. 171-182

26. Schnepper U., Müller R-P., Schnepper E. (1991): Funktionelle Störungen der Mikrozirkulation in der Rattenniere nach Kobalt-60-Bestrahlung, gemessen anhand des Gewebe-pO_2. Strahlenther.Onkol. 167, 120-123

27. Thomlinson R.H., Gray L.H. (1955): The histologic structure of some human lung cancers and the possible implications for radiotherapy. Br.J.Cancer 9, 537-549

28. Thomlinson R.H. (1968): Changes of oxygenation in tumors in relation to irradiation. Front.Radiat.Ther.Oncol. 3, 109-121

29. Thomlinson R.H. (1969): Reoxygenation as a function of tumor size and histopathological type. In Proceedings of the Carmel Conference on time and dose relationships in radiation biology as applied to radiotherapy. BNL Report 50203 (C-57) Ed: Bond V.P., Upton, New York, pp. 242-245

30. Tucker S.L., Thames H.D. (1989): The effect of patient to patient variability on the accuracy of predictive assays of tumor response to radiotherapy: a theoretical evaluation. Int.J.Radiat.Oncol.Biol.Phys. 17, 145-147

31. van Putten L.M., Kallmann R.F. (1968): Oxygenation status of a transplantable tumor during fractionated radiotherapy. J.Natl.Cancer Inst. 40, 441-451

32. Vaupel P. Kallinowski F., Okunieff P. (1989): Blood flow, oxygen and nutrient supply, and metabolic microenvironment of human tumors: a review. Cancer Res. 49, 6449-6465

33. Vaupel P., Schlenger K., Knoop C., Hoeckel M. (1991): Oxygenation of human tumors: evaluation of tissue oxygen distribution in breast cancer by computerized O_2 tension measurements. Cancer Res. 52, 3316-3322
34. Vaupel P. (1993): Oxygenation in solid tumors. In Drug Resistance in Oncology. Ed. Teicher B.A., Dekker M., New York, pp. 53-85

64

OXYGEN SUPPLY DURING CARDIOPULMONARY BYPASS (CPB) IN CYANOTIC PATIENTS

Haldun Tekinalp, Semih Barlas, Emin Tireli, Hasibe Çavuşodlu, and Cemil Barlas

Department of Cardiovascular Surgery
Istanbul Faculty of Medicine
TR-34280 Çapa-Istanbul, Turkey

INTRODUCTION

Today, repair of congenital heart defects has become a routine operation with low mortality and morbidity in many subgroups of anomalies[1]. The knowledge that the ventricular myocardium continues to develop and conditions itself according to the loads present and that this mechanism can influence the long-term outcome in many cardiac anomalies, stress the operations towards infantile and neonatal age groups[2,3]. This, in turn, stresses the importance of the possible damage to the other organs and systems still developing at the time of cardiopulmonary bypass (CPB). While myocardial protection attempted by lowering the energy demands through topical cold and chemical cardioplegia[1,4,5], the protection of other systems, including central nervous system (CNS), remains dependent on the oxygenation and the local wash out of metabolites by means of CPB.

Oxygenation and oxygen consumption in CPB are the easiest and cheapest parameter to monitor continuously. Enzymatic studies not only give delayed results, thus making an up to date monitoring giving the perfusionist the possibility to react and correct the problem impossible, but are also generally too expensive to repeat frequently. We control the adequacy of perfusion by our patients on CPB by means of oxygen uptake ratio and oxygen consumption with arterial and venous continuous oxygen saturation monitoring.

In our survey of the literature we found that all the studies concerning the total body oxygenation during CPB are made on animals or in adult patient groups[1,6,7,8]. This led us to review the data obtained on pediatric patients during cardiac operations on CPB. Since it is well known that even simple deep hypothermia is able to prevent damage to neural tissue in adults for 30 min[9], we focused on cyanotic patients with CPB times longer than 90 minutes.

Oxygen Transport to Tissue XVII, Edited by Ince et al.
Plenum Press, New York, 1996

Table 1. Correlation of Consumption and standardised cons.values to temperature and time

	T5	T15	T30	T45	T60	T75	T90	T120
Consumptyon	0.517279	0.654873	0.532763	0.526494	0.59774	0.719538	0.80694	0.559199
Cons/kg	0.416703	0.509095	0.502669	0.650193	0.638952	0.581549	0.497	0.477442
Cons/bsa	0.444479	0.515034	0.468292	0.461204	0.544431	0.655598	0.491716	0.55891
log of Cons	0.502052	0.691947	0.621612	0.437441	0.772651	0.731955	0.329935	0.671352

MATERIAL AND METHOD

In this retrospective study, we reviewed the perfusion data of 54 consecutive patients operated for correction of cyanotic heart defects in our clinic in 1993-1994. The patient population consisted of 21 male and 33 female pediatric patients with a median age of 5 years (± 3.86 years); an average Body Surface Area (BSA) of 0.71m^2 (± 0.28 m^2) and an average weight of 17.78 kg (± 10.02 kg). The average cross clamp time was 75 ± 15 minutes and the average perfusion time was 110 ± 17 minutes.

William Harvey HF4700 membrane oxygenator and Biomedicus centrifugal pump primed with albumin 1 g/kg body weight + crystalloid and banked CPD blood to achieve a Htc 20% if needed were used on all patients. Bypass was begun with cooled (26°C) perfusate to allow a fast core cooling. Mean rectal temperature was kept at 28°C during the perfusion. The pH-Stat acid-base management protocol was used.

Arterial and venous blood samples for blood gas analyses were withdrawn on 5, 15, 30, 45th, 60, 75, 90 and 120 minutes of perfusion and analyzed using Gem-6-Stat Mallincrodt Inc. blood gas analyzer. Simultaneously pump flow, arterial and venous mean pressures, rectal and esophageal temperatures were recorded. Arterial and venous oxygen contents, oxygen supply, oxygen consumption and systemic vascular resistance were calculated using this data. Oxygen consumption values are standardized to BSA and body weight. Log10 of oxygen consumption is also used in assessing the correlation between consumption and temperatures according to the *Arrhenius's theory*, stating that the logarithm of a chemical reaction is inversely proportional to the reciprocal of the absolute temperature[10].

The statistical analyses were made using Excel 5.0 analysis tool pack, Microsoft Inc. and GB-STAT, Dynamic Microsystems Inc.

RESULTS

The evaluation of measured and calculated values show that there is a strong positive relation between the temperature and oxygen consumption, irrespective of sampling time (Table 1.).

The negative correlation between esophageal temperature and systemic vascular resistance was a strong at 15, 30, 45, 75 and 120 min (p < 0.01) and a weak at 60 and 90 min

Table 1. Correlation of SVR and pH to temperature and time

	T5	T15	T30	T45	T60	T75	T90	T120
SVR	0.071002	-0.56386	-0.57889	-0.67194	-0.23028	-0.5727	-0.26694	-0.6019
pH	0.052422	0.210751	0.369322	0.091758	-0.05901	0.136887	0.140813	0.168866

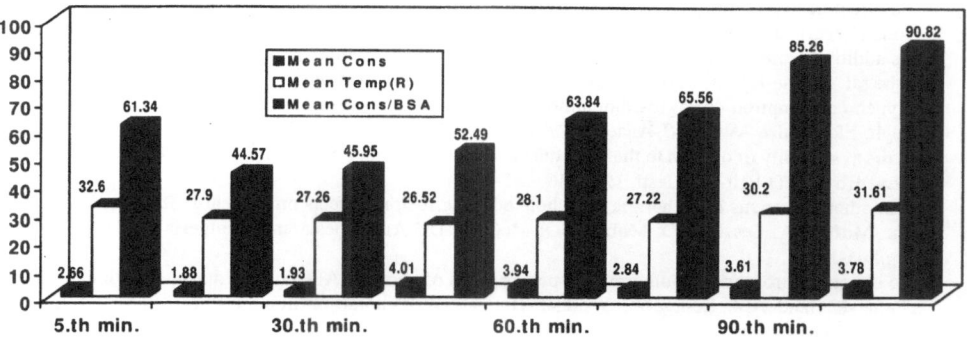

Figure 1. Relation of oxygen consumption to rectal temperatures and perfusion time.

(p < 0.05), but the "t" test found no statistical difference between variances. There was a weak negative correlation between rectal temperature and systemic vascular resistance in all data sets (Table 2.).

Oxygen consumption was noted to increase with the perfusion time, even in the period 15 to 90 minutes, where the temperature remained relatively unchanged. The Kruskal-Wallis analysis of variance, which was used due to small sample volume, revealed that there is a statistically significant rise in oxygen consumption related to the perfusion time (p < 0.001) (Figure 1).

CONCLUSION

Today, the average age of patients operated for congenital cyanotic heart diseases tends towards less than 1 year. However, the regulation of perfusion in these patients is based on studies on adults and on animal experiments. We found in this retrospective study based on our cyanotic pediatric patients a strong relation between the oxygen consumption and rectal temperature; an inverse relation between temperature and systemic vascular resistance as anticipated[7,8]. A direct relation between consumption and perfusion time was also observed, which is not explained by the rise of temperature towards the end of the perfusion. We also found that, the esophageal temperature has a greater correlation to the oxygen consumption than does the rectal temperature.

The physiological basis of direct relation of oxygen consumption to perfusion time, needs to be evaluated. Our results have encouraged us to speculate that in cyanotic patients, the oxygen supply during CPB should be increased with the perfusion time as well as temperature and that the use of esophageal temperature monitoring is more appropriate than the use of rectal temperature for monitoring of rate of metabolism.

REFERENCES

1 Cardiac Surgery Ed.by Kirklin-JW and Barrat-Boyes-BG Second Ed. Churchill Livingstone 1993

2 Simple transposition of great arteries: The arterial switch operation in neonates. Castenada-AR in Pediatric Cardiac Operations Ed. by Jacobs-ML and Norwood-WI. Butterworth-Heinemann 1992 p.1-12

3 Physiologic repair of aortic atresia-hypoplastic left heart syndrome. Norwood-WI; Lang-P; Hansen-D N-Engl-J-Med. 1983; 308: 23

4 The effect of hypothermic ischemia on recovery of left ventricular function and preload reserve in the neonatal heart. Bove-EL; Gallagher-KP; Drake-DH et al. J-Thorac-Cardiovasc-Surg. 1988; 95:814

5 The additive protective effects of hypothermia and chemical cardioplegia during ischemic cardiac arrest in the rat. Hearse-DJ; Stewart-DA; Braimbridge-MV J-Thorac-Cardiovasc-Surg. 1980; 79:39

6 Oxygen consumption during cardiopulmonary bypass with moderate hypothermia in man. Harris-EA; Seeyle-ER; Squire-AW Brit-J-Anaesth. 1971; 43 :1113-20

7 On the availability of oxygen to the body during cardiopulmonary bypass in man. Harris-EA; Seeyle-ER; Barrat-Boyes-BG Brit-J-Anaesth 1974; 46 :425-31

8 Normothermia versus hypothermia for whole body perfusion: Effects on myocardial and body metabolism. Moffit-EA; Sessler-AD; Molnar-GD; McGoon-DC Anaesthesia and Analgesia 1971; 50 (4) Jul-Aug:505-16

9 The effects of profound hypothermia on preservation of cerebral ATP content during circulatory arrest. Kramer-R; Sanders-A; Lesage-A; Woodhall-B; Sealy-W J-Thorac-Cardiovasc-Surg. 1968; 56:699

10 Oxygen consumption of animals and tissues as a function of temperatures. Fuhrman-GJ;Fuhrman-FA J-Gen-Physiol. 1959; 42:715

65

THE EFFICACY OF PULSE OXYMETRY IN THE POSTOPERATIVE RESPIRATORY MANAGEMENT OF CARDIAC PATIENTS

Haldun Tekinalp, Semih Barlas, Ufuk Alpagut, Rasim Sarioğlu, and Cemil Barlas

Department of Cardiovascular Surgery
Medical Faculty of the University of Istanbul
TR-34280 Çapa-Istanbul, Turkey

INTRODUCTION

Today, early postoperative respiratory support is routine for patients with prolonged general anesthesia, especially after cardiopulmonary bypass. The adequacy of respiration during this period is evaluated by serial blood gas samplings and the appropriate changes according to the patients' status are made on this basis.

Blood gas analyzers used in our clinic require 1-3 ml of heparinized blood for an analysis, taking approximately 5 minutes and every cartouche needs to be changed after maximally 45 analyses. This procedure takes approximately another 90 minutes. These time limitations and the high cost of every sample analyzed have led us to investigate the effectiveness and reliability of transcutaneous O_2 saturation measurings in the follow-up of postoperative respiratory support compared to the invasive blood gas samplings.

MATERIALS AND METHODS

In this clinical trial 100 consecutive patients (45 male, 55 female) operated for cardiopulmonary bypass are evaluated. The following surgical procedures are done: Coronary Artery Bypass Grafting (CABG): 35 patients; Valve replacement/Repair: 20 patients; Repair for non-cyanotic Congenital Heart Defects: 32 patients and Repair of cyanotic Congenital Heart Defects: 13 patients. The mean age was 24.4 years (1-72 years). In the Intensive Care Unit (ICU) Transcutaneous O_2 saturation (SO_2t), arterial and central venous pressures, ECG and rectal temperature are continuously monitored using a Datascope 3000 monitor. At the entry to ICU all patients were connected to a Servo 900C ventilator in Pressure Control mode using FiO_2: 1.0; Positive End Expiratory Pressure: 4 mmHg and a tidal volume of 10 ml/kg body weight. Inotrops are used if necessary. FiO_2 is tapered to 0.4 and the respiration mode is changed to Continuous Positive Airway Pressure (CPAP) through

Oxygen Transport to Tissue XVII, Edited by Ince et al.
Plenum Press, New York, 1996

509

Pressure Support Ventilation. After 2 hours of CPAP ventilation, patients with stable hemodynamics and without postoperative bleeding problems were extubated. The mean extubation time was 8.4 Hours (4-17 hours) for the CABG and Valve Replacement/Repair group, 4.3 Hours (2-12 hours) for Repair of non-cyanotic Heart Defects and 13.2 Hours (10-22 hours) for patients undergoing surgery for Repair of Cyanotic Heart Defects. In patients 6 years or older Cardiac Output and systemic and pulmonary resistance's are measured at the entry to ICU, before extubation and 3 hours later using a 8F Swan Ganz Thermodilution Catheter and Cardiac Computer.

In the first 15 patients, the effect of probe placement in transcutaneous SO_2 monitoring is evaluated by using both thumbs with inversely placed probes.

Data analyses are made with the Analysis tool pack of Excel 5.0, Microsoft Corp.

RESULTS

A set of 600 invasive/non invasive SO_2 measuring couples and 135 Cardiac Output measurings are analyzed for the differences in measuring couples and their relation to rectal temperature and systemic vascular resistance.

The statistical analysis revealed that there is a strong positive correlation between invasive and non invasive SO_2 measurements if all data sets are considered (r = 0.54).The correlation coefficient improves further to r = 0.97 (P < 0.0001 compared to the whole group) in the subgroup of measurements with rectal temperature over 37°C. Also a strong correlation between positive inotrops (Dopamine > 7 µg/min/kg or Dobutamin > 5 µg/min/kg) and the difference of invasive-non invasive SO_2 measurements is found. No statistical differences were found between age groups.

When comparing the peripheral vascular resistance obtained by cardiac output measurings with the saturation differences at the same time, the direct relation was strong (r = 0.72). There is also a strong positive relation between positive inotropic dosage and peripheral vascular resistance.

In the first 15 patients with dual transcutaneus probes, the wrong positioned probes have lead to a lesser degree of correlation to the invasive measurings than the properly placed group (r = 0.36 versus r = 0.54).

CONCLUSION

Today, continuous transcutaneous non invasive SO_2 monitoring is a routine procedure in the ICU[1,2,3,4]. The reliability of this method depends as also seen in our results, on the adequacy of the peripheral circulation of the patient as well as the proper positioning of the probe[5,6].

During the postoperative respiratory support, the adequacy of ventilation is generally monitored by serial blood gas samplings[7,8,9]. Our study has revealed that the transcutaneous SO_2 monitoring is a safe and accurate method for assessing oxygenation within some limits in a wide age and patient group[1,9,10]. Lowering the number of blood gas analyses helps to lower the cost of ICU stays and, more pronounced in infants and pediatric patients, reduces significantly the amount of blood withdrawn for analysis.

In the follow-up of a patient with transcutaneous SO_2 monitoring it must be remembered that lower rectal temperatures than 37°C and higher inotropic doses of dopamine and dobutamine 7 g/min/kg and 5 g/min/kg respectively can lead to lower readouts than actual. In our clinic we no longer draw blood gas samples in patients fulfilling that criterion unless there is major change in the clinical picture of the patient or major respiratory changes. This policy allowed us to reduce the blood gas sampling frequency from 18 analyses per day per patient to 7.5 analyses per day per patient.

REFERENCES

1. Pulse Oxymetry: an alternative method for the assessment of oxygenation in newborn infants. Jennis-MS; Peabody-JL Pediatrics. 1987 Apr; 79 (4): 524-8

2. Noninvasive monitoring of cardiorespiratory parameters. Abraham-E. Emerg-Med-Clin-North-Am. 1986 Nov; 4(4): 791-807

3. Evaluation of noninvasive measurements of oxygenation in stable infants. Mok-J; Pintar-M Benson-L; McLaughlin-FJ; Levison-H Crit-Care-Med. 1986 Nov; 14(11): 960-3

4. Pulse oximetry in the accident and emergency department. Holburn-CJ; Allen-MJ. Arch-Emerg-Med. 1989 Jun; 6(2): 137-42

5. The current status of pulse oximetry. Clinical value of continuous noninvasive oxygen saturation monitoring. Taylor-MB; Whitwam-JG. Anaesthesia. 1986 Sep; 41(9): 943-9

6. Pulse oximetry for continuous oxygen monitoring in sick newborn infants. Durand-M; Ramanathan-R. J-Pediatr. 1986 Dec; 109(6): 1052-6

7. Cardiopulmonary interactions in the cardiac patient in the intensive care unit. Marini-CE; Lodato-RF; Guiterrez-G Crit-Care-Clin. July 1989; Vol 5(3): 533-49

8. Postoperative Care in Cardiac Surgery Eds. Kirklin-JW and Barrat-Boyes-BG Vol 1 195-248 Second Edition Curchill Livingstone Inc. 1993

9. Development of a method of continuous monitoring of spontaneous respiration in the postoperative phase. 1. Normal values for cutaneous pO2- and pCO2-partial pressure together with pulse oximetry determined oxygen saturation in healthy young volunteers. Lehmann-KA; Asoklis-S; Ground-S; Huttarsch-H; Anaesthesist. 1992 Mar; 41(3): 121-9

10. Continuous pulse oximetry and the diagnosis of pulmonary embolism in critically ill trauma patients. Brathwaite-CE; O'Malley-KF; Ross-SE; Pappas-P; Alexander-J; Spence-RK. J-Trauma. 1992 Oct; 33(4): 528-30

66

EXTRACORPOREAL OXYGENATION

F. Esen

Department of Anesthesiology
Medical Faculty of the University of Istanbul
Turkey

INTRODUCTION

Adult respiratory distress syndrome still causes a substantial mortality due to the injury of the lung parenchyma. Since ARDS may complicate various disease processes, effective treatment of the syndrome requires the resolution of the original underlying disease. However, survival depends upon supportive measures aimed at providing adequate gas exchange, while avoiding further damage to the lungs and other organ systems. Mechanical ventilation in its various forms is presently the most common treatment for ARDS. However one must consider the possible "iatrogenic cost" of the respiratory support, hypothesizing that the support, when applied to severely injured lungs, may induce further damage both to the lung structure and possibly to the other organs. Extracorporeal support has emerged as an alternative to more conventional ventilatory managements.

The artificial lung was first introduced in cardiac surgery in the 1950's as a "bubble oxygenator" in which blood oxygenation was performed through direct contact between oxygen and blood. However this method leads to hemolysis in a few hours, preventing the long-term use of these devices. In 1956 a new artificial lung was introduced, in which an artificial membrane, permeable to oxygen and CO_2, separates the blood and gas phases. The artificial membrane lung due to increased bio-compatibility, had potential as a long-term substitute for alveolar capillary function. In 1972, Hill et al (1) reported the first successful long-term treatment using an artificial membrane lung in a patient with severe posttraumatic acute respiratory failure.

Subsequently, several centers, both in the United States and in Europe, applied this new technique as supportive therapy in acute respiratory failure, aiming to secure time for lung healing. In 1974, the National Heart Lung and Blood Institute founded a prospective, multicenter randomized trial to compare ECMO and continuous positive-pressure ventilation with continuous positive ventilation alone in the adult respiratory distress syndrome. The results of the study showed 90% mortality rates in both techniques, and these results dampened enthusiasm for extracorporeal support in adults (2,3). However, it has been thought that the potential of extracorporeal support was not fully exploited during the ECMO trial for several reasons:

Oxygen Transport to Tissue XVII, Edited by Ince et al.
Plenum Press, New York, 1996

a. the continuous positive pressure ventilation in the ECMO-treated patients was essentially the same as in the control group

b. the bypass was venous-arterial, with possible lung hypoperfusion and

c. the complication rate (mainly bleeding) was unacceptably high.

The following studies showed that ECMO was very successful in term and near term infants suffering respiratory failure.

In 1977, Kolobow and colleagues (4) developed a new kind of spiral coil membrane lung especially designed for CO_2 removal. The original goal was to use the CO_2 membrane lung for CO_2 dialysis in chronic lung diseases. So with Gattinoni (5), they introduced the concept of extracorporeal CO_2 removal as opposed to extracorporeal oxygenation. Since $PaCO_2$ could be removed by the extracorporeal lung, ventilation could be set at any desired combination of tidal volume and frequency, down to complete apnea. The rationale for extracorporeal support lies on the assumptions that:

a. mechanical ventilation is disadvantageous in the ARDS lung because of the anomalies in structure-function relationship

b. lung rest provides a better environment for healing.

ARDS, since its original description was considered a syndrome diffusely affecting the lung parenchyma. Furthermore, there is increasing evidence indicating that ARDS is an uncontrolled diffuse inflammation of the lung parenchyma. However, in their study Gattinoni et al (6) showed by using computed tomography that lung lesions are primarily located in the dependent lung suggesting a nonhomogeneous distribution of the disease. It has also been shown that the dimensions of the normally inflated lung, which are directly estimated by the lung compliance measurement, may be as small as 20-30% of the dimension of a normal adult lung, and may assume the size of a "baby lung".

The structural modifications of the ARDS lung are strictly associated with changes in function (7). Compliance estimates the dimensions of the baby lung, while compliance estimates the open lung, the gas exchange impairment is associated with the non-inflated lung. The greater the non-inflated compartment, the greater the oxygenation deficit and the higher the shunt fraction. However, the CO_2 clearance occurs only in the baby lung. Consequently, in order to ventilate patients with severe ARDS, high pressures and large tidal volumes, relative to baby lung size, are required, which is in turn associated with increased frequency of "barotrauma" and morphological changes in the lung tissues.

It has not been proved directly that lung rest provides a better environment for healing. However, in the last few years, there has been a wide spread tendency toward the lung rest strategy, target the mechanical ventilation to increased $PaCO_2$ values like permissive hypercapnia (8).

TECHNIQUE

The term "extracorporeal support" includes different techniques.

ECMO refers to high-flow venoarterial bypass as used in the multicenter ECMO study. The main objective of ECMO is to provide an immediate improvement in the oxygenation of the arterial blood. The choice was thus high flow venoarterial bypass. The term "extracorporeal CO_2 removal" refers mainly to venovenous bypass at relatively low flow (20-30% of the cardiac output), and this technique is intended to provide lung rest by reducing the ventilation of diseased lung. During ECMO, some CO_2 is cleaned by the extracorporeal circuit, and during extracorporeal CO_2 removal, some oxygen is provided by the artificial lung. The amounts are roughly in proportion to the ratio of extracorporeal blood flow to the cardiac output.

The term "partial extracorporeal CO_2 removal" indicates a very low-flow veno-venous bypass with partial removal of the CO_2 produced.

Whatever terminology is used, a correct description of the extracorporeal support system should include the following:

a. type of bypass (venoarterial or venovenous)

b. the ratio of the extracorporeal blood flow to the cardiac output.

c. the ventilatory management of the natural lungs (low frequency, positive pressure ventilation, high-frequency jet ventilation, continuous positive pressure ventilation etc.)

Here I would like to focus on the technique of extracorporeal CO_2 removal, which we are more familiar with in the clinical setting of Adult RDS.

For oxygenation purposes, an adequate pulmonary blood flow, and inflated lung are necessary. No ventilation is required, as the oxygenation may be provided by supplying oxygen equal to consumption.

For CO_2 removal, the blood flow requirement is low, due to high mixed venous CO_2 content, but high ventilation mandatory. During routine extracorporeal CO_2 removal, the two respiratory functions are dissociated.

70-80% oxygenation-natural lungs. The amount of Oxygen provided by the membrane lung depends on the extracorporeal blood flow, the hemoglobin concentration, and the FiO_2 of the gas mixture ventilating the artificial lung. CO_2 clearance is less dependent on extracorporeal blood flow. Here, the key factor for extracorporeal CO_2 removal is the surface area of the membrane lung and its adequate ventilation. A 9 m^2 surface is usually required to provide adequate extracorporeal CO_2 removal in adults.

Vascular Access

Until 1989, access to the bypass veins was performed at the bedside by a vascular surgeon. Percutaneous cannulation of the femoral or jugular vein with a single double lumen catheter (34Fr) was being used. Thereafter, with this method, bleeding from the cannulation sites has been reported to reduce virtually to zero.

Extracorporeal Apparatus

Extracorporeal apparatus consists of a blood section (artificial lung, blood pump, blood circuit). Gas section (flowmeter, gas line, humidifiers). Monitoring apparatus (oxygen saturation of the input blood, differential procedures across the membrane lung, temperature).

Two kinds of artificial lungs are available in Europe for long-term bypass. The Kolobow membrane lung (Sci-med) and the microphourus heparinized membrane lung (Medtronic)(9). Kolobow membrane lung were reported to be safer and more suitable for long-term use.

Until 1988-89, roller pumps were continued to be used as a device for blood pumping. With a heparinized pump head, centrifugal pump allowed for simplification of the circuit, as the blood reservoir is not required.

CLINICAL DATA

For the extracorporeal CO_2 removal application, an update entry criteria is reported as follows (10):

a. severe hypoxemia:

- fast entry: $PaO_2 < 50$ mmHg with $FiO_2 = 1.0$, PEEP $> = 5$ cmH_2O for 2 hrs.
- slow entry: $PaO_2 < 50$ mmHg with $FiO_2 > = 0.6$, PEEP $> = 5$ cmH_2O for > 12 hrs

b. low respiratory compliance:$< 30ml/cmH_2O$.

c. lack of positive response to PEEP response, tested from 5 to 15 cm H_2O (no recruitment phenomena)

After connecting the patient to bypass, extracorporeal blood flow is started with close attention to body temperature and hemodynamic variables. After 20-30 min, it is usually possible to set the blood flow at the maintenance rate of 30% CO, which will be adequate for CO_2 removal.

Ventilation is then decreased to 2-4 breaths/min, while PEEP is progressively increased to maintain mean airway pressure at the same level as during previous continuous positive pressure ventilation period, at limited peak pressure up to 35-45 cmH_2O). Pure oxygen is then delivered into the carina through a small catheter to provide for the oxygen consumed during the long expiratory pause (apneic oxygenation).

When the target PaO_2 of 80-100 mmHg is maintained, weaning from pressure starts by decreasing PEEP value. The entire procedure may take several days and disconnection from bypass is accomplished when the patient is able to tolerate PSV or CPAP for 6 to 12 hrs.

Bleeding is the main complication during long-term bypass, since systemic antico-agulation needs to be performed during the procedure. (15-30 u/kg/hr). A complete coagu-lation assessment must be performed at least twice day. However since 1992, heparin coated circuits with heparinized artificial lungs, provided an adequate blood flow with minimal risk of major bleeding.

CLINICAL RESULTS

Before discussing outcome in adult patients treated with extracorporeal support, the high rate of success obtained in newborns is worth mentioning. Data updated in October 1991 from the extracorporeal life support organization shows an overall survival rate of 83% in > 5000 neonatal cases (11). This survival rate decreased to 47% in 285 pediatric patients.

The experience in Europe in more than 300 adult patients, shows an overall survival rate of 47%.

Milan-Monza group treated 94 patients of whom 89 patients had ARDS. In these patients, the most frequent etiology was pneumonia with 63% (11).

Our experience with the $ECCO_2R$ technique started on an experimental basis in ARDS induced pigs (12).

We evaluated and compared respiratory and hemodynamic effects of conventional volume controlled PEEP ventilation, pressure regulated volume controlled ventilation with an I/E ratio of 4:1 and LFPPV-$ECCO_2R$ in pigs with acute respiratory failure due to surfactant depletion. All three modes, randomly applied to achieve

$PaO_2 > 350$mmHg and $PaCO_2$ about 40mmHg, were as follows.

Table 1. Respiratory parameters during the administration of ventilation modes (n = 15)

	CM1	CM2	M1	M2	M3
pPAW	16.8 ± 4.4	32.3 ± 8.7	45.4 ± 8.5	31.5 ± 7.7	0.7 ± 6.0
mPAW	6.1 ± 2.2	8.8 ± 4.1	23.9 ± 3.4	25.7 ± 6.7	20.5 ± 3.3
PEEP	4	4	14.1 ± 3.9	15.0 ± 3.3	16.9 ± 2.9
AP	13.1 ± 3.0	27.8 ± 3.4	31.2 ± 3.3	15.7 ± 3.4	15.4 ± 3.2
TSLC	47.3 ± 5.8	25.7 ± 4.4	30.9 ± 3.8	32.2 ± 3.0	34.8 ± 3.7
MV	5.0 ± 1.1	5.0 ± 0.9	4.8 ± 0.9	4.9 ± 1.0	1.0 ± 0.4

pPAW:Peak airway pressure, mPAW:Mean airway pressure, PEEP:Positive end expiratory pressure,
AP:Pressure amplitude

M_1: VCV with measured best-PEEP, V_T8-12 ml/kg fr 12/min and I/E 1:2.

M_2: PRVCV with PEEPs 4 cmH_2O, F 12 b/min, pP_{aw} adjusted to get V_T 8-12 ml/kg and I:E ratio 4:1. Around 45 cmH_2O pP_{aw} was applied during the first 5 min to open the alveoli.

M_3: LFPPV with measured best-PEEP, tidal volume (V_T) 5 ml/kg, fr 5/b/min I:E ratio 1:2 1-2 L/min O_2 given through a catheter (ID/mm) inserted through the tracheal tube and advanced to the level of carina; $ECCO_2R$ with a pump speed of 20-30% of the CO measured during CM_2 and membrane lungs with a surface area of 7 m^2. V_T was adjusted to achieve pP_{aw} of 45 cmH_2O during the first 5 min to open the alveoli.

Adequate oxygenation was provided by VCV with best PEEP, however, the safety of this method is not convening due to the resulting high pP_{aw}. On the other hand PRVCV with I:E ratio 4:1 and LFPPV-$ECCO_2R$ have comparable results with respect to gas exchange, airway pressures and hemodynamic parameters with low pP_{aw} (Table 1).

In another experimental study (13) we aimed to show the changes in right ventricular performance with these three modes of ventilation in ARDS induced pigs. Again the results supported the data of our previous work, since better gas exchange was observed with no cardiocirculatory depression and no impairment of right ventricle function (Table 2, Figures 1-2).

Table 2. Hemodynamic parameters during the administration of ventilation modes

	CM1	CM2	M1	M2	M3
CO	5.5 ± 1.6	5.0 ± 1.6	4.7 ± 1.9	4.7 ± 1.9	5.0 ± 1.4
RVEF	53.4 ± 5.7	33.6 ± 11.1	35.2 ± 9.2	37.3 ± 8.2	39.9 ± 6.8
HR	164 ± 23	156 ± 34	160 ± 15	154 ± 20	150 ± 22
SV	38.0 ± 8.7	39.2 ± 11.4	38.0 ± 13	38.3 ± 13.7	39.9 ± 17.3
RVESV	36.1 ± 12	60.6 ± 17.6	59.0 ± 18.8	56.1 ± 18.8	63.2 ± 23.2
RVEDV	73.3 ± 21	127 ± 47	123 ± 60	117.5 ± 45	103 ± 33
RVSWI	7.8 ± 2.0	12.5 ± 4.3	11.8 ± 3.8	11.4 ± 3.8	10.4 ± 2.7
LVSW	166.4 ± 22	60.8 ± 17.8	58.8 ± 23	56.0 ± 18	54.6 ± 18
MPAP	25.8 ± 4.4	35.1 ± 4.1	32.2 ± 5.1	35.4 ± 4.6	34.8 ± 4.2
PCWP	12.7 ± 5.9	14.9 ± 5.2	17.9 ± 4.6	16.6 ± 4.2	16.8 ± 4.4
CVP	9.8 ± 2.9	10.1 ± 2.3	11.0 ± 2.2	11.7 ± 2.1	12.5 ± 2.7
MAP	130 ± 17	133 ± 22	125 ± 19	120 ± 22	121 ± 15
PVR	208 ± 165	374 ± 220	406 ± 245	400 ± 245	394 ± 246
SVR	1757 ± 488	1711 ± 551	1623 ± 488	1600 ± 499	1590 ± 506

CO:Cardiac output, RVEF:Right ventricle ejection fraction, HR:Heart rate, SV:Stroke volume,
RVESV:Right ventricle end systolic volume, RVEDV:Right ventricle end diastolic volume, RVSWI:Right
ventricle stroke work index, LVSW:Left ventricle stroke work, MPAP:Mean pulmonary artery pressure,
PCWP:Pulmonary capillary wedge pressure, CVP:Central venous pressure, MAP:Mean arterial pressure,
PVR:Pulmonary vascular resistance, SVR:Systemic vascular resistance.

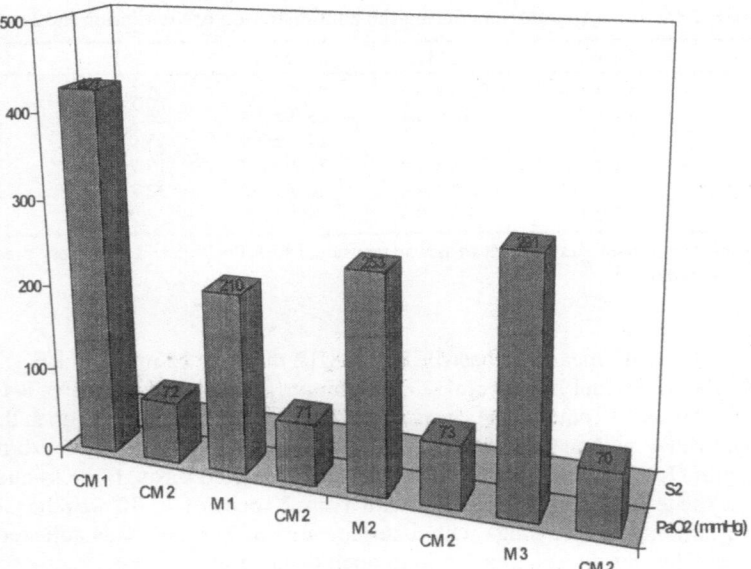

Figure 1. PaO$_2$ during each mode trial.

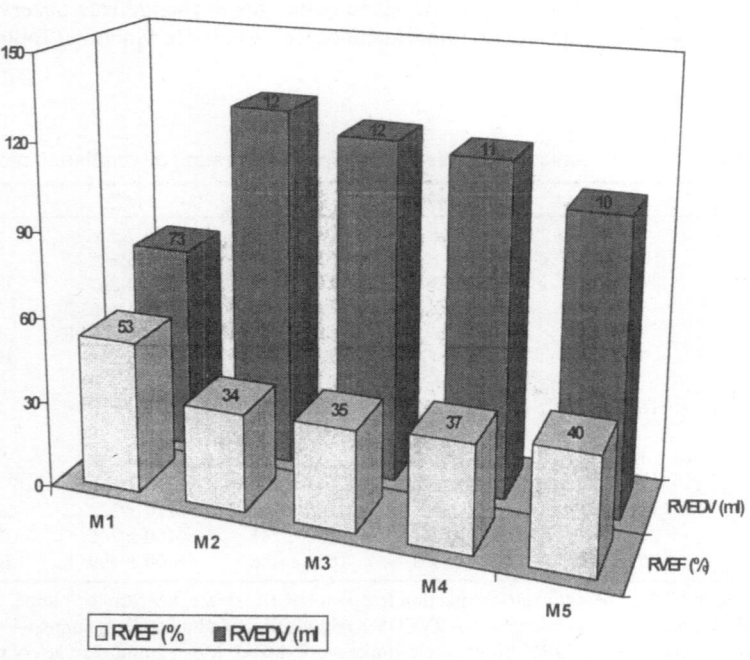

Figure 2. Right ventricle ejection fraction (RVEF), end diastolic volume (EDV) during each mode trial.

However, we all know that LFPPV-ECCO$_2$R is more than just a ventilation mode. It is invasive, expensive, needs additional equipment, and involves special training and attention by the medical staff.

After 20 years of experience with this technique in adults, its actual place in the treatment of severe ARDS is still being discussed. The damages induced by mechanical ventilation in severe ARDS are well documented and justify the research for alternative therapeutic strategies. Permissive hypercapnia is a treatment along this line and may decrease the need for extracorporeal support. In adults, extracorporeal technique is still being considered as an experimental technique to be used in special centers with sufficient motivation, technical skills and manpower. Only when technical improvement assure safe conduction of long-term bypass, this technique should be tested as an alternative to more conventional treatment of ARDS.

REFERENCES

1. Hill JB, O Brian TG, Murray IT: Prolonged extracorporeal oxygenation for acute post-traumatic respiratory failure (shock lung syndrome). N England J Med 1972; 286: 629-634.
2. Bathesda MD, Protocol for extracorporeal support for respiratory insufficiency collaborative program. National Heart, Lung and Blood Institute Division of Lung Disease, May 1974.
3. Zapol W, Snider MT, Hill JD, et al: Extracorporeal membrane oxygenation in severe acute respiratory failure. A randomized prospective study JAMA 1979; 242: 2193-96.
4. Kolobow T, Gattinoni L, Tomlinson T, et al: The carbondioxide membrane lung (CDMC) A new concept. Trans. Am. Soc. Artif. Intern. Organs 1977; 23: 17-21.
5. Kolobow T, Gattinoni L, Tomlinson T, et al: Control of breathing using an extracorporeal membrane lung. Anesthesiology 1977; 46: 138-141.
6. Gattinoni L, Mascheroni D, Torresin A, et al: Morphology response to positive end- expiratory pressure in acute respiratory failure. Computerized tomography study. Intensive Care Med 1986; 12: 137-142.
7. Gattinoni L, Pesenti A, Baglioni S, et al: Inflammatory pulmonary edema and positive end-expiratory pressure. Correlations between imaging and physiologic studies. J Thorac Imaging 1988; 3: 59-64.
8. Hickling KG, Henderson SJ, Jackson R: Low mortality associated with low volume-pressure limited ventilation with permissive hypercapnia in severe adult respiratory distress syndrome. Intensive Care Med 1990; 16: 372-77.
9. Bindslev L, Eklund J, Worlander W, et al: Treatment of acute respiratory failure by extracorporeal CO$_2$ elimination performed with a surface heparinized artificial lung. Anesthesiology 1987; 67: 117-120.
10. Gattinoni L, Pesenti A, Mascheroni D, et al: Low frequency positive-pressure ventilation with extracorporeal CO$_2$ removal in severe acute respiratory failure. JAMA 1986; 256: 881-886.
11. Gattinoni L, Pesenti A, Bambino M, Pelosi P, Brazzi L: Role of extracorporeal circulation in Adult Respiratory Distress Syndrome Management. New Horizons 1990. Vol 1; 4: 603-612.
12. Kesecioğlu J, Telci L, Denkel T, et al: Comparison of different modes of Artificial Ventilation with extracorporeal CO$_2$ elimination on gas exchange in an animal model of acute respiratory failure. Adv Exp Med Biol Vol 317: 893-899.
13. Telci L, Kesecioğlu J, Esen F, et al: Effects of CPPV, PC-IRV, and LFPPV-ECCO$_2$R on right ventricular functions in pigs with ARDS. Adv Exp Med Biol Vol 317: 599-604.

67

COMPARISON OF GASTRIC INTRAMUCOSAL pH MEASUREMENTS WITH OXYGEN SUPPLY, OXYGEN CONSUMPTION AND ARTERIAL LACTATE IN PATIENTS WITH SEVERE SEPSIS

Figen Esen,[1] Lütfi Telci,[1] Nahit Çakar,[1] Ahmet Tütüncü,[1]
Jozef Kesecioğlu,[2] and Kutay Akpir[1]

[1] Department of Anesthesiology and Intensive Care
University of Istanbul, Medical Faculty
Istanbul, Turkey
[2] Department of Intensive Care
Academic Medical Center- University of Amsterdam
Amsterdam, The Netherlands

INTRODUCTION

Tissue hypoxia caused by the imbalance between the oxygen demand and the oxygen uptake, is considered to be the most important factor to the mortality and morbidity in patients with severe sepsis. However, the assessment of tissue oxygenation is still controversial, since direct measurement of the adequacy of tissue oxygenation has not yet been available in the clinical setting.

Lactate levels have been considered to reflect anaerobic metabolism of the tissue in septic patients (1). Several studies have shown that blood lactate levels are closely related to outcome from septic shock (2-5). In their clinical work, Bakker et al. concluded that blood lactate levels have a strong relation to survival from septic shock and can thus be a reliable clinical guide to therapy (6). However the involved metabolic process of lactate can be complex since severe alteration in the clearance mechanism of lactate by the liver and other organs could contribute to the maintenance of high blood lactate levels.

Studies have indicated that a pathologic relationship between oxygen delivery and oxygen consumption can reflect the presence of global oxygen debt of the tissues (7-10). This phenomenon has been associated with increased mortality and its recognition has been thought to have important clinical consequences (11-14). Pinsky concluded that the relationship between increasing DO_2 and VO_2 in patients in whom maximal tissue O_2 extraction is already occurring would identify those patients with an oxygen debt and patients increasing

Oxygen Transport to Tissue XVII, Edited by Ince et al.
Plenum Press, New York, 1996

521

DO_2 by whatever means should increase survival (15). Nevertheless, the need for additional organ specific parameters to establish the adequacy of tissue oxygenation still exists.

Recent studies have shown that (non-invasive) gastric intramucosal pH (pHi) measurements as first focused by Fiddian–Green is a good indicator of gut ischemia (16-18). These measurements are attractive because they are minimally invasive and attempt to assess the adequacy of tissue oxygenation in an organ system that may play a pivotal role in sepsis. Showing gastrointestinal mucosal ischemia as one of the earliest manifestations of impaired tissue perfusion in the critically ill, investigators suggested pHi measurements to be a sensitive indicator of gut ischemia (19-21). In this study we aimed to evaluate tonometric measurements of gastric intramucosal pH with oxygen supply dependency states and lactic acid values as systemic indices of tissue oxygenation in patients with severe sepsis.

METHODS

18 patients meeting criteria for severe sepsis as defined by Bone et al. (22), were included in the study.

1. temperature > 38 C or < 36 C
2. heart rate > 90/min
3. respiratory rate > 20 breaths/min or $PaCO_2$ < 32 mmHg
4. white blood cell count > 12.000/mm, < 4.000/mm or > 10% immature band forms

(Systemic response to infection manifested by two or more of the conditions above which is associated with organ dysfunction, hypoperfusion or hypotension.)

In all patients hypovolemia, anemia (hemoglobin < 9 g/dl), and hypoxemia (PaO_2 < 60 mmHg) had been previously corrected. Patients with hemodynamic instability under vasopressors or inotropes, coagulation disorders, massive gastrointestinal bleeding, and esophageal varices were not included. A pulmonary artery catheter was placed percutaneously to evaluate hemodynamic variables. All patients were artificially ventilated with a Servo 900C (Siemens–Elema) on a pressure controlled mode set to maintain normal blood gas values. FiO_2 was below 0.6 in all patients in order to limit error in measuring direct VO_2.

A gastric tonometer (TRIP tonometric, Worchester, MA) was introduced in place of a standard nasogastric tube. Correct positioning was determined roentgenographically. Measures of pHi were obtained as recommended by Fiddian-Green et al (18). Normal saline solution of 2.5ml was used to fill the balloon of the tonometer. 60 min. was allowed for the equilibration of CO_2 between gastric lumen and the saline.

Fluid withdrawn anaerobically was sent for determination of PCO_2. Sample of arterial blood was drawn for the measurement of arterial HCO_3 concentration. Gastric pHi was calculated using the Handerson–Hasselbach equation. Each patient had received Histamine receptor blocker (ranitidine) in order to minimize the contribution of intraluminal CO_2 production to gastric intramucosal acidosis.

After baseline measurements, dobutamine infusion was started at 5 µgr/kg/min and increased to 10 and 15 µgr/kg/min at 60 minutes intervals. Measurements were repeated at the end of 60 min infusion of each dose of dobutamine. No changes in the therapy, including the rate of infusions, level of inspired oxygen concentration and ventilation mode were allowed during the transitional period.

Intravascular pressures were measured in the supine position with a Viggo spectramed transducer on a Horizon 2000 (Mennen–Medical) monitor. Mean arterial pressure (MAP) was obtained from an arterial catheter inserted to the radial artery. Cardiac output (CO) was measured by thermodilution technique (Hemopro-1 Spectramed), and the cold bolus of injection was started at the end of inspiratory phase. Five measurements were averaged to obtain each cardiac output value. Arterial and mixed venous blood gas determinations as well as the gastric PCO_2 measurements were done by ABL 300 (Radiometer-Copenhagen). To avoid erroneous measurements of PCO_2 from saline , blood gas analyzer was standardized for measuring PCO_2 in saline. Hemoglobin concentration (Hb), arterial and mixed venous saturation and arterial O_2 content were measured with Hemoximeter (OSM3, Radiometer). Arterial lactate concentration was determined using an enzymatic method. Oxygen delivery (DO_2) was calculated by the simplified formula; $DO_2 = CO \times Hb \times SaO_2 \times 1.39$. Oxygen consumption (VO_2 ml/min) was determined by direct analysis of expired gas (mVO_2) (Datex-multicap Helsinki), and by indirect calculation of the product of cardiac index and arteriovenous oxygen content difference (cVO_2).

Data Analysis

All data were expressed as mean ± SEM. Wilcoxon pairs test was used to compare the differences between variables measured during baseline and each dose of dobutamine

Table 1. Characteristics and outcome
of the patients included in the study

Patient	Age	Diagnosis	Outcome
1	35	Polytrauma,sepsis	Survived
2	67	Pneumonia,ARDS	Survived
3	65	Sepsis, ARDS	Survived
4	18	Polytrauma, sepsis, ARDS	Survived
5	45	Liver transplant, sepsis	Survived
6	47	Liver transplant, sepsis	Survived
7	22	Pneumonia, ARDS	Survived
8	43	Pneumonia	Survived
9	27	Pneumonia, sepsis	Survived
10	67	Polytrauma, sepsis	Survived
11	23	Sepsis, ARDS	Died
12	39	Polytrauma,sepsis	Died
13	18	Sepsis, ARDS	Died
14	62	Sepsis, ARDS	Died
15	21	Polytrauma, sepsis, ARDS	Died
16	27	Renal transplant, pneumonia	Died
17	66	Polytrauma, sepsis, ARDS	Died
18	35	Renal transplant, pneumonia	Died

Table 2. Hemodynamic parameters measured before and after each dose dobutamine (Db)

	Baseline	Db5	Db10	Db15
HR (beats/min)	118.520.4	123.517.5	12718.0	126.312.9
MAP (mmHg)	73.414.9	76.114.3	78.414.0	77.210.7
RAP (mmHg)	13.41.9	13.22.4	12.82.1	12.32.4
MAP (mmHg)	32.710.9	32.18.7	30.710.4	29.810.1
PCWP (mmHg)	18.85.6	18.35.4	17.45.6	17.2w.6
CI (l/min/m^2)	5.21.2	5.91.4[a]	6.51.21[a]	6.61.2
SI (ml/m^2)	43.09.6	50.212.7[a]	53.611.6	54.813.1
SVR (dyne sec/cm^5)	514240	555315	520257	515227
PVR (dyne sec/cm^5)	11056	12177	11674	11678

[a] p from the previous dose
HR-heart rate, MAP-mean arterial pressure, RAP-right atrial pressure, MPAP-mean pulmonary arterial pressure, PCWP-pulmonary arterial wedge pressure, Cl-cardiac index, Sl-stroke index, SVR-systemic vascular resistance, PVR-pulmonary vascular resistance.

infusion. Data were examined in each patient to establish the presence of oxygen supply dependency, defined as a linear relationship between DO_2 and VO_2 having a positive slope. To determine the linearity, a linear function ($y = a + bx$) were fitted to the data from each patient measured values of DO_2 (x) and VO_2 (y). Relationships between the pooled data of all the measurements of a given variable (VO_2, lactate, and pHi) and DO_2 were analyzed graphically using linear (least square) regression analysis.

A level of significance of $p < 0.05$ was used in the study.

RESULTS

Clinical characteristics and outcome of the patients are summarized in table 1. Each dose of dobutamine infusion was well tolerated in all patients. Increasing dose of dobutamine showed a consistent increase in cardiac index (CI l/min m^2) and stroke index (SI ml/m^2), but no significant change in cardiac pressures were noted (table 2). No significant differences were noted in blood gasses, whereas slight increase was seen in mixed venous O_2 saturation (SvO_2) (table 3). Mean values of both mVO_2, cVO_2 and DO_2 showed a parallel increase with each dose of dobutamine administration. Lactate levels which were elevated at baseline decreased significantly with the increase in O_2 supply. Mean values of pHi also increased significantly (table3).

In each patient DO_2 / VO_2 relationships showed oxygen supply dependency by a linear function (figure 1). When all the measured mVO_2 and cVO_2 were plotted against the corresponding level of DO_2, a linear correlation between both mVO_2-DO_2 ($y = 169 + 0.16x$, $p < 0.0001$) and cVO_2-DO_2 ($y = 50 + 0.29x$, $p < 0.0001$) was found (figures 2 and 3). The relation between mVO_2 and cVO_2 ($y = 127 + 0.6x$, $p < 0.0001$) is shown in figure 7.

The relationships of each measured DO_2 to the corresponding measures of arterial lactate and pHi are shown in figures 4 and 5 respectively. There was a clear correlation between DO_2 and lactate levels ($y = 7-0.004x$, $p < 0.0001$), but no significant linearity between DO_2 and pHi ($y = 7.2 + 0.0001x$, $p = 0.051$).

When measurements of pHi and lactate were related to survival, initially significant lower levels of pHi and higher levels of lactate were observed in nonsurvivors, however

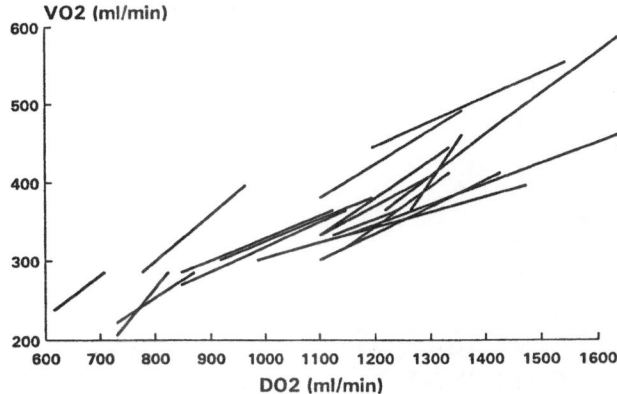

Figure 1. VO_2 - DO_2 relationship. Lines of best fit as determined by simple linear regression in 18 patients.

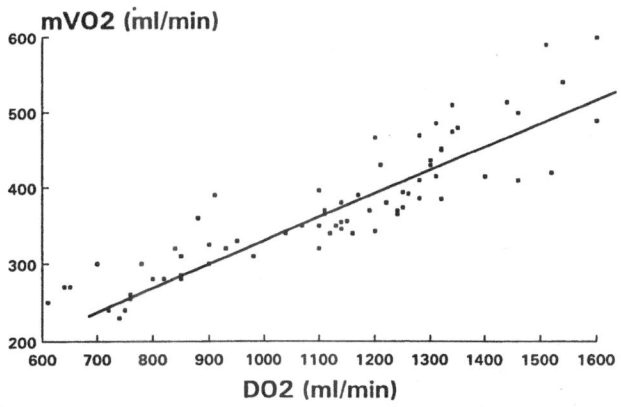

Figure 2. Correlation between DO_2 and mVO_2.

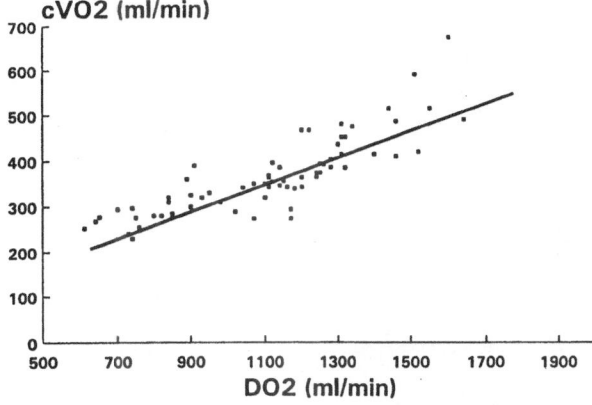

Figure 3. Correlation between DO_2 and cVO_2.

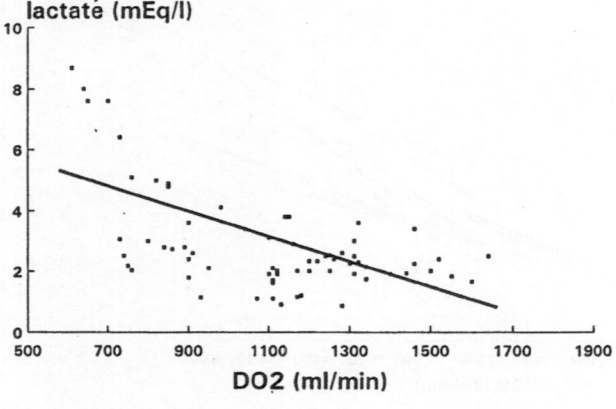

Figure 4. Correlation between DO_2 and lactate.

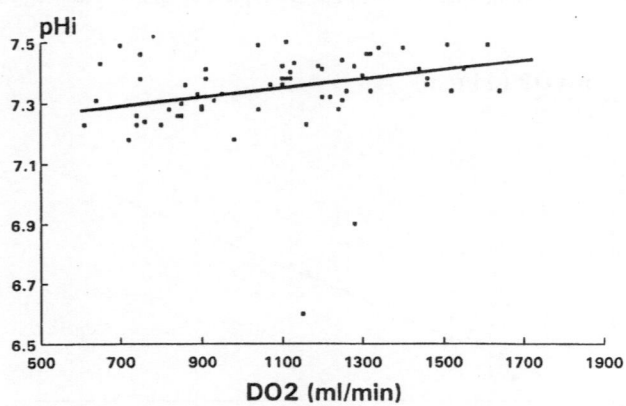

Figure 5. Correlation between DO_2 and pHi.

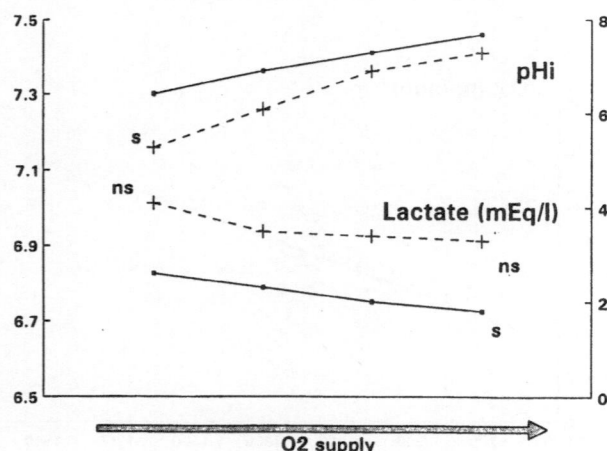

Figure 6. Arterial lactate and pHi values with the increase of oxygen supply in survivors (s) and nonsurvivors (ns).

Table 3. Respiratory parameters, pHi and lactate values measured before and after each dose dobutamine (Db)

	Baseline	Db5	Db10	Db15
pH	7.380.09	7.350.08	7.370.06	7.380.03
PaO_2 (mmHg)	11136	11630	10927	11128
$PaCO_2$ (mmHg)	40.46.8	40.58.9	40.35.6	40.18.6
PvO_2 (mmHg)	37.54.5	37.75.1	37.65.6	385.9
$PvCO_2$ (mmHg)	43.57.7	43.78.7	42.77.2	41.57.0
SvO_2 (%)	67.06.7	68.18.8	67.79.7	70.28.6
mVO_2 (ml/min)	33067	35071[b]	37674[b]	408105[b]
cVO_2 (ml/min)	31264	33874[a]	35779[a]	381110[a]
DO_2 (ml)	1001205	1070234[b]	1152254[b]	1197282[b]
lactate (mEq/l)	3.31.8	2.81.6[b]	2.51.6[b]	2.41.7[a]
pHi	7.240.17	7.310.11[b]	7.390.07[b]	7.430.06[b]

[a] p 0.05 from the previous dose
[b] p 0.01 from the previous dose
mVO_2- oxygen consumption (direct), cVO_2-oxygen consumption (calculated), DO_2-oxygen delivery, pHi- gastric intramucosal pH.

increases in oxygen supply caused similar changes both for survivors and nonsurvivors (figure 6).

DISCUSSION

Occult cellular hypoxia may develop in sepsis, and contribute to significant morbidity and mortality. Proper assessment of tissue oxygenation is of paramount importance to monitor the effects of therapy. Clinical evaluation of tissue perfusion and oxygenation by noninvasive means has always been the primary goal in the management of the critically ill. Studies indicate that decreases in tissue oxygenation are associated with various pathologic processes (23-25), and if not reversed the resultant tissue hypoxia may lead to multiple organ dysfunctions and death.

Investigators have considered that lactate levels identify tissue hypoxia in septic patients. Although blood lactate can be easily measured in critically ill patients, its interpretation can be complex since a blood concentration of any substance reflect not only its production, but also its elimination. Lactate clearance, which takes place primarily in the liver, can be altered in acute circulatory failure. However lactic acidosis can also be associated with disease processes other than circulatory failure. Lactic acidosis, as a consequence of an acute decrease in oxygen delivery has been shown in some (3,4), but not in all studies (26). This has been thought as a result of variations in oxygen extraction capabilities with the type and severity of the disease (27,28). Hence, blood lactate levels which reflect an imbalance between the metabolic requirements and the oxygen supply has been shown to be closely related to outcome in septic shock (4,5,6,29,30).

In clinical studies where blood lactate levels were measured, the oxygen uptake supply dependency has been observed in patients with increased lactate levels (8,10,13,31). Based on observations that there is a close relationship between changes in DO_2 and VO_2 over a wider than normal DO_2 range in patients with ARDS, sepsis, chronic congestive heart failure, acute liver failure or after cardiopulmonary by-pass

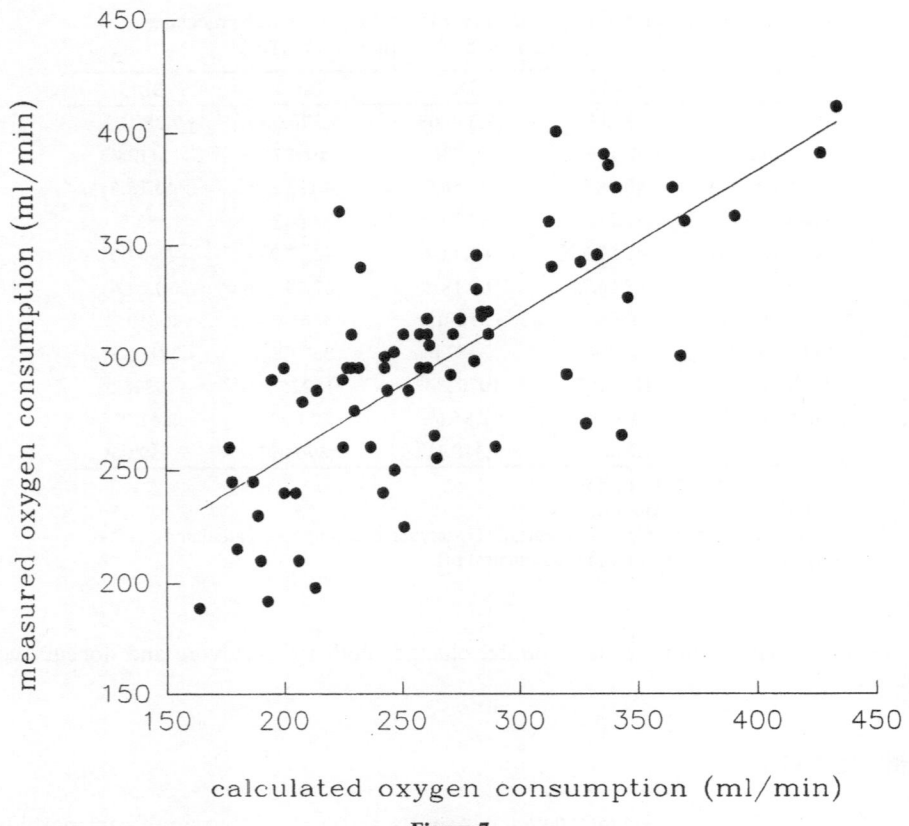

Figure 7.

operation, it has been concluded that in these disease states VO_2 is limited by DO_2. These observations suggested that this phenomenon of pathological relationship between VO_2 and DO_2 is the sign of inadequacy of tissue oxygenation. The recognition of this pathologic supply dependency which could be related to various factors, including local hypoxia, abnormal distribution of blood flow, development of interstitial edema, formation of microemboli has been thought to have important clinical consequences. However the clinical application of this phenomenon is still a matter of controversy, since recent studies concluded that the measurements of VO_2 and DO_2 and their relationship were not related to outcome from septic shock (6,32).

Experimental studies have shown that intramucosal pH of the gut decreases with hypoxia (16), therefore measurements of gastric mucosal acidosis by tonometry which is relatively noninvasive has been thought to provide a monitor of systemic oxygenation in critically ill patients. Low pHi values were shown to be associated with reduced end organ (gut) perfusion in animals (16), and also associated with high morbidity and mortality in patients (20). Clinical studies showed measurements of pHi values as a therapeutic index of tissue oxygenation in critically ill patients and support the hypothesis that a substantial proportion of ICU patients may benefit from pHi monitoring (20,21,33) since gastrointestinal tract is the first organ to be affected by inadequate oxygenation resulting from decreases in arterial partial pressure of oxygen or from hypoperfusion (33-35).

Our study was designed to compare these three techniques of monitorization of tissue oxygenation status, and to evaluate the measurements of pHi with increasing levels of oxygen delivery in patients with severe sepsis. The oxygen uptake supply dependency is usually detected by measuring oxygen consumption before and after acute changes in oxygen delivery. Fluid or blood administration has been used (8,13,23), however a pharmacological intervention such as dobutamine which induces minimal hemodynamic and metabolic effects other than an increase in cardiac output, was thought to represent a more reliable and a reversible therapeutic trial. So we used dobutamine test, as indicated by Vincent and coworkers (10), up to 15 μgr/kg/min since its effects on oxygen consumption has been found minimal in patients with heart failure (36).

All of our patients demonstrated the phenomenon of O_2 supply dependency with elevated lactate levels. There was a clear linear function between all the values of DO_2 and VO_2 in every patient. In most of the studies reporting oxygen supply dependency, VO_2 was calculated as the product of cardiac output and arteriovenous oxygen content difference. CO and CaO_2 are used in the calculation of both VO_2 and DO_2 which may result in a high correlation even when no functional relationship between these variables exists. Vermeij et al., evaluating the relationship between VO_2 and DO_2, found significant correlation between VO_2 and DO_2 measured from Fick equation but found no correlation between DO_2 and directly measured VO_2 (38). They considered that this relationship might be because of the mathematical coupling of the shared variable. However, the effect of coupled error has been found small in another study looking at the relationship between oxygen uptake and supply (40). In this study, in order to exclude the shared variable in calculating VO_2, we measured VO_2 by two different methods, and found a good correlation between calculated and measured oxygen consumption.

In a series of studies, gastrointestinal tract has been shown to be particularly susceptible to alterations in oxygen supply. Significant correlation found between the levels of splanchnic oxygen supply and gastric intramucosal pH demonstrated the direct relationship between local oxygen supply and pHi (16,41). However, parallel to the results of the study by Gutierrez et al (39), in our results, although there seemed increases in mean pHi values by the increases of mean values of oxygen supply, no significant correlation was found when pHi values were plotted against the corresponding levels of DO_2. As interpreted in their study by Gutierrez et al (39), this poor correlation may be due to DO_2 representing an overall measurement of oxygen supply to the systemic tissues, whereas pHi values only reflects the balance between oxygen demand and DO_2 in the gastric mucosa .

Silverman et al, in their study showed the effectiveness of dobutamine in improving low levels of pHi in septic patients (21). In the same study, they suggested that lactic acid levels may not be a sensitive marker of splanchnic tissue oxygenation since their results showed poor correlation between lactate levels and pHi. However our study demonstrated high levels of lactic acid with low pHi before dobutamine infusion while a significant decrease in lactate levels were noticed by increasing levels of DO_2. There was good correlation between arterial lactate values, as well as between lactate and pHi.

Gutierrez et al (20), in their study of evaluating pHi as a therapeutic index of tissue oxygenation, demonstrated poor outcome in patients admitted with low pHi values . Doglio et al (37) determined the effect of therapeutic interventions in the early evaluation of gastric intramucosal pH, and found significant differences in patient mortality according to the admission and 12 h pHi values. In their study, they concluded that pHi monitoring may be useful as an index to guide therapeutic interventions in critically ill patients. In our study protocol, we didn't evaluate pHi values of the patients on admission. However, we found significantly low pHi and higher levels of lactate in nonsurvivors compared to survivors during the study period. On the other hand when DO_2 and VO_2 were related to survival, neither the DO_2 and VO_2 values nor the pathologic supply dependency could separate the survivors from nonsurvivors.

In summary, our results showed that oxygen supply dependency is a common finding in patients with severe sepsis, however neither the value of DO_2 nor the presence of supply dependency was predictive of mortality. On the other hand arterial lactate levels and gastric intramucosal pH measurements were closely associated with poor prognosis. The results of this study are of clinical importance that measurements of pHi as a noninvasive procedure is a valuable adjunct to lactic acid as a prognostic indicator showing the severity of oxygen debt of the tissues in patients with severe sepsis. We think that this beneficial monitoring device may lead to early therapeutic interventions including fluid replacement or inotropic support to increase oxygen supply to the tissues.

REFERENCES

1. Vincent JL, Dufaye P, Barre J, Leeman M, Degaute JP, Kahn RJ. Serial lactate determinations during circulatory shock. Crit Care Med 1983; 11 : 449-451.
2. Groneveld ABJ, Kester ADM, Nauta JJP. Relation of arterial blood lactate to oxygen delivery and hemodynamic variables in human shock states. Circ Shock 1987; 22:35-53.
3. Rashkin M, Boxkin C, Baughman R. Oxygen delivery in critically ill patients: relationship to blood lactate and survival. Chest 1985; 87:580-84.
4. Tuchschmidt J, Fried J, Swinney R, Sharma P. Early hemodynamic correlates of survival in patients with septic shock. Crit Care Med 1989; 17:719-23.
5. Azimi G, Vincent JL. Ultimate survival from septic shock. Resuscitation 1986; 14:245-53.
6. Bakker J, Coffernils M, Leon M, Gris P, Vincent JL. Blood lactate levels are superior to oxygen-derived variables in predicting outcome in human septic shock. Chest 1991; 99:956-62.
7. Kaufman BS, Rackow MD, Falk JL. The relationship between O_2 delivery and consumption during fluid recusitation of hypovolemic and septic shock. Chest 1984; 85:336-40.
8. Gilbert EM, Haupt MT, Mardaras RY, Huaringen AJ. The effect of fluid loading , blood transfusions, and catecholamine infusion on oxygen delivery and consumption in patients with sepsis. Am Rev. Respir Dis. 1986; 134:873-78.
9. Astiz M, Rackow E, Falk JL et al. O_2 delivery and consumption in patients with hyperdynamic septic shock. Crit Care Med. 1987; 15:26-29.
10. Vincent JL, Roman A, DeBacker D, Kahn RJ. O_2 uptake/supply dependency: effects of short term dobutamine infusion. Am Rev Respir Dis. 1990; 141:2-7.
11. Gutierrez G, Dohil RJ. Oxygen consumption is nearly related to O_2 supply in critically ill patients. J Crit Care. 1986; 1: 45-53.
12. Gilbert EM, Haupt MT, Mardanas RY, Huaringa AJ, Carlson RW. The effects of fluid loading, blood transfusion, and catecholamin infusions on oxygen delivery and consumption in patients with sepsis. Am. Rev Respir Dis. 1986; 134:873-8.
13. Haupt MT, Gilbert EM, Carlson RW. Fluid loading increases oxygen consumption in septic patients with lactic acidosis. Am Rev Respir Dis. 1985; 131:912-6.
14. Bihari D, Smithies M, Gimson A, Tinker J. The effects of vasodilatation with prostacylin on oxygen delivery and uptake in critically patients. N Engl. J. Med. 1987; 317: 397-403.
15. Pinsky MR. Oxygen delivery and uptake in septic patients. Yearbook of Intensive Care and Emergency Medicine. Ed. JL Vincent. Springer Verlag. Berlin, Heidelberg, Newyork.
16. Grum CM, Fiddian-Green RG, Pittenger GI, Grant BJB, Rothman ED, Dantzker DR. Adequacy of tissue oxygenation in intact dog intestine. J Appl Physiol. 1984; 56: 1065-69.
17. Hartmann M, Montgomery A, Johnsonn K, Haglung U. Tissue oxygenation in hemorrhagic shock measured as transcutaneous oxygen tension, and gastrointestinal mucosal pH in pigs. Crit Care Med. 1991; 19:205-10.
18. Fiddian–Green RG, Pittenger G, et al. Back diffusion of CO_2 and its influence on the intramural pH in gastric mucosa. J Surg Res. 1982; 33:39.
19. Russell JA. O_2 delivery and consumption in adult respiratory distress syndrome and sepsis. Update in Intensive Care and Emergency Medicine. (Ed. by JL Vincent) Springer–Verlag Berlin, Heidelberg; 1991:175-181.

20. Gutierrez G, Palizas F, Doglio G, et al. Gastric intramucosal pH as a of tissue oxygenation in critically ill patients. Lancet 1992; 339: 195-99.

21. Silverman HJ, Tuma P. Gastric tonometry in patients with sepsis. Chest. 1992; 102: 184-188.

22. Bone RC, Balk AR, Cerra CB, Dellinger RP, Fein AM, Knaus WA, Schein RMH, Sibbald WJ. Definition for sepsis and organ failure and guidelines for the use of innovative therapies in sepsis. Chest 1992; 101:1644-55.

23. Gutierrez G, Pohil RJ. Oxygen consumption is linearly related to O_2 supply in critically ill patients. Journal of Crit Care. 1986; 1:45-53.

24. Haglung U, Fiddian-Green RG. Assesment of adequate tissue oxygenation in shock and critical illness: oxygen transport in sepsis. Intensive Care Med. 1989; 15:475.

25. Kuttila K, Niinikoski J, Haglung U. Visceral and peripheral tissue perfusion after cardiac surgery. J Thorac Cardiovasc Surg. 1989; 2: 389-99.

26. Astiz ME, Rackow EC, Kaufman B, Falk JL, Weil MH. Relationship of oxygen delivery and mixed venous oxygenation to lactic acidosis in patients with sepsis and acute myocardial infarction. Crit Care Med. 1988; 16: 655-8.

27. Kreisberg RA. Pathogenesis and management of lactic acidosis. Annu Rev Med. 1984; 35: 181-93.

28. Vincent JL, Weil MH, Puri V, Carlson RW. Circulatory shock associated with purulent peritonitis. Am J Surg. 1981; 142:262-70.

29. Groeneveld AB, Bronsveld W, Thijs LG. Hemodynamic determinants of mortality in human septic shock. Surgery 1986; 99: 140-52.

30. Vincent JL, Dufaye PH, Berre J, Leeman M, Degaute JP, Kahn RJ. Serial lactate determinations during circulatory shock. Crit Care Med. 1982; 11:449-51.

31. Annat C, Viale JP, Percival C, Froment M, Motin J. Oxygen delivery and uptake in the adult respiratory syndrome : lack of relationship when measured independently in patients with normal blood lactate concentrations. Am Rev Respir Dis. 1986; 133: 999-1001.

32. Parker MM, Shelhamer JH, Natanson C. Serial cardiovascular variables in survivpors and nonsurvivors of human septic shock: heart rate as an early predictor of prognosis. Crit Care Med. 1987; 15:923-29.

33. Fiddian-Green RG. Splancnic ischemia and multipl organ failure in the critically ill. Ann R Coll Surg Engl. 1988; 70: 128-154.

34. Shoemaker WC. Tissue perfusion and oxygenation : a primary problem in acute circulatory failure and shock. Crit Care Med. 1991; 19:595-596.

35. Clark CH, Gutierrez G. Gastric intramucosal pH: a noninvasive method for the indirect measurement of tissue oxygenation. Am J of Crit Care. 1992; 2: 53-60.

36. Cohn JN, Levine TB, Olivari MM, et al. Plasma norepinephrine as a guide to prognosis in patients with chronic congestive failure. N Engl J Med. 1984; 311: 819-23.

37. Doglio GR, Pusajo JF, et al. Gastric mucosal pH as a prognostic index of mortality in critically ill patients. Crit Care Med. 1991; 19: 1037-40.

38. Vermeij CG, Feenstra BWA, Bruining HA. Oxygen delivery and oxygen uptake in postoperative and septic patients. Chest. 1990; 98:415-20.

39. Gutierrez G, Bismar H, Dantzker DR, Silva N. Comparison of gastric intramucosal pH with measures of oxygen transport and consumption in critically ill patients. Crit Care Med. 1992; 20:451-57.

40. Stratton HH, Fenstel PJ, Newell JC. Regression ao calculated variables in the presence of shared measurement error. J Appl Physiol. 1987; 62: 2083-93.

41. Nelson DP, Sansel RW, Wood LDH, et al. Pathological supply dependence of systemic and intestinal O_2 uptake during endotoxemia. J Appl. Physiol. 1988; 64: 2410-19.

68

MECHANICAL VENTILATION IN ARDS

J. Kesecioğlu

Department of Intensive Care
Academic Medical Center
Amsterdam, The Netherlands

INTRODUCTION

Adult respiratory distress syndrome (ARDS) is a type of acute respiratory failure resulting from altered ventilation-perfusion ratios leading to hypoxemia, hypo- or hypercapnia and decreased lung compliance. Asbaugh and Petty (1) used the term ARDS for the first time in 1967 to describe the clinical picture of diffuse pulmonary infiltration on chest X-ray, impairment of oxygenation and pulmonary congestion. However, the existence of this syndrome was already reported after the first World War by Weed and McAfee (2) who discussed the respiratory failure as a sequela of wound shock.

Other names such as congestive atelectasis, wet lung syndrome, shock lung, pump lung, transfusion lung, post-traumatic pulmonary insufficiency and Danang lung have also been used to identify this entity. Due to the increased lung water in ARDS, the syndrome has also been referred to as non-cardiogenic pulmonary edema or interstitial pulmonary edema.

Although lung injury associated with shock was recognized many years ago, it is only during the last 25 years that ARDS has been recognized as a serious problem in intensive care patients. Various disorders such as shock, chest trauma, non-thoracic trauma, abdominal or extra-abdominal sepsis, acute pancreatitis, bacterial or viral pneumonia, aspiration pneumonia, head injury, smoke inhalation, drug overdose, radiation pneumonitis etc. have been identified as the initiating events of this pathological entity. Serious hypoxemia and pulmonary edema without pulmonary venous hypertension are uniformly characteristic, regardless of the etiology.

PATHOPHYSIOLOGY OF ARDS

Respiratory failure follows either a primary lung disease or extra-thoracic injury of many origins. Primary lung injury damages the alveolar cells and results in impaired surfactant activity. Alveolar macrophages are activated and they liberate chemotactic substances which promote granulocyte micro-aggregation in the lung capillaries. Secondary lung damage is due to extra-thoracic injuries where damaged tissues liberate blood-borne

Oxygen Transport to Tissue XVII, Edited by Ince et al.
Plenum Press, New York, 1996

533

mediators which in turn lead to micro-aggregate formation in the pulmonary micro-vasculature (polymorphonuclear cells, platelets, etc.). An interaction occurs between micro-aggregation and humoral substances (blood-borne but also produced by the micro-aggregates) in the pulmonary capillaries. This interaction causes capillary endothelial injury in the lungs, particularly of capillary basement membranes resulting in increased permeability of these membranes. This situation is also called "capillary leak syndrome". Accumulation of fluid and cells in the interstitium and in the alveoli leads to loss of surfactant activity and hypoxemia and ARDS (3-5).

PATHOLOGICAL CHANGES IN THE LUNG

In order to have an efficient gas exchange, a minimum diffusion distance between alveolar air and blood, and a large surface area of well-perfused micro-vasculature exposed to air is required. This blood-gas barrier is extremely delicate and highly vulnerable; damage to this barrier in the acute stage of ARDS leads to increased permeability and flooding of the alveolar space with blood constituents.

Pathologically, ARDS is divided into an early exudative phase which, after about one week, develops into a proliferative phase and eventually into fibrosis (6,7). Changes can be grouped into three stages:

- acute ARDS: alveolar edema with fibrin and leukocyte debris, damage to type I pneumocytes and endothelial cells;
- sub-acute ARDS: persistent capillary endothelial damage, type II cell proliferation and the beginning of squamous transformation;
- chronic ARDS: thickening of the interstitium by an increased number of type II cells and fibrosis (8,9).

Early ultrastructural changes may occur during the course of ARDS (10-13). Based on a series of 200 human autopsies, Blaisdell and Lewis (14) reported the timing of pathological events in ARDS as follows:

1. First 6 hours: minimal congestion of capillaries and fibrin with platelet and leukocyte aggregates in pulmonary arteries 25-250 microns in diameter.
2. At 12 hours: marked capillary congestion.
3. Between 16-19 hours: periarterial hemorrhage.
4. At 24 hours: severe congestion and interstitial hemorrhage.
5. At 72 hours: generalized consolidation with thrombi and emboli and the appearance of hyaline membranes.
6. Later in the first week: inflammatory cell changes resembling bronchopneumonia.
7. Between the second and sixth week: progressive fibrosis.

CLINICAL PRESENTATION AND DIAGNOSIS OF ARDS

The clinical presentation of patients with ARDS has hardly changed since its original description in 1967 and consists of a history of preceding noxious event; a symptom-free interval lasting hours to days; the subsequent progression of severe hypoxemia, decreased lung compliance and diffuse non-cardiogenic pulmonary edema. An accurate definition of the syndrome is essential for the standardization of the entry criteria into various clinical stages. Variable diagnostic criteria are used by different investigators during the last 25 years. These criteria include chest X-ray, presence or absence of positive end expiratory pressure

(PEEP), specific values for the alveolar-arterial oxygen tension gradient, pulmonary shunt fraction, FiO_2 and pulmonary compliance. However, these criteria are not universally accepted and vary widely among investigators (15).

THERAPEUTIC STRATEGIES

During recent years, there has been a growing tendency to consider mechanical ventilation as a supportive measure rather than therapeutic in the treatment of ARDS. The lack of prospective randomized studies comparing the mortality in ARDS patients ventilated with different modes of mechanical ventilation is probably the main reason for this concept. The immediate therapeutic aim in ARDS is to treat hypoxemia, which in most cases is marked by sophisticated ventilatory support. Of concern is the possibility that ARDS may be a response to barotrauma and that the use of high airway pressures to ventilate patients with ARDS may itself result in injury and morphological changes similar to ARDS, in formerly uninvolved parts of the lung (16-20).

To prevent further barotrauma, techniques such as pressure controlled inverse ratio ventilation (PC-IRV) (16), high frequency ventilation (HFV) (21-24) and low frequency positive pressure ventilation with extracorporeal carbondioxide removal (LFPPV-ECCO$_2$R) (25) have been used during the last 15 years. The aims of these techniques are to treat hypoxemia, but simultaneously avoid high peak inspiratory pressure (PIP) and intrapulmonary pressure swings while keeping the alveoli open. Furthermore, remembering the sequence of the pathological changes in the ARDS lungs it is important to apply a suitable mode of ventilation as early as possible to provide maximum opening of the alveoli and to keep them open, as consolidation is reported to occur already at 72 hours after the initiation of the respiratory failure (14).

An understanding of the biochemical alterations that take place in the lungs of patients with ARDS has led to evaluation of the therapeutic use of cyclo-oxygenase inhibitors, antibodies directed against endotoxin or C5a, surfactant replacement therapy and antiprotease and antioxidant therapy. However, no specific therapy for ARDS is yet available other than ventilatory and, if necessary, circulatory support measures.

CHOICE OF VENTILATION MODE

After 30 years of experimental studies and clinical experience, the application of the best mechanical ventilation mode in ARDS is still a subject of controversy. The aims of full or partial ventilatory support in ARDS can be summarized as follows:

1. Achievement of optimal arterial oxygenation and adequate alveolar ventilation;
2. Avoidance of damage to pulmonary parenchymal structures by high inflation pressures;
3. Prevention of cardiovascular system depression by high intrapulmonary airway pressures.

Considering the above mentioned pathological changes occurring in the lung, maximum recruitment of alveoli will be possible only in acute ARDS. At this stage application of almost any type of mechanical ventilation will recruit the atelectic lung regions.

Conventional volume controlled ventilation (VCV) with PEEP has been successfully used for years to treat hypoxemia in ARDS. In this mode, tidal volume (V_T) maintains ventilation and PEEP prevents airway closure. The level of PIP is regulated by the ventilator depending on the lung compliance, to deliver the pre-set tidal volume. This method results in high intrapulmonary pressure amplitude to recruit atelectatic parts in ARDS lungs. Moreover, clinically used levels of external PEEP do not stabilize all the alveoli as there is no homogeneous distribution of the lung damage in ARDS (26,27). In an editorial Lachmann (28) stressed that, a preset external PEEP will only balance the increased retractive forces of only a part of the damaged lungs preventing their collapse during the expiration of the ventilatory cycle, but will still not be sufficient to keep all parts of the lungs open. Highly damaged lung regions will only be aereated at the end of inspiration and gas exchange may be highly decreased due to the short time of insufflation, whereas during the same period capillary perfusion is reduced by the high intra-alveolar pressure. Moreover, the applied external PEEP may probably lead to capillary compression and ventilation/perfusion mismatching in the healthy regions of the lungs, which will be more prominent during the inspiration.

Alternatively, PC-IRV has shown to provide good oxygenation, normocapnia with low PIP and intrapulmonary pressure amplitude. In accordance with the pathophysiology of ARDS, ventilatory support should provide a persistent positive alveolar pressure to balance the high retractive forces due to diminished surfactant system. To maintain this balance it would be ideal to maintain a constant alveolar pressure. Therefore, in ARDS, when artificial ventilation is required, it is logical to control pressure rather than volume since the physical and therapeutic effects of intermittent positive pressure ventilation are closely linked to positive airway pressures. This concept implies that insufflation pressure is responsible for stabilization of lung units while release of this pressure during the expiratory phase is responsible for ventilation. In ARDS, the ventilatory support should open up closed units and keep them open and should avoid local hyperinflation. Once the alveoli are opened at PC-IRV with initially applied high PIP, the PIP can be set at the level that balances elastic recoil caused by the increased surface forces. Ventilation is then achieved by briefly releasing that pressure so that expiration takes place without leaving time for alveolar collapse. By applying this concept the lung will be kept optimally open and be ventilated with a smaller intrapulmonary pressure amplitude compared to VCV-PEEP. Kesecioglu and colleagues (29) demonstrated in an ARDS model that PC-IRV with inspiration/expiration (I/E) ratio of 4:1 provided significantly more homogeneous ventilation and lower intrapulmonary pressure amplitude almost equal to the control values of the healthy lungs. These results confirm the beneficial effect of prolonged inspiration in stabilizing the airways and the alveoli, resulting in homogeneous ventilation.

On the other hand, in later stages of ARDS the pathological changes develop into a situation where recruitment of alveoli are no longer possible by the application of pressure into the lungs. At this stage pressure limited ventilation with permissive hypercapnia would be a logical alternative mode of mechanical ventilation (30). In this mode, low V_T and PEEP are used aiming to ventilate the remaining healthy lung units only. The overdistension of these units by standard V_T are avoided. The rise in arterial carbon dioxide tension is permitted unless the patient suffers from high intracranial pressure, ischemic heart disease or hypertension.

In sophisticated centers, LFPPV-ECCO$_2$R can be used alternatively at later stages of ARDS (25). This mode allows a reduction in the minute ventilation and PIP by applying low tidal volume and frequency and prevents further damage to the lungs. The carbon dioxide is removed by the extracorporal circulation. Oxygenation is maintained by the ventilation of the membrane lungs with O$_2$ in addition to the intrapulmonary administration of oxygen via a catheter. Although it is effective, this method is still invasive, expensive, needs additional equipment and involves special training and attention from the attending medical staff.

CONCLUDING REMARKS

The results of various studies summarized in this article indicate that ARDS is caused by multiple factors consisting of primary or secondary lung damage. A complex mediator cascade is activated leading to the "capillary leak syndrome", pathological changes in the lungs varying from alveolar edema with fibrin and leucocyte debris, to fibrosis and clinical symptoms such as severe hypoxemia and decreased lung compliance. In spite of extensive work done in this field, successful therapeutic measures are still limited to mechanical ventilation. However, ventilatory support, being the treatment of hypoxemia in ARDS, may itself cause lung damage and pathological changes similar to ARDS in healthy parts of the lungs, if applied with high intrapulmonary pressure swings. Considering the pathological changes occurring in ARDS, a ventilation mode opening the lung and keeping it open with the lowest possible pressure swings should be applied as early as possible in the course of ARDS.

Almost all modes of mechanical ventilation provide adequate oxygenation in recruitable lungs with ARDS. However, PC-IRV, pressure limited ventilation with permissive hypercapnia and LFPPV-ECCO$_2$R avoid high PIP and intrapulmonary pressure amplitude reducing the risk of barotrauma. In the early phase of ARDS in which most of the alveoli are still recruitable, PC-IRV should be an adequate form of treatment without the use of further invasive methods. However, at a later stage of the disease when only marginal alveolar recruitment is possible, in which no form of ventilation would be effective to provide gas exchange, pressure limited ventilation with permissive hypercapnia or if the facilities exist LFPPV-ECCO$_2$R should be considered as the treatment of choice.

REFERENCES

1. Asbaugh D.G., Bigolow DB, Petty TL, Levine BE 1967. Acute respiratory distress in the adult, *Lancet* 2: 319-323.
2. Weed F.W., McAfee L. 1927. Wound shock, *The Medical Department of the United States Army in the World War II*: 185-213.
3. Zapol W.M., Kobayashi K., Snider M.T., Greene R., Laver M.B., 1977, Vascular obstruction causes pulmonary hypertension in severe acute respiratory failure, *Chest* 71:306-307.
4. Bone R.C., Francis P.B., Pierce A.K., 1976, Intravascular coagulation associated with the adult respiratory distress syndrome, *Am J Med* 61:585-589.
5. Ashbaugh D.G., Petty T.L.,1972, Sepsis complicating the acute respiratory distress syndrome, *Surg Gynecol Obstet* 135:865-869.
6. Katzenstein A.L., Bloor C., Leibow A.A., 1976, Diffuse alveolar damage, the role of oxygen shock and related factors, *Am J Path* 85:210-228.
7. Nash G., Blennerhassett J.B., Pontoppidan H., 1967, Pulmonary lesions associated with oxygen therapy and artificial ventilation, *New Engl J Med* 276:368-374.
8. National Heart, Lung and Blood Institute, Division of Lung Diseases, 1971, Extracorporeal support for respiratory insufficiency, *National Institute of Health, Bethesda MD* :243-245.
9. Murray J.F., 1977, Conference report - Mechanisms of acute respiratory failure, *Am Rev Resp Dis* 115:1071-1078.
10. Bachofen M., Weibel E.R., 1974, Basic pattern of tissue repair in human lungs following unspecific injury, *Chest* 65S:14S-19S.
11. Bachofen M., Weibel E.R., 1977, Alterations of the gas exchange in adult respiratory insufficiency associated with septicemia, *Am Rev Resp Dis* 116:589-615.
12. Bachofen M., Weibel E.R., 1982, Structural alterations of of lung parenchyma in the adult respiratory distress syndrome, *Clin Chest Med* 3:35-56.
13. Pratt P.C., Vollmer R.T., Shelburne J.D., Crapo J.D., 1979, Pulmonary morphology in a multihospital collaborative extracorporeal membrane oxygenation project, I.Light microscopy, *Am J Path* 95:191-214.
14. Blaisdell F.W., Lewis F.R., 1977, Respiratory distress syndrome of shock and trauma: post-traumatic respiratory failure. *Major Problems in Clinical Surgery* XXI. W.B. Saunders, Philadelphia.
15. Artigas A., Carlet J., Chastang C.l., Le Gall J.R., Blanch L., Fernandez R., 1992, Adult respiratory distress syndrome: clinical presentation, prognostic factors and outcome in Adult Respiratory Distress Syndrome. Artigas A., Lemaire F., Suter P.M., Zapol W.M. (eds.), *Churchill Livingstone*: 509-525.

16. Lachmann B., Danzmann E., Haendly B., Jonson B., 1982, Ventilator settings and gas exchange in respiratory distress syndrome. In: Prakash O (ed) Applied physiology in clinical respiratory care. *Martinus Nijhoff Publishers, The Hague*:141-176.
17. Greenfield L.J., Ebert P.A., Benson D.W., 1964, Effect of positive pressure ventilation on surface tension properties of lung extracts. *Anesthesiology* 25: 312-316.
18. Reynolds E.O.R., Taghizadeh A., 1974, Improved prognosis of infants mechanically ventilated for hyaline membrane disease, *Arch Dis Child* 49:505-515.
19. Lachmann B., Jonson B., Lindroth M., Robertson B., 1982, Modes of artificial ventilation in severe respiratory distress syndrome. Lung function and morphology in rabbits after washout of alveolar surfactant, *Crit Care Med* 10:724-732.
20. Mead J., Collier C., 1959, Relation of volume history of lungs to respiratory mechanics in anaesthetized dogs, *J Appl Physiol* 14:669-678.
21. Hamilton P.P., Onayemi A., Smith J.A., Gillan J.E., Cutz E., Froese A.B., Bryan A.C., 1983, Comparison of conventional and high frequency ventilation: oxygenation and lung pathology, *J Appl Physiol* 55:131-138.
22. MacIntyre N., Follet J., Deitz J., 1986, Jet ventilation at 100 breaths per minute in adult respiratory failure, *Am Rev Respir Dis* 134:897.
23. Sjöstrand U., 1980, High-Frequency Positive-Pressure Ventilation (HFPPV): A Review, *Crit Care Med* 8:345-364.
24. Schuster D.P., Klain M., Snyder J.V., 1982, Comparison of high frequency jet ventilation to conventional ventilation during severe acute respiratory failure in humans, *Crit Care Med* 10:625-630.
25. Gattinoni L., Pesenti A., Caspani M.L., Pelizzola A., Mascheroni D., Marcolin R., Fumagali R., Rossi F., Iapichino G., Romagnoli B., Uziel L., Agostoni A., Kolobow T., Damia G., 1986, Low frequency positive-pressure ventilation with extracorporeal CO_2 removal in severe acute respiratory failure, *JAMA* 256:881-886.
26. Dantzker D., 1982, Gas exchange in the adult respiratory distress syndrome, *Clin Chest Med* 3:57-67.
27. Gattinoni L., Pesenti A., Bombino M. et al, 1988, Relationships between lung computed tomographic density, gas exchange, and PEEP in acute respiratory failure, *Anesthesiology* 69:824-832.
28. Lachmann B., 1992, Open up the lung and keep the lung open, *Intensive Care Med* 18:319-321.
29. Kesecioglu J., Gültuna I., Pompe J.C., Hop W.C.J., Ince C., Erdmann W., Bruining H.A., Assessment of Ventilation Inhomogeneity and Gas Excange with Volume Controlled Ventilation and Pressure Regulated Volume Controlled Ventilation on Pigs with Surfactant Depleted Lungs, *Adv.Exp.Med.Biol.* (in press).
30. Hickling K.G., Henderson S.J., Jackson R., Low mortality associated with pressure limited ventilation with permissive hypercapnia in severe adult respiratory distress syndrome, *Intensive Care Med. 16:372-377.*

69

ASSESSMENT OF VENTILATION INHOMOGENEITY AND GAS EXCHANGE WITH VOLUME CONTROLLED VENTILATION AND PRESSURE REGULATED VOLUME CONTROLLED VENTILATION ON PIGS WITH SURFACTANT DEPLETED LUNGS

J. Kesecioğlu,[1,3] I. Gültuna,[1] J. C. Pompe,[2] W. C. J. Hop,[5] C. Ince,[4] W. Erdmann,[1] and H. A. Bruining[2].

[1] Department of Anesthesiology
[2] Department of Surgery
 University Hospital, Rotterdam
 The Netherlands
[3] Department of Intensive Care
[4] Department of Anesthesiology
 Academic Medical Center
 Amsterdam, The Netherlands
[5] Department of Epidemiology Biostatistics
 University Hospital, Rotterdam
 The Netherlands

INTRODUCTION

In adult respiratory distress syndrome (ARDS), whilst a great fraction of the lung is affected by the disease, some parts still function normally creating a non-homogeneous distribution of lung pathology. This unequal distribution results in increased differences in time constants of different lung units (1-3). Conventional volume controlled ventilation (VCV) with positive end expiratory pressure (PEEP) with a standard inspiration/expiration (I/E) ratio of 1:2 may be inadequate to ventilate these affected lung units equally, as they tend to open later and close earlier compared to healthy alveoli due to high retractive forces. Prolonging the inspiration time of the respiratory cycle is suggested to overcome these time-constant inequalities within the ARDS lungs, ventilate all alveoli despite different retraction forces in the airways and improve the gas exchange (4,5). Furthermore, the application of pressure controlled inverse ratio ventilation (PC-IRV) is shown to provide better gas exchange with significantly lower peak inspiratory pressures (PIP) (4). On the

Oxygen Transport to Tissue XVII, Edited by Ince et al.
Plenum Press, New York, 1996

contrary, high PIP is necessary to provide adequate gas exchange with VCV with PEEP, which is reported to lead to barotrauma or progressive injuries in the lungs (6-10). In previous studies, investigators used commercial ventilators for the assessment of ventilation inhomogeneity in ARDS patients (11,12). A new method was recently developed to measure ventilation inhomogeneity and end expiratory volume (EEV) with a multibreath indicator gas washout technique during mechanical ventilation (13).

In the present study lung lavage was used to create surfactant depletion in pig lungs, aiming to assess the ventilation inhomogeneity caused by this acute respiratory failure model. Keeping PaO_2 values above 450 mmHg with equal expired minute volumes (V_E), it was thereafter investigated whether pressure regulated volume controlled ventilation (PRVCV) with I/E ratio of 4:1 led to a more homogeneous ventilation compared to VCV with PEEP in pigs with surfactant depleted lungs.

METHODS

This study was performed at the University Hospital of Rotterdam, Department of Surgery. Approval of the local Animal Investigation Committee was obtained.

Nine pigs, 21.1 ± 0.5 kg (range 19-22 kg) were investigated in this study. Anesthesia was induced with 30 mg/kg ketamine given im. The trachea was intubated with a portex tube with an internal diameter of 7 mm and the cuff was inflated to avoid air leakage. The lungs were ventilated with a Servo 900C (Siemens–Elema Solna, Sweden) ventilator, fitted with an indicator gas injector (13). Anesthesia was maintained by infusion of midazolam (0.02 mg/kg/min) and ketamine (0.05 mg/kg/min). Pancuronium bromide infusion (0.08 mg/kg/min) was administered after a bolus dose of 4 mg for muscle relaxation. A catheter for fluid replacement and a 7F Swan–Ganz thermodilution catheter for hemodynamic monitoring were inserted through the right and left internal jugular veins, respectively. The carotid artery was cannulated for blood sampling and invasive blood pressure monitoring. An 18F Foley urine catheter was placed into the bladder by cystostomy to monitor urine output.

Arterial blood gases were determined by AVL 945 automatic bloodgas system. PIP, mean airway pressure (mPaw), static-PEEP [$PEEP_S$=preset PEEP in mode 1 (M1)] and total PEEP [$PEEP_T$= $PEEP_S$ + auto-PEEP in mode 2(M2)] displayed by the ventilator (by pressing end-expiratory hold button) were recorded. Intrapulmonary pressure amplitude (ΔP) was expressed as PIP - $PEEP_T$ (or $PEEP_S$).

A multiple breath indicator gas washout test was used to calculate end expiratory volume (EEV) and an index for ventilation inhomogeneity (S) was achieved as described by Huygen and colleagues (13). The blood insoluble gas indicator SF6 was used in a small fraction (2%) to minimize the effects on the metabolic gases. Gas fractions were continuously measured by a mass-spectrometer (Airspec MGA 3000, UK) and information about EEV and ventilation inhomogeneity was obtained.

The experimental animals were subjected to lung lavage to establish a model of ARDS (14). After completing baseline measurements at the control mode before lung lavage (BLL), lung lavage was performed with 50 ml/kg warm saline solution and repeated in 5-10 min intervals until a PaO_2 < 100 mmHg was achieved.

The following modes of ventilation were used during the study, where either M1 or M2 was applied following the control mode. FiO_2 was kept as 1.0.

BLL   =   VCV with $PEEP_S$ 4 cm H_2O, frequency (f) 12/min, tidal volume (V_T) 8-12 ml/kg and I/E ratio of 1:2 (25% I, 10% pause).

After lung lavage=   same as BLL, before M1.
(ALL)1
    M1   =   VCV with $PEEP_S$ to produce a PaO_2 above 450 mm Hg, f
                          12/min, V_T 8-12 ml/kg and I/E ratio of 1:2 25% I, 10% pause).
    ALL2   =   same as BLL, before M2.
    M2   =   PRVCV with $PEEP_T$ to produce a PaO_2 above 450 mm Hg, f
                          12/min, PIP adjusted to obtain a V_T of 8-12 ml/kg and I/E ratio
                          of 4:1 (80% I). Around 45 cm H_2O PIP was applied during the
                          first 5 min to open the alveoli.

Data are expressed as mean ± SEM. For all parameters, carry on and period effects were controlled. As such effects were not present data were analyzed by the paired t-test. Significance was accepted at $p \leq 0.05$.

RESULTS

EEV and S values are shown in figure 1. Lung lavage produced a significant decrease in EEV. M1 and M2 had significantly higher EEV values compared to BLL and their control

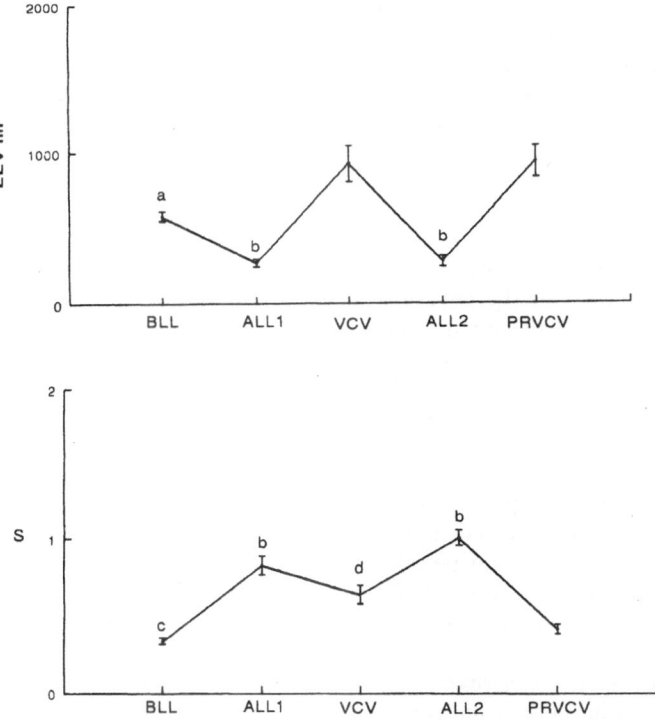

Figure 1. Changes in end expiratory volume (EEV) and ventilation inhomogeneity index (S) during the application of the modes. BLL= before lung lavage; ALL= after lung lavage; VCV= volume controlled ventilation with PEEP; PRVCV= pressure regulated volume controlled ventilation; a= significantly different from ALL1, VCV, ALL2 and PRVCV; b= significantly different from VCV and PRVCV; c= significantly different from ALL1, VCV and ALL2; d= significantly different from PRVCV.

Table 1. Respiratory data measured during the study

n=9	BLL	ALL1	M1	ALL2	M2
mPaw (cmH$_2$O)	6.8±0.3[a]	11.4±0.7[b]	21.6±0.9[c]	11.4±0.7[c]	29.7±2.1
PIP (cmH$_2$O)	19.4±0.8[a]	38.6±1.9[b]	44.6±2.1[c]	38.9±1.5	34.8±2.1
PEEP$_T$ (cmH$_2$O)	4[d]	4[d]	16.7±0.7	4[d]	16.6±0.6
ΔP (cmH$_2$O)	15.4±0.8[e]	34.6±1.9[b]	27.3±2[c]	34.9±1.5[c]	17.6±1.2
PaO$_2$ (mmHg)	489±24[f]	66±5[b]	474±26	66±5[c]	493±25
PaCO$_2$ (mmHg)	41.8±1.4[f]	48.7±0.5[b]	43.4±1	48.9±1[c]	43.2±1.4

Values are presented as mean ± SEM. M=mode; BLL=before lung lavage; ALL1=after lung lavage, before M1; ALL2=after lung lavage, before M2; mPaw=mean airway pressure; PIP=peak inspiratory pressure; PEEP$_T$=total PEEP [values shown in BLL, ALL1, M1 and ALL2 are static-PEEP(PEEP$_S$) and in M2 is PEEP$_S$+auto-PEEP];ΔP=intrapulmonary pressure amplitude;
a=significantly different from ALL1, M1, ALL2 and M2 (P=<0.05); b=significantly different from M1 (P<0.05); c=significantly different from M2 (p=<0.05); d=significantly different from M1 and M2 (P<0.05); e=significantly different from ALL1, M1 and ALL2 (P<0.05); f= significantly different from ALL1 and ALL2 (P<0.05).

modes ALL1 and ALL2. No significant difference was found between M1 and M2. ALL1 and ALL2 had significantly higher S values compared to BLL. ALL2 had also significantly higher ventilation inhomogeneity compared to ALL1. Ventilation inhomogeneity was significantly improved with the application of M1 and M2. However, M2 had a significantly lower ventilation inhomogeneity compared to M1. V$_E$ was kept around 4.5 l/min in all modes.

Respiratory values obtained are shown in Table 1. M2 had significantly higher mPaw compared to M1. PIP and ΔP were significantly lower in M2 compared to M1. No difference were observed between M1 and M2 concerning PaO$_2$ and PaCO$_2$ values.

DISCUSSION

Lung lavage is an acute respiratory failure model which is known to be stable for several hours to evaluate different modes in a single animal model. Our results confirm the stability of this model with similar results obtained with the control modes ALL1 and ALL2, in all respiratory parameters. This model was introduced by Lachmann and colleagues (14) and is known to produce surfactant deficiency and alveolar collapse, resulting in hypoxemia and imitating the early phase of ARDS. The results of this study confirm the great reduction of EEV after lung lavage. Moreover, two to three fold increases in S values proves this method as a valuable acute respiratory failure model providing inhomogeneous distribution of ventilation as well as impaired gas exchange.

In this study no difference in arterial oxygenation was observed between VCV with PEEP and PRVCV. This might be related to the use of almost equal PEEP levels which provide PaO$_2$ values above 470 mmHg and would indicate adequate alveolar recruitment without overdistension. VCV with PEEP and PRVCV also produced similar results considering the changes in EEV. However, PRVCV provided a significant improvement in ventilation inhomogeneity. Furthermore, S values measured with PRVCV were similar to the values obtained with normal lungs before lung lavage. In VCV with PEEP, due to short inspiration time healthier alveoli receive the volume and pressure earlier at the inspiratory phase. At the end of inspiration, diseased alveoli with lower compliance tend to open later and may still be less inflated compared to healthier units. At the expiratory phase the diseased alveoli close earlier. In VCV with PEEP I/E ratio 1:2, due to long expiration time, the diseased alveoli which have already not been totally opened may contain lower volume and pressure at the end of the expiration compared to healthier lung units. These differences in opening and closing times between different lung units affect the washin and washout procedures of SF6 resulting in increased S values.

On the contrary, in PRVCV with I/E ratio 4:1, the alveoli are opened with high PIP during the first 5 min. The PIP is reduced thereafter to a level to provide the desired V_T. Increased inspiration time causes the even distribution of the V_T and pressure in already opened lung units in spite of the difference in resistances and compliances. The short expiration time prevents the collapse of the diseased alveoli at this phase resulting in improved ventilation inhomogeneity. Gattinoni and co-workers (5) also discussed the possibility of a more even distribution of gas due to prolonged inspiration, overcoming the scattered time constants of different lung units and improving the ventilation perfusion mismatch and arterial oxygenation.

PIP necessary to ventilate the lungs was about 35 cm H_2O in PRVCV compared to 45 cm H_2O in VCV with PEEP. PIP around 40-50 cm H_2O is known to be associated with a higher risk of barotrauma and morphological damage of the lungs (6-10). Recently, in a review article Hickling (15) also stressed the importance of high PIP as a cause of barotrauma and pulmonary interstitial emphysema which shows the importance of low PIP achieved during PRVCV for the prevention of iatogenic lung damage. Moreover, once the lungs were opened with high PIP, ΔP value necessary to ventilate the lungs was 17 cm H_2O in PRVCV and was comparable to the control values. ΔP was around 27 cm H_2O in VCV with PEEP which could result in high shear forces, accompanying the nonhomogeneous expansion of the lungs (16) which could be a reason for damage of bronchiolar and alveolar epithelium, as in ARDS, causing eventually the formation of hyaline membranes.

In conclusion, according to the multi-breath indicator gas washout test used in this study, lung lavage proved to be a good model of acute respiratory failure to investigate ventilation inhomogeneity. Although no change in EEV was observed, PRVCV with I/E 4:1 was related with a significant improvement of ventilation inhomogeneity compared to VCV with PEEP, allowing more even distribution of pressure and volume between the alveolar units with different time constants and therefore avoiding unequal ventilation of the alveoli. Furthermore, lower PIP and ΔP were needed during PRVCV therefore avoiding potentially damaging shear forces at each closure and opening of the lung units.

REFERENCES

1. Gattinoni L., Pesenti A., Avalli L., Rossi F., and Bombina M., 1987, Pressure-volume curve of total respiratory system in acute respiratory failure: computed tomographic scan study, *Am.Rev.Respir.Dis.* 136:730-736.
2. Gattinoni L., Pesenti A., Bombino M., Baglioni S., Rivolta M., et al., 1988, Relationships between lung computed tomographic density, gas exchange, and PEEP in acute respiratory failure, *Anesthesiology* 69:824-832.
3. Dantzker D., 1982, Gas exchange in the adult respiratory distress syndrome, *Clin.Chest.Med.* 3:57-67.
4. Lachmann B., Danzmann E., Haendly B., and Jonson B., 1982, Ventilator settings and gas exchange in respiratory distress syndrome, In: Prakash, O., (ed): Applied physiology in clinical respiratory care, Martinus Nijhoff Publishers, The Hague, 141-176.
5. Gattinoni L., Pesenti A., Caspani M.L., Pelizzola A., Mascheroni D., Marcolin R., Iapichino G., Langer M., Agostonia., Kolobow T., Melrose D.G., and Damia G., 1984, The role of total static lung compliance in the management of severe ARDS unresponsive to conventional treatment, *Intensive Care Med.* 10:121-126.
6. Greenfield L.J., Ebert P.A., and Benson D.W., 1964, Effect of positive pressure ventilation on surface tension properties of lung extracts, *Anesthesiology* 25:312-316.
7. Reynolds E.O.R., and Taghizadeh A., 1974, Improved prognosis of infants mechanically ventilated for hyaline membrane disease, *Arch.Dis.Child.* 49:505-515.

8. Lachmann B., Jonson B., Lindroth M., and Robertson B., 1982, Modes of artificial ventilation in severe respiratory distress syndrome. Lung function and morphology in rabits after washout of alveolar surfactant, *Crit.Care Med.* 10:724-732.

9. Mead J., and Collier C., 1959, Relation of volume history of lungs to respiratory mechanics in anaesthetized dogs, *J.Appl.Physiol.* 14:669-678.

10. Webb H.H., and Tierney D.F., 1974, Experimental pulmonary edema due to intermittent positive pressure ventilation with high inflation pressures. Protection by positive end-expiratory pressure, *Am.Rev.Respir.Dis.* 110:556-563.

11. Rossi A., Gottfried S.B., Higgs B.D., Zocchi L., Grassino A., and Milic-Emili J., 1985, Respiratory mechanics in mechanically ventilated patients with respiratory failure, *J.Appl.Physiol.* 58:1849-1858.

12. Broseghini C., Brandolese R., Poggi R., Polese G., Manzin E., Milic-Emili J., and Rossi A., 1988, Respiratory mechanics during the first day of mechanical ventilation in patients with pulmonary edema and chronic airway obstruction, *Am.Rev.Respir.Dis.* 138:355-361.

13. Huygen P.E., Gültuna I., Ince C, Zwart A., Bogaard J.M., Feenstra B.W., and Bruining H.A., 1993, A new ventilation inhomogeneity index from multiple breath indicator gas wash-out tests in mechanically ventilated patients, *Crit.Care Med.* 21:1149-1158.

14. Lachmann B., Robertson B., and Vogel J., 1980, In-vivo lung lavage as an experimental model of the respiratory distress syndrome, *Acta Anaesthesiol. Scand.* 24:231-236.

15. Hickling K.G., 1990, Ventilatory management of ARDS: can it affect the outcome? *Intensive Care Med.* 16:219-226.

16. Mead J., Takishima T., and Leith D., 1970, Stress distribution in lungs: a model of pulmonary elasticity, *J.Appl.Physiol.* 28:596-608.

70

OPERATIVE LUNG CONTINUOUS POSITIVE AIRWAY PRESSURE VIA THE LUMEN OF A MOVABLE BRONCHIAL BLOCKER

M. Tuğrul, Ĝ. Köprülü, H. Karpat, and K. Pembeci

Department of Anesthesiology and Intensive Care
Medical Faculty, University of Istanbul
Istanbul, Turkey

INTRODUCTION

The primary aim of an anesthesiologist is to ventilate both lungs of the anesthetised patient. But isolation of a lung might be necessary under certain conditions such as bleeding, infection, air cysts or bronchial rupture, creating problems for the anesthesiologist. Besides these obligatory conditions, prevention of tidal movements of the surgical area is known to have many technical advantages for the surgeon and it has been requested frequently during thoracic surgery.

One lung ventilation (OLV) is an important addition to the armamentarium of the anesthesiologist for managing thoracic surgical cases. But it has a risk of developing hypoxemia and hypercarbia. Inspired oxygen concentrations (FiO_2) of 1.0 and keeping the same minute volume are commonly employed during OLV to minimize these risks. High frequency jet ventilation (HFJV), continuous airway pressure (CPAP) applied to the non-ventilated lung (nondependent lung-NDL) and PEEP or high tidal volumes to ventilated lung (dependent lung - DL) or combinations of these methods were used widely for the treatment of hypoxemia without interrupting surgery and reinflating the NDL.

In our study, we investigated the effects of CPAP to NDL, applied via the narrow lumen of a movable bronchial blocker with a special device modified in our clinic, on pulmonary shunting and oxygenation as well as surgical interference.

MATERIAL AND METHOD

Fifteen patients (age 18-71, mean 49.1±9.7 years, weight 47-82, mean 67.6±10.3 kg) scheduled for thoracotomy were studied after informed consent had been obtained. They had no other pathologies except thoracic surgical indication; their forced vital capacity and forced expiratory volumes at the first second were greater than 60% (FVC, FEV1 > 60%).

After premedication with morphine sulphate (0.15 mg/kg IM), anesthesia was induced by fentanyl (3 µg/kg), na-thiopentone (5-7 mg/kg). Endotracheal intubation was

Oxygen Transport to Tissue XVII, Edited by Ince et al.
Plenum Press, New York, 1996

545

facilitated by vecuronium (0.1 mg/kg) and an endotracheal tube (size 7.5, 8, 8.5) with a movable bronchial blocker (Univent, Fuji Corp., Japan) was placed and the blocker was advanced to the mainstem bronchus to be occluded using a fiberoptic bronchoscope.

50% O_2/N_2O and isoflurane 0.5-1.5% were used for the maintenance of anesthesia. During the OLV period, N_2O was stopped and fentanyl infusion (0.075 µg/kg) was used for analgesia. Patients were ventilated with a tidal volume of 10 ml/kg for both OLV and TLV **(TLV - step 1)**. Respiratory rate was adjusted to maintain a PETCO$_2$ between 30-35 mmHg. FiO$_2$ was adjusted as 0.5 for TLV and 1.0 for OLV. After thoracotomy OLV **(step 2)** was commenced and 5 cm H_2O CPAP to NDL, 5 cm H_2O PEEP to DL **(step 3)** was applied. Then CPAP was increased to 10 cm H_2O **(step 4)**. There was a 20-30 minute stabilization period before measurements in each step. Servo 900-C ventilator (Siemens, Elema, Sweden) and Criticare 1100 hemodynamic monitor were used during the procedures.

Pulmonary capillary wedge pressure **(PCWP)**, central venous pressure **(CVP)**, mean arterial pressure **(MAP)**, heart rate **(HR)**, arterial and mixed venous blood gases **(PaO$_2$, PaCO$_2$, PvO$_2$)**, pulmonary shunt **(Qs/Qt)** were measured and calculated at the end of each step.

Five cm H_2O PEEP to DL was applied by the ventilator. 5 and 10 cm H_2O PEEP valve (Vital Signs), and a continuous O_2 flow source was connected by stopcocks to perform CPAP via the blocker's lumen and the pressure in the lumen was monitored by a transducer (Fig. 1).

Surgical interference of the NDL expansion due to CPAP was evaluated by the surgeon giving a value between 1 and 5 (1- perfect conditions for surgery, 2- good for surgery, 3- sufficient for surgery, 4- partially limiting surgery, 5- seriously limiting surgical procedure).

Statistical analyses were performed by Student's t test and p<0.05 was regarded as significant.

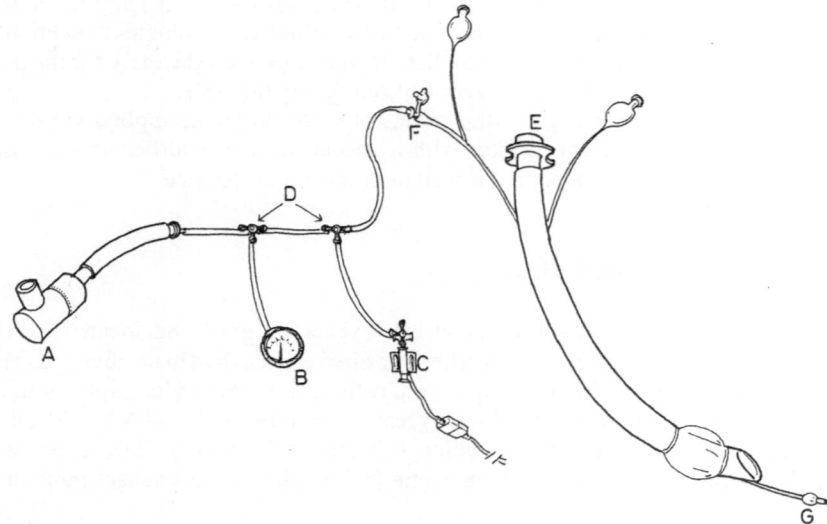

Figure 1. The device that was modified in our clinic: A. PEEP valve; B. continuous O_2 flow source; C. transducer for monitoring; D. stopcocks; E. univent tube; F. bronchial blocker's lumen; G. bronchial blocker.

Table 1. Results of the study

	TLV	OLV	OLV + CPAP 5, NDL PEEP 5, DL	OLV + CPAP 10, NDL PEEP 5, DL
HR (beat/m)	79.8±11	78.8±13.6	79.6±13.6	79.4±13.9
MAP (mmHg)	93±10.5	91.5±10.4	92.4±8	94±8
PCWP (mmHg)	16.5±4	16.6±3.7	16.9±3.3	17.2±3.4
CVP (mmHg)	9.2±4	8.6±3.5	9.5±3.4	9.8±3
PaO_2 (mmHg)	194±67	175±121	217±96	269±100
$PaCO_2$(mmHg)	34±6.7	38.9±9	38.8±9	36.4±7.4
Qs/Qt (%)	13.95±7.4	27±8	24.3±9.7	21±7.4

RESULTS

The results were summarized in Table 1. There were no significant differences in hemodynamic parameters (MAP, HR, CVP, PCWP) between each step.

Although there was no significant difference between step 1 and 2 for PaO_2 (p=0.16), it should be remembered that FiO_2 was increased to 1.0 in step 2. The differences between step 2, 3 and 4 were significant for PaO_2 (2,3-p=0.006, 2,4-p=0.00003, 3,4-p=0.00001).

The differences between the values of $PaCO_2$ for step 1-2, 2-4, 3-4 were significant (p=0.0002, p=0.01, p=0.008, respectively).

For Qs/Qt ratio the significance was valid for the differences between step 1-2, 2-3, 2-4, 3-4 (p=0.00003, p=0.02, p=0.000007, p=0.005, respectively).

Surgical conditions during step 2 were designated perfect by the surgeons. Surgical conditions were evaluated as 50% perfect, 50% good for step 3, and 80% good, 20% sufficient for step 4.

DISCUSSION

OLV is often used as a method for isolating a diseased lung from a normal lung and improving operating conditions. But hypoxemia can frequently be a consequence of OLV.

Alveolar hypoxia, whether caused by a low ventilation to perfusion ratio or atelectasis, causes pulmonary vasoconstriction, and the phenomenon is called hypoxic pulmonary vasoconstriction (HPV). This is an important compensatory mechanism for hypoxia. There is an increase in pulmonary vascular resistance that causes a diversion of blood flow away from the hypoxic lung to better ventilated normoxic lung. Blood flow diversion decreases the amount of shunt flow that can occur by the existence of a collapsed lung.

It was seen clearly in our study that expected 50% shunt fraction was decreased to 25% as a result of the compensatory mechanism explained above.

There are various factors that were contributing or counteracting hypoxia during thoracic operations. Surgical manipulation might increase HPV, whereas mediators released from surgical area might decrease HPV. Major factors contributing to hypoxia include: reduced lung volume of the dependent lung due to position, induction of general anesthesia, lung compression by mediastinal contents, absorption atelectasis due to high FiO_2, fluid transudation into DL and secretions (1).

Maintaining adequate oxygenation during OLV is often a problem. Hypoxia can easily occur in spite of high FiO_2 when the NDL is allowed to collapse. The optimal way to increase PaO_2 during OLV is to apply moderate levels of CPAP or HFJV to NDL and PEEP to DL.

Double lumen endotracheal tubes, Carlens or Robert Shaw, were preferred for managing OLV. Continuous oxygen flow via the nonventilated lung's lumen was used to improve oxygenation earlier, but later because of the unsatisfactory results an expiratory resistance was added to the system (2).

CPAP to NDL was applied via a modified Mapleson-D circuit by Slinger (3). Hensley used a Bain circuit with 15 cm H_2O pop-off valve (4) and Brown (5) and Hannenberg (6) preferred a T piece with 10 cm H_2O valve for applying CPAP . All authors used an oxygen flow of 1-10 l/min., and the level of CPAP in the circuit was monitored by cuff pressure gauge.

We used the Univent tube (Fuji Sys.Corp.,Tokyo, Japan) that was first introduced by Inoue et al. (7) just over a decade ago in our clinic for OLV. It is an endotracheal tube with a movable bronchial blocker integrated into the side wall of the tube and is technically easy to place properly.

Benumof et al. (8) have shown that CPAP can be applied easily and effectively through the blocker lumen of the Univent tube. The blocker lumen, that has a 2 mm diameter, may also be used for oxygen insufflation, HFJV, suction and irrigation. Benumof used a tube connected to the blocker lumen that has a length of 15 cm and a diameter of 1.5 mm instead of a CPAP valve that was used in our device for applying CPAP. He used an oxygen flow of 1.6-3.3 l/min. for maintaining 5-20 cm H_2O CPAP. By connecting a CPAP valve to system we maintained 10 cm H_2O CPAP with an oxygen flow less than 0.5 l/min.

Low levels of CPAP to NDL simply maintain the patency of NDL airways, allowing some oxygen distention of gas exchanging alveolar space in the NDL without significantly affecting the pulmonary vasculature. High levels of nonventilated lung CPAP improve oxygenation by permitting oxygen uptake in the nonventilated lung as well as by causing blood flow diversion to the ventilated lung where oxygen and carbon dioxide exchange can take place (9). This phenomenon was observed in our study, and the increase in CPAP resulted in a reduction of shunt fraction and CO_2 and an improvement in oxygenation.

PEEP to DL is the other component of treating hypoxia that occurs during OLV. The impaired dependent lung ventilation appears to result from a decrease in functional residual capacity (FRC), which may promote airway closure in DL. The decrease in FRC can be counteracted by the application of PEEP to DL. However, increase in thoracic pressure in DL caused by PEEP increase the blood flow through the NDL resulting an increase in shunt fraction; this can be balanced by applying CPAP to NDL. For this reason, we preferred to use PEEP and CPAP together.

As a result, we agree with Benumof(9) that CPAP can be easily applied via the narrow lumen of the Univent tube. We believe that connecting a CPAP valve to system will reduce the oxygen flow that is necessary for maintaining adequate oxygenation. Low levels of NDL CPAP are as efficacious as high levels and have less surgical interference; it is logical to use low levels of NDL CPAP in the beginning and increase CPAP levels if necessary.

REFERENCES

1. Gayes JM: Pro: One-Lung ventilation is best accomplished with the Univent Endotracheal Tube. J Cardiothoracic Vascular Anesth 7:103-7, 1993
2. Capan LM, et al: Optimization of arterial oxygenation during one-lung anesthesia. Anesth Analg 59: 847-51, 1980.
3. Slinger P, Triolet V, Wilson J: Improving arterial oxygenation during one-lung ventilation. Anesthesiology, 68: 291, 1988.

4. Hensley FA, Martin F, Skeehan TM: High pressure pop-off safety device when using the Bain circuit for CPAP oxygenation during one-lung ventilation. Anesthesiology, 67: 863-4, 1987.

5. Brown DL, Davis RF: A simple device for oxygen insufflation with CPAP during one-lung ventilation. Anesthesiology, 61:481-2, 1984.

6. Hannenberg AA, Satwicz PR, Dienes RS, O'Brien J: A device for applying CPAP to nonventilated upper lung during one-lung ventilation. Anesthesiology, 60:254-5, 1984.

7. Inoue H, Shotsu A, et al: Endotracheal tube with movable blocker to prevent aspiration of intratracheal bleeding. Ann Thor Surg, 37:497-9, 1984

8. Benumof JL, Gaughan S, Ozaki GT: Operative lung constant positive airway pressure with the Univent bronchial blocker tube. Anesth Analg 74: 406-10, 1992

9. Alfery DD, Benumof JL, Trousdale FR: Improving oxygenation during one-lung ventilation in dogs: The effects of PEEP and blood flow restriction to the nonventilated lung. Anesthesiology 55: 382-5, 1981.

71

SEPSIS AND NITRIC OXIDE

J. A. M. Avontuur,[1] H. A. Bruining,[1] and C. Ince[2]

[1.] Department of Surgery
University Hospital Rotterdam Dijkzigt
Dr. Molewaterplein 40, 3015 GD Rotterdam, The Netherlands
[2.] Department of Anaesthesiology
Academic Medical Center
University of Amsterdam
Meibergdreef 9, 1105 AZ Amsterdam, The Netherlands

INTRODUCTION

Sepsis or the septic syndrome is the number one cause of mortality in todays intensive care. Overall mortality of sepsis is estimated to be 40 to 60% and when shock or organ failure is present mortality rate is even higher despite recent progress in antibiotic and vasopressor therapy (Bone et al. 1991). The initial cardiovascular changes during hyperdynamic sepsis are characterized by massive vasodilatation with normal to high cardiac output, low peripheral vascular resistance and severe hypotension (Astiz et al. 1987, Groeneveld et al. 1985). In a large number of patients the hypotension is unresponsive to treatment with fluid substitution or vasopressors. This causes hypotension to remain present leading to higher mortality rates. Unresponsive hypotension is present in 50% of patients that die of septic shock (Parker et al. 1987). In the first week, unresponsive hypotension is the primary cause of mortality of septic patients. In a recent retrospective study with one hundred intensive care patients with sepsis, 80% of mortality in the first week was caused by severe hypotension (Ruokonen et al. 1991). For mortality in the second week multiple organ failure (MOF) was the primary cause.

Most important in the pathophysiology of sepsis are macro- and microvascular disturbances in combination with myocardial depression resulting from the complex interaction between bacterial cell products like lipopolysaccharides (LPS) and host defense mechanisms leading to the release of several cytokines [e.g., tumour necrosis factor (TNF), interleukines (IL-1, IL-2 and IL-6)], myocardial depressant factor (MDF), platelet activating factor (PAF), kinins, arachidonic acid metabolites and several other mediators. At macrocirculatory level the disturbances lead to increased cardiac output, reduced vascular resistance and hypotension (Parillo 1990). At microcirculatory level the disturbances lead to increased permeability of capillaries and maldistribution of blood flow causing inadequate tissue perfusion and tissue damage (Groeneveld et al. 1991, Rackow et al. 1988, Gosh et al. 1993). Myocardial depression is often present in septic patients and may aggravate the

Oxygen Transport to Tissue XVII, Edited by Ince et al.
Plenum Press, New York, 1996

551

macro- and microvascular defects (Quezado et al. 1992). The cardiovascular derangements will eventually lead to inadequate oxygen transport to tissue, causing MOF and death.

ROLE OF NITRIC OXIDE IN VASCULAR RELAXATION

In 1987 it was demonstrated that the NO radical was the same molecule as the earlier discovered but never isolated Endothelium Derived Relaxing Factor (EDRF; Palmer et al. 1987, Ignarro et al. 1987). We can say that this was the beginning of a whole new area of cardiovascular research. Nitric Oxide (NO) released from the endothelium is an important endogenous vasodilator and plays an important role in the regulation of tissue perfusion (Moncada et al. 1991, 1993). Other biological roles of NO include a role as neurotransmitter in the brain and peripheral nerves (Knowles et al. 1989), a role in host defense by being cytotoxic to bacteria (Marletta et al. 1988) and tumour cells (Hibbs et al. 1987), and a role in preventing platelet aggregation (Radomski et al. 1990). NO is released from the aminoacid L-arginine, but not D-arginine, by the enzyme NO synthase, that is located in a number of different cell types (Moncada et al. 1991). The NO radical is synthesized from the terminal guanidino nitrogen atom of L-arginine and forms L-citrulline as a co-product. Evidence from studies using $^{18}O_2$ and mass spectrometry indicates that in this reaction molecular oxygen is incorporated into both NO and citrulline by NO synthase which shows that the enzyme is a dioxygenase (Leone et al. 1991). Endothelial cells react to physical and chemical stimuli like shear stress, hypoxia or endothelial dependent vasodilators such as bradykinine, acetyl-choline, serotonine and adenosinediphosphate which can elicit Ca^{2+} release from intracellu-lar stores, with the release of NO in small amounts and for short periods (Rubanyi et al. 1986, Palmer et al. 1988, Whittle et al. 1989). The enzyme responsible for NO production in endothelial cells is called constitutive NO synthase (c-NOS). NO synthase is a heme containing enzyme which belongs to the P-450 class of enzymes (Wang et al. 1993). The constitutive NO synthase is activated by Ca^{2+}, which binds to calmodulin, forming a complex that is a crucial cofactor for enzyme activity (Busse et al. 1990). Furthermore nicotinamide adenine dinucleotide phosphate (NADPH) and tetrahydrobiopterin (BH$_4$) are required as cofactors (Marletta et al. 1988, Mayer et al. 1991). Following its production, the NO radical diffuses readily to the vascular smooth muscle cells where, by binding to iron in the heme at the active site of soluble guanylate-cyclase, the enzyme is activated to convert guanyl-tri-phosphate (GTP) to cyclic-3'5'guanyl-mono-phosphate (c-GMP). The intracellular rise of c-GMP will lead to relaxation of the vascular smooth muscle cell which causes vasodilation (Ignarro et al. 1985). NO can be considered as an endogenous nitrovasodilator since the same mechanism is responsible for vasodilation upon infusion of nitrovasodilator drugs as glyceryltrinitrate or nitroprusside that act as exogenous donors of NO (Ignarro et al. 1981). Organic nitrovasodilators, such as glyceryltrinitrate, undergo intermediate metabolism, an enzymatic step requiring thiol groups whereas the inorganic nitrovasodilators, such as sodiumnitroprusside, release NO readily when exposed to vascular tissue by a mechanism that is not fully understood. The half-life of the NO radical is only a couple of seconds because following production the NO radical reacts easily with sulphydryl groups in aminoacids or proteins, metal ions in redox reactions, hemoglobin, oxygen and radicals like superoxide. Nitrate and nitrite are the inactive and stable endproducts of NO production (Moncada et al. 1991). Although cultured endothelial cells depend on the presence of L-arginine in their medium for production of NO for longer periods (Palmer et al. 1988), intracellular concentrations of L-arginine *in vitro* and *in vivo* are under normal conditions not rate limiting in the production of NO so that administration of L-arginine will not lead to vasodilation (Palmer et al. 1988, Rees et al. 1989a, Kilbourn et al 1990a).

ROLE OF NITRIC OXIDE IN EXPERIMENTAL SEPSIS

Recently much interest has been focused on the role of nitric oxide in septic shock. Following discovery of the NO radical it was clear that NO could play a role in the vascular relaxation and hypotension during sepsis and endotoxemia. *In vitro* studies showed that isolated macrophages produced increased amounts of nitrate and nitrite upon stimulation with lipopolysaccharides (LPS=endotoxin) (Marletta et al. 1988, Stuehr et al. 1989). This was shown to be caused by the increased production of NO that depended on the presence of L-arginine and could be inhibited by analogues of L-arginine. The enzyme responsible for production of NO in macrophages differs from the c-NOS in endothelial cells. The c-NOS is calcium and calmodulin dependent and releases small amounts of NO whereas the enzyme in activated macrophages is calcium and calmodulin independent and releases high amounts of NO for a prolonged period of time independent of any stimuli. The same enzyme shown in macrophages could also be found in neutrophils, hepatocytes, myocardial cells and vascular smooth muscle stimulated with endotoxin (LPS) or cytokines and is called inducible NOS (i-NOS) (Brady et al. 1992, Geller et al. 1993, Moncada et al. 1991). The i-NOS is expressed under influence of endotoxin (LPS) and several cytokines like Tumor Necrosis Factor (TNF), interleukin-1β (IL-1β) and interferon-γ (Geller et al. 1993, Kilbourn et al. 1990b). The expression of i-NOS can be inhibited by pretreatment with glucocorticoids and inhibitors of protein synthesis (Knowles et al. 1990, Rees et al. 1990a). Once the i-NOS is expressed glucocorticoids are without effect. Both c-NOS and i-NOS depend on the presence of NADPH and tetrahydrobiopterin as a cofactor (Mayer et al. 1991, Gross et al. 1992). There is some evidence that when i-NOS is present the availability of the substrate L-arginine can be a limiting factor of NO production (Gonzales et al. 1992) .

A schematic representation of the different types of NO synthase and their role in the circulation are displayed in Fig 1. In general one can say that the constitutive NO synthase plays a physiological role in maintaining organ perfusion, whereas the expression of the inducible form (as in sepsis and endotoxemia) plays a more pathological role leading to the

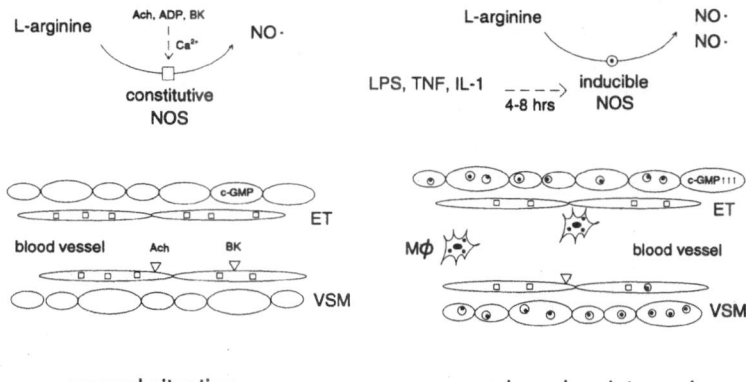

normal situation   **sepsis and endotoxemia**

Figure 1. Schematic representation of blood vessels with the different types of NO synthase in the normal situation (*left*) and during sepsis and endotoxemia (*right*). In the normal situation only constitutive NO synthase is present in endothelial cells (ET) that can be stimulated by acetylcholine (Ach), adenosinediphosphate (ADP), or bradykinine (BK). During sepsis and endotoxemia a second type of NO synthase is induced in macrophages (MΦ) and vascular smooth muscle cells (VSM) by stimulation with cytokines [e.g., interleukin-1 (IL-1) and tumor necrosis factor (TNF)] and endotoxin (LPS) with a lag-phase of 4-8 hours. The inducible NO synthase produces large amounts of NO for long periods of time resulting in massive vasodilation and tissue damage.

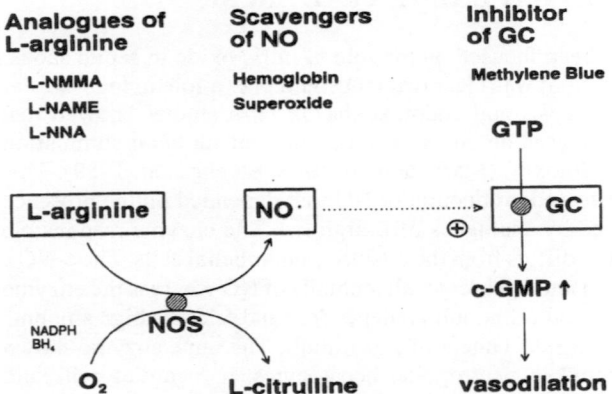

Figure 2. Schematic representation of the chemical reaction leading to formation of NO and respective vasodilation and the different pharmacological agents that intervene with the arginine/NO/c-GMP pathway at their specific sites. L-NMMA, N^G-monomethyl-L-arginine; L-NAME, N^G-nitro-L-arginine methyl ester; L-NNA, N^G-nitro-L-arginine; NO, nitric oxide; GC, guanylate-cyclase; GTP, guanyltriphosphate; c-GMP, cyclic-guanylmonophosphate; BH_4, tetrahydrobiopterin.

production of excessive amounts of NO resulting in vascular relaxation and tissue damage (Wright et al. 1992).

INHIBITORS OF NO SYNTHESIS

There are several ways to intervene with the L-arginine/NO/c-GMP pathway. Nitro-substituted or methylated analogues of L-arginine are able to competitively and stereospecific inhibit the production of NO by the constitutive and inducible NO synthase (Fig. 2). Analogues that are often used are N^G-monomethyl-L-arginine (L-NMMA; Palmer et al. 1988, Rees et al. 1989b), N^ω-Nitro-L-arginine (L-NNA; Moore et al. 1990) en N^ω-Nitro-L-arginine methylester (L-NAME; Rees et al. 1990b). These analogues all inhibit NO synthesis by competition with L-arginine, however, differ in their specific actions. The order of potency in inhibiting endothelial derived NO synthesis seems to be L-NNA>L-NAME>L-NMMA whereas in macrophages their order of potency is L-NMMA>L-NAME>L-NNA (Rees et al. 1990b, Mc Call et al. 1991). The pharmacokinetics of these NO inhibitors are largely unknown, although some differences have been shown in uptake and degradation. L-NMMA is converted to L-citrulline by endothelial cells whereas L-NNA is not (Hecker et al. 1990). In dogs and rabbits L-NAME is rapidly converted to L-NNA which seems to be the final degradation product (Schwarzacher et al. 1992, Kreycy et al. 1993). L-NAME however easily enters the intracellular compartment whereas L-NNA does not. These three

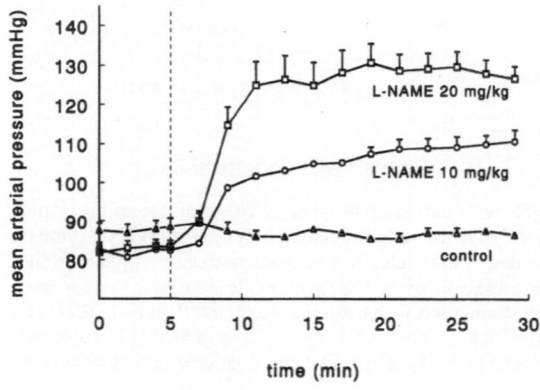

Figure 3. Effect of L-NAME on mean arterial blood pressure (±SEM) in anaesthetized male Wistar rats. Following anaesthesia with pentobarbitone blood pressure was measured by inserting a canula in the carotid artery. At t=5 min rats were injected with a bolus injection (0.3 ml) of either L-NAME 10 mg/kg (n=4) or L-NAME 20 mg/kg (n=4). The control animals only received NaCl 0.9% (n=3). Note that following injection of a single bolus L-NAME mean arterial blood pressure remains elevated for at least 30 min.

analogues of L-arginine are orally active. *In vivo* these analogues of L-arginine can raise blood pressure by inhibition of NO synthesis. Figure 3 shows the dose-dependent rise of blood pressure in the anaethesized rat following inhibition of NO synthesis with a bolus injection of L-NAME.

Scavengers of NO production can be used to inactivate the NO radical itself preventing interaction of NO with the ferrous heme-containing receptor site on soluble guanylate cyclase. Two potent scavengers of NO are the superoxide anion (O_2^-) and free hemoglobin (Martin et al. 1985). The superoxide anion can inactivate NO/EDRF in the vascular endothelium thus causing local vasoconstriction and selective loss of endothelium dependent vasodilation (Gryglewski et al. 1986). The chemical reaction that could be involved is that the superoxide anion reacts with NO to form peroxynitrite (NO_3^-) (Blough et al. 1985). The enzyme superoxide dismutase (SOD) specifically catalyses the reaction $2O_2^- + 2H^+ \rightarrow H_2O_2 + O_2$, and thus decreases the local concentration of the superoxide anion. Since with SOD less O_2^- is present to inactivate NO, the half-life of NO is increased by SOD. This may explain the vasodilating effect of superoxide dismutase in the microcirculation. Hemoglobin is a scavenger that directly binds NO (Martin et al. 1986). Since hemoglobin is a large protein that does not enter the intracellular space, the binding of NO is confined to the extracellular space. However, because NO readily crosses cell membranes, binding by hemoglobin in the extracellular space may increase the concentration gradient across the cell membranes, increase the net rate of diffusion out of the cells and thereby reduce the magnitude of relaxation. Hemoglobin from lysed erythrocytes has been implicated in the profound cerebral vasospasm which often follows cerebral hemorrhage (Osaka et al. 1977). The NO scavenging properties of hemoglobin could play a role in this.

Another way to interfere with NO metabolism is to inhibit the effector enzyme of NO synthesis with methylene blue (MB) which inhibits the enzyme soluble guanylate cyclase (GC) in the vascular smooth muscle cell (Martin et al. 1985). Methylene blue easily enters cells and inhibits soluble guanylate cyclase probably by oxidizing the ferrous form of the hemoprotein component of the enzyme to the ferric form. However, other mechanisms of action by MB, independent of guanylate cyclase inhibition, have also been suggested. MB can act as a direct inhibitor of NOS (Mayer et al. 1993) and by generation of a superoxide anion which directly inactivates NO (Wolin et al. 1990).

NO SYNTHASE INHIBITORS IN ANIMAL STUDIES

Studies with isolated vessels of septic rats show that the contractile response to noradrenaline is reduced (McKenna et al. 1986). The same is seen in vessels of endotoxemic animals and in vessels of healthy animals incubated with LPS (Flemming et al. 1990, Julou-Schaeffer et al. 1990). The reduced response to noradrenaline is accompanied by a rise in c-GMP content of the vessel wall (Flemming et al. 1991). Inhibitors of NO synthesis like L-NMMA or L-NNA and methylene blue, an inhibitor of soluble guanyl cyclase, can normalise the vasodilatation and the reduced response to noradrenaline (Flemming et al. 1991a, Gray et al. 1990). Furthermore it was found that following endotoxin injection the i-NOS is present primarily in the vascular smooth muscle cells and not in the vascular endothelium (Julou–Schaeffer et al.1990, Flemming et al. 1991b). These *in vitro* studies are underscored by *in vivo* experiments. Rats injected with LPS have reduced blood pressure and a diminished response to noradrenaline together with increased serum levels of nitrate and nitrite. Inhibition of NO synthesis in these endotoxemic rats increases blood pressure and restores the response to noradrenaline (Thiemermann et al. 1990, Gray et al. 1991, Guc et al. 1992). Continuous infusion of endotoxin for 24 hours in conscious sheep resulted in a hyperdynamic circulation with systemic vasodilatation, hypotension and increased cardiac output (Meyer et al. 1992). This model resembles the cardiovascular findings of human sepsis. Infusion of L-NAME could normalise the hemodynamic changes despite continuation of endotoxin infusion. Although oxygen delivery was reduced following L-NAME by reduction of cardiac output, total oxygen consumption was unaffected. It was concluded that L-NAME caused reduced perfusion of tissues with low metabolic demand. In the same model the effect of L-NAME on blood perfusion of the separate organs was studied (Meyer et al. 1993). Twenty-four hours following endotoxin infusion increased blood flow to brain, heart, intestines, liver and kidneys was seen. L-NAME reduced all regional flows. No detrimental effects of inhibition of NO synthesis were seen in this model of endotoxemia. However, a direct negative inotropic effect on the myocardium causing the reduction in cardiac output could not be excluded.

These findings suggest that NO plays an important role in the pathophysiology of sepsis and that inhibition of NO synthesis may offer a novel therapeutic target for treatment of septic patients with hypotension. Many studies have therefore been undertaken to evaluate the effect of inhibitors of NO synthesis during experimental sepsis and endotoxemia. However some controversy has arisen in the different animal studies on the use of NO synthase inhibitors during sepsis and endotoxemia.

CONTROVERSY IN ANIMAL STUDIES USING NO SYNTHASE INHIBITORS

In endotoxemic dogs two hours following injection of LPS, inhibition of NO synthesis resulted in correction of hypotension (Kilbourn et al. 1990a). Similar results were seen when TNF (Kilbourn et al 1990b) or IL-1 (Kilbourn et al. 1992) were used for induction of hypotension. It was concluded that increased release of NO was responsible for the hypotension found in these models. However studies with isolated cells show that induction of i-NOS takes a lag-phase of at least 4 to 8 hours needed for *de novo* synthesis of protein (Kilbourn et al. 1990c). Hypotension seen early following injection of LPS or TNF can therefore not be caused by increased production of NO by i-NOS but rather by other mechanisms. Recently it was found that increased NO synthesis by the c-NOS produces the early (within 60 min) hypotension and hyporeactivity to noradrenaline following LPS injection whereas the delayed hyporeactivity to noradrenaline and hypotension is caused by increased NO synthesis of i-NOS (Szabo et al. 1993). When interpreting results of studies on the increased NO release during endotoxemia or sepsis, the elapsed time between the induction of endotoxemia or sepsis and the time of NO inhibition should be mentioned. This may indicate which subtype of NO synthase is responsible for the changes in hemodynamics.

Although inhibition of NO synthesis may be of help in restoring blood pressure many have reported negative effects upon inhibition of NO synthesis during sepsis and endotoxemia. Inhibition of NO synthesis has been shown to raise mortality in a canine model of endotoxin shock (Cobb et al. 1992). Several studies have shown increased damage to the liver (Billiar et al. 1990), gastrointestinal tract (Hutcheson et al. 1990) and kidney (Schulz et al. 1992b) when inhibitors of NO synthesis are given during endotoxin shock probably because of a combined effect of organ hypoperfusion and trombosis. S-nitroso-N-acetyl-penicillamine (SNAP), an exogenous NO donor, could protect against the acute intestinal damage induced by LPS infusion in the rat (Boughton-Smith et al. 1990). Also many have reported reductions in cardiac output following inhibition of NO synthesis during sepsis and endotoxemia (Kilbourn et al. 1990, Klabunde et al. 1991, Nava et al. 1991). The cause for this is still unknown, however hypoperfusion of the myocardium has been hypothesised. A reduction in cardiac output may be detrimental by compromising tissue perfusion especially when cardiac output is already reduced, e.g. a hypodynamic circulation exists. Many studies on inhibition of NO synthesis however use models of sepsis and endotoxemia that cause a hypodynamic circulation with reduced cardiac output (Cain et al. 1991, Fink et al. 1990). In this situation there is already an underperfusion of the organs and not an overperfusion as in hyperdynamic septic shock (Lang et al. 1983). Additional vasoconstriction by inhibition of NO synthesis can, in this situation, lead to even further reductions in cardiac output and organ perfusion. Therefore inhibition of NO synthesis might be less effective in hypodynamic shock. When interpreting results of studies on the efficacy of NO inhibition during sepsis and endotoxemia one should include whether a hypo- or hyperdynamic model of shock was used. Since correction of hypotension at the cost of cardiac output could act counter productive, the effect on cardiac output should be mentioned

The increased release of NO by activated macrophages seems to play an important role in the killing of bacteria and fungi in several animal species (Marletta et al. 1988, Stuehr et al 1989). Arginine-supplemented diets could improve survival in septic rats probably by a beneficial effect on immune defense and bactericidal mechanisms via the arginine/NO pathway (Gianotti et al 1993). Theoretically inhibition of NO synthesis during sepsis could therefore reduce the immune response of macrophages to live bacteria. However in rats that were intraperitoneally injected with live E. Coli a higher survival rate was seen in the group that received a combination of imipenem-celastatine and L-NMMA as compared to imipenem-cilastatine or L-NMMA alone (Teale et al. 1992). This suggests a minimal effect of NO inhibition on immune response *in vivo*. It is unclear whether human monocytes or macrophages have the same NO dependent mechanism of bacterial killing as in rodents. Inhibition of NO synthesis with L-NMMA reversed the antimicrobial effects of activated mouse peritoneal macrophages but in human activated macrophages L-NMMA had no effect (Murray et al. 1992). Recently human macrophages failed to secrete NO even after activation with endotoxin, interferon-γ, granulocyte-macrophage colony-stimulating factor, tumour necrosis factor-α, bacteria or proliferating lymphocytes (Schneeman et al. 1993). Exogenous tetrahydrobiopterin, an essential cofactor of NO synthase not synthesized by human macrophages, did not stimulate NO production in human macrophages.

The effect of NO inhibition depends on the concentration of the inhibitor used. In a study of Nava et al. (1992) infusion of L-NMMA 10 mg/kg to endotoxemic rats resulted in a recovery of hypotension where all animals survived while in the control group of endotoxemic rats 40% of the animals died. When however a very high dose of 300 mg/kg L-NMMA was used the initial rise in blood pressure was followed by a steep fall in blood pressure while all animals died. A similar pattern of rise and fall in blood pressure with increased mortality was seen in endotoxemic rabbits using L-NMMA 300 mg/kg (Wright et al. 1992). The negative effects of NO inhibitors seem therefore dose related and are mostly seen in studies using extremely high doses of NO inhibitors.

SELECTIVE INHIBITORS OF INDUCIBLE NO SYNTHASE

The dose related toxicity of NO synthase inhibitors may be also related to the nonspecificity of the inhibitors used i.e. inhibition of both the inducible and constitutive NO synthase. In most studies inhibitors are used that affect both the constitutive and the inducible NO synthase. The inhibition of only the i-NOS may attenuate the excessive vasodilation, correct hypotension and prevent the toxic effects of increased NO synthesis without interfering in physiologic NO mediated processes. However, inhibition of the constitutive enzyme will lead to the disturbance of physiologic processes regulated by NO and may lead to tissue hypoperfusion and ischemia. That some NO is required for normal vascular function was shown in a study where co-administration of S-nitroso-N-penicillamine (SNAP), an exogenous nitric oxide donor drug, protected against the deleterious effects of a large dose of L-NMMA (300 mg/kg) in endotoxic shock in the rabbit (Wright et al. 1992). To prevent tissue hypoperfusion, total inhibition of NO synthesis combined with an exogenous NO donor could therefore be helpful. There is some evidence that aminoguanidine is an inhibitor of NO synthesis that has selectivity for the inducible enzyme (Griffiths et al. 1993). *In vitro* aminoguanidine caused a dose-dependent increase in phenylephrine-induced tension of intact and endothelium denuded vascular rings from endotoxin-treated rats, but had no effect on sham-treated controls. In endotoxin-stimulated cultured macrophages aminoguanidine was approximately seven times more potent than L-NMMA in inhibiting the i-NOS (Tilton et al. 1993). In contrast L-NMMA was 15 times more potent than aminoguanidine at inhibiting constitutive NO synthase isolated from rat brain. *In vivo* aminoguanidine was 40 fold less potent in increasing mean arterial pressure in anaesthetized rats (Corbett et al. 1992). Whether aminoguanidine can raise blood pressure in animals during sepsis or endotoxemia is unknown. L-canavanine is another inhibitor which was found to have some selectivity against i-NOS although the concentration range in which selectivity exists is narrow (Umans et al. 1992). Tetrahydrobiopterin (BH_4) is an essential cofactor for all isoforms of NO synthase and an absolute requirement for NO synthesis in macrophages, endothelial cells and vascular smooth muscle cells. (Tayeh et al. 1989, Marletta et al. 1989, Gross et al. 1992). Recently pretreatment with N-acetylserotonin, an inhibitor of tetrahydrobiopterin biosynthesis, could prevent the fall in blood pressure and the increase in plasma nitrite levels induced by E.Coli lipopolysaccharides in the rat (Klemm et al. 1993). In control rats receiving only N-acetylserotonin no effects on blood pressure or nitrite levels were seen. Although inhibitors of BH_4 synthesis may provide a novel approach for therapy in pathological conditions with excess NO, these BH_4 inhibitors have to be given before induction of BH_4, much like dexamethasone has to be given before induction of i-NOS to prevent excess NO production. Further studies using selective inhibitors of i-NOS *in vivo* during sepsis or endotoxemia must be awaited; such selective inhibitors may be of special value in the future management of septic patients.

NITRIC OXIDE IN THE MYOCARDIUM

In the coronary vessels NO plays an important role in the regulation of myocardial blood flow. In the isolated perfused rabbit heart there is a basal release of NO and increased release of NO accounts for the vasodilatation induced by acetylcholine (Amezcua et al. 1988) and bradykinine (Kelm et al. 1988). In the same heart preparation inhibition of NO synthesis with L-NMMA increases coronary resistance and prevents the vasodilator response to acetylcholine (Amezcua et al. 1989). In the awake dog L-NMMA causes a dose-dependent reduction in epicardial coronary diameter (Chu et al. 1991). In humans intracoronary infusion of L-NMMA causes a reduction in epicardial calibre and coronary blood flow (Lefroy et al. 1993).

Common features of sepsis are myocardial depression with ventricular dilatation, decreased stroke work and reduced left ventricular and right ventricular ejection fractions (Parillo et al. 1990). Similar changes are seen in healthy volunteers when given endotoxin (Suffedrini et al. 1989) and in animal models of sepsis and endotoxemia (Abel et al. 1989, McDonough et al. 1984, Goldfarb et al. 1986). The myocardial depression does not depend on the presence of myocardial ischemia (Adams et al. 1989). The human coronary circulation during sepsis is often characterized by inappropriately high coronary flow with decreased oxygen extraction, closely mimicking the peripheral circulatory anomalies of vascular relaxation (Cunnion et al. 1986, Dhainaut et al. 1987). This pattern of disordered coronary flow regulation could also be seen in models of hyperdynamic sepsis in the rat both *in vivo* and *in vitro* and was suggested to result from the release of a putative vasodilator substance (Lang et al. 1983, Rumsey et al. 1988). In the isolated rabbit heart preparation the coronary vasodilation induced by endotoxin was found to be nitric oxide dependent and could be inhibited by dexamethasone suggesting that i-NOS plays a role (Smith et al. 1991). Recently Baydoun et al. (1993) showed that in the isolated rat heart endotoxin rapidly stimulates endothelium dependent vasodilatation which suggest that c-NOS plays a role. Both isoforms of NO synthase may therefore play a role in the coronary vasodilatation during sepsis. Not only the coronary vessels express NO synthase, but also the endocardium and myocardium itself have the capacity to express the constitutive and inducible NO synthase in both animal and human. The constitutive form of NO synthase has been shown to be present in cultured endocardial cells of the pig (Schulz et al. 1991) and in normal rat myocardium (Schulz et al. 1992a). The inducible NO synthase is expressed following endotoxemia and isolated rat myocytes express the inducible NO synthase after stimulation with cytokines which can be inhibited with dexamethasone (Schulz et al. 1992a). The increased release of NO by i-NOS within cardiomyocytes itself could play a role in myocardial depression during sepsis. In the isolated papillary muscle of pig and ferret, endocardial NO release reduces the duration of contraction and accelerates relaxation, an effect that is mediated by c-GMP (Smith et al. 1991, Shah et al. 1991). In a study of Brady et al. (1992), isolated ventricular myocytes of the guinea pig heart following endotoxemia showed reduced contractility to electrical stimulation. The depression in myocyte contraction could be corrected with L-NAME, L-NMMA, MB or dexamethasone pretreatment which suggests a causative role for i-NOS. In this study myocytes exposed to MB showed signs characteristic of calcium overload with hypercontractility and fibrillation. This suggests that MB has direct toxic effects on cardiomyocytes as reported by others (Baydoun et al. 1990). Human ventricular tissue of patients with dilated cardiomyopathy shows a significant activity of the inducible NO synthase and increased levels of c-GMP with low activity of the constitutive enzyme (De Belder et al. 1993). This could suggest that the inducible enzyme plays a pathological role in the human myocardium, whereas the constitutive enzyme plays a more physiological role.

In theory inhibition of NO synthesis during sepsis or endotoxemia could prevent the overperfusion of myocardial tissue and correct the contractile dysfunction caused by over-production of NO. However, in contrast to this, many investigators have reported reductions in heart frequency and cardiac output following inhibition of NO production during sepsis and endotoxemia *in vivo* (Petros et al. 1991, 1994, Schneider et al. 1992, Wright et al. 1992, Cobb et al. 1992, Klabunde et al. 1991). Combined with the increased vascular resistance seen after inhibition of NO formation such reductions in cardiac output may compromise myocardial and tissue perfusion. It is not clear whether the fall in cardiac output seen after administration of inhibitors of NO synthesis during sepsis and endotoxemia is secondary to increased blood pressure and increased afterload or a direct effect of NO synthase inhibition in the heart. A direct negative effect on cardiomyocytes or decreased coronary flow causing myocardial ischemia have both been hypothesized (Cobb et al. 1992, Hotchkiss et al. 1992). A direct negative effect on isolated cardiomyocytes was however shown not to be present using the NO inhibitor L-NMMA (Amrani et al. 1992) and L-NAME (Brady et al. 1992). There is however some evidence that inhibition of NO synthesis can cause myocardial ischemia. In a recent study by Duncker et al. (1994) using radioactive microspheres, L-NNA was shown to exacerbate myocardial hypoperfusion during exercise in the presence of a coronary stenosis in awake dogs. In the isolated perfused rat heart L-NMMA caused a dose dependent reduction in coronary flow together with a reduction in cardiac performance and increased release of lactate in the effluent which is suggestive for anaerobic metabolism and myocardial ischemia (Amrani et al. 1992). Administration of L-NMMA has been reported to exacerbate global myocardial ischemia *in vivo* in a rabbit model of endotoxemia as indicated by changes on the electrocardiogram (Wright et al.1992). However, in this model of endotoxemia, ischemic changes were already present before administration of L-NMMA which suggest that a hypodynamic circulation was present with already reduced coronary flow rates. We recently showed that L-NNA or MB can cause local myocardial ischemia in a hyperdynamic model of endotoxemia in the isolated perfused rat heart using the fluorescent properties of reduced nicotinamide adenine dinucleotide (NADH) and a videoanalyzer to detect local ischemia (Avontuur et al. 1994). Endotoxin treated hearts showed increased coronary flow and oxygen consumption which could be corrected with L-NNA or MB. However, NADH videofluorimetry revealed areas of high NADH fluorescence, indicative of myocardial ischemia, in endotoxin treated hearts but not normal hearts following L-NNA or MB (Fig. 4). This may suggest that endotoxemia promotes myocardial ischemia when NO synthesis is inhibited.

ROLE OF NITRIC OXIDE IN HUMAN SEPSIS

Plasma levels of nitrate (NO_3^-) and nitrite (NO_2^-) represent the stable endproducts of NO synthesis and can be used as a marker of NO synthase activity (Palmer et al. 1987, Green et al. 1982). In a study by Ochoa et al. (1991) in 39 intensive care patients it was found that septic patients had increased levels of nitrate and nitrite whereas trauma patients had not. Systemic vascular resistance showed an inverse correlation to the levels of nitrate and nitrite in the plasma. High plasma levels of c-GMP during sepsis correlate with low systemic vascular resistance (Schneider et al. 1993). In another clinical study plasma levels of nitrate were evaluated in 12 patients treated with IL-2 as antitumour therapy (Hibbs et al. 1992). An important complication of this therapy is the presence of a hyperdynamic circulation with hypotension and low systemic vascular resistance. Patients under treatment showed an eight-fold raise in plasma nitrate levels and urine nitrate excretion. This increased production of nitrate was derived from the guanidino nitrogen atom of L-arginine. There was a significant negative correlation between mean arterial pressure and maximum increase in plasma nitrate levels.

Another way to indirectly estimate NO production is to determine consumption of L-arginine that delivers the nitrogen atom in the NO radical. Freund et al. (1978, 1979) showed that arginine levels in the plasma of septic patients were reduced 24% as compared

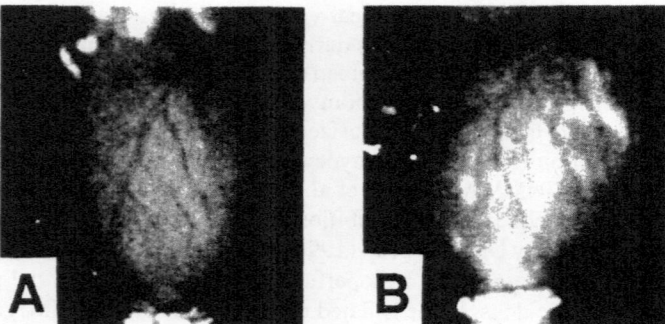

Figure 4. NADH fluorescence pictures of the isolated rat heart following inhibition of soluble guanylate-cyclase with methylene blue (MB). MB could correct the massive vasodilatation in the endotoxin-treated heart. However, following MB, the control heart (A) showed no signs of ischemia whereas in the endotoxin-treated heart (B) areas of high NADH fluorescence are present, indicative of local myocardial ischemia.

to normal. Non-survivors had lower serum levels of arginine than survivors although a causative relation is difficult to establish.

NO plays a physiological role in human microcirculation (Moncada et al. 1993). Intra-arterial infusion of L-NMMA to healthy volunteers resulted in a rise in vascular resistance in the forearm and a reduced response to acetylcholine (Vallance et al. 1991). Local intracoronary infusion of L-NMMA showed a reduction in vascular diameter without the presence of myocardial ischemia and prevented the response to achethylcholine (Lefroy et al. 1993). This shows that in humans NO plays a role in both the coronary vasculature and the peripheral arterial system however that under normal conditions NO is not vital to keep flow to the tissues adequate for demand. Possibly NO gives the vasculature some reserve in conditions of increased demand as during exercise. In patients with chronical renal failure endogenous inhibitors of NO synthase are present in the serum (Vallance et al. 1991). It is not clear whether these endogenous inhibitors of NO synthesis play a role in the hypertension seen in these patients.

INHIBITORS OF NO SYNTHESIS IN HUMAN SEPSIS

Inhibitors of NO synthesis have been used in the treatment of patients with septic shock. The number of patients in these studies is small and no toxicologic data are currently available on inhibitors of NO synthesis; also all of the patients in these studies were seriously ill and did not respond to conventional therapy. Conclusions should therefore be drawn with caution.

Schneider et al. (1992) used methylene blue, an inhibitor of soluble guanylate -cyclase, the effector enzyme of nitric oxide, in two patients with septic shock. Methylene blue has been used for years in the treatment of patients with NO and nitrate poisoning (Clutton-Brock et al. 1967). Furthermore the oxidizing properties of MB have been used clinically to induce methemoglobinemia in the treatment of cyanide poisoning (Goodman et al. 1975). Infusion of MB in a dose of 1-3 mg/kg caused an early rise in blood pressure and systemic vascular resistance with a small reduction in cardiac output. Also the dose of noradrenaline could be reduced. *In vitro* studies have shown that MB has direct toxic effects on cardiomyocytes (Baydoun et al. 1990, Brady et al. 1992). Although there was a small reduction in cardiac output, no negative side effects were reported in this study. Despite the hemodynamic improvements both patients died 4-6 days later because of multiple organ failure (MOF).

Petros et al. (1991) describes the use of NO synthase inhibitors in two patients with unresponsive septic shock. In the first patient L-NMMA (0.3-1.0 mg/kg) increased blood pressure and systemic vascular resistance. In the this patient the infusion of noradrenaline could be stopped. Because the effect of a bolus injected was only short (10-15 min at a dose of 1 mg/kg) in the second patient L-NAME was given in a bolus of 0.15 mg/kg continued with continuous infusion of 5 µg/kg/min. Bolus injection increased mean arterial pressure from 84 to 102 mmHg after 5 min. This effect was present for 10-15 min and was caused by increased systemic vascular resistance (cardiac output and pulmonary wedge pressure were unchanged). During continuous administration the noradrenaline infusion could be reduced from 0.6 to 0.2 µg/kg/min. Within 48 hours the infusion of noradrenaline and L-NAME could be stopped and blood pressure and systemic vascular resistance returned to normal. The first patient eventually became independent of intensive care support. The second patient died two days later because of the combination of repeated intra-abdominal sepsis, adult respiratory distress syndrome (ARDS), and diffuse intravascular coagulation (DIC).

Schilling et al. (1993) used L-NMMA in the treatment of unresponsive hypotension in a septic patient that had two episodes of severe hypotension within a period of four weeks.

Treatment with this inhibitor of NO synthesis resulted in an increase in mean arterial blood pressure, a reduction in cardiac output and an increase in arterial oxygen pressure. During infusion of L-NMMA infusion of catacholamines could be reduced. The initial response to a bolus of L-NMMA lasted only 5-10 min so that bolus injection had to be repeated several times.

In a recent randomized, double blind, placebo controlled study of Petros et al. (1994) the effects of L-NMMA was investigated in 12 patients with severe sepsis. L-NMMA (0.3 and 1.0 mg/kg) resulted in a dose-dependent increase in mean arterial pressure, systemic vascular resistance, pulmonary vascular resistance, central venous pressure, and pulmonary artery occlusion pressure, and a decrease in cardiac output and heart rate. Continuous infusion of L-NMMA (1 mg/kg/h) produced sustained hemodynamic changes. In the L-NMMA group 2/5 patients survived the 28 day study period whereas in the control group 1/6 patients survived which was not significant. One patient was withdrawn from the study because of infection with HIV.

In a prospective intervention study in septic patients by Lorente et al. (1993) the effect of a bolus injection of L-NNA was evaluated followed by a bolus injection of L-arginine. Injection of L-NNA in a dose of 20 mg/kg resulted in a generalised vasoconstriction of the systemic and pulmonary circulation with hypertension and a strong reduction of cardiac output. These effects could be reversed by administration of L-arginine. Administration of L-arginine in a separate group of septic patients resulted in transient hypotension and an increase in cardiac index together with an increase in oxygen consumption. This could be due to increased formation of NO, however, both L-arginine and D-arginine can cause vasodilation when administered into the brachial artery of healthy volunteers which suggests that this effect is not mediated by NO (Calver et al. 1990). Compared to the studies mentioned above the dose of L-NNA used was very high. Furthermore in vitro studies have shown that L-NNA is a more potent inhibitor of constitutive NO synthase than L-NMMA and L-NAME. This could explain the generalised vasoconstriction seen in this study.

From the experiments mentioned above one can say that inhibition of NO synthesis in human sepsis leads to increased blood pressure and systemic vascular resistance. This allowed the dose of vasopressors to be reduced. Inhibition of NO synthesis could therefore be of help in patients with hyperdynamic septic shock where hypotension does not respond to conventional therapy. Furthermore in most studies a reduction in cardiac output, decreased heart frequency and an increase in pulmonary vascular resistance is seen. This makes inhibition of NO synthesis less desirable in a situation of hypodynamic septic shock. In the end stages of sepsis when MOF is present, underperfusion of organs is a common feature. In this situation inhibition of NO synthesis may aggrevate tissue hypoperfusion. Larger studies are required to determine the effect of NO inhibition on survival in patients with sepsis.

CONCLUSIONS

Nitric oxide plays a central role in the cardiovascular derangements of sepsis and endotoxemia in animals and humans. Inhibition of NO synthesis in animal models of endotoxemia and sepsis may be either beneficial or counterproductive depending on the specific model used. Inhibition of NO synthesis in humans provides a new approach in the treatment of severe hyperdynamic sepsis with unresponsive hypotension. However, cardiac monitoring is desired since myocardial perfusion may be compromised. Selective inhibition of the different isoenzymes of NO synthase may provide new tools for therapy in the future.

REFERENCES

Abel FL: Myocardial function in sepsis and endotoxin shock. *Am J Physiol* 1989;257:R1265-R1281.

Adams HR, Parker JL, Laughlin MH: Intrinsic Myocardial dysfunction during endotoxemia: Dependent or independent of myocardial ischemia. *Circ Shock* 1990;30:63-76.

Amezcua JL, Dusting GJ, Palmer RM, Moncada S: Acetylcholine induces vasodilatation in the rabbit isolated heart through the release of nitric oxide, the endogenous nitrovasodilator. *Br J Pharmacol* 1988 95;830-834.

Amezcua JL, Palmer RMJ, de Souza BM, Moncada S: Nitric oxide synthesized from L-arginine regulates vascular tone in the coronary circulation of the rabbit. *Br J Pharmacol* 1989;97:1119-1124.

Amrani M, O'shea J, Allen NJ, Harding SE, Jayakumar J, Pepper JR, Moncada S, Yacoub MH: Role of basal release of nitric oxide on coronary flow and mechanical performance of the isolated rat heart. *J Physiol* 1992;456:681-678.

Astiz ME, Rackow EC, Falk JL, Kaufman BS, Weil MH: Oxygen delivery and consumption in patients with hyperdynamic septic shock. *Crit Care Med* 1987;15(1):26-28.

Avontuur JA, Bruining HA, Ince C: Inhibition of nitric oxide synthesis causes myocardial ischemia in endotoxemic rats. *Circ Res* 1995;76:418-425.

Baydoun AR, Peers SH, Cirino G, Woodward B: Vasodilator action of endothelin-1 in the perfused rat heart. *J Cardiovasc Pharmacol* 1990;15:759-763.

Baydoun AR, Foale RD, Mann GE: Bacterial endotoxin rapidly stimulates prolonged endothelium-dependent vasodilatation in the rat isolated perfused heart. *Br J Pharmacol* 1993;109:987-991.

Belder De AJ, Radomski MR, Why HJ, Richardson PJ, Bucknall CA, Salas E, Martin JF, Moncada S: Nitric oxide synthase activities in human myocardium. *Lancet* 1993;341:84-85.

Billiar TC, Curran RD, Harbrecht BG, et al: Modulation of nitrogen oxide synthesis *in vivo*:: NG-monomethyl-L-arginine inhibits endotoxin-induced nitrite/nitrate biosynthesis while promoting hepatic damage. *J Leukoc Biol* 1990;48:565-569.

Blough NV, Zafiriou OC: Reaction of superoxide with nitric oxide to form peroxonitrite in alkaline aqueous solution. *Inorg Chem* 1985;24:3502-3504.

Bone RC: The pathogenesis of sepsis. *Ann Int Med* 1991;115: 457-469.

Boughton–Smith NK, Hutcheson IR, Deakin AM, Whittle BJR, Moncada S: Protective effect of S-nitroso-N-acetyl-penicillamine in endotoxin-induced acute intestinal damage in the rat. *Eur J Pharmacol* 1990;191:485-488.

Brady JB, Poole–Wilson, Harding SE, Warren JB: Nitric Oxide production within cardiac myocytes reduces their contractility in endotoxemia. *Am J Physiol* 1992;263 (*Heart Circ Physiol* 32): H1963-H1966.

Busse R, Mülsch A: Calcium-dependent nitric oxide synthesis in endothelial cytosol is mediated by calmodulin. *FEBS Lett* 1990;265:133-136.

Cain MS, Curtis SE: Experimental models of pathologic oxygen supply dependency. *Crit Care Med* 1991;19: 603-612.

Calver A, Collier J, Vallance P: L-arginine-induced hypotension. *Lancet* 1990;ii:1016-1017.

Chu A, Chambers DE, Lin CC, Kuehl WD, Palmer RM: Effects of inhibition of Nitric Oxide formation on basal vasomotion and endothelium-dependent responses of the coronary arteries in awake dogs. *J Clin Invest* 1991;87: 1964-1968

Clutton–Brock J: Two cases of poisoning by contamination of nitrous oxide with higher oxides of nitrogen during anaesthesia. *J Anaesth* 1967;39:388-392.

Cobb JP, Natanson C, Hoffman WD, Lodato RF, Banks S, Koev CA, Solomon MA, Elin RJ, Hosseini JM, Danner RL: N$^{\omega}$-Amino-L-arginine, an inhibitor of Nitric Oxide Synthase, raises vascular resisistance but increases mortality rates in awake canines challenged with endotoxin. *J Exp Med* 1992;176:1175-1182

Corbett JA, Tilton RG, Chang K, Hasan KS, Ido Y, Wang JL, Sweetland MA, Lancaster JR, Williamson JR, McDanie ML: Aminoguanidine, a novel inhibitor of nitric oxide formation, prevents diabetic vascular dysfunction. *Diabetes* 1992;41,552-556.

Cunnion RE, Schaer GL, Parker MM, Natanson C, Parillo JE: The coronary circulation in human septic shock. *Circulation* 1986;73(4):637-644.

Dhainaut JF, Huygebaert MF, Monsalier JF, Lefevre G, Ava-Santucci J, Brunet F, Villemant D, Carli A, Raichvarg D: Coronary hemodynamics and myocardial metabolism of lactate, free fatty acids, glucose, and ketones in patients with septic shock. *Circulation* 1987;75(3):533-541.

Duncker DJ, Bache RJ: Inhibition of nitric oxide production aggrevates myocardial hypoperfusion during exercise in the presence of a coronary artery stenosis. *Circ Res* 1994;74:629-640.

Fink MP, Heard SO: Laboratory models of sepsis and septic shock. *J Surg Res* 1990;49:186-196.

Flemming I, Gray GA, Julou-schaeffer G, Parret JR, Stoclet JC: Incubation with endotoxin activates the L-arginine pathway in vascular tissue. *Biochem Biophys Res Commun* 1990;171:562-568.

Flemming I, Julou–Schaeffer, Gray GA, Parret JR, Stoclet JC: Evidence that an L-arginine/nitric oxide dependent elevation of tissue cyclic GMP content is involved in depression of vascular reactivity by endotoxin. *Br J Pharmacol* 1991a;103:1047-1052

Flemming I, Gray GA, Schott C, Stoclet JC: Inducible but not constitutive production of nitric oxide by vascular smooth muscle cells. *Eur J Pharmacol* 1991b;200:375-376.

Freund H, Atamian S, Holroyde J, Fischer JE: Plasma amino acids as predictors of the severity and outcome of sepsis. *Ann Surg* 1979;190:571-576

Freund H, Ryan JA, Fischer4 JE: Aminoacids derangements in patients with sepsis: Treatment with branched chain aminoacids rich infusions. *Ann Surg* 1978;188:423-429

Geller DA, Nussler AK, Di Silvio M, Lowenstein CJ, Shapiro RA, Wang SC, Simmons RL, Billiar TR: Cytokines, endotoxin, and glucocorticoids regulate the expression of inducible nitric oxide synthase in hepatocytes. *Proc Natl Acad Sci USA* 1993;90;522-526

Gianotti L, Alexander W, Pyles T, Fukushima R: Arginine-supplemented diets improve survival in gut-derived sepsis and peritonitis by modulating bacterial clearance. *Ann Surg* 1993;217:644-654.

Goldfarb RD, Nightingale LM, Kish P, Weber PB, Loegering DJ: Left ventricular function during lethal and sublethal endotoxemia in the swine. *Am J Physiol* 1986;251(20): H364-H373.

Goodman LS and Gilman A: The pharmacological basis of therapeutics. *Macmillan, New York* 1975; pp. 1003-1004.

Gonzales C, Fernandez A, Martin C, Moncada S, Estrada C: Nitric oxide from endothelium and smooth muscle modulates resposes to sympathetic nerve stimulation: Implications for endotoxin shock. *Biochem Biophys Res Commun* 1992;86:151-156.

Gosh S, Latimer RD, Gray BM, Harwood RJ, Oduro A: Endotoxin induced organ injury. *Crit Care Med* 1993;21:S19-S24.

Gray GA, Julou-schaeffer G, Oury K, Flemming I, Parret JR, Stoclet JC: A L-arginine-derived factor mediates endotoxine-induced vascular hyposensivety to calcium. *Eur J Pharmacol* 1990;191:89-92

Gray GA, Schott C, Juluo–Schaeffer G, Fleming I, Parratt JR, Stoclet JC: The effect of inhibitors of the L-arginine/nitric oxide pathway on endotoxin induced loss of vascular responsiveness in anaesthetized rats. *Br J Pharmacol* 1991;103:1218-1224

Green LC, Wagner DA, Glogowski J, Skipper PL, Wishnok JS, Tannenbaum SR: Analysis of nitrate, nitrite, and [15N]nitrate in biological fluids. *Anal Biochem* 1982:126:131-138

Griffiths MJ, Messent M, MacAllister RJ, Evans TW: Aminoguanidine selectively inhibits inducible nitric oxide synthase. *Br J Pharmacol* 1993:110;963-968.

Groeneveld AB, Bronsveld W, Thijs LG: Hemodynamic determinants of mortality in human septic shock. *Surgery* 1985;140-152.

Groeneveld AB, Lambalgen van AA, Bos van den GC, Bronsveld W, Nauta JP, Lambertus GT: Maldistribution of heterogeneous coronary blood flow during canine endotoxin shock. *Cardiovasc Res* 1991;25:80-88.

Gross SS and Levi R: Tetrahydrobiopterin Synthesis. *J Biol Chem* 1992;267:25;722-25729.

Gryglewski RJ, Palmer RM, Moncada S: Superoxide anion is involved in the breakdown of endothelium-derived relaxing factor. *Nature* 1986;320:454-456.

Guc MO, Furman BL, Parratt JR: Modification of α-adrenoceptor-mediated pressor reponses by N^G-nitro-L-arginine methyl ester and vasopressin in endotoxin-treated pitched rats. *Eur J Pharmacol* 1992;224:63-69.

Hecker M, Mitchell JA, Harris HJ, Katsura M, Thiemermann C, Vane JR: NG-monomethyl-L-arginine is metabolized by endothelial cells to L-citrulline. *Biochem Biophys Res Commun* 1990;167:1037-1043.

Hibbs JB, Vavrin Z, Taintor RR: L-arginine is required for expression of the activated macrophage effector mechanism causing selective metabolic inhibition in target cells. *J Immunol* 1987;138:550-565.

Hibbs JB, Westenfelder C, Taintor R, Vavrin Z, Kablitz C, Baranowski RL, Ward JH, Menlove RL, MCCurry MP, Kusher JP: Evidence for cytokine-inducible nitric-oxide synthesis from L-arginine in patients receiving IL-2 therapy. *J Clin Invest* 1992;89:867-877

Hotchkiss RS, Karl IE, Parker JL, Adams HR: Letter to the editor. *Lancet* 1992;339(15):434-435.

Hutcheson IR, Whittle BJ, Boughton-Smith NK: Role of nitric oxide in maintainining vascular integrety in endotoxin-induced acute intestinal damage in the rat. *Br J Pharmacol* 1990;101:815-820.

Ignarro LJ, Byrns RE, Buga GM, Wood KS: Endothelium derived relaxing factor from pulmonary artery and vein possesses pharmacologic and chemical properties identical to those of nitric oxide radical. *Circ Res* 1987:61:866-879.

Ignarro LJ, Lippton H, Edwards JC, Baricos WH, Hyman AL, Kadowitz PJ, Gruetter CA: Mechanism of vascular smooth muscle relaxation by organic nitrates, nitrites, nitroprusside and nitric oxide:

evidence for the involvement of S-nitrosothiols as active intermediates. *J Pharmacol Exp Ther* 1981;218:739-749

Ignarro LJ, Lippton, Kadowitz PJ: The pharmacological and physiological role of cyclic GMP in vascular smooth muscle. *Ann Rev Pharmacol Toxicol* 1985;25:171-191.

Julou-Schaeffer G, Gray GA, Fleming I, Scott C, Parrat JR, Stoclet JC: Loss of vascular respon-siveness induced by endotoxin involves L-arginine pathway. *Am J Physiol* 1990;259 (*Heart Circ Physiol* 28): H1038-H1043.

Kelm M, Schrader J: Control of coronary vascular tone by Nitric Oxide. *Circ Res* 1990;66:1561-1575

Kilbourn RG, Jubran A, Gross SS, Griffith OW, Levi R, Adams J, Lodato RF: Reversal of endotoxin-mediated shock by N^G-methyl-L-arginine, an inhibitor of nitric oxide synthesis. *Biochem Biophys Res Comm* 1990a;172(3):1132-1138

Kilbourn RG, Jubran A, Gross SS, Griffith OW, Levi R, Adams J, Lodato F: N^G-methyl-L-arginine inhibits tumor necrosis factor-induced hypotension: Implications for the involvement of nitric oxide. *Proc Natl Acad Sci USA* 1990b;87:3629-3632.

Kilbourn RG, Belloni P: Endothelial cell production of nitrogen oxides in response to interferon gamma in combination with tumor necrosis factor, interleukine-1, or endotoxin. *J Natl Cancer Inst* 1990c;82:772-776

Kilbourn RG, Gross SS, Lodato RF, Levi R, Miller LL, Lachman LB, Griffith OW: Inhibition of interleukin-1-alpha-induced nitric oxide synthase in vascular smooth muscle and full reversal of interleukin-1-alpha-induced hypotension by N-omega-amino-L-arginine. *J Natl Cancer Inst* 1992;84:1008-1016

Klabunde RE, Ritger RC: N^G-monomethyl-L-arginine (NMA) restores arterial blood pressure but reduces cardiac output in a canine model of endotoxic shock. *Biochem Biophys Res Comm* 1991;178 (3):1135-1140

Klemm P, Ostrowski J, Morath T, Gruber C, Martorana PA, Henning R: *N*-Acetylserotonine prevents the hypotension induced by bacterial lipopolysaccharides in the rat. *Eur J Pharmacol* 1993;250:R9-R10.

Knowles RG, Palacios M, Palmer RM, Moncada S: Formation of nitric oxide from L-arginine in the central nervous system: a transduction mechanism for stimulation of the soluble guanyl-cyclase. *Proc Natl Acad Sci USA* 1989;86:5159-5162.

Knowles RG, Salter M, Brooks SL, Moncada S: Anti-inflammatory glucocorticoids inhibit the induction by endotoxin of nitric oxide synthase in the lung, liver and aorta of the rat. *Biochem Biophys Res Comm* 1990;172(3):1042-1048.

Kreycy K, Schwarzacher S, Raberger G: Distribution and metabolism of N^G-nitro-L-Arginine and N^G-nitro-L-Arginine methylester in canine blood in vitro. *Arch Pharmacol* 1993;347:342-345.

Lang CH, Bagby GJ, Ferguson JL, Spitzer JJ: Cardiac output and redistribution of blood flow in hypermetabolic sepsis. *Am J Physiol* 1983;246:R331-R337

Lefroy DC, Crake T, Uren NG, Davies GJ, Maseri A: Effect of inhibition of Nitric Oxide synthesis on epicardial caliber and coronary blood flow in humans. *Circulation* 1993;88:43-54

Leone AM, Palmer RMJ, Knowles RG, Francis PL, Ashton DS, Moncada S: Constitutive and inducible nitric oxide synthases incorporate molecular oxygen into both nitric oxide and citrulline. *J Biol Chem* 1991;266:23790-23795.

Lorente JA, Landin L, Pablo de R, Renes E, Liste D: L-Arginine pathway in the sepsis syndrome. *Crit Care Med* 1993;21:1287-1295.

Martin W, Smith JA, White DG: The mechanisms by which haemoglobin inhibits the relaxation of rabbit aorta induced by nitrovasodilators, nitric oxide, or bovine retractor penis inhibitory factor. *Br J Pharmacol* 1986;89:563-571.

Martin W, Villani GM, Jothianandan D, Furchgott RF: Selective blockade of endothelium dependent and glyceryl trinitrate-induced relaxation by hemoglobin and by methylene blue in the rabbit aorta. *J Pharmacol Exp Ther* 1985;232:708-716.

Marletta MA, Yoon PS, Iyengar R, Leaf CD, Wishnok JS: Macrophage oxidation of L-arginine to nitrite and nitrate: Nitric oxide is an intermediate. *Biochemistry* 1988;27:8706-8711.

Mayer B, Brunner F, Schmidt K: Inhibition of nitric oxide synthesis by methylene blue. *Biochem Pharmacol* 1993;45:367-374.

Mayer B, John M, Heinzel B: Brain nitric oxide synthase is a biopterin and flavin containing multi-functional oxido-reductase. *FEBS Lett* 1991;288:187.

McCall TB, Feelisch M, Palmer RM, Moncada S: Identification of N-iminoethyl-L-ornithine as an irreversible inhibitor of nitric oxide synthase in phagocytic cells. *Br J Pharmacol* 1991;102:234-238.

McDonough KH, Lang CH, Spitzer JJ: Depressed function of isolated hearts from hyperdynamic septic rats. *Circ Shock* 1984;12:241-251.

McKenna TC,Martin FM, Chernow B, Briglia FA: Vascular endothelium contributes to decreased aorctic contractility in experimental sepsis. *Circ Shock* 1986;19:267-273.

Meyer J, Stothert JC, Pollard V, Hinder F, Herndon DN, Traber DL: Increased organ bloodflow in sepsis and its reversal with the nitric oxide synthesis inhibitor L-NAME. *Crit Care Med* 1993;21:S280(Abst)

Meyer J, Traber LD, Nelson S, Nelson S, Lentz CW, Nakazawa H, Herndon DN, Noda H, Traber DL: Reversal of hyperdynamic responses to continuous endotoxin administration by inhibition of NO synthesis. *J Appl Physiol* 1992;73:324-328

Moncada S, Higgs EA: The L-arginine - Nitric Oxide pathway. *N Engl J Med* 1993;30:2002-2012

Moncada S, Palmer RM, Higgs EA: Nitric Oxide: Physiology, Pathophysiology, and Pharmacolo-gy. *Pharmacol Rev* 1991;43:109-142.

Moore PK, al-Swayeh OA, Chong NWS, Evans RA, Gibson A: L-N^G-nitro-arginine (L-NOARG), a novel, L-arginine-reversible inhibitor of endothelium-dependent vasodilation in vitro. *Br J Pharmacol* 1990;99:408-412.

Murray HW and Teitelbaum RF: L-arginine-dependent reactive nitrogen intermediates and the antimicrobial effect of activated human mononuclear phagocytes. *J Infect Dis* 1992;165:513-517.

Nava E, Palmer RM, Moncada S: Inhibition of nitric oxide synthesis in septic shock: how much is beneficial? *Lancet* 1991;338:1555-1557

Nava E, Palmer RM, Moncada S: The role of nitric oxide in endotoxic shock: Effects of N^G-monomethyl-L-arginine. *J Cardiovasc Pharmacol* 1992;20:S132-134.

Ochoa JB, Udekwu AO, Billiar TR, Curran RD, Cerra FB, Simmons RL, Peitzman AB: Nitrogen oxide levels in patients after trauma and during sepsis. *Ann Surg* 1991;214:621-626.

Osaka K: Prolonged vasospasm produced by the breakdown products of erythrocytes. *J Neurosurg* 1977;47:403-410.

Palmer RM, Ferrige AG, Moncada S: Nitric oxide release accounts for the biological activity of endothelium-derived relaxing factor. *Nature* 1987;327:524-526.

Palmer RM, Ashton DS, Moncada S: Vascular endothelial cells synthesize nitric oxide from L-arginine. *Nature* 1988;333:664-666.

Parillo JE: Septic shock in humans. *Ann Intern Med* 1990;113:227-242.

Parker MM, Shelhamer JH, Natanson C, Alling DW, Parillo JE: Serial cardiovascular variables in survivors and nonsurvivors of human septic shock: Heart rate as an early predictor of prognosis. *Crit Care Med* 1987;15:923-929.

Petros A, Bennet D, Vallance P: Effect of nitric oxide synthase inhibitors on hypotension in patients with septic shock. *Lancet* 1991;338:1557-1558.

Petros A, Lamb G, Leone A, Moncada S, Bennet D, Vallance P: Effects of a nitric oxide synthase inhibitor in humans with septic shock. *Cardiovasc Res* 1994;28:34-39.

Quezado ZM, Natanson C: Systemic hemodynamic abnormalities and vasopressor therapy in sepsis ans septic shock. *Am J Kidney Dis* 1992;20:214-222.

Rackow EC, Astiz ME, Weil MH: Cellular oxygen metabolism during sepsis and shock. *JAMA* 1988;259 (13):1989-1993.

Radomski MW, Palmer RM, Moncada S: An L-arginine:nitric oxide pathway present in human platelets regulates aggregation. *Proc Natl Acad Sci USA* 1990;87:5193-5197.

Rees DD, Palmer RM, Moncada S: Role of endothelium derived nitric oxide in the regulation of blood pressure. *Proc Natl Acad Sci USA* 1989a;86:3375-3378.

Rees DD, Palmer RMJ, Hodson HF, Moncada S: A specific inhibitor of nitric oxide formation from L-arginine attenuates endothelium dependent relaxation. *Br J Pharmacol* 1989b;96:418-424.

Rees DD, Cellek S, Palmer RM, Moncada S: Dexamethasone prevents the induction by endotoxin of a nitric oxide synthase and the associated effects on vascular tone: an insight into endotoxin shock. *Biochem Biophys Res Comm* 1990a;173 (2):541-547

Rees DD, Palmer RM, Schulz R, Hodson HF, Moncada S: Characterisation of three inhibitors of endothelial nitric oxide synthase in vitro and in vivo. *Br J Pharmacol* 1990b;101:746-752.

Rubanyi GM, Romero JC, Vanhoutte PM: Flow-induced release of endothelium-derived relaxing factor. *Am J Physiol* 1986;250(*Heart Circ Physiol* 19):H1145-H1149.

Rumsey WL, Kilpatrick L, Wilson DF, Erecinska M: Myocardial metabolism and coronary flow: effects of endotoxemia. *Am J Physiol* 1988;255(*Heart Circ Physiol* 24): H1295-H1304.

Ruokonen E, Takala J, Kari A, Alhava E: Septic shock and multiple organ failure. *Crit Care Med* 1991;19: 1146-1151

Schilling J, Cakmakci M, Bättig U, Geroulanos S: A new approach in the treatment of hypotension in human septic shock by N^G-monomethyl-L-arginine, an inhibitor of the nitric oxide synthethase. *Intensive Care Med* 1993;19:227-231

Schneider F, Lutun P, Hasselmann M, Stoclet JC, Tempe JD: Methylene blue increases vascular systemic resistance in human septic shock. *Intensive Care Med* 1992;18:309-311.

Schneider F, Lutun P, Couchot A, Bilbault P, Tempe JD: Plasma cyclic guanosine 3'-5'monophosphate concentrations and low vascular resistance in human septic shock. *Intensive Care Med* 1993;19:99-104

Schneemann M, Schoedon G, Hofer S, Blau N, Guerrero L, Schaffner A: Nitric oxide synthase is not a constituent of the antimicrobial armature of human mononuclear phagocytes. *J Infect Dis* 1993;167:1358-1363.

Schulz R, Smith JA, Lewis MJ, Moncada S: Nitric oxide synthase in cultured endocardial cells of the pig. *Br J Pharmacol* 1991;104:21-14.

Schulz R, Nava E, Moncada S: Induction and potential biological relevance of a Ca^{2+} -independent nitric oxide synthase in the myocardium. *Br J Pharmacol* 1992a;105:575-580

Schultz PJ, Raij L: Endogenously synthesized nitric oxide prevents endotoxin-induced glomerular trombosis. *J Clin Investig* 1992b;90:1718-1725.

Schwarzacher S, Raberger G: L-N^G-Nitro-Arginine Methyl Ester in the anaesthetized rabbit: venous vasomotion and plasma levels. *J Vasc Res* 1992;29:290-292.

Shah AM, Lewis MJ, Henderson AH: Effects of 8-bromo-cyclic GMP on contraction and on inotropic response of ferret cardiac muscle. *J Mol Cell Cardiol* 1991;23:55-64.

Smith RE, Palmer RM, Moncada S: Coronary vasodilatation induced by endotoxin in the rabbit isolated perfused heart is nitric oxide-dependent and inhibited by dexamethasone. *Br J Pharmacol* 1991;104:5-6.

Smith JA, Shah AM, Lewis MJ: Factors released from endocardium of the ferret and pig modulate myocardial contraction. *J Physiol* 1991;439;1-14.

Stuehr DJ, Gross SS, Sakuma I, Levi R, Nathan C: Activated murine macrophages secrete a metabolite of arginine with the bioactivity of endothelium-derived relaxing factor and the chemical reactivity of nitric oxide. *J Exp Med* 1989;169:1011-1020.

Suffredini AF, Fromm RE, Parker MM, Brenner M, Kovacs JA, Wesley RA, Parillo JE: The cardiovascular response of normal humans to the administration of endotoxin. *N Engl J Med* 1989;321: 280- 287.

Szabo C, Mitchell JA, Thiemermann C, Vane JR: Nitric-oxide-mediated hyporeactivity to noradrenaline preceeds the induction of nitric oxide synthase in endotoxin shock. *Br J Pharmacol* 1993;108;786-792

Tayeh and Marletta: Macrophage oxidation of L-arginine to nitric oxide, nitrite, and nitrate. *J Biol Chem* 1989;264:19654-19658.

Teale DM, Akinson AM: Inhibition of nitric oxide synthesis improves survival in a murine peritonitis model of sepsis and that is not cured by antibiotics alone. *J Antimicrob Chemother* 1992;30:839-842

Thiemermann C, Vane JR: Inhibition of nitric oxide synthesis reduces the hypotension induced by bacterial lipopolysacharides in the rat in vivo. *Eur J Pharmacol* 1990;182:591-595

Tilton RG, Chang K, Hasan KS, Smith SR, Petrash JM, Misko TP, Moore WM, Currie MG, Corbett JA, McDaniel ML, Williamson JR: Prevention of diabetic vascular function by guanides. Inhibition of nitric oxide synthase versus advanced glycation end-product formation. *Diabetes* 1993;42:221-232.

Umans GJ, Samsel RW: L-Canavanine selectively augments contraction in aortas from endotoxemic rats. *Eur J Pharmacol* 1992;210:343-346.

Vallance P, Collier J, Moncada S: Effects of endothelium-derived nitric oxide on peripheral arteriolar tone in man. *Lancet* 1991;997-998

Vallance P, Leone A, Calver A, Collier J, Moncada S: Accumulation of an endogenous inhibitor of nitric oxide synthesis in chronic renal failure. *Lancet* 1992;339:572-575.

Wang J, Stuehr DJ, Ikeda-Saito M, Rousseau DL: Heme coordination and structure of the catalytic site in nitric oxide synthase. *J Biol Chem* 1993;268:22255-22258.

Whittle BJ, Lopez–Belmonte J, Rees DD: Modulation of the vasopressor actions of acetylcholine, bradykinine, substance P and endothelin in the rat by a specific inhibitor of nitric oxide formation. *Br J Pharmacol* 1989;98:646-652.

Wolin MS, Cherry PD, Rodenburg JM, Messina EJ, Kaley G: Methylene blue inhibits vasodilation of skeletal muscle arterioles to acetylcholine and nitric oxide via the extracellular generation of superoxide anion. *J Pharmacol Exp Ther* 1990;254:872-876.

Wright CE, Rees DD, Moncada: Protective and pathological roles of nitric oxide in endotoxin shock. *Cardiovasc Res* 1992;26:48-57.

72

ASSESSMENT OF PRESSURE-VOLUME CURVE OF THE RESPIRATORY SYSTEM IN MECHANICALLY VENTILATED PATIENTS WITH ARDS

A. S. Tütüncü, N. Çakar, Ğ. Köprülü, F. Esen, and L. Telci

Department of Anesthesiology
Medical Faculty of University of Istanbul, Turkey

INTRODUCTION

Measurement of total dynamic compliance of the respiratory system is routinely obtained in mechanically ventilated patients with ARDS. Determination of the pressure-volume (P-V) curve of the total respiratory system is recommended to be more useful as a diagnostic and therapeutic means (1-3).

Because compliance is volume dependent, a single value of static compliance is not as informative as a P-V curve of the respiratory system. The P-V curve can be constructed with several techniques (4-7), and the conventional method commonly used is the super-syringe technique (6,7). Despite the useful clinical information provided by a P-V curve, its performance has not become a part of routine practice for patient assessment in the ICU in clinical conditions because of the need for technical facilities. More recently, Levy and colleagues described a new method for P-V curve determination without disconnecting the patient from the ventilator (8). In this method, volume and pressure recordings were made from the mechanical ventilator without necessitating extra technical requirements. By using the same technique, in this study, we aimed to evaluate the assessment of P-V curves of patients with severe ARDS.

METHODS

We studied 13 patients (mean age 42 ± 18 years) with severe ARDS of different etiologies (6 patients with polytrauma, 5 patients with pneumonia and 2 patients with pancreatitis). All met the criteria of having PaO_2/FiO_2 ratio less than 150 and having bilateral lung infiltration without a history of previous lung disease or clinical suspicion of left heart failure.

Oxygen Transport to Tissue XVII, Edited by Ince et al.
Plenum Press, New York, 1996

Initially the patients were mechanically ventilated with pressure controlled ventilation, using a tidal volume of 10 ml/kg, respiratory frequency of 15 breaths per minute and PEEP of 5 cm H_2O. For determination of the P-V curve, patients were sedated, paralyzed and switched to volume-controlled mode under Servo 900C (Siemens–Elema, Sweden) to get a constant inspiratory flow waveform. After setting a pre-determined minute ventilation volume, PEEP was removed and FiO_2 was set to 1. During the procedure, arterial pressure and arterial oxygen saturation (pulse-oximeter) were continuously monitored.

Technique

After an end-expiratory hold for determining auto-PEEP, inflation P-V curve was performed by intermittently changing the respiratory frequency to a desired value to correspond to pre-determined tidal volume. From the initial ventilator setting, the first step is to perform an end-expiratory hold. After reaching a plateau pressure, the frequency knob is rapidly set at a desired value, and the expiratory pause hold knob is released. At this point, the tidal volume will be equal to the minute ventilation divided by the new respiratory frequency. Thereafter, the inspiratory pause hold knob is immediately pressed and tidal volume and the plateau pressure shown in the display of the ventilator are manually recorded. During the end-inspiratory pause, the frequency is set to a new value and the hold knob is released for expiration. The procedure is repeated with another frequency for pre-determined tidal volumes, and a P-V curve is constructed for each patient at the end of the procedure.

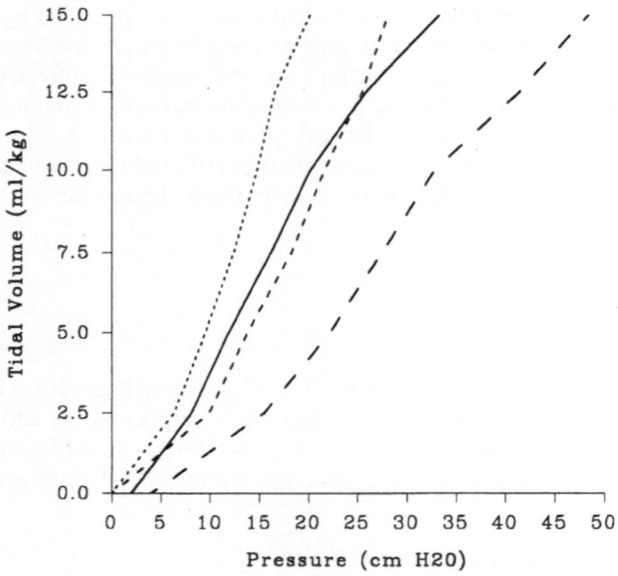

Figure 1. Representative P-V curves of 4 patients, demonstrating an auto-PEEP in 2 patients.

Table 1. Static compliance of the respiratory system (ml/cm H_2O) at different levels of tidal volume (VT) in each patient

Patient no.	VT		
	7.5 ml/kg	10 ml/kg	12.5 ml/kg
1	41.3	39.7	30.4
2	21.2	22.6	21.8
3	43.4	42.4	47.0
4	22.6	22.0	25.6
5	24.3	25.3	27.7
6	38.3	40.8	40.8
7	53.6	58.1	61.7
8	30.9	31.4	35.3
9	23.2	24.5	25.9
10	36.1	40.6	43.9
11	46.5	47.6	44.8
12	35.5	39.0	37.5
13	38.7	40.1	43.8
Range	21.2 - 53.6	22.0 - 58.1	21.8 - 61.7

RESULTS

All patients tolerated the procedure which lasted 60 to 90 seconds, and they were clinically stable during the procedure with unchanged arterial oxygen saturation. An auto-PEEP, ranging between 2 to 7 cm H_2O, was detected in 7 patients.

The P-V curve of some patients are depicted in Figure 1, demonstrating a decrease in slope of the curve at high tidal volumes, whereas some patients demonstrated a significantly high respiratory system compliance.

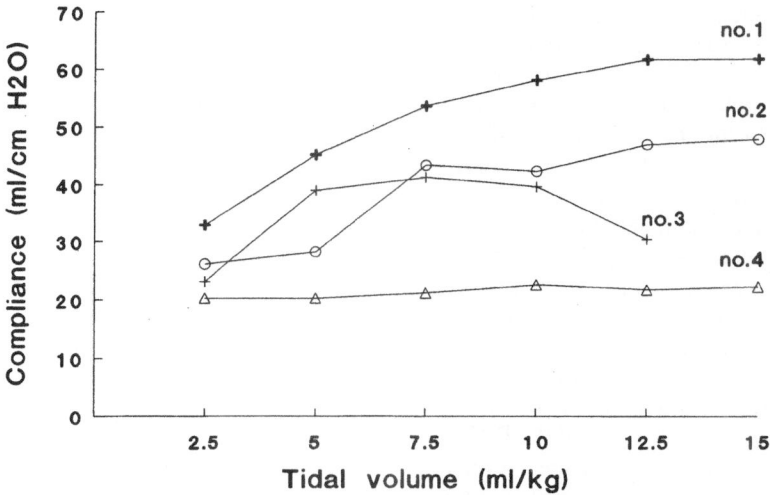

Figure 2. The respiratory system compliance exhibited a great variability at the same lung volume among the patients despite the same clinical severity level of the lung disease. Additionally, the response in compliance to increasing lung volumes were different between the patients.

The static compliance of the respiratory system as calculated by inspiratory tidal volume divided by end-inspiratory pause airway pressure exhibited a high variability in patients with ARDS, despite the same level of clinical severity of the disease (Table 1). The compliance of the respiratory system corresponding to increasing tidal volumes are depicted for some patients in Figure 2.

DISCUSSION

The mechanical properties of the respiratory system is known to be adversely affected by a reduction in functional residual capacity volume of the lung in patients with ARDS. Since the lung pathology is nonhomogeneous and, the severity of the disease and the clinical course of impairment of respiratory mechanics vary widely in ARDS, ventilatory management should be adjusted according to the mechanical status of the lung in each patient. This study demonstrated that this technique of determination of the mechanical status of the respiratory system can be performed easily at the bedside without technical requirements.

This method of P-V curve determination during mechanical ventilation without disconnecting the patient from the ventilator is described by Levy et al. as an alternative to the super-syringe technique (4). In their technique, volume and pressure signals from the ventilator were recorded on a computerized system during which intermittent end-inspiratory airway occlusions at different inflation volumes were performed. This technique was further modified and tested by Fernandez et al. (9). The authors used the pressure data on the digital display of the ventilator, thereby eliminating the computerized system and making this test more simple and available on the bedside. The comparison of the data from manual recordings to the computer-derived data showed a good correlation and, therefore, demonstrated the accuracy of the data collection directly from the ventilator.

With this technique, only the inflation part of the P-V curve is performed. Although the deflation P-V curve is theoretically possible to perform with the same technique, the measurements during deflation carries the risk for erroneous recordings since the expiratory pressure is not displayed digitally on the ventilator. On the other hand, the importance of the deflation part of P-V curve is questioned by various investigators (10,11). The potential problem leading to calculation errors is the lung volume changes due to continuing gas exchange depending on the duration of the procedure (usually 60 to 90 sec). However, the end-inspiratory occlusion period lasts only 3 to 5 sec and, the effect of continuing gas exchange during this short period of end-inspiratory airway occlusion at each lung volume should be negligible with this technique (4).

Another point for consideration during this technique is the possible pressure leaks in the ventilator-patient system. Therefore, special care for the control of leaks must be given prior to the measurements for the accuracy of the P-V determinations.

This study demonstrated that in some patients with ARDS, the static P-V curve exhibited upward concavity which reflected progressive alveolar recruitment with increasing inflation (lung) volumes. In other patients, the static P-V curve exhibited a progressive decrease in slope as an upward convexity, suggesting that there was no progressive recruitment with increasing inflation volume and the functional alveoli were close to their maximal volume (e.g, in the flat part of the P-V curve).

Assessment of mechanical properties of the respiratory system in patients with acute respiratory failure is of clinical use for providing information in order to optimize the ventilatory support. The shape of the static P-V curve has been suggested to identify the patients who are likely to benefit from PEEP therapy and, thus, respond with alveolar recruitment (12).

The conventional technique for P-V determination is the super-syringe method. This technique of super-syringe involves some points for discussion. A temporary disconnection leading to changes in lung volume, and prolonged time for performance leading to errors because of continuing gas exchange; these are the main points that may lead to calculation errors. Therefore, these points have been subjected to several studies (10,11). Furthermore, this method requires a higher FiO_2 than that used during respiratory support.

Others have proposed that the P-V curve may reflect the therapeutic level of PEEP in patients with ARDS (5,13). The inflection point on the P-V curve has been suggested to correspond to the best PEEP because this was believed to reflect the critical opening pressure of collapsed alveoli and airways (5).

Static compliance of the respiratory system has been used for staging and determining the prognosis of patients with ARDS. Static compliance is calculated from the slope of the linear part of the P-V curve. Since the inflation volume has a determining effect on the value of lung compliance, the measurement is not limited with a single point analysis with this technique. In contrast to the compliance calculation as displayed in some ventilators, the calculation of compliance from the P-V curve data is more accurate and reliable for assessing the severity of the lung pathology.

In the present investigation, we determined intrinsic PEEP (iPEEP) in the majority of patients (7 of 13 patients). The presence of iPEEP has been demonstrated in mechanically ventilated patients (14) and has also been termed as auto PEEP (15). The implications of iPEEP on measurement of respiratory system compliance have been clearly defined by Rossi et al. (16). In this study, iPEEP was taken into account during measurement of compliance to avoid errors in the measurement (underestimation) of true compliance of respiratory system.

As conclusion, together with the study of Faradez et al. (9), our data showed that the elastic properties of the respiratory system can be assessed at the bedside easily, rapidly, and the measurements do not require special devices. Moreover, the assessment of the P-V curve during inflation provides clinically useful information during mechanical ventilation in patients with ARDS.

REFERENCES

1. Suter PM, Fairly HB, Isenberg MD. Effect of tidal volume and positive end-expiratory pressure on compliance during mechanical ventilation, Chest 1978; 73:158-162
2. Bone R.C. Compliance and dynamic characteristic curves in acute respiratory failure. Crit Care Med 1976; 4:173-179
3. Mancebo J, Benito S, Martin M, Net A. Value of static pulmonary compliance in predicting mortality in patients with acute respiratory failure. Intensive Care Med 1988; 14:110-114
4. Suter P, Fairley B, Isenberg M. Optimum end-expiratory airway pressure in patients with acute pulmonary failure. N Eng J Med 1975; 292:284-289
5. Mankikian B, Lemaire F, Benito S, Brun-Buisson C, Harf A, Maillot JP, Becker J. A new device for measurement of pulmonary pressure-volume curves in patients on mechanical ventilation. Crit Care Med 1983; 11:897-901
6. Gattinoni L, Pesenti A, Caspani ML et al. The role of the total static lung compliance in the management of severe ARDS unresponsive to conventional treatment. Intensive Care Med 1984; 10:121-126
7. Matamis D, Lemaire F, Harf A, Brun-Buisson C, Ansquer JC, Atlan G. Total respiratory pressure-volume curves in the adult respiratory distress syndrome. Chest 1984; 86:58-66
8. Levy P, Similowski T, Cobeil C, Albala M, Pariente R, Milic-Emili J, Jonson B. A method for studying the static volume-pressure curves of the respiratory system during mechanical ventilation. J Crit Care 1989; 4:83-89
9. Fernandez R, Blanch L, Artigas A. Inflation static pressure-volume curves of the total respiratory system determined without any instrumentation other than the mechanical ventilator. Intensive Care Med 1993; 19:33-38

10. Gattinoni L, Mascheroni D, Basilico E, Foti G, Pesenti A, Avalli L. Volume-pressure curve of total respiratory system in paralysed patients: artifacts and correction factors. Intensive Care Med 1987; 13:19-25

11. Dall'Ava-Santucci J, Armagandis A, Brunet F, Dhainaut JF, Chelucci GL, Monsallier JF, Lockhart A. Causes of error of pressure-volume curves in paralyzed subjects. J Appl Physiol 1988; 64:42-49

12. Ranieri VM, Giuliani R, Fiore T, Dambrosio M, Milic-Emili J. Volume-pressure curve of the respiratory system predicts effects of PEEP in ARDS: occlusion versus constant flow technique. Am J Respir Crit Care Med 1994; 149:19-27

13. Benito S, Lemaire F. Pulmonary pressure-volume relationship in acute respiratory distress syndrome in adults: role of positive end-expiratory pressure. J Crit Care 1990; 5:27-34

14. Jonson B, Nordstrom L, Olsson SC et al. Monitoring of ventilation and lung mechanics during automatic ventilation: a new device. Bull Eur Physiopath Respir 1975; 11:729

15. Pepe PE, Marini JJ. Occult positive end-expiratory in mechanically ventilated patients with airflow obstruction. Am Rev Respir Dis 1982; 126:166-170

16. Rossi A, Gottfried SB, Zocchi L, Higgs BD, Lennox S, Calverley PMA, Begin P, Grassino A, Milic-Emili J. Measurement of static compliance of the total respiratory system in patients with acute respiratory failure during mechanical ventilation. Am Rev Respir Dis 1985; 131:672-677

73

TITRATING PEEP THERAPY IN PATIENTS WITH ACUTE RESPIRATORY FAILURE

A. S. Tütüncü,[1] N. Çakar,[1] F. Esen,[1] J. Kesecioğlu,[2] L. Telci,[1] and K. Akpir[1]

[1] Department of Anesthesiology
  Medical Faculty of University of Istanbul, Turkey
[2] Department of Intensive Care
  Academic Medical Center
  Amsterdam The Netherlands

INTRODUCTION

The use of PEEP has become a well-established method to improve arterial oxygenation under mechanical ventilatory support in patients with adult respiratory distress syndrome (ARDS). In clinical conditions, it is difficult to determine the level of PEEP which can be considered optimal in order to achieve a satisfactory PaO_2 associated with the least cardiocirculatory compromise. Various end-points to determine such an optimal PEEP level have been defined (1-3). The aim in this study was to define the variability of optimal PEEP level in patients at the same clinical severity level of ARDS.

METHODS

Thirty adult patients (48±15 years) with acute respiratory failure due to various etiologies (e.g. trauma, sepsis, pneumonia, acute pancreatitis) were studied on the first day of admission to the ICU at the Medical Faculty of Istanbul. The study entry criteria were bilateral lung infiltrates on chest X-ray, no history of chronic lung disease or clinical suspicion of left heart failure, and PaO_2 to inspired O_2 fraction (FiO_2) ratio being less than 200 with the following initial ventilatory settings: pressure-controlled ventilation, tidal volume of 10 ml/kg, respiratory frequency of 15/min, inspiratory to expiratory time ratio of 1/2, FiO_2 of 1.0 and PEEP of 5 cm H_2O.

All patients were orotracheally intubated and mechanical ventilation was provided by a Servo 900C ventilator (Siemens–Elema, Sweden). All patients were paralyzed with vecuronium (0.1 mg/kg/h) and sedated with midazolam (0.3 mg/kg/h). A radial artery was catheterized for monitoring arterial pressure (Viggo–Spectramed, USA) and blood sampling for blood gas analyses (ABL-500, Radiometer, Copenhagen, Denmark). In addition, all

Oxygen Transport to Tissue XVII, Edited by Ince et al.
Plenum Press, New York, 1996

patients had a central venous catheter for pressure monitoring which was normalized in all patients prior to the trial.

According to the study protocol, patients fulfilling the above criteria received 3 cm H_2O increments of PEEP until a PEEP of 15 cm H_2O was reached or peak airway pressure exceeded 45 cm H_2O, or a significant drop in blood pressure occurred. The above ventilator settings were kept constant at each PEEP level in all patients, and gas exchange parameters were recorded at each PEEP level after a stabilization period of 20-30 min.

Data were analyzed using the Wilcoxon signed-rank test within the groups and a p value of less than 0.05 was accepted as statistically significant.

RESULTS

Of the 30 patients, 14 received PEEP up to 15 cm H_2O, while the trial had to be stopped at PEEP level of 12 cm H_2O in the remaining (due to high peak airway pressure in 6 patients and decrease in blood pressure in 10 patients).

According to the response of PaO_2 to increasing PEEP levels, patients were separated into 3 groups (Table 1). In the first group of 16 patients, PaO_2 increased in parallel to increased PEEP and was significantly different at PEEP levels of 9, 12 and 15 cm H_2O when compared to PEEP of 6 cm H_2O. In the second group of 7 patients, no significant PaO_2 changes occurred at increasing PEEP levels while in the remaining group initial increase at PaO_2 was followed by a significant decrease above PEEP level of 12 cm H_2O.

DISCUSSION

This study demonstrated that there is a great variability in patients with ARDS in terms of the response in arterial oxygenation to different levels of PEEP and, therefore, an optimum level of PEEP for each patient should be applied instead of a standard PEEP during mechanical ventilation.

Although there has been no controlled studies comparing PEEP with no PEEP in patients with acute respiratory failure, there is general agreement that PEEP should be used to treat respiratory failure due to diffuse lung parenchymal inflammation. The proposed mechanisms responsible for alterations in gas exchange both beneficial and detrimental are recruitment of collapsed alveolar units, reduction of cardiac output and overinflation of lung structures (1).

The present data suggest that not all patients with ARDS benefit from PEEP. In ARDS, lung injury is nonhomogeneous in a sense that a healthy lung zone is coexisting together with poorly inflated lung zone and atelectatic (dependent) lung zone (4). All these

Table 1. Gas exchange parameters in treatment groups (G1,G2,G3) at incremental PEEP levels (Mean ± SD)

		6 cmH$_2$O	9 cmH$_2$O	12 cmH$_2$O	15 cmH$_2$O
PaO$_2$ (mmHg)	G1	155 ± 70	236 ± 95	282 ± 109	294 ± 82
	G2	154 ± 78	161 ± 86	163 ± 99	154 ± 36
	G3	163 ± 74	177 ± 63	147 ± 43	139 ± 20
PaCO$_2$ (mmHg)	G1	40 ± 9	39 ± 9	38 ± 10	37 ± 13
	G2	42 ± 8	42 ± 8	40 ± 10	49 ± 14
	G3	38 ± 7	40 ± 8	39 ± 5	44 ± 3

zones respond to PEEP in a different way. The progressive improvement of oxygenation with increasing PEEP levels suggests that there was progressive recruitment with increasing inflation pressure and/or the previously recruited lung units could be prevented from end-expiratory collapse.

In the absence of significant alveolar recruitment, the application of PEEP may result in hyperinflation of functional lung units and increased risk of pulmonary barotrauma (5). In Group 3 patients, the increased arterial oxygenation followed by a reduction in PaO_2 at PEEP> 9 cmH_2O is highly suggestive of a lung overdistension impeding alveolar perfusion. This may also explain the reduced blood pressure (cardiac output) despite the additional intravascular volume support as observed in many of the patients in this group. The impaired cardiovascular stability suggested that the inflation pressures at such high PEEP levels were almost at the flat part of the pressure-volume curve of the respiratory system.

The interpretation of optimal PEEP without evidence of hemodynamic stability has no validity because a low cardiac output due to either primarily cardiac dysfunction or in response to decreased preload may essentially lead to a low mixed venous oxygen saturation. This eventually may result in lower PaO_2 and the optimal PEEP may be underestimated. Therefore, close care was given in this study to maintain the hemodynamic stability during the procedure and, in 10 patients out of 30, there was a tendency for decreased blood pressure despite the intravascular volume support and the procedure had to be stopped at PEEP level of 12 without further increasing the PEEP in order not to increase the pulmonary shunt fraction.

In conclusion, it is worth to emphasize once more that the most important goal during mechanical ventilation in ARDS patients should be the assessment of adequate oxygen delivery and tissue oxygenation rather than a simple analysis of arterial blood gases. Thus, the aim of determining optimal PEEP during ventilatory support should be considered as a part of this concept and optimal PEEP has to be determined in each patient instead of applying on an empirical ground.

REFERENCES

1. Suter PM, Fairley HB, Isenberg MD. Optimum end-expiratory airway pressure in patients with acute pulmonary failure. N Engl J Med 1975; 292:284-289
2. Nelson LD, Civetta JM, Hudson-Civetta J. Titrating positive end-expiratory pressure therapy in patients with early, moderate arterial hypoxemia. Crit Care Med 1987; 15:14-19
3. Murray IP, Modell JH, Gallagher TJ, Banner MJ. Titration of PEEP by the arterial minus end-tidal carbon dioxide gradient. Chest 1984; 85:100-104
4. Gattinoni L, Pesenti A, Bombino M, et al. Relationships between lung computed tomographic density, gas exchange, and PEEP in acute respiratory failure. Anesthesiology 1988; 69:824-832
5. Ranieri VM, Eissa NT, Corbeil C, Chasse M, Braidy J, Matar N, Milic-Emili J. Effects of positive end-expiratory pressure on alveolar recruitment and gas exchange in patients with the adult respiratory distress syndrome. Am Rev Respir Dis 1991; 144:544-551

74

EFFECTS OF DIFFERENT CPAP SYSTEMS ON WEANING PARAMETERS IN PATIENTS RECOVERING FROM ACUTE RESPIRATORY FAILURE

A. S. Tütüncü,[1] N. Çakar,[1] F. Esen,[1] J. Kesecioğlu,[2] and L. Telci [1]

[1] Department of Anesthesiology
Medical Faculty of University of Istanbul
Istanbul, Turkey
[2] Department of Intensive Care
Academic Medical Center
Amsterdam The Netherlands

INTRODUCTION

Weaning from mechanical ventilatory support involves a transition from assisted to spontaneous ventilation. There are several ventilatory techniques to wean patients from mechanical ventilation, one of which is continuous positive airway pressure (CPAP). One strategy to optimize weaning is to decrease the workload of respiratory muscles. However, a common problem with some ventilatory modalities is the presence of demand valve which may induce extra work of breathing as, in the CPAP mode. Therefore, different CPAP techniques have been introduced in clinical practice.

Predicting patient responses during weaning procedure is based on some bedside measurements of respiratory mechanics and the clinical status of the patient. Although various tests have been proposed as indicators of successful weaning, no single parameter has proven to be superior in terms of predicting weaning accurately (1-3).

The aim of this study was to compare three different CPAP systems in terms of the effects on weaning parameters measured by a portable bedside respiratory monitor and on pulmonary gas exchange in patients recovering from acute respiratory failure.

METHODS

Fourteen adult patients (4 female and 10 male), with a mean age of 44 ± 17 years, who were diagnosed with acute respiratory failure from various etiologies were included in this study (Table 1). All patients were orotracheally intubated and they underwent mechanical

Oxygen Transport to Tissue XVII, Edited by Ince et al.
Plenum Press, New York, 1996

Table 1. Demographic data

Patient no.	Diagnosis	Age (years) sex	Duration of mechanical ventilation (days)
1	pneumonia	57, M	11
2	polytrauma	66, F	9
3	polytrauma	60, F	17
4	polytrauma	24, F	9
5	polytrauma	60, M	20
6	m. gravis	16, F	7
7	fat emboli	22, M	12
8	polytrauma	37, M	14
9	polytrauma	45, M	8
10	polytrauma	37, M	16
11	polytrauma	42, M	8
12	polytrauma	29, M	9
13	polytrauma	58, M	6
14	pneumonia	63, M	3
Mean ± SD		44 ± 17	11 ± 5

ventilation for an average of 11 ± 5 days. All patients were studied using a Servo 900C ventilator (Siemens-Elema, Sweden); the patient trigger sensitivity was set at -0.5 cmH$_2$O and FiO$_2$ was kept as 0.5 throughout the study.

For measurement of weaning parameters, an esophageal balloon catheter was inserted to the lower third part of the esophagus and connected to a respiratory monitor (CP-100 Bicore Monitoring Systems, Irvine, USA) together with a flow-pressure transducer positioned between the Y-piece of the breathing circuit and the endotracheal tube (4). The correct placement of the catheter was checked by the Baydur method (5). This monitor provided real time measurements of pressure-volume and calculations of weaning parameters. Patients were studied in three different CPAP modes at 10 cmH$_2$O for 30 min in random order and, pressure support ventilation was applied between the CPAP trials for 30 min. For each parameter, at least 3 measurements were performed at intervals not less than 15 sec. The nutritional intake was kept constant throughout the study in all patients. Each patient was kept in semi-recumbent position and no sedatives or narcotics were administered during the study.

Modes

Two modes of CPAP with demand valves and one with continuous flow system were applied.

CPAP-1. CPAP in pressure support mode of the Servo 900C ventilator with zero inspiratory pressure level (demand flow).

CPAP-2. CPAP mode of the Servo 900C ventilator (demand flow).

CPAP-3. CPAP with continuous flow balloon system. This system is routinely used in our ICU and, has a latex bag reservoir of 25 liter. The flow is adjusted to 40 L/min during application.

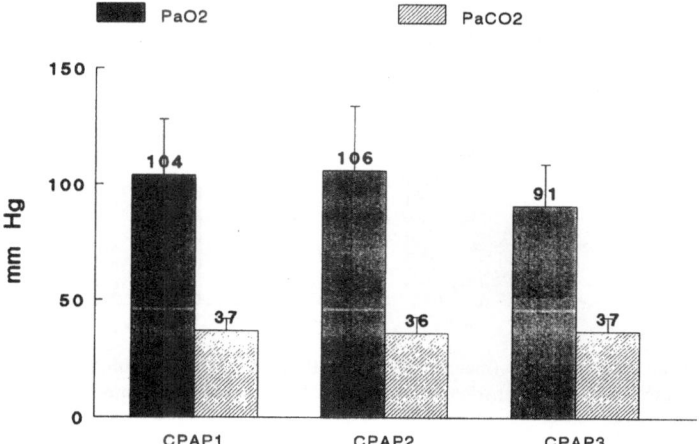

Figure 1. Arterial blood gases (mean ± SD) during CPAP modes.

Parameters Measured and/or Calculated

Minute ventilation volume (VE), tidal volume (VT), respiratory frequency (f), work of breathing (WOB), tracheal occlusion pressure (P0.1), rapid shallow breathing index (f/VT), inspiratory time to total respiratory cycle ratio (TI/TTOT), pressure time index (PTI), arterial blood gases and hemodynamics.

RESULTS

All three CPAP modes were successfully completed in all patients with stable heart rate and arterial blood pressures. Arterial blood gases did not change significantly between the three CPAP systems (Figure 1), while minute ventilation volume was significantly (p < 0.05) less during CPAP-3 when compared to the other modes.

WOB in CPAP-1, CPAP-2 and CPAP-3 were as follows: 1.29 ± 0.50, 1.41 ± 0.71 and 1.0 ± 0.27 Joule/L, respectively. It follows that WOB in CPAP-1 mode was less than in CPAP-2 mode but this finding was statistically insignificant. On the other hand, WOB in CPAP-3 was significantly (p < 0.05) less when compared to CPAP-1 and CPAP-2 (Figure 2).

There was a significant (p < 0.05) difference for P0.1 values between CPAP-3 mode and other CPAP modes (Table 2). The other weaning parameters (f/VT, TI/TTOT, PTI and f) remained unchanged during the CPAP modes (Table 2).

The hemodynamic variables were comparable during the three CPAP modes. The mean data for systolic arterial blood pressure during the CPAP modes were 125 ± 72, 135 ± 78 and 127 ± 72 mmHg, respectively. The mean data for heart rate (beats/min) were as follows: 115 ± 17 in CPAP-1, 111 ± 18 in CPAP-2 and 107 ± 16 in CPAP-3.

DISCUSSION

The results of this study demonstrated that despite the unchanged blood gases and hemodynamic parameters, CPAP system with continuous flow provided better weaning

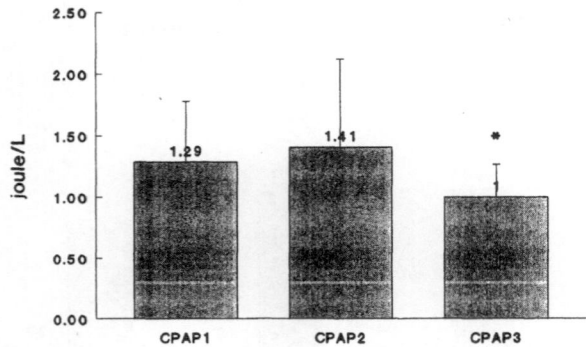

Figure 2. The mean data for WOB (mean ± SD) after 30 min of CPAP breathing with different systems.
* = statistically significant (p < 0.05) when compared to CPAP-1 and CPAP-2 modes.

parameters in the short term in comparison to CPAP systems with demand valve as assessed by work of breathing and tracheal occlusion pressure parameters in patients recovering from acute respiratory failure.

Increased WOB indicated that the patient is performing more work on CPAP-1 and CPAP-2, and, thus, spending more oxygen during breathing (6). The decreased WOB in CPAP-3 is primarily due to the continuous flow from the elastic reservoir bag which can suddenly augment high base-line flows. In this way, high inspiratory flow demands can be supplied with this system. When these demands are not met, pressures within the central airway and trachea will fall dramatically during forceful inspiration as the patients peak demand exceeds the rate at which the gas flows continuously through the circuit. Additional energy is then expended against the resistance in the ventilator circuit (7).

Inspiratory demand valve resistance is likely to be an important contributing factor to the increased work of breathing during CPAP-1 and CPAP-2 modes (8). The inspiratory effort of the patient has to trigger the preset pressure to open the demand valve and must overcome the time delay of the demand valve. The expiratory valve resistance in the ventilator is also suggested to play a role in increased WOB during CPAP (9).

Although the mean values for WOB were statistically comparable, less work was done in CPAP-2 mode when compared to CPAP-1 mode. The slightly better performance on CPAP-2 was most likely due to the small amount of pressure support in CPAP-2 mode which may be already present even though the pressure support option is not activated (10).

The tracheal occlusion pressure is a parameter to reflect the neuromuscular drive of the patients as assessed by the airway occlusion pressure at 0.1 ms from the onset of the inspiration. In our study, the tracheal occlusion pressures in CPAP-1 and CPAP-2 modes were greater than in CPAP-3 mode, indicating an increased respiratory center output and increased inspiratory neuromuscular activity. Inappropriately elevated drive places unnecessary stress on the respiratory muscles and predisposes to fatigue (7). The increase in P0.1

Table 2. Weaning parameters as measured by Bicore Respiratory Monitor

	VE	f	f/VT	TI/TTOT	PTI	$P_{0.1}$
CPAP-1	12.9 ± 3.5	31 ± 8	74 ± 26	0.41 ± 0.08	0.08 ± 0.03	4.4 ± 3.1
CPAP-2	13.0 ± 3.6	31 ± 9	76 ± 28	0.39 ± 0.07	0.09 ± 0.04	4.4 ± 3.6
CPAP-3	11.1 ± 3.2	30 ± 6	89 ± 25	0.40 ± 0.06	0.11 ± 0.04	1.9 ± 1.0
p	< 0.05	NS	NS	NS	NS	< 0.05

in these modes may be due to the same factors in the ventilatory circuit, leading to high inspiratory mechanical load as mentioned above.

Similar to our findings, Gibney et al. have demonstrated that CPAP breathing on demand valve system resulted in a high WOB compared to high flow CPAP circuits (11). On the other hand, Samodelow et al. have shown that, on the electromechanical lung model, the demand valve devices performed similar results as compared to continuous flow CPAP system; that is the work performed was similar in both systems. However, these results may not reflect the situation in clinical conditions (10).

Although we have obtained similar values for the other weaning parameters (f/VT, T_i/T_{tot}, PTI, and f) in all CPAP modes in this study, this may not reflect the situation for long-term application. During long-term application, the patient may react by reducing the tidal volume and increasing the respiratory rate to compensate for the increased work and, thus, this may increase f/VT values during CPAP breathing.

The P0.1 and WOB are both used as weaning predictors. Unsuccessful weaning is indicated for P0.1 values more than 2 cmH$_2$O and for WOB values more than 0.4-0.6 Joule/L during spontaneous breathing. It seems advantageous to keep these values in normal ranges during CPAP breathing, and this will increase the chance of successful weaning. According to the results of this study, at least for short-term application, by using continuous flow CPAP systems this aim could be reached.

In conclusion, although this study does not provide data on the predictability of successful weaning, the short-term results indicated that CPAP system with continuous flow technique may be superior to others by reducing the workload of respiratory muscles and, this may be an advantageous approach for long-term use during weaning patients from mechanical ventilation.

REFERENCES

1. Sahn SA, Lakshminarayan S. Bedside criteria for discontinuation of mechanical ventilation. Chest 1973; 63:1002-1005
2. Hodgkin JE, Bowser MA, Burton GG. Respiratory weaning. Crit Care Med 1974; 2:96-102
3. Sassoon CSH, Light RW, Lodia R, Sieck GC, Mahutte CK. Pressure-time products on T-piece, continuous positive airway pressure and pressure support ventilation during weaning from mechanical ventilation. Am Rev Respir Dis 1991; 143:469-475
4. Petros AJ, Lamond CT, Bennett D. The Bicore pulmonary monitor. A device to assess the work of breathing while weaning from mechanical ventilation. Anaesthesia 1993; 48:985-988
5. Baydur AD, Behrakis WZ, Jeager MA. A simple method for assessing the validity of the esophageal balloon technique. Am Rev Respir Dis 1982; 126:788-791
6. Marini JJ. Strategies to minimize breathing effort during mechanical ventilation. In: Mechanical Ventilation. Tobin MJ (ed); Critical Care Clinics 1990, Vol 6, No3, pp 635-659
7. Tobin MJ, Yang K. Weaning from mechanical ventilation. In: Mechanical Ventilation. Tobin MJ (ed); Critical Care Clinics 1990, Vol 6, No3, pp 725-747
8. Brochard L, Pluskwa F, Lemaire F. Improved efficiency of spontaneous breathing with inspiratory pressure support. Am Rev Respir Dis 1987; 136:411-415
9. Sassoon CSH, Giron AE, Ely EA, Light RW. Inspiratory work of breathing on flow-by and demand flow continuous positive airway pressure. Crit Care Med 1989; 17:1108-1114.
10. Samodelov LF, Falke KJ. Total inspiratory work with modern demand valve devices compared to continuous flow CPAP. Intensive Care Med 1988; 14:632-639.
11. Gibney NRT, Wilson RS, Pontoppidon H. Comparison of work of breathing on high gas flow and demand valve continuous positive airway pressure systems. Chest 1982; 82:692-695.

75

NEGATIVE SLOPE OF EXHALED CO$_2$ PROFILE

Implications for Ventilation Heterogeneity during Partial Liquid Ventilation

E. A. Mates,[1] P. Tarczy–Hornoch,[2] J. Hildebrandt,[1,3] J. C. Jackson,[2] and M. P. Hlastala[1,3]

[1] Department of Physiology and Biophysics, SJ-40
[2] Division of Neonatal and Respiratory Disease, RD-20
[3] Division of Pulmonary and Critical Care Medicine, RM-12
University of Washington
Seattle, Washington 98195

INTRODUCTION

In the course of studying gas exchange during partial liquid ventilation (PLV) in healthy and injured piglets, we noted a reversal in the profile of exhaled CO$_2$ (P$_E$CO$_2$) versus time. Rather than a positive slope, the CO$_2$ expirogram often reached a peak early in expiration and fell toward the end of the breath. In addition to a change in sign, the absolute value of the slope was very large. The change in profile led us to question the generally accepted practice of using "end-tidal" P$_E$CO$_2$ to represent average alveolar P$_A$CO$_2$ during PLV. Given the steep rate of change of P$_E$CO$_2$ over a breath, the use of a single point on the CO$_2$ expirogram to represent alveolar gas during PLV seemed in error. We hypothesized that a combination of increased ventilation heterogeneity and diffusion limitation could account for the reversal in sign and exaggeration of the slopes. To further investigate this problem we explicitly measured P$_E$CO$_2$ vs. exhaled volume in two additional experiments and compared these findings to 40 previously studied animals.

The slope of phase III of exhaled PCO$_2$ has been used by many authors as an index of ventilation heterogeneity[5,8,12]. A plot of P$_E$CO$_2$ vs. expired volume (V$_E$) in healthy lungs results in a curve similar to Figure 1. There is a steep upslope in P$_E$CO$_2$ (phase II) during the transition from non-gas exchanging airways (phase I) to the alveolar plateau phase or (phase III). The dead space can be defined as the midpoint of phase II. Phase III is typically characterized by varying degrees of upward slope with the last gas exiting the lung associated with the highest P$_E$CO$_2$. This "end tidal" P$_E$CO$_2$ is often used to represent mixed alveolar PCO$_2$ since in normal lungs the gas originates from distal regions of the lung. The positive slope of phase III may most simply be attributed to accumulation of CO$_2$ in alveoli during

Oxygen Transport to Tissue XVII, Edited by Ince et al.
Plenum Press, New York, 1996

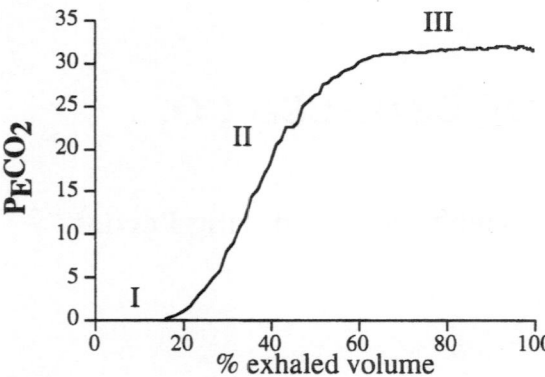

Figure 1. Normal gas ventilated CO_2 expirogram illustrating the phases of exhalation.

the course of exhalation. Grønlund et.al.[5] showed, however, that a continuous source model without ventilation heterogeneity could not fully explain the positive slope of exhaled PCO_2 in healthy animals. They concluded some portion of the sloping alveolar plateau is due to ventilation heterogeneity in addition to accumulation.

During PLV the lung commonly contains a volume of liquid perfluorochemical (PFC) equal to the animal's functional residual capacity (FRC). Conventional mechanical ventilation with O_2-enriched gas is employed and positive end-expiratory pressure (PEEP) applied to maintain a gas volume in the lung at all times. The resultant mixture of gas and fluid of low surface energy dramatically alters lung mechanics and gas exchange as compared to the gas-filled lung. The pattern with which PFC distributes in the lung and spreads in alveoli during PLV is only partially understood. There is probably a spectrum ranging from completely gas-filled, to part liquid and part gas, to completely liquid-filled alveoli. Heterogeneity of fluid distribution would lead to heterogeneity of ventilation and gas exchange. During PLV in healthy animals, increases in shunt, diffusion limitation, and ventilation/perfusion ($\dot{V}_A/\dot{Q}$) mismatch in the lung result in widening ((a-A)D's) of O_2 and CO_2 partial pressures[7]. Increased $\dot{V}_A/\dot{Q}$ mismatch can be due to increased heterogeneity in $\dot{Q}$, $\dot{V}_A$ or both. We hypothesized that PFC in the airspace of the lung would increase ventilation heterogeneity by altering regional time constants and would increase heterogeneity of P_ACO_2 due to diffusion limitation in the fluid.

METHODS

Group I

Forty piglets, 7-14 days old, 2-4 kg were studied previously to assess gas exchange during various applications of PLV. PFC volume- and time-dependent changes in gas exchange were investigated in healthy animals and in lung injury. Exhaled PCO_2 vs. time profiles were systematically re-analyzed to determine the timing and frequency of negative slopes of P_ECO_2. Sedation was with ketamine/xylazine (24 and 2.75 mg•kg^{-1} IM, respectively) and anesthesia was maintained with pentobarbital (1 mg•kg^{-1}•min^{-1} IV, supplemented with 15-30 mg•kg^{-1} hourly), or pentobarbital and ketamine (3 and 5 mg•kg^{-1}•hr^{-1} IV respectively). Pancuronium bromide was administered in 0.2 mg•kg^{-1} IV doses as needed to prevent respiratory efforts. Intravenous access was established via the left jugular vein and the right carotid artery was cannulated for blood gas sampling and arterial pressure measurements. Tracheotomies were performed using a metal Y-shaped cannula with sideport for administration of perfluorochemcials.

A Harvard single piston volume ventilator was used to deliver tidal breaths of 100% O_2 to animals in the supine position. Tidal volume and frequency were set to maintain arterial P_aCO_2 less than 40 mmHg prior to PLV with LiquiVent™ (perflubron [$C_8F_{17}Br$], Alliance Pharmaceutical Corp., San Diego, CA) and were not changed. Positive end-expiratory pressure (PEEP) was applied by immersion of the distal expiratory port in 5 centimeters water.

Arterial and mixed venous blood samples were drawn simultaneously for each condition. Blood gas measurements were made using a Corning model 170 pH/blood gas analyzer which showed linear PO_2 response to tonometered blood up to 700 mmHg with and without perflubron vapor present. Exhaled PCO_2 was continuously monitored using a Novametrix model 7000 infrared analyzer situated in-line between the piglet and solenoid valve. The analyzer showed close linear correlation with a Perkin-Elmer medical mass spectrometer with respect to PCO_2 in the presence of water and perflubron vapor. Warmed, non-preoxygenated perflubron was instilled in the lungs via the sideport of the endotracheal tube. Each dose was administered by depositing small volumes into the tube at end-expiration during mechanical ventilation. Animals were rotated into right and left lateral and supine positions during dosing to improve fluid distribution. Several dosing protocols were followed. In 11 animals, three doses of 10 ml•kg^{-1} PFC were given at hourly intervals and 10 ml•kg^{-1} was drained for the final hour of PLV. Blood gases and P_ECO_2 profiles were recorded 45-60 minutes after each dose. In 5 animals a single dose of 20 ml•kg^{-1}, and in 9 animals 30 ml•kg^{-1} (roughly equal to FRC volume in healthy piglets) was given at the start of PLV. Blood samples and CO_2 expirograms were recorded every 60 minutes for 3 hours. Hourly doses of 2 ml•kg^{-1} were given to replace evaporative losses of PFC.

In 15 piglets, lung injury was induced prior to PLV by IV infusion of 0.1 to 0.14 ml•kg^{-1} oleic acid over 30 minutes. Eleven were given hourly doses of 10 ml•kg^{-1} up to the maximum each could retain at zero airway pressure (a volume equal to FRC). Four were given a single FRC-volume dose of perflubron post-injury and followed for three hours. Blood gas analysis and CO_2 expirograms were recorded after each dose or at hourly intervals as described for the healthy animals.

Group II

Two animals were studied to obtain P_ECO_2 vs. exhaled volume data from which the slope of phase III could be measured. Animals were sedated, anesthetized, and instrumented as described above. Exhaled CO_2 was continuously monitored using a Novametrix model 7000 infrared analyzer situated at the proximal end of the endotracheal tube. A pneumo-tachometer was placed between the ventilator and IR adapter to obtain instantaneous flow. A Novametrix NVM volume monitor provided continuous outputs of minute ventilation, frequency, and tidal volume. Flow and P_ECO_2 signals were collected on a Macintosh computer for slope analysis.

A pressure-limited time-cycled ventilator was used to deliver 100% O_2 at a frequency of 20 breaths per minute and pressures were adjusted to maintain constant tidal volume. Minute ventilation was chosen to achieve P_aCO_2 in the range 30-35 ml•kg^{-1} in the healthy animal. P_ECO_2 and flow were recorded for each of five conditions at a respiratory rate of twenty, with and without PEEP. Eight to ten breaths of P_ECO_2 and flow were acquired for each condition. The first set of measurements were made during conventional gas ventilation in the healthy anesthetized, paralyzed piglet. The animals were subsequently injured by IV infusion of 0.1 to 0.14 ml•kg^{-1} oleic acid over 30 minutes. Once injury was established (P_aO_2 < 250 mmHg), P_ECO_2 and flow data were recorded on a computer with and without PEEP.

Measurements were repeated during PLV in the injured animal with 10, 20, and 30 ml•kg^{-1} perflubron in the lung. A minimum of ten to fifteen minutes elapsed between recording data at different PEEP levels while twenty to thirty minutes separated recordings following changes in PFC dose.

To characterize the response time of the instrumentation, a step change in flow and PCO$_2$ was applied simultaneously to the CO$_2$ monitor and pneumotachometer and their response times were compared. The CO$_2$ monitor lagged the flow signal by approximately 20 milliseconds therefore CO$_2$ signals were shifted in time by 20 milliseconds prior to plotting P$_E$CO$_2$ vs. exhaled volume. Volume was obtained from flow in one of two ways. In one experiment the volume signal was obtained from flow via an analog signal integrator, then digitized and stored on the computer along with signals for flow and P$_E$CO$_2$. In the other experiment, the digitized flow signal was numerically integrated on the computer to derive volume.

P$_E$CO$_2$ vs. volume curves were determined by ensemble averaging of 8-10 breaths for every condition. Individual breaths were lined up in time according to the instant at which the volume (or flow) curve changed significantly from zero. Resultant P$_E$CO$_2$ and volume curves are therefore point-by-point averages of sequential exhaled breaths. Exhaled volume was converted to % exhaled volume for slope calculations. Graphs of P$_E$CO$_2$ vs. % exhaled volume were created for each condition and the slope of phase III determined by linear fit to the curve between 75% and 95% exhaled volume.

RESULTS

Group I

Negative slopes were seen during PLV in all but two animals in this group but were never seen during GV in the same group of animals. Sustained periods (40-180 min) of negative slopes of exhaled CO$_2$ vs. time (Figure 2) were noted in roughly half of the healthy animals and in all injured animals. Figure 3 depicts the percentage of healthy and injured animals exhibiting sustained negative slopes during GV and PLV. It also indicates the percent which showed periodic, unsustained negative slopes, usually following administration of PFC or upon removal of PEEP. Where exhaled volume was recorded along with exhaled PCO$_2$, peak P$_E$CO$_2$ occurred at approximately 60% of exhaled volume and the last 40% exhaled volume was associated with lower PCO$_2$. Seven of eleven healthy animals given graded doses of PFC showed sustained negative slopes and of the seven, the negative slopes appeared 1/7 times at 10 ml•kg^{-1} PFC, 3/7 at 20 ml•kg^{-1}, and 7/7 at 30 ml•kg^{-1} suggesting a dose-dependent effect. Of fourteen healthy animals given a single dose of PFC for three hours, only five showed sustained negative slopes. Of six animals receiving 20 ml•kg^{-1} in

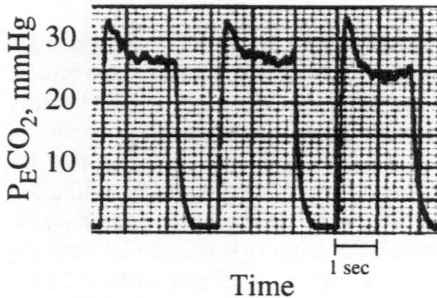

Time   1 sec

Figure 2. P$_E$CO$_2$ vs. time measured at the mouth during PLV.

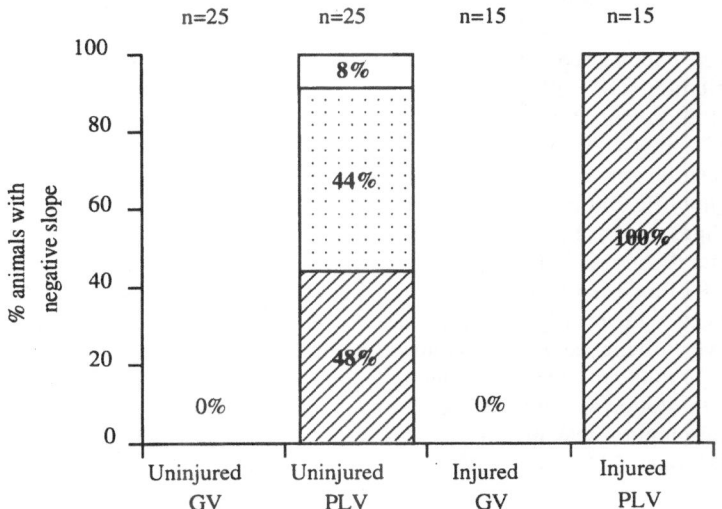

Figure 3. Percent animals exhibiting negative slope of P_ECO_2 vs. time. GV is conventional gas ventilation and PLV is partial liquid ventilation at any dose of PFC. Hatched regions represent animals with sustained (40-180 min) negative slopes, dotted regions show animals with periodic (5-40 min) negative slopes, and open bar regions show animals never exhibiting negative slopes during PLV.

this group, one showed sustained negative slopes. The remaining eight received 30 ml•kg⁻¹ PFC of which four showed sustained negative slopes. Animals with lung injury all showed sustained negative slopes during PLV independent of dose. In all animals, removal of PEEP accentuated the negative slopes, in some instances converting a phase III slope from positive to negative. These changes reversed upon re-application of PEEP.

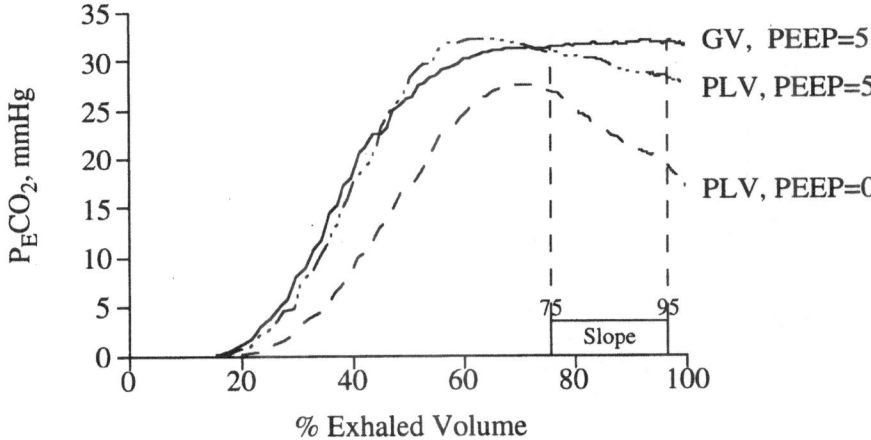

Figure 4. P_ECO_2 vs. % exhaled volume in piglet #2. Curves are an average of eight breaths. Solid line is healthy animal during GV. Dotted lines are post injury PLV with 30 ml•kg⁻¹ PFC at two levels of PEEP. The region between 75 and 95% exhalation from which phase III slope is calculated is illustrated.

Group II

Figure 4 shows typical P_ECO_2 vs. volume curves from animal #2. CO_2 expirograms of gas ventilation (GV), PLV with 30 ml•kg^{-1} PFC and 5 cmH$_2$O PEEP, and PLV with 30 ml•kg^{-1} PFC and no PEEP are shown. The slope of phase III becomes negative with PLV at a PEEP of 5 cmH$_2$O. Removing PEEP causes the slope to become even more negative and decreases the total amount of CO_2 eliminated in the same tidal volume. Negative slopes were seen throughout PLV in both animals at all doses of PFC.

Changes in phase III slope (mmHg per % volume) with experimental condition and with PEEP are illustrated in Figure 5. Slopes were zero or positive during baseline and injury GV for PEEP of 5 cmH$_2$O and no PEEP. The introduction of 10 ml•kg^{-1} PFC to the injured lungs produced significantly negative phase III slopes in both animals. The slopes became more negative with increasing PFC volume (20 ml•kg^{-1} and 30 ml•kg^{-1}). Reduction of PEEP from 5 cmH$_2$O to 0 cmH$_2$O accentuated the slope for all doses of PFC in both animals.

P_aCO_2, plotted in Figure 6, increases with lung injury and again with the addition of PFC. CO_2 elimination was most efficient (i.e. P_aCO_2 was lowest) with PEEP of 5 cmH$_2$O for all conditions: GV, injury GV, PLV with 10 ml•kg^{-1}, 20 ml•kg^{-1}, and 30 ml•kg^{-1} perflubron. P_aCO_2 increased with lung injury in pig 1 but not in pig 2 and the decrease in P_aO_2 was much greater in pig 1 than pig 2. Removal of PEEP caused significant CO_2 retention in pig 1 and slight increase in pig 2.

DISCUSSION

In this paper, the novel appearance of exhaled CO_2 profile during PLV is evaluated in terms of its implications for ventilation heterogeneity. The CO_2 expirogram reflects both lung mechanics and regional gas exchange of which little is known during partial liquid ventilation. Interpretation of expired CO_2 vs. exhaled volume during PLV provides unique insight into the perturbations which occur in ventilation and gas exchange when a low surface energy fluid fills the lung. In order for the slope of phase III to be negative, 3 conditions

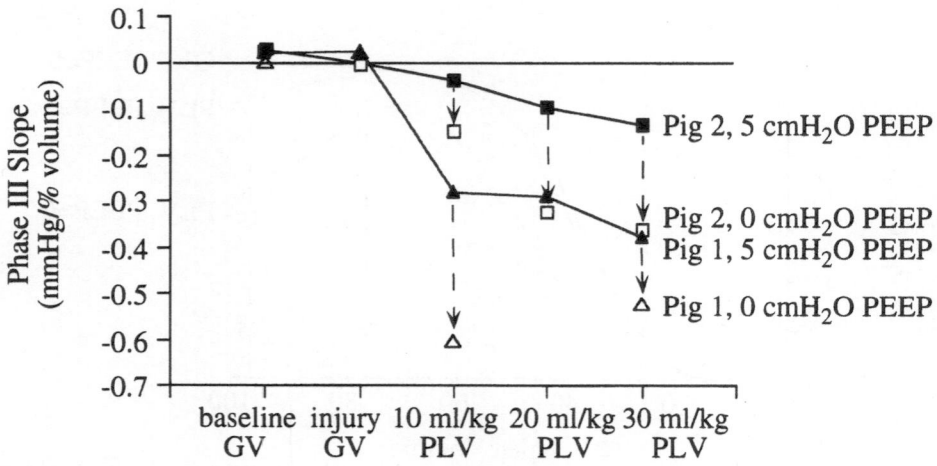

Figure 5. Phase III slope vs. experimental condition. GV is conventional gas ventilation, PLV is partial liquid ventilation. Solid symbols are for PEEP of 5, triangles represent pig #1, and squares pig #2. Removal of PEEP during PLV results in more steeply negative slopes at all doses (arrows to open symbols).

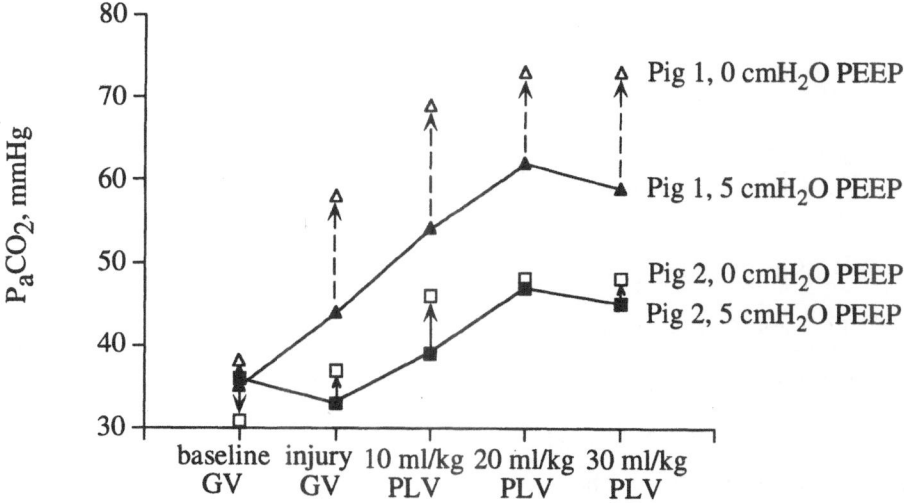

Figure 6. P_aCO_2 in group II animals. GV is conventional gas ventilation and PLV partial liquid ventilation at different doses of PFC in the lung. Solid symbols represent PEEP of 5, triangles represent pig #1 and squares represent pig #2. Removal of PEEP results in increased P_aCO_2 during GV in injury and during PLV at all doses (arrows to open symbols).

must be met: 1) lung regions must empty asynchronously; 2) alveolar P_ACO_2 must be distributed heterogeneously throughout the lung; and 3) the sequence with which the lung empties must be such that regions with low P_ACO_2 exit later than those with high P_ACO_2.

In an ideal homogeneous lung, arterial P_ACO_2 equals alveolar P_ACO_2 with which it equilibrated across the alveolar-capillary membrane. Arterial-alveolar differences of CO_2 ((a–A)D CO_2) result from inefficiencies in lung function. Heterogeneity of P_ACO_2 arises from regional differences in (a–A)D CO_2 in the non-ideal lung. The causes of (a–A)D CO_2 are threefold: shunt (blood flow to regions which are not ventilated), diffusion limitation (incomplete equilibration between blood and alveolar gas), and ventilation perfusion ($\dot{V}_A/\dot{Q}$) mismatch. Shunt does not typically affect CO_2 elimination unless it is quite large (50% or more) such that regional (a–A)D CO_2 is determined by diffusion limitation and $\dot{V}_A/\dot{Q}$ mismatch. $\dot{V}_A/\dot{Q}$ mismatch is worsened by increased heterogeneity in $\dot{V}_A$ or in $\dot{Q}$. The steep slope of P_ECO_2 vs. volume is indicative of increased $\dot{V}_A$ heterogeneity during PLV. Little is known about changes in the distribution of $\dot{Q}$ during PLV. Shaffer et al.[13] found a modest increase in heterogeneity of $\dot{Q}$ after draining PFC from a filled lung however it is difficult to extrapolate this work to PLV. Heterogeneity of P_ACO_2 during PLV is most likely due to increased $\dot{V}_A$ heterogeneity in combination with diffusion limited gas exchange between capillary blood and alveolar gas.

The distribution of ventilation is dependent on static properties of the lung and dynamic distribution of gas volume. Figure 7 depicts the lung as parallel compartments united by a common airway. Each compartment has unique resistance (R), and compliance (C) determined by static properties such as regional airway resistance, tissue elastance, and surface tension (γ). The dynamic distribution of flow among units is dependent on regional time constants, τ, which are equal to the product of local R and C. Time constants describe the time it takes to fill or empty a compartment by 63% of the total volume change at a given pressure. Figure 7 illustrates how heterogeneity of τ leads to non-uniform or asynchronous emptying. Branch 1 has low C_1 and normal R_1 hence shorter than average τ and empties

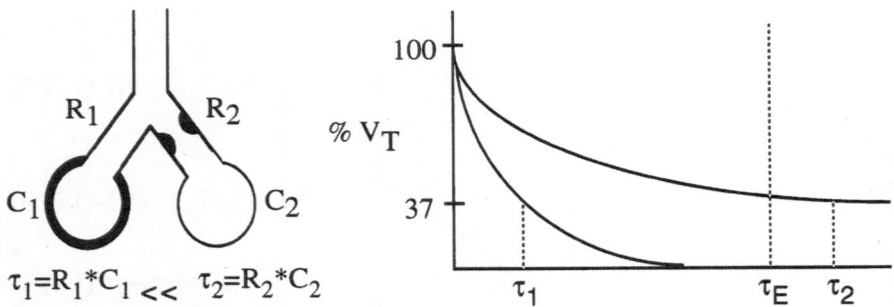

Figure 7. RC model of ventilation distribution which shows one airway with increased R due to obstruction and one with decreased C due to increased surface tension (γ). $\tau_1 < \tau_2$ due to low C_1 and high R_2. The left unit empties quickly and entirely, while right unit empties slowly and incompletely.

faster and completely. Branch 2 has normal C_2 and large R_2 hence longer than average τ and empties slowly and incompletely. Healthy, gas ventilated lungs have fairly uniform regional time constants and empty well before end expiration. Perturbations in local R and C due to disease or the presence of surface tension altering fluid change local τ's. Non-uniform distribution of a disease process or of PFC leads to heterogeneity in τ and asynchronous emptying of lung regions hence increased ventilation heterogeneity.

The dynamics of CO_2 exchange between blood and alveolar gas is altered when fluid is present in alveoli. Convective flow in the capillaries delivers blood laden with CO_2 to the alveolar-capillary membrane where diffusive exchange with alveolar gas strips the blood of CO_2. In a healthy, gas-filled lung the membrane is thin and diffusive conductance is high such that partial pressure equilibration between blood and alveolar gas is complete before red cells exit alveolar capillaries. In a fluorocarbon filled the lung, the alveolar membrane separates fluid from fluid. Diffusion times of CO_2 in PFC are approximately five orders of magnitude longer than in the gas phase[15]. The kinetics of partial pressure equilibration between blood and alveolar gas depends on the solubility of CO_2 in PFC, its diffusion coefficient in PFC, and on the thickness of the fluid layer separating blood from gas.

We developed a mathematical model of gas exchange in a fluid-filled alveolus to explore the contribution of diffusion limitation to (a–A)D CO_2. The model is described by a system of three differential equations of mass flux in three compartments: capillary, PFC, and alveolar gas (see Appendix). The equations were solved for steady state conditions such that the ratio of blood:PFC, blood:alveolar gas, and PFC:alveolar gas CO_2 partial pressures are constant in time. The compartments are assumed spherical in shape with the capillary compartment being outermost, the PFC compartment "lining" the blood compartment, and the gas compartment is a spherical "hole" in the center of the PFC. Figure 8 schematically describes the model and graphically displays solutions of (a–A)D CO_2 vs. PFC layer thickness. The size of the gas compartment changes over the respiratory cycle, being small (3 ml•kg^{-1}) at end-expiration, increasing greatly by the end of inspiration (15 ml•kg^{-1} + 3 ml•kg^{-1}). Gas exchange will differ in these two cases due to the large change in PFC:alveolar gas surface area. Figure 8 shows model solutions of (a–A)D CO_2 at each of the three doses of PFC used in these experiments (10, 20, and 30 ml•kg^{-1}) for inspiration and expiration. The value of P_ECO_2 measured at the mouth will likely lie somewhere in between the predicted P_ECO_2 for inspiration and expiration at a given P_aCO_2. These solutions predict the degree (a-A)D CO_2 which would be created by PFC in the alveoli. It will underestimate (a-A)D CO_2 in oleic acid injured piglets where edema fluid adds to diffusion limitation.

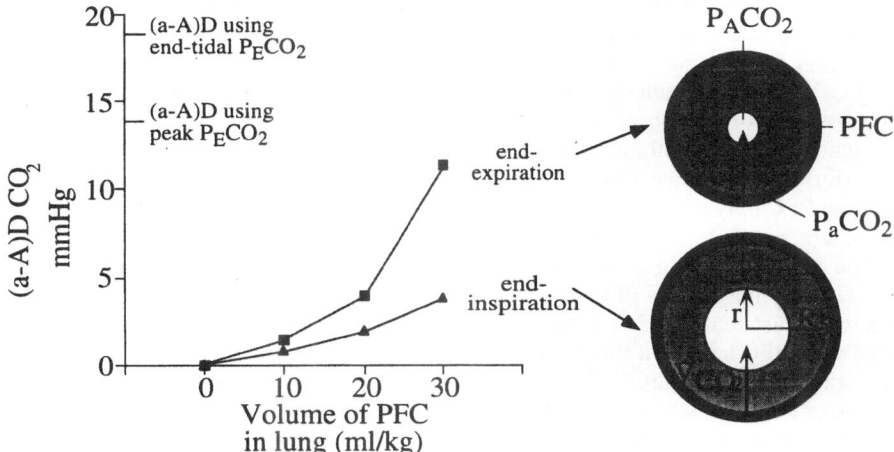

Figure 8. Schematic of gas exchange model and solutions for (a-A)DCO$_2$ vs. PFC layer thickness at end-inspiration and end-expiration. Values for peak and end-tidal PCO$_2$ from piglet #2 during PLV at 30 ml•kg^{-1} are labeled for comparison.

These solutions show that fairly large (a-A)D CO$_2$ can be created by PFC and that P$_A$CO$_2$ can vary to quite a degree throughout the lung due to variation in PFC thickness and differences in the degree of gas inflation.

Figure 9. Schematic depicting four conditions of the lung in this study. GV is gas ventilation (clear alveoli), PLV is partial liquid ventilation (dotted regions), injury is depicted by hatched regions, nl refers to normal GV values. A labels alveolus 1 and 2, R is resistance, C is compliance, τ is time constant.

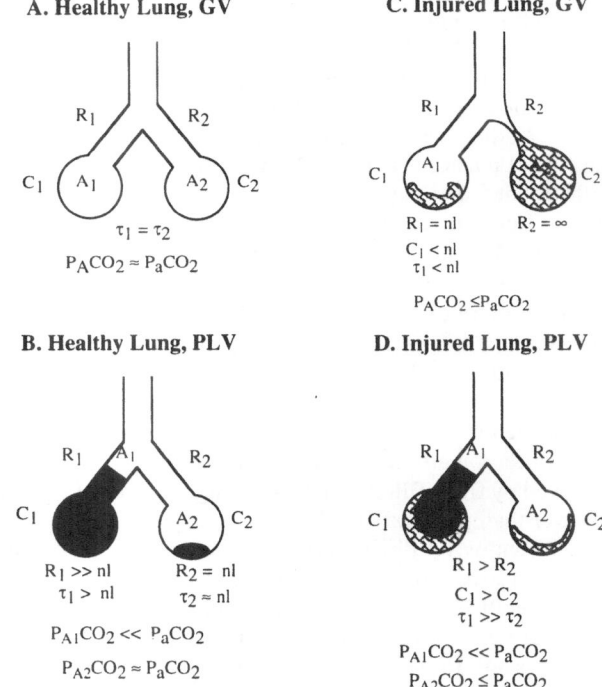

Figure 9 schematically illustrates the conditions which may exist in each of the 4 experimental preparations described which could lead to negative sloping alveolar plateaus of P_ECO_2. A healthy gas ventilated lung (Figure 9a) will have uniform τ and negligible (a-A)D CO_2. On expiration, lung units empty synchronously and at the same P_ACO_2. The slight positive slope of phase III in healthy lungs (Figure 1) is attributable to mild $\dot{V}_A/\dot{Q}$ heterogeneity and accumulation of CO_2 in alveoli over the course of a breath[5]. Healthy lungs with uniformly distributed PFC will behave similarly to that of Figure 9a. A healthy lung containing PFC may produce a positive slope of phase III of the CO_2 washout if changes in R and C are small. Lung emptying would still be uniform and heterogeneities of P_ACO_2 would be averaged during exhalation if they existed.

Healthy lungs with PFC distributed heterogeneously among acini (Figure 9b) will have a wider distribution of τ and areas of increased (a-A)D CO_2. Some acinar units will be flooded with PFC such that the gas:PFC interface moves in and out of small airways with the respiratory cycle (Figure 9b, A_1). This greatly increases R_1 which lengthens τ_1 while units with little or no PFC such as Figure 9b, A_2 have similar R_2 to the healthy gas ventilated lung.

The slower emptying compartments (those with larger PFC volumes) would have lower P_ACO_2 secondary to diffusion resistance (Figure 8). Upon exhalation areas with less fluid, shorter τ, and higher P_ACO_2 (little or no (a-A)D CO_2) empty first followed by obstructed regions with low P_ACO_2 producing a negative slope. This model of CO_2 elimination is supported by our findings that 1/9 healthy animals given graded doses of PFC showed negative phase III slope, 3/9 had sustained negative slope at 20 ml•kg^{-1}, and 7/9 exhibited the negative slope at 30 ml•kg^{-1} PFC. Larger doses of PFC lead to increased flooding of acinar units and small airway obstruction, increasing R and diffusion limitation.

Oleic acid injured piglets suffer patchy disease characterized by bloody alveolar exudate which inactivates surfactant and elevates surface tension. Figure 9c illustrates heterogeneously distributed disease with areas of atelectasis (9c, A_2) mixed with normal or mildly injured regions (9c, A_1). Atelectatic regions do not contribute to exhaled volume and the remaining regions are the least injured and empty fairly uniformly. Overall (a-A)D CO_2 increased in many of our injured animals. Figure 6 shows an increase in P_aCO_2 post injury in piglet #1, while P_ACO_2 was unchanged (data not presented). This increase is largely due to shunt (>50% in pig 1) created by atelectasis. The absence of increased P_aCO_2 in piglet #2 is explained by the fact that shunt fraction was only 15% post-injury in this case.

Adding PFC to oleic acid injured lung opens areas of atelectasis by lowering interfacial tension[14] (Figure 9d, A_1). Fluid distributes heterogeneously throughout the lung with flooded and partially filled acinar units (Figure 9d, A_1 and A_2). PFC lowers γ in the injured lung, increasing compliance (C_1) in proportion to the surface area of the PFC:lung interface. The presence of PFC in small airways increases R_1 and τ_1 is lengthened due increases in both R_1 and C_1. "Dry" (non-PFC filled) injured regions have lower than normal C_2 and short time constant, τ_2. The combination of longer time constants in fluid filled regions and shorter than normal τ's for dry regions leads to an even wider distribution of time constants during PLV compared to healthy animals. This may explain why negative slopes of phase III are seen more consistently in injured than healthy animals (Figure 3). PFC-filled areas have lower P_ACO_2 due to diffusion resistance in the fluid (Figure 8). Volume exhaled early in the breath comes from dry alveoli (Figure 9d, A_2) with low C_2 and high P_ACO_2 followed by fluid-filled regions with higher R_1 and C_1 and low P_ACO_2 so that end-tidal PCO_2 is lower than peak exhaled PCO_2.

Removing PEEP during PLV decreases gas volume, airway pressure, and alveolar volume at FRC while fluid volume remains constant. This results in mouthward movement of the gas:PFC interface, increasing the percent of small airways filled with fluid (Figures 9b, A_1 & 9d, A_1). In health and injury this results in increased heterogeneity of $\dot{V}_A$ with a greater proportion of units emptying later in the breath due to increased R. (a-A)D CO_2

widens with increased diffusion distances in fluid-filled regions, further dropping the end-tidal PCO_2. Figure 4 compares CO_2 expirograms from the same injured animal at 30 ml•kg^{-1} with PEEP of 5 cmH₂O and no PEEP. Exhaled PCO_2 is significantly lower throughout exhalation with no PEEP and the slope between 75% and 95% exhalation is more negative than the companion curve at 30 ml•kg^{-1}. Figure 6 shows increased P_aCO_2 for piglet 2, no PEEP vs. piglet 2, PEEP of 5 cmH₂O, consistent with our prediction that a larger proportion of lung regions with long τ and low P_ACO_2 result in decreased CO_2 excretion and delayed emptying.

In conclusion, partial liquid ventilation results in CO_2 retention and increased $(a-A)D\ CO_2$ at constant respiratory rate, tidal volume, and positive end-expiratory pressure. This is a consequence of increased ventilation heterogeneity as well as heterogeneously distributed diffusion limitation creating uneven P_ACO_2. Negative phase III slopes of CO_2 expirograms occur during PLV when PFC is heterogeneously distributed and flooded regions characterized by prolonged emptying times and low P_ACO_2 empty late in expiration. Larger volumes of PFC and the absence of positive end-expiratory pressure accentuate ventilation heterogeneity and diffusion limitation resulting in greater CO_2 retention. Higher PEEP, lower respiratory rates, and larger tidal volumes may improve CO_2 removal during PLV. However, the disadvantages of increased mean airway pressure must be weighed against advantages of lower P_aCO_2 and markedly improved oxygenation[6,13,16] in individuals with lung injury. A practical implication of the negative sloping CO_2 profile is that capnography must be used with caution in estimating arterial PCO_2. Peak expired CO_2 is a closer approximation to arterial PCO_2 than end-tidal values and both may be significantly lower than arterial PCO_2.

APPENDIX

Mass flux $(dM/dt = d(\beta \bullet P \bullet V)/dt)$ in the capillary (c), fluorocarbon (PFC) and alveolar gas (g) compartments, respectively, are described by the following equations:

$$\frac{dP_c}{dt} = \frac{\dot{Q}}{V_c} \cdot (P_{\bar{v}} - P_a) + (\frac{D_{pfc} \cdot \beta_{pfc} \cdot 4\pi \cdot R^2}{\beta_b \cdot V_c}) \frac{dP_{pfc}}{dr}\Big|_{r=R} \tag{1}$$

$$\frac{\partial P_{pfc}}{\partial t} = D_{pfc} \cdot (\frac{\partial^2 P_{pfc}}{\partial r^2} + \frac{2}{r}\frac{\partial P_{pfc}}{\partial r}) \tag{2}$$

$$\frac{dP_g}{dt} = \frac{D_{pfc} \cdot \beta_{pfc} \cdot 4\pi \cdot (R-r)^2}{\beta_g \cdot V_g} - \frac{\dot{V}_A}{V_g} \cdot P_g \tag{3}$$

Where $\dot{Q}$ is blood flow (mL•sec^{-1}), $\dot{V}_A$ is alveolar ventilation (ml•sec^{-1}), t is time (sec), V's are compartment volumes (mL), P's are CO_2 partial pressures (mmHg), D_{pfc} is molecular diffusion coefficient in PFC (cm²•sec^{-1}), β's are solubility coefficients (mL CO_2•100 mL solvent^{-1}•mmHg^{-1}), and r is the spatial coordinate in the radial direction with r=R being the capillary:PFC boundary, r=r the PFC:alveolar gas boundary, and r=0 being the center of the sphere (see figure 8).

By assuming steady state gas exchange conditions, the set of 3 differential equations (1-3) reduces to two linear equations by setting all dP/dt = 0 and setting mass flux through each compartments equal:

$$\dot{Q} \cdot \beta_b \cdot P_{\bar{v}} = (\dot{Q} \cdot \beta_b + D_{pfc} \cdot \beta_{pfc} \cdot 4\pi \frac{R \cdot r}{R-r}) \cdot P_c - (D_{pfc} \cdot \beta_{pfc} \cdot 4\pi \frac{R \cdot r}{R-r}) \cdot P_g \qquad (4)$$

$$0 = (D_{pfc} \cdot \beta_{pfc} \cdot 4\pi \frac{R \cdot r}{R-r}) \cdot P_c - (\dot{V}_A \cdot \beta_g + D_{pfc} \cdot \beta_{pfc} \cdot 4\pi \frac{R \cdot r}{R-r}) \cdot P_g \qquad (5)$$

Where $P_{\bar{v}}$ is mixed venous blood partial pressure (mmHg), R is radius of the gas exchange unit, and r is the radius of the gas compartment. The two unknowns, P_g and P_c can be solved for using these equations:

$$P_c = P_{\bar{v}} \cdot \frac{\dfrac{\dot{V}_A \cdot \beta_b \cdot (R-r)}{D_{pfc} \cdot \beta_{pfc} \cdot 4\pi \cdot R \cdot r} + \dfrac{\beta_b}{\beta_g}}{\dfrac{\dot{V}_A \cdot \beta_b \cdot (R-r)}{D_{pfc} \cdot \beta_{pfc} \cdot 4\pi \cdot R \cdot r} + \dfrac{\beta_b}{\beta_g} + \dfrac{\dot{V}_A}{\dot{Q}}} \qquad (6)$$

$$P_g = P_{\bar{v}} \cdot \frac{\dfrac{\beta_b}{\beta_g}}{\dfrac{\dot{V}_A \cdot \beta_b \cdot (R-r)}{D_{pfc} \cdot \beta_{pfc} \cdot 4\pi \cdot R \cdot r} + \dfrac{\beta_b}{\beta_g} + \dfrac{\dot{V}_A}{\dot{Q}}} \qquad (7)$$

Parameters for model solutions in Figure 8 included matched $\dot{V}_A$ and $\dot{Q}$ similar to values typically measured in piglets. R was chosen as 300 μm, the approximate size of a terminal sac, and r is variable and based on the degree of inflation (3 ml•kg^{-1} at end-expiration and 18 ml•kg^{-1} at end-inspiration) and the volume of PFC in the lung. β_{blood} for CO_2 was assumed linear in the range of PCO_2 30-50 mmHg and calculated for pig blood using blood gas subroutines of Olszowka et.al.[9]. CO_2 molecular diffusion coefficient in PFC was estimated from Tham et.al.[15], $D_{CO_2} = 4.35*10^{-5}$ cm^2•sec^{-1}. β_{pfc} provided by Alliance Pharmaceutical Corp as 0.276 ml CO_2•100 ml PFC^{-1}•mmHg^{-1}. β_g was calculated from the Ideal Gas Law at BTPS: $\beta_g = 0.132$ ml•100 ml^{-1}•mmHg^{-1}.

ACKNOWLEDGMENTS

The expert technical assistance of Wayne Lamm and Dr. T. Standaert was invaluable in successfully completing these experiments. Alliance Pharmaceutical Corp. (San Diego, CA) provided the perflubron used in this study. This research was supported in part by NIH grant R37HL12174.

REFERENCES

1. Crank, J. Diffusion in a sphere. In: *The Mathematics of Diffusion*. Oxford: The Clarendon Press, 1970, chapt. 6.
2. Curtis, S. Perfluorocarbon-associated gas exchange: A hybrid approach to mechanical ventilation. Crit. Care Med. 19: 600-601, 1991.
3. Curtis, S. E. and J. T. Peek. Perfluorocarbon associated gas exchange in a canine model of acute lung injury. American Thoracic Society. A886, 1993.
4. Fuhrman, B. P., P. R. Paczan and M. DeFrancisis. Perfluorocarbon-associated gas exchange. Crit. Care Med. 19: 712-22, 1991.

5. Grønlund, J., E. R. Swenson, J. Ohlsson and M. P. Hlastala. Contribution of continuing gas exchange to phase III exhaled PCO_2 and PO_2 profiles. J. Appl. Physiol. 62: 2467-2476, 1987.

6. Leach, C. L., B. P. Fuhrman, F. C. Morin and M. G. Rath. Perfluorocarbon-associated gas exchange (partial liquid ventilation) in respiratory distress syndrome: A prospective, randomized, controlled study. Crit. Care Med. 21: 1270-1278, 1993.

7. Mates, E.A., J.C. Jackson, J. Hildebrandt, W.E. Truog, T.A. Standaert, and M.P. Hlastala. Respiratory gas exchange and inert gas retention during partial liquid ventilation. Oxygen Transport To Tissue XVI, in press.

8. Neufeld, G. R., S. Gobran, J. E. Baumgardner, S. J. Aukburg, M. Schreiner and P. W. Scherer. Diffusivity, respiratory rate and tidal volume influence inert gas expirograms. Respir. Physiol. 84: 31-47, 1991.

9. Olszowka, A.J., and L.E. Farhi. A system of digital computer subroutines for blood gas calculations. Respir. Physiol. 4: 270-280, 1968.

10. Otis, A. B., C. B. McKerrow, R. A. Bartlett, J. Mead, M. B. McIlroy, N. J. Selverstone and E. P. Radford. Mechanical factors in distribution of pulmonary ventilation. J. Appl. Physiol. 8: 427-443, 1956.

11. Sargent, J. W. and R. J. Seffl. Properties of perfluorinated liquids. Fed. Proc. 29: 1699-1703, 1970.

12. Scherer, P. W., S. Gobran, S. J. Aukburg, J. E. Baumgardner, R. Bartkowski and G. R. Neufeld. Numerical and experimental study of steady-state CO_2 and inert gas washout. J. Appl. Physiol. 64: 1022-1029, 1988.

13. Shaffer, T. H., C. A. Lowe, V. K. Bhutani and P. R. Douglas. Liquid ventilation: effects on pulmonary function in distressed meconium-stained lambs. Ped. Res. 18: 47-52, 1984.

14. Tarczy-Hornoch, P., J. Hildebrandt, T. Standaert, W. Lamm and J. C. Jackson. In-situ interfacial tension during liquid ventilation. Ped. Res. 35: 355a, 1994.

15. Tham, M. K., R. D. Walker and J. H. Modell. Diffusion coefficients of O_2, N_2, and CO_2 in fluorinated ethers. J. Chem. Eng. Data. 18: 411-412, 1973.

16. Tutuncu, A. S., N. S. Faithfull and B. Lachmann. Intratracheal perfluorocarbon administration combined with mechanical ventilation in experimental respiratory distress syndrome: dose-dependent improvement of gas exchange. Crit. Care Med. 21: 962-969, 1993.

76

NEBULIZATION OF ENDOTOXIN DURING MECHANICAL VENTILATION

An Experimental Model of ARDS in Pigs

J. C. Pompe,[1] J. Kesecioğlu,[2] H. A. Bruining,[1] and C. Ince[3]

[1] Departments of Surgery
[2] Department of Anesthesiology
   University Hospital Rotterdam, The Netherlands
[3] Department of Anesthesiology
   Academic Medical Center
   University of Amsterdam
   The Netherlands

INTRODUCTION

Gram-negative sepsis is the most common setting in the ICU in which abnormalities in lung function known as the Adult Respiratory distress Syndrome (ARDS) develops[1]. Lipo-polysaccharide molecules residing in the outer membrane of gram-negative bacteria (endotoxins), are the bacterial components presumed to elicit the inflammatory response of the host that underlie the etiology of ARDS. ARDS is associated with significant abnormalities of lung mechanics[2,3]. Decreased compliance with resultant rapid shallow breathing and increased deadspace ventilation, not refractory hypoxemia, are often rate-limiting factors preventing weaning from mechanical ventilation in patients with esthablished ARDS[4]. Decreases in compliance are also associated with increased airway pressure in the mechanically ventilated patient and with the development of macroscopic barotrauma[5]. The pathophysiology of lung injury induced by endotoxin has been studied extensively in vivo after intravenous or intraperitoneal infusion of endotoxin in a wide variety of animal models, using different methods and species, including the rat[6], pig[7], dog[8] and sheep[9]. However, responses to intravenous infusion nearly always include symptoms of systemic involvement, with varying degrees of hemodynamic instability as a result. The same problem exist in other models of ARDS such as lung-lavage and oleic acid infusion. In order to produce a stable model in these instances, aggressive volume and inotropic support is usually necessary. As a result, treatment strategies under investigation are difficult to evaluate.

The aim of our study was to develop an animal model of endotoxin induced acute lung injury in mechanically ventilated pigs, that does not suffer from this limitation. We hypothesized that by direct application of endotoxin as aerosol to the lungs, the cardiovas-

Oxygen Transport to Tissue XVII, Edited by Ince et al.
Plenum Press, New York, 1996

cular instability normally seen during intra-venous administration of endotoxin could be prevented in order to optimize the pulmonary response.

MATERIALS AND METHODS

Thirteen healthy domestic pigs (weight: 28.1±4.0 kg) were anesthetized (midazolam/ke-tamine) and paralyzed (pancuronium bromide). All animals were ventilated with a Siemens Servo 900C ventilator in the volume controlled mode (VT=15 ml/kg, FiO_2=0.6, PEEP=4 cmH_2O). An ultrasonic nebulizer Hico–Ultrasonat 760E) was incorporated in the inspiratory part of the breathing circuit. Seven pigs (Endotoxin group) were challenged with aerosolized E. Coli endotoxin (serotype 0127:B8, Sigma) until PaO_2 < 90 mmHg and six pigs (Control group) were challenged with an equal amount aerosolized distilled water. Arterial, central venous and pulmonary artery pressures were measured by intravascular catheters. Arterial and mixedvenous blood samples were analyzed by a blood gas analyzer (AVL 945).

After a stabilization period of 1 hour, baseline measurements were accepted when $PaCO_2$ was between 35 and 45 mmHg. After baseline ventilatory settings remained unchanged.

After the nebulization period, measurement were done at 1 hour intervals for 4 hours.

RESULTS

Baseline values (n=13): PaO_2 353±32 mmHg, mean arterial pressure (MAP) 98±20 mmHg and mean pulmonary pressure (MPP) 20±2 mmHg. In 5 pigs of the endotoxin group, severe respiratory distress developed after 161±33 minutes of nebulization during which period 118.5±20.6 mg endotoxin was administered. PaO_2 was 62±20 mmHg, MAP 99±8 mmHg and MPP 29±9 mmHg in these 5 pigs. The model was stable for 4 hours and there were no signs of hemodynamic instability (figures 1 and 2). Measurement of pulmonary mechanics (peak airway pressure, figure 3) were in close agreement with those found clinically in ARDS. Chest X-ray pictures showed bilateral infiltrates. Histopathologic changes consisted of pulmonary oedema and a diffuse neutrophilic alveolitis. In 2 pigs of the endotoxin group, the nebulization period was nearly 4 times longer (436±46 minutes in which 196.5±54.5 mg endotoxin

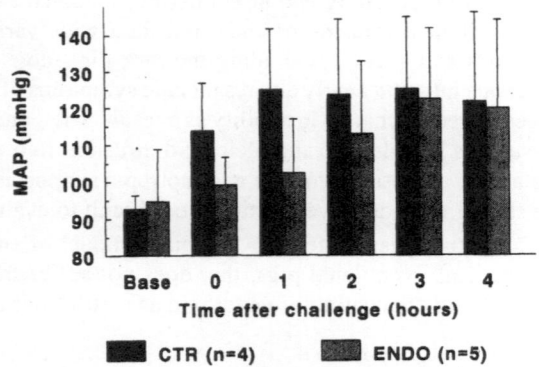

Figure 1. Mean systemic arterial pressure.

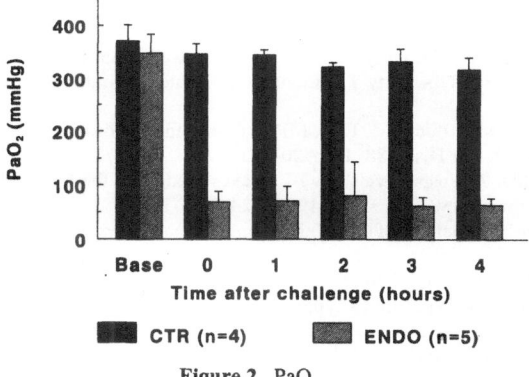

Figure 2. PaO$_2$.

was nebulized). This group was evaluated separately and was characterized as having a different sensitivity to nebulized endotoxin. In 4 pigs of the control group there were no significant changes from baseline measurements (PaO$_2$ 346±19 mmHg, MAP 114±13 mmHg and MPP 18±2 mmHg). However, in 2 pigs of the control group severe (instable) respiratory distress developed 2 hours after the nebulization period (PaO$_2$ 77±20 mmHg).

CONCLUSIONS

We conclude that this model can be used as an experimental analog of endotoxin induced ARDS. The lack of any systemic response with this method of endotoxin administration, offers the opportunity to investigate and compare the effects of therapeutic interventions on pulmonary parameters. However, differences in sensitivity and specificity of the model need to be anticipated.

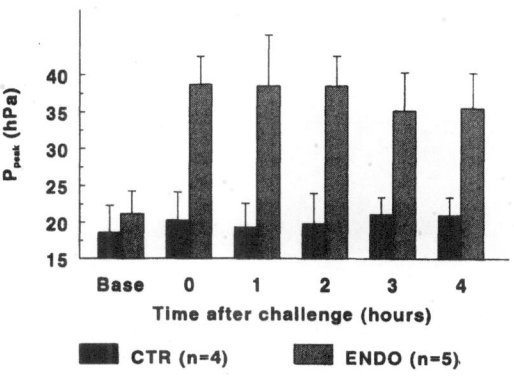

Figure 3. Peak airway pressure.

REFERENCES

1. Newman JH. Sepsis and pulmonary edema. Clin Chest Med 1985; 6:371-91
2. Ashbaugh DG, Bigelow DB, Petty TL, Levine BE. Acute respiratory distress in adults. Lancet 1967; 2:319-323.
3. Murray JF, Matthay MA, Luce JM, Flick MR. An expanded definition of the adult respiratory distress syndrome. Am Rev Respir Dis 1988; 138:720-723.
4. Yang KL, Tobin MJ. A prospective study of indexes predicting the outcome of trials of weaning from mechanical ventilation. New Engl J Med 1991; 324:1445-1450.
5. Hinson JR, Marini JJ. Principles of mechanical ventilator use in respiratory failure. Annu Rev Med 1992; 43:341-361.
6. Rinaldo JR, Dauber MH, Christman J, Rogers RM. Neutrophil alveolitis following endotoxemia: enhancement by previous exposure to hyperoxia. Am Rev Respir Dis 1984; 130:1065-1071.
7. Steinberg SM, Dehring DJ, Gower JR, Vento JM, Lowery FF, Cloutier CT. Prostacyclin in experimental septic acute respiratory failure. J Surg Res 1983; 34:298-302.
8. Krausz MM, Utsunomiya T, Feuerstein G, Wolfe JHN, Shepiro C, Hechtman HP. Prostacyclin reversal of lethal endotoxemia in dogs. J Clin Invest 1981; 67:1118-1125.
9. Ebenshade AM, Newman J, Lams P, Jolles H, Brigham K. Respiratory failure after endotoxin infusion in sheep: lung mechanics and fluid balance. J Appl Physiol 1979; 53:967-976.

77

SUPPORTING TISSUE OXYGENATION DURING ACUTE SURGICAL BLEEDING USING A PERFLUOROCHEMICAL-BASED OXYGEN CARRIER

Peter E. Keipert, N. Simon Faithfull, Duane J. Roth, JoAnn D. Bradley, Sanjay Batra, Philip Jochelson, and Kathryn E. Flaim

Alliance Pharmaceutical Corporation
3040 Science Park Road
San Diego, CA 92121

INTRODUCTION

Attempts to develop so-called "blood substitutes" have historically focused on three approaches: 1) acellular hemoglobin (Hb) solutions, 2) encapsulation of Hb, and 3) perfluorochemical emulsions. Work on Hb solutions first began over 100 years ago,[1] by lysing red blood cells to extract the oxygen-carrying Hb molecules. For the past 60 years, efforts have concentrated on purifying the Hb from the contaminating stromal lipid,[2] and on developing ways to crosslink or polymerize the Hb molecules[3] to prevent them from splitting into dimers (i.e., half-molecules) which are rapidly filtered by the kidney. Based on the fact that free Hb has a very short intravascular persistence and is potentially toxic to the kidney,[4] some investigators have used microencapsulation as a means to package the Hb inside sub-micron lipid-coated vesicles.[5,6]

Perfluorochemicals (PFCs), in contrast, are completely synthetic liquids developed originally as part of the Manhattan Project during World War II.[7] It was not until the pioneering work of Drs. Leland Clark and Robert Geyer in the 1960's that the potential of PFCs as oxygen-carriers was appreciated.[8,9] Since PFCs are not miscible with water, they must be mixed with water and a surfactant (e.g., egg yolk phospholipid, Pluronic) in order to produce injectable emulsions.[10] The first commercial PFC emulsion, *Fluosol*, was made by Green Cross Corporation in Japan, and consisted of two PFCs (F-decalin and F-tripropylamine) emulsified with Pluronic-F68 and egg yolk phospholipid.[11] *Fluosol* represents the only temporary oxygen carrier approved by the U.S. Food and Drug Administration to date.

More recently, development efforts have investigated different PFCs and surfactants in order to improve emulsion characteristics and overcome the limitations of *Fluosol* (i.e., *Fluosol* must be frozen, thawed and mixed with two annex solutions prior to use, and is only stable for 8 hours once prepared). One of these approaches is based on emulsifying

Oxygen Transport to Tissue XVII, Edited by Ince et al.
Plenum Press, New York, 1996

perflubron (perfluorooctyl bromide; $C_8F_{17}Br$) with egg yolk phospholipid (Alliance Pharmaceutical Corp.) to produce a concentrated emulsion (*Oxygent*™) which is stable and can be terminally heat-sterilized.[10] One of the clinical indications for this product will be as a temporary oxygen carrier during surgical procedures involving significant blood loss, as a means to reduce or eliminate the need for allogeneic (donor) blood transfusions.[12] It is anticipated that *Oxygent* will function as an "anti-hypoxic" agent and thereby maintain tissue oxygenation during acute progressive surgical anemia.

MATERIALS AND METHODS

Computer Modeling and PFC-Based Oxygen Transport

A computer model has been developed to predict the mixed venous blood PO_2 (PvO_2) level and the physiological responses to both acute normovolemic hemodilution (ANH) and volume-compensated surgical blood loss.[13] Inputs include those determining the oxygen content (e.g., total Hb, blood PO_2) and the position of the Hb dissociation curve (Kelman constants, pH, PCO_2, temperature); those determining total oxygen delivery (cardiac output, cardiac output response to ANH, rate of bleeding) and consumption (VO_2 - measured or calculated); and those related to the properties of the oxygen carrier (either Hb- or PFC-based) being administered (e.g., dose, solubility for oxygen, blood half-life). When the arterial and venous blood gases, Hb concentrations, and oxygen consumption are entered, all of the oxygen transport and VO_2 variables can be determined for any of the 3 compartments, i.e., the red cells, the plasma, and the oxygen carrier.

Canine Model of Hemodilution and Surgical Bleeding

A dog model was designed to mimic normovolemic hemodilution and surgical bleeding, using isoflurane-anesthetized mechanically-ventilated beagle dogs. Surgical preparation consisted of a midline laparotomy to tie off splenic blood vessels (to prevent release of sequestered red cells during ANH and bleeding), bladder catheterization, and placement of arterial, venous, and pulmonary artery catheters. Heart rate (from ECG) and blood pressures were monitored continuously. Following ANH to a target Hb level of 8.0 g/dL with 1:1 (vol/vol) *Hespan* (6% hydroxyethyl starch solution, Dupont) replacement, the FiO₂ was increased to 1.0 to maximize the oxygen-carrying capacity of the PFC and to establish the PvO_2 on 100% oxygen ventilation. Animals were then randomized to a treatment group that received a single dose of perflubron emulsion (1.35 g PFC/kg), or to a control group that received an equal volume of a balanced electrolyte solution (1.5 mL/kg of *Plasma-Lyte*). All dogs were then subjected to a volume-compensated blood loss at 0.75 mL/kg/min until Hb concentrations reached 2.0 g/dL.

Clinical Studies

To date, in excess of 145 human subjects have been tested with perflubron emulsions in various European and U.S. clinical studies at doses ranging from 0.45 to 3.0 g PFC/kg). Clinical testing in the U.S. has consisted of a saline-controlled Phase I dose-escalation safety study in 57 conscious healthy volunteers; and a multi-center randomized, Ringer's lactate-controlled dose-escalation Phase I/II safety study in 30 consenting adult surgical patients undergoing low-blood-loss elective surgery under general anesthesia. More recently, a multi-center single-dose Phase IIa study to evaluate safety and drug activity was completed in 7 consenting adult patients undergoing high blood loss surgical procedures.

RESULTS

From the computer model used in this study to simulate the addition of a PFC-based oxygen carrier to a hemodiluted patient, it is possible to calculate the predicted PvO_2 and the relative contribution of each compartment (i.e., the red cells, plasma, and PFC) to total oxygen consumption (VO_2). Under normal air-breathing conditions, plasma-dissolved oxygen does not contribute significantly to VO_2, but following hemodilution the relative plasma volume is increased (as red cells are removed) and therefore can contribute about 6% of VO_2 when Hb levels are reduced to 8.0 g/dL. Breathing 100% oxygen, however, significantly increases the amount of dissolved oxygen, such that the plasma alone can provide about 30% of VO_2 at normal Hb levels and up to approximately 50% of VO_2 when hemodiluted to Hb levels of 8 g/dL. (Model assumptions for these calculations include: $PaO_2 = 500$ mmHg breathing 100% oxygen, $VO_2 = 3.0$ mL/min/kg at 37°C, and a cardiac output increase of 0.5 L/min/g Hb decrease for a 70 kg adult under anesthesia.) The effect of injecting a PFC emulsion is to increase the net oxygen solubility of the plasma compartment. The relative contribution of each of these compartments to VO_2 is illustrated in Figure 1, following a 1.8 g PFC/kg dose of perflubron emulsion administered at various levels of hemodilution. The PFC-dissolved oxygen is able to provide about 15% of VO_2 and can effectively replace almost one third of the total contribution required from the remaining red cell mass.

The computer model also permits a calculation of how much blood loss can occur before the mixed venous oxygen tension returns to pre-dosing levels. Since PvO_2 is generally accepted as the best overall indicator for whole body oxygenation status,[14] it can be monitored during active bleeding in surgical patients with pulmonary artery catheters as a guide (together with all the other parameters normally monitored such as blood loss, ECG changes, hemodynamics, etc.) to when a blood transfusion is indicated. If a dose of PFC emulsion is administered in place of an intraoperative blood transfusion, the PvO_2 response in the face of ongoing blood loss can be predicted by the computer model, as shown in Figure 2. These results demonstrated that a single dose of perflubron emulsion (1.35 g PFC/kg) could maintain global tissue oxygenation (i.e., based on keeping PvO_2 at or above predosing

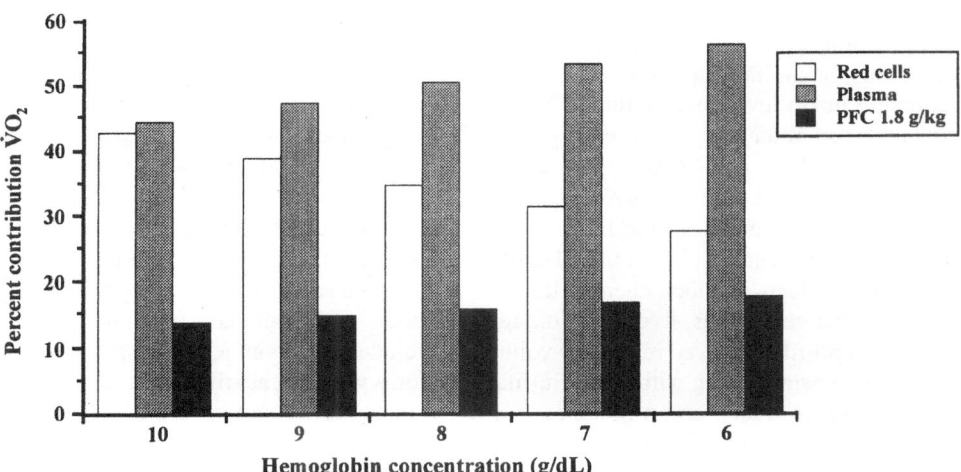

Figure 1. Relative contribution of red cells, plasma, and PFC to total oxygen consumption when breathing 100% oxygen. Each group compares the effects when the PFC emulsion dose is injected at different levels of hemodilution (indicated by Hb levels).

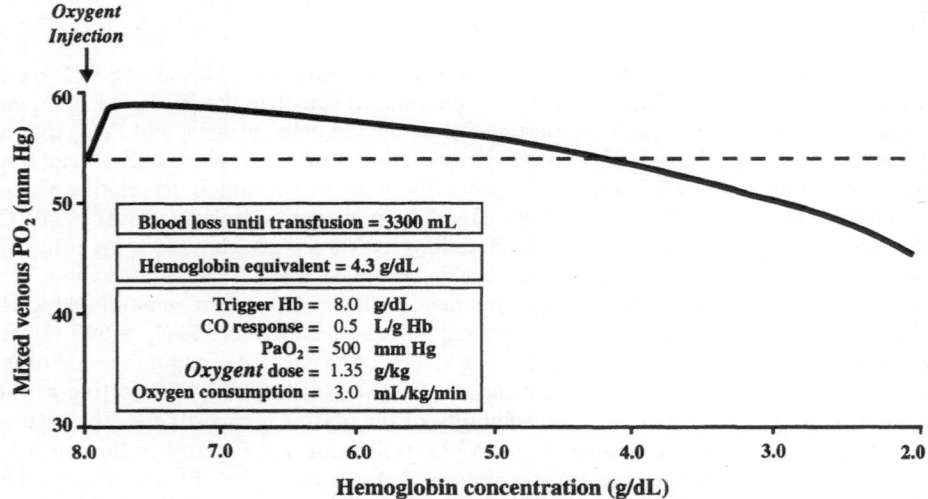

Figure 2. Computer simulation of the predicted changes in mixed venous oxygen tension following administration of a PFC emulsion at a Hb of 8 g/dL. The hypothetical patient is then subjected to "surgical bleeding" at 1.0 L/hour. (Model assumptions and conditions for this simulation are indicated in the figure.)

levels) while allowing blood loss to lower total Hb levels from about 8.0 to 4.0 g/dL (corresponding to a blood loss of approximately 3 Liters).

Results from the canine hemodilution study demonstrated that there were no appreciable changes in mean arterial pressures, heart rate, or pulmonary wedge pressures throughout the protocol. In response to the reduced blood viscosity caused by removal of red blood cells, cardiac outputs increased in both groups as expected. Total oxygen consumption was maintained closer to baseline levels in treated dogs despite the severe anemia induced by progressive volume-compensated blood loss. The increase in PvO_2 predicted by the computer model after administering PFC emulsion was also seen in these dog studies.[15] A significant increase of ~15 mmHg above the pretreatment (oxygen-breathing) PvO_2BL was observed, versus only a 2 mmHg increase in the control group. The magnitude of the reduction in Hb levels before the PvO_2 returned to pre-dosing PvO_2BL level for each dog was recorded and averaged for each group. These results are displayed in Figure 3. Thus, treated dogs were able to withstand significantly greater blood loss (almost 70 mL/kg) compared to 10 mL/kg in controls.

To date, in excess of 145 human subjects have been tested with perflubron emulsions in various European and U.S. clinical safety and efficacy studies. There were no clinically meaningful effects on blood chemistries, coagulation values (PT, PTT, fibrinogen), red cell counts, white cell counts, liver function, renal function, and pulmonary function tests. The side effect profile observed in healthy volunteers included a delayed febrile response (at 6-8 hours post-dosing) along with some flu-like symptoms which generally resolved spontaneously by 12-24 hours, and a moderate reduction in platelet count with a nadir of about 170,000/μL at 3 days post-dosing with full recovery by 7 days. Similar side effects were subsequently encountered in surgical patients, but relative to the control group were only observed at the highest dose studied (1.8 g PFC/kg).[16] These included a transient febrile response (onset at ~ 4 hours post-dosing; peak temperature of 38-39°C at 6-8 hours; full resolution by 12-14 hours) and a transient reduction in platelet count (mean nadir

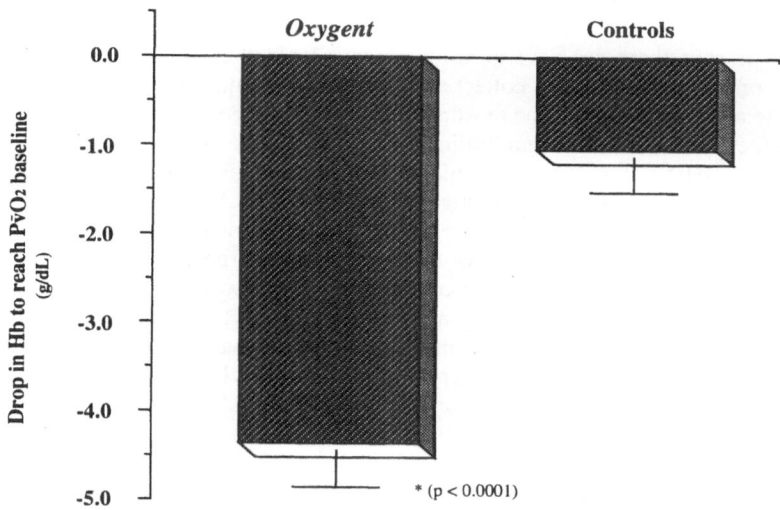

Figure 3. Reduction in total Hb concentration in dogs subjected to rapid blood loss (0.75 mL/kg/min) following administration of a single dose of perflubron emulsion. The PvO$_2$ levels achieved on 100% oxygen breathing alone were used as the lower limit to determine how much blood loss the dogs could tolerate after being dosed with perflubron emulsion (1.35 g PFC/kg).

~130,000/µL on day 3; with return to normal by 7 days). Patients in the control group experienced similar responses of slightly less magnitude.

Preliminary data from the pilot Phase II efficacy study in hemodiluted patients undergoing high-blood-loss surgical procedures demonstrated increases in PvO$_2$ levels that remained elevated above predosing levels (achieved on 100% oxygen-breathing) while ongoing surgical bleeding caused total Hb to decrease from about 8.3 g/dL (the Hb level where perflubron emulsion was dosed instead of initiating blood transfusion) to 6.5 g/dL (the point where blood transfusion was initiated).[17] Four of the seven patients did not require any allogeneic blood. These preliminary data suggest that the equivalent of about 1-2 units of blood were avoided with a single dose (0.9 g PFC/kg) of perflubron emulsion.

DISCUSSION

Traditionally, transfusions with allogeneic (donor) blood were regarded as an effective therapeutic intervention to correct the anemia resulting from surgical blood loss, and most physicians adhered to the "10/30" rule, i.e., a Hb of 10 g/dL (or a 30% hematocrit) was the level at which a blood transfusion was given. In the mid-1980s, however, the safety of the blood supply was seriously questioned when it was recognized that the human immunodeficiency virus (HIV) which causes AIDS could be transmitted via a blood transfusion. This threat of AIDS and other infectious diseases such as hepatitis, along with the known compatibility risks of allogeneic blood, have generated considerable publicity in both the medical press and in news reports.[18] Patients therefore became very concerned about the risks for disease transmission, and clinicians started re-evaluating existing transfusion guidelines and clinical practise in order to reduce patient exposure to allogeneic blood. The American College of Physicians recently published a *Clinical Guideline* which now recommends that elective transfusion with allogeneic blood be regarded as an outcome to be avoided.[19]

At present there are several strategies available that are being employed, independently or in combination, to collect autologous blood and thereby reduce the patient's exposure to allogeneic blood. The most common include preoperative autologous donation (PAD), acute normovolemic hemodilution (ANH), and intra- or post-operative autologous blood salvage (ABS). Each single technique, however, has limitations which often prevent the patient from having sufficient autologous blood available.[20] PAD is a technique whereby patients, in the 6 weeks prior to elective surgery, can deposit several units of their own autologous blood to be stored for later use during their surgery. PAD, however, requires access to an established predonation program and is not always feasible when the patient does not have enough lead time, or may not be healthy enough to donate sufficient blood. A related technique, ANH, involves collecting a portion of the patient's blood (i.e., 2 to 4 units) in the operating room just prior to surgery and replacing it with a crystalloid and/or colloid plasma expander to maintain blood volume. Perioperative ANH is somewhat limited in its application due to organizational logistics, inexperience with the technique, and the amount of blood that can be collected safely. Blood salvage devices used in ABS are designed to suction blood from the surgical field during active bleeding and collect the recovered red cells for reinfusion into the patient.

Avoidance of allogeneic blood in surgical patients, however, has been shown to be highly dependent on combining several autologous blood collection techniques. A novel approach has been developed to combine the use of a perflubron-based emulsion with intraoperative autologous blood conservation strategies (since the plasma expanders used in surgery to maintain constant circulating blood volume do not replace the oxygen-carrying capacity of the missing red blood cells.[12]) It is anticipated that administration of perflubron emulsion will allow for safer hemodilution by preventing decreases in tissue oxygenation, and ultimately can reduce or eliminate intraoperative allogeneic blood transfusion requirements.[21]

A relatively small dose of a PFC emulsion provides a significant contribution to VO_2 based on the very high efficiency of oxygen extraction from the PFC and plasma. PFCs obey Henry's law and dissolve increasing amounts of oxygen in direct proportion to the surrounding PO_2.[13] Thus, overall extraction of PFC-dissolved oxygen is about 90% when going from arterial blood to mixed venous blood (PaO_2 = 500 mmHg, PvO_2 = 50 mmHg while breathing 100% oxygen). In contrast, oxygen extraction from whole blood under these same conditions would only be about 25-30% (with about 20% of this oxygen extraction coming from plasma and 80% from the red cell Hb).[22] Also, the dissolved oxygen carried by the PFC is present in the plasma compartment (i.e., does not require diffusion across the red cell membrane), thereby increasing the effective surface area for diffusion of oxygen to the tissues.

Thus, PAD and/or ANH supplemented with perflubron emulsion (@FiO_2=1.0), and ABS during surgery has the potential to safely allow more profound ANH and might safely permit surgery at lower Hb levels. The net savings would be fewer red cells lost during active surgical bleeding at lower Hb levels and more fresh autologous whole blood available for reinfusion once hemostasis has been achieved, thereby reducing or potentially eliminating exposure to allogeneic blood.

REFERENCES

1. Ponfick, E., 1875, Experimentelle Beitrage auf Lehre von der Transfusion. *Virchows Arch.* 62:273-335.
2. Rabiner, S. F., Helbert, J. R., Lopas, H., and Friedman, L. H., 1967, Evaluation of a stroma-free hemoglobin solution. *J. Exp. Med.* 126:1127-1142.

3. Mok, W., Chen, D., and Mazur, A., 1975, Cross-linked hemoglobin as potential plasma extenders. *Fed. Proc.* 34:1458-1460.
4. Savitsky, J. P., Doczi, J., Black, J., and Arnold, J. D., 1978, A clinical safety trial of stroma-free hemoglobin. *Clin. Pharmacol. Ther.* 23:73-80.
5. Djordjevich, L., and Miller, I. F., 1980, Synthetic erythrocytes from lipid encapsulated hemoglobin. *Exp. Hematol.* 8:584-592.
6. Hunt, C. A., Burnette, R. R., MacGregor, R. D., Strubbe, A. E., Lau, D. T., Taylor, N., and Kawada, H., 1985, Synthesis and evaluation of a prototypal artificial red cell. *Science* 230:1165-1168.
7. Tremper, K. K., 1985, Preface. In: *Perfluorochemical Oxygen Transport,* Tremper, K. K., (ed), Little, Brown, & Co., Boston, p. xv.
8. Clark, L. C., and Gollan, R., 1966, Survival of mammals breathing organic liquids equilibrated with oxygen at atmospheric pressure. *Science* 152:1755-1756.
9. Geyer, R. P., Monroe, R. G., and Taylor, K., 1968, Survival of rats totally perfused with a perfluorocarbon-detergent preparation. In: *Organ Perfusion and Preservation*, Norman, J. V., Folkman, J., Hardison, L. E., Ridolf, L. E., and Veith, F. J., (eds), Appleton-Century-Crofts, New York, pp. 85-95.
10. Riess, J. G., 1991, Fluorocarbon-based *in vivo* oxygen transport and delivery systems. *Vox Sang.* 61:225-239.
11. Naito, R., and Yokoyama, K., 1978, Perfluorochemical blood substitutes Fluosol-43, Fluosol-DA 20% and 35%. *Technical Information Bulletin Series* No. 5, The Green Cross Corporation, Osaka, Japan.
12. Roth, D. J., Keipert, P. E., Faithfull, N. S., Zuck, T. F., and Riess, J. G., 1994, Facilitating oxygen delivery in conjunction with hemodilution. *United States Patent 5,344,393.*
13. Faithfull, N. S., 1994, Mechanisms and efficacy of fluorochemical oxygen transport and delivery. *Art. Cells, Blood Substit., Immob. Biotech.* 22:181-197.
14. Snyder, J. Y., 1987, Assessment of systemic oxygen transport. In: *Oxygen Transport in the Critically Ill*, Snyder, J. V., Pinsky, M. R., (eds), Year Book Medical Publishers, Chicago, pp. 179-198.
15. Keipert, P. E., Bradley, J. D., Faithfull, N. S., Hazard, D. Y., Spooner, M., Mackley, K. M., Rusheen, P. D., Batra, S., and Flaim, S. F., 1994, Effects of *Oxygent*™ (perflubron emulsion) on O_2 transport in a dog model of hemodilution and surgical bleeding. *FASEB J.* 8:A901.
16. Cernaianu, A. C., Spence, R. K., Vasilidze, T. V., DelRossi, A. J., Carrig, T., White, P. F., Nathanson, M., Okonkwo, N., Wahr, J. A., Faithfull, N. S., Keipert, P. E., and Flaim, K. E., 1994, A safety study of a perfluorochemical emulsion, *Oxygent*™, in anesthetized surgical patients. *Anesthesiology* 81: A397.
17. Wahr, J. A., Trouwborst, A., Spence, R. K., Henney, C. P., Tremper, K. K., Cernaianu, A. C., Lau, W., Vasilidze, T. V., Faithfull, N. S., Flaim, K. E., and Keipert, P. E., 1994, A pilot study of the efficacy of an oxygen carrying emulsion, *Oxygent*™, in patients undergoing surgical blood loss. *Anesthesiology* 81: A313.
18. Newman, R. J., Podolsky, D., and Loeb, P., 1994, Bad Blood. *U.S. News*, June, pp. 68-78.
19. American College of Physicians Clinical Guideline, 1992, Practice strategies for elective red blood cell transfusion. *Annals Intern. Med.* 116:403-406.
20. Mercuriali, F., Inghilleri, G., Biffi, E., Colotti, M. T., and Vinci, A., 1991, *Autologous Blood. A safe alternative for surgical patients.* Istituto Ortopedico "Gaetano Pini", Immunology & Transfusion Service, TransMedica Europe Ltd., pp. 1-30.
21. Keipert, P. E., 1995, Use of *Oxygent*, a perfluorochemical-based oxygen carrier, as an alternative to intraoperative blood transfusion. *Artif. Cells, Blood Substit., Immob. Biotech.* 23: (in press).
22. Keipert, P. E., Faithfull, N. S., Bradley, J. D., Hazard, D. Y., Hogan, J., Levisetti, M. S., and Peters, R. M., 1994, Enhanced oxygen delivery by perflubron emulsion during acute hemodilution. *Artif. Cells, Blood Substit., Immob. Biotech.* 22: 1161-1167.

78

THE ROLE OF AMINO GROUPS IN HEMOGLOBIN OXYGEN BINDING

P. J. Anderson,[1] J. P. Vaccani,[1] and G. P. Biro[2]

[1] Departments of Biochemistry
[2] Departments of Physiology
Faculty of Medicine
University of Ottawa
Ottawa, Ontario Canada K1H 8M5

INTRODUCTION

The role of amino groups in hemoglobin function and the effects of chemical modification of these on the function of the protein have recently been reviewed (1). Amino groups are contributed by either the N-terminal residues of the globin chains or the lysine residues of the polypeptides. Depending on the reagents used and the conditions chosen, a variable degree of specificity towards amino groups can be achieved. Such studies can be of value in determining the role of individual functional groups in biological function and may be of use in the preparation of derivatives that have modified functional properties desirable in material to be used in a hemoglobin based oxygen carrier. We have recently developed a new method for hemoglobin amino group modification which has a much greater rate of reaction with α amino groups than with the ε amino groups of lysines at physiological pH (2). This method utilizes the activation of exogenous carboxyl containing compounds with water soluble carbodiimides and subsequent reaction with the amino groups of proteins such as hemoglobin. We report here the effects on oxygen binding observed as a function of the extent and conditions of hemoglobin modification with citrate used as the exogenous carboxyl containing compound.

METHODS

Hemoglobin prepared from outdated human blood provided by the Canadian Red Cross was used throughout these studies. It was purified by ultrafiltration and was stored frozen until use. Methemoglobin levels were monitored using a Co-oximeter and were less than 5% in all materials used. The water soluble carbodiimide, 1-ethyl-3 (3-dimethyl aminopropyl)-carbodiimide (EDC) was obtained from Pierce Chemical Company. It was used to activate the carboxyl groups of citric acid in 0.1M MOPS buffer, pH 7.2. The

Oxygen Transport to Tissue XVII, Edited by Ince et al.
Plenum Press, New York, 1996

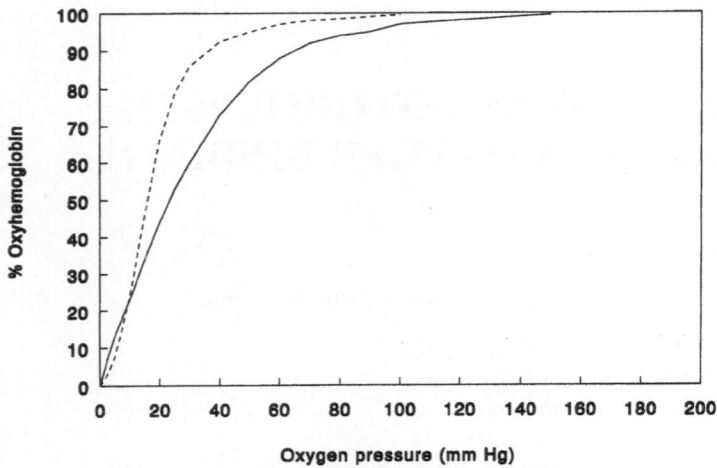

Figure 1. The effect of amino group modification restricted to the N-terminal residues of hemoglobin, on oxygen binding. –, untreated hemoglobin; ÄÄ, modification with EDC activated citrate for 10 minutes under conditions described in the text.

activated carboxyls were allowed to react at 0° with hemoglobin. Reactions were stopped by dilution and subsequent dialysis. Incorporation of citrate was monitored using [1,5-^{14}C] citric acid obtained from Amersham. Peptide maps of tryptic digests of radiolabelled material were prepared as previously described (2). Levels of incorporation of radioactivity were determined by scintillation counting. Oxygen binding properties were determined using a Hem-O-Scan dissociation curve analyzer (SLM Aminco) on solutions concentrated using Amicon Centricon Concentrators. All measurements were carried out in Krebs-Ringer bicarbonate and all gas mixtures contained 5.7% carbon dioxide.

RESULTS

At pH values near neutrality the α amino groups of hemoglobin react much more rapidly with carbodiimide activated citric acid than do the ε amino groups of lysine residues. Thus, conditions can be obtained in which modification is essentially restricted to N-terminal residues. At hemoglobin concentrations of 0.33 mM, in the first few minutes of reaction at 0° in the presence of 50 mM citrate and 25 mM EDC, it can be demonstrated by subsequent peptide mapping of tryptic digests that radioactive citrate (specific activity 100 μCi/mmol) is located to a significant extent only in two peptides which have previously been shown to correspond to the tryptic peptides derived from the N-terminal regions of α globin and β globin (2).

Figure 1 shows the oxygen binding curve obtained for hemoglobin in which amino group modification is restricted to the N-terminal residues. Under these conditions the incorporation of radioactivity when ^{14}C-citrate is used indicates that very close to one mole of citrate per globin chain is incorporated and that this is largely restricted to the α amino groups of the N-terminal residues of α and β globin. It can be seen that significant alterations of oxygen binding properties accompanying the modification.

Figure 2 shows that oxygen binding properties depend on the extent of modification. Oxygen affinity increases as additional amino groups of lysine residues are derivatized.

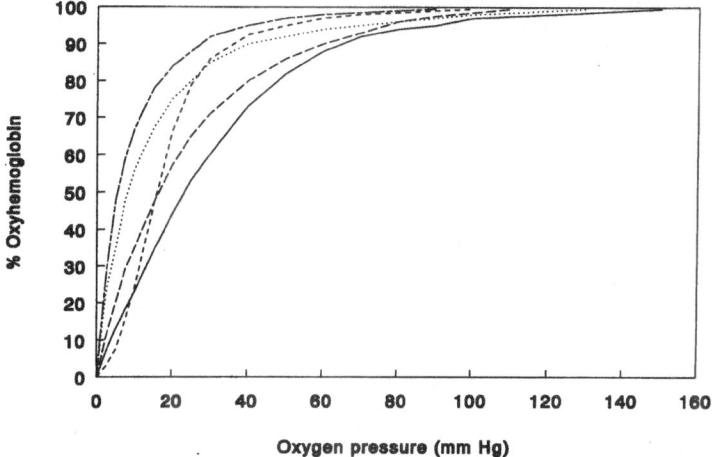

Figure 2. The effect of increasing the extent of amino group modification on oxygen binding by hemoglobin. —, untreated hemoglobin; – – – reaction for 10 minutes; — —, 20 minutes; ...., 60 minutes; — —, 16 hours under conditions described in the text.

From the Hill plots of binding data shown in Figure 3, it can be seen that a series of parallel lines can be generated from oxygen measurements of oxygen binding as a function of extent of amino group modification.

Table 1 summarizes oxygen binding parameters (p50 and Hill coefficients) calculated from the data shown in Figures 1 to 3. N-terminal amino group modification by citrate is accompanied by a decrease in oxygen affinity as indicated by an increased p50 compared to untreated hemoglobin. As more amino groups are modified the oxygen affinity increases

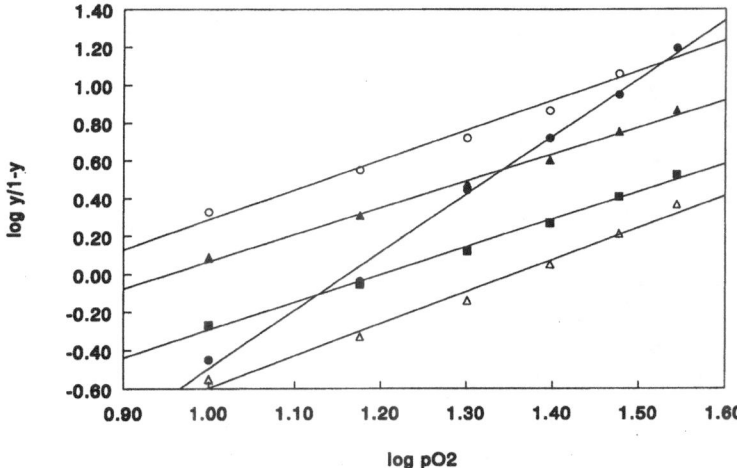

Figure 3. Hill plots of the oxygen binding data shown in Figure 2. y represents the fractional saturation with oxygen as a function of oxygen tension. ●, untreated hemoglobin. Hemoglobin reacted with citrate for △, 10 minutes; ■, twenty minutes; ▲, 1 hour and ○, 16 hours.

Table 1. Oxygen binding parameters for citrate modified hemoglobin

Duration of reaction	p50 (mm Hg)	Hill coefficient (n)
0	16	2.14
10 minutes	24	1.58
20 minutes	16	1.37
1 hour	8	1.40
16 hours	5.5	1.54

Oxygen binding parameters for citrate modified hemoglobin. Values were obtained from the data shown in Figures 1, 2 and 3. The p50 is the oxygen pressure at which 50% of the hemoglobin is in the oxyhemoglobin form. The Hill coefficient, the slope of the lines in Figure 3, indicates the degree of co-operativity of binding.

significantly as seen by the substantial decreases in p50 with increasing extent of modification. The Hill coefficients indicate that the co-operative nature of oxygen binding is quickly lost and that increasing the extent of modification has no further effect on this.

The reaction of hemoglobin amino groups with reagents containing activated carboxyls such as that described for citrate in the presence of EDC can be used to examine effects of physiological modifiers of hemoglobin oxygen binding properties.

Figure 4 illustrates kinetic data for citrate incorporation in the presence and absence of sodium chloride (0.2M) obtained by measuring the incorporation of radioactivity as a function of time of reaction. The overall first order rate constant in the absence of sodium chloride is .032 min^{-1}. The presence of the salt protects amino groups decreasing the overall rate constant to .011 min^{-1}. Preliminary experiments using peptide mapping of tryptic digests of radiolabelled indicate a degree of specificity associated with this protection. It would be

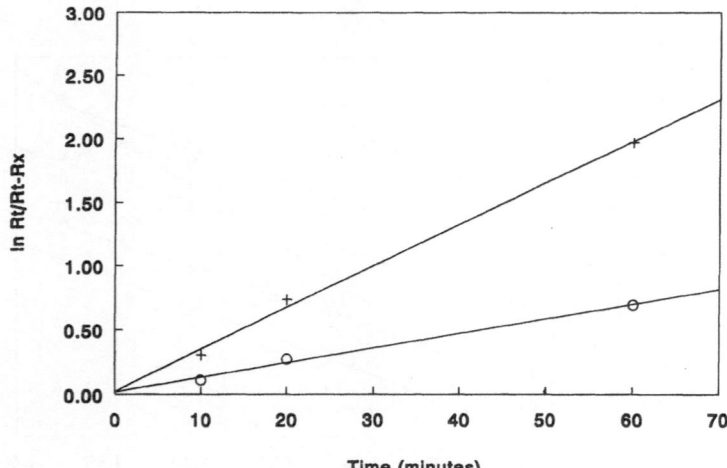

Figure 4. Kinetic data for citrate incorporation in the presence (+) and absence (o) of 0.2M sodium chloride. The incorporation of radioactivity was assumed to be complete after 16 hours. The natural log of the ratio of radioactivity incorporated after 16 hours (R_T) to the radioactivity incorporated after 16 hours minus the radioactivity at specific time intervals ($R_T - R_x$) is plotted as a function of the time intervals at which incorporation of radioactivity was measured. The slope of the line is the first order rate constant.

of value to determine if effects on oxygen binding properties accompanying amino group protection by chloride.

DISCUSSION

The effects on oxygen binding accompanying hemoglobin amino group modification by citrate in the presence of the carboxyl activating carbodiimide EDC are different than previously reported effects of amino group modification. Carboxymethylation of N-terminal amino groups of hemoglobin using the aldehyde glyoxalate and sodium cyanoborohydrate to reduce the Schiff base has been reported (3). As observed for citrate modification of N-terminal residues in the present study, this results in a decrease in oxygen affinity as indicated by an increase of p50 from 13 mm Hg to 30 mm Hg in 5% carbon dioxide. This compares with an increase of p50 from 16 mm to 24 mm when citrate modification is restricted to the N-terminal residues. Unlike N-carboxymethylation (3), further modification of amino groups by citrate increases the oxygen affinity. In addition, there is a large difference between the two derivatives with respect to cooperativity of binding as indicated by the Hill coefficient. Hemoglobin carboxymethylated on the N-terminal residues (3) has an identical Hill coefficient as unmodified hemoglobin (n=2.3). In the present study, the Hill coefficient decreased from 2.14 to 1.58 on N-terminal amino group modification and subsequent further lysine amino group modification had no additional effect. This indicates that, in addition to the loss of the modified functional group, the nature of the derivative formed effects the properties.

There are potential uses for the activation of carboxyl groups in reagents to be subsequently used for hemoglobin modification. The method could provide the basis for studying the effects of physiological regulation of hemoglobin by modifying agents which act in a non covalent way such as chloride. The present study shows that the reactivity of amino groups with activated citrate is greatly reduced in the presence of chloride. It could be of interest to determine whether specific amino groups are protected differentially. Carbodiimides have been used to modify hemoglobin by activating of the carboxyls of the hemoglobin molecule and subsequent reaction with an exogenous nucleophile (4). The new method we describe provides an alternate means of modification in which the nucleophilic amino groups of hemoglobin react with exogenous activated carboxyls. In the experiments described, the compound used to provide activated carboxyls was citric acid. However, a variety of compounds containing one or more carboxyls could be used to modify hemoglobin or other proteins. These could be of value in inter and intra molecular crosslinking. A range of sizes of crosslinkers can be examined using the methodology described. These could be of value in understanding the structural basis of hemoglobin function as well as in developing ways to make hemoglobin based oxygen delivering compounds of possible clinical value.

ACKNOWLEDGMENTS

This work was supported by a grant from Miles Canada / CRCS.

REFERENCES

1. Manning, J.M., 1994, Methods in Enzymology, 231:225-246.
2. Anderson, P.J., 1993, J. Biol. Chem. 268:15504-15509.
3. Manning, L.R. and Manning, J.M., 1988, Biochemistry 27:6640-6644.
4. Rao, M.J. and Acharya, S.S., 1992, Biochemistry 31:7231-7236.

79

EFFECTS OF OXYGENATOR AND PUMPING DEVICES ON BLOOD PARAMETERS IN OPEN HEART SURGERY

A Clinical Study

Semih Barlas, Emin Tireli, Haldun Tekinalp, Enver Dayioğlu, Leyla Sevgenay, and Cemil Barlas

Department of Cardiovascular Surgery
Medical Faculty, University of Istanbul
Turkey

INTRODUCTION

Despite the recent technological advances in open heart surgery, an oxygenator being absolutely harmless to blood cells and providing excellent perfusion to organ tissues, especially the brain, via an extra corporeal pump is yet to be developed.

Today three different types of cardiopulmonary bypass (CPB) pumps, roller, centrifugal and pulsatile are in worldwide use with bubble and membrane oxygenators. It's now common knowledge that the roller pump leads to severe hemolysis and postoperative bleeding, especially following long-standing extra corporeal bleeding. The only reason for its present use is economical, its price being 50% of the others.

In this prospective study, centrifugal and pulsatile pumps have been used in our department in various combinations with bubble and membrane oxygenators and the combinations have been compared with respect to their physiological consequences.

MATERIAL AND METHODS

This study was done during 1992 - 1993 in Istanbul University, Istanbul Medical Faculty, Thoracic and Cardiovascular Department. 80 patients undergoing open heart surgery, 50 due to a congenital heart disease and 30 due to either valvular or ischemic heart disease were included in the study. The patients were randomly allocated to four equal groups. In Group 1 pulsatile pump (Sarns Ltd., USA) + membrane oxygenator (HF- 5700 Bard, USA); in Group 2 centrifugal pump (Biomedicus, USA) + membrane oxygenator; in Group 3 pulsatile pump + bubble oxygenator (HF - 1300 Bard, USA); and in Group 4 centrifugal pump + bubble oxygenator was used.

Oxygen Transport to Tissue XVII, Edited by Ince et al.
Plenum Press, New York, 1996

Patients with preoperative hematological disorders (severe anemia, thrombocy-topenia, severe cyanotic lesions) and hepatic dysfunction, as well as those undergoing a redo open heart procedure and those in whom Aprotinine was administered peroperatively were not included in the study.

White blood cell (WBC), erythrocyte (RBC), thrombocyte (PLT) and hemoglobin were obtained from Medonic CA 470, whereas Mallincrodt Gem 6 Plus blood gas analyzer was used for pH, PO2 and K^+ values. Systemic vascular resistance (SVR) and cardiac output (CO) was measured with Spectramed Oxymetrix cardiac output device. In order to reduce the risk of hemolysis caused by bank blood, peroperative auto transfusion with Elec-tromedics AT 750 cell saver machine was performed and administration of conserved blood was avoided if possible.

All patients were monitored with the Cerebrotrac (SRD Medical Ltd., Shorashim, Israel), a computerized electroencephalogram, for continuous evaluation of brain function. This device works with 5 electrodes located on the patient's scalp and generates brain electrical activity as the spontaneous EEG trace along with isochrone wave form reflections. The mean wave width/frequency ('Spectral edge frequency - SEF') and height ('Amplitude - AMP') describes the patient status as "Sleepy" / "Cerebral hypothermia" at SEF 0 - 4 Hz and AMP 5μv; "Stupor" / "General anesthesia" at SEF 5 - 7 Hz and AMP 20 - 65 μv; "Relaxed state" at SEF 8 - 12 Hz; "Awake state" at SEF >13 Hz and AMP 50 μv; "Cerebral ischemia" / "Hypoxia" at SEF 0 - 4 Hz and AMP >65 μv; and "CPB" at SEF 7 Hz and AMP 10 - 15 μv.

The first hematological, blood gas and cardiac output data were obtained when the patients were taken to theater. SEF and AMP values obtained with the Cerebrotrac on arrival to theater and at the induction of general anesthesia were recorded.

For all patients, the priming solution for the oxygenator and the extracorporeal pump circuits consisted of crystalloid Ringer's solution and fresh donor blood with a resulting HCT of 25% according to the formula:

$$HCTpm = \frac{Body\ weight\ (kg) \times f \times 1000 \times HCTpt}{Body\ weight\ (kg) \times f \times 1000 \times Pump\ blood\ volume}$$

HCTpm: Hematocrit of the priming solution; HCTpt: Hematocrit of the patient prior to CPB; f: 0.08 (constant for infant - 12 years-of-age), 0.065 (constant for >12 years-of-age). Heparin 3 mg/kg were given to all patients. Anterograde aortic root cardioplegia was done with the St. Thomas solution containing 16 mmol/l potassium chloride, 16 mmol/l magnesium sulphate, 2 mmol/l sodium bicarbonate and 55 mmol/l glucose at a temperature of 4°C and a pH of 7.45. Initially 2 ml/kg of the cardioplegic solution was given and 1 ml/kg boluses were repeated every 30 minutes. The heart was externally cooled peroperatively by "ice slush" and "Shumway" techniques using cold Isolyte solution. Systemic hypothermia was provided with a hypothermia device, reducing the temperature down to 25 - 32°C depending on the type and length of the operation and the age of the patient. Any arterial filter known to lead to hemolysis and platelet destruction was avoided during the CPB.

Cerebrotrac, hematological, cardiac output and blood gas analysis data were obtained at 30 min and at the end of CPB. Amount of bleeding during the patient's postoperative stay in the Intensive Care Unit was recorded.

The groups were first analysed with the paired Student's t-test with respect to the significance of the data obtained before, at 30 min and after CPB. Comparison between groups were done with the unpaired Student's t-test. $p < 0.05$ was considered statistically significant.

Table 1. Mean Values of Patient Data (G1-4. Groups 1-4; Pre. PreCPB; 30dk. 30.nth Minute of CPB; Post. at the End of CPB. RBC x 106, PLT x 102, HB. % and Drenage in ml)

	G 1			G 2			G 3			G 4		
	Pre	30dk	Post	Pre	30dk	Post	Pre	30dk	Post	Pre	30dk	Post
WBC	5700	5200	4976	5814	5625	5270	5978	5200	4690	5216	5150	4605
RBC	4.15	3.20	3.06	3.50	3.13	2.96	4.07	2.80	2.50	3.64	3.08	3.03
PLT	1423	1015	856	1278	1062	909	1396	865	574	1380	1026	777
HCT	43	28	25	37	27	26	39	27	25	43	29	27
K$^+$	4.1	4.3	4.5	4.6	4.8	4.7	3.6	4.2	4.4	3.9	4.1	3.9
HB	13.8	6.9	14.6	14.8	7.3	16.9	15.6	9.5	13.6	16.6	6.7	23.4
pH	7.46	7.43	7.46	7.42	7.38	7.38	7.39	7.35	7.37	7.44	7.39	7.38
PO$_2$	205	280	390	250	311	340	198	268	304	290	350	380
SVR	1650	1730	1798	1745	1854	2394	1940	2237	2446	1860	1976	2545
CO	4.10	4.80	5.10	3.65	3.87	3.98	3.10	3.15	3.17	3.60	3.75	3.83
Dren			350			320			710			630

Table 2. Percent changes at 30th minute and at the end of CPB against preCPB values

	PREOP - 30th Minute of CPB				PREOP - End of CPB			
	G 1 %	G 2 %	G 3 %	G 4 %	G 1 %	G 2 %	G 3 %	G 4 %
WBC	8.77	3.25	13.01	1.26	12.7	9.35	21.54	11.71
RBC	22.89	10.57	31.20	15.38	26.26	15.42	38.57	16.75
PLT	28.67	16.90	38.03	25.65	39.84	28.87	58.88	43.69
HCT	34.88	27.02	30.76	32.55	41.86	29.72	35.89	37.20
K^+	4.87	4.34	16.66	5.12	9.75	2.17	22.22	0
HB	50	50.67	39.10	59.63	5.79	14.18	12.82	40.96
PH	0.4	0.5	0.5	0.6	0	0.5	0.2	0.8
PO_2	36.58	24.4	35.35	20.68	90.24	36	53.53	31.03
SVR	4.84	6.24	15.30	6.23	8.96	37.19	26.08	36.82
CO	17.07	6.02	1.61	4.16	24.39	9.04	2.25	6.38

RESULTS

The mean values of the four groups obtained at pre, at 30 min, and post CPB are presented in Table 1.

WBC, RBC, PLT and HCT counts were found to decrease in all groups after going onto CPB; the decrease in Groups 1, 2, and 3 were statistically significant ($p<0.05$) whereas it was found to be insignificant in Group 4. The decrease in RBC count in Groups 1, 2, and 4 was statistically significant ($p<0.05$), the significance was greater in Group 3 ($p<0.01$). The decrease in PLT and HCT counts were all statistically significant ($p<0.05$).

Blood K^+ levels were found to increase in all groups after going onto CPB; the increase was statistically significant in Groups 1 and 4 ($p < 0.05$), highly significant in Group 3 ($p<0.01$) and insignificant in Group 2.

A statistically significant decrease ($p<0.05$) was observed in blood free hemoglobin levels in all groups. There was no significant change in blood pH values in any of the groups. The increase in blood PO2 values in all groups were statistically significant ($p<0.05$). Systemic vascular resistance was increased in all groups, this increase being statistically significant ($p<0.05$) in Groups 2 and 4, highly significant in Group 3 ($p<0.01$), and

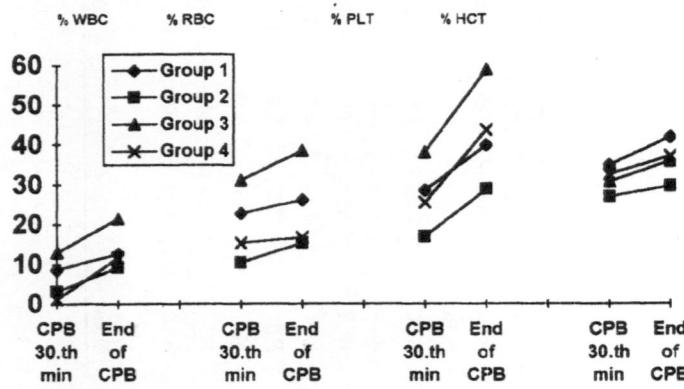

Figure 1. Percent changes in WBC, RBC, PLT and HCT on CPB at 30' and at the end of CPB compared to preCPB values.

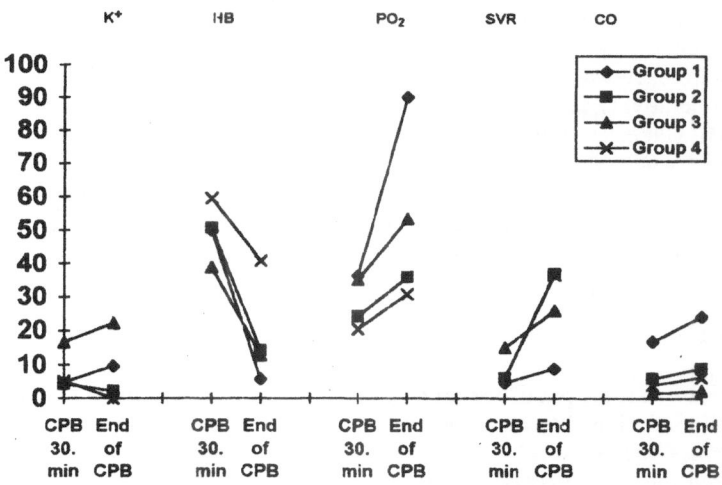

Figure 2. Percent changes of K+, HB, PO2, SVR and CO on CPB 30' and at the end of CPB compared to preCPB Values.

insignificant in Group 1. CO increases were insignificant in Groups 2, 3, and 4, whereas it was found to be statistically significant ($p < 0.05$) in Group 1.

Percent changes of the variables at 30 min of CPB when compared to preCPB values are shown in Table 2 and Figures 1 and 2.

Percent changes of all variables other than pH were found to be statistically significant ($p < 0.05$) when Group 1 was compared with Groups 2 and 3. % changes of all variables other than HB and pH were statistically significant ($p < 0.05$) when Group 2 and 3 were compared. % changes in WBC, RBC, PLT, HCT and HB values were found to be statistically significant ($p < 0.05$) between Groups 2 and 4. % changes of all variables except pH were found to be statistically significant ($p < 0.05$) when Groups 3 and 4 were compared.

When % changes of the variables at the end of CPB were compared to preoperative values, statistical significance ($p < 0.05$) was found between Groups 1 and 2 for all variables other than K+, HB, and pH; between Groups 1 and 3 for all variables other than pH and PO_2; between Groups 1 and 4 for RBC, HB, PO2, SVR and CO; between Groups 2 and 3 for all variables other than HCT and pH; between Groups 2 and 4 for WBC, RBC, PLT, HCT and HB; and between Groups 3 and 4 for all variables other than HCT, pH and CO.

Table 3. Cerebrotrac values

	G 1		G 2		G 3		G 4	
	SEF	AMP	SEF	AMP	SEF	AMP	SEF	AMP
PREOP	14	52	13	50	14	66	15	58
GA	5	55	7	60	5	65	6	57
30.DK	0	5	7	11	3	9	8	13
POST	5	30	5	35	7	40	6	20

SEF. Spectral edge frequency, AMP. Amplitude Preop. Preoperative, GA. General Anaesthesia Induction, 30.DK. CPB 30'., POST. at the end of CPB , G1. Group 1, G2. Group 2, G3. Group 3,G4. Group 4.

Mean SEF and AMP values obtained from the Cerebrotrac preoperatively, at the induction of general anesthesia, at 30 min and at the end of CPB are shown in Table 3.

No difference was observed between groups concerning the SEF and AMP values obtained preoperatively, at the induction of general anesthesia or at the end of CPB; however, SEF and AMP values at 30 min of CPB were found to be significantly higher (p<0.05) in Groups 2 and 4 when compared to Groups 1 and 3.

When postoperative bleeding was observed, there was no difference between Groups 1 and 2, yet the amount of bleeding was found to be significantly less (p<0.05) in these groups when compared to Groups 3 and 4.

DISCUSSION

There are various published articles documenting the beneficial effects of pulsatile pumping devices during CPB on myocardium, brain and kidneys. However, the majority of these studies has been done as a comparison of roller vs. pulsatile pumps, and no one has ever compared the pulsatile and centrifugal pumps[1-4]. On the other hand, no significant explanation has been given about the hemolysis after long open heart procedures demonstrating whether the problem is related with the oxygenators or the pumping devices.

Our results have shown us that both devices cause hemolysis and thus lower the WBC, RBC, PLT, and HCT levels. This fact correlates with the literature findings[1-5]. Parallel with time WBC, RBC and PLT were found to decrease mostly in Group 3, whereas HCT decreased in Group 1. Free K+ after hemolysis was noted to be highest in Group 3. Free HB level was the highest in Group 4, but this level decreased parallel with time.

It was noted that the pulsatile pumping + bubble oxygenator was the most deleterious combination to blood cells. When the pumping device was kept constant, bubble oxygenator caused most hemolysis; when the oxygenator type was kept constant, pulsatile pump caused most hemolysis. Postoperative blood loss was the highest in bubble oxygenator groups without any significance on pump types.

No system caused a notable difference in pH values. Correlated with time, PO2 and CO increased in Group 1. Increase in the SVR was the lowest in Group 1 and highest in Groups 2 and 4. This fact leads to the conclusion that pulsatile pumping preserves myocardial functions and tissue oxygenation, whereas centrifugal pumping increases the SVR as the duration of CPB increases.

The studies of Habbal[6] and Maddoux[7] demonstrate the increase in left ventricular ejection fraction with the pulsatile perfusion in the early postoperative period after CPB. Some authors have also speculated that this technique reduces the incidence of perioperative AMI rates[8,9].

EEG monitoring has shown that during CPB, pulsatile pumping causes better cerebral cooling. However, normal cerebral activity was noted after CPB in all groups. The literature findings[10] and our data have clearly demonstrated that to keep the cerebral injury at minimum, surgeons and perfusionists have to keep the CPB time as short as possible and use membrane oxygenators with pulsatile pumps.

The cost ineffectiveness of membrane oxygenators is the main reason for the limited usage of these devices. Classical extracorporeal circuit tubings can be used with pulsatile pumps whereas centrifugal pumps necessitate a very expensive pump head.

Another feature of membrane oxygenators is that the durability is 72 hours whereas it is only 6 hours in bubble oxygenators.

CONCLUSION

1. Pulsatile pumps + bubble oxygenators cause greater hemolysis.
2. Pulsatile pumps cause optimal PO2 levels.
3. Pulsatile pumps increase SVR minimally.
4. Pulsatile pumps cause CO increase.
5. Pulsatile pumps cause optimal cerebral hypothermia.
6. Bubble oxygenators cause greater postoperative bleeding.
7. The durability of membrane oxygenators is higher.
8. Centrifugal pumps and membrane oxygenators are more expensive.

Pulsatile and centrifugal pumps, as well as membrane oxygenators have beneficial and deleterious effects in open heart procedures. Bubble oxygenators have no favorable effect other than the cost.

REFERENCES

1. Mori F, Ivey TD, Itoh T, Thomas R, Breazeale DG, Misbach G: Effects of pulsatile perfusion on postischemic recovery of myocardial function after global hypothermic cardiac arrest. J Thorac Cardiovasc Surg 1987; 93:719.
2. Williams GD, Seifen AB, Lawson NW, Norton JB, Readinger RI, Dungan TW, Callaway JK: Pulsatile perfusion versus conventional high-flow nonpulsatile perfusion for rapid core cooling and rewarming of infants for circulatory arrest in cardiac operation. J Thorac Cardiovasc Surg 1979; 78:667.
3. Singh RKK, Barratt–Boyes BG, Harris EA: Does pulsatile flow improve perfusion during hypothermic cardiopulmonary bypass. J Thorac Cardiovasc Surg 1980; 79:827.
4. Trinkle JK, Helton NE, Wood RE, Bryant LR: Metabolic comparison of a new pulsatile pump and a roller pump for cardiopulmonary bypass. J Thorac Cardiovasc Surg 1969; 58:562.
5. Vorhees ME, Elgas R: Membrane and bubble oxygenators in Techniques in extracorporeal circulation. 3rd ed. Ed. Kay P. Butterworth–Heinemann Ltd. 1992; 42.
6. Habal SM, Weiss MB, Spontiz HM: Effects of pulsatile and nonpulsatile coronary perfusion on performance of the canine left ventricle. J Thorac Cardiovasc Surg 1976; 72: 742.
7. Maddoux G, Pappas G, Jenkins M: Effect of pulsatile and non pulsatile flow during cardiopulmonary bypass on left ventricular ejection fraction early after aortocoronary bypass surgery. Am J Cardiol 1976; 37:1000.
8. Pappas G, Winter SD, Kopriva CJ, Steel PP: Improvement of myocardial and other vital organ functions and metabolism with a simple method of pulsatile flow (IABP) during clinical cardiopulmonary bypass. Surgery 1975; 34.
9. Laschinger JC, Cunningham JN, Catinella FP: Pulsatile left atrial-femoral artery bypass. a new method of preventing extension of myocardial infaction. Arc Surg 1983; 118:965.
10. Smith P: The neurological sequelae of cardiopulmonary bypass in Techniques in extracorporeal circulation. 3rd ed. Ed. Kay P. Butterworth–Heinemann Ltd. 1992; 183.

80

CBV DURING EXCHANGE TRANSFUSION IN INFANTS WITH ERYTHROBLASTOSIS

Hubert Fahnenstich, Martin Günther, and Nikolaus Kau

Department of Neonatology
University Childrens Hospital
Adenauerallee 119
D-53113 Bonn, Germany

Since the introduction of postpartal and prepartal Anti-D-prophylaxis (Bowman et al., 1978; Thornton et al., 1989), the rate of sensibilisation to the rhesus system has decreased in Germany from 1:20.000 before prophylaxis to 1:1.000 today (Maas et al., 1991). There is, however, little chance that the future could hold an even lower incidence of infants with isoimmunisation caused by Rhesus incompatibility, since prophylaxis has a constant failure rate. In addition, there are other types of isoimmunisations caused by other blood groups, especially anti-c, Kell and Duffy, where no prophylaxis is yet possible. Therefore exchange transfusion (ET) will continue to be a part of the postnatal management in neonates suffering from Morbus haemolyticus.

Although ET has a relatively low complication rate of about 1-4% in larger studies (Panagopoulus et al., 1969; Kitchen, 1970), there are some severe, known cardio-vascular complications that arise when this procedure is used. The morbidity in newborns of a gestational age below 30 weeks rose to 4% (Keenan et al., 1985). No information is available in the literature on how cerebral hemodynamics is influenced during ET. Therefore we studied variations in cerebral blood volume (CBV) in ET by means of near-infrared spectroscopy (NIRS) in order to obtain initial information.

PATIENTS AND METHODS

We studied 6 patients with a mean birth weight of 2.847 ± 222 g and a mean gestational age of 36.5 ± 1.1 weeks. Patients were not sedated and this fact caused a high rate of artefacts in 4 patients, whose data have been excluded from the evaluation. During ET we were able to evaluate changes in CBV 50 times while withdrawing blood from the patients and 51 times while transfusing blood using a syringe. The exchange volume was 20 ml. One single withdrawal and one transfusion each took about 90 to 180 seconds. ET was done over an umbilical vessel catheter. The tip was placed in the right atrium or in the upper part of the Vena cava inferior.

Oxygen Transport to Tissue XVII, Edited by Ince et al.
Plenum Press, New York, 1996

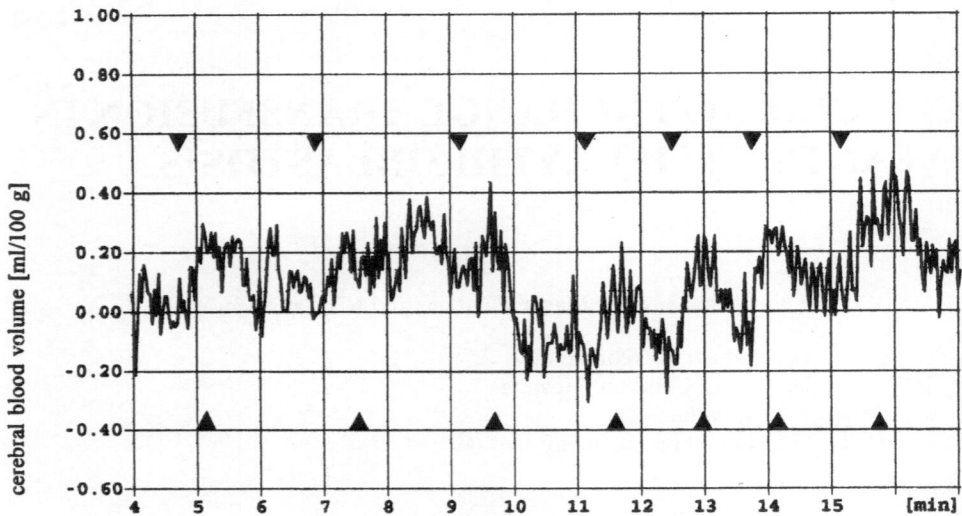

Figure 1. Short-time changes in cerebral blood volume during exchange transfusion.

To measure changes in CBV we used the near-infrared spectroscopy supplied by Radiometer®, Copenhagen, Denmark. The method, which allows non-invasive bed-sited monitoring, is described elsewhere (Reynolds et al., 1988). Changes in CBV were calculated according to changes in total hemoglobin (Wyatt et al., 1990). Heart frequency, respiration rate and blood pressure were also monitored.

RESULTS

CBV during ET differed between "short-time changes," which took place immediatly after withdrawing and transfusing the blood, and "long-time changes", which took place over the whole period of ET.

During the short-time changes, the withdrawal and transfusion stages could be easily differentiated: CBV decreased and increased 5 to 20 sec after onset of the procedure. Mean blood volume decreased during blood withdrawal by 0.39 ± 0.2 ml/100 g brain weight (range 0.1-1.0 ml/100 g). The increase during blood transfusion was somewhat higher with 0.46 ± 0.2 ml/100 g (range 0.1-1.2 ml/100 g). Figure 1 demonstrates the cyclic variations in blood volume according to the kind of practical management during ET.

In spite of equivalent volumes during ET we also found an additional absolute change in CBV in 4 out of 6 patients. The maximum increase in CBV during ET reached a value of 1.4 ml/100 g at the end of near-infrared monitoring, lasting up to one hour (Fig. 2). During ET all the patients were in a very stable condition; heart frequency and blood pressure did not show any changes that were relevant for changes in CBV.

DISCUSSION AND CONCLUSIONS

Our findings allow the conclusion that ET has an effect on cerebral haemodynamics: the variations in CBV are induced by blood volume changes while withdrawing and

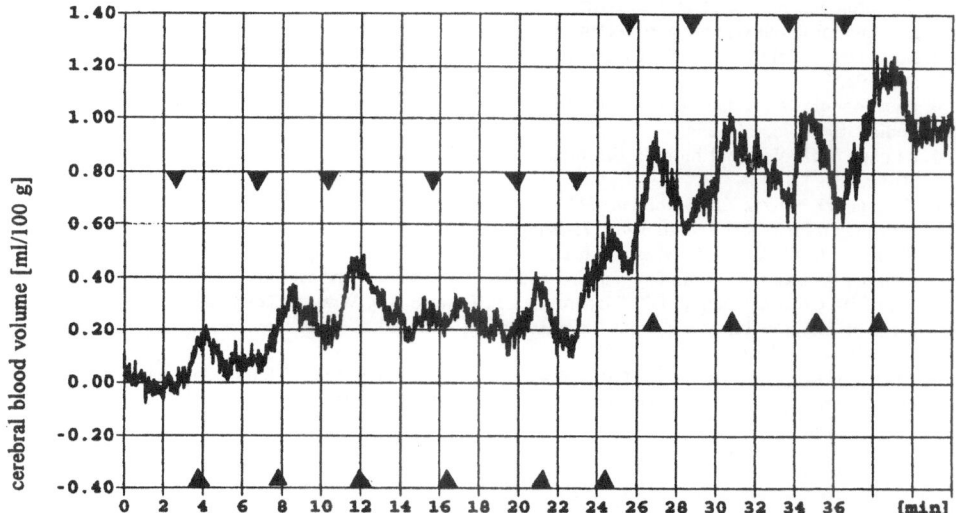

Figure 2. Increase in cerebral blood volume (long-time changes) during exchange transfusion.

transfusing blood through the umbilical catheter and these are directly transmitted to cerebral vessels. The short-time changes of CBV are probably associated with changes in the cerebral perfusion pressure. The group under study with a relatively high gestational age were able to tolerate these changes in cerebral volume and pressure: no side effects of any clinical relevance were found in these newborns.

Two hypotheses are possible to interpret the long-time changes in CBV during ET: the repeated strain on the cerebral vessels by the exchange of blood volume leads to changes in the vessels. At the venous site the vessels dilate and the increase in CBV is then caused by a venous pooling of the blood. That peripheral venous vessels can react to the increase in volume is a known factor. At the arterial site of the cerebral perfusion the repeated strain can lead to a loss of vascular resistance; the vessels are consequently able to store more blood.

Viewed from a methodological standpoint, an alternative explanation could be that the cerebral hematocrit is altered by the transfused blood. In the newborns under study we found a hematocrit of between 32% and 45%. The transfused blood was adjusted to a hematocrit of 45% to 50%. Changes in the blood hematocrit are important for the absorbtion of light and for the calculation of CBV by means of NIRS (Cope et al., 1988). In consequence the long-time changes registered could be based on an error in the total measured hemoglobin and the calculated CBV.

In general, changes in CBV might not be clinically relevant for newborns with an intact cerebral autoregulation, and this fact is supported by the overall low complication rate in ET. It remains open why it is that prematures with a lower gestational age and/or newborns with a delayed cerebral autoregulation are sometimes not able to cope with the variations during ET.

REFERENCES

1. Bowman JM, Chown B, Lewis M, Pollock JM. Rhesus isoimmunization during pregnancy: Antenatal prophylaxis. Can Med Assoc J 1978; 118: 623-627.

2. Thornton JG, Page C, Foote G, Arthur GR, Tovey LAD, Scott JS. Efficacy and long-term effects of antenatal prophylaxis with anti-D immunoglobulin. Brit Med J 1989; 298: 1671- 1673.

3. Maas DHA, Ünlü C, Schneider J. Bedeutung der antenatalen Rhesusprophylaxe. Pädiat Prax 1991; 42: 593-598.

4. Panagopoulos G, Valaes T, Doxiades SA. Morbidity and morality related to exchange transfusions. J Pediatr 1969; 75: 247-254.

5. Kitchen WH. Neonatal mortality in infants receiving an exchange transfusion. Aus Paediatr 1970; 6: 30-40.

6. Keenan WJ, Nowak KK, Sutherland JM, Bryla DA, Fetterly KL. Morbidity and mortality associated with exchange transfusion. Pediatr 1985; 75 (suppl.): 417-421.

7. Reynolds EOR, Wyatt JS, Azzopardi D, Delpy DT, Cady EB, Cope M, Wray S. New non- invasive methods for assessing brain oxygenation and haemodynamics. Brit Med Bull 1988; 44: 1052-1075.

8. Wyatt JS, Cope M, Delphy DT, Richardson CE, Edward AD, Wray S, Reynolds EOR. Quantitation of cerebral blood volume in human infants by near-infrared spectroscopy. J Appl Physiol 1990; 68: 1086-1091.

9. Cope M, Delpy DT, Reynolds EOR, Wray S, Wyatt JS, van der Zee P. Methods of quantitating cerebral near infrared spectroscopy data. Adv Exp Med Biol 1988; 222: 183- 189.

81

EFFECTS OF VOLUME CONTROLLED VENTILATION WITH PEEP, PRESSURE REGULATED VOLUME CONTROLLED VENTILATION AND LOW FREQUENCY POSITIVE PRESSURE VENTILATION WITH EXTRACORPOREAL CARBON DIOXIDE REMOVAL ON TOTAL STATIC LUNG COMPLIANCE AND OXYGENATION IN PIGS WITH ARDS

J. Kesecioğlu,[1] L. Telci,[2] A. S. Tütüncü,[2] F. Esen,[2] T. Denkel,[2]
W. Erdmann,[3] K. Akpir,[2] and B. Lachmann[3]

[1] Department of Intensive Care
   Academic Medical Center, Amsterdam, The Netherlands
[2] Department of Anesthesiology and Intensive Care
   University of Istanbul, Faculty of Medicine, Istanbul, Turkey
[3] Department of Anesthesiology
   Erasmus University Hospital Dijkzigt
   Rotterdam, The Netherlands

INTRODUCTION

Adult respiratory distress syndrome (ARDS) is characterised by decreased lung compliance and functional residual capacity (FRC) and increased intrapulmonary shunting resulting in hypoxemia. The immediate treatment of this critical situation is respiratory therapy of one form or the other and various modes have been recommended since the description of the disease. Positive end expiratory pressure (PEEP) with large tidal volume (V_T), which recruits atelectic areas and increases FRC, was once suggested as the treatment of ARDS.[1-6] However, this mode of ventilation may cause barotrauma and/or morphological changes due to high peak inspiratory pressures (PIP).

Alveolar recruitment, being the aim during the initial phase of the disease, is shown by Lachmann[7] to be achieved successfully by pressure controlled inverse ratio ventilation

Oxygen Transport to Tissue XVII, Edited by Ince et al.
Plenum Press, New York, 1996

(PC-IRV) with inspiration expiration (I/E) ratio of 4:1. This mode of ventilation is reported to facilitate the opening of the lung units and to prevent their collapse due to short expiration time. Furthermore, high PEEP and pressure related complications are avoided by the application of this mode.

Low frequency positive pressure ventilation with extracorporeal carbon dioxide removal (LFPPV-ECCO$_2$R) is suggested by Gattinoni and co-workers[8] as an alternative treatment of severe acute respiratory failure, unresponsive to other forms of mechanical ventilation. In this method, while oxygen is delivered into the lungs by apneic diffusion, CO$_2$ is removed extracorporeally through the artificial lung. This form of treatment, beside providing lung rest, also avoids high PIP and overdistension of the ventilated alveoli.

Total static lung compliance (TSLC) is an important parameter of the pressure volume relationship of the total respiratory system and is useful in monitoring of the status and progress of mechanically ventilated patients.[9-12] Furthermore, it has been used in guiding the respiratory therapy in ARDS patients.[13] Suter and colleagues[14] suggested the use of compliance as an indicator of the level of PEEP to be applied to produce optimum oxygen delivery. They further related the increase seen with TSLC, during mechanical ventilation with PEEP, to the re-expansion of the atelectatic lung areas. They have also shown reduction of lung compliance with inappropriately higher V$_T$ and PEEP levels during VCV.[11]

Although TSLC is reported to give no information of poorly or non-areated tissues,[15] it might be important to observe its trend during different modes of ventilation. Being a parameter reflecting the amount of normally areated tissues,[15] its fast increase within few minutes could be used as an indicator of recruitment and its fast decrease as a loss of ventilated lung units, under stable conditions in experimental ARDS. However, it is difficult to differentiate between the improvement of lung distensibility and recruitment over longer periods. This differentiation is only possible with measurements of FRC changes in the lung.

In this study VCV with PEEP, pressure regulated volume controlled ventilation (PRVCV) with I/E of 4:1 and LFPPV-ECCO$_2$R were compared in pigs with surfactant depleted lungs. The aim being to achieve similar PaCO$_2$ tensions, their effects on TSLC, oxygenation and airway pressures were investigated.

METHODS

This study was approved by the Animal Committee of the Medical Faculty of the University of Istanbul.

Ten male pigs, 46.6±3.1 kg (range 40-50 kg) were premedicated with midazolam (0.5 mg/kg) intramuscularly. Anesthesia was induced with thiopentone (2-4 mg/kg), given through a 20 G cannula placed into an ear vein. Tracheostomy was performed and a portex tube with an internal diameter of 7 mm was inserted. The tube was fixed to the trachea with stitches and the cuff was inflated to avoid air leakage. Lungs were ventilated with a Servo 900C (Siemens–Elema Solna, Sweden) ventilator thereafter. Anesthesia was maintained by infusion of midazolam (0.2 mg/kg/min) and fentanyl (2 µg/kg/min). Pancuronium bromide (0.08 mg/kg/min) infusion was administered after a bolus of 0.2 mg/kg for muscle relaxation.

A three lumen catheter (Abbott Critical Systems, 7F) for fluid replacement to avoid hypovolemia and a Swan–Ganz thermodilution catheter (Spectramed Model SP 5537-H,7F) for hemodynamic monitoring were inserted to the right and left internal jugular veins, respectively.

Two cannulae (Cook, 20F) were inserted to the right and left femoral veins for the administration of extracorporeal circulation. Femoral artery was cannulated for blood sampling and invasive blood pressure monitoring. An 18F Foley urine catheter was placed into the bladder by cystostomy to monitor urine output.

Hemodynamic parameters were evaluated by Viggo Spectramed transducers. Cardiac output (CO) and heart rate (HR) measurements were made on a Horizon 2000 (Mennen, Medical). Arterial blood gases were determined by ABL 300 and OSM Hemoximeter (Radiometer-Copenhagen). Best-PEEP and TSLC values were measured with Compli 80 System (Kontron). In this system, a pressure-volume curve was determined after disconnecting the pigs from the ventilator (zero PEEP). Then, an automatically driven syringe (1.5 L) inflated the lungs by stepwise movements of 100 ml air, to a total volume of 10 ml/kg body weight. TSLC was calculated by the computer as the ratio between the inspiratory volume and the pressure difference. Best-PEEP was calculated by the computer as the minimal pressure at which the pressure-volume curve became linear and corresponded to the pressure at which PEEP must be set to increase functional residual capacity. A centrifugal pump (Biomedicus) and membrane lungs (SciMed) were used for $ECCO_2R$. The membrane lungs were each ventilated with 10 L of humidified pure O_2.

Pigs were subjected to lung lavage to establish a model of ARDS which is reported to be stable to evaluate different modes in a single animal.[16,17] After completing baseline measurements at control mode (CM) 1, lung lavage was performed with 50 ml/kg of warm saline solution, to induce severe respiratory failure. PaO_2 < 100 mm Hg was accepted as ARDS.

Randomly applied modes of ventilation were as follows:

CM1:   VCV with a preset static PEEP ($PEEP_S$) of 4 cm H_2O, V_T of 10 ml/kg, frequency (f) of 12 breaths/min and I/E ratio of 1:2 (25% I, 10% pause).

CM2:   Same as CM 1. After lung lavage.

Mode (M) 1: VCV with measured best-PEEP, V_T of 10 ml/kg, f of 12 breaths/min and I/E ratio of 1:2 (25% I, 10% pause).

M2:   PRVCV with PEEP of 4 cm H_2O, PIP adjusted to get a V_T of 10 ml/kg, f of 12 breaths/min and I/E ratio of 4:1 (80% I). Around 45 cm H_2O PIP was applied during the first 5 min to open the alveoli.

M3:   LFPPV with best PEEP, V_T of 5 ml/kg, f of 5 breaths/min and I/E ratio of 1:2 (25% I, 10% pause) 1-2 L/min O_2, given through a catheter (ID 1 mm), inserted through the tracheal tube and advanced to the level of the carina; $ECCO_2R$ with a pump speed of 20-30% of the CO measured during CM 2 and membrane lungs with a surface area of 7 m[2] [18]. V_T was adjusted to achieve a PIP of 45 cm H_2O during the first 5 min to open the alveoli.

Fluid was administered to keep CVP between 7-14 mmHg. Measurements were made 60 min after changing the ventilatory mode. $PEEP_T$ ($PEEP_S$+auto-PEEP) of M2 displayed by pressing the end-expiratory hold button, PIP and mean airway pressure (mPaw) of all modes displayed by the ventilator were recorded. Intrapulmonary pressure amplitude (ΔP), defined as the difference between $PEEP_T$ or $PEEP_S$ and PIP, were recorded. Before the use of different ventilatory settings, CM2 was applied for 30 min to investigate whether any changes in gas exchange had occurred over time.

Data were compared by two-way analysis of variance (ANOVA) test. All data are expressed as mean±s.d. Student's t-test was used for pair-wise comparison. Significance was considered at P<0.01.

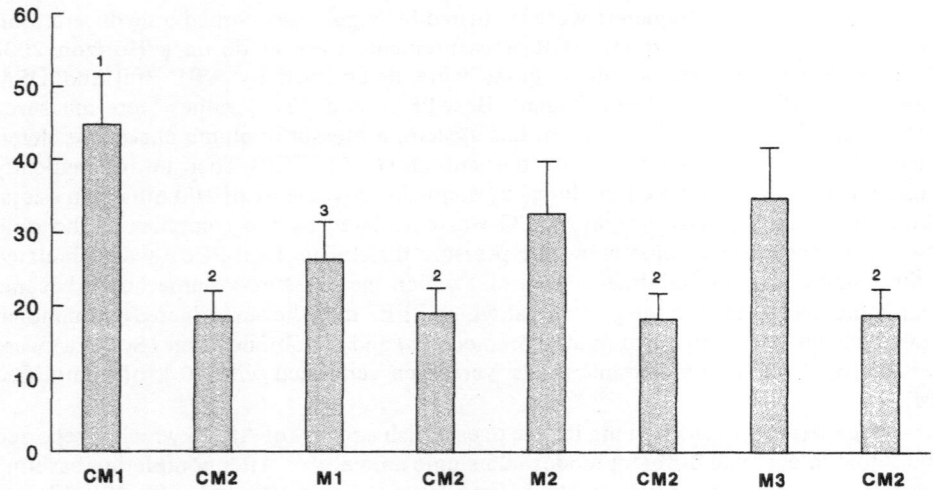

Figure 1. Changes in total static lung compliance (TSLC) during the application of modes. CM= control mode; M= mode; 1= significantly different from CM2, M1, M2 and M3 (P<0.01); 2= significantly different from M1, M2 and M3 (P<0.01); 3= significantly different from M2 and M3 (P<0.01).

RESULTS

Lung lavage was followed with a significant drop in TSLC. All the treatment modes provided increases in this parameter. Changes observed with the application of M2 and M3 were significantly higher than M1 (figure 1).

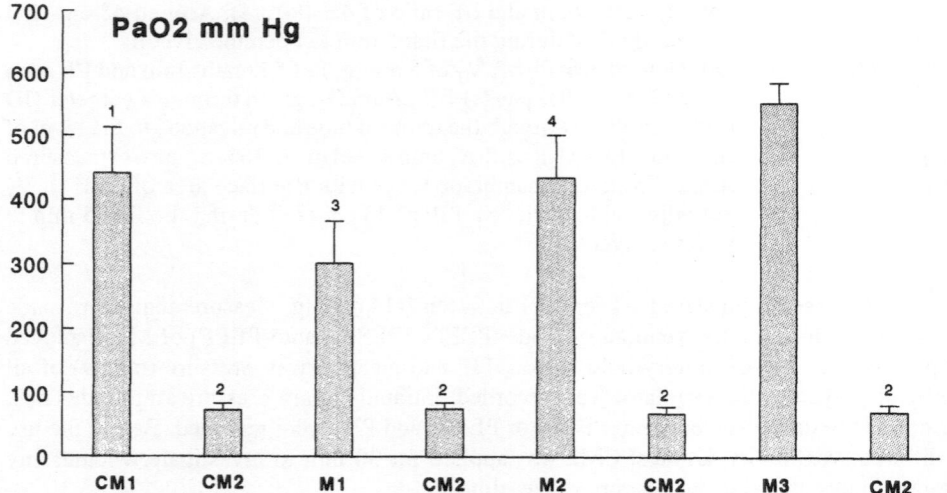

Figure 2. Changes in PaO_2 during the application of modes. CM= control mode; M= mode; 1= significantly different from CM2 and M1 (P<0.01); 2= significantly different from M1, M2 and M3 (P<0.01); 3= significantly different from M2 and M3 (P<0.01); 4= significantly different from M3 (P<0.01).

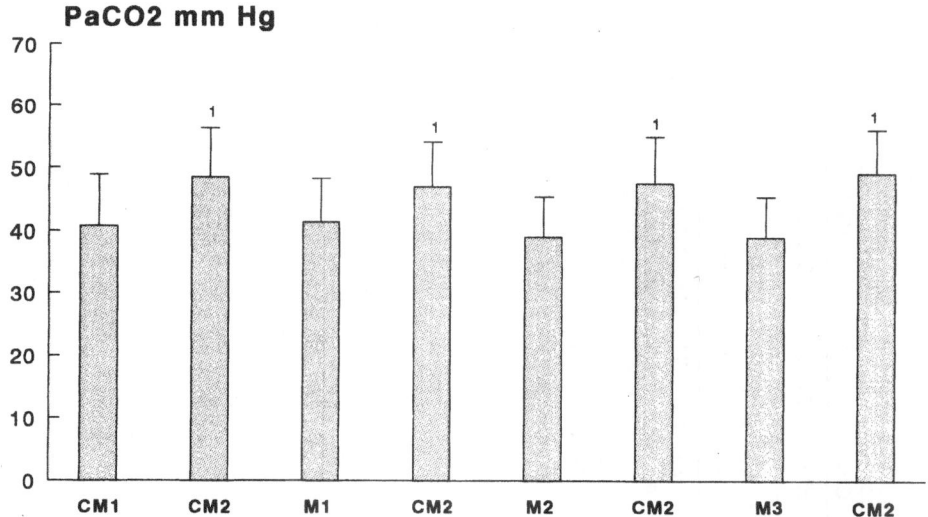

Figure 3. Changes in $PaCO_2$ during the application of modes. CM= control mode; M= mode; 1= significantly different from CM1, M1, M2 and M3 (P<0.01).

Similarly, significantly higher PaO_2 were achieved with M2 and M3 compared to M1. M3 provided the highest values (figure 2). No statistical significance was observed between the $PaCO_2$ values achieved by the treatment modes (figure 3).

Pump flow rate was around 1.1 ± 0.1 l/min. PIP, Mpaw, best-PEEP and $PEEP_T$ values measured during the study, along with the baseline values are listed in Table 1. M1 had significantly higher PIP and significantly lower Mpaw values compared to the other modes. $PEEP_T$ values obtained were similar in all treatment modes. ΔP values were around 17 cm H_2O in M2 and M3. This was significantly higher with M1 (29 cm H_2O).

Cardiac output (CO), mean pulmonary artery pressure (MPAP), pulmonary capillary wedge pressure (PCWP), mean arterial blood pressure (MABP), central venous pressure (CVP) and intrapulmonary shunt (Qs/Qt) values and levels of significance are shown in Table 2. A significant increase in MPAP was observed after the lung lavage. This increase persisted throughout the whole study. Similarly, Qs/Qt increased significantly after lung lavage. Application of all the treatment modes decreased Qs/Qt significantly.

DISCUSSION

TSLC might be a useful index to indicate aereated functioning lung units. Gattinoni and co-workers[15] showed the existence of three zones in the lungs with acute respiratory failure and named these as healthy, recruitable and diseased zones. They suggested that P-V curve investigated only the first two of these zones. However, an ideal assessment of the lungs should also include the measurements of functional residual capacity (FRC). We did not have this possibility in our experimental conditions.

On the other hand, lung lavage, introduced by Lachmann and colleagues,[16,17] and used in this study, is an acute respiratory failure model where surfactant is removed from the alveoli causing their collapse. In this model, most of the alveoli are recruitable and the early stage of clinical ARDS, where no consolidation is yet seen, is imitated. Therefore, in

Table 1. Ventilatory parameters during the application of trial modes (mean ±SD)

	CM1	CM2	M1	CM2	M2	CM2	M3	CM2
PIP	17.4±5.5[1]	29.2±5.3[2]	43.8±4.3[3]	28.7±5.3[2]	27.9±5.2	29.0±4.8[2]	30.5±5.1	28.5±5.0[2]
mPaw	6.7±1.6[1]	9.6±3.0[4]	16.9±4.8[3]	9.7±3.1[4]	26.1±5.3	10.0±3.3[4]	21.2±6.0	9.4±3.1[4]
PEEP$_T$	4[4]	4[4]	13.1±2.1	4[4]	13.4±1.9	4[4]	14.3±1.6	4[4]
ΔP	13.9±3.6[5]	24.5±3.1[3]	29.3±4.6[3]	24.5±4.1[3]	15±2.7	26.2±3.6[3]	17.6±3.5	25.1±4.5[3]

PIP= peak inspiratory pressure (cm H_2O); mPaw= mean airway pressure (cm H_2O); Values shown in CM1, CM2, M1 ans M3 are static-PEEP (PEEP$_S$) and in M2 is PEEP$_S$ + auto-PEEP; PEEP$_T$= total PEEP (cm H_2O); ΔP= intrapulmonary pressure amplitude.
[1] Significantly different from CM2, M1, M2, and M3 ($p<0.05$).
[2] Significantly different from M1 ($p<0.05$).
[3] Significantly different from M2 and M3 ($p<0.05$).
[4] Significantly different from M1, M2 and M3 ($p<0.05$).
[5] Significantly different from CM2 and M1 ($p<0.05$).

Table 2. Parameters measured and calculated during the application of trial modes (mean ±SD)

	CM1	CM2	M1	CM2	M2	CM2	M3	CM2
CO	5.2±1.7	4.9±2.2	4.8±1.2	4.8±2.0	4.7±1.8	4.8±1.6	4.8±1.8	4.7±1.9
MPAP	22.1±6.1[1]	30.0±4.7	30.9±4.9	30.1±4.4	33.2±6.1	31.1±4.9	33.4±5.9	32.0±5.3
PCWP	13.1±5.3	13.9±5.4	16.4±4.9	14.1±5.1	17.1±4.7	14.0±4.3	16.5±5.4	14.3±4.9
MABP	141±22	147±23	137±20	148±21	136±25	149±19	138±19	145±19
CVP	10±5	11±5	12±5	10±5	12±5	11±5	12±5	11±5
Qs/Qt	18.3±11.7	46.5±15.3[2]	25.9±14.1	44.7±16.2[2]	21.0±10.2	43.9±15.9[2]	19.7±14.6	44.1±14.2[2]

CO= cardiac output (L/min); MPAP= mean pulmonary artery pressure (mm Hg); PCWP= pulmonary capillary wedge pressure (mm Hg); MABP= mean arterial blood pressure (mm Hg); CVP= central venous pressure (mm Hg); Qs/Qt= intrapulmonary shunt (%).
[1] Significantly different from all CM2 modes, M1, M2, and M3 ($p<0.05$).
[2] Significantly different from CM1, M1, M2, and M3 ($p<0.01$).

this study, a trend of PaO_2 and TSLC was observed within a limited period and increases in both of these parameters were accepted as an indication of recruitment.

The measurement and application of best-PEEP in M1, improved oxygenation and caused some increase in TSLC after several hours. But, these increases were considerably lower compared to M2 and M3. Mpaw obtained were also lower in this mode. More importantly, PIP and ΔP were higher, indicating great changes in intrapulmonary pressures and a possibility for increased shear forces. These findings were in accordance with Lachmann and co-workers[17] who stressed that adequate gas exchange with VCV with PEEP was only possible when dangerous levels of inspiration pressures were reached.

In VCV with PEEP a preset external PEEP will only balance the increased retractive forces of a part of the damaged lungs preventing their collapse during the expiration of the ventilatory cycle, but will still not be sufficient to keep all parts of the lungs open. Highly damaged lung regions will only be aereated at the end of inspiration and gas exchange may be highly decreased due to the short time of insufflation, whereas during the same period capillary perfusion is reduced by the high intra-alveolar pressure. Moreover, the applied external PEEP may probably lead to local alveolar hyperdistension and capillary compression in the healthy regions of the lungs, which will be more prominent during the inspiration. This situation may lead to low TSLC values.

On the other hand, prolonging of the inspiration time to 80% of the respiratory cycle in PRVCV seems to overcome this problem of non-homogeneous distribution of volume and pressures applied if the lungs are first opened with high PIP for the re-aeration of the collapsed alveoli. The PIP can be reduced thereafter to the lowest possible level to keep the alveoli open. With this mode, the lung units fill quickly to a stable volume and recruitment is achieved when a constant and adequate pressure is applied long enough. High mPaw achieved with this mode which is also another important factor to keep the alveoli open. This was confirmed with increased PaO_2 and TSLC values with this mode of ventilation compared to M1. Moreover, PIP and ΔP were much lower than conventional techniques proving PRVCV as a reliable technique in acute respiratory failure to prevent barotrauma.

Another mode which produced superior gas exchange was LFPPV-ECCO$_2$R. The application of high Mpaw, PEEP, avoidance of deflation of the lungs and apneic oxygenation probably allowed even gas distribution in the alveoli. PaO_2 values obtained were even higher than M3, probably due to the additional gas exchange provided by the oxygenators. In this mode, PIP was also considerably low compared to M1 due to low V_T used. The model of respiratory failure used in this study consists of collapsed but recruitable alveoli and no consolidation. In M3 V_T was adjusted to a higher value initially to recruit the alveoli which might have caused considerable amount of alveolar recruitment. High levels of PEEP combined with even low V_T ventilation applied thereafter has probably kept the recruited alveoli open therefore causing an increase TSLC. However the lack of FRC measurements prevents us from further speculation on this matter.

Alveolar shunting was decreased in LFPPV-ECCO$_2$R as in other treatment modes, compared to CM2. Although this effect was expected with other modes this data concerning M3 was in contradiction with the study of Gattinoni and collegues[19] where more shunting was observed with V_T of 3 ml/kg and f of 4. This shunting was decreased with higher V_T of 10 and 15 ml/kg. The higher level of PEEP used in this study compared to 5 cm H_2O of PEEP in the former, has possibly caused an increase in FRC and clarifies the reduction of Qs/Qt in spite of low V_T used.

In conclusion, PV-IRV with I/E ratio of 4:1 and LFPPV-ECCO$_2$R provided higher levels of oxygenation and increased TSLC compared to VCV with best PEEP suggesting optimal alveolar recruitment. Moreover, airway pressures obtained were within safe levels.

On the other hand, VCV with best-PEEP gave the impression of being a dangerous mode in producing barotrauma due to high PIP and ΔP.

REFERENCES

1. Kumar A., Falke K.J., Geffin B., et al., 1970, Continuous positive-pressure ventilation in acute respiratory failure: effects on hemodynamics and lung function, *N. Engl.J.Med.* 283:1430-1436.
2. Falke K.J., Pontoppidan H., Kumar A., et al., 1972, Ventilation with end expiratory pressure in acute lung disease, *J.Clin.Invest.* 51:2315-2323.
3. Hedley–Whyte J., Pontoppidan H., and Morris M.J., 1966, The response of patients with respiratory failure and cardiopulmonary disease to different levels of constant volume ventilation, *J.Clin.Invest.* 45:1543-1554.
4. Ashbaugh D.G., Bigelow D.B., Petty T.L., et al., 1967, Acute respiratory distress in adults, *Lancet* 2:319-323.
5. McIntyre R.W., Lwas A.K., and Ramachandran P.R., 1969, Positive expiratory pressure plateau; improved gas exchange during mechanical ventilation, *Can.Anaesth.Soc.J.* 16:477-486.
6. Levine M., Gilbert R., and Auchincloss J.H. Jr., 1972, A comparison of the effects of sighs, large tidal volumes, and positive end expiratory pressure in assisted ventilation, *Scand.J.Respir.Dis.* 53:101-108.
7. Lachmann B., Danzmann E., Haendly B., and Jonson B., 1982, Ventilator settings and gas exchange in respiratory distress syndrome. In: Prakash, O, (ed): Applied physiology in clinical respiratory care, Martinus Nijhoff Publishers, The Hague, Boston, London.
8. Gattinoni L., Agostoni A., Pesenti A., Pelizzola A., Rossi G.P., Langer M., Vesconi S., Uziel L., Fox U., Longoni F., Kolobow T., and Damia G., 1980, Treatment of acute respiratory failure with low-frequency-positive-pressure ventilation and extracorporeal removal of CO_2, *Lancet* 2:292.
9. Bone R.C., 1976, Compliance and dynamic characteristics curves in acute respiratory failure, *Crit.Care Med.* 4:173-179.
10. Wilson R.S., 1976, Monitoring the lung: mechanics and volume, *Anaesthesiology* 45:135-145.
11. Suter P.M., Fairley H.B., and Isenberg M.D., 1978, Effect of tidal volume and positive end expiratory pressure on compliance during mechanical ventilation, *Chest* 73:158-162.
12. Bone R.C., 1983, Monitoring ventilatory mechanics in acute respiratory failure, *Respir.Care* 28:597-604.
13. Gattinoni L., Pesenti A., Caspani M.L., Pelizzola A., Mascheroni D., Marcolin R., Iapichino G., Langer M., Agostoni A., Kolobow T., Melrose D.G., and Damia G., 1984, The role of total static lung compliance in the management of severe ARDS unresponsive to conventional treatment, *Intensive Care Med.* 10:121-126.
14. Suter P.M., Fairley H.B., and Isenberg M.D., 1975, Optimum end-expiratory airway pressure in patients with acute pulmonary failure, *N.Engl.J.Med.* 292:284-289.
15. Gattinoni L., Pesenti A., Avalli L., Rossi F., and Bombino M., 1987, Pressure-volume curve of total respiratory system in acute respiratory failure: computed tomographic scan study, *Am.Rev.Respir.Dis.* 136:730-736.
16. Lachmann B., Robertson B., and Vogel J., 1980, In-vivo lung lavage as an experimental model of the respiratory distress syndrome, *Acta Anaesthesiol.Scan.* 24:231-236.
17. Lachmann B., Jonson B., Lindroth M., and Robertson B., 1982, Modes of artificial ventilation in severe respiratory distress syndrome: Lung function and morphology in rabbits after wash-out of alveolar surfactant. *Crit.Care Med.* 10:724-732.
18. Gattinoni L., Pesenti A., Mascheroni D., Marcolin R., Fumagalli R., Rossi F., Iapichino G., Romagnoli G., Uziel L., Agostoni A., Kolobow T., and Damia G., 1986, Low-frequency positive-pressure ventilation with extracorporeal CO_2 removal in severe acute respiratory failure, *JAMA* 256:881-886.
19. Gattinoni L., Kolobow T., Tomlinson T., Iapichino G., Samaja M., White D., and Pierce J., 1978, Low-frequency positive pressure ventilation with extracorporeal carbon dioxide removal (LFPPV-$ECCO_2R$): an experimental study, *Anesth.Analg.* 57:470-477.

82

EFFECTS OF DIFFERENT MODES OF VENTILATION ON RIGHT VENTRICULAR FUNCTION IN PATIENTS WITH ARDS

G. Köprülü, F. Esen, A. Tütüncü, N. Çakar, L. Telci, and K. Akpir

Department of Anesthesiology and Intensive Care
Medical Faculty
University of Istanbul
Turkey

INTRODUCTION

Adult respiratory distress syndrome **(ARDS),** is defined as an acute lung injury that results in widespread bilateral pulmonary infiltrates, severe refractory hypoxemia, and a marked reduction in lung compliance.

It is a common pathway resulting from a wide variety of insults at the cellular and molecular level that compromise the integrity of alveolar capillary barrier function. ARDS is characterized by acute pulmonary hypertension and pulmonary edema arising from increased lung microvascular permeability. Although ARDS may be caused or associated with a large variety of conditions or diseases, most patients demonstrate similar clinical and pathologic features irrespective of the cause of acute lung injury.

Treatment of underlying etiology (if possible) is essential in the management. Unfortunately, no therapeutic modality exists at this time that will halt or reverse either the capillary leak or the pulmonary fibrosis, so that the therapy must be supportive.

The major basic component of supportive care is ventilator therapy. Providing adequate gas exchange, while avoiding complications such as further lung injury and hemodynamic depression, is the aim of ventilatory therapy.

Besides pulmonary pathology, hemodynamic changes are likely to have considerable effects on oxygen transport and delivery. While increasing oxygen content, mechanical ventilation may worsen oxygen transport by reducing cardiac output (CO). The changes in intrathoracic pressure caused by positive pressure ventilation may influence the performance of the heart and especially right ventricle (RV).

Because of the possible benefits of ventilation modes in ARDS, it is important to extend our knowledge of hemodynamic effects, especially to RV. In this study, we tried to evaluate effects of different modes of ventilation (volume controlled ventilation, {VCV}, pressure controlled ventilation, {PCV, I:E-1:2}, pressured controlled inverse ratio ventilation, {PC-IRV, I:E: 1:1, 2:1, 3:1}) on right ventricular function and on pulmonary mechanics.

Oxygen Transport to Tissue XVII, Edited by Ince et al.
Plenum Press, New York, 1996

MATERIAL AND METHOD

Twelve patients, 4 females and 8 males (age 17-58, mean 35±12 years), hospitalized in intensive care unit of our department for ARDS were studied after informed consent of the closest relative had been obtained. The following criteria were accepted for inclusion of the patients to the study: 1) Arterial hypoxemia with a $PaO_2<100$ mmHg with FiO_2 of 1.0 in combination with 4-6 cm H_2O PEEP, 2) Bilateral diffuse infiltrates on chest roentgenogram, 3) Pulmonary capillary artery wedge pressure <18 mmHg, 4) Lung injury score (LIS)>2.5, 5) No evidence of intracranial hypertension.

All patients received sedation with fentanyl and muscle relaxation with vecuronium at appropriate doses as infusion during whole procedure. The endotracheal tube was checked for leakage before the trial.

The assessment of right ventricular function was made by thermodilution technique using modified pulmonary artery catheters (Edward's Lab. USA) equipped with fast response thermistors. Since pulmonary artery catheters are routinely used in the management of ARDS, this adaptation may represent an uncomplicated and free adjunct to monitorization.

Servo 900C (Siemens, Elema, Sweden) ventilator and Horizon 2000 (Mennen Medical) hemodynamic monitor were used during whole procedure.

Right ventricle ejection fraction (RVEF) is used as an index of right ventricular function, which is influenced by contractility, afterload, and to a lesser extent by preload. The following hemodynamic, respiratory, and laboratory parameters were measured and calculated: heart rate (HR), mean arterial pressure (MAP), mean pulmonary arterial pressure (PMAP), pulmonary capillary wedge pressure (PCWP), central venous pressure (CVP), cardiac output and cardiac index (CO, CI), stroke volume and index (SV, SI), (RVEF), right ventricle end diastolic volume (RVEDV), right ventricle end systolic volume (RVESV), pulmonary vascular resistance (PVR), peak airway pressure (PIP), mean airway pressure (MAWP), PEEP, and arterial blood gases (PaO_2, $PaCO_2$).

As base line measurements, hemodynamic and pulmonary parameters first measured with VCV at a respiratory rate (RR) of 14/min., using a tidal volume of 10 ml/kg in combination with 4-6 cm H_2O PEEP at a constant FiO_2 of 1.0. Then best PEEP titration was performed during PCV 1:2 in order to assess the optimal PEEP set throughout the trials. Assessed best PEEP was applied during VCV and PCV 1:2 as external PEEP and it was set according to autoPEEP during PC-IRV ventilation as to keep total PEEP constant. Then the trial was started with VCV with best PEEP. PCV with best PEEP was used secondly and it was repeated for 30 minutes before each mode as a control mode. PC-IRV modes were applied randomly. Minute volume, respiratory rate, total PEEP and FiO_2 were kept constant during the trial. There was a 45 minute stabilization period before each measurement.

Statistical analyses were performed by Student's t-test and $p<0.05$ was regarded as significant.

RESULTS

The results were summarized in Table 1.

Changing ventilation mode from VCV to PCV and increasing I:E ratios resulted in a significant increase in PaO_2 values. Also there was a trend to lower $PaCO_2$ at high I:E ratios, only the difference between VCV and PC-IRV 3:1 was significant.

PIP showed a remarkable decrease with PCV and inverse ratio modes. At the same time MAwP increased significantly.

Table 1. Hemodynamic and respiratory parameters measured during each trial

	G1 VCV 1:2	G2 PCV 1:2	G3 PC-IRV 1:1	G4 PC-IRV 2:1	G5 PC-IRV 3:1
Peak P	43.8±3.7	37.2±3.1	36.0±4.0	35.9±4.0	33.6±4.8
Mean P	15.5±0.8	16.6±0.9	20.9±4.0	23.8±4.5	25.9±4.1
PEEP	10.1±2.0	10.1±2.0	10.1±2.3	10.2±4.2	12.4±4.6
PaO_2	92±20	103±35	168±88	204±106	229±133
$PaCO_2$	39.8±5.6	37.9±5.9	37.9±6.5	36.2±6.7	35.4±6.7
MAP	79.6±16	83.8±16	83.9±15	84.8±13	78.8±10
PMAP	34.0±6.4	32.6±5.8	35.8±6.6	38.4±5.8	40.2±5.6
CVP	11.4±1.5	10.8±1.4	12.2±1.7	13.2±2.7	14.2±2.9
PCWP	16.0±1.3	15.4±0.8	16.8±1.5	18.0±1.5	19.4±2.4
HR	115±18	110±22	109±24	107±24	111±26
RVEF	38.2±7.0	45.5±9.0	44.2±8.0	43.4±8.0	42.7±8.0
RVEDV	182±18	202±47	205±45	184±22	177±26
RVESV	112±25	113±33	117±28	111±14	110±18
SV	69.6±12	92.0±24	92.6±27	79.8±22	75.0±20
SI	36.8±6	49.0±12	49.2±13	41.6±11	40.0±9
CO	8.6±1.8	10.3±2.3	10.5±2.5	8.8±2.0	8.2±1.8
CI	4.5±0.9	5.5±1.4	5.6±1.2	4.6±0.9	4.3±0.9
PVR	168±85	137±59	140±52	178±78	198±79
Qs/Qt	19.2±7.7	8.5±5.5	5.9±4.0	4.5±3.9	4.0±3.3

Peak pressure: G1 significantly different from G2, G3, G4, G5; G2 significantly different from G4, G5; G3 significantly different from G5.

Mean pressure: G1 significantly different from G3, G4, G5; G2 significantly different from G3,G4,G5; G3 significantly different from G5

PaO_2 : G1 significantly different from G2, G3, G4, G5; G2 significantly different from G3, G4, G5; G3 significantly different from G4, G5; G4 significantly different from G5

$PaCO_2$: G1 significantly different from G5

SV - CO: G1, G4, G5 significantly different from G2, G3

PVR: G2, G3 significantly different from G4, G5; G1 significantly different from G5

CVP-PCWP-MPAP: G1, G2 significantly different from G4, G5

Qs/Qt: G1 significantly different from G2, G3, G4, G5; G2 significantly different from G4, G5

There were no significant differences in hemodynamic parameters (HR, MAP, RVEF, ESV, and EDV) between each group. Only SV and CO improved while PCV and PC-IRV 1:1 respect to other modes. There was no difference between VCV and PC-IRV 2:1, 3:1 for the same parameters. Increases in CVP, PCWP, PMAP and PVR were observed with increasing MAWP during PC-IRV 2:1, 3:1. But increases in MAWP had no clinical effect on other hemodynamic paramaters and never reached critical values.

Shunt fraction decreased significantly due to the increasing I:E ratios.

DISCUSSION

The overall goal in treatment of ARDS has been to recruit and stabilize closed, potentially functional alveolar units while minimizing inhomogeneity of expansion, shear forces, and barotrauma. Ideally, this should be accomplished in such a way that keeps the PIP to the minumum acceptable, provides adequate oxygen transport, allows for CO2 excretion and an appropriate cardiac output with acceptable vascular pressures. Thus it should provide a positive alveolar pressure to balance the retractive forces and should open up the closed units and keep them open while avoiding local or general overinflation (1). If a pressure more than that balancing the surface tension is applied, overinflation and lung injury will be easily seen. Then to maintain the patency of alveoli, controlling pressure rather than volume must be essential. This is in contradiction to VCV-PEEP, where PEEP prevents airway closure and tidal volume maintains ventilation without controlling the pressure (2-3).

But in PC-IRV, insufflation pressure is responsible for stabilization of lung units while releasing of this pressure during the expiratory phase is for ventilation. That means that adequate ventilation is produced by releasing the pressure for time periods just long enough to allow the escape of tidal volume, but shorter than the time needed for closure of alveoli (4).

The results of this study show that, patients in pressure controlled modes, especially PC- IRV modes, had higher PaO_2 compared to that in VCV. Also increasing I:E ratios at PC modes resulted in significant improvement in oxygenation, moreover it was maintained with low PIP. This may be due to the pattern of inspiratory gas flow, which is a function of both the I:E ratio and pressure control mode (5). The PC-IRV produces an initially high inspiratory flow followed by a rapidly decelerating flow pattern compared to constant flow pattern seen in volume control ventilation. A constant inspiratory flow rate, with gradually increasing airway pressure, is used with VCV to attain the preset volume (6).

An additional benefit from the flow pattern is that high inspiratory flow rates have been shown to improve gas exchange and provide a more even distribution of ventilation to lung units. In other words, mechanism of this ventilation mode seems to be related to the alveolar time constant that is known to be short in healthy alveoli and long in diseased alveoli. By shortening the expiration time, a gas trapping effect is obtained, which creates autoPEEP, and results in recruitment of collapsed alveoli with long alveolar time constant (7).

Since no significant change occurred in PaO_2 with significant decreases in PIP, this implies some combination of a reduction in dead space to tidal volume ratio (VD/VT), an improvement in ventilation perfusion matching (VA/Q) possibly secondary to more efficient recruitment of ventilation perfusion surface area and improvement in shunt (Qs/Qt) that was showed clearly in our study (8-9).

High airway pressures are associated with increased lung damage in both human and animal studies (8). High PIP is often cited as the major determinant of risk for barotrauma and may result in shear forces at the alveoli (10). The PC-IRV may reduce these shear forces by producing a more even inflation and stabilization of the terminal respiratory structures (11).

The significant reductions in PIP seen with PC-IRV were accompanied by a small but significant increase in MAWP. Several investigators have suggested a role between mean airway pressure and oxygenation (12) while others failed (5). The relation between MIP and oxygenation appears complex. But it seems an oversimplification to attribute the improvement in oxygenation to that increase (5).

Necessity of heavy sedation and neuromuscular blockade have been a concern as a disadvantage of PC-IRV (4). However this mode of ventilation is reserved for a population of patients that have severe respiratory failure, that means that they need heavy sedation. Survivors who underwent PC-IRV were amnestic for the period of time PC-IRV were maintained in our study. We believe this is necessary for protecting the patient from a serious psychological trauma

Mortality rate was specifically not utilized as an end point in our study. Our rate was about 75% (3 survivors) due to the selection for severe and refractory ARDS and confirm the data in literature (13). In one study in which patients were selected for severe ARDS, mortality rate was as high as 92% (14). Furthermore, our study size is too small for making any mortality comparison.

Any mode of mechanical ventilation is likely to cause a decrease in CO, when high intrapulmonary pressures are used (4). Changes in right ventricle function plays a major role in possible decrease in CO. This can be explained by different mechanisms:

a) Interdependence between the left and right ventricles; afterloaded right ventricle may cause restriction in left ventricle filling b) Decrease in RV preload, inducing a secondary decrease in left ventricular filling c) Real dysfunction of RV d) Increased afterload (15-16).

Assessment of RV function in ARDS might be more attractive for two reasons. First RV dysfunction can be amplified by an increase in RV afterload, represented by the pulmonary hypertension frequently observed in severe ARDS and because of high inspiratory pressures during mechanical ventilation. Secondly, in clinical practice when RV preload (best appreciated by {RVEDV}) is decreased, modification in CO can be treated by blood volume expansion. In case of true RV dysfunction (an increase of RVEDV) fluid challenge is not indicated for treatment. And also initial RV ejection fraction (RVEF) already has a prognostic value in ARDS and septic shock. This has not been observed with measurements of LVEF, probably due to the peripheral compensatory mechanisms that alter its function (17).

Several investigators have reported adverse hemodynamic effects of PEEP in animals (18) and ARDS patients (19-20). Others found no significant changes (21-22). Similar controversies exist for the hemodynamic effects of PC-IRV. Some authors observed a decrease in CO (23) and the others observed an increase (24). We did not observe any deleterious effect of PC-IRV with a proper fluid management. There were no significant changes in HR, MAP, RVEF, ESV, and EDV between VCV+PEEP and PC-IRV.

Although CVP, PCWP, MPAP, and PVR were increased, it did not reach dangerous levels due to that the high retractive forces in ARDS lungs may limit the airway pressures being transmitted to the pulmonary vasculature, which normally would result in a decreased preload for LV and an increased afterload for RV. There was no increase in ESV during PC-IRV suggesting an increase of afterload of RV. This implies that if cardiac performance was not effected by primary pathology or inotropic support was maintained, the increase in afterload would be easily balanced by the patient.

In short, PC-IRV was hemodynamically well tolerated by patients. Proper fluid management combined with positive inotropic drugs will be sufficient for obtaining adequate CO, if any adverse effect occurs (25).

As conclusion, these results suggest that PC-IRV may be a beneficial ventilatory modality in the treatment of ARDS since it results in improvement in arterial oxygenation at significantly reduced airway pressures without any concomitant adverse effects on other hemodynamic parameters including right ventricular performance. Besides the equipment and technology required to implement PC-IRV is present or easily obtainable by most hospitals. Also patients can be transported while on PC-IRV, and invasive vascular access is not required.

REFERENCES

1. Lachmann B Open up the lung and keep the lung open. Intensive Care Med 1992 18:319-321
2. Dantzker D Gas exchange in the adult respiratory distress syndrome.Clin Chest 1982 3:57-67
3. Gattinoni L, Pesenti A Relationships between lung computed tomographic density, gas exchange and PEEP in acute respiratory failure. Anaesthesiology 1988 69:824-832
4. Kesecio∂lu J , L.Telci Evaluatyon of oxygenation with different modes of ventilation in patients with adult respiratory distress syndrome. Adv Exp Med Biol 1992;317 :901-906
5. Tharratt R.S, Allen R.P. Pressure controlled inverse ratio ventilation in severe adult respiratory failure Chest 94; 4: 755-761
6. Jansson L, Jonson B. A theoretical study of flow patterns of ventilators. Scan J. Repir Dis 1972; 53: 237-246
7. Baum M, Benzer H Inverse ratio ventilation (IRV) Anaesthesist 1980; 29: 592 596

8. Lachmann B, Danzmann E. Ventilator settings and gas exchange in respiratory distress syndrome. In: Prakash O, ed. Applied physiology in clinical respiratory care. Boston: Martinus Nijhoff, 1982; 141-176
9. Cole AG, Weller SF. Inverse ratio ventilation compared with PEEP in adult respiratory failure. Intensive Care Med 1984; 10: 227-232
10. Kolobow T, Moretti MP Severe impairment in lung function induced by high peak airway pressure during mechanical ventilation : An experimental study. Am Rev Respir Dis 1987 ; 135: 312-315
11. Kesecioðlu J, Tibboel D. Advantages and rationale for pressure controlled ventilation. Year book of intensive care and emergency medicine 1994 : Edited by Vincent JL.
12. Boros SJ. Variations in inspiratory:expiratory ratio and airway pressure wave form during mechanical ventilation: the significance of mean airway pressure. J Pediatr 1979; 94: 114-117
13. Taylor Rw. The adult respiratory distress synrome. In : Kirby RR, Taylor Rv ed. Respiratory failure . Chicago: Year Book Medical Publishers 1986:208-244
14. Zapol WM, Snider MT. Extracorporeal membrane oxygenation in severe acute respiratory failure. JAMA 1979; 242: 2193-2196
15. Martin C, Saux P . Right ventricular function during PEEP. Chest 1987; 92: 999-1003
16. Vincent J.L., Lenaers A. Right heart function and its evaluation. Perspectives in Critical Care 1989: Vol:2 Num:1 141-156
17. Vincent J.L., Reuse C. Right ventricular dysfunction in septic shock: assessment by measurements of right ventricular ejection fraction using the thermodilution technique. Acta Anaesthesiol Scand 1989:33: 34-38
18. Qvist J, Pontoppidan H. Hemodynamic responses to mechanical ventilation with PEEP; the effects of hypervolemia. Anesthesiology 1975; 42: 45-55
19. Suter PM, Fairlay HB. Optimum end-expiratory airway pressure in patients with acute pulmonary failure. N Engl J Med 1975; 292: 284-289
20. Jardin F, Farcot JC. Influence of positive end-expiratory pressure on left ventriculer performance. N Engl J Med 1981; 304: 387-392
21. Pinsky MR, Desmet JM. Effect of positive end-expiratory pressure on right ventricular function in humans. Am Rev Respir Dis 1992; 146: 681-687
22. Ellman H, Dembin H. Lack of adverse hemodynamic effects of PEEP in patients with acute respiratory failure. Crit Care Med 1982; 10: 706-711
23. Cole AG, Weller SF. Inverse ratio ventilation compared with PEEP in adult respiratory failure. Intensive Care Med 1984; 10: 227-232
24. Poelaert JI, Vogalers DP. Evaluation of hemodynamic and respiratory effects of inverse ratio ventilation with a right ventricular ejection fraction catheter. Chest 1991; 99: 1445-1449
25. Kesecioğlu J, L.Telci PRVCV with different ratios in comparision with conventional volume controlled PEEP ventilation in patients suffering from ARDS. Acta Anaesthesial Scand 1994; 38: 879-884

83

PULMONARY MECHANICS DURING LAPAROSCOPIC SURGERY

G. Köprülü, F. Esen, K. Pembeci, and T. Denkel

Department of Anesthesiology and Intensive Care
Medical Faculty
University of Istanbul
Turkey

INTRODUCTION

The primary aim of an anesthesiologist is to maintain optimum conditions per and postoperatively for the patient. Laparoscopic surgery is a technique that is getting popular because of its rapid postoperative recovery and short-term hospital stay. But CO_2 insufflation for visualization of the peritoneal cavity may cause undesired conditions, such as hypercarbia or high inspiratory pressures, for anesthesiologists. This study was performed prospectively to compare the effects of laparoscopic and traditional surgical techniques on pulmonary mechanics, arterial blood gases, and routinely used hemodynamic parameters.

MATERIAL AND METHODS

Fourty adult patients, ASA physical status I or II, scheduled for elective abdominal surgical procedure (open or laparoscopic cholecystectomy) were studied after institutional approval. Patients were randomly assigned to one of the two groups: laparoscopic or traditional techniques.

Each group consisted of 20 patients: 6 males, 14 females (age 47±6 years, range 29-64 years) in the laparoscopic group (L); and 8 males, 12 females (age 42±9 years, range 26-59) in the traditional surgery group, as a control (C) group. After insertion of an intravenous canula, all patients received 10 ml/kg bolus of 0.9% NaCl solution intravenously prior to induction followed by 4 ml/kg/hour infusion of the same solution during surgery. Anesthesia was induced by fentanyl (2 microg/kg), Na-thiopenton (3-5 mg/kg) and endotracheal entubation was facilitated by vecuronium (0.1 mg/kg). 50% O_2/N_2O and isoflurane 1-1.5% were used for the maintenance.

Patients were ventilated with a tidal volume of 10 ml/kg, applying zero external PEEP with a respiratory rate of 15 breaths per minute in volume controlled mode using a Servo 900 C ventilator (Siemens, Elema, Sweden). They were insufflated with carbon dioxide gas

Oxygen Transport to Tissue XVII, Edited by Ince et al.
Plenum Press, New York, 1996

643

intraperitoneally by an automatic gas insufflator that kept intraperitoneal pressure constant at 14 mmHg after induction of anesthesia. A 45 degree of reverse Trendelenburg and 25 degree of left lateral position were used in order to get a better visualization of the right upper region during surgical procedure. Parameters including mean arterial pressure (MAP), heart rate (HR), arterial blood gases (PaO_2, $PaCO_2$), pulmonary pressures (peak {PIP}, pause, and end expiratory pressure) were measured and static compliance was calculated by dividing the tidal volume by the difference between pause and end expiratory pressures. Positive end expiratory pressure was measured by means of an end-expiratory occlusion of at least 5 seconds. For the analysis of mechanical ventilation parameters, five consecutive cycles were recorded, and averaged. Hemodynamic parameters were measured and recorded by a Horizon 2000 hemodynamic monitor (Mennen Medical). Arterial blood gas samples were analyzed by an ABL 505 (Radiometer, Copenhagen, Denmark).

Measurements were done at the following steps for each group:

L1. After induction and intubation
L2. When the maximal intraabdominal pressure was reached (14mmHg)
L3. Peroperatively 30 minutes
L4. When whole abdomen was deflated

C1. After induction and intubation
C2. At skin incision
C3. Peroperatively 30 minutes
C4. At the end of the operation

Statistical analyses were performed by Student's t-test regarding $p < 0.05$ significant.

RESULTS

There were no significant differences between the two groups with regard to age and other demographic data. Surgical operations were open or laparoscopic cholecystectomy in both groups, that means equal surgical conditions.

MAP, HR, PaO_2, $PaCO_2$, PIP and compliance did not change in the control group during the surgical procedure.

Although there were no differences in PIP and compliance in the control group, peak pressure increased 50% and compliance decreased 50% when the maximal intraabdominal pressure was reached and at peroperatory 30 minutes in the laparoscopic technique. Arterial CO_2 increased significantly in step 4, but it did not reach pathological values in group L. On the other hand there were no significant changes in HR, MAP, PaO_2 in group L.

Results of the study are summarized in Table 1.

DISCUSSION

Patients are insufflated with CO_2 intraperitoneally at constant pressure for clear visualization of the cavity during laparoscopic surgery. Intraperitoneal pressure (IPP) is increased by creation of carbon dioxide pneumoperitoneum. Increased IPP leads to a decrease in static compliance (1) and functional residual capacity (2,3) due to its limitation effect on diaphragmatic excursion (4). The result of this phenomenon is an increase in airway pressures (4,5,6,7). This was shown clearly in our study. When IPP reached its highest level, PIP increased 50% compared to baseline level, and at the same time compliance decreased 50%. High airway pressure continued during the whole surgical procedure. Pulmonary

Table 1. Hemodynamic and respiratory parameters measured during each step

	MAP	HR	PaO$_2$	PaCO$_2$	PEAK P	COMPL.
L1	94±10	79±8	211±35	34±4	17.2±5	58.5±19
L2	95±16	69±11	205±29	36±5	25.1±6*	31.8±8*
L3	104±13	75±8	188±27	37±4	22.5±3*	37.2±7*
L4	101±14	77±8	184±33	44±5*	18.0±3	46.4±7
C1	100±5	77±6	214±18	35±4	13.7±2	58.5±7
C2	111±11	82±5	201±11	33±4	14.3±2	53.9±7
C3	103±9	75±3	180±22	31±6	14.5±2	52.4±7
C4	107±3	87±4	178±12	33±5	15.8±2	54.0±94

(*) P<0.05

pressures and compliance return to normal values at the end of the operation due to deflation of the abdominal cavity.

On the other hand, CO_2 is continuously absorbed into the circulation transperitoneally and may lead to hypercarbia. Development of hypercarbia and cardiovascular compromise have been reported during laparoscopy (8-12). The cardiovascular compromise was caused by mechanical factors directly related to increased IPP affecting ventilation and venous return as well as by the absorption of CO_2, leading to acidosis and further depression of the cardiovascular system (10,11,12). Laparoscopic technique was hemodynamically well tolerated in patients in our study. Proper fluid management prior to induction and maintaining of normocapnia during surgery are the possible reasons for this. Although CO_2 increased in step 4, it did not reach pathological values. This implies that our model of ventilation was suitable for this group of patients.

The aim of ventilation during surgical operation is to maintain anesthesia and also to provide gas exchange, while avoiding complications such as progressive lung injury and hemodynamic depression due to high inspiratory pressures. These goals can be achieved by keeping PIP to the minimal acceptable level. It was clearly shown that a PIP level over 40 cm H_2O is associated with an increased risk of barotrauma and morphological changes in airways (13). Because the risk of barotrauma increases with rising peak airway pressures, strategies decreasing or limiting the increase must be instituted.

Although the increases in the pulmonary pressures were about 50%, it did not reach barotrauma values in our study due to that our patients had no previous pulmonary problems. But the laparoscopic technique is claimed to be preferred in patients who have pulmonary problems because of its rapid postoperatory recovery (14). We do not believe that laparoscopic technique is contraindicated in patients with pulmonary pathologies but the ventilation of such patients deserves special attention, especially after the abdominal cavity is inflated (15).

As a result, the anesthesia method must include the use of techniques that can be modified according to the pulmonary mechanics and adequate monitoring must be provided during laparoscopic surgical technique for evaluating pulmonary mechanics, oxygenation and hemodynamics.

REFERENCES

1. Irwin SR. Mechanical ventilation Part I: Initiation. In: Rippe JM, Irwin SR., Intensive Care Medicine, 2nd Ed. Boston-Toronto-London: Little Brown Company p: 563-575
2. Puri GD., Singh H. Ventilatory effects of laparoscopy under general anaesthesia. Br J Anaesth 1992: 68(2): 211-213
3. Fitzgerald SD., Andrus CH. Hypercarbia during carbon dioxide pneumoperitoneum. Am J Surg 1992: 163(1): 186-190

4. Leighton T., Planim N. Effectors of hypercarbia during experimental pneumoperitoneum. Am Surg 1992: 58(12): 717-721

5. Lister DR, et al. Carbon dioxide absorption is not linearly related to intraperitoneal carbon dioxide insufflation pressure in pigs. Anesthesiology 1994: 80: 129-136

6. Wittgen CM., Andrus CH. Analysis of hemodynamic and ventilatory effects of laparoscopic cholecystectomy. Arch Surg Aug 1991: 126(8): 997-1000

7. Brantley JC., Riley PM. Cardiovascular collapse during laparoscopy: A report of two cases. Am J Obstet Gynecol 1988:15: 735-737

8. Shanta TR., Harden J. Laparoscopic cholecystectomy: Anesthesia related complications and guidelines. Surg Laproscopy Endoscopy 1991: 1: 173-178

9. Kent RB. Subcutaneous emphysema and hypercarbia following laparoscopic cholecystectomy Arch Surg 1991: 126: 1154-1156

10. Scott DB., Julian DG. Observations on cardiac arrhythmias during laparoscopy BMJ 1972: 1: 411-413

11. Holzman M., Sharp K. Hypercarbia during carbon dioxide gas insufflation for therapeutic laparoscopy. Surg Laproscopy Endoscopy 1992: 2(1): 11-14

12. Tan PL., Lee TL. Carbon dioxide absorption and gas exchange during pelvic laparoscopy. Can J Anaesth 1992: 39(7): 677-681

13. Kolobow T, Moretti MP. Severe impairment in lung function induced by high peak airway pressure during mechanical ventilation: An experimental study. Am Rev Respir Dis 1987; 135: 312-315

14. Frazee RC, Roberts JW. Open versus laparoscopic cholecystectomy. Ann Surg 1991: 213: 651

15. Marco AP., Yeo CJ. Anaesthesia for a patient undergoing laparoscopic cholecystectomy. Anesthesiology 1990: 73: 1268

84

SHORT TERM EFFECTS OF SIMV AND PSV ON WORK OF BREATHING BY PULMONARY MONITOR CP-100 (BICORE)

N. Çakar,[1] A. Tütüncü,[1] Ĝ. Köprülü,[1] F. Esen,[1] L. Telci,[1] and J. Keseciğlu[2]

[1] Department of Anesthesiology and Intensive Care
  Medical Faculty of University of Istanbul, Turkey
[2] Department of Intensive Care
  Academic Medical Center
  Amsterdam, The Netherlands

INTRODUCTION

Synchronized Intermittent Mandatory Ventilation (SIMV) and Pressure Support Ventilation (PSV) are used as partial ventilatory support or weaning modes.(1) SIMV is the "old" and widely used weaning mode and is a combination of spontaneous and assisted ventilation. At intervals determined by the SIMV frequency setting, the ventilator becomes "sensitized" to the patients inspiratory effort and responds to that effort by delivering a mechanical "assisted" breath. Between these cycles, the patient breaths spontaneously at a rate and depth of his or her own choosing. PSV is the "new" mode of weaning in which a physician-selected fixed amount of pressure is applied throughout inspiration on each breath to augment the patient's own respiratory rate, inspiratory time, inspiratory flow rate, and tidal volume [2-5].

During weaning or partial ventilatory support work sparing efficiency and work characteristics of the mode may be important for the muscle reconditioning, relieving the sensation of dyspnea, and increasing patient comfort [6].

Direct measures of the work performed by the respiratory system include the rate of oxygen consumption of the ventilatory muscles, quantitative electromyography, mechanical (external) work of breathing and the product of pressure developed by the inspiratory muscles and duration their contraction [7]. In the clinical setting, the quantitative assessment of breathing activity can be most precisely and easily quantitated by measures of pressure development and work of breathing [7].

A recently developed pulmonary monitor, the Bicore CP-100, could measure work of breathing by separating machine and patient work in each spontaneous and machine aided cycle [8]. The aim of this study was to evaluate short-term effects of SIMV and PSV on work of breathing by using the Bicore CP-100.

Oxygen Transport to Tissue XVII, Edited by Ince et al.
Plenum Press, New York, 1996

Table 1. Demographic data of the study population

NO/SEX	Age (years)	Weight (kg)	Diagnosis	Mech. vent.	ICU stay
1 M	21	85	SIRS+Abd.surg. postop	24	38
2 M	69	72	SIRS+Abd.surg. postop	19	26
3 F	21	68	Thoracal surg. postop	1	3
4 F	24	55	Polytrauma	10	13
5 F	56	63	Abd.surg. postop	1	3
6 F	23	71	Abd.surg. postop	7	9
7 M	65	76	Thoracal surg. postop	1	3
8 M	53	68	Thoracal surg. postop	2	4
9 M	23	74	Myasthenia gravis	14	17
10 M	26	78	Myasthenia gravis	13	15
11 M	51	63	Polytrauma	5	7
12 F	68	67	COPD	6	8
13 M	61	62	Polytrauma	9	11
14 M	67	73	Thoracal surg. postop	1	3
15 M	67	76	Polytrauma	7	9
Mean±SD	46±20	70±8		8±7	11±10

M = male; F = female; Mech.vent = Total mechanical ventilation duration; SIRS = systemic inflammatory response syndrome; Abd. surg = Abdominal surgery.

METHODS

Patients

Fifteen mechanically ventilated adult patients (5 female, 10 male) with a mean duration of 8±7 days recovering from acute respiratory failure due to different etiologies were studied (Table 1). The age of the group ranged between 21 and 69 years (mean 46±8 years).

All patients were awake, alert and diagnosed to be weaned off the ventilator. Inclusion criteria were: stable or improving chest radiography, $PaO_2 > 60$ mmHg with a $FiO_2 < 0.4$, $PaCO_2 < 50$ mmHg, PEEP < 5 cm H_2O, negative inspiratory effort <-2 cm H_2O, hemodynamic stability as evidenced by a regular cardiac rhythm and a mean arterial pressure (MAP) higher than 70 mmHg without any vasopressor administration.

Protocol

The patients were randomly ventilated for 30 min with SIMV and PSV on the same day with a Servo 900C ventilator (Siemens-Elema). Both modes were set to provide 150-180 ml/kg minute volume with same FiO_2 and PEEP. This was done by adjusting peak pressure in PSV and tidal volume with frequency in SIMV. The nutritional intake was kept constant throughout the study in all patients. Each patient was placed in semi-recumbent position which remained constant throughout the study. No sedatives or narcotics were administered.

WOB of the patient (WOBp), WOB of the ventilator (WOBv) and the other respiratory parameters were measured by the pulmonary monitor CP-100 (BICORE).

For the measurement of Bicore parameters, an esophageal balloon catheter was placed in the lower third part of the esophagus, and a flow transducer was attached in between the intubation tube and Y-piece of ventilatory lines. The correct placement of the catheter was checked by the Baydur method [9].

At the end of each ventilation mode hemodynamic parameters, blood gas analysis, WOBp, WOBv (for SIMV WOBps = the work done by the patient during spontaneous respiration phase, WOBpm = the work done by the patient during mandatory phase, WOBvm = the work done by the respirator during mandatory ventilation phase, WOBvs = the work done by the ventilator during spontaneous ventilation phase), tidal volume (VT), tidal volume for SIMV mandatory cycle (VTm), tidal volume for SIMV spontaneous cycle (VTs), peak airway pressure (Paw peak), mean airway pressure (Paw mean), auto-PEEP (aPEEP), expiratory minute volume (VE), respiratory rate (RR), dynamic compliance (Cdyn), expiratory airway resistance (Rawe), ratio of inspiratory cycle time to total cycle time (Ti/Ttot), rapid shallow breathing index (f/VT), tracheal occlusion pressure (P0.1), delta esophageal pressure (delta PES), for spontaneous cycle of SIMV (delta PES SIMVs) and for mandatory cycle of SIMV (delta PES SIMVm) were recorded.

By using the measured parameters, the total work of breathing of the patient in one minute (WOBminp), the work of breathing done by the ventilator in one minute (WOBminv) and the sum of these (WOB min tot) were calculated according to the following equations:
For PSV mode

$$WOBminp = (WOBp)x(VE)$$

$$WOBminv = (WOBv)x(VE)$$

For SIMV mode

$$WOBminp = (WOBps)x(VEs)+(WOBpm)x(VEm)$$

$$WOBminv = (WOBvs)x(VEs)+(WOBvm)x(VEm)$$

$$VEs = (spontaneous\ respiratory\ rate)\ x\ (spontaneous\ tidal\ volume),$$

$$VEm = (mandatory\ respiratory\ rate)\ x\ (mandatory\ tidal\ volume)$$

For both modes

$$WOBmintot = (WOBminp)+(WOBminv)$$

Results were evaluated by Student's t-test. Significance was considered at $p < 00.5$.

RESULTS

All patients successfully completed the study. Four patients complained of dyspnea at the end of the SIMV period. Both ventilation modes provided similar blood gas-pH values. VE, Cdyn and RR were similar; blood pressure and heart rate remained stable throughout the study in both groups.

The mean values for gas exchange, pH, ventilation parameters and hemodynamic parameters are listed in Table 2.

Although similar results were obtained in both modes for VE, RR, SaO_2, pH, PaO_2, $PaCO_2$, MAP, and HR, there was great difference between the mean values of WOBp and WOBminp. Mean values for WOBminp were 3.2 ± 2.1 joule/min in PSV group and 16.2 ± 7.5 joule/min in SIMV group. The work done by the ventilator in one minute (WOBminv) is greater in the PSV group than in the SIMV group and mean values were 20.9 ± 7.5 and 12.3 ± 8.4 joule/min, respectively. These differences were statistically significant. Although WOBmintot was greater in the SIMV group this was statistically nonsignificant and, the

Table 2. Mean values for gas exchange, pH, ventilation, and hemodynamic parameters

	PSV	SIMV
VE (ml/kg/min)	155±34	158±36
VT (ml)	640±210	625±150 (SIMVm)
		420±123 (SIMVs)
f (/min)	20±5	22±5
Peak Paw (cmH$_2$O)	21.7±4.4	21.7±5.1 (SIMVm)
		9.0±2.6 (SIMVs)
Mean Paw (cmH$_2$O)	10.0±2.4	6.9±1.7
SaO$_2$ (%)	98.5±1.0	98.3±1.5
pH	7.4±0.05	7.4±0.09
PaO$_2$ (mmHg)	140.5±46.2	143.8±55.4
PaCO$_2$ (mmHg)	39.3±9.3	43.2±12.7
HR (beat/min)	109±17	108±19
MAP (mmHg)	86±10	82±12

PSV = pressure support ventilation; SIMV = synchronized intermittent mandatory ventilation; VE = minute volume; VT = tidal volume; f = respiratory frequency; Peak Paw = peak airway pressure; Mean Paw = mean airway pressure; HR = heart rate; MAP = mean arterial pressure; SIMVm = for mandatory cycle of SIMV; SIMVs = for spontaneous cycle of SIMV.

mean values were 24.2±9.1 joule/min for PSV and 28.4±12.7 joule/min for the SIMV group. The mean values for WOB in both modes are listed in Tables 3 and 4.

As shown in Table 4, mean WOBp value was greater during spontaneous breathing cycles in SIMV mode (WOBps) than the mandatory breaths in SIMV (WOBpm) and PSV (WOBp). It is worth noting that the lowest work is performed in PSV mode per breath.

The comparisons for the mean values of aPEEP, Cdyn, Ti/Ttot, f/VT, P0.1 for SIMV and PSV mode are shown in Table 5. P0.1 value was greater in SIMV mode (3.6±2.7 cmH$_2$O) when compared to PSV mode (1.4±0.1 cmH$_2$O); f/VT was also greater in SIMV group (51.8±18.3) than PSV group (35.6±16.9). These differences were statistically significant.

Delta PES SIMVm 13.2±7.4 cmH$_2$O, and delta PES SIMVs 11.9±7.4 cmH$_2$O were both greater than delta PES in PSV 4.1±3.6 cmH$_2$O. The values for aPEEP in SIMV group and PSV group were 1.1±1.1 cmH$_2$O, 0.2±0.4cmH$_2$O, respectively; this difference was statistically significant.The mean values of dynamic compliance were similar in SIMV group (64.6±27.6 ml/cmH$_2$O) and PSV group (64.2±30.3ml/cmH$_2$O).

Table 3. Mean values for work of breathing

	PSV	SIMV
WOB min tot (j/min)	24.2±9.1	28.4±12.7
WOB min p (j/min)	3.2±2.1	16.2±7.5
WOB min v (j/min)	20.9±7.5	12.3±8.4

WOB min tot = Total work of breathing in one minute; WOB min p = Work of breathing of the patient in one minute; WOB min v = Work of breathing of the ventilator in one minute.

Table 4. Mean values for work of breathing

PSV		SIMV	
WOB p (j/ml)	0.3	WOB ps (j/min)	1.20
WOB v (j/ml)	1.7	WOB vs (j/min)	0.4
		WOB pm (j/min)	0.7
		WOB vm (j/min)	1.5

WOB p = Work of breathing of patient per liter; WOB v = Work of breathing of ventilator per liter; WOB ps = Work of breathing of patient per minute for spontaneous cycles of SIMV; WOB vs = Work of breathing of ventilator per minute for spontaneous cycles of SIMV; WOB pm = Work of breathing of patient per minute for mandatory cycles of SIMV; WOB vm = Work of breathing of ventilator per minute for mandatory cycles of SIMV.

DISCUSSION

The results of this study demonstrated that PSV reduces ventilatory muscle work when compared with SIMV in the ventilatory settings which provide similar oxygenation and ventilation in the short term. Since there is only the need for an effort of triggering the set level of the ventilator by the patient in PSV, once triggered the ventilator delivers tidal volumes with a flow, rate and duration he or she needs depending on the both preset airway pressure level and lung compliance. In short PSV is a pressure assist form of mechanical ventilatory support that augments the patients spontaneous inspiratory efforts with a clinician-selected level of positive airway pressure[3].

But in SIMV the tidal volume, flow rate and the duration of flow are determined by the physician on the ventilatory settings. Observation has shown that clinician controlled volume and flows of the volume assisted breath (SIMV) may not always be synchronous with the patient ventilatory drive [10]. Several investigators have demonstrated that the

Table 5. Comparison of mean values in both modes

	PSV	SIMV
Cdyn (ml/cmH$_2$O)	64.2±30.3	64.6±27.6
Rawe (cmH$_2$O/L/sn)	9.5±10.5	11.2±11.1 (SIMVm)
		7.5±11.2 (SIMVs)
Ti/Ttot	0.3±0.5	0.4±0.9 (SIMVm)
		0.4±0.1 (SIMVs)
f/VT	36.6±16.9	51.8±18.3
P0.1 (cmH$_2$O)	1.4±0.1	3.6±2.7
aPEEP	0.2±0.4	1.1±1.1
Delta PES	4.1±3.6	13.2±7.4 (SMIVm)
		11.9±7.4 (SIMVs)

C dyn = Dynamic compliance; Rawe = expiratory airway resistance; Ti/Ttot = ratio of inspiratory cycle time to total cycle time; f = respiratory frequency; VT = tidal volume; P0.1 = tracheal occlusion pressure; aPEEP = auto-PEEP; delta PES = delta esophageal pressure; SIMVm = for mandatory cycle of SIMV; SIMVs = for spontaneous cycle of SIMV.

spontaneous inspiratory effort which triggers a mechanical assisted breath does not cease with delivery of that breath [11]. Thus the potential for significant imposed ventilatory muscle loads exists if the mechanical assisted flow and volumes do not match the patient's desired flow pattern and tidal volume. Subjective dyspnea, as well as increased muscle energy demand, would appear to be a consequence of such imposed loads [12].

This dsysynchrony has been shown to be an important source of imposed load when using volume assisted breaths because of the fixed flow and volume characteristics of such breaths [11]. Moreover, if these volume assisted breaths are given only intermittently, there appears to be a consequent in patient ventilatory drive that can further increase this imposed load.

In our study the ventilatory drive (P0.1) in SIMV group (3.6±2.7) is greater than the PSV group (1.4±0.7). The difference was statistically significant and it is known that increased P0.1 indicates increased respiratory center output and increased neuromuscular activity [12].

In the meantime for the intermittent spontaneous breathing cycles during SIMV the patient must overcome the deleterious effects of the ventilatory tubings and endotracheal tubes besides the demand valves. In these cycles the contribution of the mechanical ventilator is minimal. In Table 4, we can see that the WOBp is 0.3±0.1 j/min for PSV, but in SIMV the WOBp is 0.7±0.4 j/min for mandatory cycles and 1.2±0.4 j/min for spontaneous cycles of SIMV.

According to J.J. Marini, the connection between perceived effort and external work accomplished in ventilation is a very loose one, depending on the neural stimulation to contraction, the force reserves, the coupling efficiency of the neural signal to muscular force generation, the structural integrity of the ventilatory pump, the pattern (coordination) of muscular contraction and the impedance to airflow. However, for any given individual, if all these remain intact and unchanging (so that overall efficiency of the system is unchanged) then the mechanical work of breathing correlates quite well with the oxygen consumed by the ventilatory task and perceived effort [7].

If we accept these factors are stable in our patients then the increased WOBp would be parallel with increased oxygen consumption of respiratory muscles.

According to MacIntyre, PSV reduces ventilatory muscle work but perhaps more importantly altered the pressure/volume change characteristics of this work [6].

In summary, we have found that in the short term PSV reduces WOBp more than SIMV and this finding suggests that SIMV may impose increased potential for weaning failure in long-term application. This point may also be important for patients with limited cardiopulmonary reserve; such as cardiac patients on mechanical ventilation who are at high risk of myocardial damage, or patients in acute exarbation of COPD whom may have respiratory muscle fatigue and in who increased work of breathing and thereby increased total oxygen consumption will deteriorate the patient's clinical status.

REFERENCES

1. Irwin, S.R. Mechanical ventilation. Part I Initiation. Intensive Care Medicine. Second Edition. Rippe, J.M.(Ed). pp562-583, 1991.
2. Downs, J.B., Stock, M.C., Tanbeling, B.: Intermittent mandatory ventilation (IMV). A primary Ventilatory Support Mode. Ann Chir Gynaecol 196(suppl):57-63, 1982.
3. Luce, J.M., Pierson, D.J., Hudson, L.D.: Intermittent Mandatory Ventilation. Chest 79:678-685, 1981.
4. Weisman, I.M., Rinaldo, J.E., Rogers, R.M.: Intermittent Mandatory Ventilation. Am Rev Respir Dis 127:641-647, 1983
5. Sassoon, S.H.C., Ckoes, M., Light, W.R. Ventilator modes Old and New .In: Mechanical Ventilation. Tobin, J.M.(Ed). Critical Care Clinics, Vol.6, No.3, pp.605, 1990
6. Mac Intyre, N.R.: Respiratory Function During Pressure Support Ventilation. Chest 89:677-683, 1986.

7. Marini, J.J.: Strategies to minimize breathing effort during mechanical ventilation. In Mechanical Ventilation. Tobin, J.M.(Ed). Critical Care Clinics, Vol.6, No.3, pp635-659, 1990.

8. Petros, A.J., Lamond, C.T., Bennett, D.: The Bicore Pulmonary Monitor. Anaesthesia 48:985-988, 1993.

9. Baydur, D.A., Behrakis, W.Z., Jaeger, M.: A Simple method for assessing the validity of the esophageal balloon technique. Am Rev Respir Dis 126:788-791, 1982.

10. Marini, J.J., Caps, S.S., Culver, B.H.: The inspiratory work of breathing during assisted mechanical ventilation. Chest 87:612-618, 1985.

11. MacIntyre, N.R.: An Li Ingho Effects of initial flow rate once breath termination criteria on pressure support ventilation. Chest 99:134-138, 1991.

12. Tobin, J.M., Yang, K.: Weaning from mechanical ventilation. In: Mechanical Ventilation. Tobin, J.M.(Ed). Critical Care Clinics, Vol.6, No.3, p724, 1990.

AUTHOR INDEX

655

SUBJECT INDEX

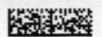